Undergraduate Texts in Physics

AF166838

Series Editors

Kurt H. Becker, NYU Polytechnic School of Engineering, Brooklyn, USA

Jean-Marc Di Meglio, Matière et Systèmes Complexes, Université Paris Diderot, Bâtiment Condorcet, Paris, France

Sadri D. Hassani, Department of Physics, Loomis Laboratory, University of Illinois at Urbana-Champaign, Urbana, USA

Morten Hjorth-Jensen, Department of Physics, Blindern, University of Oslo, Oslo, Norway

Michael Inglis, Patchogue, USA

Bill Munro, NTT Basic Research Laboratories, Optical Science Laboratories, Atsugi, Kanagawa, Japan

Susan Scott, Department of Quantum Science, Australian National University, Acton, Australia

Martin Stutzmann, Walter Schottky Institute, Technical University of Munich, Garching, Germany

Undergraduate Texts in Physics (UTP) publishes authoritative texts covering topics encountered in a physics undergraduate syllabus. Each title in the series is suitable as an adopted text for undergraduate courses, typically containing practice problems, worked examples, chapter summaries, and suggestions for further reading. UTP titles should provide an exceptionally clear and concise treatment of a subject at undergraduate level, usually based on a successful lecture course. Core and elective subjects are considered for inclusion in UTP.

UTP books will be ideal candidates for course adoption, providing lecturers with a firm basis for development of lecture series, and students with an essential reference for their studies and beyond.

David S. Simon

Introduction to Quantum Science and Technology

 Springer

David S. Simon
Department of Physics and Astronomy
Stonehill College
North Easton, MA, USA

ISSN 2510-411X   ISSN 2510-4128   (electronic)
Undergraduate Texts in Physics
ISBN 978-3-031-81314-6   ISBN 978-3-031-81315-3   (eBook)
https://doi.org/10.1007/978-3-031-81315-3

© The Editor(s) (if applicable) and The Author(s), under exclusive license to Springer Nature Switzerland AG 2025

This work is subject to copyright. All rights are solely and exclusively licensed by the Publisher, whether the whole or part of the material is concerned, specifically the rights of translation, reprinting, reuse of illustrations, recitation, broadcasting, reproduction on microfilms or in any other physical way, and transmission or information storage and retrieval, electronic adaptation, computer software, or by similar or dissimilar methodology now known or hereafter developed.
The use of general descriptive names, registered names, trademarks, service marks, etc. in this publication does not imply, even in the absence of a specific statement, that such names are exempt from the relevant protective laws and regulations and therefore free for general use.
The publisher, the authors and the editors are safe to assume that the advice and information in this book are believed to be true and accurate at the date of publication. Neither the publisher nor the authors or the editors give a warranty, expressed or implied, with respect to the material contained herein or for any errors or omissions that may have been made. The publisher remains neutral with regard to jurisdictional claims in published maps and institutional affiliations.

This Springer imprint is published by the registered company Springer Nature Switzerland AG
The registered company address is: Gewerbestrasse 11, 6330 Cham, Switzerland

If disposing of this product, please recycle the paper.

Dedicated to Michael A. Horne, a great friend and a great colleague, greatly missed by all who knew him.

Preface

Many aspects of quantum mechanics, such as superposition, entanglement, and non-locality, have always been mystifying to our classical intuition. But over the past few decades, these very same quantum properties have come to be viewed as resources that can produce new scientific and technological effects that would once have seemed impossible. The goal of this text is to provide a path of entry into understanding many of these new quantum mechanical applications.

As is discussed in more detail in Chap. 1, there are many reviews of quantum computing and quantum information processing, but finding introductory reviews at the advanced undergraduate level for other applications, such as quantum sensing and quantum communication, is harder. The goal of this book is to span an unfilled gap by providing an undergraduate-level survey of these applications, bundled together with introductions to the experimental and theoretical background required to understand them. The hope is that students who have assimilated the material here will be well-equipped, when entering graduate school or industry, to quickly jump into research in related areas. The book will also hopefully be a useful introduction to these areas for graduate students and researchers coming from other fields.

We mostly assume here that the reader has at least a basic background in electromagnetism, quantum mechanics, calculus, and linear algebra, at the level typical of a junior or senior physics major. But the early chapters cover enough background on the basics of quantum mechanics for engineering and computer science students without any prior quantum background to quickly acquire the prerequisites needed. For physics students, the chapter on basic quantum mechanics (Chap. 3) should provide a quick review and fix the notation (primarily Dirac bra-ket notation) to be used in the remainder of the book, but it may also introduce a few new topics they likely have not seen before. Additional reviews of basic probability, matrices, and other topics are included in the appendices.

Plan of the Book This book is basically an introduction to modern applications of quantum mechanics in the areas of computation, communication, and measurement. Some of the applications to be presented are already in use in real-world situations, but many are still in development. The book is divided into four parts:

- Part I covers background material on basic quantum mechanics and on other topics such as optics, electromagnetism, and classical computation. The purpose of this first part is to make sure that the reader has all of the necessary conceptual tools needed to follow the rest of the book. Most readers will already have been exposed to many of the topics presented in Part I, but the majority of readers will not be familiar with all of them. So the goal of Part I is to get everyone up to speed and fill in any gaps in their background, in order to provide a level field for everyone as they enter the later parts of the book. Obviously, depending on the background of the reader, some of this material may be just quickly skimmed.
- Part II provides a more in-depth dive into modern aspects of quantum mechanics. Although some of these topics are beginning to work their way into newer introductory quantum textbooks, the majority of students who have taken standard modern physics or introductory quantum mechanics courses will not be familiar with them. The emphasis in this part is on the aspects of quantum mechanics that will be useful for the applications in Part III.
- Part III uses the background from Parts I and II to present a range of applications. These applications include quantum-enhanced versions of optical sensors, computers and algorithms, imaging systems, communication and cryptography systems, and precision measuring devices. Some of these applications have already made real-world impact, such as super-resolution microscopes in biomedical settings or quantum-enhanced interferometry in the detection of gravitational waves emitted by colliding black holes. A different but equally important strand of recent research is the development of novel, exotic materials in which quantum mechanics allows unprecedented new effects such as topologically protected states (states which propagate without scattering or loss due to conserved topological quantum numbers), and states that seem to have fractional charges and exotic (i.e., neither Bose nor Fermi) statistics; these "topological materials" will be discussed in Chap. 21.
- Finally, in Part IV, short introductions to some of the hardware and physical platforms used to actually implement quantum states for modern applications is provided. These include topics such as laser cooling and optical traps, optical detectors, sources of entangled photon pairs, MRI systems, and solid-state qubit systems. These are the physical systems that will embody the qubits discussed more abstractly in the previous parts and that will carry out the quantum computations and measurements outlined schematically in Part III. The final chapter briefly mentions additional topics that have not been discussed in the main portion of the book, but that are likely to be of increasing importance in the near future, such as quantum walks, the quantum internet, and quantum methods in biology.

User's Guide There is enough material here to fill a two-semester course. In the more likely event that this is used as a basis for a single-semester course, the instructor can pick and choose from among the chapters to produce a number of different courses, tailored to the goals and interests of the students and instructor. A few suggested courses are listed in the table below; each of these would consist of a set of common core chapters all tracks share, followed by a cluster of additional, more specialized chapters. The instructor should choose 4–6 of the more specialized chapters from the given track, based on their goals and on the time available in the course.

When planning a course, it should be kept in mind that the chapters vary considerably in length: Chaps. 2–5, 8, 21, 24 are longer (30 to 45 pages), while the remainder are shorter (mostly around 20 pages); the longer chapters would likely take 5–6 class days to cover in their entirety, but the shorter chapters might be comfortably covered in two to three class meetings. The instructor may choose to omit some of the sections in the core chapters at their own discretion. For example, if pursuing an optical track, the section on graphene can be omitted in Chap. 4, but for a solid state or atomic track the sections on nonlinear optics and metamaterials might be omitted from Chap. 5.

Because of the possibility of instructors following different tracks, some redundancy is built into the text, with some definitions being made multiple times in different chapters. But making sure that required background material is covered prior to its use will still imply a few constraints. For example, anyone planning on covering Sects. 23.4, 24.7–24.8, 26.3, or 25.2 should cover Chap. 16 first, since Chap. 16 covers material on quantum gates that is frequently mentioned in those other sections. Similarly, Sect. 23.1, which introduces the concepts of π and $\frac{\pi}{2}$ pulses, is a prerequisite for Sects. 24.7–24.8 and 25.3, and for Chap. 26.

Acknowledgments I would like to thank all of my friends and colleagues for their help and support over the years, including Ed Deveney, Francesca Fornasini, Ruby Gu, Gregg Jaeger, Anthony Manni, Alessandro Massarotti, Olga Minaeva, Abdoulaye Ndao, Shuto Osawa, Chris Schwarze, Alexander Sergienko. Thanks also to Nemul Khan and Tina Infanta and all of the editors at Springer Nature for their help and guidance during the preparation of this book. This book would also never have existed without the support of Marcia, Jo, and Alee, and of course the cats.

Whether a student, instructor, or researcher, I hope the reader finds this book to be interesting and useful. Please feel free to report any errors found to me at dsimon@stonehill.edu.

North Easton, MA, USA David S. Simon
August 2024

Suggested Chapters for Potential Course Tracks

Focus of course	Core chapters	Additional chapters: choose 3–5
Survey of modern scientific applications (More science, less technology)	1–3 4 (Sections 1–3, 5) 5, 8, 9	7, 10, 11, 12, 13, 14, 15, 21
Optical measurement, imaging, and sensing	1–3 4 (Sections 1–3, 5) 5, 8, 9	6, 7, 10, 11, 19, 20, 22
Optical info. processing/communication applications	1–6, 8, 9	6, 12, 13, 14, 16, 17, 18, 22
Applications to atomic, ionic, NMR systems	1 2 (Sections 1,2,4–7,9) 5 (Sections 1–7) 8, 9, 12	10, 13, 14, 16–19, 23–25
Applications to superconducting and solid-state systems	1 2 (Sections 1,2,4–7,9) 5 (Sections 1–7) 8, 9, 12	7, 10, 13, 15–17, 19, 21, 26
Quantum computing	1–4 5 (Sections 1–7)	18, 23–26

Contents

Part IV   Physical Implementations

Part I
Background and Basics

Part I
Background and Basics

Chapter 1
Introduction and Preliminaries

In this chapter, a brief overview is given, describing how quantum effects such as superposition and entanglement have stimulated advances in many areas of science and technology. In addition, some fundamental concepts such as the ideas of qubits and uncertainty are introduced.

1.1 Overview of Quantum Technology

During the first three decades or so of the twentieth century, quantum mechanics was developed and then gradually coalesced into its modern form. The new quantum formalism was revolutionary, in the sense that it removed the strict determinism that had been one of the main pillars of physics for centuries and introduced an intrinsic probabilistic underpinning into physics that could not be removed simply by making more accurate measurements. Despite the radical changes that quantum theory wrought, it was quickly accepted as a result of its repeated experimental verification and due to the fact that it gave simple, logically coherent explanations of a broad swath of previously unexplained physical phenomena. In very short order, the new quantum theory explained the photoelectric effect, the discrete spectral lines in atoms, and even the stability of atoms themselves. Within a few more decades, it led to the band theory of solids, explanations of a wide range of astrophysical and chemical phenomena, and consistent theories of superconductivity and superfluidity. When combined with relativity, it grew into a comprehensive theory of fundamental particle physics, known as the standard model. By the end of the twentieth century, quantum mechanics had become the most useful and most successful scientific theory in the history of science.

Toward the end of the twentieth century, roughly beginning in the late 1980s or early 1990s, we saw the beginning of what has sometimes been called the second quantum revolution, in which weird, nonintuitive aspects of quantum mechanics,

© The Author(s), under exclusive license to Springer Nature Switzerland AG 2025
D. S. Simon, *Introduction to Quantum Science and Technology*, Undergraduate Texts in Physics, https://doi.org/10.1007/978-3-031-81315-3_1

such as superposition and entanglement, began to be viewed as useful assets, rather than as curiosities. In particular, many of these new developments involved the idea that quantum methods could be used as a tool for improved processing of information. This allowed a wide vista of new potential effects that could be harnessed to exponentially speed up computations, to break through the classical limits on optical resolution, and to manipulate matter at the atomic level. Inherent in this new approach is the idea that quantum entanglement could be viewed as a resource that could be quantified and manipulated.

Just as the first quantum revolution had a wide range of technological spin-offs that have become ubiquitous in modern life (lasers, transistors, and microprocessors, to name just a few), the second quantum revolution has promised a new range of technologies that are likely to be equally wide-ranging. A short sampling of these technologies (some already in use, some still in development) would include the following:

- **Quantum computing** uses the ability of quantum bits (or qubits) to be in superpositions of logical 0 and 1 states, can exponentially speed up some computations, and can sometimes carry out computations that are simply not practical at all on classical devices.
- **Quantum lithography** uses entangled-photon pairs to etch microcircuitry at size scales unattainable with classical light.
- **Quantum cryptography** (**quantum key distribution**) uses the rules of quantum mechanics to ensure that secret communications can be carried out securely, making any eavesdropping attempt clearly apparent to the legitimate users of the system.
- **Quantum "ghost" imaging** allows images to be constructed from photons that never interacted with the object being viewed. This is done by monitoring their *correlations* with photons that did interact with the object, but which cannot by themselves form an image. These correlated imaging processes allow imaging beyond the classical limit and can reduce the image distortions produced by turbulence.
- **Quantum microscopy** with entangled photons allows super-resolution imaging of microscopic objects at sub-wavelength scales and can produce images through intervening turgid media.
- **Quantum sensors**: a wide variety of quantum-based sensors can provide unprecedented resolution and sensitivity in measurements of a wide range of variables, including phase differences, accelerations, rotation rates, magnetic fields, and temperature. Quantum sensors designed to detect minute variations in the gravitational field, thereby allowing detection of underground structures, are already on the commercial market, as are sensors for noninvasively mapping brain function via the measurement of tiny magnetic fields at the surface of the skull.

Quantum computing and quantum information processing are widely reviewed in a number of textbooks and review articles at a variety of levels [1–12], but finding introductory reviews at the advanced undergraduate level is more difficult for many of the other topics listed above. Exceptions include a few books such as [13] for optical applications and [14] for topological phases of matter and a variety of review articles in journals; but these are of narrower scope and a little more advanced in level, and they expect that the reader already has a strong background in the necessary fields. The goal of this book is to span an unfilled gap by providing, at the undergraduate level, a survey of these applications. To make the treatment as self-contained as possible, introductions to much of the background on required experimental and theoretical topics are also included.

A wide range of physics disciplines are covered in the following chapters, primarily because qubits and entangled states can be implemented in a number of different physical platforms, including photonic and atomic systems, as well as in multiple condensed matter systems, such as superconducting materials and crystal lattice defects. As a result, the first part of the book will review background in areas such as optics, quantum mechanics, and solid state physics, to make sure all readers are up to speed for the later portions of the book. Theoretical background is largely in Parts I and II, with more detail on experimental platforms in Part IV, while Part III discusses applications.

1.2 Measurement and Uncertainty

One of the recurring themes in the following chapters is that the use of specialized quantum states can allow measurements at higher levels of resolution and sensitivity than is possible with purely classical measurements. So before proceeding, it is useful to give a brief review of ideas and definitions related to measurement and uncertainty.

Consider a measurement of a physical variable, x. Regardless of whether x is a position, a voltage, an optical intensity, or any other physical quantity, we can make some general statements related to its measurement.

First, if the same quantity is measured multiple times under seemingly identical conditions, it is still likely that each measurement will give a slightly different value. The results are likely to be scattered over a small range of values as shown in Fig. 1.1. So, when we report our results, we really need to report two numbers. The first number gives our best estimate of the "true" value of the measured quantity, which most often is taken to be the mean $\bar{x}$ of the measurements. The second number gives a measure of the size of the spread of the data around the mean, or in other words, of the **uncertainty**, Δx. This uncertainty can come from many sources. Just to list a few:

- **Thermal fluctuations:** as molecules in a gas randomly collide or molecules in a solid exchange energy via vibrations of the crystal lattice, each individual molecule fluctuates in speed and energy. These fluctuations increase with temperature and can affect things like electrical current measurements or the phases of emitted photons.

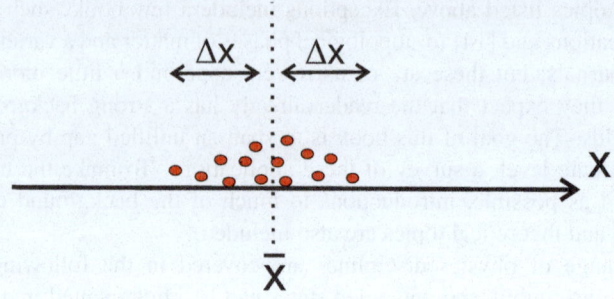

Fig. 1.1 Repeated measurement of the same variable, x, will produce a scatter of values. The mean value, $\bar{x}$, gives a measure of where the center of the distribution, while the uncertainty Δx measures the width of the region over which the points are spread

- **Shot noise:** the electrons in an electric current or the photons in a light beam are detected as discrete particles. They will arrive at a detector or other measuring device at randomly varying intervals, leading to fluctuations in measured currents or intensities.
- **Heisenberg uncertainty:** thermal noise and shot noise are universal effects that occur in both classical and quantum systems. They can be diminished by various means. But eventually, as the classical noise is removed, the purely quantum mechanical fluctuations required by the Heisenberg uncertainty principle become more important. These are of a fundamental nature and cannot be removed by using better measuring devices.

The uncertainty is most often measured by the standard deviation of the probability distribution from which the data is drawn, $\Delta x = \sigma$. (See Appendix A.3 for some basic statistics, including the definitions of the mean and standard deviation.) The results of the experiment would now be reported in the form of a **confidence interval**:

$$\bar{x} \pm \Delta x = \bar{x} \pm \sigma, \tag{1.1}$$

meaning that the "true" value will most likely fall in the range

$$\bar{x} - \sigma \leq \mu \leq \bar{x} + \sigma. \tag{1.2}$$

To be a little more precise, for most cases (for normally distributed data), the probability of being inside the confidence interval is about 68%. (The probabilities of being within two or three standard deviations of the mean are 95% and 99.7%, respectively.)

Another important quantity experimentally is often the **signal-to-noise ratio (SNR)**, which is the ratio of the uncertainty to the mean value of some measured quantity, $SNR = \frac{\bar{x}}{\Delta x}$. This is useful because it is always dimensionless, and so it gives a measure of signal quality that is independent of the type of variable being measured or of the units used to measure it.

The identification $\Delta x = \sigma$ assumed that the values of x are measured one at a time. On the other hand, we could make measurements of averages $\bar{x}$ on each of a sequence of samples of size n and then take the mean of those sample averages. We would expect the average of those sample means to vary less than single values would, since inside each sample errors of opposite sign would partially cancel. So the sample means $\bar{x}$ should have smaller uncertainties than individual values. In statistics books, the distribution of $\bar{x}$ is known as the **sampling distribution of the mean**, and it is shown that samples of size n have a standard deviation $\sigma_{\bar{x}}$ that is smaller than the single-value standard deviation σ by a factor of $\sqrt{n}$: $\sigma_{\bar{x}} = \frac{\sigma}{\sqrt{n}}$. So the confidence interval is now of the form

$$\bar{x} \pm \Delta x = \bar{x} \pm \frac{\sigma}{\sqrt{n}}. \tag{1.3}$$

As might be expected, increasing the number of measurements decreases the uncertainty.

In quantum mechanics, the act of measuring a variable causes a back reaction on the system that can affect other variables; these then, in turn, affect the measured variable, leading to a reduction in the accuracy of the measurement (Sect. 11.2). This back reaction leads to what is known as the **standard quantum limit (SQL)** [15]. In laser physics and other areas, this often coincides with the **shot noise limit**, a limit on the minimum quantum noise caused by the random arrival times of discrete photons or electrons. (Note that naming conventions of different authors are not consistent with each other, so terminology may vary.) The shot noise leads to a scaling factor of $\frac{1}{\sqrt{n}}$. For conventional measurement methods, increasing the number of measurements cannot decrease the measurement uncertainty by more than this $\frac{1}{\sqrt{n}}$ factor.

However, this standard quantum limit is not fundamental and can be evaded by clever measurement strategies. As will be seen repeatedly in later chapters, it is sometimes possible to use characteristic features of quantum mechanics (such as superposition and entanglement) to create situations where this limit is violated. These quantum-enhanced measurements can evade the standard quantum limit and achieve measurements beyond the resolution allowed by classical physics. However, at some point, the Heisenberg uncertainty principle will prevent further improvements, leading to what is known as the **Heisenberg limit**, which says that the uncertainty can't decrease with n faster than $\frac{1}{n}$. Unlike the standard quantum limit, the Heisenberg limit is a fundamental result that cannot be overcome. It provides an ultimate limit on the quality of a measurement.

As an example, consider an optical interferometer. A typical interferometer (Fig. 1.2a) splits a beam of light into two beams, sends the beams through two different paths to the same endpoint, and then recombines them. One of the beams may gain an extra phase shift ϕ relative to the other. This phase shift could be due to one beam traveling a longer distance or one beam passing through a different

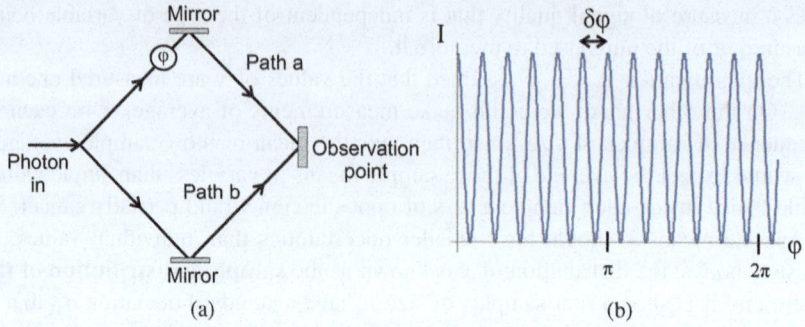

Fig. 1.2 (**a**) Schematic of a basic interferometer. Incoming light has two (or more) paths it can take to the observation point (which could consist of a viewing screen, a camera, or an optical detector). In general, the light will gain a different phase along each path, leading to interference when the beams are recombined at the observation point. (**b**) Intensity versus phase shift in an interference pattern. The minimum resolution $\delta\phi$ with which the phase can be measured is essentially determined by the distance between the intensity peaks, which act like the marks on a ruler. To achieve higher resolution, the interference pattern has to be made to oscillate more rapidly with phase

material. But when the beams are recombined, an interference pattern arises, with intensity maxima (bright bands) at points where ϕ is a multiple of 2π and minima (dark bands) at points where ϕ is an odd multiple of π. The goal is to measure the phase shift with the highest possible resolution, discerning as small a phase difference between the beams as possible.

The intensity of the interference pattern oscillates rapidly as the phase changes (Fig. 1.2b). The peaks in the oscillating pattern serve the same role as the markings on a ruler; the more closely spaced the peaks are (as a function of ϕ), the more precisely the phase can be measured.

Interferometers can be built which can operate at the individual photon level, interfering amplitudes for multiple different paths of a single photon. Increasing the number of photons in each measurement gives an improvement proportional to $\frac{1}{\sqrt{n}}$ as expected, since each photon detection is essentially an independent measurement. But n-photon entangled states can be constructed in which the photons act cooperatively, as if the n photons formed a single super-particle. In this case, the wavefunction of the n-photon states oscillates n times faster than the wavefunction of any of the individual photons. The interference pattern then oscillates faster as well, allowing the phase resolution of the measurement to improve, scaling now as $\frac{1}{n}$, and potentially reaching the Heisenberg limit. The important point is that the improvement is due to the entanglement of the photons, so this is an inherently quantum mechanical effect.

An alternate approach to evading classical limits is to squeeze the uncertainty in a variable of interest, in a sense that will be made clear in Sect. 9.2. This will allow that variable to be measured to a higher precision, but it comes at a cost: the Heisenberg principle forces the uncertainty in another variable to become larger. So, for example, the position can be measured to as high a precision as desired, but at the expense of losing any possibility of determining the momentum.

It is important to keep in mind how the uncertainty in a variable propagates through functions. Suppose $f(x)$ is some function of the variable x and that x has an uncertainty Δx. Then the uncertainty in the value of the function is found by using the chain rule from calculus:

$$\Delta f(x) = \frac{df}{dx} \Delta x. \tag{1.4}$$

If there are several measured variables x, y, z, ... with independent uncertainties Δx, Δy, Δz, ..., then the uncertainties add according to the Pythagorean theorem, like components of a vector. This is referred to as **addition of quadratures**:

$$\Delta f(x) = \sqrt{\left(\frac{\partial f}{\partial x}\Delta x\right)^2 + \left(\frac{\partial f}{\partial y}\Delta y\right)^2 + \left(\frac{\partial f}{\partial z}\Delta z\right)^2 + \ldots}. \tag{1.5}$$

Before concluding this section, we should define some terms that are often used in relation to measurement uncertainty (see [16–18] for more detail):

- **Resolution** is the smallest detail of a signal or object that can be seen distinctly. Any detail smaller than the resolution size is simply a blur. Two values of a measured variable that are closer together than the resolution size cannot be distinguished as different values, since the blurred images they give will blend into each other.
- **Sensitivity** is the smallest *change* in a variable that can be unambiguously detected with a given measurement setup.
- **Accuracy** is a measure of how close a measurement is to a "true" or accepted value.
- **Precision** is a measure of repeatability; it measures how much the same measurement is likely to vary when it is repeated multiple times.

In Fig. 1.2b, the resolution corresponds to the spacing of the peaks, while the sensitivity corresponds to the maximum slope. Clearly, resolution and sensitivity are closely related, since the higher the slope, the more peaks can be fit into a given phase range.

1.3   Qubits

One key idea in quantum mechanics is that of superposition: if two states $|\psi_1\rangle$ and $|\psi_2\rangle$ are both solutions to the Schrödinger equation, then any linear superposition of those two states, $|\psi\rangle = \alpha|\psi_1\rangle + \beta|\psi_2\rangle$, is also a legitimate solution. In a sense, the system of interest can be viewed as being in both states at the same time. The wavefunction of the system does not settle into one state or the other until interactions (a measurement, for example, or thermal interactions with the environment) force it to make a choice. Until that happens, both possibilities have to be included in all calculations.

The system can be in a superposition of any number of different mutually orthogonal states (even infinitely many), but the simplest and most common case is often when the superposition contains just two terms. For example, the system might be an electron in a superposition of spin-up and spin-down states, or it could be a photon in a superposition of horizontal and vertical polarizations. In such cases of two-state systems, we will often label the two possible states as $|0\rangle$ and $|1\rangle$, so that the superposition state will be of the form

$$|\psi\rangle = \alpha|0\rangle + \beta|1\rangle. \tag{1.6}$$

The coefficients (which may be complex) are usually chosen so that the state satisfies the normalization condition $\langle\psi|\psi\rangle = 1$; this then implies that $|\alpha|^2 + |\beta|^2 = 1$. The states $|0\rangle$ and $|1\rangle$ represent a basis for a two-dimensional Hilbert space.

Although the states $|0\rangle$ and $|1\rangle$ can represent values of any binary state variable, the choice to label them by "0" and "1" is common because it allows an analogy between quantum systems and computational systems. In computers, 0 and 1 are used as labels for the two binary states (or **bits**) that can be stored in a given memory unit. As a result, the $|0\rangle$, $|1\rangle$ basis is often called the **computational basis**. But in a classical system, the value of the bit is always well-defined: it is always one value or the other. Even if the *value* of the specific bit is not known by the user of the computer, the fact that it *has* a definite value is certain. However, in a quantum system, this is not the case. As the Bell inequalities (Chap. 14) will make clear, the notion that a quantum system is in a definite (even if unknown) state is often untenable. Superposition states are unavoidable in quantum mechanics, and so whereas classical computers operate by performing operations on classical bits, quantum computers will perform operations on quantum **qubits**, i.e., superpositions of the form of Eq. 1.6. We will see that this is what leads to many of the advantages of quantum systems. In quantum computers, for example, operations are performed on both bits of the qubit simultaneously, allowing multiple computations to effectively be carried out at the same time. Although the result of only one of these computations can be read out at the end, this set of simultaneously conducted computations offers a new possibility not present in classical systems: interference can occur between the different terms in the computer's superposition state. By judiciously measuring the resulting interference effects, new information can often be extracted from the quantum calculation that cannot be obtained from a classical computer. Examples of using these quantum interference effects for computation will be the focus of Chap. 17.

There are many physical two-state systems that can provide embodiments of the two states labeled by 0 and 1 in the qubit, so there will be many ways to physically implement the abstract qubit states. Just to list a few:

- The two basis states can be spin-up and spin-down states of a fermion: $|0\rangle = |\uparrow\rangle$ and $|1\rangle = |\downarrow\rangle$, so that the qubits are superpositions of the form $\alpha|\uparrow\rangle + \beta|\downarrow\rangle$. The fermion itself could be an electron in a crystal lattice, or it could be a nuclear spin in a nuclear magnetic resonance system.
- The qubit could also be photon polarization. The two states in the superposition could be horizontal or vertical linear polarizations, left- and right-circular polarizations, or linear polarizations along any arbitrary pair of axes at right angles to each other. Note that optical polarization is essentially a consequence of the *spin* angular momentum of the photon; qubits can also be stored in the *orbital* angular momentum states of light that will be discussed in Chap. 6.
- The two states could represent the ground state $|0\rangle$ and the first excited state $|1\rangle$ an atom or of a harmonic oscillator potential.
- In a Mach-Zehnder interferometer (Sect. 2.5), a photon can be sent into one of two input ports of the interferometer on the left, and then it will exit at one of two output ports on the right. A phase shift inserted into one arm of the interferometer can control the amplitudes for the light to exit at each of the output ports. In this case, the two possible input ports represent the basis states: the photon can enter the upper port (state $|0\rangle$) or the lower port (state $|1\rangle$), or it can be in a superposition of both. Similarly, the output state can be a superposition of the two possible exit ports, which will again represent the $|0\rangle$ and $|1\rangle$ states of the output qubit.
- Solid-state implementations of qubits also exist. For example, nitrogen-vacancy or NV centers are defects in the crystal lattices of diamond, in which one lattice site has nitrogen instead of carbon and an adjacent site is empty. This lattice defect can be either electrically neutral or electrically charged, providing the required two states needed to form a qubit. NV centers will be discussed in Sect. 26.2.

The above discussion implies a close connection between quantum computation and interferometry. Interference effects will be essential in nearly all of the applications presented in this book. For example, in Chaps. 19 and 20, methods will be presented to make optical interference patterns oscillate faster than would be the case for the corresponding classical interference patterns using the same wavelength of light; this allows resolution beyond the classical diffraction limits and forms the basis of a wide variety of high-resolution imaging and sensing systems.

Other superposition states beyond the qubit state will be important. The qubit is a superposition of two states of a *single* particle. But superpositions of *pairs* of particles can also occur and lead to additional effects. These two-particle superpositions are **entangled states**. Entanglement will be one of the key tools required for applications going beyond the capabilities of purely classical systems. Entanglement will be discussed in detail in Chaps. 3 and 13.

References

1. M. Le Bellac, *A Short Introduction to Quantum Computer Information and Quantum Computation* (Cambridge University Press, Cambridge, 2006)
2. V. Vedral, *Introduction to Quantum Information Science* (Oxford University Press, Oxford, 2007)
3. G. Jaeger, *Quantum Information: An Overview* (Springer, Berlin, 2007)
4. N.D. Mermin, *Quantum Information Science: An Introduction* (Cambridege University Press, Cambridege, 2007)
5. N.S. Yanofsky, M.A. Mannucci, *Quantum Computing for Computer Scientists* (Cambridge University Press, Cambridge, 2008)
6. S.M. Barnett, *Quantum Information* (Oxford University Press, Oxford, 2009)
7. M.A. Nielsen, I.L. Chuang, *Quantum Computation and Quantum Information: 10th Anniversary Edition* (Cambridge University Press, Cambridge, 2011)
8. J.A. Jones, D. Jaksch, *Quantum Information, Computation, and Communication* (Cambridge University Press, Cambridge, 2012)
9. M.M. Wilde, *Quantum Information Theory*, 2nd edn. (Cambridge University Press, Cambridge, 2017)
10. J.D. Hidary, *Quantum Computing: An Applied Approach*, 2nd edn. (Springer, Berlin, 2021)
11. J.A. Bergou, M. Hillery, M. Saffman, *Quantum Information Processing: Theory and Implementation*, 2nd edn. (Springer, Berlin, 2021)
12. T.G. Wong, *Introduction to Classical and Quantum Computing* (Rooted Grove Press, New York City, 2022)
13. D.S. Simon, G. Jaeger, A.V. Sergienko, *Quantum Metrology, Imaging, and Communication* (Springer, Berlin, 2016)
14. R. Moessner, J.E. Moore, *Topological Phases of Matter* (Cambridge University Press, Cambridge, 2021)
15. V.B. Braginsky, F.Ya. Khalili, *Quantum Measurement* (Cambridge University Press, Cambridge, 1992)
16. I. Hughes, T. Hase, *Measurements and their Uncertainties: A Practical Guide to Modern Error Analysis* (Oxford University Press, Oxford, 2010)
17. S.V. Gupta, *Measurement Uncertainties: Physical Parameters and Calibration of Instruments* (Springer, Berlin, 2012)
18. A. Possolo, J. Meija, *Measurement Uncertainty: A Reintroduction* (National Research Council of Canada, Ottawa, 2022)

Chapter 2
Classical Optics and Optical Devices

Light plays a central role in many applications of physics to real-world tasks. In particular, it is one of the chief tools for quantum metrology and quantum communication, and for reasons to be discussed in later chapters, it is a promising physical mechanism for implementing various quantum information processing tasks. So, this chapter gives an overview of classical optics, with particular emphasis to aspects of classical light which will have close analogs in quantum mechanics; since both quantum mechanics and optics are based on linear wave equations, there are many classical optical phenomena that can be used to simulate quantum-like behavior. Furthermore, polarization in a classical optical beam is mathematically identical to spin in a spin-$\frac{1}{2}$ quantum system. So, both spin and polarization are two-state systems that can serve as physical implementations of quantum qubits. Similarly, the classical optical law of Malus can be viewed as the result of quantum state projections induced by measurements. For these and other reasons, it is profitable to treat quantum mechanics and optics in parallel.

2.1 Electromagnetic Waves and Light

Classically, light is described as an electromagnetic wave: it is composed of an electric field and a magnetic field, with both fields oscillating in orthogonal directions with a common frequency. Let ρ and J be the charge density (charge per volume) and current density (current per area). Then recall that Maxwell's equations in matter can be written in differential form as follows:

$$\nabla \cdot D = \frac{\rho}{\epsilon_0} \qquad \text{(Gauss' law)} \qquad (2.1)$$

$$\nabla \cdot B = 0 \qquad \text{(Absence of magnetic monopoles)} \qquad (2.2)$$

© The Author(s), under exclusive license to Springer Nature Switzerland AG 2025
D. S. Simon, *Introduction to Quantum Science and Technology*, Undergraduate
Texts in Physics, https://doi.org/10.1007/978-3-031-81315-3_2

$$\nabla \times E = -\frac{\partial B}{\partial t} \qquad \text{(Faraday's law)} \tag{2.3}$$

$$\nabla \times H = J + \frac{\partial D}{\partial t} \qquad \text{(Ampere-Maxwell law).} \tag{2.4}$$

The magnetic field and the electric displacement, H and D, are related to the magnetic inductance and electric field, B and E, by

$$D = \epsilon E, \qquad B = \mu H, \tag{2.5}$$

where ϵ and μ are the permittivity and permeability of the material. ϵ and μ in the material are related to the corresponding vacuum values by the relative permittivity (or dielectric constant) ϵ_r and relative permeability μ_r:

$$\epsilon = \epsilon_r \epsilon_0 \qquad \text{and} \qquad \mu = \mu_r \mu_0. \tag{2.6}$$

Consider vacuum in a region of space where there are no charges or currents, $J = \rho = 0$. Maxwell's equations then reduce to

$$\nabla \cdot E = 0 \tag{2.7}$$

$$\nabla \cdot B = 0 \tag{2.8}$$

$$\nabla \times E = -\frac{\partial B}{\partial t} \tag{2.9}$$

$$\nabla \times B = \mu_0 \epsilon_0 \frac{\partial E}{\partial t}. \tag{2.10}$$

Then by combining these equations pairwise (see Problem 1), wave equations can be derived for each component of the electric and magnetic fields:

$$\left(\nabla^2 - \epsilon_0 \mu_0 \frac{\partial^2}{\partial t^2} \right) E_j = 0 \tag{2.11}$$

$$\left(\nabla^2 - \epsilon_0 \mu_0 \frac{\partial^2}{\partial t^2} \right) B_j = 0. \tag{2.12}$$

Comparison to the standard wave equation given in every introductory physics textbook, $\left(\nabla^2 - \frac{1}{c^2} \frac{\partial^2}{\partial t^2} \right) \psi(x, t) = 0$, where c is the speed of the wave, indicates that all components of E and B travel at the same speed, namely, at

$$c = \frac{1}{\sqrt{\epsilon_0 \mu_0}}. \tag{2.13}$$

When the numerical values of ϵ_0 and μ_0 are inserted, this gives the result

$$c = 299792458 \frac{m}{s} \approx 3 \times 10^8 \frac{m}{s},$$

which is exactly the speed of light! This fact was the first indication that light was in fact an electromagnetic wave.

Inside a material, the vacuum values ϵ_0 and μ_0 must be replaced by the values in the material, giving a reduced optical speed:

$$v = \frac{1}{\sqrt{\epsilon\mu}} = \frac{1}{\sqrt{\epsilon_r\epsilon_0\mu_r\mu_0}} = \frac{c}{n}, \tag{2.14}$$

where the index of refraction is defined by

$$n = \sqrt{\epsilon_r\mu_r}. \tag{2.15}$$

In nonmagnetic materials, $\mu_r \approx 1$, so that we may simply write

$$n = \sqrt{\epsilon_r}. \tag{2.16}$$

The fact that $n \geq 1$ ensures that $v \leq c$ inside matter. But note two caveats to this. The first is that inside a conductor, the refractive index will be complex-valued at some frequencies, $n = n_r + in_i$. But the wave has a phase factor of the form e^{ikx}, where the wavenumber $k = nk_0$ is proportional to the refractive index and the wavenumber in vacuum. So while the real part n_r of n gives an oscillatory contribution, $e^{ik_r x} = e^{in_r k_0 x}$, the imaginary part n_i leads to an exponentially decaying factor, $e^{i(ik_i)x} = e^{-n_i k_0 x}$, indicating absorption. For example, this is why metals are always opaque at optical frequencies where the imaginary part is nonzero; at higher frequencies (usually with a threshold in the ultraviolet), the index becomes real and the metal will allow the waves to pass. Complex n is a complication we won't worry about any further here; see any standard text on solid-state physics or on the optical properties of materials. The second caveat can be found in the box below.

Box 2.1 Negative Refractive Index Material

A **metamaterial** [1, 2] is an artificially engineered material, designed to have a desired set of optical or conductive properties. Metamaterials are often constructed from nanoscale resonators or arrays of nanoscale dielectric scattering elements. One particular category of metamaterials that will have significant technological impact in coming years is **negative refractive index** metamaterials.

The possibility of negative index materials was explored by Veselago in 1967 [3], but began to be taken seriously after further theoretical develop-

(continued)

Box 2.1 (continued)

ments and experiments more than 30 years later. Pendry et al. [4] and Shelby et al. [5] showed that practical implementations are possible.

Recall that the refractive index is $n = \sqrt{\epsilon_r \mu_r}$. Normally, ϵ_r and μ_r are taken to be positive, but Veselago considered the possibility of taking them both negative. Consider $\epsilon_r = \mu_r = -1$. Care has to be taken with signs of imaginary parts to get sensible outcomes. So write ϵ_r and μ_r in exponential form. This can be done with two possible signs in the exponent:

$$\epsilon_r = \mu_r = e^{\pm i\pi}. \tag{2.17}$$

The correct prescription requires all imaginary parts to be positive, so that the plus signs must be used. The result then is a negative refractive index:

$$n = \sqrt{\epsilon_r}\sqrt{\mu_r} = e^{i\frac{\pi}{2}} \cdot e^{i\frac{\pi}{2}} = (i)^2 = -1. \tag{2.18}$$

As a result of the negative index, light bends the "wrong" way in Snell's law (Fig. 2.1a). Among other implications, this means that lenses can be made from flat slabs of negative-index material (Fig. 2.1b). Without the need for curved surfaces, lenses can be made very compact for integrated optical circuits and other applications.

One other important implication of negative-index materials is that inside them evanescent waves are amplified over distance, rather than decaying. This opens up new possibilities for near-field imaging and allows sub-wavelength resolution. For an introductory review of negative-index metamaterials and their applications, see [6]. See also Sect. 5.10 of this book.

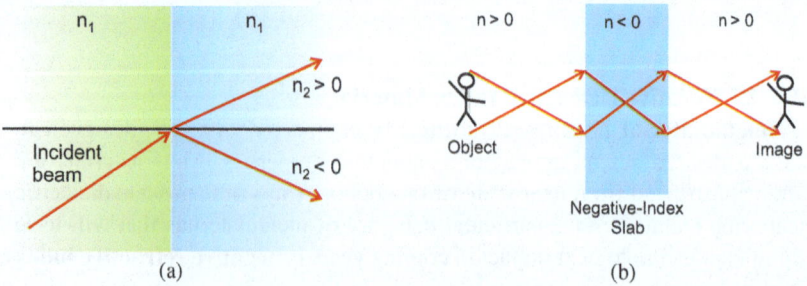

(a)   (b)

Fig. 2.1 (**a**) Snell's law: the refracted light bends in opposite directions for $n > 0$ and $n < 0$. (**b**) The refraction of light in the "wrong" direction in a negative-index slab allows perfect images to be created, with no aberration or loss of small-scale detail

2.2 Light Propagation

The wave equations 2.11 and 2.12 apply to each component of E and B separately. Let us focus on the electric field. Further, let us assume for simplicity that the light is linearly polarized, so that the direction of E is constant. Then 2.11 will also apply to the magnitude or **field strength**, $E_{phys}(r, t) = \sqrt{E_x^2 + E_y^2 + E_z^2}$,

$$\nabla^2 E_{phys} - \frac{1}{c^2} \frac{\partial^2 E_{phys}}{\partial t^2} = 0. \tag{2.19}$$

It will be convenient to allow the electric field to be treated as complex, so that the real, physically measurable electric field strength E_{phys} will be the real part of the complex field E: $E_{phys} = \mathrm{Re}(E)$. If the light is monochromatic (a single frequency), then the complex field will have a simple time dependence:

$$E(r, t) = E(r)e^{-i\omega t}; \tag{2.20}$$

the wave equation then reduces to the **Helmholtz equation**, which controls the spatial dependence:

$$\left(\nabla^2 + k^2\right) E(r) = 0, \tag{2.21}$$

where $k = \frac{\omega}{v}$. The intensity of the light at a given point is then given by $I(r) = |E(r)|^2$.

Two particularly useful solutions to the Helmholtz equation are those for plane waves:

$$E(r) = E_0\, e^{\pm i k \cdot r}, \tag{2.22}$$

and spherical waves:

$$E(r) = E_0\, \frac{e^{\pm ikr}}{kr}. \tag{2.23}$$

(Notice that for the spherical wave, the exponent is the product of the magnitudes of k and r, not the dot product of the vectors.) The plus and minus signs in the exponents represent right- and left-moving waves, respectively, for the plane wave, and they represent waves moving outward or inward from a source at the origin in the spherical case.

Recall that the time-independent Schrödinger equation is $H\psi(x) = E\psi(x)$. Split the Hamiltonian (the energy operator) into kinetic and potential energy parts, $H = \frac{\hat{p}^2}{2m}\nabla^2 + V(r) = -\frac{\hbar^2}{2m} + V(r)$, where the three-dimensional momentum operator is $\hat{p} = -i\hbar\nabla$. Using this Hamiltonian, the Schrödinger equation can then be rearranged into the form

$$\left(\nabla^2 + \frac{2m}{\hbar^2}\left(E - V(r)\right)\right)\psi(r) = 0, \tag{2.24}$$

or equivalently,

$$\left(\nabla^2 + k^2\right)\psi(r) = 0, \tag{2.25}$$

which is the same form as Eq. 2.21. Here, the wavenumber is given by $k = \sqrt{\frac{2m}{\hbar^2}\left(E - V(r)\right)}$. So the Helmholtz equation is often used in experiments to provide an optical simulation of quantum systems, with the electric field playing the role of the wavefunction. Since k is proportional to the refractive index,

$$k = \frac{2\pi}{\lambda} = \frac{2\pi}{\lambda_0}n, \tag{2.26}$$

where λ_0 is the wavelength in vacuum, a spatially varying index of refraction $n(r)$ essentially plays the role of the spatially dependent potential $V(r)$.

If only light staying close to the axis is measured, then we can simplify the Helmholtz equation by considering only **paraxial solutions**. The paraxial approximation is valid if both the field and its derivative along the z-axis vary sufficiently slowly. To be more precise, write the field in terms of an envelope function $A(r)$:

$$E(r) = A(r)e^{ik_z z}. \tag{2.27}$$

Then the paraxial Helmholtz equation,

$$\nabla_\perp^2 A + 2ki\,\partial_z A = 0, \tag{2.28}$$

will hold provided that

$$\frac{\partial A}{\partial z} << kA, \quad \frac{\partial^2 A}{\partial z^2} << k^2 A. \tag{2.29}$$

(Here, the transverse Laplacian is defined by $\nabla_\perp^2 = \frac{\partial^2}{\partial x^2} + \frac{\partial^2}{\partial y^2}$.)

Suppose we know the distribution of light in some initial object plane (taken to be the xy-plane at $z = 0$), and we want to know the distribution of light in some image plane (taken to be the xy-plane intersecting the z-axis at some nonzero z). We assume the light is propagating freely through a medium of refractive index n (Fig. 2.2). Suppose the light has wavelength λ inside the material. Then the magnitude of the **wavevector k** is called the **wavenumber** and is given by

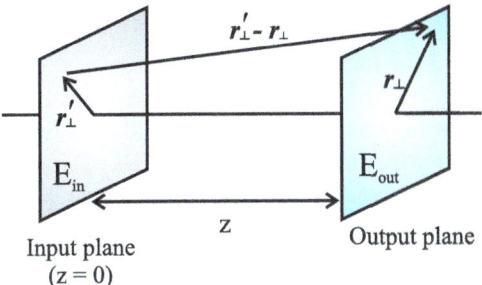

Fig. 2.2 Free propagation of light from an input plane (or object) to an output (or image) plane. z is the longitudinal distance between the planes, while $\boldsymbol{r}'_\perp$ and $\boldsymbol{r}_\perp$ are the initial and final position vectors in the transverse (xy) plane. Each point of the final plane receives light from every point in the initial plane, so that the output consists of the propagated light emitted from each input point, integrated over all points in the initial plane (Eq. 2.35)

$$k = \frac{\omega}{v} = \frac{2\pi}{\lambda} = \frac{2\pi n}{\lambda_0} = nk_0, \tag{2.30}$$

where λ_0 and k_0 are the values in free space $(n = 1)$: $\lambda = \frac{\lambda_0}{n}$, $k = nk_0$. Let $\boldsymbol{r}_\perp$ and $\boldsymbol{k}_\perp$ denote the position and wavevector in the transverse (xy) plane, with magnitudes $r_\perp = \sqrt{x^2 + y^2}$ and $k_\perp = \sqrt{k_x^2 + k_y^2}$. Propagation is simpler in momentum space than in coordinate space, so we start with the momentum-space propagation first.

The z-component of the wavevector, k_z (often called the **propagation constant** and written as β), can be written as

$$k_z = \sqrt{k^2 - k_x^2 - k_y^2} = \sqrt{k^2 - k_\perp^2}, \tag{2.31}$$

so for fixed wavelength (fixed magnitude k), the wavevector really only has two independent degrees of freedom, k_x and k_y. Let $E_{in}(\boldsymbol{k})$ be the Fourier transform of the electric field in the initial plane:

$$E_{in}(\boldsymbol{k}) = \frac{1}{\sqrt{2\pi}} \int E(\boldsymbol{r}_\perp) e^{-i\boldsymbol{k}\cdot\boldsymbol{r}_\perp} d^2 r_\perp. \tag{2.32}$$

Similarly, the Fourier-transformed field in the final plane is $E_{out}(\boldsymbol{k})$. Each point in the object plane sends out a Huygens wavelet that propagates along the z-axis, with each momentum component traveling in a straight line, gaining a phase proportional to the distance traveled. So the fields in the two planes are related by

$$E_{out}(\boldsymbol{k}) = H(\boldsymbol{k}) E_{in}(\boldsymbol{k}), \tag{2.33}$$

where the free-space **transfer function** $H(\boldsymbol{k})$ is

$$H(k) = e^{izk_z} = e^{iz\sqrt{k^2 - k_x^2 - k_y^2}}. \tag{2.34}$$

One important thing to notice here is that not all waves will propagate to the final plane. If $k_\perp^2 < k^2$, then the exponential is an oscillatory function and the waves will propagate to arbitrarily large distances. But if $k_\perp^2 > k^2$, the square root in the exponent becomes imaginary, and $H(k)$ becomes an exponentially decaying function of z. These **evanescent waves** don't propagate, but are important in various settings, such as near-field imaging and during the coupling of light into integrated optical circuits.

Undoing the Fourier transforms, we have the coordinate-space relation

$$E_{\text{out}}(r_\perp) = \int h(z, r_\perp - r'_\perp) E_{\text{in}}(r'_\perp) d^2 r'_\perp, \tag{2.35}$$

where r here is used for the two-dimensional position vector in the transverse plane, $r_\perp = (x, y)$. The function $h(z, r_\perp, r'_\perp)$ is called the **impulse response function** or the **point spread function (PSF)** and is the Fourier transform of the transfer function:

$$h(z, r_\perp) = \frac{1}{\sqrt{2\pi}} \int H(k) e^{ik \cdot r_\perp} d^2 k. \tag{2.36}$$

Note that Eq. 2.35 is saying that the final image is simply the convolution of the initial image with the PSF. We have taken advantage of the fact that free space is translation invariant, which means that h is only a function of position *differences*, $h(r, r') = h(r - r')$: the physics is completely unchanged if all the position vectors are shifted by a fixed constant.

The basic idea of Eq. 2.35 is that the value of something (in this case, electric field) at a final point can be found by seeing how much of it (represented by h) propagates to the final point from every possible initial point, and then summing or integrating over all those initial points. This method is used in many areas of science and engineering, and in particular, it is often useful in quantum mechanics and quantum field theory. See, for example, the action of the time-evolution operator on a quantum wavefunction in Eq. 3.52. Because similar methods are used in so many fields, there is wide variation in terminology. What we are referring to as the impulse response function may, in other contexts, be called a *propagation function*, a *propagator*, an *integration kernel*, or a *Green's function*. Here we are discussing propagation through free space, but we can define response functions of optical instruments as well, which take input fields to output fields. For instance, we will see the impulse response function for a lens in the next section.

The exact form of the PSF is given by the Huygens-Fresnel principle, which says that each point in the object emits a spherical wave that propagates outward in all directions; the field at any image point is then the sum over the spherical waves emitted by all the object points. Letting r now represent the full three-dimensional position vector, $r = (r_\perp, z)$, then

$$h(\boldsymbol{R}) = \frac{i}{\lambda R} e^{-ikR}, \tag{2.37}$$

where

$$R = |\boldsymbol{r} - \boldsymbol{r}'| = \sqrt{z^2 + (x - x')^2 + (y - y')^2}. \tag{2.38}$$

The square root in R often makes it impossible to perform the integral in Eq. 2.35 exactly. So approximations are usually necessary.

Assume z is much larger than any of the transverse distances, and then start with a binomial expansion of R:

$$R = z + \frac{(x - x')^2 + (y - y')^2}{2z} - \frac{((x - x')^3 + (y - y')^3)}{8z^3} + \dots, \tag{2.39}$$

dropping terms higher than quadratic:

$$R \approx z + \frac{(x - x')^2 + (y - y')^2}{2z} \tag{2.40}$$

$$= z + \frac{x^2 + y^2}{2z} - \frac{x \cdot x' + y\, y'}{z} + \frac{x'^2 + y'^2}{2z}. \tag{2.41}$$

Keeping just the first three of the four terms on the right leads to the **Fraunhofer or far-field approximation**, while keeping all four gives the **Fresnel or near-field approximation**. The Fraunhofer approximation holds when the observation point is far enough from the object that the wavefronts reaching it are approximately planar. To be more quantitative, define the Fresnel number:

$$F = \frac{a^2}{\lambda z}, \tag{2.42}$$

where a^2, the maximum value of $x'^2 + y'^2$ in the input plane, is essentially the size of the object being viewed. Then the near field corresponds to $F \geq 1$, while regions with $F << 1$ lie squarely in the far field.

The resulting output fields are then given by

$$E_{\text{out}}^{(\text{Fraun})}(\boldsymbol{r}) = \frac{1}{i\lambda z} e^{ik\left(z + \frac{x^2 + y^2}{2z}\right)} \int E_{\text{in}}(\boldsymbol{r}') e^{-\frac{ik}{z}(x\, x' + y\, y')} dx' dy', \tag{2.43}$$

$$E_{\text{out}}^{(\text{Fresnel})}(\boldsymbol{r}) = \frac{1}{i\lambda z} e^{ikz} \int E_{\text{in}}(\boldsymbol{r}')\, e^{\frac{ik}{2z}[(x-x')^2 + (y-y')^2]} dx' dy'. \tag{2.44}$$

Notice that, aside from some overall constants, the Fraunhofer output is simply the Fourier transform of the input:

$$E_{\text{out}}^{(\text{Fraun})}(r) = C\tilde{E}_{\text{in}}\left(\frac{kr}{2\pi z}\right) = C\tilde{E}_{\text{in}}\left(\frac{r}{\lambda z}\right), \tag{2.45}$$

where $C = \frac{\sqrt{2\pi}}{i\lambda z}e^{ik\left(z+\frac{x^2+y^2}{2z}\right)}$. Light carries out an analog computation of the Fourier transform, simply by propagating to the far field!

Unfortunately, it takes propagation to a fairly large distance to reach the far field, on the order of 1000 m for visible light when passing through an aperture of a few centimeters. However, the far field can be reached in a much shorter distance by placing a converging lens in the light path, as will be seen in the next section.

2.3 Lenses and Imaging

A lens is simply a piece of transparent material with two curved surfaces. Light passing through it refracts at each surface in a manner determined by the curvature of the surfaces. Consider light rays entering the lens parallel to the axis. For a converging lens, all the rays will pass through the axis at a single point, called the **focal point**, F (Fig. 2.3, top). The distance f from the lens to the focal point is the **focal length** and will be taken positive for a converging lens. For a diverging lens, one must trace the rays backward to find the focal point (Fig. 2.3, bottom), and the focal length is then taken to be negative. Here, for simplicity, we consider only the case where the lens is thin compared to its focal length, and where the two curved surfaces are spherical, with radius of curvature R. We will also assume for simplicity that $|R|$ is the same for both surfaces of the lens. In the following, we attach a sign to R, which will be equal to the sign of f.

Consider light passing through a transparent slab of material of refractive index n and of variable thickness, $d(x, y)$. Then it is not hard to show (see, e.g., [7, 8]) that the impulse response function in the paraxial case can be written as

Fig. 2.3 Top: A converging lens bends light inward, producing a focal point F on the side of the outgoing light. The focal length f is positive. Bottom: A diverging lens bends light outward, so the rays must be traced backward (the dashed segments) to find the focal point, which occurs on the incoming side of the lens. The focal length is negative

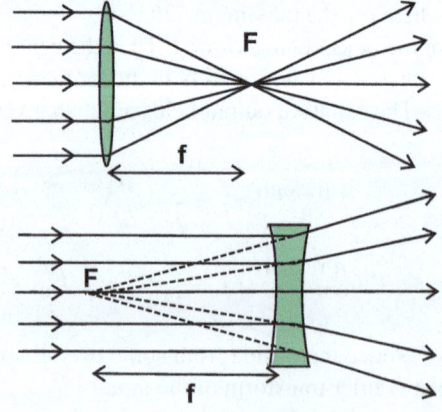

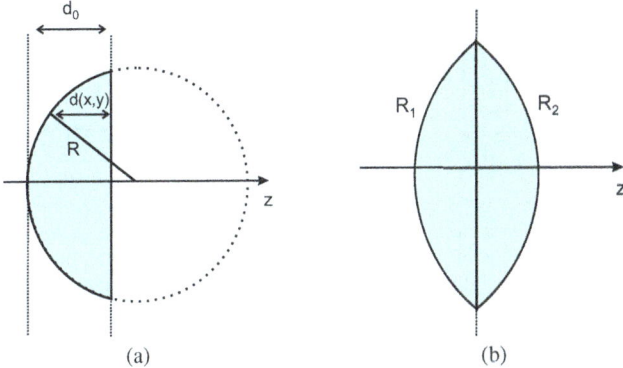

Fig. 2.4 (**a**) A plano-convex lens, with radius of curvature R. (**b**) A doubly convex lens with radii R_1 and R_2

$$h(x, y) = e^{ik_0 d_0} e^{i(n-1)k_0 d(x,y)}, \tag{2.46}$$

where d_0 is the maximum width of the lens at its thickest point and $k_0 = \frac{2\pi}{\lambda_0}$ is the wavenumber in the surrounding air.

For the plano-convex lens shown in Fig. 2.4a, assume that the radius of the spherical surface R is large compared to the lens d_0. Then the function $d(x, y)$ can be approximated (see Problem 3) as

$$d(x, y) \approx d_0 - \frac{x^2 + y^2}{2R}. \tag{2.47}$$

Plugging this into Eq. 2.46, we find that the impulse response of the lens is

$$h(x, y) = e^{ink_0 d_0} e^{-\frac{ik_0}{2f}(x^2 + y^2)}, \tag{2.48}$$

where the focal length is given by

$$f = \frac{R}{n - 1}. \tag{2.49}$$

Similarly, for a lens with two spherical surfaces of radii R_1 and R_2 (Fig. 2.4b), Eq. 2.48 holds with the focal length now given by

$$\frac{1}{f} = (n - 1) \left(\frac{1}{R_1} - \frac{1}{R_2} \right). \tag{2.50}$$

Consider the system shown in Fig. 2.5a, where the object and image planes are both one focal length from a lens. This so-called **2f system** produces the Fourier

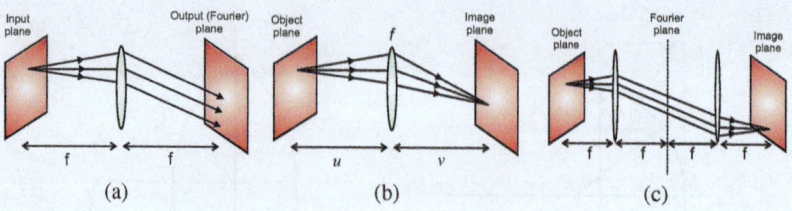

Fig. 2.5 (**a**) A 2f optical system produces a Fourier-transformed version of the input at the final focal plane. (**b**) A single lens of focal length f produces an image when the image and object distances obey the imaging condition $\frac{1}{u} + \frac{1}{v} = \frac{1}{f}$. The image is magnified by a factor of $m = -\frac{v}{u}$. If $m < 0$, the image is inverted; otherwise, it is upright. (**c**) A 4f system carries out two successive Fourier transforms, resulting in an inverted image of the input

transform of the object as output at the image plane; in other words, it acts like a Fraunhofer system but without requiring the distance to the final plane to be in the far field. By multiplying the two free-field response functions (before and after the lens) with the response function of the lens and integrating over the input plane, we find that the output of the 2f system is

$$E_{\text{out}}(x, y) = -\frac{i}{\lambda f} e^{ik(d+f)} e^{2ikf} \tilde{E}_{\text{in}}\left(\frac{x}{\lambda f}, \frac{y}{\lambda f}\right), \tag{2.51}$$

or more compactly,

$$E_{\text{out}}(\boldsymbol{r}_\perp) = -\frac{i}{\lambda f} e^{ik(d+f)} e^{2ikf} \tilde{E}_{\text{in}}\left(\frac{\boldsymbol{r}_\perp}{\lambda f}\right). \tag{2.52}$$

Aside from an overall phase and a decrease in amplitude of $\frac{1}{\lambda f}$, the result is just a Fourier transform.

One of the principal uses of a lens is to form images. One way of forming an image is with a single lens. Consider the arrangement shown in Fig. 2.5b, where the object is a distance u from a lens of focal length f and an image is formed at some distance v. As shown in every introductory physics book, the image and object distances for the single lens system are related by the **imaging condition**:

$$\frac{1}{u} + \frac{1}{v} = \frac{1}{f}. \tag{2.53}$$

The image is magnified relative to the object by a factor

$$m = -\frac{v}{u}. \tag{2.54}$$

Another way of forming lenses is with a **4f system**. This just consists of two concatenated 2f systems with the input plane at one end and the output at the other

(Fig. 2.5c). The system performs two consecutive Fourier transforms, which simply gives back the original input, but (as verified in Problem 6) spatially inverted and decreased in amplitude:

$$E_{\text{out}}(\boldsymbol{r}) = \left(\frac{1}{\lambda f}\right)^2 E_{\text{in}}(-\boldsymbol{r}). \qquad (2.55)$$

One advantage of such a system is that optical filters can be placed in the Fourier plane in order to provide processing of the image. For example, high spatial-frequency noise can be filtered out.

2.4 Resolution

One of the most important measures of quality for an optical system is its **resolution**, or its ability to accurately distinguish small structures. Resolution can be quantified by giving a minimum resolvable size scale or, more commonly, a minimum resolvable angular size.

The reason resolution is limited in optical devices is that light has to be collected through some aperture, which will have a finite size (Fig. 2.6). As a result, interference between light waves that diffracted at different parts of the aperture will interfere, and the resulting interference will smear out the smaller details of an image. Considering only the case of a circular aperture, which is by far the most common case, the diffraction pattern looks as shown in Fig. 2.7a, b. Let the radius of the aperture be a. The bright spot in the center is called the **Airy disk**, and the intensity at distance r from the center of the aperture axis can be given in terms of the first Bessel function, $J_1(r)$,

$$I = I_0 \left(\frac{2J_1(ka \sin \alpha)}{ka \sin \alpha}\right)^2, \qquad (2.56)$$

Fig. 2.6 Light passing through an aperture of radius a strikes an observation plane, forming a diffraction pattern whose intensity at distance r from the axis depends on the distance R from the aperture and the opening angle α

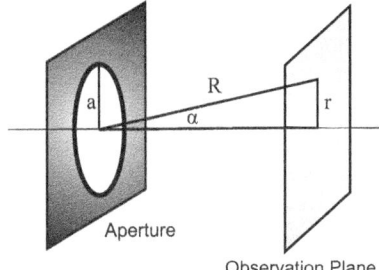

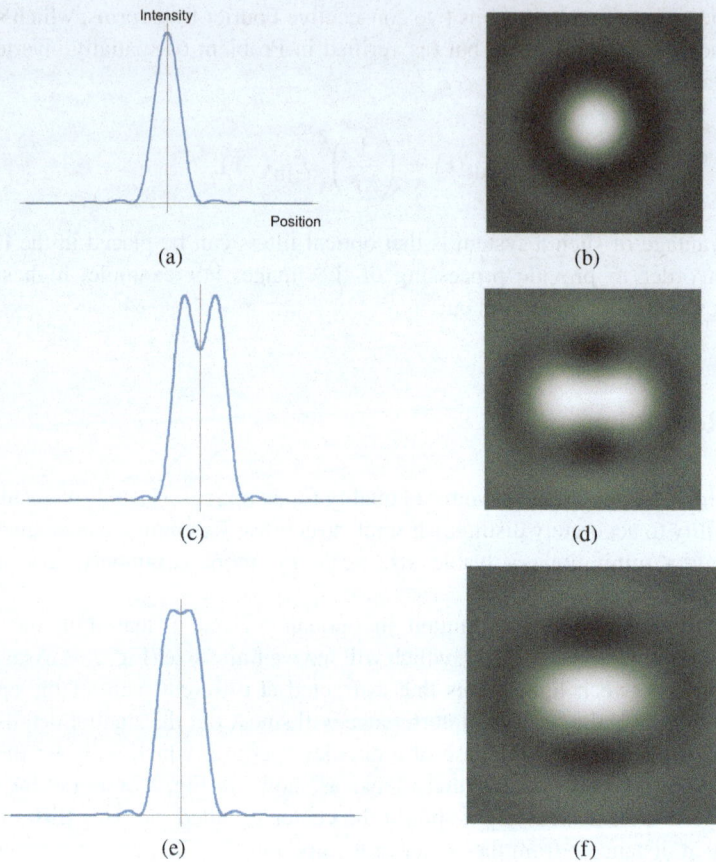

Fig. 2.7 (**a**) and (**b**): An Airy disk. (**c**) and (**d**): Two Airy disks near the Rayleigh resolution limit. (**e**) and (**f**): Two Airy disks near the Abbé resolution limit. In the right column, the darker areas represent lower intensities

where I_0 is the maximum intensity at the center of the Airy disk and $k = \frac{2\pi}{\lambda}$ is the wavenumber. The wavelength in the material, λ, is related to the wavelength in vacuum, λ_0, by $\lambda = \frac{\lambda_0}{n}$.

To discuss resolution, begin by making a few definitions. Consider an optical instrument (a lens, microscope, telescope, etc.) with light entering it through some aperture, and imagine that the instrument is viewing an object that is small compared to the distance between it and the viewing device. Then define an **angular radius** α as the $\frac{1}{2}$-angle subtended by the device as viewed from the object position (Fig. 2.6). It is often convenient to write $\sin \alpha = \frac{r}{R}$, where R is the distance from the aperture to the observation plane. Then the **numerical aperture** NA is defined by

$$NA = n \sin \alpha, \tag{2.57}$$

where n is the refractive index of the medium that the light is traversing. The resolution should improve as either of two conditions change:

(i) As the aperture size increases, diffractive effects become less important, leading to an improved resolution.
(ii) As the index n increases, the wavelength of the light becomes smaller for fixed frequency, again leading to improved resolution.

But both of these changes also increase the numerical aperture as well. So in general, we expect better resolution for higher values of NA. Typical values for a microscope are roughly $1.0 \lesssim NA \lesssim 1.5$. Often oil is used as the propagation medium in high-power microscopes in order to increase n and NA. For devices in which the light is collected through an optical fiber, the numerical aperture is usually much smaller, $0.05 \lesssim NA \lesssim 0.4$.

Both Rayleigh (1896) and Abbé (1873) gave quantitative estimates of resolution in terms of NA. They arrived at slightly different results, because they used different criteria for what counts as "resolvable," but the difference is small enough that it is often irrelevant. Imagine two point sources of light, such as two distant stars, or two fluorescing molecules. Each produces an Airy diffraction pattern as in Fig. 2.7b. As the two point sources move closer together, the Airy patterns begin to overlap, and eventually they overlap so much that it is impossible for the observer to tell whether they are viewing a single Airy disk or if they are viewing two that have begun to merge. In terms of NA, the intensity pattern of Eq. 2.56 for a single source can be rewritten as

$$I = I_0 \left(\frac{2 J_1 \left(\frac{2\pi NA\, r}{\lambda_0} \right)}{\left(\frac{2\pi NA\, r}{\lambda_0} \right)} \right)^2. \tag{2.58}$$

Then the **Rayleigh criterion** says that the limit of resolution occurs when the distance between the Airy disks is small enough that the first dark rings in the two interference patterns coincide or, equivalently, that the Airy disks are just beginning to overlap. In terms of the angular distance between the points, this means that the minimum resolvable angular distance is

$$\theta_{min} \approx \frac{1.22\lambda_0}{2\,a}. \tag{2.59}$$

The factor 1.22 occurs because the first minimum of the Bessel function $J_1(\frac{kar}{R}) \approx J_1(ka \sin\theta)$ occurs at $ka \sin\theta \approx 3.83$, so that

$$\theta \approx \sin\theta \approx \frac{3.83}{ka} = \frac{3.83\lambda_0}{2\pi a} = \frac{1.22\lambda_0}{2a}. \tag{2.60}$$

In terms of minimum resolvable distance between the objects, $\Delta x = 2R \sin\theta$, the Rayleigh limit becomes

$$\Delta x_{min} \approx \frac{1.22\lambda_0}{NA}. \tag{2.61}$$

The **Abbé criterion** is a little less restrictive and considers the two Airy disks to be distinguishable until the intensity dip between the two maxima becomes unobservable. According to this criterion, the minimum resolvable angle and distance are

$$\theta_{min} \approx \frac{1.0\lambda_0}{2a}, \qquad\qquad \Delta x_{min} \approx \frac{1.0\lambda_0}{2NA}. \tag{2.62}$$

One other resolution criterion that is often used in astronomy is the **Sparrow criterion**, which gives the resolution limit

$$\Delta x_{min} \approx \frac{0.47\lambda_0}{NA}. \tag{2.63}$$

These classical resolution limits make several assumptions, including plane wave illumination and simple intensity measurements at the detection side. In recent decades, many methods have been developed that violate these assumptions. As a result, there are now a number of methods to carry out super-resolution imaging, in which resolution surpasses the Rayleigh and Abbé limits. See [9] for a review. As will be seen in Chap. 20, the use of entangled (quantum-correlated) states is one way to achieve super-resolution imaging.

Example 2.4.1 The Hubble Space Telescope collects light using a mirror of diameter of $D = 2.40$ m. Here, the radius of the mirror plays the role of the aperture, so $a = \frac{D}{2} = 1.20$ m. The minimum angular separation at which the telescope can resolve two distinct stars using light of wavelength $\lambda = 550$ nm is then given by

$$\theta_{min} \approx \frac{1.22\lambda}{2a} = \frac{(1.22)(5.5 \times 10^{-7} \text{ m})}{2.4 \text{ m}} = 2.8 \times 10^{-7} \text{rad} \approx 1.6 \times 10^{-5} \text{ degrees.} \tag{2.64}$$

Therefore, when viewing a pair of objects at a distance of $r = 10 \ ly = 9.461 \times 10^{16} m$ away from the earth, the objects can be seen to be distinct as long as they are not within a distance

$$\Delta x = r\theta_{min} = (9.461 \times 10^{16} \text{m})(2.8 \times 10^{-7} \text{ rad}) = 2.65 \times 10^{10} \text{m}. \tag{2.65}$$

of each other.

For comparison, under ideal conditions, ground-based telescopes have resolutions of $\theta_{min} \sim 1$ arcsec $\approx 4.85 \times 10^{-6}$rad or a bit less, giving roughly an order of magnitude less resolution. In the presence of significant atmospheric turbulence, the advantage of space-based telescopes relative to ground-based telescopes increases further. ∎

2.5 Beam Splitters

One of the most useful devices in optics is the **beam splitter (BS)**. Light incident on the beam splitter has some probability T to be transmitted through the beam splitter, and some probability R of being reflected, as shown in Fig. 2.8. Assuming that negligible light is lost in the beam splitter, the transmission and reflection probabilities have to sum to 100%, $R + T = 1$. As a result, it is often convenient to parameterize the beam splitter by an angle θ, defined so that $R = \cos^2 \theta$ and $T = \sin^2 \theta$. One important aspect of beam splitters is that they can be used even at the single photon level: a single photon entering the BS has fixed amplitudes to transmit or reflect, and it exits in a *superposition* of both states. Such superposition states will be discussed in more detail in Chap. 3.

Beam splitters can be made various ways. For example, the dielectric beam splitter cubes often found in optics labs are formed by gluing two glass or clear plastic prisms together, with a thin layer of adhesive resin between their joined surfaces. Another type of beam splitter is the half-silvered mirror, consisting of a thin, partially transparent layer of aluminum or other metal on a substrate of glass or clear plastic.

If the glass prisms in the dielectric beam splitter are replaced by birefringent crystals, a **polarizing beam splitter (PBS)** can be formed. Here, the reflection and transmission probabilities depend on polarization. In the most common case, one of the polarizations (e.g., horizontally polarized) transmits with near-100% probability, while the orthogonal polarization (vertical) reflects with near certainty.

To describe the action of the beam splitter on incoming light, label the four sides of the beams splitter by the numbers 1–4, with the light entering from sides 1 and 2, and exiting at sides 3 and 4. Then a generic input state can be written in column vector form as

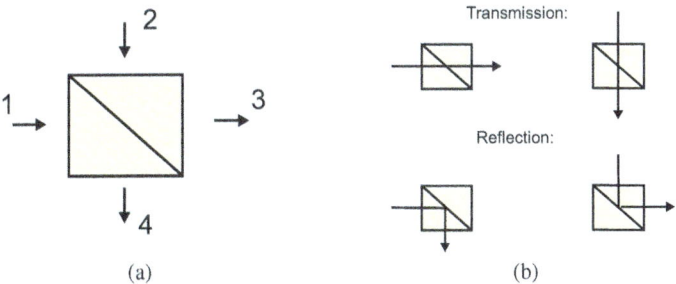

(a)   (b)

Fig. 2.8 (**a**) A beam splitter has two entrance ports (labeled 1 and 2 here) and two exit ports (3 and 4). Light entering each input port has fixed proportions exiting the two outputs. If light enters both inputs, then the light at each exit port will be a combination of the two inputs. (The light can also be reversed, entering at 3 and 4, and exiting at 1 and 2.) (**b**) Each photon entering an input port has an amplitude to transmit or reflect. In general, the output will be a superposition of both possibilities

$$E_{\text{in}} = \begin{pmatrix} E_1 \\ E_2 \end{pmatrix}, \tag{2.66}$$

where $E_{1,2}$ are the field amplitudes entering input ports 1 and 2. Similarly, the state of the field exiting the BS can be written in terms of the amplitudes at output ports 3 and 4 as

$$E_{\text{out}} = \begin{pmatrix} E_3 \\ E_4 \end{pmatrix}. \tag{2.67}$$

A 2×2 matrix can then be defined that describes the action of the beam splitter and tells how the input is transformed into the output:

$$E_{\text{out}} = U_{BS} E_{\text{in}}. \tag{2.68}$$

The matrix U_{BS} has to be unitary (see Sect. 3.2.3 for the definition of unitarity), which puts constraints on the form of its entries. The most general expression for the beam splitter matrix is

$$U_{BS} = e^{i\phi_0} \begin{pmatrix} \sin\theta \ e^{i\phi_1} & \cos\theta \ e^{-i\phi_2} \\ \cos\theta \ e^{i\phi_2} & -\sin\theta \ e^{-i\phi_1} \end{pmatrix}, \tag{2.69}$$

where θ is the same parameter defined above. Of particular interest are 50–50 beam splitters, with have equal transmission and reflection probabilities, $T = R = 50\%$; in this case, $\theta = \frac{\pi}{4}$, so that U_{BS} reduces to

$$U_{BS} = \frac{1}{\sqrt{2}} e^{i\phi_0} \begin{pmatrix} e^{i\phi_1} & e^{-i\phi_2} \\ e^{i\phi_2} & -e^{-i\phi_1} \end{pmatrix}. \tag{2.70}$$

Two common special cases can be obtained by making particular choices of the phase shifts. The dielectric beam splitter formed from glass wedges has $\phi_0 = \phi_1 = \phi_2 = 0$, which leads to

$$U_{BS} = \frac{1}{\sqrt{2}} \begin{pmatrix} 1 & 1 \\ 1 & -1 \end{pmatrix}. \tag{2.71}$$

On the other hand, the *symmetric* beam splitter has $\phi_0 = -\phi_1 = \frac{\pi}{2}$ and $\phi_2 = 0$, leading to

$$U_{BS} = \frac{1}{\sqrt{2}} \begin{pmatrix} 1 & i \\ i & 1 \end{pmatrix}. \tag{2.72}$$

The beam splitter action above was written in terms of the classical field E. But in the single-photon case, E gets replaced by the single-photon creation operator $\hat{a}^\dagger$

(Chap. 5). The input-output relation, Eq. 2.68, remains unchanged, with the same forms for U_{BS}. It should be noted that the beam splitter mixes the two inputs and produces outputs that have those inputs superposed in proportions determined by the parameter θ. Conversely, if there is input at only one port, the output consists of a superposition of amplitudes at *both* outputs. It should also be noted for future use that when the light is reflected, it experiences an extra phase shift of $i = e^{\frac{i\pi}{2}}$, relative to the transmitted beam.

Example 2.5.1 In practice, it usually doesn't matter which form you use for the beam splitter matrix, since the different beam splitter matrices can be transformed into each other by adding phase shifts at the input and output ports. For example, the dielectric version can be transformed into the symmetric version simply by inserting phase shifters with phase $e^{\frac{i\pi}{2}} = i$ at input 2 and output 4:

$$U_{BS}^{\text{diel}} = \begin{pmatrix} 1 & 1 \\ 1 & -1 \end{pmatrix} \rightarrow \begin{pmatrix} 1 & 0 \\ 0 & i \end{pmatrix} \begin{pmatrix} 1 & 1 \\ 1 & -1 \end{pmatrix} \begin{pmatrix} 1 & 0 \\ 0 & i \end{pmatrix}$$

$$= \begin{pmatrix} 1 & i \\ i & 1 \end{pmatrix} = U_{BS}^{\text{symm}}. \quad \blacksquare \qquad (2.73)$$

Beam splitters are *directionally biased* in the sense that light that enters in one direction cannot turn around and exit from the same port. So, despite having four ports, beam splitters are really two-dimensional devices: they take two input modes to two output modes. It is possible to devise directionally unbiased devices, which allow exit from any port, including the input port. Fabry-Perot and Michelson interferometers are both examples of two-port directionally unbiased devices. Three-port and four-port unbiased devices have recently been investigated [10–12]; these have been found to produce new quantum interference effects [13, 14], to provide control over the distribution of entanglement in optical networks [15], and (by replacing beam splitters with unbiased multiports inside the interferometer) to produce enhanced capabilities of standard interferometers [16, 17].

2.6 Simple Interferometers

Recall that the absolute phase of a field can't be measured. But phase *differences* are measurable, and one major application of beam splitters is to form interferometers, which can be used to measure such differences. One example is the **Mach-Zehnder (MZ) interferometer** shown in Fig. 2.9. Light is input to ports 1 and 2. The beam splitter sends out linear combinations of the two inputs at ports 3 and 4. The light at port 3 propagates unchanged through the lower path to point 5. Meanwhile, the light at point 4 is passed through a phase shifter in the upper branch (e.g., a thin piece of glass), before entering the second BS at port 6. The second BS then mixes the beams at 5 and 6 to produce the final output at ports 7 and 8. We can easily trace

Fig. 2.9 A Mach-Zehnder interferometer. Light enters ports 1 and 2, exiting at ports 7 and 8. Two input beams mix at the beam splitters, and the phase shift ϕ of the upper branch relative to the lower branch determines the fraction of light exiting each of the two output ports

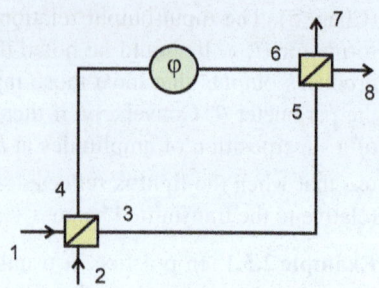

through the interferometer to find the output by associating a matrix to each step in the process. We already know the beam splitter matrix (we will use the symmetric version, Eq. 2.72, and assume the beam splitters are both 50–50). The phase shift acts only on the upper branch, leaving anything on the lower branch unchanged, and so the phase shift can be expressed as

$$\begin{pmatrix} e^{i\phi} & 0 \\ 0 & 1 \end{pmatrix}. \tag{2.74}$$

So, the output can be written in terms of the input fields by

$$\begin{pmatrix} E_8 \\ E_7 \end{pmatrix} = \left[\frac{1}{\sqrt{2}} \begin{pmatrix} 1 & i \\ i & 1 \end{pmatrix} \right] \begin{pmatrix} e^{i\phi} & 0 \\ 0 & 1 \end{pmatrix} \left[\frac{1}{\sqrt{2}} \begin{pmatrix} 1 & i \\ i & 1 \end{pmatrix} \right] \begin{pmatrix} E_1 \\ E_2 \end{pmatrix} \tag{2.75}$$

$$= \frac{1}{2} \begin{pmatrix} (e^{i\phi} - 1)E_1 + i(e^{i\phi} + 1)E_2 \\ i(e^{i\phi} + 1)E_1 - (e^{i\phi} - 1)E_2 \end{pmatrix}. \tag{2.76}$$

This can be rewritten in a more suggestive form:

$$\begin{pmatrix} E_8 \\ E_7 \end{pmatrix} = i\, e^{\frac{i\phi}{2}} \begin{pmatrix} \sin\frac{\phi}{2} & \cos\frac{\phi}{2} \\ \cos\frac{\phi}{2} & -\sin\frac{\phi}{2} \end{pmatrix} \begin{pmatrix} E_1 \\ E_2 \end{pmatrix}. \tag{2.77}$$

In this form, it is clear that (up to an overall phase shift) the interferometer is performing a rotation in the space of electric fields.

Notice that if light is only entering through port 1 (so $E_2 = 0$), then the output becomes

$$\begin{pmatrix} E_8 \\ E_7 \end{pmatrix} = \frac{1}{2} \begin{pmatrix} (e^{i\phi} - 1)E_1 \\ i(e^{i\phi} + 1)E_1 \end{pmatrix} = i\, e^{\frac{i\phi}{2}} \begin{pmatrix} \sin\frac{\phi}{2} \\ \cos\frac{\phi}{2} \end{pmatrix} E_1. \tag{2.78}$$

For $\phi = 0$, all of the light then exits at port 7:

$$\begin{pmatrix} E_8 \\ E_7 \end{pmatrix} = i \begin{pmatrix} 0 \\ E_1 \end{pmatrix}, \tag{2.79}$$

while for $\phi = \pi$, all the light exits at port 8:

$$\begin{pmatrix} E_8 \\ E_7 \end{pmatrix} = - \begin{pmatrix} E_1 \\ 0 \end{pmatrix}. \tag{2.80}$$

So the Mach-Zehnder acts as a phase-controlled switch: we can direct the outgoing light to a desired output direction, simply by choosing the appropriate value for ϕ. Or, more generally, we can send any desired fraction of the incoming light out each of the two exit ports, so that we can also think of the MZ interferometer as a sort of controlled beam splitter. This point of view has applications in areas such as the design of optical neural networks or the implementation of general unitary transformations with meshes of MZ interferometers.

We can also go in the opposite direction, taking ϕ as an unknown variable, instead of as a controllable parameter. Then, by measuring the amount of output at each of the two ports, we can invert the above equations to determine the unknown phase. This gives a means of measuring small phase shifts to a high degree of accuracy. As will become apparent in later chapters, the MZ interferometer plays a central role in optical quantum information processing.

The **Michelson interferometer**, which was instrumental in disproving the existence of the luminiferous ether theory around the turn of the twentieth century, is another important example of interferometry that is still widely used today; for example, it was used in the first successful detection of gravitational waves in 2015 (see Sect. 9.3). The goal of the Michelson interferometer is to measure length differences between two paths. The basic setup is shown in Fig. 2.10. Light is input at point A, and split at the beam splitter, so that half the light travels toward mirror M_1 and the other half heads toward mirror M_2. The distances from the BS to the two mirrors are L_1 and L_2. The two light beams reflect back to be mixed with each other at the beam splitter. Part of the light exits back out port A and part exits at B.

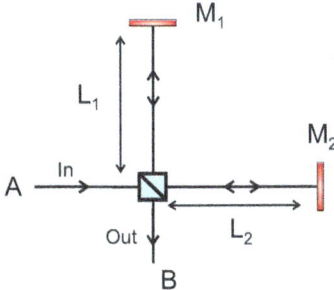

Fig. 2.10 Michelson interferometer. A beam splitter sends portions of the light input at A into each arm. After reflection at the mirrors M_1 and M_2, the two portions are recombined at the beam splitter. Measurement of the interference at port B then allows the path length difference $\Delta L = L_2 - L_1$ to be measured

Output at either port could be measured, but usually the output at B is used, since the interference fringes there have greater visibility than those at port A [18].

Suppose light of amplitude E_0 is sent into port A. The beam splitter sends amplitude $\frac{1}{\sqrt{2}}E_0$ toward mirror M_1 and amplitude $\frac{i}{\sqrt{2}}E_0$ toward M_2. The round trips back to the beam splitter add phases to these, so that the amplitudes returning to the BS are $\frac{1}{\sqrt{2}}E_0 e^{2ikL_2}$ and $\frac{i}{\sqrt{2}}E_0 e^{2ikL_1}$. Feeding these both back into the beam splitter, the output at port B is then

$$E_B = \frac{i}{\sqrt{2}}\left(\frac{1}{\sqrt{2}}E_0 e^{2ikL_2}\right) + \frac{1}{\sqrt{2}}\left(\frac{i}{\sqrt{2}}E_0 e^{2ikL_1}\right) \tag{2.81}$$

$$= \frac{i}{2}E_0\left(e^{2ikL_2} + e^{2ikL_1}\right) \tag{2.82}$$

$$= \frac{i}{2}E_0 e^{2ik(L_1+L_2)}\left(e^{ik(L_2-L_1)} + e^{ik(L_1-L_2)}\right) \tag{2.83}$$

$$= iE_0 e^{2ik(L_1+L_2)}\cos k\Delta L, \tag{2.84}$$

where $\Delta L = L_2 - L_1$. The intensity at B is proportional to $\cos^2 k\Delta L$, so the output provides a means of measuring small path length differences ΔL. This can be generalized to allow for different indices of refraction in each branch (so that the k's in the two arms are different) or additional phase shifts ϕ_1 and ϕ_2 in each arm. Then small index differences and phase differences, Δn and $\Delta\phi$, can be measured with the same arrangement.

Another useful example of an interferometer is the **Sagnac interferometer** [19, 20], which is used to measure rotations. Consider the apparatus shown schematically in Fig. 2.11. Imagine that the whole setup is allowed to rotate about an axis perpendicular to the page. The incident beam is split into two parts by BS_2, and these parts propagate around the loop in opposite directions. If the loop is rotating

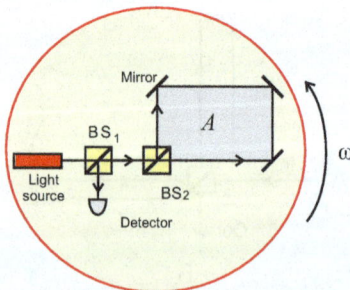

Fig. 2.11 A Sagnac interferometer. The second beam splitter splits the incident beam into two parts, which propagate in opposite directions around the loop. These beams then are recombined and directed to the detector at the beam splitter BS_1. When the system is rotating at angular speed ω, there will be a phase shift between the two interfering beams. This shift will be proportional to the rotation rate, ω

at angular speed ω, then the beams in the two directions take different times to make a complete loop and reach the detector. The interferometer loop is drawn as a rectangle in Fig. 2.11; the behavior is independent of the shape of the loop, so for simplicity, assume that the optical loop is actually circular, of radius r. Then the time difference between the clockwise and counterclockwise paths can be shown to be

$$\delta t = \frac{4A\omega}{c^2 - r^2\omega^2} \approx \frac{4A\omega}{c^2}, \qquad (2.85)$$

where A is the area enclosed by the loop. BS_2 then recombines the two beams, while BS_1 directs the recombined beam to the detector. (Half of the intensity is lost at each passage through BS_1, but as long as the beam is still intense enough to detect, that doesn't matter.) This time difference between the counter-propagating beams then leads to a phase difference at the detector of

$$\delta\phi = \omega\,\delta t \approx \frac{8\pi A\omega}{\lambda c}. \qquad (2.86)$$

The phase difference can be measured in various ways (such as measuring the displacement of interference fringes or by putting a MZ interferometer at the output). Since $\delta\phi$ is proportional to ω, measurement of the phase shift amounts to a measurement of the rotation rate about the axis perpendicular to the plane of the interferometer. By taking the area of the loop to be larger, the phase can be made larger for the same ω, increasing the sensitivity to small rotation rates. The Sagnac effect has a number of applications to fiber-optical gyroscopes [21], measurement of gravitational effects [22], and quantum sensing [23].

Interference can only occur if the beams being combined are **coherent**. Coherence means that the beams have a stable phase relationship; if the relative phase of the two interfering beams is rapidly fluctuating, then the fluctuations will destroy the interference pattern. In a thermal light source, like the sun or a heated gas, each atom emits photons independently of the other, at random times. Because of this randomness, the coherence of thermal light sources tends to be low. In contrast, the photons in a laser are emitted in a coordinated fashion that gives the emitted beam high coherence.

Generally, interferometers work by splitting a beam, making the two daughter beams take different paths, and recombining them on a screen, a camera, or other viewing device. If the paths are of different lengths, then the light in the two beams could have been produced at different times. We find that if the time difference between the emissions of the two interfering beam segments is too large, the interference pattern decays and becomes impossible to see. The maximum time difference that allows visible interference is called the **coherence time**, τ_c, of the light source. This time defines a distance along the beam, called the **longitudinal coherence length**, $L_c = c\tau_c$.

In a similar manner, if the optical wavefronts that are interfering came from parts of the beam that were too far apart in the transverse direction, the interference will also be destroyed. The maximum transverse separation is called the **transverse coherence length**, Δx_c. For example, consider the Young two-slit experiment. If two slits separated by distance D are illuminated with light of transverse coherence length Δx_c, interference will only be seen if $D \lesssim \Delta x_c$

A more quantitative discussion of coherence in terms of correlation functions will be given in Chap. 10.

2.7 Linear Polarization and Jones Vectors

From a quantum mechanical point of view, light is composed of a collection of **photons**, discrete quanta, or bundles of energy carried by excitations of the electromagnetic field. As will be discussed in more detail in Chap. 6, photons are spin-1 particles, carrying one unit of spin angular momentum (in units of $\hbar$). In general, a spin S particle would be expected to have $2S + 1$ possible spin states, with z-components running from $m_z = -S\hbar$ to $m_z = +S\hbar$. But photons are massless, $m = 0$, as they must be in order to move at the speed of light. For reasons that would take us too far afield to explain here (see any introductory book on quantum field theory or gauge theory, such as [24–26]), massless particles are missing one spin component. So the photon has only two orthogonal spin states. These are called **polarization** states and the polarization direction is defined to be the direction of the electric field vector. Because photons have two polarization states and electrons have two spin states relative to the z-axis, there is a direct analogy or isomorphism between photon polarization and electron spin which allows them to be treated in the same manner mathematically; we will often take advantage of this fact in later chapters.

For propagating light, the allowed polarization states are always perpendicular to the direction of motion of the light. The direction of motion is conventionally called z (the **longitudinal** direction), which means the polarization vector is in the **transverse** or xy-plane. If the light is linearly polarized, we can take any two orthogonal directions in the xy-plane as a basis for the polarization states. So for example, the x- and y-axes themselves are often used. Then light with electric field along the x-axis is called **horizontally polarized**, and the orthogonal polarization along the y-axis is called **vertically polarized**. An alternative basis could be the set of directions in the transverse plane at $45°$ to the x- and y-axes. In that case, light polarized along the line at $+45°$ above the x-axis is called **diagonally polarized**, while light with electric field along the line at $-45°$ is called **anti-diagonally polarized**. The electric fields in the xy-basis are equal superpositions of the two polarizations in the diagonal/anti-diagonal basis (and vice versa):

$$E_H = E_x = \frac{1}{\sqrt{2}}(E_D + E_A) \qquad\qquad E_D = \frac{1}{\sqrt{2}}(E_H + E_V) \quad (2.87)$$

$$E_V = E_y = \frac{1}{\sqrt{2}}(E_D - E_A) \qquad\qquad E_A = \frac{1}{\sqrt{2}}(E_H - E_V). \quad (2.88)$$

More generally, if $\hat{n}$ is a unit vector at angle θ from the x-axis, the directions parallel and perpendicular to $\hat{n}$ can be used as a basis for the polarization states and are related to the fields in the xy-basis by

$$E_\| = E_H \cos\theta + E_V \sin\theta \qquad\qquad (2.89)$$

$$E_\perp = E_H \sin\theta - E_V \cos\theta. \qquad\qquad (2.90)$$

The two spin states of the photon actually correspond to states of **circular polarization**, in which the electric field rotates about the z-axis as it propagates. The two circular polarization states are called **left-circular** and **right-circular**; if the light is heading toward you, then these are, respectively, the states in which you see the field rotating clockwise or counterclockwise. Once again circular polarization and linear polarization states can be written as linear combinations of each other:

$$E_R = \frac{1}{\sqrt{2}}(E_H + iE_V) \qquad\qquad E_L = \frac{1}{\sqrt{2}}(E_H - iE_V) \qquad (2.91)$$

In a linearly polarized state, the two basis components E_x and E_y are oscillating in phase with each other so that they both reach a minimum or a maximum at the same time. The direction of field stays constant over time. But in a circularly polarized state, the E_x and E_y are out of phase by $\frac{\pi}{2}$, with one leading the other. As a result, one component will reach a maximum, while the other is passing through zero; the ratio of the two components will not be constant over time, and so the angle from the x-axis, $\tan\theta = \frac{E_y}{E_x} = \frac{E_V}{E_H}$, will rotate. More generally, if the amplitudes of the components along the two axes are unequal, then the light will be **elliptically polarized**.

Most natural light sources produce unpolarized light, with the direction of the field fluctuating randomly, whereas laser light is generally polarized. Polarized light can also be produced from unpolarized light either by reflection at the Brewster angle or by passing the light through a material that selectively absorbs one polarization and transmits the other. We'll come back to this below.

For a monochromatic plane wave traveling along the z-axis, we can always write the electric field in the form $\boldsymbol{E} = E_x\hat{x} + E_y\hat{y}$, where

$$E_x(z, t) = A_x \cos(\omega t - kz + \phi_x), \qquad\qquad (2.92)$$

$$E_y(z, t) = A_y \cos(\omega t - kz + \phi_y), \qquad\qquad (2.93)$$

where A_x and A_y are the amplitudes, while ϕ_x and ϕ_y are the phases of the components at $t = z = 0$. Going over to complex fields, these become

$$E_x(z, t) = A_x e^{i(\omega t - kz + \phi_x)}, \qquad\qquad (2.94)$$

$$E_y(z, t) = A_y e^{i(\omega t - kz + \phi_x)}. \tag{2.95}$$

Physically, only the difference $\phi_y - \phi_x$ matters, not ϕ_x and ϕ_y separately. If this difference fluctuates randomly, then the light will be unpolarized. The light will be polarized if this phase difference is stable.

To describe the polarization state, it is common to use **Jones vectors**. Discarding the $e^{i(\omega t - kz)}$, which is the same for both field components and therefore irrelevant for our current purposes, the Jones vector is given by

$$J = \begin{pmatrix} J_x \\ J_y \end{pmatrix} = \frac{1}{N} \begin{pmatrix} A_x e^{i\phi_x} \\ A_y e^{i\phi_y} \end{pmatrix}, \tag{2.96}$$

where the constant N is a normalization factor given by the square root of the intensity, $N = \sqrt{I}$. N is chosen so that $J^\dagger \cdot J = |J_x|^2 + |J_y|^2 = 1$. (Some authors use an unnormalized version with $N = 1$, so that $J^\dagger \cdot J = I$.)

Then linear polarization corresponds to the case where the initial phases are equal, $\phi_x = \phi_y$. For monochromatic light, this guarantees that the total phases of the components, $\omega t - kz + \phi$, are equal for all time. Then the light is linearly polarized at an angle $\theta = \tan^{-1}\left(\frac{A_y}{A_x}\right)$, and the electric field simply becomes periodically larger and smaller along this direction over time (Fig. 2.12). The Jones vector for a linearly polarized state at angle θ is given by

$$J_{\text{lin}} = \begin{pmatrix} \cos\theta \\ \sin\theta \end{pmatrix}. \tag{2.97}$$

As special cases, $A_y = 0$, $A_x = 0$, $A_x = A_y$, and $A_x = -A_y$ lead to horizontally and vertically polarized, diagonal, and anti-diagonal cases, respectively, with Jones

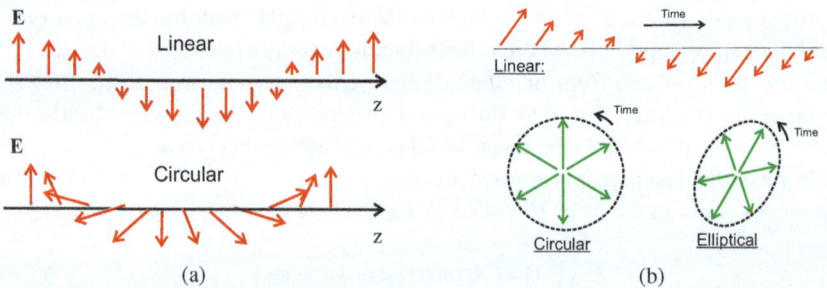

(a)   (b)

Fig. 2.12 (**a**) For linearly polarized light, the electric field becomes larger and smaller, and then flips polarity, periodically and along a single line, without rotating at all. Circular and elliptical polarizations have electric fields that rotate about the propagation axis. (**b**) The fields as viewed with the light propagating directly out of the page. For linear polarization, the field is drawn at a single point but at multiple times, with the time running horizontally. For circular and elliptical cases, the field at a point rotates over time, as indicated by the black arrows

vectors:

$$A_y = 0 \; (\theta = 0) \longrightarrow J_H = \begin{pmatrix} 1 \\ 0 \end{pmatrix} \tag{2.98}$$

$$A_x = 0 \; \left(\theta = \frac{\pi}{2}\right) \longrightarrow J_V = \begin{pmatrix} 0 \\ 1 \end{pmatrix} \tag{2.99}$$

$$A_x = A_y \; \left(\theta = \frac{\pi}{4}\right) \longrightarrow J_D = \frac{1}{\sqrt{2}} \begin{pmatrix} 1 \\ 1 \end{pmatrix} \tag{2.100}$$

$$A_y = -A_y \; \left(\theta = -\frac{\pi}{4}\right) \longrightarrow J_A = \frac{1}{\sqrt{2}} \begin{pmatrix} 1 \\ -1 \end{pmatrix}. \tag{2.101}$$

For elliptical polarization, $\phi_x - \phi_y = \pm\frac{\pi}{2}$, so that the two components are $1/4$ of a cycle out of phase, causing one component to be a minimum when the other is at a maximum. So as z and t advance with propagation of the light, the angle of the electric field from the x-axis increases linearly. For $\phi_x - \phi_y = -\frac{\pi}{2}$, the E rotates counterclockwise when looking toward the light source (right-handed elliptical), while for $\phi_x - \phi_y = +\frac{\pi}{2}$, it rotates clockwise (left-handed elliptical). (Be aware that many authors use the opposite definitions for right- and left-elliptical polarizations.) For the special case $A_x = A_y$, the elliptical polarization becomes circular. Jones vectors for right- and left-circular polarization are

$$J_R = \frac{1}{\sqrt{2}} \begin{pmatrix} 1 \\ i \end{pmatrix} \qquad J_L = \frac{1}{\sqrt{2}} \begin{pmatrix} 1 \\ -i \end{pmatrix}. \tag{2.102}$$

Note that the Jones vectors for the three pairs of polarizations $\{H, V\}$, $\{D, A\}$, and $\{R, L\}$ each define bases for two-dimensional spaces of polarization states. The first two bases are related by a simple 45° rotation (an orthogonal transformation), while they are each related to the third basis by a unitary transformation (see Sect. 3.2.3 for definitions of orthogonal and unitary transformations.) All three bases are orthonormal; in terms of the Jones matrices, the orthogonality can be expressed as

$$J_H^\dagger \cdot J_V = J_D^\dagger \cdot J_A = J_R^\dagger \cdot J_L = 0. \tag{2.103}$$

Suppose linearly polarized light is incident on a surface that serves as a boundary between two media of refractive indices n_1 (on the incident side) and n_2. The fraction of the incident field and intensity that is reflected and the fraction that is transmitted into the second medium is given by the Fresnel equations (see Appendix B). Given these equations, it can be shown starting from Snell's law that when the incident and transmitted angles obey the relation $\theta_1 + \theta_2 = \frac{\pi}{2}$, then all of the reflected light is polarized perpendicular to the incident plane. The incident angle at which this happens is called the **Brewster angle** or **polarizing angle**:

Fig. 2.13 At the Brewster angle,

$\theta_1 = \theta_B = \tan^{-1}\left(\frac{n_2}{n_1}\right)$,

incident and refracted angles obey $\theta_1 + \theta_2 = \frac{\pi}{2}$, and the reflected light is completely polarized, with its electric field perpendicular to the incident plane

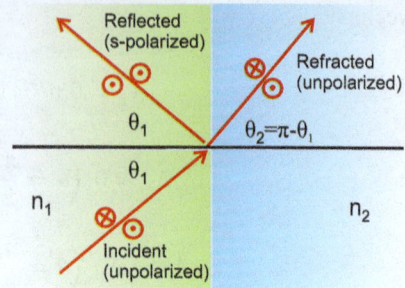

$$\tan \theta_B = \frac{n_2}{n_1}, \tag{2.104}$$

as in Fig. 2.13. This provides a simple means of polarizing initially unpolarized light. Note that the refracted light still contains both polarizations. As the incident angle moves away from the Brewster angle, the proportion of reflected light that is p-polarized smoothly increases.

The other means of polarizing unpolarized light is by transmission through a medium that is opaque to one polarization and transparent to the perpendicular polarization. To see how this works, suppose a material has molecules in which electrons can move easily in one direction (the y-direction, say), but have much more limited mobility in the x-direction. Then when light that is polarized in the y-direction passes through a collection of these molecules, the electrons can easily follow the oscillations of the electric field. Therefore, the energy of the light is converted to kinetic energy of the electrons and the y-polarized light is absorbed. On the other hand, light polarized in the x-direction can only cause a small amount of electron movement and so can only lose a small amount of energy. Thus, the x-polarized light is transmitted. Initially, unpolarized light entering the material becomes x-polarized during passage through it. Such a material is referred to as **dichroic**. Many naturally occurring materials, such as tourmaline, exhibit this type of behavior, as do artificially produced materials such as polaroid filters.

A sheet of dichroic material can be used as a **polarizer** or **polarizing filter**, selectively passing one polarization component and blocking the orthogonal component. An ideal polarizer would perfectly block 100% of the undesired polarization and pass 100% of the desired one. Real-world polarizers never quite achieve this ideal, but we will henceforth assume that all polarizers mentioned are close approximations to the ideal case.

Consider unpolarized light of intensity I_0 incident on a polarizing filter that is oriented to pass horizontally polarized light (Fig. 2.14). Since on average perfectly unpolarized light should be equally likely to be vertically or horizontally polarized, the light exiting the filter should have half the initial intensity, $I_1 = \frac{1}{2}I_0$. This exiting light should all be horizontally polarized. Now suppose that the light encounters a second polarizer, oriented to pass light polarized at angle θ from the horizontal. Then the **law of Malus** tells us that the intensity of light exiting the second filter

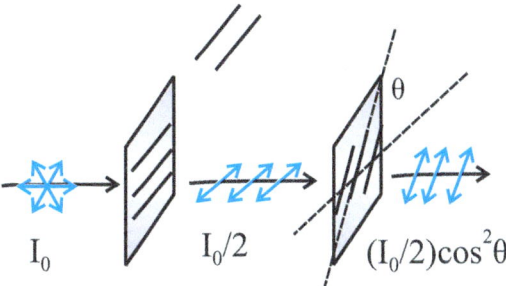

Fig. 2.14 Initially, unpolarized light has its intensity cut in half by a polarizing filter. Light exiting the filter is now polarized in the direction singled out by the filter. Subsequent polarizers will each cut the intensity by a factor of $\cos^2 \theta$, where θ is the angle between the incoming light's polarization vector and the direction of the filter. At the end, the surviving light will be polarized in the direction of the last filter

should be

$$I_2 = I_1 \cos^2 \theta. \tag{2.105}$$

The surviving light will now be polarized at θ from the horizontal. Each additional polarizing filter will again decrease the intensity by another factor of $\cos^2 \theta$, where at each step, θ is the angle between the light striking the filter and the preferred polarization direction passed by the filter. Although the law of Malus dates back to the early nineteenth century, we will see later that it is essentially a consequence of the laws of quantum mechanics.

We can describe the action of the polarizing filter by a two-by-two **Jones matrix** T that acts on the incoming Jones vector, $\boldsymbol{J}_{in}$,

$$\boldsymbol{J}_{\text{out}} = T \, \boldsymbol{J}_{\text{in}}. \tag{2.106}$$

Example 2.7.1 For example, the matrices

$$T_H = \begin{pmatrix} 1 & 0 \\ 0 & 0 \end{pmatrix}, \qquad T_V = \begin{pmatrix} 0 & 0 \\ 0 & 1 \end{pmatrix} \tag{2.107}$$

represent linear polarization filters that allow passage of horizontal and vertical polarizations, respectively.

More general polarization filters can be arrived at by rotating one of the matrices given above. For example, consider a filter polarized at angle θ from the horizontal. Then the Jones matrix of the polarizer is given by applying rotation matrices to the horizontal filter:

$$T_\theta = R(\theta) T_H R^{-1}(\theta)$$

$$= \begin{pmatrix} \cos\theta & -\sin\theta \\ \sin\theta & \cos\theta \end{pmatrix} \begin{pmatrix} 1 & 0 \\ 0 & 0 \end{pmatrix} \begin{pmatrix} \cos\theta & \sin\theta \\ -\sin\theta & \cos\theta \end{pmatrix}$$

$$= \begin{pmatrix} \cos^2\theta & \sin\theta\cos\theta \\ \sin\theta\cos\theta & \sin^2\theta \end{pmatrix}. \tag{2.108}$$

If the polarizer is diagonal, $\theta = 45°$, then the Jones matrix becomes

$$T_D = \frac{1}{2} \begin{pmatrix} 1 & 1 \\ 1 & 1 \end{pmatrix}. \quad \blacksquare \tag{2.109}$$

Example 2.7.2 The law of Malus easily follows from the Jones matrix of Eq. 2.108. Suppose the initial light is polarized along the x-axis with amplitude E_1, so that the incident *unnormalized* Jones vector (dropping the normalization factor in Eq. 2.96) and the intensity are

$$J_{in} = \begin{pmatrix} E_{in} \\ 0 \end{pmatrix}, \qquad I_{in} = J^\dagger J = |E_{in}|^2. \tag{2.110}$$

Then the output is

$$J_{out} = T J_{in} = \begin{pmatrix} \cos^2\theta & \sin\theta\cos\theta \\ \sin\theta\cos\theta & \sin^2\theta \end{pmatrix} \begin{pmatrix} E_{in} \\ 0 \end{pmatrix} \tag{2.111}$$

$$= E_1 \begin{pmatrix} \cos^2\theta \\ \sin\theta\cos\theta \end{pmatrix}, \tag{2.112}$$

which gives the output intensity

$$I_{out} = |E_0|^2 \left(\cos^4\theta + \sin^2\theta\cos^2\theta \right) \tag{2.113}$$

$$= I_0 \cos^2\theta \left(\cos^2\theta + \sin^2\theta \right) \tag{2.114}$$

$$= I_0 \cos^2\theta, \tag{2.115}$$

in agreement with Eq. 2.105. $\blacksquare$

In addition to being dichroic, it is possible for a material to be **birefringent** or **doubly refractive**. Birefringence was discovered as early as 1669 by Danish scientist Rasmus Bartholin, who first observed it in calcite. Birefringent means that the index of refraction depends on the polarization of the light. This implies that different polarization states of light propagate at different speeds through the material. The best-known example of a birefringent material is calcite (CaO_3), also known as Iceland spar. If a small object is illuminated with unpolarized light and viewed through a calcite crystal, two images will appear. Since the two orthogonal polarization states will have different refractive indices, then by Snell's law, they will refract at different angles when entering the crystal and emerge from the opposite side of the crystal at different points, thus producing two images. That each of the images is produced by a different polarization is easily demonstrated: place a polarizing filter between your eye and the calcite and slowly rotate it. At some particular orientation, one of the images will disappear. Then rotate by another 90°: the other image should now vanish.

Another interesting property is easy to demonstrate. Try rotating the crystal now. As you do, one image should stay fixed, while the other image rotates around the fixed one. The light forming the fixed image is said to be formed by the **ordinary ray** or **o-ray**, while the image that rotates is formed by the **extraordinary ray** or **e-ray**. To describe what these rays are, we first must define the optic axis of the crystal. Birefringence can only occur in crystalline minerals in which there exists an axis of symmetry; rotating the crystal about this axis does not change the behavior of the light. Light propagating parallel to this axis has a single refractive index, n_o, regardless of its polarization. (Note that the light is a transverse wave, so the polarization of light propagating along this axis is perpendicular to the axis.) This axis is called the **optic axis**, and the light propagating parallel to it constitutes the o-ray.

Light traveling in any other direction will have polarization components both parallel and perpendicular to the optic axis, which results in the e-rays. While the wavefronts of the o-rays are spherical, the wavefronts of the e-rays are elliptical because the rays propagating in different directions, at different angles from the optic axis, see different indices of refraction and therefore move at different speeds. Just as n_o was the refractive index for light propagating parallel to the optic axis, define n_e to be the index for line propagating perpendicular to the axis. For light that is propagating at any angle θ other than 0° or 90° from the optic axis, the refractive index $n_{eff}(\theta)$ will be a combination of n_o and n_e, namely:

$$\frac{1}{n_{eff}(\theta)} = \frac{\cos^2\theta}{n_o^2} + \frac{\sin^2\theta}{n_e^2}. \tag{2.116}$$

So when unpolarized light enters a birefringent medium from an initial medium of index n_1, the o- and e-rays refract at different angles θ_o and θ_e given by two copies of Snell's law:

$$n_1 \sin\theta_1 = n_o \sin\theta_o \tag{2.117}$$

$$n_1 \sin\theta_1 = n_{eff} \sin\theta_e, \tag{2.118}$$

leading to the double refraction effect (Fig. 2.15).

In general, a crystal can have up to two optic axes, but here we have restricted ourselves to the simplest case of a single optic axis, i.e., to the case of a **uniaxial** crystal. Some additional terminology is sometimes useful: a crystal is said to be **positive uniaxial** if $n_o > n_e$ and **negative uniaxial** if $n_e > n_o$. Calcite is an example

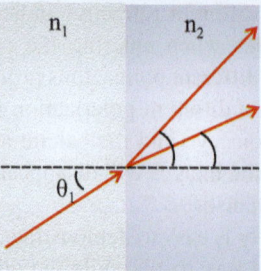

Fig. 2.15 Double refraction: different polarization components see different effective refractive indices, so Snell's law refracts each component at a different angle. The original ray is split into two outgoing rays, each exiting at a different point on the right side, and each in different polarization states

of a negative uniaxial crystal. The polarization components with the lower and higher refractive indices are referred to, respectively, as the **fast and slow waves**.

In a sufficiently strong electric field, birefringence can be induced in a normally non-birefringent material. The field effectively stretches and distorts the molecules to artificially induce anisotropy, leading to appearance of an optic axis. This occurs via the **Kerr or quadratic opto-electric effect** (Chap. 5).

2.8 Circular Polarization

Recall that for circularly polarized light, the electric field rotates about the propagation axis as the light moves through space. Depending on the direction of rotation, the light is said to be left- or right-circularly polarized. The circular polarizations are linear combinations of the two linear polarizations, but with complex coefficients, indicating that one component leads the other in time by a quarter of a cycle, $\delta\phi = \phi_x - \phi_y = \pm\frac{\pi}{2}$.

The corresponding Jones vectors for circular polarization are

$$J_R = \frac{1}{\sqrt{2}} \begin{pmatrix} 1 \\ i \end{pmatrix} \qquad J_L = \frac{1}{\sqrt{2}} \begin{pmatrix} 1 \\ -i \end{pmatrix}. \tag{2.119}$$

More generally, elliptical polarizations are given by

$$J_R = \frac{1}{\sqrt{A_x^2 + A_y^2}} \begin{pmatrix} A_x \\ i A_y \end{pmatrix} \qquad J_L = \frac{1}{\sqrt{A_x^2 + A_y^2}} \begin{pmatrix} A_x \\ -i A_y \end{pmatrix}. \tag{2.120}$$

Just as different linear polarizations can behave differently and see different refractive indices in a birefringent material, different circular polarizations can also experience different refractive indices and move at different speeds. Materials in which this occurs are called **chiral** or **optically active**. Because the linear polarizations are linear combinations of these differently moving circular polarizations, any linearly polarized light entering a chiral material will see its polarization vector rotate.

The amount of rotation of the linear polarization will depend on the distance that the light has propagated through the chiral material. Let the rotation angle be ϕ and let z be the propagation distance in the material. Then the degree of chirality of the material can be quantified by defining the **rotatory power** in radians per meter:

$$\rho = \frac{d\phi}{dz} = \frac{\pi}{\lambda_0} \left(n_L - n_R \right), \tag{2.121}$$

where λ_0 is the wavelength of the light in air and n_L and n_R are the refractive indices of the medium for left- and right-circular polarizations.

The rotatory power is strongly dependent on the frequency of the light. If $\rho < 0$, the linear polarization rotates in the same direction as the electric field of a left-hand circularly polarized beam; the material is then called **left-chiral** or **levarotatory**. Similarly, if $\rho > 0$, a **right-chiral** or **dextrorotatory** material causes linear polarizations to rotate in the same direction as a right-circularly polarized beam. Most organic molecules, including amino acids, proteins, and sugars, are chiral, as are many inorganic materials such as selenium, quartz, and cinnabar (HgS).

The rotatory power can be easily measured by a polarimeter. A polarizing filter of known orientation is placed between the light source and the material, so the polarization direction of the ingoing light is known. Then a second polarizer is placed in the path of the outgoing light on the other side of the material. Rotating the second polarizer until the intensity it transmits is maximum, the orientation angle of this filter relative to the first one then gives the rotation angle $\Delta\phi$ in the material. So for material of thickness Δz, the rotatory power is $\rho = \frac{\Delta\phi}{\Delta z}$.

Example 2.8.1 Consider a 4 mm-thick slice of quartz. For light at wavelength $\lambda_0 = 500$ nm, the rotatory power is $\rho \approx 30°/\text{mm} = 0.2618 \frac{\text{rad}}{\text{mm}}$. This translates to a difference

$$n_L - n_R = \frac{\lambda_0 \rho}{\pi} = \frac{500 \text{ nm}}{\pi} \left(0.2618 \frac{\text{rad}}{\text{mm}} \right) = 4.167 \times 10^{-5}. \tag{2.122}$$

In the process of traversing the material, the linear polarization vector will rotate through an angle of

$$\Delta\phi = \rho \Delta z = \left(30°/\text{mm} \right) \cdot (4 \text{ mm}) = 120°. \tag{2.123}$$

In contrast, if the light is of wavelength $\lambda_0 = 600$ nm, then $\rho \approx 22°/\text{mm}$, so the rotation angle will be only $\Delta\phi = 88°$. ∎

Just as birefringence can be induced in a material by a strong electric field, a strong magnetic field can induce optical activity. The resulting rotatory power will be proportional to the size of the applied magnetic field:

$$\rho = V B, \tag{2.124}$$

where V is called the **Verdet constant** of the material. This induced chirality effect is known as the **Faraday effect** and was discovered in 1845 by Michael Faraday. The Verdet constant is in units of $\frac{rad}{T \cdot m}$ and is frequency-dependent. The Faraday effect provides a way to construct a controllable polarization rotator. An optically active material with large V is used to rotate linearly polarized light. The angle of rotation is controlled by the strength of the applied magnetic field.

If a material rotates the polarization by angle θ, then it can be described by a Jones matrix of the form

$$T_{\text{rot}}(\theta) = \begin{pmatrix} \cos\theta & -\sin\theta \\ \sin\theta & \cos\theta \end{pmatrix} = R(\theta), \tag{2.125}$$

where $R(\theta)$ is the two-dimensional rotation matrix.

One simple application of the polarization rotator is in the construction of an **optical isolator**. An optical isolator enforces one-way motion in an optical system: light can pass through it once, but can't reflect back through it in the other direction. But it does this at the expense of losing some of the initial intensity.

Consider the situation shown in Fig. 2.16. Light entering from the left passes first through a vertical polarizer. It then passes through a Faraday rotator, arranged to rotate the polarization by 45°, followed by another linear polarizer, also at 45° to the initial direction, so that all of the light exiting the rotator passes through the second filter. Now, if light reflected backward, it will pass through the rotator again. After this passage, the light will have been rotated by 45° twice, making it orthogonal to the first polarizer. The light is therefore blocked from reaching the left side of the apparatus. Note that the setup works in both directions: the isolator allows a first

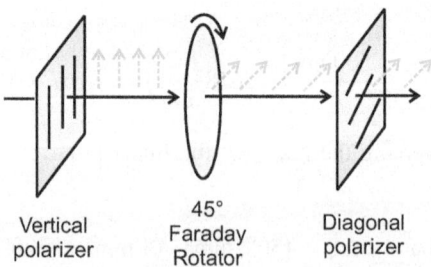

Vertical
polarizer

45°
Faraday
Rotator

Diagonal
polarizer

Fig. 2.16 An optical isolator. The light passing the first, vertical polarizing filter is rotated 45° so that it can pass the second, diagonal polarizer, allowing it to continue propagating to the right. However, if any of the light reflects back, it will be rotated again so that it will be blocked by the vertical polarizer. So no light can reflect back to the left side of the system

passage in either direction, but does not allow back reflection, or turning around, of the light.

Such an isolator is useful, for example, in stopping laser light from reflecting back into a laser, which can lead to unstable oscillations in the laser cavity. More generally, it allows control over the flow of light through an optical system.

2.9 The Poincaré Sphere and Stokes Parameters

The polarization states of light can be conveniently represented pictorially in terms of points on a sphere, called the **Poincaré sphere**. This picture is especially useful because an essentially identical representation (the **Bloch sphere**) can be used to describe spin-$\frac{1}{2}$ states, or two-states systems more generally (see Chap. 6).

Let A_x and A_y be the amplitudes in Eqs. 2.94 and 2.95, with phases ϕ_x and ϕ_y. Then define the complex polarization ratio,

$$\mathcal{R} = \frac{A_x}{A_y} = |\mathcal{R}| e^{i\,\delta\phi}, \tag{2.126}$$

where $\delta\phi = \phi_x - \phi_y$.

The **Stokes parameters** are defined by

$$S_0 = |A_x|^2 + |A_y|^2 \tag{2.127}$$

$$S_1 = |A_x|^2 - |A_y|^2 \tag{2.128}$$

$$S_2 = 2\,\mathrm{Re}\left(A_x^* A_y\right) \tag{2.129}$$

$$S_3 = 2\,\mathrm{Im}\left(A_x^* A_y\right). \tag{2.130}$$

These define the components of the four-dimensional **Stokes vector**:

$$S = \begin{pmatrix} S_0 \\ S_1 \\ S_2 \\ S_3 \end{pmatrix}. \tag{2.131}$$

The definitions imply that

$$S_0^2 \geq S_1^2 + S_2^2 + S_3^2. \tag{2.132}$$

The Stokes vector is more general than the Jones vector, in the sense that Jones vectors can only be defined for fully polarized light. The Stokes vector can be defined for unpolarized or partially polarized light as well. The Stokes vectors for some common polarization states are given in Table 2.1.

Table 2.1 Stokes vectors for common polarization states. The intensity has been normalized so that $S_0 = 1$

Unpolarized	$\begin{pmatrix} 1 \\ 0 \\ 0 \\ 0 \end{pmatrix}$	Left-circular	$\begin{pmatrix} 1 \\ 0 \\ 0 \\ -1 \end{pmatrix}$
Horizontal (0°)	$\begin{pmatrix} 1 \\ 1 \\ 0 \\ 0 \end{pmatrix}$	Diagonal (+45°)	$\begin{pmatrix} 1 \\ 0 \\ 1 \\ 0 \end{pmatrix}$
Vertical (+90°)	$\begin{pmatrix} 1 \\ -1 \\ 0 \\ 0 \end{pmatrix}$	Anti-diagonal (−45°)	$\begin{pmatrix} 1 \\ 0 \\ -1 \\ 0 \end{pmatrix}$
Right-circular	$\begin{pmatrix} 1 \\ 0 \\ 0 \\ 1 \end{pmatrix}$		

S_0 is simply the total intensity of the light. The other three parameters are related to the polarization. S_1 measures the excess of horizontal polarization over vertical, S_2 measures the excess of diagonal over anti-diagonal, and S_3 measures the excess of right-circular over left-circular. The overall **degree of polarization** p of the light,

$$p = \frac{\sqrt{S_1^2 + S_2^2 + S_3^2}}{S_0}, \tag{2.133}$$

satisfies $0 \le p \le 1$, where $p = 0$ corresponds to unpolarized light and $p = 1$ is the complete polarization, with values in between representing partial polarization.

Using the Stokes parameters, we can define a pair of angles, χ and ψ, according to the following:

$$S_1 = S_0 \cos 2\chi \cos 2\psi \tag{2.134}$$

$$S_2 = S_0 \cos 2\chi \sin 2\psi \tag{2.135}$$

$$S_3 = S_0 \sin 2\chi. \tag{2.136}$$

Inverting these definitions, we find that

$$2\psi = \tan^{-1}\left(\frac{S_2}{S_1}\right) \tag{2.137}$$

$$2\chi = \tan^{-1}\left(\frac{S_3}{\sqrt{S_1^2 + S_2^2}}\right). \tag{2.138}$$

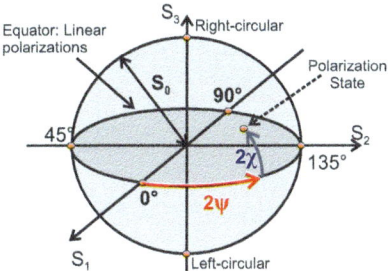

Fig. 2.17 The Poincaré sphere, used to represent polarization states of light. Right- and left-circular polarizations are represented by the north and south poles, respectively, while linear polarization states lie on the equator. The angles on the equator represent the angle of the polarization vector from the horizontal. Fully polarized states lie on the surface of the sphere, while partially polarized light lies in the interior. Completely unpolarized light is represented by the center of the sphere. 2ψ and 2χ are the angles, respectively, from the S_1 axis and the equator

Then the **Poincaré sphere** is the sphere of radius S_0 shown in Fig. 2.17, with 2ψ and 2χ representing, respectively, to the angles from the S_1 axis and from the equator. Henceforth, we can choose units to normalize $S_0 = 1$, so that this becomes a unit sphere. Fully polarized light lies on the surface of the sphere at $r = 1$, while partially polarized light of polarization degree p is in the interior, at radius $r = p$.

Two fully polarized states are orthogonal to each other if they are antipodal points on the Poincaré sphere (i.e., 180° apart from each other on opposite sides of a great circle). More generally, the overlap or inner product of two fully polarized states will be proportion to $\pi - \theta$, where θ is the angle of the circular arc joining them on the sphere's surface.

The angles ψ and χ giving the location of a fully polarized state on the Poincaré sphere are related to the **polarization ellipse**. As the light propagates, the tip of the electric field vector traces out an ellipse. ψ and χ determine the shape of the ellipse, as shown in Fig. 2.18. Let a and b be the lengths of the semi-major and semi-minor axes. Then ψ gives the inclination of the major axis from the x-axis (the horizontal), while χ is called the **ellipticity angle**, $\chi = \tan^{-1} \frac{b}{a}$. The **eccentricity** is defined by

$$e = \sqrt{1 - \frac{b^2}{a^2}}. \tag{2.139}$$

For $e \to 0$, the ellipse becomes a circle (circular polarization), and for $e \to 1$, the ellipse collapses to a line segment (linear polarization).

Fig. 2.18 The polarization ellipse of an optical state can be determined by the angles specifying the state's location on the Poincaré sphere, as shown. The ellipticity angle χ is related to the semi-major and semi-minor axes, a and b, by $\tan \chi = \frac{b}{a}$, while the angle ψ gives the tilt of the major axis relative to the horizontal

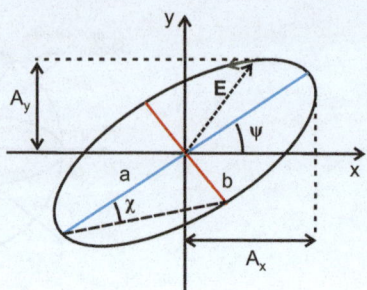

2.10 Wave Plates

A **wave plate** or **phase retarder** is an optical element that provides different phase shifts to two orthogonal polarization components in order to change the polarization state of the outgoing light relative to that of the incoming light. Recall that light of free-space wavelength λ_0 passing a distance d through a material of index n gains a phase shift of

$$\phi = 2\pi \frac{d}{\lambda} = \frac{2\pi}{\lambda_0} nd, \tag{2.140}$$

where $\lambda = \frac{\lambda_0}{n}$ is the wavelength in the material. So now consider a thin slice of calcite or some other birefringent material. The components of the light polarized parallel and perpendicular to the optic axis travel at different speeds and see different refractive indices. As a result they gain different phase shifts. So when the light exits the material, there is a relative phase shift between the ordinary and extraordinary components, given by

$$\delta\phi = \phi_e - \phi_o = \frac{2\pi}{\lambda_0} (n_e - n_o) d = 2\rho d, \tag{2.141}$$

where ρ is the rotatory power. To see what effect this has on the polarization state of the light, recall that for linearly polarized light, the two orthogonal polarization states are in phase with each other ($\phi_H - \phi_V = 0$), but for circularly (or elliptically) polarized light, the phase difference is $\pm \frac{\pi}{2}$.

The crystal is cut so that the surface is perpendicular to the propagation axis of the light. If $n_e < n_o$ (as in calcite), then the direction of the optic axis is called the **fast axis**, while the direction perpendicular both the optic axis and the propagation direction is the **slow axis**. If $n_o < n_e$, then these definitions are reversed. In the following, we take the x-axis along the fast direction and the y-axis along the slow axis.

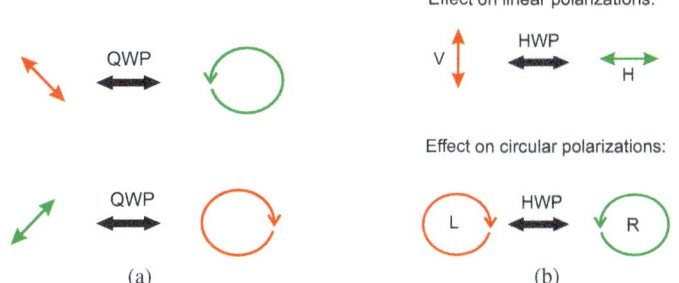

Fig. 2.19 (**a**) A quarter-wave plate converts linearly polarized light into circularly polarized light and vice versa. (**b**) A half-wave plate reflects linear polarizations across the optic axis, interchanging the two polarizations that are at 45° from that axis. It also interchanges left- and right-circular polarizations

The Jones matrix for the wave plate is of the form

$$T = \begin{pmatrix} 1 & 0 \\ 0 & e^{i\delta\phi} \end{pmatrix}. \tag{2.142}$$

Consider two special cases:

- A **quarter-wave plate (QWP)** is a wave plate with $|\delta\phi| = \frac{\pi}{2}$. This will convert linearly polarized light into circularly or elliptically polarized light and vice versa (Fig. 2.19a).
- A **half-wave plate (HWP)** has $|\delta\phi| = \pi$. Its effect is to convert the two orthogonal states in a given polarization basis into each other: if the light is linearly polarized, it interchanges horizontal and vertical polarizations, but if the light is circularly polarized, it interchanges left-circular with right-circular polarizations (Fig. 2.19a). (Here, horizontal and vertical are defined to be the slow and fast directions of the wave plate.)

2.11 Spatial Light Modulators

A **spatial light modulator (SLM)** consists of an array of elements (cells or pixels) that impose phase and/or amplitude modulations onto a beam of light reflecting off or transmitting through them. In this way, a desired spatial structure can be imprinted on the wavefront or phase profile of the outgoing light. Typically, each cell is individually controlled electronically, so that a specific modulation pattern can be programmed onto the SLM surface, allowing complex structures to be imprinted onto the reflected light. The modulation can be performed either by using a deformable mirror for each cell or by using cells composed of liquid crystals. LCD

SLMs can have on the order of a million pixels per square centimeter, leading to very high resolution, making them a popular choice for television screens, computer monitors, and projectors. But they have also become essential for a wide variety of scientific uses, such as super-resolution microscopy [27], the generation of high OAM light beams [28, 29], optical manipulation of nanoparticles, and spatial-frequency filtering.

Although most SLMs are electronically addressable, meaning that the phase and amplitude change imparted by each pixel is controlled by applying appropriate voltages, some use optical addressing schemes.

Problems

1. Consider light traveling through a vacuum. Use the vacuum Maxwell's equations, Eq. 2.7, to 2.10 derive the two wave equations 2.11 and 2.12.
2. By explicitly doing the Fourier transforms, show that Eq. 2.35 leads to Eq. 2.33.
3. (a) Consider a converging lens formed by two spherical surfaces of radius R. Take z along the axis of the lens, with x and y in the transverse plane, perpendicular to the axis. Show that the thickness of the lens as a function of x and y is given by Eq. 2.47, where d is the maximum thickness of the lens (the thickness at the center) and that the focal length is given by 2.49.
   (b) For a lens formed from two spherical surfaces of radii R_1 and R_2, show that the focal length is given by Eq. 2.50.
4. (a) Starting from Eq. 2.75, verify that 2.77 is correct. (b) Show that the MZ interferometer conserves total intensity: $|E_1|^2 + |E_2|^2 = |E_7|^2 + |E_8|^2$. (This indicates that the interferometer transformation is unitary.)
5. Consider a circular Sagnac interferometer with radius r and area A. Derive the difference between the light arrival times for the clockwise- and counterclockwise-propagating beams of Eq. 2.85, and from there show that the phase difference of Eq. 2.86 is correct.
6. Show that two successive Fourier transforms of the form of Eq. 2.52 gives back an inverted copy of the field, as in Eq. 2.55.
7. Derive Brewster's law from Snell's law by assuming that the light is refracted back into the original medium, i.e., that $\theta_2 > 90°$.
8. Prove the inequality given in Eq. 2.132.
9. Suppose both input ports of a Mach-Zehnder are simultaneously illuminated with light of intensity $\frac{I}{2}$. What is the intensity at each output port for $\theta = 0$, $\theta = \frac{\pi}{2}$, and $\theta = \pi$?
10. Consider vertically polarized light incident on a horizontally oriented polarizing filter. Clearly, it would be expected that no light will be transmitted.

   (a) *Show* that no light is transmitted using the Jones calculus formalism.
   (b) Now place a second filter oriented at 45° in front of the first one (so that the light strikes the new filter first). Using Jones matrices, show that the presence of the additional filter actually *increases* the output intensity. Find the new output.

References

1. L. Solymar, E. Shamonina, *Waves in Metamaterials* (Oxford University Press, Oxford, 2009)
2. C. Simovski, S. Tretyakov, *An Introduction to Metamaterials and Nanophotonics* (Cambridge University Press, Cambridge, 2020)
3. V.G. Veselago, Sov. Phys. Usp. **10**, 509 (1968)
4. J.B. Pendry, A.J. Holden, D.J. Robbins, W.J. Stewart, IEEE Trans. Microwave Theory Tech. **47**, 2075 (1999)
5. R.A. Shelby, D.R. Smith, S. Schultz, Science **292**, 77 (2001)
6. J.B Pendry, D.R. Smith, Phys. Today **57**, 37 (2004)
7. B.E.H. Saleh, M.C. Teich, *Fundamentals of Photonics*, 3rd edn. (Wiley, Hoboken, 2019)
8. J.W. Goodman, *Introduction to Fourier Optics*, 4th edn. (W. H. Freeman, New York City, 2017)
9. J. Vangindertael, R. Camacho, W. Sempels, H. Mizuno, P. Dedecker, K.P.F. Janssen, Methods Appl. Fluoresc. **6**, 022003 (2018)
10. D.S. Simon, C.A. Fitzpatrick, A.V. Sergienko, Phys. Rev. A **93**, 043845 (2016)
11. S. Osawa, D.S. Simon, A.V. Sergienko, Opt. Exp. **26**, 27201 (2018)
12. I. Kim, et al., Opt. Exp. **29**, 29527–29540 (2021)
13. S. Osawa, D.S. Simon, A.V. Sergienko, Phys. Rev. A **102**, 063712 (2020)
14. D.S. Simon, S. Osawa, A.V. Sergienko, Phys. Rev. A **101**, 032118 (2020)
15. D.S. Osawa, D.S. Simon, A.V. Sergienko, Phys. Rev. A **104**, 012617 (2021)
16. D.S. Simon, C. Schwarze, A.V. Sergienko, Phys. Rev. A **106**, 033706 (2022)
17. C.R. Schwarze, D.S. Simon, A.V. Sergienko, Phys. Rev. A **107**, 052615 (2023)
18. A. Labeyrie, S.G. Lipson, P. Nisenson, *An Introduction to Optical Stellar Interferometry* (Cambridge University Press, Cambridge, 2006)
19. E.J. Post, Rev. Mod. Phys. **39**, 475 (1967)
20. P. Hariharan, *Optical Interferometry*, 2nd edn. (Academic Press, San Francisco, 2003)
21. H.C. Lefevre, *The Fiber-Optic Gyroscope*, 2nd edn. (Artech House, Boston, 2014)
22. K.-X. Sun, M.M. Fejer, E. Gustafson, R.L. Byer, Phys. Rev. Lett. **76**, 3053 (1996)
23. I.L. Paiva, R. Lenny, E. Cohen, arXiv:2110.05824 [quant-ph] (2021)
24. K. Moriyasu, *An Elementary Primer for Gauge Theory* (World Scientific, Singapore, 1983)
25. L.H. Ryder, *Quantum Field Theory*, 2nd edn. (Cambridge University Press, Cambridge, 1996)
26. A. Zee, *Quantum Field Theory in a Nutshell*, 2nd edn. (Princeton University Press, Princeton, 2010)
27. C. Maurer, A. Jesacher, S. Bernet, M. Ritsch-Marte, Laser Phot. Rev. **5**, 81 (2011)
28. A. Forbes, A. Dudley, M. McLaren, Adv. Opt. Photon. **8**, 200 (2016)
29. A.M. Yao, M.J. Padgett, Adv. Opt. Photonics **3**, 161 (2011)

Chapter 3
Quantum Mechanics: The Essentials

Building on the background in optics and waves covered in the last chapter, this chapter gives a brief review of quantum mechanics, focusing on those aspects that will be important in later parts in this book. It is assumed that the reader has at least some prior familiarity with quantum mechanics, so the goal here is partly to refresh the reader's memory and fill in any gaps. Along the way, a few new ideas such as fidelity and decoherence, which are not normally taught in introductory quantum courses, will be introduced for later use. Along with the previous chapter, this one establishes a foundation for the remainder of the book.

3.1 Particles and Waves

The twentieth century began with a radical revision of our view of nature, as it was realized that the localized, continuous-energy particles that populate our classical, intuitive picture of the world had to be replaced with delocalized wavefunctions describing particles carrying energy in discrete bundles. In this new picture, an electron doesn't exist at a precise location but is spread over all of space. As a result, the position is, to some extent, indeterminate and has a margin of error or width Δx, the position **uncertainty**. Not only do other variables, such as momentum and energy, come equipped with their own uncertainties, but it turns out that in some cases, these uncertainties were intrinsically related to each other and subject to constraints formalized by Heisenberg as his famous uncertainty relations.

The uncertainty relations are built into nature and can't be evaded by improved measurement devices, even in principle. This inherent measurement limitation was just one of the novel features that quantum mechanics introduced. Given that a particle has wavelike properties, it is necessary to introduce a wave amplitude, or **wavefunction**, to describe the system, which in turn must obey a wave equation, the Schrödinger equation. Being linear, multiple known solutions to the wave equation

© The Author(s), under exclusive license to Springer Nature Switzerland AG 2025
D. S. Simon, *Introduction to Quantum Science and Technology*, Undergraduate
Texts in Physics, https://doi.org/10.1007/978-3-031-81315-3_3

can be added to each other to form new solutions. Since the wavefunction is in some sense describing the state of the system, this addition or **superposition** of wavefunctions seems to imply that the particle can be in two separate states at the same time!

Other novel features quickly emerged: for example, the wavefunctions were complex-valued, and yet somehow real-valued physical variables had to emerge from them. In addition, the absolute square of these wavefunctions turned out to describe the probabilities of states. The resulting probabilistic nature of quantum mechanics, in stark contrast to the strictly deterministic laws of classical mechanics, was probably the greatest psychological shock to the classically trained physicists of the day. Even some of the founders of quantum theory, most famously Einstein, never accepted this inherently probabilistic aspect and spent the rest of their lives trying to find a way to banish it.

These new features, though jarring to many at the time, came to be accepted within a few years as they became thoroughly corroborated by overwhelming experimental evidence. Now, in the twenty-first century, many of these new features, present only in quantum theory and often difficult for our classical intuition to understand, have begun to be harnessed for practical use. These new uses, mostly still in the development stages, range from ultrahigh precision measurements that evade classical resolution limits to quantum computers that can solve problems that are intractable on classical devices.

In this chapter, the basics of quantum mechanics are reviewed. Along with standard introductory quantum mechanics textbooks such as [1–3], books such as [4] (or [5] at a more conceptually advanced level) may be useful as extra references for the reader.

3.2   States and Operators

Introductory quantum mechanics textbooks usually focus on the wavefunction $\psi(x)$ of a quantum system. Here we will be more general and work primarily with **state vectors**, expressed in Dirac's bra-ket notation. A state vector or **ket** contains all possible information about the system and may most easily be thought of as a column vector with an infinite number of components. State vectors will normally be denoted generically as $|\psi\rangle$ or $|\phi\rangle$. They may also be distinguished by some continuous or discrete label; for example, the state of a system currently residing in its jth energy level, E_j, may be denoted as either $|E_j\rangle$ or more compactly as $|j\rangle$. In a similar manner, a particle known to be currently at position x (in other words, its wavefunction is only nonzero at that one point) may be denoted as $|x\rangle$, while the state of a particle, with a definite value of momentum p, may be written as $|p\rangle$. (For a more detailed discussion of state vectors, starting by analogy to three-dimensional vectors, see Appendix C.)

Every ket vector $|\psi\rangle$ has a corresponding Hermitian adjoint **bra** vector, $\langle\psi|$. If $|\psi\rangle$ is thought of as a column vector, then $\langle\psi|$ would be the complex conjugate of the corresponding row vector. Inner products between a bra and a ket (a "bra-ket" or "bracket"), $\langle \phi | \psi \rangle$, are then just a generalized version of the usual matrix multiplication between row and column vectors.

The position-space wavefunction $\psi(x)$ is the projection of $|\psi\rangle$ onto position x. In other words,

$$\psi(x) = \langle x | \psi \rangle. \tag{3.1}$$

Taking the complex conjugate of both sides, we find that

$$\psi^*(x) = \langle \psi | x \rangle; \tag{3.2}$$

note that complex conjugation interchanges the bra and ket. The description of the system in terms of wavefunctions is not unique. Instead of characterizing the system as a function of x, we could, for example, think of it as a function of momentum by carrying out a different projection:

$$\psi(p) = \langle p | \psi \rangle. \tag{3.3}$$

The transition from the position-space wavefunction $\psi(x)$ to the momentum-space wavefunction should simply be thought of as a generalized change of basis in the infinite-dimensional space of states. Like any change of basis, the transition may make the system look different, but physically measurable quantities like energy or angular momentum should come out to be the same in either description.

Since state vectors are generally infinite-dimensional, we can't write down explicit expressions for them. Nonetheless, we can manipulate them mathematically and extract information from them by acting on them with **operators**. If states are viewed as infinite-dimensional column vectors, then operators can be thought of as infinitely large square matrices that transform state vectors into other state vectors.

Recall that in quantum mechanics, we are usually concerned with two main classes of operators: Hermitian and unitary. So, we turn to these next.

3.2.1 Hermitian Operators

Hermitian or **self-adjoint** operators represent physical observables, or in other words, variables whose values can be physically measured. A Hermitian operator $\hat{O}$ is one that equals its own Hermitian adjoint:

$$\hat{O} = \hat{O}^\dagger, \tag{3.4}$$

where the Hermitian adjoint of the matrix representing the operator is found by taking the complex conjugate of the transpose, $\hat{O}^\dagger = \left(\hat{O}^T\right)^*$. An equivalent definition is that

$$\langle \hat{O}^\dagger \phi | \psi \rangle = \langle \phi | \hat{O} \psi \rangle, \tag{3.5}$$

for any states $|\phi\rangle$ and $|\psi\rangle$. Common examples of Hermitian operators are the position and momentum, the Hamiltonian or energy operator, and angular momentum operators.

Hermitian operators have real eigenvalues, so these eigenvalues can represent the values of observables. If a state has a definite value for an observable, then it is an eigenfunction of the corresponding Hermitian operator. For example, the states $|x\rangle$ and $|p\rangle$ given above are eigenstates of the position and momentum operators, respectively:

$$\hat{x}|x\rangle = x|x\rangle \quad \text{and} \quad \hat{p}|p\rangle = p|p\rangle, \tag{3.6}$$

where the quantities with the hats over them are the operators and those without are the c-number eigenvalues of the observables.

In contrast, $|p\rangle$ is not an $\hat{x}$ eigenstate, so the momentum eigenstate does not have a definite position. $|p\rangle$ is in a superposition of different positions; if its position is measured, any one of these position values could be obtained, with some fixed probability distribution. Similarly, states of definite energy are eigenstates of the Hamiltonian, $\hat{H}$,

$$\hat{H}|E\rangle = E|E\rangle, \tag{3.7}$$

but are not (in general) eigenstates of $\hat{x}$ or $\hat{p}$.

To be more general, suppose some physically observable quantity (energy, momentum, etc.) is represented by a Hermitian operator $\hat{O}$. If $\hat{O}$ is measured, the only values that can be obtained are the eigenvalues of $\hat{O}$. After the measurement, the original state $|\psi\rangle$ "collapses" to the eigenstate of $\hat{O}$ corresponding to the measured eigenvalue. So if the eigenvalues and eigenvectors of $\hat{O}$ are λ_j and $|\psi_j\rangle$, with $j = 1, 2, 3, \ldots$, then measuring the value of $\hat{O}$ in state $|\psi\rangle$ gives one of the results λ_j, where the probability of the discrete eigenvalue λ_j occurring is

$$p_j = |\langle \psi_j | \psi \rangle|^2. \qquad \text{(Born rule).} \tag{3.8}$$

The state itself is then changed by the measurement: the initial state, which may have been a *superposition* of eigenstates, is now projected onto *one* of the eigenstates. If the initial $|\psi\rangle$ was already an eigenstate, then this implies that the corresponding eigenvalue appears with certainty and that the state remains unchanged.

In the case where the spectrum is continuous, a similar result holds, except that Eq. 3.8 gives the probability *density* (probability per unit distance); for example, the probability density to measure position value x is

$$p(x) = |\langle x|\psi\rangle|^2. \tag{3.9}$$

(See Appendix A.3 for a review of probabilities and probability densities.)

Box 3.1 Non-Hermitian Systems

Often a quantum system (call it S) is coupled to the surrounding environment (E), so that energy can be lost to or gained from the surroundings. Then the time evolution of the system can be non-unitary (unitarity is defined below, Sect. 3.2.3), and physical observables may fail to be Hermitian. This apparent Hermiticity failure is due to the fact that we are only collecting information from part (S) of the complete system while ignoring the degrees of freedom in the other part (E). Hermiticity is restored if measurements are made on the full physical system $S + E$. Commonly, though, the environment has too many degrees of freedom to keep track of, so it is advantageous to treat the observable as a non-Hermitian operator acting on S alone.

In a non-Hermitian system, the eigenvalues of the Hamiltonian are inherently complex, which allows new types of singularities to occur, such as branch cuts in the complex plane. In particular, singularities called **exceptional points (EPs)** or **non-Hermitian degeneracies** occur. Normally, degeneracy in a Hermitian system means that two or more eigenvalues become equal, while the corresponding eigenstates remain orthogonal. However, at an exceptional point, the eigenstates also become equal (they become parallel in the Hilbert space), meaning that the dimension of the Hilbert space decreases. The existence of exceptional points can lead to nontrivial topological states in the system (see Chap. 21).

It was pointed out in [6] that non-Hermitian systems will still have real eigenvalues if they are PT-symmetric and that non-Hermitian but PT-symmetric systems have interesting physical properties near exceptional points. P stands for **parity** or spatial-reflection symmetry. A Hamiltonian $H(r)$ is P-symmetric if it is invariant under $P : H(r, t) \to H(-r, t)$. Similarly, a T-symmetric or **time-reversal** symmetric system has a Hamiltonian that is unchanged under reversal of the direction of time, $T : H(r, t) \to H(r, -t)$. A PT-symmetric theory may not necessarily be invariant under P or T operations separately, but is invariant under the combined application of both, $H(r, t) \to H(-r, -t)$. In mechanical or field-theoretical systems, a necessary and sufficient condition for PT-symmetry is that the potential function obey $V(r) = V^*(-r)$, where $*$ represents complex conjugation. The imaginary parts of the energy correspond to gain (e.g., pumping in a laser system) or loss.

It was quickly realized that non-Hermitian but PT-invariant systems can be readily realized in optics, since loss is inherent in all optical systems, while

(continued)

Box 3.1 (continued)

gain is necessary for lasers and other active optical systems. By engineering a balance between gain and loss, PT-symmetry can be arranged, with the PT-symmetric condition becoming a requirement that the index of refraction be complex and position-dependent, with real and imaginary parts being, respectively, symmetric and antisymmetric under $r \rightarrow -r$. Such optical systems have recently become a focus of interest, because they can lead to large increases in the sensitivity of optical sensors when the system is near an exceptional point. Such sensors have been constructed using coupled optical cavities with controlled amounts of gain and loss. Non-Hermitian, PT-symmetric Hamiltonians have also been implemented in acoustic, electronic, and metamaterial systems.

Use of exceptional points has also allowed the fabrication of single-mode lasers in microcavities [7–9]. For reviews of non-Hermitian physics, see [10]. Non-Hermitian optics is reviewed in [11, 12] and applications to sensing in [13].

3.2.2   Projection Operators and Complete Sets of States

Consider a set of vectors, $\{|e_j\rangle\}$, where j is some discrete label, possibly ranging over an infinite number of values. Then the set is said to be **complete** if any other vector can be built as a linear combination of the vectors in the set; in other words, it is possible to write any vector $|\psi\rangle$ in the form

$$|\psi\rangle = \sum_j c_j |e_j\rangle, \tag{3.10}$$

for some set of constants c_j. Another way of saying the same thing is that the $|e_j\rangle$ form a **basis** for the space of states. Usually, the basis vectors are required to be orthonormal,

$$\langle e_i | e_j \rangle = \delta_{ij}, \tag{3.11}$$

where δ_{ij} is the Kronecker delta.

The components of a vector in the orthonormal e_j basis can be easily found. Suppose we wish to know the kth component, c_k, in the expansion $|\psi\rangle = \sum_j c_j |e_j\rangle$. Then take the inner product of $|\psi\rangle$ with $\langle e_k|$:

$$\langle e_k | \psi \rangle = \sum_j c_j \langle e_k | e_j \rangle \tag{3.12}$$

$$= \sum_j c_j \delta_{jk} \tag{3.13}$$

$$= c_k. \tag{3.14}$$

This simply expresses the fact that the component c_k is the projection of the vector $|\psi\rangle$ onto the kth direction.

The completeness of a discrete basis can be expressed in the following useful form. If the basis formed by the vectors $|e_j\rangle$ is complete, then the identity operator $\hat{I}$ can be written in the form

$$\hat{I} = \sum_j |e_j\rangle\langle e_j|, \tag{3.15}$$

which is referred to as the **completeness relation**. This simply says that the sum of the projections onto all the basis directions gives back all of the components needed to reconstruct the original vector:

$$|\psi\rangle = \hat{I}|\psi\rangle = \sum_j |e_j\rangle\langle e_j|\psi\rangle = \sum_j c_j|e_j\rangle. \tag{3.16}$$

Once an orthonormal basis is defined, then inner products are easily given a more explicit expression. Given two vectors of the form

$$|\psi\rangle = \sum_j a_j|e_j\rangle \qquad \text{and} \qquad |\phi\rangle = \sum_j b_j|e_j\rangle, \tag{3.17}$$

their inner product becomes

$$\langle\psi|\phi\rangle = \sum_j a_j^* b_j. \tag{3.18}$$

This is clearly a straightforward generalization of the dot product defined for ordinary vectors in three-dimensional space.

Instead of a discrete basis labeled by an integer like j, the basis vectors may form a continuous set, labeled by continuous variables such as x or p. For example, the state of all position eigenstates will form a complete set; the only difference for the continuous case is that the sum over discrete labels becomes an integral over the continuous variable. So any state $|\psi\rangle$ can be written in a form like

$$|\psi\rangle = \int_{-\infty}^{\infty} \psi(x)|x\rangle dx. \tag{3.19}$$

The orthonormality of the basis vectors (the position eigenstates) is then given in terms of a Dirac delta function instead of a Kronecker delta:

$$\langle x|x'\rangle = \delta(x - x'). \tag{3.20}$$

Notice that these equations are consistent with Eq. 3.1:

$$\langle x|\psi\rangle = \langle x| \left(\int \psi(x')|x'\rangle dx' \right) \tag{3.21}$$

$$= \int \psi(x')\langle x|x'\rangle dx' \tag{3.22}$$

$$= \int \psi(x')\delta(x - x') dx' \tag{3.23}$$

$$= \psi(x). \tag{3.24}$$

Similar expressions also hold for the basis of momentum eigenstates or any other continuous basis.

The continuous version of the discrete completeness relation Eq. 3.15 is

$$\hat{I} = \int |x\rangle\langle x| dx, \tag{3.25}$$

and the inner product becomes

$$\langle \psi|\phi\rangle = \int \psi^*(x)\phi(x) dx. \tag{3.26}$$

The **norm** or length of a state vector is given by

$$|\psi|^2 = \langle \psi|\psi\rangle = \int_{-\infty}^{\infty} \psi^*(x)\psi(x) dx. \tag{3.27}$$

We require that all physically reasonable wavefunctions be square integrable: the expression $\int_{-\infty}^{\infty} \psi^*(x)\psi(x) dx$ must be finite. The space of such square integrable functions is usually denoted L_2 and is called a **Hilbert space**. More precisely, a Hilbert space is a space of functions that are square integrable and complete, where completeness here means that the limiting value of any sequence in the space is also contained in the space. The completeness is required in order to do calculus on the set of functions and will always hold in physically realistic systems.

Since we want to maintain the interpretation of squared wavefunctions as probability densities, we usually require that state vectors be normalized:

$$|\psi|^2 = \langle \psi|\psi\rangle = 1. \tag{3.28}$$

Factors of the form $|e_j\rangle\langle e_j|$ in Eq. 3.15 and $|x\rangle\langle x|$ in Eq. 3.25 are examples of **projection operators**. When acting on a vector $|\phi\rangle$, $|e_j\rangle\langle e_j|$ projects out the part of $|\phi\rangle$ parallel to the jth basis direction, discarding the rest of the vector, while $|x\rangle\langle x|$

projects out the part of $|\phi\rangle$ located at position x. In general, a projection operator can always be written in the form

$$\hat{\mathcal{P}}_\psi = |\psi\rangle\langle\psi| \tag{3.29}$$

and acts on state vectors by projecting onto the direction of $|\psi\rangle$ in the Hilbert space. So, for example, $\hat{\mathcal{P}}_\psi|\phi\rangle = |\psi\rangle\langle\psi|\phi\rangle$ is a new (un-normalized) vector parallel to $|\psi\rangle$. In general, projection operators must have the following properties:

$$\hat{\mathcal{P}}_\psi \circ \hat{\mathcal{P}}_\psi = \hat{\mathcal{P}}_\psi \tag{3.30}$$

$$\hat{\mathcal{P}}_\psi \circ \hat{\mathcal{P}}_\phi = 0 \quad \text{if } \langle\psi|\phi\rangle = 0 \tag{3.31}$$

$$\sum_j \hat{\mathcal{P}}_{e_j} = \hat{I} \quad \text{if the } |e_j\rangle \text{ form an orthonormal basis.} \tag{3.32}$$

The last line is simply the completeness relation, Eq. 3.16, for the basis.

Measurements often correspond to making projections in a given basis, as in the following examples.

Example 3.2.1 One of the most important categories of quantum system is the **two-state system**. Examples of such systems are spin-$\frac{1}{2}$ particles that can be spin-up or spin-down, a photon that can be vertically or horizontally polarized, or a particle that can take two possible paths through an interferometer. All two-state systems, regardless of their physical origin, are equivalent to each other mathematically. So, as a representation of these systems, consider an electron with two basis states $|\uparrow\rangle$ and $|\downarrow\rangle$, representing spin-up and spin-down along the z-axis. These two basis vectors can be represented as column matrices:

$$|\uparrow\rangle = \begin{pmatrix} 1 \\ 0 \end{pmatrix}, \qquad |\downarrow\rangle = \begin{pmatrix} 0 \\ 1 \end{pmatrix}. \tag{3.33}$$

Clearly, these two basis vectors are orthogonal, and any normalized spin-$\frac{1}{2}$ vector can be expanded in this basis,

$$|\hat{n}\rangle = \cos\frac{\theta}{2}|\uparrow\rangle + \sin\frac{\theta}{2}a|\downarrow\rangle = \begin{pmatrix} \cos\frac{\theta}{2} \\ \sin\frac{\theta}{2} \end{pmatrix}, \tag{3.34}$$

where θ is the angle of $\hat{n}$ from the z-axis. If the spin is oriented along $|\hat{n}\rangle$, but a measurement is made along z, the probabilities of finding the spin pointing up or down along the z-axis are $p_\uparrow = \cos^2\frac{\theta}{2}$ and $p_\downarrow = \sin^2\frac{\theta}{2}$.

The projection operators onto the spin-up and spin-down states are given by

$$\mathcal{P}_\uparrow = |\uparrow\rangle\langle\uparrow| = \begin{pmatrix} 1 \\ 0 \end{pmatrix}\begin{pmatrix} 1 & 0 \end{pmatrix} = \begin{pmatrix} 1 & 0 \\ 0 & 0 \end{pmatrix} \tag{3.35}$$

$$\mathcal{P}_\downarrow = |\downarrow\rangle\langle\downarrow| = \begin{pmatrix} 0 \\ 1 \end{pmatrix}\begin{pmatrix} 0 & 1 \end{pmatrix} = \begin{pmatrix} 0 & 0 \\ 0 & 1 \end{pmatrix}. \tag{3.36}$$

Notice that the projections,

$$\mathcal{P}_\uparrow|\hat{n}\rangle = \langle\uparrow|\hat{n}\rangle|\uparrow\rangle = \cos\frac{\theta}{2}|\uparrow\rangle, \qquad \mathcal{P}_\downarrow|\hat{n}\rangle = \langle\downarrow|\hat{n}\rangle|\downarrow\rangle = \sin\frac{\theta}{2}|\downarrow\rangle, \tag{3.37}$$

give the coefficients in the basis expansion of Eq. 3.34. ∎

Example 3.2.2 As another example, consider two one-dimensional Gaussian wavefunctions, one centered at the origin and one centered at $x = x_0$:

$$\psi(x) = Ne^{-\left(\frac{x}{a}\right)^2}, \qquad\qquad \phi(x) = Ne^{-\left(\frac{x-x_0}{a}\right)^2}, \tag{3.38}$$

where a is a measure of the width of the wavefunction, and the normalization constant in front is given by $N^2 = \sqrt{\frac{2}{\pi a^2}}$. (Verify that the normalization is correct!) We can ask what is the overlap between these two states, or in other words, what is the amplitude for one state to convert to the other? Using the Gaussian integrals in Appendix A.4, this can easily be computed:

$$\langle\psi|\phi\rangle = N^2 \int \psi^*(x)\phi(x)\,dx = N^2 \int e^{-\left(\frac{x}{a}\right)^2}e^{-\left(\frac{(x-x_0)}{a}\right)^2}$$

$$= N^2 \int e^{-\frac{1}{a^2}\left[2x^2 - 2xx_0 + x_0^2\right]}dx. \tag{3.39}$$

Complete a square and then pull x-independent pieces out of the integral:

$$\langle\psi|\phi\rangle = N^2 \int e^{-\frac{2}{a^2}\left[x^2 - 2\left(\frac{x_0}{2}\right)x + \left(\frac{x_0}{2}\right)^2\right]}e^{-\frac{1}{2}\left(\frac{x_0}{a}\right)^2}dx$$

$$= N^2 e^{-\frac{1}{2}\left(\frac{x_0}{a}\right)^2} \int_{-\infty}^{\infty} e^{-\frac{2}{a^2}\left(x - \frac{x_0}{2}\right)^2}dx. \tag{3.40}$$

Shifting the integration variable $x \to x + \frac{x_0}{2}$ and doing the Gaussian integral, we find

$$\langle\psi|\phi\rangle = N^2\sqrt{\frac{a^2\pi}{2}}\,e^{-\frac{1}{2}\left(\frac{x_0}{a}\right)^2} = e^{-\frac{1}{2}\left(\frac{x_0}{a}\right)^2}. \tag{3.41}$$

So if a particle is placed in state $|\phi\rangle$, its transition probability to state $|\psi\rangle$ is proportional to

$$|\hat{P}_\psi |\phi\rangle|^2 = |\langle\psi|\phi\rangle|^2 = e^{-\left(\frac{x_0}{a}\right)^2}. \tag{3.42}$$

This should seem reasonable: the probability of a transition starts out equal to 1 when the states are equal ($x_0 = 0$) and then decreases exponentially with distance, going to zero asymptotically as x_0 becomes much larger than the width a. ■

Any operator can always be decomposed in the following manner. Given an operator $\hat{O}$ with eigenvalues $\{\lambda\}$, the corresponding eigenvectors $\{|\lambda\rangle\}$ form an orthonormal basis. Then $\hat{O}$ can be written as

$$\hat{O} = \sum_\lambda \lambda |\lambda\rangle\langle\lambda| = \sum_\lambda \lambda \hat{P}_\lambda. \tag{3.43}$$

This just says that $\hat{O}$ acts on each eigenspace (the space spanned by each eigenvector) independently, and on each of these eigenspaces, $\hat{O}$ simply acts by multiplying the vectors in that subspace by the corresponding eigenvalue. Again, this relation can also generalized to the case where the eigenvalues form a continuous set:

$$\hat{O} = \int \lambda |\lambda\rangle\langle\lambda|\, d\lambda = \int \lambda \hat{P}_\lambda d\lambda. \tag{3.44}$$

Henceforth, we will only write out only one version (either the discrete or the continuous expression) for each result; given one, the form of the other version should always be obvious.

Using Eq. 3.44, the expectation value of an operator $\hat{O}$ in state $|\psi\rangle$ is

$$\langle\hat{O}\rangle = \langle\psi|\hat{O}|\psi\rangle \tag{3.45}$$

$$= \int d\lambda\, \lambda \langle\psi|\lambda\rangle\langle\lambda|\psi\rangle \tag{3.46}$$

$$= \int d\lambda\, \lambda\, |\langle\psi|\lambda\rangle|^2 \tag{3.47}$$

$$= \int d\lambda\, \lambda\, p(\lambda). \tag{3.48}$$

Here $p(\lambda) = |\langle\psi|\lambda\rangle|^2$ is the probability density of being in the eigenspace with eigenvalue λ. Equivalently, it is the probability density (probability per unit lambda) of finding the value λ when the physical observable O corresponding to operator $\hat{O}$ is measured. Equation 3.48 simply expresses the fact that the expectation value gives the average or statistical mean value of λ when a large sample of measurements is made.

3.2.3 Unitary Operators and Pauli Matrices

While Hermitian operators are used to describe physical observables like energy and momentum, unitary operators are used to describe how states change under transformations.

Consider an operator $\hat{U}$. This operator is **unitary** if it obeys condition

$$\hat{U}^{\dagger}\hat{U} = \hat{I}, \quad \text{or equivalently} \quad \hat{U}^{-1} = \hat{U}^{\dagger}. \tag{3.49}$$

The eigenvalues of a unitary operator are always of unit norm, $|\lambda| = 1$, and so they lie on the unit circle in the complex plane. This reflects the fact that the action $\hat{U}$ on a state, $|\psi'\rangle = \hat{U}|\psi\rangle$, cannot change the magnitude or norm of the state ($|\psi'|^2 = |\psi|^2$), but can only perform a rotation of the complex-valued wavefunction in the Hilbert space.

These complex rotations are the generalizations of the real-valued rotations familiar from classical mechanics. Rotation matrices on a real-valued space are given by orthogonal operators, which are defined to be operators or matrices that obey the condition

$$\hat{O}^T\hat{O} = \hat{I}, \quad \text{or equivalently} \quad \hat{O}^{-1} = \hat{O}^T. \tag{3.50}$$

Such an operator preserves the lengths of real-valued vectors: if $v' = \hat{O}v$ then $|v'|^2 = |v|^2$. Clearly, orthogonal or rotation matrices are special cases of unitary matrices: they are unitary matrices with real entries.

There is an intimate connection between Hermitian and unitary operators. Recall that Hermitian operators have real eigenvalues. So if $\hat{O}$ is a Hermitian operator, then the exponential $\hat{U} = e^{-\frac{i}{\hbar}\hat{O}}$ will have eigenvalues on the unit circle, suggesting that $\hat{U}$ is unitary. This is in fact correct: you should be able to easily show using 3.4 and 3.49 that if $\hat{O}$ is Hermitian, then $\hat{U} = e^{-\frac{i}{\hbar}\hat{O}}$ is unitary.

One especially important unitary operator is the **time-evolution operator**, $\hat{U}(T) = e^{-\frac{i}{\hbar}\hat{H}T}$, which acts by translating states forward in time:

$$|\psi(t + T)\rangle = \hat{U}(T)|\psi(t)\rangle. \tag{3.51}$$

Here, $\hat{H}$ is the Hamiltonian operator; we say that the Hamiltonian is the **generator** of time translations. Written in terms of wavefunctions instead of state vectors, this becomes

$$\psi(x, t + T) = \int dy\, u(x - y, T)\psi(y, t), \tag{3.52}$$

where the propagation function is $u(x - y, T) = \langle x|\hat{U}(T)|y\rangle = \langle x|e^{-\frac{i}{\hbar}\hat{H}T}|y\rangle$. Equation 3.51 expresses the fact that the wavefunction at point x at a given time is formed from all of the previous amplitudes that have propagated to x from all other points in space during time T. Note that 3.51 has the same mathematical form as the optical propagation formula 2.35; the main difference is simply that here we are using time to parameterize the evolution, while in Eq. 2.35 the evolution is with respect to the spatial position z.

In a similar manner, other Hermitian operators can be used to generate additional unitary transformations. For example, the Hermitian momentum operator $\hat{p}$ generates unitary translations in space: if

$$\hat{U}(v) = \int d^3r |r + v\rangle\langle r| = e^{-\frac{i}{\hbar}v \cdot \hat{p}} \tag{3.53}$$

for some vector v, then

$$\hat{U}(v)|x\rangle = |x + v\rangle \tag{3.54}$$

and

$$\langle x|\hat{U}(v) = \langle x - v|. \tag{3.55}$$

In terms of wavefunctions, this becomes

$$\hat{U}(v)\psi(x) = \psi(x - v). \tag{3.56}$$

The latter can be verified by writing $\hat{p} = -i\hbar\nabla$ and expanding the exponential on the left side of the equation and then separately expanding the right side in a Taylor series about x. When this is done, it is immediately seen that all of the expansion terms on the two sides are equal. It can also be seen as follows:

$$\langle x|U(v)|\psi\rangle = \int d^3r \, \langle x|r + v\rangle\langle r|\psi\rangle \tag{3.57}$$

$$= \int d^3r \delta(r + v - x)\langle r|\psi\rangle \tag{3.58}$$

$$= \langle x - v|\psi\rangle \tag{3.59}$$

$$= \psi(x - v) \tag{3.60}$$

Other examples of Hermitian operators that are commonly used to generate unitary transformations are the orbital angular momentum L and the spin angular momentum S, whose corresponding unitary operators $e^{-\frac{i}{\hbar}\hat{L}\theta}$ and $e^{-\frac{i}{\hbar}\hat{S}\theta}$ generate rotations of spatial and spin wavefunctions. The examples given so far represent transformations in space-time, but similar operations can be generated for internal variables. For example, isospin rotations (converting protons and neutrons into each other) can be generated by the isospin-space analogs of angular momentum (see the example below).

A special set of operators are that will be of central importance in the coming chapters are those described by the **Pauli matrices:**

$$
\sigma_x = \begin{pmatrix} 0 & 1 \\ 1 & 0 \end{pmatrix}, \qquad \sigma_y = \begin{pmatrix} 0 & -i \\ i & 0 \end{pmatrix}, \qquad \sigma_z = \begin{pmatrix} 1 & 0 \\ 0 & 1 \end{pmatrix}. \tag{3.61}
$$

In the quantum information literature, these are often abbreviated as with single-letter names: $\hat{X} = \sigma_x, \hat{Y} = \sigma_y, \hat{Z} = \sigma_z$. The Pauli matrices are *both* Hermitian *and* unitary, signaling their dual role of being proportional to the observable spin operators, and of serving as transformation operators acting on those spins.

For a spin-$\frac{1}{2}$ particle, the spin operator is given by

$$
S = \frac{\hbar}{2}\sigma, \tag{3.62}
$$

where

$$
\sigma = \sigma_x \hat{x} + \sigma_y \hat{y} + \sigma_z \hat{z} = \begin{pmatrix} \hat{z} & \hat{x} - i\hat{y} \\ \hat{x} + i\hat{y} & -\hat{z}. \end{pmatrix} \tag{3.63}
$$

is a matrix-valued vector. For a unit vector $\hat{n}$, the operator $\hat{n} \cdot \sigma$ gives the spin along the direction of $\hat{n}$.

The use of Pauli matrices as unitary transformation operators follows from the identity (whose proof will be left as an exercise at the end of the chapter):

$$
e^{i\theta\hat{n}\cdot\sigma} = \hat{I}\cos\theta + i\hat{n} \cdot \sigma \sin\theta. \tag{3.64}
$$

So, exponentiating a Hermitian operator built from Pauli matrices just gives back a unitary transformation operator, also built from Pauli matrices. The spin-up and spin-down states along direction $\hat{n}$ are given by

$$
|\hat{n}, \uparrow\rangle = \begin{pmatrix} \cos\frac{\theta}{2} e^{-i\phi/2} \\ \sin\frac{\theta}{2} e^{+i\phi/2} \end{pmatrix}, \qquad |\hat{n}, \downarrow\rangle = \begin{pmatrix} -\sin\frac{\theta}{2} e^{-i\phi/2} \\ \cos\frac{\theta}{2} e^{+i\phi/2} \end{pmatrix}. \tag{3.65}
$$

For a spin along the z-axis ($\theta = 0$), the states above reduce to

$$
|\uparrow\rangle_z = \begin{pmatrix} 1 \\ 0 \end{pmatrix}, \qquad |\downarrow\rangle_z = \begin{pmatrix} 0 \\ 1 \end{pmatrix}. \tag{3.66}
$$

Some of the most useful properties of the Pauli matrices are given below. In the following, $\left[\hat{A}, \hat{B}\right] = \hat{A}\hat{B} - \hat{B}\hat{A}$ and $\left\{\hat{A}, \hat{B}\right\} = \hat{A}\hat{B} + \hat{B}\hat{A}$ are, respectively, the **commutator** and **anti-commutator** of the operators $\hat{A}$ and $\hat{B}$, while the totally antisymmetric Levi-Civita tensor ϵ_{ijk} is defined by $\epsilon_{ijk} = +1$ if $\{ijk\}$ is a cyclic permutation of $\{ijk\}$ is an anti-cyclic permutation, and $\epsilon_{ijk} = 0$ if two of the indices are equal. Explicitly, $\epsilon_{xyz} = \epsilon_{yzx} = \epsilon_{zxy} = +1$, but $\epsilon_{zyx} = \epsilon_{yxz} = \epsilon_{xzy} = -1$.

Properties of Pauli Matrices

1. They obey the anticommution relations

$$\{\sigma_i, \sigma_j\} = 2\delta_{ij}\hat{I}. \tag{3.67}$$

2. Their commutators obey the same relations as angular momentum operators:

$$[\sigma_i, \sigma_j] = 2i \sum_k \epsilon_{ijk}\sigma_k. \tag{3.68}$$

If $i \neq j$, all of the terms in the sum will vanish, except for one. So, for example, $[\sigma_x, \sigma_y] = 2i\sigma_z$, but $[\sigma_x, \sigma_z] = -2i\sigma_y$. Products are similar, but without the factors of 2:

$$\sigma_i\sigma_j = i \sum_k \epsilon_{ijk}\sigma_k. \tag{3.69}$$

3. They are traceless and square to the identity matrix: for any j,

$$Tr[\sigma_j] = 0, \quad \text{and} \quad \sigma_j^2 = \hat{I}. \tag{3.70}$$

(The trace is defined in the next subsection.)

4. Generalizing the last result, define a unit vector $\hat{n}$. Then:

$$(\hat{n} \cdot \sigma)^2 = \hat{I}. \tag{3.71}$$

5. The trace of two Pauli matrices is given by

$$Tr(\sigma_i\sigma_j) = 2\delta_{ij}. \tag{3.72}$$

6. Any unitary 2×2 matrix can be built as a linear combination of the identity matrix and the Pauli matrices. In other words, for any unitary 2×2 matrix, there exist complex numbers $\alpha, \beta_x, \beta_y, \beta_z$ such that

$$U = \alpha\hat{I} + \beta_x\sigma_x + \beta_y\sigma_y + \beta_z\sigma_z = \alpha\hat{I} + \beta \cdot \sigma. \tag{3.73}$$

(Equation 3.64 is an illustration of this fact.) Another way to state this result is that the identity matrix and the Pauli matrices together form a basis for the group $SU(2)$; see Appendix E.

Example 3.2.3 Consider an isospin state of a nucleon inside a nucleus. The basis for the nucleon states is given by states representing a proton or a neutron:

$$|p\rangle = \begin{pmatrix} 1 \\ 0 \end{pmatrix}, \qquad |n\rangle = \begin{pmatrix} 0 \\ 1 \end{pmatrix}. \tag{3.74}$$

Notice that this two-state system is formally identical to the spin-up/spin-down system of Example 3.2.1, with the proton and neutron representing isospin-up and isospin-down states in some abstract "internal" space of the nucleon. What is the transformation that converts a proton into a neutron? Clearly, a rotation by π about either the x- or y-axis can convert isospin-up along the z-axis to isospin-down. So consider an isospin analog of the unitary rotation operator about the y-axis:

$$\hat{U} = e^{-\frac{i}{\hbar}\pi S_y} = e^{-i\frac{\pi}{2}\sigma_y} \tag{3.75}$$

$$= \hat{I}\cos\frac{\pi}{2} - i\sigma_y\sin\frac{\pi}{2} \tag{3.76}$$

$$= \begin{pmatrix} 0 & -1 \\ 1 & 0 \end{pmatrix}, \tag{3.77}$$

where Eq. 3.64 has been used. The action of this operator is exactly what we wanted:

$$\hat{U}|p\rangle = |n\rangle, \qquad \hat{U}|n\rangle = -|p\rangle \quad \blacksquare \tag{3.78}$$

More generally, rotation of a spinor by angle α about unit vector

$$\hat{n} = \sin\theta\cos\phi\hat{x} + \sin\theta\sin\phi\hat{y} + \cos\theta\hat{z} \tag{3.79}$$

in a spin or isospin space can be expressed as

$$U = e^{-\frac{i}{\hbar}\alpha\hat{n}\cdot S} \tag{3.80}$$

$$= \hat{I}\cos\frac{\alpha}{2} - i\hat{n}\cdot S\sin\frac{\alpha}{2}, \tag{3.81}$$

where

$$\hat{n}\cdot S = \frac{\hbar}{2}\begin{pmatrix} \cos\theta & \sin\theta\,e^{-i\phi/2} \\ \sin\theta\,e^{i\phi/2} & -\cos\theta \end{pmatrix}. \tag{3.82}$$

3.2.4   Traces of Operators

For a matrix, the trace is simply defined to be the sum of the diagonal entries. For example,

$$\hat{A} = \begin{pmatrix} a & b \\ c & d \end{pmatrix} \implies \text{Tr } \hat{A} = a + d. \tag{3.83}$$

The trace is independent of basis:

$$\text{Tr}(\hat{U} \hat{A} \hat{U}^\dagger) = \text{Tr } \hat{A}, \tag{3.84}$$

for any unitary matrix $\hat{U}$. In particular, by going to a basis in which the matrix is diagonal, it follows that the trace must equal to the sum of eigenvalues of $\hat{A}$:

$$\text{Tr } \hat{A} = \sum_i \lambda_i. \tag{3.85}$$

Another useful property of the trace is that matrices can be permuted cyclically inside it:

$$\text{Tr}(\hat{A}\hat{B}\hat{C}) = \text{Tr}(\hat{C}\hat{A}\hat{B}) = \text{Tr}(\hat{B}\hat{C}\hat{A}). \tag{3.86}$$

These facts are well-known for finite-dimensional matrices, but for quantum mechanics, we need to know how to extend them to (possibly infinite-dimensional) operators. To see how to do this, consider the two-by-two matrix $\hat{A} = \begin{pmatrix} a & b \\ c & d \end{pmatrix}$. The two vectors

$$|0\rangle = \begin{pmatrix} 1 \\ 0 \end{pmatrix}, \qquad |1\rangle = \begin{pmatrix} 0 \\ 1 \end{pmatrix}$$

form an orthonormal basis for the space $\hat{A}$ acts upon. So the trace can be rewritten as

$$\text{Tr}\hat{A} = \langle 0|\hat{A}|0\rangle + \langle 1|\hat{A}|1\rangle = \sum_{j=0}^{1} \langle j|\hat{A}|j\rangle. \tag{3.87}$$

The reader can verify that the diagonal basis

$$|\psi_1\rangle = |+\rangle = \frac{1}{\sqrt{2}} \begin{pmatrix} 1 \\ 1 \end{pmatrix}, \qquad |\psi_2\rangle = |-\rangle = \frac{1}{\sqrt{2}} \begin{pmatrix} 1 \\ -1 \end{pmatrix},$$

which is also orthonormal, gives the same trace:

$$\text{Tr}\hat{A} = \langle +|\hat{A}|+\rangle + \langle -|\hat{A}|-\rangle = \sum_{j=0}^{1} \langle j|\hat{A}|j\rangle. \tag{3.88}$$

The last equation generalizes from matrices to operators: for any operator $\hat{A}$ acting on an N-dimensional system with a discrete orthonormal basis $|n\rangle$ with $n = 1, 2, \ldots N$, The trace can be written as

$$\text{Tr}\hat{A} = \sum_{n=1}^{N} \langle n|\hat{A}|n\rangle. \tag{3.89}$$

Here, N can be finite or countably infinite.

For the case where the eigenstates are labeled by continuous, rather than discrete, variables, the sum is simply replaced by an integral:

$$\text{Tr}\hat{A} = \int d\chi \, \langle \chi|\hat{A}|\chi\rangle. \tag{3.90}$$

In this manner, traces over position or momentum space can be carried out.

Example 3.2.4 Consider the operators

$$(a) \quad \hat{A} = a|0\rangle\langle 1| + b|1\rangle\langle 1| + c|0\rangle\langle 0| \tag{3.91}$$

$$(b) \quad \hat{B} = \sum_{n=0}^{\infty} \left(2^{-n}|n\rangle\langle n| + 3^{-n}|n\rangle\langle n-1|\right) \tag{3.92}$$

$$(c) \quad \hat{C} = e^{-a\hat{x}^2}, \tag{3.93}$$

where $\hat{x}$ is the position operator. Find the traces.

(a) For the first operator, we have

$$\text{Tr}\hat{A} = \sum_{j=0}^{1} \left\langle j \left| \left[\left(a|0\rangle\langle 1| + b|1\rangle\langle 1| + c|0\rangle\langle 0|\right) \right] \right| j \right\rangle \tag{3.94}$$

$$= c\langle 0|0\rangle\langle 0|0\rangle + b\langle 1|1\rangle\langle 1|1\rangle \tag{3.95}$$

$$= c + b. \tag{3.96}$$

Of course, the same result could be obtained simply by writing the operator in matrix form:

$$\hat{A} = \begin{pmatrix} c & a \\ 0 & b \end{pmatrix}, \tag{3.97}$$

and summing the diagonal elements.

(b) For the second operator, assuming the $|m\rangle$ states form a complete orthonormal basis,

$$\mathrm{Tr}\hat{B} = \sum_{m,n=0}^{\infty} \langle m| \left(\left(2^{-n}|n\rangle\langle n| + 3^{-n}|n\rangle\langle n-1|\right)\right)|m\rangle \tag{3.98}$$

$$= \sum_{m,n=0}^{\infty} \left(\left(2^{-n}\delta_{mn} + 3^{-n}\delta_{m,n-1}\delta_{mn}\right)\right) \tag{3.99}$$

$$= \sum_{n=0}^{\infty} 2^{-n} \tag{3.100}$$

$$= \frac{1}{1 - \frac{1}{2}} \tag{3.101}$$

$$= 2, \tag{3.102}$$

summing the geometric series.

(c) Inserting the complete position basis $|x\rangle$ and integrating over the continuous x variable,

$$\mathrm{Tr}\,\hat{C} = \int_{-\infty}^{\infty} dx \langle x|e^{-a\hat{x}^2}|x\rangle \tag{3.103}$$

$$= \int_{-\infty}^{\infty} dx \langle x|e^{-ax^2}|x\rangle \tag{3.104}$$

$$= \int_{-\infty}^{\infty} dx\, e^{-ax^2} \tag{3.105}$$

$$= \sqrt{\frac{\pi}{a}}, \tag{3.106}$$

where $\hat{x}|x\rangle = x|x\rangle$ was used in the second line, and the Gaussian integral was performed in the last line.

3.2.5 Density Operators

So far, we have taken it for granted that the quantum system of interest can be described by a specific state vector $|\psi\rangle$. But this assumes that the experimenter is certain that the one particular state described by $|\psi\rangle$ was prepared. Unfortunately, in real life, this is usually not the case. More often, there is some range of possible states being produced, with some probability distribution describing the likelihoods of each one. For example, a thermally excited electron in an atom may have several lower-energy states it could decay into (Fig. 3.1). In that case, the photon produced

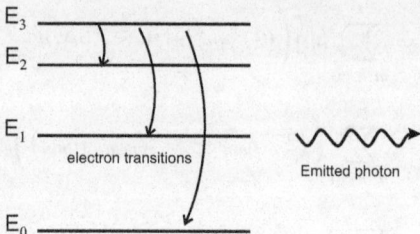

Fig. 3.1 An electron in an excited state may decay to various lower-energy levels, with varying probabilities. As a result, the emitted photon may have any one of several different frequencies. If the frequency is not measured, then the photon will be in a mixed state

in the decay could have several different frequencies, with some transitions being more likely than others.

To address cases like this, we introduce the **density operator** or **density matrix**, $\hat{\rho}$. Suppose several possible states $|\psi_j\rangle$, with $j = 1, 2, 3, \ldots, n$ (where n could be infinite), can be produced, each with corresponding probability p_j. Then as long as we don't make a measurement to see which $|\psi_j\rangle$ is present, the full state of the system has to include all of the possible $|\psi_j\rangle$ and cannot be described by a single-state vector or wavefunction. A meaningful description of the full state is instead given by the operator

$$\hat{\rho} = \sum_{j=1}^{n} p_j |\psi_j\rangle\langle\psi_j|. \tag{3.107}$$

Note that since the p_j are probabilities, they have to satisfy the conditions

$$0 \le p_j \le 1 \text{ for all } j, \qquad \text{and} \qquad \sum_{j=1}^{n} p_j = 1. \tag{3.108}$$

A result called **Gleason's theorem** assures us that for any set of probabilities satisfying Eq. 3.108, we can find a density matrix that will reproduce them.

$\hat{\rho}$ is simply a convex sum of projection operators $\mathcal{P}_j = |\psi_j\rangle\langle\psi_j|$, where convex here means that the coefficients of the operators sum to unity.

If the density operator can be written in a form with only a single term present in the sum of Eq. 3.107, $\hat{\rho} = |\psi\rangle\langle\psi|$, then $\hat{\rho}$ is called a **pure state**, and it can simply be described by the state vector $|\psi\rangle$ or the corresponding wavefunction $\psi(x)$. If the state is *not* pure, then the system is said to be in a **mixed state**. Mixed states represent statistical distributions of pure states; for example, if a cathode ray tube is producing a stream of electrons with random spins, then the coefficients of the $|\uparrow\rangle\langle\uparrow|$ and $|\downarrow\rangle\langle\downarrow|$ terms in $\hat{\rho}$ give probabilities of a random electron being spin-up or spin-down, respectively. The density matrix of a mixed state cannot be written in the single-state form $\hat{\rho} = |\psi\rangle\langle\psi|$, but will need to be expressed as a *sum* over projectors that cannot be factored into a single product.

Example 3.2.5 If the experimenter is producing state $|\psi\rangle = |0\rangle$ with 100% probability, then the resulting pure state can be expressed as a density operator:

$$\hat{\rho} = |0\rangle\langle 0| = \begin{pmatrix} 1 \\ 0 \end{pmatrix} (1\ 0) = \begin{pmatrix} 1 & 0 \\ 0 & 0 \end{pmatrix}. \tag{3.109}$$

In contrast, if the source of the states produces $|0\rangle$ with probability p and state $|1\rangle$ with probability $1 - p$, then the resulting mixed state is described by operator

$$\hat{\rho} = p|0\rangle\langle 0| + (1 - p)|1\rangle\langle 1| = \begin{pmatrix} p & 0 \\ 0 & 1 - p \end{pmatrix}. \tag{3.110}$$

The fact that $\hat{\rho}$ is a linear combination of two pure density operators is a signal of the mixed nature of the state.

However, whether a state is mixed or pure is not always immediately obvious from looking at the density operator. For example, consider a density matrix of the form

$$\hat{\rho} = p|0\rangle\langle 0| + (1 - p)|1\rangle\langle 1| + q|0\rangle\langle 1| + q^*|1\rangle\langle 0| = \begin{pmatrix} p & q \\ q^* & 1 - p \end{pmatrix} \tag{3.111}$$

which is a perfectly legitimate density matrix. In general, it is a mixed state. But if there are complex numbers α and β such that $p = |\alpha|^2$, $\alpha\beta^* = q$, and $|\alpha|^2 + |\beta|^2 = 1$, then the state becomes

$$\hat{\rho} = |\alpha|^2|0\rangle\langle 0| + |\beta|^2|1\rangle\langle 1| + \alpha\beta^*|0\rangle\langle 1| + \alpha^*\beta|1\rangle\langle 0| \tag{3.112}$$

$$= \left(\alpha|0\rangle + \beta|1\rangle\right)\left(\alpha^*\langle 0| + \beta^*\langle 1|\right) \tag{3.113}$$

$$= |\psi\rangle\langle\psi|, \tag{3.114}$$

which is clearly a pure state. This again highlights that the state only need be factorable in *one* Hilbert space basis in order to be pure and may only *look* non-factorable in the basis in which it is given.   ∎

Aside from being a positive operator (meaning that the p_j, or equivalently the eigenvalues of $\hat{\rho}$, are all non-negative), the other fundamental property of the density operator is that it has unit trace:

$$\text{Tr}\ \hat{\rho} = 1. \tag{3.115}$$

Before showing this, recall that the trace can be computed by choosing an orthonormal basis $|n\rangle$, $n = 1, 2, \ldots$, and then for any operator $\hat{O}$, the trace is

$$\text{Tr } \hat{O} = \sum_n \langle n|\hat{O}|n\rangle. \tag{3.116}$$

The trace is basis-independent, so any other orthonormal basis will give the same result.

Applying this to the density operator, we find that

$$\text{Tr } \hat{\rho} = \sum_{n,j} \left\langle n \left| \left(p_j |\psi_j\rangle\langle\psi_j| \right) \right| n \right\rangle$$

$$= \sum_{n,j} p_j \langle\psi_j|n\rangle\langle n|\psi_j\rangle \qquad \text{(rearranging)}$$

$$= \sum_j p_j \langle\psi_j|\psi_j\rangle \qquad \left(\text{using completeness, } \sum_n |n\rangle\langle n| = \hat{I} \right)$$

$$= \sum_j p_j \qquad \left(\text{since } |\psi_j\rangle \text{ is normalized, } \langle\psi_j|\psi_j\rangle = 1 \right),$$

$$= 1. \tag{3.117}$$

Although the trace of $\hat{\rho}$ is always the same, that is not true of its square, which must obey the following:

$$0 < \text{Tr } \hat{\rho}^2 \leq 1. \tag{3.118}$$

For a pure state $\hat{\rho} = |\psi\rangle\langle\psi|$, we find that

$$\text{Tr } \hat{\rho}^2 = \sum_n \langle n|\hat{\rho}^2|n\rangle \tag{3.119}$$

$$= \sum_n \langle n|\psi\rangle\langle\psi|\psi\rangle\langle\psi|n\rangle \tag{3.120}$$

$$= \sum_n \langle\psi|n\rangle\langle n|\psi\rangle \tag{3.121}$$

$$= \langle\psi|\psi\rangle \tag{3.122}$$

$$= 1. \tag{3.123}$$

On the other hand, a mixed state $\hat{\rho} = \sum p_j |\psi_j\rangle\langle\psi_j|$ must always have $\text{Tr } \hat{\rho}^2 < 1$, since

$$\text{Tr}\hat{\rho}^2 = \sum_{njk} \langle n|p_j p_k|\psi_j\rangle\langle\psi_j|\psi_k\rangle\langle\psi_k|n\rangle \tag{3.124}$$

$$= \sum_{njk} p_j p_k \langle n|\psi_j\rangle \delta_{jk} \langle \psi_k|n\rangle \tag{3.125}$$

$$= \sum_{nj} p_j^2 \langle \psi_j|n\rangle \langle n|\psi_j\rangle \tag{3.126}$$

$$= \sum_{j} p_j^2 \langle \psi_j|\psi_j\rangle \tag{3.127}$$

$$= \sum_{j} p_j^2 \tag{3.128}$$

$$< \sum_{j} p_j \tag{3.129}$$

$$= 1. \tag{3.130}$$

In the next to last line, we used the fact that $p_j^2 < p_j$ for $0 < p_j < 1$ and that for a mixed state, all of the p_j must be *strictly* less than one (not less than or equal to). Therefore, it is common to define a quantity called the **purity**,

$$P = \text{Tr } \hat{\rho}^2, \tag{3.131}$$

so that $P = 1$ for pure states, and $0 < P < 1$ for mixed states. Computing the purity of the operators in the previous example immediately tells us that the states of Eqs. 3.109 and 3.110 are pure and mixed, respectively. Similarly, applying the condition $\text{Tr } \hat{\rho}^2 = 1$ tells us that the state of Eq. 3.111 is pure if and only if $p^2 = |q|^2$, or equivalently, if $|\alpha|^2 + |\beta|^2 = 1$.

Expectation values for operators in states described by density operators are found by multiplying by $\hat{\rho}$ and taking the trace:

$$\langle \hat{O}\rangle = \text{Tr}\left(\hat{\rho}\hat{O}\right) = \text{Tr}\left(\hat{O}\hat{\rho}\right). \tag{3.132}$$

We can easily see that this is consistent with how we defined the expectation operator for a pure state. For pure state $|\psi\rangle$,

$$\text{Tr}\left(\hat{\rho}\hat{O}\right) = \text{Tr}\left(|\psi\rangle\langle\psi|\hat{O}\right) \tag{3.133}$$

$$= \sum_{n} \langle n|\psi\rangle\langle\psi|\hat{O}|n\rangle \tag{3.134}$$

$$= \sum_{n} \langle\psi|\hat{O}|n\rangle\langle n|\psi\rangle \tag{3.135}$$

$$= \langle\psi|\hat{O}|\psi\rangle \tag{3.136}$$

$$= \langle\hat{O}\rangle, \tag{3.137}$$

where we have again used completeness. More generally, for any state (mixed or pure),

$$\text{Tr}\left(\hat{\rho}\hat{O}\right) = \text{Tr}\left(\sum_j p_j |\psi_j\rangle\langle\psi_j|\hat{O}\right) \tag{3.138}$$

$$= \sum_{n,j} p_j \langle n|\psi_j\rangle\langle\psi_j|\hat{O}|n\rangle \tag{3.139}$$

$$= \sum_{n,j} \langle\psi_j|\hat{O}|n\rangle\langle n|\psi_j\rangle \tag{3.140}$$

$$= \sum_j p_j \langle\psi_j|\hat{O}|\psi_j\rangle \tag{3.141}$$

$$= \langle\hat{O}\rangle. \tag{3.142}$$

3.2.6   Comparing States: Fidelity and Trace Distance

The **fidelity** is a measure of the overlap between two states, or equivalently, it is a measure of the probability of mistaking one state for the other. For pure states, the fidelity between $|\phi\rangle$ and $|\psi\rangle$ is defined to be

$$\mathcal{F}(|\phi\rangle, |\psi\rangle) = |\langle\phi|\psi\rangle|^2. \tag{3.143}$$

For mixed states with density operators $\hat{\rho}$ and $\hat{\sigma}$, this generalizes to

$$\mathcal{F}(\hat{\rho}, \hat{\sigma}) = Tr(\hat{\rho}\hat{\sigma}). \tag{3.144}$$

The latter definition clearly reduces back to the previous one if the states are, in fact, pure; for $\hat{\rho} = |\psi\rangle\langle\psi|$ and $\hat{\sigma} = |\phi\rangle\langle\phi|$, we find that

$$\mathcal{F}(\hat{\rho}, \hat{\sigma}) = Tr(\hat{\rho}\hat{\sigma}) = Tr\left(|\psi\rangle\langle\psi|\phi\rangle\langle\phi|\right) \tag{3.145}$$

$$= Tr\left(\langle\phi|\psi\rangle\langle\psi|\phi\rangle\right) \tag{3.146}$$

$$= |\langle\phi|\psi\rangle|^2. \tag{3.147}$$

In the second line, the cyclic property of the trace was used.

In the case of a mixed state ρ and a pure state $|\psi\rangle$, the fidelity can also be written as

$$\mathcal{F} = \langle\psi|\hat{\rho}|\psi\rangle. \tag{3.148}$$

The reader should be aware that there are multiple slightly different definitions of the fidelity in use in the literature. Some authors (e.g., reference [14]) define the fidelity as the square of what we have defined here. Another variation arises for the following reason. If two pure states are equal, then according to Eq. 3.143, their fidelity equals 1. But according to Definition 3.144, this is not true for identical mixed states: if $\hat{\rho} = \hat{\sigma}$, then $\mathcal{F}(\hat{\rho}, \hat{\sigma}) = Tr(\hat{\rho}^2) < 1$. Because of this, some authors prefer to use the definition

$$\mathcal{F}(\hat{\rho}, \hat{\sigma}) = Tr\sqrt{\hat{\rho}\hat{\sigma}}, \tag{3.149}$$

or

$$\mathcal{F}(\hat{\rho}, \hat{\sigma}) = \left(Tr\sqrt{\hat{\rho}\hat{\sigma}}\right)^2, \tag{3.150}$$

which both have the advantage that when $\hat{\rho} = \hat{\sigma}$, then $\mathcal{F}(\hat{\rho}, \hat{\sigma}) = Tr(\hat{\rho}) = 1$, for both mixed and pure states.

A second measure of similarity between two states is the **trace distance**. For two density operators $\hat{\rho}$ and $\hat{\sigma}$, this is defined to be

$$D(\hat{\rho}, \hat{\sigma}) = \frac{1}{2}\mathrm{Tr}\left(|\hat{\rho} - \hat{\sigma}|\right). \tag{3.151}$$

The absolute value signs mean to take the absolute value of each matrix entry. This satisfies all of the necessary conditions for a true distance measure:

$D(\hat{\rho}, \hat{\sigma}) = D(\hat{\sigma}, \hat{\rho})$ (symmetric)

$D(\hat{\rho}, \hat{\sigma}) \geq 0$ (non-negative)

$D(\hat{\rho}, \hat{\sigma}) = 0$ if and only if $\hat{\rho} = \hat{\sigma}$ (non-degeneracy)

$D(\hat{\rho}, \hat{\sigma}) \leq D(\hat{\lambda}, \hat{\rho}) + D(\hat{\lambda}, \hat{\sigma})$ (for any density $\hat{\lambda}$).

The fidelity and trace distance will appear in Chap. 13, where they will be used to quantify entanglement. They can be shown to obey the inequality:

$$1 - \sqrt{\mathcal{F}} \leq D \leq \sqrt{1 - \mathcal{F}}. \tag{3.152}$$

Example 3.2.6 Consider a pair of pure photon polarization states:

$$|\psi\rangle = |H\rangle, \qquad |\phi\rangle = \cos\theta|H\rangle + e^{i\phi}\sin\theta|V\rangle.$$

The density matrices of these states are, respectively,

$$\hat{\rho} = |H\rangle\langle H|$$

$$\hat{\sigma} = \left(\cos\theta|H\rangle + e^{i\phi}\sin\theta|V\rangle \right)\left(\cos\theta\langle H| + e^{-i\phi}\sin\theta\langle V| \right)$$

$$= \cos^2\theta|H\rangle\langle H| + \sin^2\theta|V\rangle\langle V| + \sin\theta\cos\theta\left(e^{i\phi} + e^{-i\phi} \right)\left(|H\rangle\langle V| + |V\rangle\langle H| \right)$$

$$= \cos^2\theta|H\rangle\langle H| + \sin^2\theta|V\rangle\langle V| + 2\sin\theta\cos\theta\cos\phi\left(|H\rangle\langle V| + |V\rangle\langle H| \right).$$

For pure states, the fidelity can be found immediately using Eq. 3.143:

$$\mathcal{F}(|\psi\rangle, |\phi\rangle) = |\langle\psi|\phi\rangle|^2 = \cos^2\theta.$$

Alternatively, we could use Eq. 3.144, which should give the same result:

$$\mathcal{F}(\hat{\rho}, \hat{\sigma}) = \text{Tr}\left(\hat{\rho}\hat{\sigma} \right)$$

$$= \text{Tr}\left(\cos^2\theta|H\rangle\langle H| + 2\sin\theta\cos\theta\cos\phi\left(|H\rangle\langle V| + |V\rangle\langle H| \right) \right)$$

$$= \cos^2\theta. \tag{3.153}$$

The trace distance is also easily found:

$$D(\hat{\rho}, \hat{\sigma})$$

$$= \frac{1}{2}\text{Tr}\left(\left|1 - \cos^2\theta\right| |H\rangle\langle H| + \left|-\sin^2\theta\right| |V\rangle\langle V| \right.$$

$$\left. + |2\sin\theta\cos\theta\cos\phi| \left(|H\rangle\langle V| + |V\rangle\langle H| \right) \right)$$

$$= \frac{1}{2}\left(\left|1 - \cos^2\theta\right| + \left|-\sin^2\theta\right| \right)$$

$$= \sin^2\theta.$$

We see that

$$1 - \sqrt{\mathcal{F}} = 1 - |\cos\theta| \leq 1 - \cos^2\theta = \sin^2\theta = D(\hat{\rho}, \hat{\sigma})$$

$$\sqrt{1 - \mathcal{F}} = \sqrt{1 - \cos^2\theta} = |\sin\theta| \geq \sin^2\theta = D(\hat{\rho}, \hat{\sigma}), \tag{3.154}$$

so that Inequality 3.152 is satisfied. ∎

3.3 Uncertainty and Fluctuations

The aspect of quantum mechanics that probably was the hardest for physicists to accept during the early years of the theory's development was the inherently random character of many measurements. An electron, for example, is no longer a point particle with a definite position. Instead, it is described by a wavefunction that is spread over a finite region of space. If a measurement is made of the electron's position, it could be found anywhere that the wavefunction is nonzero, and which of those positions is actually measured is purely random.

Suppose a set or ensemble of identical copies of a quantum system are prepared in an identical manner, and the same variable is measured on each copy of the system. Classically, you would expect to get the same value each time (at least within the resolution of your measuring device). But quantum mechanically, you get a range of values, described by a probability distribution. For this distribution, a mean value can be obtained (the expectation value), but the measured values will be scattered randomly about the mean with some characteristic spread. This spread will be the **standard deviation** or **uncertainty** in the variable. Unlike the classical case, this uncertainty is inherent to the system and cannot be reduced by using more accurate measuring devices or by reducing the external perturbations on the system. The uncertainty will measure the characteristic size of fluctuations about the mean. As will be seen repeatedly, the fluctuations are often as important as the mean value.

A clear example of this can be seen by considering the vacuum. In quantum mechanics, the vacuum is never empty: the uncertainty principle allows the constant creation and re-annihilation of virtual particle-antiparticle pairs. To a large extent, the vacuum is like a pot of boiling water, in a constant state of agitation from the creation and bursting of vapor bubbles. On average, all fields with spin or charge (electric fields, nuclear fields, etc.) should have zero expectation value in the vacuum state. But it will be seen in Chap. 5 that the uncertainty of these fields will *not* be zero. This nonzero vacuum uncertainty in the fields caused by their constant fluctuation has experimentally observable consequences, such as the Casimir effect and the Lamb shift [15, 16].

So consider an observable represented by a Hermitian operator $\hat{A}$. The mean or **expectation value** of $\hat{A}$ for the system in state $|\psi\rangle$ is

$$\bar{A} = \langle \hat{A} \rangle = \langle \psi | \hat{A} | \psi \rangle = \int \psi^*(x) \hat{A} \psi(x) \, dx, \tag{3.155}$$

where the last expression is written in the position basis, and the others are in basis-independent form. The standard deviation or **uncertainty** is

$$\Delta A = \sqrt{\left\langle (\hat{A} - \langle \hat{A} \rangle)^2 \right\rangle} = \sqrt{\langle \hat{A}^2 \rangle - \langle \hat{A} \rangle^2}. \tag{3.156}$$

The variance is simply the square of the uncertainty:

$$\mathrm{Var}(\hat{A}) = (\Delta A)^2 = \left\langle (\hat{A} - \langle \hat{A} \rangle)^2 \right\rangle = \langle \hat{A}^2 \rangle - \langle \hat{A} \rangle^2 \tag{3.157}$$

The final equality in the last line is easy to show:

$$\left\langle (\hat{A} - \langle \hat{A} \rangle)^2 \right\rangle = \left\langle \hat{A}^2 - 2\hat{A}\langle \hat{A} \rangle + \langle \hat{A} \rangle^2 \right\rangle \tag{3.158}$$

$$= \langle \hat{A}^2 \rangle - 2\langle \hat{A} \rangle \langle \hat{A} \rangle + \langle \hat{A} \rangle^2 \tag{3.159}$$

$$= \langle \hat{A}^2 \rangle - \langle \hat{A} \rangle^2. \tag{3.160}$$

Recall that the commutator of two operators $\hat{A}$ and $\hat{B}$ is

$$\left[\hat{A}, \hat{B} \right] = \hat{A}\hat{B} - \hat{B}\hat{A}. \tag{3.161}$$

The existence of a nonzero commutator means that measurement of one operator will in some sense "disturb" the other operator and increase its uncertainty. The **Robertson uncertainty relation**, which is a generalization of the famous **Heisenberg uncertainty relation**, relates the minimum uncertainty product of two operators to their commutator:

$$\Delta A \Delta B \geq \frac{1}{2i} \langle [\hat{A}, \hat{B}] \rangle. \tag{3.162}$$

If the two operators happen to be canonically conjugate, meaning that they obey $[\hat{A}, \hat{B}] = i\hbar$, then this reduces to the Heisenberg relation:

$$\Delta A \Delta B \geq \frac{\hbar}{2}. \tag{3.163}$$

Examples of canonically conjugate pairs include position and momentum (x, p), and time and energy (t, E), leading to the uncertainty relations:

$$\Delta x \Delta p \geq \frac{\hbar}{2} \tag{3.164}$$

$$\Delta E \Delta t \geq \frac{\hbar}{2}. \tag{3.165}$$

Notice that the uncertainty products vanish as we take the classical limit $\hbar \to 0$, as they should.

There is no lower limit to the uncertainty of any single variable, only of non-commuting *pairs* of variables. So, for example, we can measure energy as precisely as we want, but only at the expense of increasing the uncertainty in time. For a concrete example, consider an atom emitting a photon. We can easily measure the photon frequency with a spectrometer. If we repeat the measurement many times on many emitted photons, we will find a set of spectral lines of finite width centered at some collection of discrete frequencies. The angular frequency ω is related to the

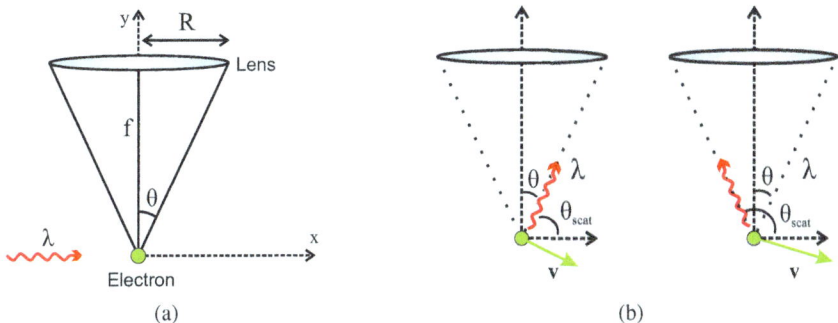

Fig. 3.2 Heisenberg microscope: a photon (**a**) strikes the electron and (**b**) leads to Compton scattering. The photon is collected in the lens in order to determine the location of the electron. The uncertainties in electron position and momentum obey an uncertainty relation

photon energy by $\omega = \frac{E}{\hbar}$, so we can make the spectral lines decrease in width (i.e., reduce the uncertainty in ω or E), by measuring for a longer time. If the detector takes a longer time to make each measurement, that amounts to increasing the uncertainty in the time at which the photon was emitted. As a result, the minimum allowable uncertainty in energy and frequency is smaller. (Here we ignore classical sources of line broadening, such as Doppler broadening or variations in energy due to collisions between the emitting atoms.)

Box 3.2 The Heisenberg Microscope

A thought experiment devised by Heisenberg gives a clear illustration of how the uncertainty principle arises in a particular case. Imagine trying to determine the position of an electron by viewing it through a powerful magnifying glass. The electron lies at the focal point. Denote the focal length and the radius of the lens by f and R. Take the x- and y-axes as shown in Fig. 3.2a. The opening half-angle of the lens is $\theta \approx \tan \theta = \frac{R}{f}$.

Imagine a photon of wavelength λ, initially traveling in the x-direction, being scattered off the electron. The photon is being used as a probe of the electron position and is to be viewed through the lens. The scattered photon has some new wavelength λ', which can be related to the scattering angle θ_{sc} by the Compton formula [1, 2],

$$\lambda' - \lambda = \frac{h}{mc}\left(1 - \cos\theta_{sc}\right). \tag{3.166}$$

But to give a rough estimate of uncertainty, we can avoid using the Compton formula by assuming that θ and $\lambda - \lambda'$ are small. The initial and final photon

(continued)

Box 3.2 (continued)
momenta in the x-direction are

$$p_{ph,x} = \frac{h}{\lambda} \quad \text{and} \quad p'_{ph,x} = \left(\frac{h}{\lambda'}\right) \cos\theta_{sc}.$$

By momentum conservation, the x-component of momentum gained by the electron is then

$$\delta p = p_{ph,x} - p'_{ph,x} = \frac{h}{\lambda} - \frac{h}{\lambda'} \cos\theta_{sc}.$$

But in order for the scattered photon to be seen in the lens, the scattering angle must obey $90° - \theta \le \theta_{scat} \le 90° + \theta$ (Fig. 3.2b). At these angles, the electron momentum takes its maximum and minimum values, so the uncertainty in the final electron momentum is

$$\begin{aligned}
\Delta p_e &= \frac{h}{\lambda'}\Big(\cos(90° - \theta) - \cos(90° + \theta)\Big) \\
&= \frac{h}{\lambda'}\Big(\sin\theta - (-\sin\theta)\Big) \\
&= \frac{2h}{\lambda'}\sin\theta \\
&\approx \frac{2h}{\lambda}\theta,
\end{aligned}$$

where in the last line, we approximated $\sin\theta \approx \theta$ and $\lambda' \approx \lambda$.

Ignoring factors of order 1, either the Rayleigh or Abbé criteria will tell us that the smallest angle that can be resolved by the lens is $\theta_{res} \approx \frac{\lambda}{2R}$, so that the position of the electron after the photon is detected could be anywhere in the range

$$-f \tan\theta_{res} \le x \le f \tan\theta_{res}.$$

This gives a position uncertainty of $\Delta x = 2f \tan\theta_{res} \approx \frac{f\lambda}{R}$. But from Fig. 3.2b, $\frac{f}{R} = \cot\theta \approx \frac{1}{\theta}$, which leads to $\Delta x \approx \frac{\lambda}{\theta}$.

Therefore, taking the product of the position and momentum uncertainty, we arrive at

$$\Delta x \Delta p = \frac{2h\theta}{\lambda} \cdot \frac{\lambda}{\theta} = 2h,$$

which is consistent (up to factors of order unity) with the Heisenberg relation.

3.4   Composite Systems and Partial Traces

Consider two particles, or more generally two quantum systems (two multiparticle systems or two quantum fields). Label the two particles A and B. If they have never interacted with each other, then each of the two will be described by some quantum state of its own: $|\psi\rangle_A$ and $|\phi\rangle_B$. As will be seen in the next section, particles that *have* interacted at some point often can't be described by individual states for each particle and will have to be described by a single *joint* state that describes the whole pair as a single unit. Even if they have never interacted, it is still often convenient to describe them as a single joint state. For example, if the particles are indistinguishable and close enough that there is overlap in their wavefunctions, then we have to take into account the symmetry (for boson) or antisymmetry (for fermions) of their joint states.

To describe joint or composite states made from multiple particles, let $\mathcal{H}_A$ and $\mathcal{H}_B$ be the Hilbert spaces for the two particles. Orthonormal bases $|\mu_j\rangle_A$ and $|\lambda_k\rangle_B$, for $j, k = 1, 2, 3, \ldots$, can be found for each Hilbert space. Then the individual single-particle states can be written in these bases:

$$|\psi\rangle_A = \sum_j a_j |\mu_j\rangle_A, \qquad |\phi\rangle_B = \sum_k b_k |\lambda_k\rangle_B,$$

for some set of complex coefficients a_j and b_k. If the two particles are unaffected by each other, then their joint two-particle state is simply the tensor product of the one-particle states:

$$|\Psi\rangle_{AB} = |\psi\rangle_A \otimes |\phi\rangle_B = \sum_{jk} a_j b_k |\mu_j\rangle_A \otimes |\lambda_k\rangle_B. \tag{3.167}$$

But more generally, any joint state formed by the two-particle composite system can be written in the form

$$|\Psi\rangle_{AB} = \sum_{jk} c_{jk} |\mu_j\rangle_A \otimes |\lambda_k\rangle_B. \tag{3.168}$$

In general, this composite state will *not* factor into a product of two single-particle states as in Eq. 3.167; in other words, there is no factorization $c_{jk} = a_j b_k$. The basis for the composite system will be the set of orthonormal vectors of the form $|\mu_j\rangle_A \otimes |\lambda_k\rangle_B$, which spans the two-particle Hilbert space $\mathcal{H}_{AB} = \mathcal{H}_A \otimes \mathcal{H}_B$.

The tensor product is often written in one of two abbreviated forms:

$$|\mu_j\rangle_A \otimes |\lambda_k\rangle_B = |\mu_j\rangle_A |\lambda_k\rangle_B = |\mu_j A \lambda_k B\rangle.$$

All three of these expressions mean exactly the same thing. When there is no ambiguity, the notation is sometimes abbreviated further by dropping the A, B

subscripts:

$$|\mu_j \lambda_k\rangle.$$

It is then understood that the first entry refers to the A subsystem and the second entry to B. In the following, the A and B labels will often be dropped on state vectors in order to unclutter notation whenever there is no danger of confusion.

To see better how the tensor product works, consider an example.

Example 3.4.1 Consider two particles A and B, where each particle can be in a superposition of two states, labeled $|0\rangle$ and $|1\rangle$:

$$|\psi\rangle_A = \alpha|0\rangle_A + \beta|1\rangle_A, \qquad\qquad |\phi\rangle_B = \gamma|0\rangle_B + \delta|1\rangle_B.$$

Then the composite system will be in the two-particle state given by

$$
\begin{aligned}
|\Psi\rangle_{AB} &= |\psi\rangle \otimes |\phi\rangle \\
&= \left(\alpha|0\rangle_A + \beta|1\rangle_A\right) \otimes \left(\gamma|0\rangle_B + \delta|1\rangle_B\right) \\
&= \alpha\gamma|0\rangle_A|0\rangle_B + \alpha\delta|0\rangle_A|1\rangle_B + \beta\gamma|1\rangle_A|0\rangle_B + \beta\delta|1\rangle_A|1\rangle_B \\
&= \alpha\gamma|00\rangle + \alpha\delta|01\rangle + \beta\gamma|10\rangle + \beta\delta|11\rangle. \qquad\qquad (3.169)
\end{aligned}
$$

Just as the single-particle state $|\psi\rangle = \alpha|0\rangle + \beta|1\rangle$ can be written in the matrix form

$$|\psi\rangle = \begin{pmatrix} \alpha \\ \beta \end{pmatrix},$$

where the two entries represent the coefficients of the basis vectors $|0\rangle = \begin{pmatrix} 1 \\ 0 \end{pmatrix}$ and $|1\rangle = \begin{pmatrix} 0 \\ 1 \end{pmatrix}$, the two-particle state can be written in the form of a column vector with a four-dimensional basis:

$$|\Psi\rangle_{AB} = \begin{pmatrix} \alpha\gamma \\ \alpha\delta \\ \beta\gamma \\ \beta\delta \end{pmatrix}, \qquad\qquad (3.170)$$

where basis vectors are

$$|00\rangle = \begin{pmatrix} 1 \\ 0 \\ 0 \\ 0 \end{pmatrix}, \quad |01\rangle = \begin{pmatrix} 0 \\ 1 \\ 0 \\ 0 \end{pmatrix}, \quad |10\rangle = \begin{pmatrix} 0 \\ 0 \\ 1 \\ 0 \end{pmatrix}, \quad |11\rangle = \begin{pmatrix} 0 \\ 0 \\ 0 \\ 1 \end{pmatrix}. \quad \blacksquare$$

(3.171)

Inner products of composite systems operate by taking the inner products of the A vectors and then separately of the B vectors:

$$\left({}_A\langle\xi| \otimes {}_B\langle\chi| \right) \cdot \left(|\psi\rangle_A \otimes |\phi\rangle_B \right) = \left(\langle\xi|\psi\rangle \right)_A \cdot \left(\langle\chi|\phi\rangle \right)_B. \qquad (3.172)$$

Clearly, all of this can be generalized in the obvious manner to composite systems formed by more than two particles, with multiparticle states of the form

$$|\Psi\rangle_{ABCD...} = |\psi\rangle_A \otimes |\phi\rangle_B \otimes |\chi\rangle_C \otimes |\xi\rangle_D \dots. \qquad (3.173)$$

Once multiparticle composite *states* have been formed, it then becomes necessary to form multiparticle *operators*. This is simple enough; given operator O_A acting on particle A and operator Q_B acting on particle B, we just form the corresponding **Kronecker product** operator:

$$\Lambda_{AB} = O_A \otimes Q_B. \qquad (3.174)$$

What this means can most easily be seen by examples.

Example 3.4.2 Consider again the state $|\Psi\rangle_{AB}$ of the previous example. Consider the operators $\sigma_{zA} \otimes \sigma_{xB}$ and $\sigma_{yA} \otimes \hat{I}_B$. They act on $|\Psi\rangle_{AB}$ according to

$$\sigma_{zA} \otimes \sigma_{xB} |\Psi\rangle_{AB} = \left[\sigma_{zA} \begin{pmatrix} \alpha \\ \beta \end{pmatrix} \right] \otimes \left[\sigma_{xB} \begin{pmatrix} \gamma \\ \delta \end{pmatrix} \right]$$

$$= \left[\begin{pmatrix} 1 & 0 \\ 0 & -1 \end{pmatrix} \begin{pmatrix} \alpha \\ \beta \end{pmatrix} \right] \otimes \left[\begin{pmatrix} 0 & 1 \\ 1 & 0 \end{pmatrix} \begin{pmatrix} \gamma \\ \delta \end{pmatrix} \right]$$

$$= \begin{pmatrix} \alpha \\ -\beta \end{pmatrix} \otimes \begin{pmatrix} \delta \\ \gamma \end{pmatrix}$$

$$= \begin{pmatrix} \alpha \begin{pmatrix} \delta \\ \gamma \end{pmatrix} \\ -\beta \begin{pmatrix} \delta \\ \gamma \end{pmatrix} \end{pmatrix}$$

$$
= \begin{pmatrix} \alpha\delta \\ \alpha\gamma \\ -\beta\gamma \\ -\beta\delta \end{pmatrix}, \tag{3.175}
$$

$$
\sigma_{yA} \otimes \hat{I}_B |\Psi\rangle_{AB} = \left[\sigma_{yA} \begin{pmatrix} \alpha \\ \beta \end{pmatrix} \right] \otimes \left[\hat{I}_B \begin{pmatrix} \gamma \\ \delta \end{pmatrix} \right]
$$

$$
= \begin{pmatrix} -i\beta \\ i\alpha \end{pmatrix} \otimes \begin{pmatrix} \gamma \\ \delta \end{pmatrix}
$$

$$
= \begin{pmatrix} -i\beta\gamma \\ -i\beta\delta \\ i\alpha\gamma \\ i\alpha\delta \end{pmatrix}. \tag{3.176}
$$

An alternate way to get the same result is to find the product operators first:

$$
\sigma_{zA} \otimes \sigma_{xB} = \begin{pmatrix} 1 & 0 \\ 0 & -1 \end{pmatrix} \otimes \begin{pmatrix} 0 & 1 \\ 1 & 0 \end{pmatrix} \tag{3.177}
$$

$$
= \begin{pmatrix} (+1)\begin{pmatrix} 0 & 1 \\ 1 & 0 \end{pmatrix} & (0)\begin{pmatrix} 0 & 1 \\ 1 & 0 \end{pmatrix} \\ (0)\begin{pmatrix} 0 & 1 \\ 1 & 0 \end{pmatrix} & (-1)\begin{pmatrix} 0 & 1 \\ 1 & 0 \end{pmatrix} \end{pmatrix} \tag{3.178}
$$

$$
= \begin{pmatrix} 0 & 1 & 0 & 0 \\ 1 & 0 & 0 & 0 \\ 0 & 0 & 0 & -1 \\ 0 & 0 & -1 & 0 \end{pmatrix}, \tag{3.179}
$$

$$
\sigma_{yA} \otimes \hat{I}_B = \begin{pmatrix} 0 & -i \\ i & 0 \end{pmatrix} \otimes \begin{pmatrix} 1 & 0 \\ 0 & 1 \end{pmatrix} \tag{3.180}
$$

$$
= \begin{pmatrix} 0\begin{pmatrix} 1 & 0 \\ 0 & 1 \end{pmatrix} & -i\begin{pmatrix} 1 & 0 \\ 0 & 1 \end{pmatrix} \\ i\begin{pmatrix} 1 & 0 \\ 0 & 1 \end{pmatrix} & 0\begin{pmatrix} 1 & 0 \\ 0 & 1 \end{pmatrix} \end{pmatrix} \tag{3.181}
$$

$$
= \begin{pmatrix} 0 & 0 & -i & 0 \\ 0 & 0 & 0 & -i \\ i & 0 & 0 & 0 \\ 0 & i & 0 & 0 \end{pmatrix} \tag{3.182}
$$

Applying these operators to the state of Eq. 3.170 gives back the results 3.175 and 3.176. ∎

Often, it is of interest to sum over all the degrees of freedom of one subsystem, system B, for instance, leaving only the A degrees of freedom as variables. This is done when we are *only* interested in subsystem A, or when we have no detailed information about B. For example, B could be the environment surrounding the system of interest, A. The environment is too complicated for its evolution to be followed in detail, but one can often still find the *average* effect of the environment on A. Another possible scenario is that Alice (A) and Bob (B) are participants on the two ends of a communication system; we can see the results of measurements in Alice's lab, but we have no knowledge of what is happening in Bob's lab. So the best we can do is take an average or expectation value over the B system's degrees of freedom.

In situations such as these, we need to take a trace over one subsystem while leaving the other subsystem untraced. The result will be to average over the unknown degrees of freedom of one system. This process is called the **partial trace**. Consider a product operator $\hat{O} = \hat{A} \otimes \hat{B}$ acting on Hilbert space $\mathcal{H}_A$ and $\mathcal{H}_B$, with respective bases $|\psi_i\rangle_A |\phi_j\rangle_B$. Then the partial trace with respect to A sums over the A degrees of freedom, leaving just the B degrees of freedom:

$$\mathrm{Tr}_A \hat{O} = \sum_i {}_A\langle \psi_i | \hat{A} \otimes \hat{B} | \psi_i \rangle_A = \sum_i {}_A\langle \psi_i | \hat{A} | \psi_i \rangle_A \, \hat{B}. \tag{3.183}$$

$\mathrm{Tr}_A \hat{O}$ is now an operator acting only on $\mathcal{H}_B$. Similarly, the partial trace over B leaves an operator that acts only on $\mathcal{H}_A$:

$$\mathrm{Tr}_B \hat{O} = \sum_j {}_B\langle \phi_j | \hat{A} \otimes \hat{B} | \phi_j \rangle_B = \sum_j {}_B\langle \phi_j | \hat{B} | \phi_j \rangle_B \, \hat{A}. \tag{3.184}$$

Example 3.4.3 Suppose a bipartite (two-particle) system is in a mixed state with density operator

$$\hat{\rho} = \sum_{ij} p_{ij} \left(|\psi_i\rangle_A \, \langle\psi_i|_A \right) \otimes \left(|\phi_j\rangle_B \, \langle\phi_j|_B \right), \tag{3.185}$$

where $|\psi_i\rangle_A$ and $|\phi_j\rangle_B$ are orthonormal bases for $\mathcal{H}_A$ and $\mathcal{H}_B$. The coefficients p_{ij} are the joint probabilities of finding the A and B systems in the states $|\psi_i\rangle_A$ and $|\phi_j\rangle_B$ simultaneously. The partial trace with respect to A is then

$$\mathrm{Tr}_A \, \hat{\rho} = \sum_k {}_A\langle \psi_k | \hat{\rho} | \psi_k \rangle_A \tag{3.186}$$

$$= \sum_{ijk} p_{ij} \Big(\langle \psi_k | \psi_i \rangle \, \langle \psi_i | \psi_k \rangle \Big)_A \Big(|\phi_j\rangle\langle\phi_j| \Big)_B \qquad (3.187)$$

$$= \sum_{ij} p_{ij} \, |\phi_j\rangle\langle\phi_j| \qquad\qquad\qquad (3.188)$$

$$= \sum_{j} p_j \, |\phi_j\rangle\langle\phi_j|, \qquad\qquad\qquad (3.189)$$

where $p_j = \sum_i p_{ij}$ is the marginal probability of state $|\phi_j\rangle_A$, regardless of the state of B. The orthonormality of the A basis was used in the third line: $\langle \psi_k | \psi_i \rangle = \delta_{ik}$. Notice that this is now of the form

$$\mathrm{Tr}_A \hat{\rho} = \sum_j p_j \big(\hat{\rho}_j \big)_B, \qquad\qquad (3.190)$$

where $\big(\hat{\rho}_j \big)_B = |\phi_j\rangle\langle\phi_j|$ are the pure states of B. So, the partial trace of a mixed state gives back another mixed state. ∎

The density matrices $\hat{\rho}_A$ and $\hat{\rho}_B$ with one subsystem traced out in this manner are called the **reduced density matrices**. Consider now a slightly different example.

Example 3.4.4 Now suppose the bipartite system is in a *pure* state given by

$$|\Phi\rangle = \alpha |0\rangle_A |0\rangle_B + \beta |1\rangle_A |1\rangle_B, \qquad\qquad (3.191)$$

with $|\alpha|^2 + |\beta|^2 = 1$. The corresponding density operator is

$$\hat{\rho} = |\Phi\rangle \otimes \langle\Phi| \qquad\qquad\qquad (3.192)$$

$$= |\alpha|^2 \Big(|0\rangle\langle0| \Big)_A \Big(|0\rangle\langle0| \Big)_B + |\beta|^2 \Big(|1\rangle\langle1| \Big)_A \Big(|1\rangle\langle1| \Big)_B \qquad (3.193)$$

$$+ \alpha\beta^* \Big(|0\rangle\langle1| \Big)_A \Big(|1\rangle\langle0| \Big)_B + \beta\alpha^* \Big(|1\rangle\langle0| \Big)_A \Big(|0\rangle\langle1| \Big)_B.$$

Tracing over the A subsystem:

$$\mathrm{Tr}_A \hat{\rho} = {}_A\langle0|\hat{\rho}|0\rangle_A + {}_A\langle1|\hat{\rho}|1\rangle_A \qquad\qquad (3.194)$$

$$= |\alpha|^2 \, {}_B|0\rangle\langle0|_B + |\beta|^2 \, {}_B|1\rangle\langle1|_B. \qquad\qquad (3.195)$$

Again, notice that even though the original state was pure, the reduced state is mixed. This will be true in general: *loss of knowledge of one subsystem reduces the purity of the other subsystem*. This is due to the influence of unknown fluctuations introduced by the unmeasured subsystem. ∎

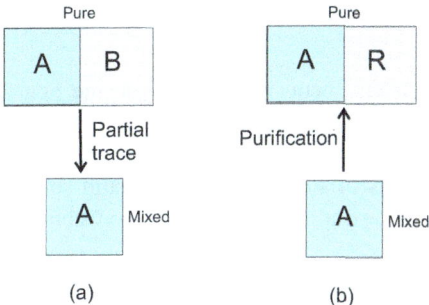

Fig. 3.3 (**a**) The partial trace of a pure system gives a mixed system, as the unaccounted-for degrees of freedom are averaged over. (**b**) In the reverse process, a mixed system can be "purified" by viewing it as a subsystem of a larger pure system, consisting of the original system A compounded with an ancillary reference system (R)

As the last example shows, the partial trace of a pure system gives a mixed system. But the process can also go the other way (Fig. 3.3): any mixed system can be viewed as a partial trace of a larger system in a pure state. This viewpoint is useful, for example, in accounting for the effects of an unknown environment on a system of interest. To see how this works, consider a mixed quantum state of a system A:

$$\hat{\rho}_A = \sum_j p_j |\psi_j\rangle\langle\psi_j|, \tag{3.196}$$

where j labels a set of basis states of A. Then if we imagine system A being coupled to a second system R (the reference system), the purification of $\hat{\rho}_A$ is given by a state

$$|\Psi_{AR}\rangle = \sum_j \sqrt{p_j} |\phi_j\rangle |\psi_j\rangle \tag{3.197}$$

of the larger system $S = A \otimes R$, where the vectors $|\phi_j\rangle$ form an orthonormal basis of R (or a subset of an orthonormal basis, if the dimension of R is larger than that of A). The original mixed state of A can be obtained as the partial trace of this pure state:

$$\hat{\rho}_A = Tr_R\big(|\Psi_{AR}\rangle\langle\Psi_{AR}|\big). \tag{3.198}$$

So the mixedness or lack of knowledge of the state of A can always be viewed as the result of entanglement with some other (possibly fictitious) system R; the partial trace takes into account our lack of knowledge of the state of the new system. The new composite system $A \otimes R$ is said to be the **purification** of A.

3.5 Superposition and Entanglement

Like the Maxwell electromagnetic wave equation, the Schrödinger equation is a
linear differential equation, meaning that there are no terms involving powers of
wavefunction ψ, except for the first power. A key property of linear wave equations
is that linear combinations of solutions are also valid solutions. So, for example,
given two state vectors $|\psi_1\rangle$ and $|\psi_2\rangle$ of a quantum system, the linear combination

$$|\psi\rangle = \alpha|\psi_1\rangle + \beta|\psi_2\rangle, \tag{3.199}$$

where α and β are complex numbers, is also a legitimate solution. As usual, the
state is assumed to be normalized, so that $|\alpha|^2 + |\beta|^2 = 1$. The system is then said
to be in a **superposition** of states $|\psi_1\rangle$ and $|\psi_2\rangle$. So, for example, a photon could be
in a superposition of vertical and horizontal polarizations:

$$|\psi\rangle = \alpha|H\rangle + \beta|V\rangle, \tag{3.200}$$

or an electron could be in a superposition of spin-up and spin-down states:

$$|\psi\rangle = \alpha|\uparrow\rangle + \beta|\downarrow\rangle. \tag{3.201}$$

In some sense, the system may be thought of as being in both of the superposed
states simultaneously. It doesn't make a choice and collapse into one of them until
a measurement is made. For example, in the photon case just cited, the photon
polarization, when measured, will be found (with probability $|\alpha|^2$) in state $|H\rangle$,
or (with probability $|\beta|^2$) in state $|V\rangle$. But until that measurement is made, all
calculations must include all available possibilities in a superposition. Two-state
superpositions such as Eqs. 3.199, 3.200, and 3.201 are called **qubits**. Rather than
two superposed states, of course, one could also imagine superpositions of many
(even infinitely many) states:

$$|\psi\rangle = \alpha|\psi_1\rangle + \beta|\psi_2\rangle + \gamma|\psi_3\rangle + \delta|\psi_4\rangle\dots. \tag{3.202}$$

From the classical point of view, this is exceedingly strange. Classical particles have well-defined values of all their variables, and these values evolve
deterministically over time. If the initial conditions are known to sufficiently high
precision, one can always (at least in principle) determine the exact outcome of any
measurement made at a later time. But in the case of a quantum superposition, it is
completely random which state the system will be found in when measured. Only
the *probabilities*, $|\alpha|^2$ and $|\beta|^2$, will evolve deterministically over time.

In the early days of quantum mechanics, Einstein and others insisted that the
system must actually be in one of the two states, $|\psi_1\rangle$ or $|\psi_2\rangle$, at each moment,
and that the superpositions simply expressed our ignorance as to which state
was the "real" one. But this position became untenable after the appearance and

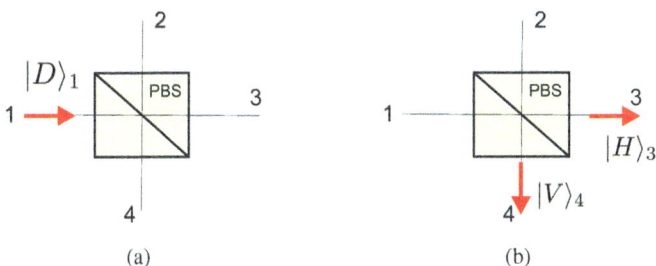

Fig. 3.4 (**a**) A diagonally polarized photon entering port 1 of a polarizing beam splitter with state vector $|D\rangle_1$. (**b**) The output is a superposition of a horizontally polarized state leaving port 3 and a vertically polarized state leaving port 4. Until a measurement is made, neither of these states can be said to be "real." Rather, only the amplitudes for each possibility have physical meaning

experimental verification of the Bell and CHSH inequalities (see Chap. 14). Until a measurement is made, we must include both possibilities and treat them as being equally real in order to avoid mathematical and physical inconsistencies.

Because of the normalization condition $|\alpha|^2 + |\beta|^2 = 1$, α and β can always be parameterized in terms of an angle θ and a pair of phases ϕ_1 and ϕ_2, meaning that they can be written in the form

$$\alpha = e^{i\phi_1} \sin\theta, \qquad \beta = e^{i\phi_2} \cos\theta. \tag{3.203}$$

The two-term superposition state 3.199 is then expressed in the form

$$|\psi\rangle = e^{i\phi_1}\left[\sin\theta|\psi_1\rangle + e^{i\phi}\cos\theta|\psi_2\rangle\right], \tag{3.204}$$

where the relative phase between the two terms is $\phi = \phi_2 - \phi_1$. For most purposes, the overall global phase $e^{i\phi_1}$ is irrelevant, and so it is often dropped.

Superposition states are easy to make. Suppose a single diagonally polarized photon is sent through a polarizing beam splitter, entering it at port 1 (Fig. 3.4). The input state is then $|\psi\rangle_{in} = |D\rangle_1$, indicating a diagonally polarized photon at port 1. Then since the diagonal polarization vector can be written as an equal linear combination of horizontal and vertical polarizations, the output state will be an equal superposition of horizontal and vertical states, which will then exit at different ports:

$$|\psi\rangle_{out} = \frac{1}{\sqrt{2}}\left(|H\rangle_3 + |V\rangle_4\right). \tag{3.205}$$

Note that this is not only a superposition of two polarization states, but it is also a superposition of two spatial modes: the photon has an amplitude to be moving to the right, out of port 3, or moving downward, out of port 4.

Every state can always be written as a superposition: even if it appears to be a single term in one Hilbert space basis, when a different basis is used, the state will decompose into a superposition of the new basis vectors. For example, an electron with spin-up or spin-down along the x-direction can always written as a sum of spin-up and spin-down terms along the z-axis:

$$|x, \uparrow\rangle = \frac{1}{\sqrt{2}} (|z, \uparrow\rangle + |z, \downarrow\rangle) = \begin{pmatrix} 1 \\ 1 \end{pmatrix} \qquad (3.206)$$

$$|x, \downarrow\rangle = \frac{1}{\sqrt{2}} (|z, \uparrow\rangle - |z, \downarrow\rangle) = \begin{pmatrix} 1 \\ -1 \end{pmatrix}. \qquad (3.207)$$

Similarly, for states that are spin up along the y-axis:

$$|y, \uparrow\rangle = \frac{1}{\sqrt{2}} (|z, \uparrow\rangle + i|z, \downarrow\rangle) = \begin{pmatrix} 1 \\ i \end{pmatrix} \qquad (3.208)$$

$$|y, \downarrow\rangle = \frac{1}{\sqrt{2}} (|z, \uparrow\rangle - i|z, \downarrow\rangle) = \begin{pmatrix} 1 \\ -i \end{pmatrix}. \qquad (3.209)$$

Notice that the electron states with spin along the x- and y-axes are analogous to the diagonal and circular polarization states of the photon (see Eq. 2.91).

Superposition states can be measured via interference. Consider two orthonormal states $|\psi_1\rangle$ and $|\psi_2\rangle$, with $\langle\psi_2|\psi_1\rangle = 0$, and consider two superpositions of these states:

$$|\psi_A\rangle = \sin\theta|\psi_1\rangle + e^{i\phi_A}\cos\theta|\psi_2\rangle \qquad (3.210)$$

$$|\psi_B\rangle = \sin\theta|\psi_1\rangle + e^{i\phi_B}\cos\theta|\psi_2\rangle. \qquad (3.211)$$

Suppose $|\psi_A\rangle$ is prepared and sent into an apparatus designed to detect the state $|\psi_B\rangle$. Then the detection probability will be

$$P = |\langle\text{Input state}|\text{output state}\rangle|^2 \qquad (3.212)$$

$$= |\langle\psi_A|\psi_B\rangle|^2 \qquad (3.213)$$

$$= |\sin^2\theta + e^{i(\phi_B - \phi_A)}\cos^2\theta|^2 \qquad (3.214)$$

$$= \sin^4\theta + \cos^4\theta + \sin^2\theta\cos^2\theta \left(e^{i(\phi_B - \phi_A)} + e^{-i(\phi_B - \phi_A)}\right) \qquad (3.215)$$

$$= \sin^4\theta + \cos^4\theta + 2\sin^2\theta\cos^2\theta\cos(\phi_B - \phi_A) \qquad (3.216)$$

The first two terms are constant background terms, independent of the phases, while the last part is an interference term that measures the phase difference between the two states. As the phase difference $\phi = \phi_B - \phi_A$ varies, the probability oscillates between the values $P_{\text{max}} = \left(\sin^2\theta + \cos^2\theta\right)^2 = 1$ and $P_{\text{min}} = \left(\sin^2\theta - \cos^2\theta\right)^2 \geq 0$; see Fig. 3.5.

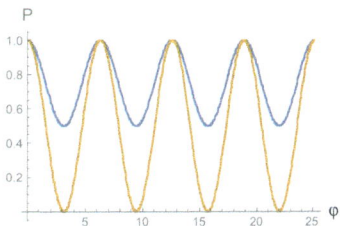

Fig. 3.5 Interference pattern described by Eq. 3.216, for $\theta = \frac{\pi}{8}$ (blue curve) and $\theta = \frac{\pi}{4}$ (yellow curve)

In order to illustrate the idea of a superposition state, Schrödinger considered the possibility of a macroscopic object in a superposition state. He imagined a cat trapped in a box with a vial of poison. The vial was arranged to open if struck by a particle created in the radioactive decay of an unstable nucleus situated near the vial. If no measurements were made on the interior of the box, it would be impossible to know if the nucleus had decayed or not, and therefore it would be impossible to know if the cat was alive or dead. Therefore, the conclusion was that Schrödinger's cat should be in a superposition state:

$$|\text{cat}\rangle = \frac{1}{\sqrt{2}} \left(|\text{alive}\rangle + e^{i\phi} |\text{dead}\rangle \right). \tag{3.217}$$

Animal cruelty issues aside, Schrödinger's cat is unrealistic because of the macroscopically large number of mutually interacting molecules present in the cat. The interactions will cause rapid **decoherence**, meaning that the relative phase ϕ between the $|\text{alive}\rangle$ and the $|\text{dead}\rangle$ terms will fluctuate randomly, washing out the interference terms of the form of that in Eq. 3.216, as will be discussed in more detail in the next section.

Nonetheless, there are special circumstances in which it is possible to put a macroscopic system into a quantum superposition. For example, superconductors can be put into macroscopic superposition states, as will be discussed in Chap. 7.

The superposition state in Eq. 3.204 describes a *single* particle that has nonzero simultaneous amplitudes for being in two different states. We can also consider superpositions of *two-particle* states. Imagine two particles (labeled 1 and 2), and suppose that each particle can be in one of two possible states (labeled A and B). Then we can define a joint two-particle state $|\psi\rangle$ describing the *pair* of particles. Suppose that the particles are either distinguishable or are very far apart, so that we don't have to worry about symmetrizing or anti-symmetrizing their joint states. Then if the particles don't interact or if they are distinguishable, the two-particle wavefunction will simply be a product of a pair of one-particle states. For example, the pair of particles might be in the state $|\psi\rangle = |\psi_A\rangle_1 |\psi_B\rangle_2$. Such a state is called **factorable** or **separable**.

But factorable states are not the only possibility. Suppose particles 1 and 2 can each be in any one of a set of states $\{A, B, C, D, \dots\}$. Then you could have a two-particle *superposition* state of the form

$$|\psi\rangle = \alpha|\psi_A\rangle_1|\psi_B\rangle_2 + \beta|\psi_C\rangle_1|\psi_D\rangle_2 + \gamma|\psi_E\rangle_1|\psi_F\rangle_2 + \ldots \qquad (3.218)$$

Rather than the general case, we will specialize to the case where there are only two terms in this sum:

$$|\psi\rangle = \alpha|\psi_A\rangle_1|\psi_B\rangle_2 + \beta|\psi_C\rangle_1|\psi_D\rangle_2, \qquad (3.219)$$

with $|\alpha|^2 + |\beta|^2 = 1$. As a further special case, it may be that $A = D$ and $B = E$, so that

$$|\psi\rangle = \alpha|\psi_A\rangle_1|\psi_B\rangle_2 + \beta|\psi_B\rangle_1|\psi_A\rangle_2. \qquad (3.220)$$

A two-particle state of this form may occur if the particles are indistinguishable, so that we can't tell which particle is in which of the two available states.

In some cases, it may be possible to go to a different basis in Hilbert space that allows states such as those in Eqs. 3.218, 3.219, or 3.220 to be factored into a product of single-particle states. If this is *not* the case, so that the system is in a superposition of two-particle (or multiparticle) states *in every possible basis*, then the state is said to be **nonseparable** or **entangled**.

Entanglement is a uniquely quantum property and is responsible for many of the nonintuitive effects that occur in quantum systems. It is a form of correlation between the two particles, which is stronger (in a sense that will be quantified in Chap. 13) than is possible in classical systems.

The non-factorability condition above defines entanglement for a *pure* state. For a *mixed* state, the state is said to be separable if it can be written as a sum of products of density matrices:

$$\hat{\rho}_{AB} = \sum_i p_i \hat{\sigma}_i \otimes \hat{\eta}_i, \qquad (3.221)$$

where σ_i and η_i are density matrices for systems A and B, and where $\sum_i p_i = 1$. If the mixed state is not separable, then it is entangled. In the special case where there is just a single term in the sum,

$$\hat{\rho}_{AB} = \hat{\sigma}_A \otimes \hat{\eta}_B, \qquad (3.222)$$

then the system is not only separable, but the subsystems A and B are in fact completely uncorrelated.

An important set of entangled states are the **Bell states**. Consider two photons, each of which can be polarized either vertically or horizontally. Then the polarization Bell states are defined as

$$|\Phi^+\rangle = \frac{1}{\sqrt{2}}\Big(|H\rangle_1|H\rangle_2 + |V\rangle_1|V\rangle_2\Big) \qquad (3.223)$$

$$|\Phi^-\rangle = \frac{1}{\sqrt{2}}\Big(|H\rangle_1|H\rangle_2 - |V\rangle_1|V\rangle_2\Big) \tag{3.224}$$

$$|\Psi^+\rangle = \frac{1}{\sqrt{2}}\Big(|H\rangle_1|V\rangle_2 + |V\rangle_1|H\rangle_2\Big) \tag{3.225}$$

$$|\Psi^+\rangle = \frac{1}{\sqrt{2}}\Big(|H\rangle_1|V\rangle_2 - |V\rangle_1|H\rangle_2\Big). \tag{3.226}$$

These form a basis for the set of all polarization-entangled two-photon states. Analogous Bell states can be defined for any two-state system, for example, by replacing H and V with spin-up and spin-down or with logical 0 and 1 states. For instance, logical Bell states are of the form

$$|\Phi^+\rangle = \frac{1}{\sqrt{2}}\Big(|0\rangle_1|0\rangle_2 + |1\rangle_1|1\rangle_2\Big) \tag{3.227}$$

$$|\Phi^-\rangle = \frac{1}{\sqrt{2}}\Big(|0\rangle_1|0\rangle_2 - |1\rangle_1|1\rangle_2\Big) \tag{3.228}$$

$$|\Psi^+\rangle = \frac{1}{\sqrt{2}}\Big(|0\rangle_1|1\rangle_2 + |1\rangle_1|0\rangle_2\Big) \tag{3.229}$$

$$|\Psi^+\rangle = \frac{1}{\sqrt{2}}\Big(|0\rangle_1|1\rangle_2 - |1\rangle_1|0\rangle_2\Big). \tag{3.230}$$

As will be seen in coming chapters, the Bell states play a major role in quantum information processing and other applications.

Box 3.3 Teleporting Energy Fluctuations

In the 1990s, the discovery that unknown quantum states could be "teleported" made news worldwide. Quantum state teleportation will be discussed in Sect. 18.2. But, more recently, it has been shown that in a certain sense, *energy* can also be teleported between labs. This is based on the fact that quantum states always have inherent fluctuations and that two particles forming an entangled state always have *correlated* fluctuations. This effect starkly demonstrates the importance of quantum fluctuations in observable phenomena.

Suppose that Alice and Bob each have half of an entangled system. Alice's system is in its ground state. Because of quantum fluctuations, the ground state energy of her system is always nonzero: there is a zero-point energy that cannot be removed. If Alice makes a measurement on her system, she will inevitably add some energy to it, raising the average energy above the ground state value. The goal is to somehow transmit this excess energy to Bob's lab.

(continued)

Box 3.3 (continued)

The idea is that Alice continuously measures the fluctuations of her system's energy. When she sees a fluctuation in one direction, she knows that Bob's energy must have a fluctuation in the opposite direction, since the total energy of the entangled system should be conserved. So when Alice sees a downward fluctuation, she informs Bob (over a classical communication channel). Bob can then extract the excess energy in that positive fluctuation; when he does so, the average energy in Alice's system decreases by a corresponding amount. If they continue with this process, all the energy Alice injected into her system through the measurement process can eventually be extracted at Bob's end.

This procedure seemed counterintuitive when it was first described by Masahiro Hota in 2008 [17], but two different experiments since 2022 have verified that it works. An experiment at the University of Waterloo [18] used the effect to teleport energy between two carbon atoms in an NMR experiment. A few months later, an experiment on IBM's Quantum Computing Platform showed similar results [19].

Both the entanglement and the communication between the observers are necessary for the procedure to work. Although the energy seems to be coming from empty space from Bob's point of view, it is not arising from nowhere: it has to be injected into the entangled system by Alice, and the total energy is being conserved at every step of the procedure. So, as amazing as this effect is, those who dream of perpetual motion machines or of extracting the zero-point energy of the vacuum to do work will still remain disappointed.

3.6 Coherence and Decoherence

In quantum mechanics, as in optics, *coherence* essentially means that the terms in a superposition have relative phases that are stable enough to produce interference. Consider, for example, a state of the form

$$|\psi\rangle = \alpha|0\rangle + \beta|1\rangle. \tag{3.231}$$

As has been mentioned, the overall phase of the state can be chosen so that the first coefficient is real, leaving the relative phase entirely in the second coefficient. α and β can then be parameterized by a pair of angles, θ and ϕ:

$$|\psi\rangle = \cos\theta|0\rangle + e^{i\phi}\sin\theta|1\rangle. \tag{3.232}$$

Then, the superposition remains coherent as long as the phase ϕ is not rapidly fluctuating.

Consider the density operator corresponding to $|\psi\rangle$. It is given by

$$\hat{\rho} = |\psi\rangle\langle\psi| \tag{3.233}$$

$$= \left(\cos\theta|0\rangle + \sin\theta e^{i\phi}|1\rangle \right)\left(\cos\theta\langle0| + \sin\theta e^{-i\phi}\langle1| \right) \tag{3.234}$$

$$= \cos^2\theta|0\rangle\langle0| + \sin^2\theta|1\rangle\langle1| \tag{3.235}$$

$$+ \sin\theta\cos\theta\left(e^{i\phi}|1\rangle\langle0| + e^{-i\phi}|0\rangle\langle1| \right)$$

$$= \begin{pmatrix} \cos^2\theta & \sin\theta\cos\theta\, e^{-i\phi} \\ \sin\theta\cos\theta\, e^{i\phi} & \sin^2\theta \end{pmatrix}. \tag{3.236}$$

Notice that the phase only appears in the off-diagonal components of the density matrix. If the phase is stable (constant or slowly varying), then the off-diagonal terms are nonzero. However, if the phase fluctuates rapidly, then the factors $e^{\pm i\phi}$ whiz around on the unit circle in the complex plane and will average to zero over time. In this case, the off-diagonal matrix elements will also wash out and average to zero, so the superposition will be incoherent.

The situation above is typical: the coherence of a system is generally determined by off-diagonal terms in $\hat{\rho}$, and if the system is incoherent, there will be a coordinate system in which the off-diagonal terms are zero. For this reason, the off-diagonal terms of $\hat{\rho}$ are sometimes called the **coherences** of the system. (Note that some care needs to be taken here: the off-diagonal terms can be zero in one coordinate system and nonzero in another, as in Example 3.2.5.)

To see briefly how decoherence works, consider the $\cos\phi$ term in Eq. 3.216. Suppose the phase is a function of time, $\phi(t)$. In particular, suppose that random thermal fluctuations and interactions between the system and its surroundings cause the phase to vary randomly over time. Then the cosine also will be fluctuating randomly. Any measurement on the system will take some finite time T. Suppose that the fluctuations are rapid enough so that their typical time scale is much shorter than T. In that case, the measurement will only see the average value of $\cos\phi(t)$:

$$\cos(\phi(t)) \to \overline{\cos(\phi(t))} = \frac{1}{T}\int_{-T/2}^{T/2} \cos(\phi(t))dt, \tag{3.237}$$

which will tend to zero as T becomes large. Thus, the system rapidly decoheres, with the interference term averaging to zero.

One way to think of the decoherence of a macroscopic system is to imagine that the environment is effectively making constant measurements of the system, rapidly forcing it to randomly collapse into one of the states $|\psi_1\rangle$ or $|\psi_2\rangle$. The resulting loss of the superposition then makes the system behave classically, with interference effects becoming unobservable.

Now consider the case where an entangled state is shared between two subsystems A and B of a composite system. To be specific, consider the state

$$|\Phi\rangle = \alpha|00\rangle + \beta|11\rangle, \tag{3.238}$$

where the first entry in each ket belongs to A and the second to B. Suppose the phases of α and β are stable, so that the entangled system is coherent. What happens if we look at one of the subsystems?

By taking the partial trace with respect to B, we arrive at the state of subsystem A:

$$\hat{\rho}_A = Tr_B |\psi\rangle\langle\psi|$$

$$= Tr_B\left(|\alpha|^2|00\rangle\langle00| + |\beta|^2|11\rangle\langle11| + \alpha\beta^*|00\rangle\langle11| + \alpha^*\beta|11\rangle\langle00|\right)$$

$$= |\alpha|^2|0\rangle\langle0| + |\beta|^2|1\rangle\langle1|$$

$$= \begin{pmatrix} |\alpha|^2 & 0 \\ 0 & |\beta|^2 \end{pmatrix}. \tag{3.239}$$

The off-diagonal terms vanish after taking the partial trace vanish, and the phase information has disappeared. Although the full entangled state is coherent, the state of the subsystem A is incoherent. (A similar result will occur for subsystem B.) This is a general result: *entanglement of one subsystem with another will lead to incoherence in each of the subsystems individually.*

Although the phase information appears to have vanished when making measurements on just one subsystem, it is actually still present; it is just hidden, lurking in the correlations between the subsystems. It can be retrieved by making **joint** measurements on the two subsystems. In other words, the phase information cannot be obtained by local measurements of A or B independently, but is distributed in a global manner over both systems.

This phenomenon can be seen, for example, in **spontaneous parametric down-conversion (SPDC)**. This is a method of producing entangled photon pairs and will be discussed in more detail in Sect. 5.8. In SPDC, an incident photon (the pump photon) entering a crystal is highly coherent, but is "split" into two outgoing photons as a result of nonlinear interactions in the crystal (the *signal* and the *idler* photons). These two new photons are each highly incoherent. Neither outgoing photon can be used to produce high-quality interference patterns. But the two photons are entangled, and the high coherence of the pump beam remains present when the two final photons are measured *jointly*. These joint measurements are usually coincidence counts, i.e., counts of the rate at which two detectors fire simultaneously (Sect. 10.4). The coincidence counting unveils the quantum correlation between the signal and the idler. The joint state of the two photons remains highly coherent, and this shows up the presence of interference arising in the coincidence rates. Consequences of this hidden coherence include the ability to produce quantum ghost images (Sect. 20.1.1) and to see two-photon interference effects such as the HOM dip (Sect. 10.6).

For decades, it was something of a mystery as to why microscopic objects like atoms, electrons, and photons showed clear signs of quantum behavior while macroscopic objects like tables, chairs, and cats did not. The distinction turns out to be one of coherence. The constituent atoms in a large object like a cat or your next-door-neighbor are in constant interaction with their environment. The atoms in your skin, for example, are constantly exchanging heat with the surrounding air, photons from the sun are being absorbed, chemical reactions are occurring with neighboring molecules, and so on. All of these interactions lead to each molecule being entangled with many other molecules in its vicinity. Even if a molecule is initially in a highly coherent state, all of this entanglement with the environment leads to rapid decoherence. So large objects cannot (at least in general) exhibit quantum behavior over a long enough time scale for it to be observed.

A decoherence time τ_D can be defined. This gives the characteristic time scale over which coherence is lost due to interactions with the environment. It is easy to see that this decoherence time will rapidly become smaller as the system size increases. Consider a common form of decoherence, the **phase flip**, in which interactions with the environment cause one term in a superposition to change sign relative to the other:

$$|\psi\rangle = \alpha|0\rangle + \beta|1\rangle \longrightarrow \sigma_z|\psi\rangle = \alpha|0\rangle - \beta|1\rangle. \tag{3.240}$$

If these flips occur for a single qubit state at a rate Γ, then the decoherence time is $\tau_D = \frac{1}{\Gamma}$ and the probability of a flip in time Δt is $p = \Gamma \Delta t = \frac{\Delta t}{\tau_D}$. But now suppose that instead of a single particle, the system consists of N entangled particles, with a state of the form $|\psi\rangle = \alpha|000\ldots0\rangle + \beta|111\ldots1\rangle$. Coherence can be destroyed if *any* of the qubits experience phase flips. If each qubit has flipping rate Γ, then the rate for the N-particle state will be $N\Gamma$, and the decoherence time will be decreased by a factor of N. So, the larger the object involved, the shorter the decoherence time. For macroscopic objects, decoherence times are typically on the order of 10^{-30} to 10^{-40} seconds or shorter, making the observation of quantum interference effects impossible.

Some exceptions to the rapid decoherence of macroscopic objects should be noted. Decoherence times can be on the order of seconds in ion traps or tens of minutes for nuclear spin-based qubits, while superconductors and superfluids can be in coherent quantum states of macroscopic size that can be maintained for indefinite periods of time. The macroscopic quantum state in a superconductor will be discussed in Chap. 7.

When two systems (say, system S and environment E) are entangled, the time evolution of the full $S + E$ system is, of course, unitary. But when only subsystem S is being viewed, ignorance of the environmental degrees of freedom leads to the appearance of non-unitary evolution in S. As a result, the eigenvalues of the

Hamiltonian may become complex, with the imaginary parts signaling the presence of gain or loss.

In some cases, a subset of states can be found that is left invariant under the non-unitary effects that lead to decoherence and loss. States in such a **decoherence-free subspace** will evolve unitarily and retain their quantum nature over time. In a typical system, the decoherence-free subspace will occupy only a minuscule fraction of the full Hilbert space; nevertheless, the use of these subspaces has been proposed for passive error-correcting codes. For a review of decoherence-free subspaces, see [20]. For more information on decoherence generally, see the articles [21, 22].

3.7 Measurement of Quantum States

Information is gained about physical systems by making measurements. Classically, this is a straightforward operation: the system is brought into contact with a measuring device and a reading is made from a dial or an electronic readout. If the system is in equilibrium, then repeated readouts will give the same result. If the system is evolving over time, then the measurements also evolve deterministically and can be accurately predicted, at least over some finite time scale set by the precision of your measuring devices and the level of sensitivity to initial conditions (in other words, the degree of chaos) in the system.

Quantum mechanically, the situation is more complicated. Even if the Hamiltonian and initial state are known perfectly, to arbitrarily high precision, the outcome can be completely random. In addition, the quantum system often must at some point be connected to a classical readout for the experimenter to have access to it; the question of where to draw the boundary between the classical and quantum behaviors in such a situation adds additional complications and connects the measurement process to the rate of concept of decoherence.

In the following subsections, we avoid the topic of coupling to a classical output and simply discuss the purely quantum part of the measurement process. This will involve defining a set of measurement operators, determining the outcomes of the measurements, and determining the effect of the operators on the initial state.

3.7.1 Projective (von Neumann) Measurements

The most commonly discussed and most idealized type of measurement is the **projective measurement** or **von Neumann measurement**, described systematically by John von Neumann and popularized in his 1932 textbook, *Mathematical Foundations of Quantum Mechanics*.

Recall that physically observable variables are represented by Hermitian operators. When a projective measurement is made of an observable, the only possible outcomes for the measurement are the eigenvalues of the operator. To be specific, consider a Hermitian operator $\hat{A}$ with eigenvalues λ_j, for $j = 1, 2, 3, \ldots$ and corresponding normalized eigenstates $|\lambda_j\rangle$. Since $\hat{A}$ is Hermitian, the λ_j are real numbers. Suppose that the observable $\hat{A}$ is measured on the pure $|\psi\rangle$. The probability of outcome λ_j is given by

$$P(\lambda_j) = |\langle \lambda_j | \psi \rangle|^2. \tag{3.241}$$

This can also be rewritten as

$$P(\lambda_j) = \langle \psi | \lambda_j \rangle \langle \lambda_j | \psi \rangle = \langle \psi | \hat{\mathcal{P}}_j | \psi \rangle, \tag{3.242}$$

where $\hat{\mathcal{P}}_j = |\lambda_j\rangle\langle\lambda_j|$ is the projection operator onto the eigenspace of eigenvalue λ_j. The observable operator itself can be written in terms of an eigenvalue decomposition using these projectors:

$$\hat{A} = \sum_j \lambda_j |\psi_j\rangle\langle\psi_j| = \sum_j \lambda_j \hat{\mathcal{P}}_j, \tag{3.243}$$

where $|\psi_j\rangle\langle\psi_j|$ project the input state onto the directions of the eigenstates.

If, instead of a pure state $|\psi\rangle$, the system is in a mixed state characterized by the density operator $\hat{\rho}$, the expression for the probability of obtaining outcome λ_j generalizes to

$$P(\lambda_j) = \mathrm{Tr}(\hat{\rho}\hat{\mathcal{P}}_j) = \mathrm{Tr}(\hat{\mathcal{P}}_j\hat{\rho}). \tag{3.244}$$

For a pure state, $\hat{\rho} = |\psi\rangle\langle\psi|$, this reduces back to Eq. 3.241.

The possibility of degenerate states can also be included. For example, if several eigenstates $|\lambda_j^{(1)}\rangle, |\lambda_j^{(2)}\rangle, \ldots |\lambda_j^{(N)}\rangle$ all share the same eigenvalue λ_j, then the relevant projection operator simply becomes

$$\hat{\mathcal{P}}_j = \sum_{k=1}^{N} |\lambda_j^{(k)}\rangle\langle\lambda_j^{(k)}|, \tag{3.245}$$

which projects onto the degenerate subspace, and then all else is the same as before.

For the case of a von Neumann measurement, the $\hat{\mathcal{P}}_j$ operators, being both Hermitian and projectors, must satisfy the following properties:

1. $\hat{\mathcal{P}}_j^\dagger = \hat{\mathcal{P}}_j$ (Hermiticity)
2. $\hat{\mathcal{P}}_j \geq 0$ (Positive operator)
3. $\sum_j \hat{\mathcal{P}}_j = 1$ (Completeness)
4. $\hat{\mathcal{P}}_i\hat{\mathcal{P}}_j = \delta_{ij}\hat{\mathcal{P}}_j$ (Orthonormality).

In the second line, a **positive operator** is one with real, non-negative eigenvalues. Equivalently, a positive operator is one for which $\langle \psi | \hat{O} | \psi \rangle \geq 0$ for all vectors $| \psi \rangle$. Positive operators are always Hermitian.

In the next subsection, we will see that for nonideal measurements, the fourth condition will need to be relaxed.

So we have seen that a projective measurement consists of projecting the initial state onto an eigenspace, and the result of the measurement will be the corresponding eigenvalues, with probabilities of the outcomes being given by Eqs. 3.241 or 3.242. After the measurement, the projection collapses the original state onto the measured eigenstate:

$$| \psi \rangle \rightarrow | \psi' \rangle = \hat{P}_j | \psi \rangle = | \lambda_j \rangle \langle \lambda_j | \psi \rangle. \tag{3.246}$$

This is not properly normalized; to normalize, we divide by the probability of the particular outcome, so that the final state is

$$| \psi' \rangle = \frac{\hat{P}_j | \psi \rangle}{| \langle \lambda_j | \psi \rangle |} = \frac{| \lambda_j \rangle \langle \lambda_j | \psi \rangle}{| \langle \lambda_j | \psi \rangle |} \tag{3.247}$$

for a pure state. Or more generally,

$$\hat{\rho}' = \frac{\hat{P}_j \, \hat{\rho} \, \hat{P}_j}{\mathrm{Tr}(\hat{P}_j \, \hat{\rho} \, \hat{P}_j)} = \frac{\hat{P}_j \hat{\rho} \hat{P}_j}{P(\lambda_j)}. \tag{3.248}$$

Because the measurement involves orthonormal projection operators, repeating the same measurement multiple times will give the same result.

Example 3.7.1 Consider the quantum state of a photon polarized horizontally, $| \psi \rangle = | H \rangle = \begin{pmatrix} 1 \\ 0 \end{pmatrix}$. We can define a set of horizontal and vertical projection operators:

$$\hat{P}_H = | H \rangle \langle H | = \begin{pmatrix} 1 & 0 \\ 0 & 0 \end{pmatrix}, \qquad \hat{P}_V = | V \rangle \langle V | = \begin{pmatrix} 0 & 0 \\ 0 & 1 \end{pmatrix}, \tag{3.249}$$

each of which has eigenvalues 0 and 1.

Now measure the polarization of the state by sending the photon through a horizontal polarizer. The possible outcomes are the eigenvalues of $\hat{P}_H$: 0 corresponds to vertical polarization and 1 to horizontal polarization. Of these, for the state $| \psi \rangle = | H \rangle$, the eigenvalue 1 will be obtained 100% of the time:

$$P(H) = \langle H | \hat{P}_H | H \rangle = 1. \tag{3.250}$$

If, instead, we send the same state through the vertical polarizer, an eigenvalue of 1 of $\hat{\mathcal{P}}_V$ now corresponds to vertical polarization and 0 to horizontal. The result will be that a vertical polarization is obtained 0% of the time:

$$P(V) = \langle H|\hat{\mathcal{P}}_V|H \rangle = 0. \tag{3.251}$$

Instead of vertical or horizontal polarizers, we could rotate the polarizer to an angle θ from the horizontal, measuring the polarization along the unit vector $|\hat{n}\rangle = \cos\theta|H\rangle + \sin\theta|V\rangle$. The corresponding projection operator will be

$$\hat{\mathcal{P}}_\theta = |\hat{n}\rangle\langle\hat{n}|$$
$$= \cos^2\theta|H\rangle\langle H| + \sin^2\theta|V\rangle\langle V| + \cos\theta\sin\theta\big(|H\rangle\langle V| + |V\rangle\langle H|\big)$$
$$= \begin{pmatrix} \cos^2\theta & \sin\theta\cos\theta \\ \sin\theta\cos\theta & \sin^2\theta \end{pmatrix}. \tag{3.252}$$

Now, the eigenvalues 1 and 0 will correspond, respectively, to the photon polarization being along the polarizer direction or perpendicular to it. The probabilities will be

$$P_\parallel(\theta) = \cos^2\theta \tag{3.253}$$
$$P_\perp(\theta) = 1 - P_\parallel(\theta)^2 = \sin^2\theta. \tag{3.254}$$

Notice that this result is simply the law of Malus, Eq. 2.105, and that this example is essentially identical to the optical polarization example of the previous chapter, Example 2.7.1.

3.7.2 Nonideal Measurements and POVM's

The projective measurements described above are highly idealized: they don't account for the possibility of measurement errors, or for the possibility that a particle may be destroyed during the measurement process. This is clearly inadequate. In real life, measurements are rife with sources of error: detectors never have 100% efficiencies, and they always have dark counts and electrical noise. Some particles can be lost on the way to the detector. Even if all sources of error can be ignored, optical detectors generally destroy the detected photon, rather than projecting it onto some eigenbasis. So the framework of the previous section needs to be generalized.

We consider the possibility of measurement errors first, which is then easily extrapolated to the most general case. To be specific, suppose a system can be in one of two states, $|0\rangle$ or $|1\rangle$. The corresponding projection operators are

$$\hat{\mathcal{P}}_0 = |0\rangle\langle 0| \tag{3.255}$$

$$\hat{\mathcal{P}}_1 = |1\rangle\langle 1|. \tag{3.256}$$

Suppose the detection system has some finite error probability p, representing the probability that state $|0\rangle$ will be mistakenly be identified as $|1\rangle$ or vice versa. Given this error, it makes sense to define a new set of operators:

$$\hat{\pi}_0 = (1-p)\hat{\mathcal{P}}_0 + p\,\hat{\mathcal{P}}_1 \tag{3.257}$$

$$\hat{\pi}_1 = (1-p)\hat{\mathcal{P}}_1 + p\,\hat{\mathcal{P}}_0. \tag{3.258}$$

In each operator, the first term accounts for correct identifications, and the second term accounts for errors. These new operators are not projection operators, because they are not orthogonal:

$$\hat{\pi}_0\hat{\pi}_1 = \hat{\pi}_1\hat{\pi}_0 = p(1-p)\big(\hat{\mathcal{P}}_0 + \hat{\mathcal{P}}_1\big) = p(1-p)\hat{I}, \tag{3.259}$$

and they don't square to themselves:

$$\hat{\pi}_0^2 = (1-p)^2\hat{\mathcal{P}}_0 + p^2\hat{\mathcal{P}}_1 \tag{3.260}$$

$$\hat{\pi}_1^2 = (1-p)^2\hat{\mathcal{P}}_1 + p^2\hat{\mathcal{P}}_0. \tag{3.261}$$

But they do correctly give the measurement probabilities:

$$P(0) = \mathrm{Tr}\big(\hat{\rho}\hat{\pi}_0\big), \qquad\qquad P(1) = \mathrm{Tr}\big(\hat{\rho}\hat{\pi}_1\big). \tag{3.262}$$

This procedure of defining new operators to account for measurement errors is a special case of the **positive operator-valued measurement (POVM)** approach. A POVM is a set of measurement operators $\{\hat{E}_j\}$ that satisfy the first three requirements for projective operators, namely:

1. $\hat{E}_j^\dagger = \hat{E}_j$ (Hermiticity)
2. $\hat{E}_j \geq 0$ (Non-negativity)
3. $\sum_j \hat{E}_j = 1$ (Completeness)

But they are allowed to violate the fourth requirement: they may not obey $\hat{E}_i\hat{E}_j \neq \hat{E}_j\delta_{ij}$, and so fail to be orthogonal projection operators. As might be expected, given a set of possible outcomes labeled by integer $m = 1, 2, 3, \ldots$, the probability of the mth outcome is given by

$$P_m = \mathrm{Tr}\big(\hat{\rho}E_m\big). \tag{3.263}$$

The operators π_j described above clearly form an example of a POVM. Projective measurements are a special case of this more general framework.

It can be shown that POVMs can be found to implement any conceivable measurement scheme, ideal or otherwise, and so they provide the most general description of the measurement process. It can also be shown that any POVM on a given system can be viewed as a projective measurement on an expanded system that contains additional ancillary particles. Essentially, POVMs are necessary when you wish to describe a system that is not closed and isolated, but that is interacting with an environment (neighboring quantum systems, optical detectors, etc.).

There can be other situations, having nothing to do with nonideal measurements, in which the measured states are not orthogonal. One example is measurements done on coherent states (Sect. 9.1); no two coherent states with finite amplitudes are ever orthogonal to each other, so POVMs need to be used to measure them. Another important application occurs when trying to identify an unknown state from among a set of possible nonorthogonal states; as will be seen in Sect. 9.6, a POVM can be constructed that will do this without error, but at the price of not always being able to reach any conclusion at all. More examples of POVMs can be found in references [5, 23].

A POVM operator can always be decomposed in terms of **Krauss operators** or **effect operators**, $\hat{K}_j$, according to $\hat{E}_j = \hat{K}_j^\dagger \hat{K}_j$. (Note that the Krauss operators may not be Hermitian, but the POVM operators are.) The completeness of the $\hat{E}_j$ implies that

$$\sum_j \hat{K}_j^\dagger \hat{K}_j = \hat{I}. \tag{3.264}$$

The result of measuring an observable has to correspond to an eigenvalue associated with one of the $\hat{E}_j$. After the measurement, the density operator of the system is changed from its initial value $\hat{\rho}$ to a final value $\hat{\rho}'$ given by

$$\hat{\rho} \rightarrow \hat{\rho}' = \frac{\hat{K}_j \hat{\rho} \hat{K}_j^\dagger}{\mathrm{Tr}(\hat{\rho} \hat{E}_j)}, \tag{3.265}$$

where the bottom of the fraction gives the probability of obtaining outcome $\hat{E}_j$:

$$P_j = \mathrm{Tr}(\hat{\rho} \hat{E}_j) = \mathrm{Tr}\hat{K}_j \hat{\rho} \hat{K}_j^\dagger. \tag{3.266}$$

The mapping 3.265 preserves the positivity of the density matrix, as well as preserving the requirement that the density operator be of unit trace: $\mathrm{Tr}\hat{\rho}' = \mathrm{Tr}\hat{\rho} = 1$.

Problems

1. Consider a single spin-$\frac{1}{2}$ particle at $x = 0$. Initially, it has spin up along the z-axis, $|\psi(0)\rangle = \begin{pmatrix} 1 \\ 0 \end{pmatrix}$. Suppose that at $t = 0$ a magnetic field $\boldsymbol{B}$ is turned on, pointing along the y-axis. The interaction with the field with the spin can be described by the Hamiltonian $\hat{H} = -\frac{\hbar\omega}{2}\hat{\sigma}_y$, where the frequency ω will be proportional to the field.

   (a) Write down the time-dependent Schrödinger equation for the state.
   (b) Find the time evolution operator $\hat{U}(t)$ and write it in matrix form. Apply it to the initial state $|\psi(0)\rangle$ in order to find $|\psi(t)\rangle = \hat{U}(t)|\psi(0)\rangle$, and show that this wave function satisfies the Schrödinger equation of part (a). What kind of motion is the spin undergoing?
   (c) Find $p_\uparrow(t)$ and $p_\downarrow(t)$, the probabilities of finding the particle in a spin-up or spin-down state along the z-axis at time t.
   (d) Rather than thinking of the state as evolving in time under the influence of a time-independent operator (the Schrödinger picture), we can instead think of the state as staying constant while the operator describing the spin evolves in time (the Heisenberg picture). Show that the time-dependent Pauli operator $\hat{\sigma}_z(t) = \hat{U}(t)^\dagger \hat{\sigma}_z \hat{U}(t)$ satisfies the Heisenberg equation of motion:

$$i\hbar \frac{\partial}{\partial t}\hat{\sigma}_z(t) = \left[\hat{\sigma}_z(t), \hat{H}\right]. \qquad (3.267)$$

2. Use the properties of the Pauli operators to show that for any pair of unit vectors $\hat{\boldsymbol{m}}$ and $\hat{\boldsymbol{n}}$, it follows that

$$\left[\hat{\boldsymbol{n}} \cdot \hat{\boldsymbol{\sigma}}, \hat{\boldsymbol{m}} \cdot \hat{\boldsymbol{\sigma}}\right] = 2i\left(\hat{\boldsymbol{n}} \times \hat{\boldsymbol{m}}\right) \cdot \hat{\boldsymbol{\sigma}}. \qquad (3.268)$$

3. Let three states be given in the computational basis by

$$|\psi\rangle = \begin{pmatrix} 1 \\ 0 \end{pmatrix}, \qquad |\phi\rangle = \frac{1}{\sqrt{2}}\begin{pmatrix} 1 \\ 1 \end{pmatrix}, \qquad |\eta\rangle = \frac{1}{\sqrt{2}}\begin{pmatrix} 1 \\ -1 \end{pmatrix}. \qquad (3.269)$$

   (a) Find the transition probabilities $|\langle\psi|\phi\rangle|^2, |\langle\psi|\eta\rangle|^2, |\langle\phi|\eta\rangle|^2$.
   (b) Find the tensor product states $|\psi\rangle \otimes |\phi\rangle$, $|\psi\rangle \otimes |\eta\rangle$, and $|\phi\rangle \otimes |\eta\rangle$, written as four-dimensional column vectors.
   (c) Find $\hat{\sigma}_x \otimes \hat{\sigma}_y$, $\hat{\sigma}_z \otimes \hat{\sigma}_x$, and $\hat{\sigma}_y \otimes \hat{\sigma}_z$ as four-by-four matrices.

4. Which of the following are product states, and which are entangled? For the product states, explicitly show how they factor into single-particle states.

(a) $|\psi\rangle = \frac{1}{2}\left(|00\rangle + |01\rangle + |10\rangle + |11\rangle\right)$

(b) $|\psi\rangle = \frac{1}{2}\left(|00\rangle + |01\rangle + |10\rangle - |11\rangle\right)$

(c) $|\psi\rangle = \frac{1}{2}\left(|00\rangle - |01\rangle - |10\rangle + |11\rangle\right)$

(d) $|\psi\rangle = \frac{1}{2}\left(|00\rangle - |01\rangle + |10\rangle - |11\rangle\right)$

(e) $|\psi\rangle = \cos^2\theta|00\rangle - \sin\theta\cos\theta|01\rangle + \sin\theta\cos\theta|10\rangle - \sin^2\theta|11\rangle$

(f) $|\psi\rangle = \frac{1}{\sqrt{2}}(|00\rangle - |11\rangle)$

5. (a) Write down the spin-up and spin-down states of Eq. 3.65 pointing along the x-axis ($\hat{n} = \hat{i}$).

   (b) Define the Bell states as in Eqs. 3.227–3.230 with $|0\rangle$ and $|1\rangle$ representing, respectively, the states that are spin-up and spin-down along the x-axis. Write these states in the z-basis of Eq. 3.66.

6. (a) Verify that the operator defined in Eq. 3.252 satisfies $\hat{P}_{\hat{n}}^2 = \hat{P}_{\hat{n}}$, as required by a projection operator.

   (b) Find a projection operator orthogonal to $\hat{P}_{\hat{n}}$.

7. Define the density matrix $\hat{\rho} = \frac{1}{2}(\hat{I} + \hat{n} \cdot \hat{\sigma})$, representing the state of a spin along unit vector $\hat{n}$.

   (a) Show that this matrix satisfies the requirements to be a density matrix.

   (b) Is the state pure or mixed?

   (c) If $\hat{m}$ is a second unit vector, at an angle θ from $\hat{n}$, find the expectation value $\langle \hat{m} \cdot \hat{n} \rangle$.

8. Find the fidelity and trace distance between the following pairs of states:

   (a) $|\psi\rangle = \begin{pmatrix} 1 \\ 0 \end{pmatrix}$ and $|\phi\rangle = \begin{pmatrix} \cos\theta \\ \sin\theta \end{pmatrix}$

   (b) $\hat{\rho} = \frac{1}{2}\begin{pmatrix} 1 & 0 \\ 0 & 1 \end{pmatrix}$ and $\hat{\sigma} = \frac{1}{2}\begin{pmatrix} 1 & 1 \\ 1 & 1 \end{pmatrix}$

   (c) $\hat{\rho}_m = \frac{1}{2}(\hat{I} + \hat{m} \cdot \hat{\sigma})$ and $\hat{\rho}_n = \frac{1}{2}(\hat{I} + \hat{n} \cdot \hat{\sigma})$, where the angle between unit vectors $\hat{m}$ and $\hat{n}$ is θ.

9. Find the partial traces Tr_A and Tr_A of density matrices for the following states of systems A and B. (Notice that A and B are each single-particle systems for parts (a)–(c), but they are two-particle systems for part (d). $\Psi^{\pm}$ and $\Phi^{\pm}$ are Bell states.)

   (a) $|\psi\rangle = |\Psi^+\rangle_{AB}$

   (b) $\hat{\rho} = \frac{1}{2}\left(_{AB}|\Psi^+\rangle\langle\Psi^+|_{AB} -_{AB}|\Phi^+\rangle\langle\Phi^+|_{AB}\right)$

   (c) $\hat{\rho} = \alpha|00\rangle\langle00| + \beta|01\rangle\langle01| + \gamma|10\rangle\langle10| + \delta|11\rangle\langle11|$, where $\alpha + \beta + \gamma + \delta = 1$.

   (d) $|\psi\rangle = |\Psi^+\rangle_A \otimes |\Psi^+\rangle_B$

10. Consider two different mixed states: one is an equal mixture of $|0\rangle$ and $|1\rangle$, while the other is an equal mixture of $|+\rangle$ and $|-\rangle$ states. Show that both of these mixed states have the same density matrix. As a result, they are experimentally indistinguishable.

References

1. R. Shankar, *Principles of Quantum Mechanics*, 2nd edn. (Plenum Press, New York, 1994)
2. R. Eisberg, R. Resnick, *Quantum Physics of Atoms, Molecules, Solids, Nuclei, and Particles*, 2nd edn. (Wiley and Sons, Hoboken, 1985)
3. D.J. Griffiths, D.F. Schroeter, *Introduction to Quantum Mechanics*, 3rd edn. (Cambridge University Press, Cambridge, 2018)
4. M. Beck, *Quantum Mechanics: Theory and Experiment* (Oxford University Press, Oxford, 2012)
5. A. Perez, *Quantum Theory, Concepts and Methods* (Kluwer Academic Publishers, Dordrecht, 1998)
6. C.M. Bender, S. Boettcher, Phys. Rev. Lett. **80**, 5243 (1998)
7. L. Feng, Z.L. Wong, R.M. Ma, Y. Wang, X. Zhang, Science **346**, 972 (2014)
8. H. Hodaei, M.A. Miri, M. Heinrich, D.N. Christodoulides, M. Khajavikhan, Science **346**, 975 (2014)
9. P. Miao, et al., Science **353**, 464 (2016)
10. Y. Ashida, Z. Gong, M. Ueda, Adv. Phys. **69**, 3 (2020)
11. R. El-Ganainy, M. Khajavikhan, D.N. Christodoulides, S.K. Ozdemir, Commun. Phys. **2**, 1 (2019)
12. M. Parto, et al., Nanophotonics **10**, 403(2020)
13. M. De Carlo, et al., Sensors **22**, 3977 (2022)
14. M.A. Nielsen, I.L. Chuang, *Quantum Computation and Quantum Information: 10th Anniversary Edition* (Cambridge University Press, Cambridge, 2011)
15. W.M.R. Simpson, U. Leonhardt, *Forces of the Quantum Vacuum: An introduction to Casimir physics* (World Scientific, Singapore, 2015)
16. A. Zee, *Quantum Field Theory in a Nutshell*, 2nd edn. (Princeton University Press, Princeton, 2010)
17. M. Hotta, Phys. Lett. A **372**, 5671 (2008)
18. N.A. Rodríguez-Briones, H. Katiyar, R. Laflamme, E. Martín-Martínez, Phys. Rev. Lett. **130**, 110801 (2023)
19. K. Ikeda, Phys. Rev. Appl. **20**, 024051 (2023)
20. D.A. Lidar, Review of decoherence-free subspaces, noiseless subsystems, and dynamical decoupling, in *Quantum Information and Computation for Chemistry*, ed. by S. Kais (Wiley, Hoboken, 2014), p. 295
21. W.H. Zurek, Phys. Today **44**(10), 36 (1991)
22. W.H. Zurek, Rev. Mod. Phys. **75**, 715 (2003)
23. S.M. Barnett, *Quantum Information* (Oxford University Press, Oxford, 2009)

Chapter 4
Elements of Solid-State and Atomic Physics

Qubits, while they were described as abstract objects in Chap. 3, in reality must always be realized by some physical system, most commonly photons, electron spin, atoms, or solid state structures. Background on photons and spin has already been covered in earlier chapters. In this chapter, we cover some basic background material in atomic, molecular, and solid-state physics in order to discuss further implementations of quantum systems in later chapters.

4.1 The Structure and Properties of Atoms

4.1.1 Atomic Spectra and the Bohr Model of Hydrogen

In the nineteenth century, the periodic table of the elements was constructed by looking at regularities in the properties of the chemical elements. However, the goal of constructing a physical explanation for these properties based on the structure of atoms remained elusive. One major clue was provided by the existence of atomic spectral lines. Newton had first seen that light passed through a prism breaks up into bands of different colors. In 1802, Wollaston found that small bands or lines of color are missing from the spectrum of light coming from the sun; soon, missing lines were found by Fraunhofer and others when white light was passed through a dilute gas, and it was realized that the pattern of lines was a characteristic of a particular chemical element of gas. These dark **absorption lines** are now known to be due to selective absorption of particular frequencies when they match the resonant frequencies of the gas atoms.

It had been known since the work of Thomas Melvill, published posthumously in 1756, that heated salts emitted light of different colors and that the patterns of colors differed depending on the species of the salt. These colors could be separated from each other with a prism or diffraction grating to create a series of brightly colored

© The Author(s), under exclusive license to Springer Nature Switzerland AG 2025
D. S. Simon, *Introduction to Quantum Science and Technology*, Undergraduate
Texts in Physics, https://doi.org/10.1007/978-3-031-81315-3_4

Fig. 4.1 The first four series
of spectral lines for hydrogen.
Each series is characterized
by the principal quantum
number of the final state

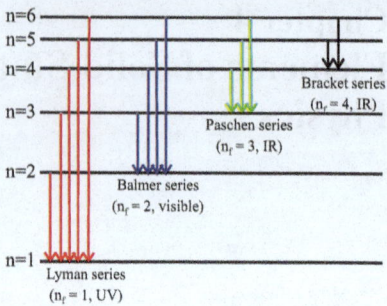

lines or bands. By 1835, Charles Wheatstone had established that the identity of a
metal could be determined from the spectrum of these **emission lines**, and by 1849
Foucault had shown that the lines corresponding to the emission and absorption
spectra coincided in frequency for a given element.

Observations of the emission spectrum of hydrogen led to the realization
that these lines could be arranged into a set of well-defined sequences, each
characterized by an integer n_f (Fig. 4.1). The first four sequences characterized by
the integers $n_f = 1, 2, 3, 4$ are called the **Lyman, Balmer, Paschen,** and **Brackett
series**, respectively. The wavelengths of the lines within each series could then
be described by a second integer $n_i > n_f$, according to the empirical **Rydberg
formula**:

$$\frac{1}{\lambda} = R \left(\frac{1}{n_f^2} - \frac{1}{n_i^2} \right), \tag{4.1}$$

where $R = 1.0973732 \times 10^7 \text{ m}^{-1}$. Any physical model of the atom would need to
explain this seemingly mysterious formula.

By the beginning of the twentieth century, the basic constituents of atoms were
all known. Electrons, initially known as **cathode rays** or **beta particles**, were first
recognized in 1897 as being emitted from the cathodes of vacuum tubes by J. J.
Thomson. These were soon realized to be the same particles Becquerel had seen
being emitted by radon and other nuclei in β-decay. Millikan and others quickly
pinned down the properties of these particles to a high degree of accuracy, showing
that they had negative electric charge and a mass about 1800 times lighter than the
hydrogen atom.

By 1911, Rutherford had established the existence of an atomic nucleus. Through
scattering experiments, it was found that these nuclei had small, hard constituents,
some positively charged and some electrically neutral. Both of these constituents
were similar in mass to the hydrogen atom. These constituents were, of course,
protons and neutrons.

How electrons, protons, and neutrons fit together to form atoms was still to be
determined. A number of models were proposed in the early days, each explaining
various properties of the atom. One possibility (the 1911 Rutherford model) was to

think of the atom as a tiny solar system with electrons orbiting a positively charged nucleus in circular or elliptical orbits, with the electrostatic attraction between protons and neutrons playing the same role that gravity plays in the actual solar system. One seemingly fatal problem with this model, though, was that such an atom would be unstable. An accelerating charge (such as the orbiting electron) necessarily radiates away energy, via electromagnetic waves. So the electron orbits should decay, leading the electron to quickly plunge into the nucleus.

Bohr provides the missing ingredient needed to make the orbiting-electron model feasible: this was the idea of quantization. It consists of two distinct assumptions:

- **Energy quantization.** The first assumption is that the allowed electron energies in the atom form a discrete sequence, rather than a continuous band. It is assumed that the frequencies v of the light in the absorption and emission spectra are proportional to differences between the allowed energy levels:

$$|E_i - E_f| = hv, \tag{4.2}$$

where i and f label the initial and final energy levels and h is Planck's constant, which was already well-known from Planck's work on the blackbody spectrum and Einstein's work on the photoelectric effect. Thus, both the electron energy levels and the emitted or absorbed photon frequencies are quantized (discrete) variables.

- **Angular momentum quantization.** The second assumption was that the angular momentum $|L| = mvr$ was also quantized, with values equal to integer multiples of a fixed constant:

$$mvr = n\hbar, \tag{4.3}$$

where $\hbar = \frac{h}{2\pi}$ and n is a non-negative integer, $n = 0, 1, 2, 3, \ldots$. In hindsight, this assumption could be seen to be equivalent to the idea that the electron's de Broglie wave (not introduced until 1924) forms a standing wave with constructive interference around the electron orbit: the assumption of angular momentum quantization is equivalent to the constructive interference condition, $n\lambda = 2\pi r$.

Restricting the energies and angular momenta to discrete values means that the electrons can only make discrete jumps between the allowed states when it is possible for the electrons and the emitted light to obey the relevant conservation laws. This stops the continuous spiral into the nucleus that doomed the original Rutherford model and guarantees the stability of the atom.

The Bohr model is simple enough that we can easily derive exact expressions for the energy levels and other variables, and we can also show that the Rydberg formula (Eq. 4.1) emerges naturally. To see this, begin with the classical, nonrelativistic energy of an electron in the Coulomb potential of the hydrogen nucleus (a single proton):

$$E = K + U \tag{4.4}$$

$$= \frac{1}{2}mv^2 - k\frac{e^2}{r}. \tag{4.5}$$

Here, $k = \frac{1}{4\pi\epsilon_0}$ is the electrostatic constant. Assuming circular orbits, Newton's second law tells us that

$$\frac{ke^2}{r^2} = m\frac{v^2}{r}, \tag{4.6}$$

where the left side is the Coulomb force and the right side is m times the centripetal acceleration. Using the last equation, the energy can be put into the form

$$E = \frac{1}{2}mv^2 - mv^2 = -\frac{1}{2}mv^2, \tag{4.7}$$

or into the form

$$E = -\frac{ke^2}{2r}. \tag{4.8}$$

Invoking angular momentum quantization, Eq. 4.3, and combining it with Eq. 4.6, we find that the allowed radii of the orbits form a discrete sequence:

$$r_n = \frac{n^2\hbar^2}{mke^2} = n^2 a_0, \tag{4.9}$$

where we have defined the **Bohr radius**,

$$a_0 = \frac{\hbar^2}{mke^2} = 0.529\,\text{Å}, \tag{4.10}$$

which gives the radius of the innermost (lowest energy) orbit in the Bohr model.

From Eq. 4.8, the allowed energies are then

$$E_n = -\frac{ke^2}{2r_n} = -\frac{ke^2}{2a_0 n^2} = \frac{E_1}{n^2}, \tag{4.11}$$

where $E_1 = -\frac{ke^2}{2a_0} = -13.6\,\text{eV}$ is the ground state energy of the hydrogen atom. These energy levels match closely with the values measured in experiments, at least in the absence of strong external electric and magnetic fields. Similarly, the quantized orbital velocities of the electrons can be found. Using the fact that the magnitude of the angular momentum may be written as $L_n = pr_n$, we find

$$v_n = \frac{p}{m} = \frac{L_n}{mr_n} = \frac{n\hbar}{mr_n} \tag{4.12}$$

$$= \frac{\hbar}{ma_0 n} = \frac{e^2}{4\pi\epsilon_0 \hbar n} \tag{4.13}$$

$$= \frac{c\alpha}{n}, \tag{4.14}$$

where the dimensionless **fine structure constant**

$$\alpha = \frac{\hbar}{ma_0 c} = \frac{e^2}{4\pi\epsilon_0 \hbar c} \approx \frac{1}{137} \tag{4.15}$$

is a measure of the intrinsic strength of the electromagnetic interaction.

Finally, consider the frequency of the light emitted as an electron decays from level i to level f:

$$\nu = \frac{E_i - E_f}{h} \tag{4.16}$$

$$= -\frac{ke^2}{2a_0 h}\left(\frac{1}{n_i^2} - \frac{1}{n_f^2}\right). \tag{4.17}$$

Using $\nu = c/\lambda$, the corresponding inverse wavelength

$$\frac{1}{\lambda} = \frac{ke^2}{2ha_0 c}\left(\frac{1}{n_f^2} - \frac{1}{n_i^2}\right). \tag{4.18}$$

If we make the identification $R = \frac{ke^2}{2ha_0 c}$, this is simply the Rydberg formula, Eq. 4.1.

Thus, starting from classical physics, combining the assumption that energy and angular momentum are quantized, the Bohr model accurately reproduces the main properties of the hydrogen atom that were known at the time.

Box 4.1 Vortex Atoms

The Bohr model was not the first attempt to model the structure of the atom. One of the earliest attempts was the vortex model created by Peter Guthrie Tait and William Thomson (soon to become Lord Kelvin) in the nineteenth century. Based on earlier ideas from Helmholtz, the vortex atom model was popular between roughly the 1860s and the 1890s.

Vortices, or whirlpools, are centered on curves inside a fluid at which some field becomes singular or undefined. The fluid circulates around the singular curve, forming a vortex that can change shape over time. Vortex lines are common in everyday life (smoke rings and tornados are both examples) and are common in superfluids and superconductors. In the days before the

(continued)

Box 4.1 (continued)

Michelson-Morley experiment and special relativity, it was a nearly universal belief that "empty" space must contain an invisible fluid, the luminiferous aether, in order to provide a medium for electromagnetic waves to propagate through. Kelvin and Tait reasoned that if vortex lines could form in this aether, then they could become knotted. Different knot configurations could therefore explain the existence of different species of atoms. The fact that different segments of the vortex lines would tend to repel each other and avoid self-crossing would keep the knots from unraveling, thereby explaining the stability of atoms. To make two segments of the knot cross would be possible, but would require large amounts of energy, so that elements could only transmute into each other at very high energies, a fact that, with our modern understanding of fission and fusion, we now know to be true. Furthermore, it was possible that different oscillation modes of the knots could explain the variety of observed atomic spectral lines.

Several years were spent by Tait and Kelvin conducting experiments on smoke rings in order to simulate the behavior of the proposed aether vortices. Kelvin's 1867 paper [1] stimulated a great deal of interest in the subject. Both Tait and J. J. Thomson published analyses of the mathematics underlying the model. Unfortunately, there was no obvious way for the model to explain the periodic properties of the atoms (embodied in the periodic table of the elements). And eventually the aether theory had to be discarded, leading to the demise of the vortex atom.

But the work of Kelvin and Tait was fruitful in the long run. Tait compiled large tables of knots of increasing complexity and investigated their properties. This became the starting point for the modern theory of knots, an important area of topology. Knots and vortices have since re-entered physics in a number of areas, including superstring theory, superconductivity (Chap. 7), and topological insulators (Chap. 21)

A nice nontechnical telling of the story of vortex atoms is given in [2].

4.1.2 The Quantum Mechanical Atom

To go beyond the Bohr model for hydrogen, we must dispense with the picture of the electron as a particle circling the nucleus in well-defined orbits. Instead, we must consider the wavelike properties of the electron by setting up the Schrödinger equation for the system and solving for the electron wavefunction. We will not work through the details here, since they can be found in every introductory quantum mechanics textbook. Instead, we summarize the result and some of the properties of the wavefunction that will be useful later.

The goal is to find the eigenvalues $\Psi_{nlm_l}(r, \theta, \phi)$ and eigenvalues E_n that will satisfy the time-independent Schrödinger equation:

$$\hat{H}\Psi_{nlm_l}(r, \theta, \phi) = E_n\Psi_{nlm_l}(r, \theta, \phi), \tag{4.19}$$

for the Coulomb Hamiltonian $\hat{H} = \frac{\hat{p}^2}{2m} - \frac{ke^2}{r}$. Recall that the Coulomb potential is a central potential: written in terms of spherical polar coordinates, $V(r, \theta, \phi)$ is in fact independent of the angular variables ϕ and θ and depends only on the radial distance r between the two charges.

For central potentials, the Schrödinger equation factorizes, and the solutions can be written in the form

$$\Psi(r, \theta, \phi) = R_{nl}(r)Y_{l,m_l}(\theta, \phi), \tag{4.20}$$

which is parameterized by three quantum numbers:

- Principal quantum number: $n = 0, 1, 2, \ldots$
- Orbital angular momentum quantum number: $l = 0, 1, 2, \ldots, n - 1$
- Magnetic quantum number: $m_l = -l, -l + 1, \ldots, 0, \ldots, l$

The radial solution $R_{nl}(r)$ depends only on the principle and orbital angular momentum quantum numbers. Its exact form involves the well-known Laguerre polynomials. We won't need the exact form of the radial functions, so we defer to standard quantum mechanics texts for a discussion of the radial wavefunction. We simply note that the number of nodes or zeros of the radial wavefunction is $n - l - 1$.

The angular part of the wavefunction, $Y_{l,m_l}(\theta, \phi)$, is known as a **spherical harmonic**. The first few spherical harmonics are given in Table 4.1. While the radial wavefunctions depend on the form of the potential, the same spherical harmonics in fact occur in the solutions for all central potentials; they are fixed by the rotational symmetry of the problem. Notice that the dependence on the azimuthal angle ϕ (the angle about the z-axis) is always given by a phase factor of the form $e^{im_l\phi}$.

Once the solutions for the wavefunction have been found, it is readily verified that the energy eigenvalues match those of the Bohr model:

Table 4.1 The spherical harmonics Y_{l,m_l} for $|l| \leq 2$.

l	m_l	Y_{l,m_l}
0	0	$\frac{1}{2\sqrt{\pi}}$
1	0	$\frac{1}{2}\sqrt{\frac{3}{\pi}}\cos\theta$
1	± 1	$\mp\frac{1}{2}\sqrt{\frac{3}{2\pi}}\sin\theta\, e^{\pm i\theta}$
2	0	$\frac{1}{4}\sqrt{\frac{5}{\pi}}(3\cos^2\theta - 1)$
2	± 1	$\mp\frac{1}{2}\sqrt{\frac{15}{2\pi}}\sin\theta\cos\theta\, e^{\pm i\phi}$
2	± 2	$\frac{1}{4}\sqrt{\frac{15}{2\pi}}\sin^2\theta\, e^{\pm 2i\phi}$

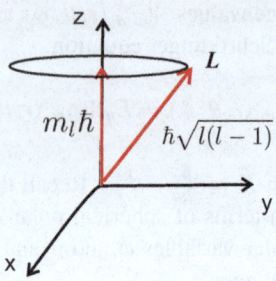

Fig. 4.2 The angular momentum vector has a length determined by the quantum number l: $|L| = \hbar\sqrt{l(l-1)}$. The z-component is quantized, with size $L_z = m_l\hbar$ determined by the integer m_l. The x- and y-components of L are indeterminate, with the tip of the L vector fluctuation randomly about the horizontal circle shown. Because m_z is quantized to a set of discrete values, the allowed angles of L from the z-axis are also quantized, $\theta_{m_l} = \tan^{-1}\left(m_l/\sqrt{l(l+1)}\right)$

$$E_n = -\frac{13.6\,\text{eV}}{n^2} \tag{4.21}$$

The significance of the l and m_l quantum numbers of course is that they determine, respectively, the magnitude and orientation (relative to the z-axis) of the orbital angular momentum vector:

$$|L| = \hbar\sqrt{l(l-1)} \qquad L_z = m_l\hbar. \tag{4.22}$$

The possible directions of the L vector are discrete, lying at angles $\theta_{m_l} = tan^{-1}\frac{L_z}{|L|} = \tan^{-1}\left(\frac{m_l}{\sqrt{l(l+1)}}\right)$, as in Fig. 4.2. In keeping with the uncertainty principle, any state with a definite value of L_z (i.e., with a definite value of m_l) must be completely uncertain in the xy-plane; the L vector can be viewed as rotating in a random manner about the z-axis, with its tip remaining at fixed angle from the axis.

Rather than specifying the integer l value of a state, it is common practice in atomic physics and other areas to use **spectroscopic notation**, using the letters s, p, d, etc. to denote the value of l:

Quantum number l	0	1	2	3	4	5
Notation:	s	p	d	f	g	h

The first three letters historically came from the words *sharp*, *principle*, and *diffuse*.

The energy levels given above depend only on the principle quantum number, n. However, in the presence of external fields, there are additional terms in the Hamiltonian that will introduce angular momentum-dependent corrections to the energies. For example, consider the **Zeeman effect**, in which the presence of an external magnetic field causes each energy level to split into multiple levels.

To describe the Zeeman effect, first recall the definition of the **magnetic dipole moment**, μ. Consider first a classical current loop, carrying a current I and

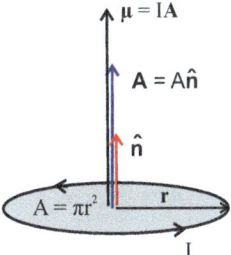

Fig. 4.3 A classical current I, when formed into a loop enclosing area $A = \pi r^2$, creates a magnetic dipole moment μ. The area is promoted to a vector by attaching a unit vector $\hat{n}$ to it to give it a direction perpendicular to the current loop. The orientation of $\hat{n}$ is fixed by using the right-hand rule on the current loop: curl the fingers of your right hand in the direction the current travels, and $\hat{n}$ is then in the direction of your thumb

enclosing an area A. Then the magnitude of the classical dipole moment is simply

$$\mu = IA. \tag{4.23}$$

By attaching a unit vector $\hat{n}$ perpendicular to the plane of the current with direction given by the right-hand rule (see Fig. 4.3), the area can be made into a vector, $\mathbf{A} = A\hat{n}$, which in turn makes the magnetic dipole into a vector:

$$\boldsymbol{\mu} = I\mathbf{A}. \tag{4.24}$$

In an external magnetic field $\mathbf{B}$, the dipole moment has potential energy

$$U = -\boldsymbol{\mu} \cdot \mathbf{B} \tag{4.25}$$

and experiences a torque

$$\boldsymbol{\tau} = \boldsymbol{\mu} \times \mathbf{B}. \tag{4.26}$$

The torque tends to align the dipole moment in the direction of the external field, which in turn minimizes the energy.

Now imagine an orbiting electron forming a circular current loop of radius r. The charge $-e$ is smeared over a circumference $2\pi r$, so that $\frac{dq}{dx} = -\frac{e}{2\pi r}$, where x is the distance along the orbit. The current can then be written as

$$I = \frac{dq}{dt} = \frac{dq}{dx} \cdot \frac{dx}{dt} = \left(\frac{-e}{2\pi r}\right) v, \tag{4.27}$$

so that

$$\mu = IA = -\frac{evr}{2} = -\frac{e}{2m}(mvr). \tag{4.28}$$

Recognizing that mvr is the magnitude of the orbital angular momentum, we arrive at an expression for μ that is more useful in quantum mechanics:

$$\mu = -\frac{e}{2m}L \tag{4.29}$$

Given that the orbital angular momentum is quantized, we find that the magnetic moment also has a quantized z-component:

$$\mu_z = -\frac{e\hbar}{2m}m_l = -\mu_B m_l, \tag{4.30}$$

where the **Bohr magneton** is defined by

$$\mu_B = \frac{e\hbar}{2m}. \tag{4.31}$$

The corresponding potential energy is then

$$U = -\mu_z B = +\mu_B m_l B. \tag{4.32}$$

This serves as a correction to the Bohr energy levels. The $2l + 1$ states of different m_l, which were previously degenerate, now split into a set of $2l + 1$ distinct energy levels, giving the Zeeman splitting. For a spin-$\frac{1}{2}$ particle, each energy level will further split into two levels, representing spin-up and spin-down states.

For μ not aligned along the z-axis, i.e., for $m_l \neq \pm l$, the torque will cause the dipole moment to precess around the z-axis (Fig. 4.4). This **Larmor precession** has frequency given by

$$\omega_L = \frac{d\phi}{dt} = \frac{1}{L\sin\theta}\frac{dL}{dt} = \frac{eB}{2m}. \tag{4.33}$$

This Larmor precession plays a role in quantum information processing, as will be seen in Chap. 23.

In addition to orbital angular momentum, an electron also has spin, with quantized magnitude and angular momentum:

$$S^2 = s(s+1)\hbar^2, \qquad S_z = m_s\hbar, \tag{4.34}$$

where, for an electron, $s = \frac{1}{2}$ and $m_s \pm \frac{1}{2}$. The spin also has a dipole moment, given by

$$\mu_s = -\frac{e}{m}S = -\frac{2\mu_B}{\hbar}S. \tag{4.35}$$

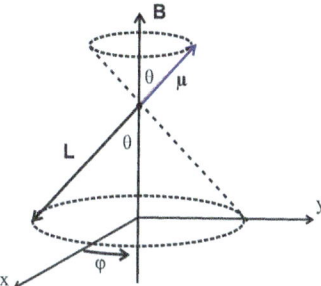

Fig. 4.4 The dipole moment is (due to the negative charge of the electron) opposite to the orbital angular momentum. Both vectors are at an angle θ from the magnetic field, which is taken along the z-axis. The magnetic field creates a torque $\tau = \mu \times B = \frac{dL}{dt}$ which is perpendicular to μ and L, causing both of these vectors to rotate about the z-axis at constant angular frequency $\omega = \frac{d\phi}{dt} = \frac{eB}{2m}$, where ϕ is the angle about the z-axis

Note that the spin enters the dipole moment with an extra factor of 2; this intrinsically relativistic effect was first noticed by Dirac. In the presence of an external B field, the interaction of the spin with the field causes energy levels to split into doublets.

We can generalize by defining the **g-factor** g, in terms of which the magnetic dipole moment may be written:

$$\mu_J = \frac{g_J}{\hbar}\mu_B J, \tag{4.36}$$

where J can represent either spin or orbital angular momentum, and where $g_s = -2$ and $g_l = -1$. (In fact, these values of g are approximate; Dirac showed that quantum effects add small corrections to them. These corrections can be both calculated theoretically and measured experimentally to extremely high precision. For example, the currently accepted value for the electron is $g_s = -2.0023193043622$.)

We can also write the last expression in the form

$$\mu = \gamma J, \tag{4.37}$$

where γ is the **gyromagnetic ratio**. For electrons, protons, and neutrons, we have

$$\gamma_e = \frac{2|\mu_e|}{\hbar} \approx 1.760 \times 10^{11}\,\mathrm{T}^{-1}s^{-1} \tag{4.38}$$

$$\gamma_p = \frac{2|\mu_p|}{\hbar} \approx 2.675 \times 10^{8}\,\mathrm{T}^{-1}s^{-1} \tag{4.39}$$

$$\gamma_n = \frac{2|\mu_n|}{\hbar} \approx 1.83 \times 10^{8}\,\mathrm{T}^{-1}s^{-1}. \tag{4.40}$$

In multi-electron atoms, new features appear. For example, the movement of electrons in the atom produces internal magnetic fields, proportional to the orbital angular momentum, that can then interact with the spin magnetic dipole moments, leading to a **spin-orbit interaction**:

$$\mu_S \cdot B_{internal} \sim S \cdot L. \tag{4.41}$$

For an electron with total angular momentum $J = L + S$, this leads to an additional energy splitting:

$$\Delta U = \mu_B \, g \, m_J \, B_{ext}, \tag{4.42}$$

where the **Landé g-factor** is given by

$$g = 1 + \frac{J(J+1) + S(S+1) - L(L+1)}{2J(J+1)}. \tag{4.43}$$

Here, S, L, and J are the spin, orbital, and total angular momentum quantum numbers for the multi-electron state (Fig. 4.5).

The full magnetic dipole of the system is the sum of those due to the orbital and spin angular momentum:

$$\mu = \mu_l + \mu_s = -\frac{e^2}{2m} (L + 2S). \tag{4.44}$$

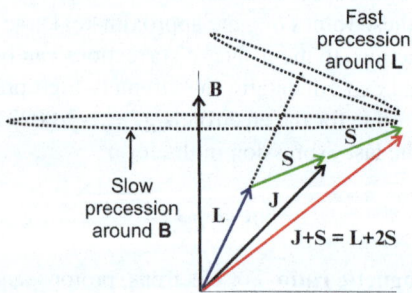

Fig. 4.5 The various angular momenta, L, S, and J, as well as the corresponding magnetic dipole moments, all precess at different rates. The precession of the entire system around the external field is much slower than the precession of total dipole moment $\sim L + 2S$ about J. (The extra factor of 2 in front of S is again due to the fact that the gyromagnetic ratio for S is double that of L)

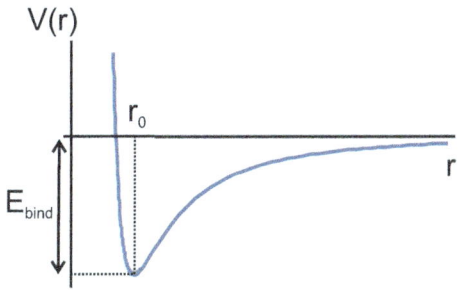

Fig. 4.6 The interaction potential between two atoms joined in a chemical bond should be roughly of the form shown, with a minimum at r_0, corresponding to the mean length of the bond. The binding energy of the molecule is given by the depth E_{bind} of the potential well

4.1.3   Molecules

As two atoms are brought close together, they may bond to form a stable molecule, if their interaction energy has a negative (attractive) minimum in its potential energy function. Often a good qualitative description of the molecule can be obtained by modeling the interaction potential in the form

$$V(r) = \frac{A}{r^n} - \frac{B}{r^m}, \qquad (4.45)$$

where A, B, m, and n are positive constants and r is the inter-atom distance. The A term leads to a repulsive force, while the B term gives an attractive force. This gives a potential of the form shown in Fig. 4.6. The position of the stable minimum determines the length of the bond, r_0, which can be arranged to match the experimental value for a given molecule by an appropriate choice of the constants. A common version of this potential is the **Lennard-Jones potential**:

$$V(r) = \epsilon_0 \left(\frac{r_0^{12}}{r^{12}} - 2\frac{r_0^6}{r^6} \right). \qquad (4.46)$$

Another variation that gives qualitatively similar behavior is the potential

$$V(r) = A\, e^{-br} - \frac{C}{r^6}. \qquad (4.47)$$

Bonds can be of a number of different kinds. For example, **ionic bonds** occur when one atom completely pulls an electron off the other, leaving one positive ion and one negative ion, which then attract electrostatically. For example, in table salt ($NaCl$), the sodium loses an electron, which attaches itself to the chlorine atom. The two resulting ions, Na^+ and Cl^-, then strongly attract, forming a stable bond. Ionic bonds are very strong; they tend to form hard, stable materials with high melting and boiling points. Because the electrons are held strongly by the atoms, these materials tend to be insulators. However, they often dissolve easily in water, since the strong electric dipole moment of the water molecule is capable of exerting a sufficiently strong attractive pull on the electron to break the ionic bond.

Covalent bonds occur when the two atoms share a pair of electrons more or less equally. One electron comes from the outermost shell of each atom. The electrons spend most of their time in the region between the atoms, creating an attractive force on each of the atoms, which can now be thought of as positively charged in the absence of the two bonding electrons. Covalent bonds, although relatively strong, are weaker in general than ionic bonds. They occur when the two atoms have similar electronegativity, i.e., when they each tend to pull equally on the electrons. Thus, covalent bonds occur, for example, in diatomic molecules formed from two atoms of the same species, such as O_2, N_2, and H_2. Large molecules, for example, complex organic molecules, tend to contain primarily covalent bonds. Conducting materials can often be considered to consist of many covalent bonds, with the electrons shared equally by all the atoms in the material; in this case, the bond is called a **metallic bond**.

In liquids, weaker **van der Waals bonds** tend to predominate. These are constantly breaking up due to thermal fluctuations and then reforming, accounting for the fluid nature of the liquid. The van der Waals force is a dipole-dipole interaction between permanent or induced electric dipole moments of the atoms. Dipole-dipole interactions decay more rapidly than electrostatic Coulomb interactions, typically having a distance dependence of the form $\sim \frac{1}{r^6}$.

On occasion, a hydrogen nucleus (a proton) can play a similar role in bonding to the role usually played by electrons. In a hydrogen bond, the H^+ attracts two negatively charged ions to it, providing a very weak bond between the two negative ions. An example occurs in the $(HF_2)^-$ ions, where the H^+ acts as a glue between the two fluorine ions F^-.

Molecular spectra tend to be complicated, because in addition to the internal energy levels of the atoms and the interaction energy in the bonds, there can also be contributions to the total energy due to rotations of the molecule or vibrations of the atoms at the ends of the molecular bonds. For example, a molecule with moment of inertia I will have rotational energies of the form

$$E_{rot} = \frac{1}{2}I\omega^2 = \frac{L^2}{2I} = \frac{\hbar^2 l(l+1)}{2I}, \qquad (4.48)$$

where the fact that $L = I\omega$ was used in the second equality and the quantization of $\hat{L}^2$ was used in the third one. These rotational energy levels tend to be on the order of $10^{-3} - 10^{-2}$ eV and so are excited by infrared and microwave photons.

Similarly, vibrations of the bonds contribute terms that can be estimated by modeling the bond as a spring undergoing simple harmonic motion. By expanding the potential in a Taylor series about its minimum r_0, the leading term should be quadratic in the displacement from the minimum, $\sim\frac{1}{2}k(r - r_0)^2$. Reading off the coefficient of this quadratic term, the effective spring constant k can be obtained. If the atoms at the ends of the bond have masses m_1 and m_2, we can also define the reduced mass:

$$\mu = \frac{m_1 m_2}{m_1 + m_2}. \tag{4.49}$$

So for a simple diatomic molecule, the resonant frequency of the bond becomes

$$\omega = \sqrt{\frac{k}{\mu}} = \sqrt{\frac{k(m_1 + m_2)}{m_1 m_2}}. \tag{4.50}$$

The vibrational energy levels are then given by the usual expression for harmonic oscillator energies:

$$E_n = \left(n + \frac{1}{2}\right) \hbar\omega. \tag{4.51}$$

For many molecules, these energies tend to be ≥ 1 eV. Since these greatly exceed the thermal energy ($kT = .025$ eV) at $T = 300°$K, these levels are not normally excited at room temperature.

Rather than looking at the molecular bonds, consider instead looking at the electron wavefunctions that arise as two atoms approach each other to form a molecule. As a concrete example, consider two hydrogen atoms. When the atoms are far apart, the electrons don't interact and each is in a standard hydrogen atom orbital. For example, they could both be in the $1s$ ground state. As the inter-atomic distance decreases, however, the electron wavefunctions begin to overlap and the Pauli exclusion principle kicks in. The two identical electrons can no longer be in the same state, and the total two-particle wavefunction must be antisymmetric under particle interchange. The original wavefunctions begin to mix to form linear combinations that respect both the spatial symmetry of the molecule and the Fermi antisymmetry. Consider just the spatial wavefunction first, neglecting spin, and assume that only one electron is present (so we have an ionic state, H_2^+). Let one atom be at the origin, and the other at position $r = R$. Consider one of the electrons, and suppose that the original wavefunction of that electron, localized at one atom, is of the form $\psi(r)$. Then at small distances, the electron can tunnel through the potential barrier between the two atoms. If the two atoms are of the same species, then the system has reflection symmetry about the midpoint between them. So the electron has no reason to spend more time near one atom than the other, and the probability density $|\psi|^2$ should be symmetric about this point. The simplest way for this to happen is for the spatial wavefunction itself to be a linear combination of wavefunctions localized at each atom and to be either symmetric

$$\psi_+(r) = \frac{1}{\sqrt{2}} (\psi(r) + \psi(r - R)) \tag{4.52}$$

or

Fig. 4.7 As the interatomic distance decreases, each of the original energy levels splits into symmetric (bonding) and antisymmetric (antibonding) states

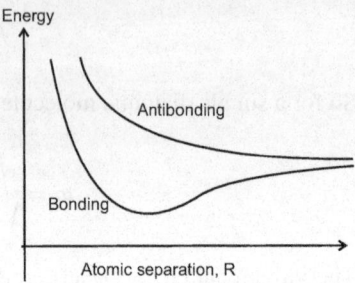

$$\psi_-(r) = \frac{1}{\sqrt{2}} \left(\psi(r) - \psi(r - R) \right) \tag{4.53}$$

antisymmetric about the midpoint, $R/2$. The antisymmetric ψ_- has higher energy than the symmetric state, ψ_+: the spatial antisymmetry implies that the wavefunction vanishes at the midpoint $R/2$, so that the electron spends more time localized near one atom or the other. As is well-known from simple examples like a quantum particle in a box (an infinite well potential), the more tightly confined and spatially localized a particle is, the higher the energy levels are. Thus, the ground state is ψ_+ and ψ_- is the first excited state. The symmetric ground state ψ_+ is called a **bonding state**, since it forms a lower energy than the two isolated atoms would have, and it therefore forms a stable bond between the atoms. Meanwhile, the antisymmetric excited state ψ_- is called an **antibonding state**, since it has higher energy than the unbound state and is therefore unstable.

Now add in the second electron to get the electrically neutral diatomic molecule H_2. Both electrons will want to be in the lower-energy bonding state. This can be done without violating the Pauli exclusion principle as long as the spin state is antisymmetric, $\sim |\uparrow\,\rangle_a |\downarrow\rangle_b - |\uparrow\,\rangle_b |\downarrow\rangle_a$. The product of the spatial and spin wavefunctions then has the required overall antisymmetry under interchange of the two electrons. The result is that the two electrons form a covalent bond, and the system has lower energy when bonded together than they would have when widely separated and noninteracting. A plot of the energy of one electron versus distance is shown in Fig. 4.7.

4.2   Band Theory and Bloch Waves

We saw above that as two atoms approach, the energy levels split into doublets: each single level splits into a higher-energy antisymmetric linear combination and a lower-energy symmetric combination of the states of the two atoms. If we now bring in a third atom, the same process repeats, with the levels splitting again to form combinations that are completely symmetric or completely symmetric in all three electrons. Continuing the process further to include N atoms, each of the original

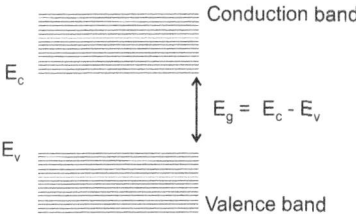

Fig. 4.8 The conduction and valence bands are separated by an energy gap, $E_g = E_c - E_v$. The conduction properties are largely determined by how large this gap is compared to the available thermal energy at a given temperature

energy levels now splits into N separate levels separated by energy gaps. As N increases, the number of levels increases, while the size of the energy splittings gets smaller. Eventually, for large enough N, the splittings are so small that we can treat the set of split levels as forming a *continuous band*. Since states of the two atoms with the same values of n and l will mix to form linear combinations, each band will have a distinct value of n and l. So, using spectroscopic notation, we can refer, for example, to $1s$, $2s$, $2p$, etc. bands. This picture of continuous energy bands that can be occupied by the electrons is the starting point of the **band theory of solids**.

In general, the set of bands can be complicated: their shapes vary with the size of the lattice spacing (the interatomic distance) and can overlap. We will ignore these complications and draw the diagrams in simple schematic form as in Fig. 4.8.

Electrons fall into the lowest unoccupied states, so that the lowest-energy bands fill first, then higher band, until all of the electrons in the system are used up. The low-energy filled bands are largely irrelevant, since the electrons are not free to really interact with their surroundings: since there are no nearby unoccupied states for the electrons to scatter into, they remain stuck in their initial states. Of particular interest, though, are the highest-energy band that is filled and the lowest-energy empty or partially empty band; these are known respectively as the **valence band** and the **conduction band**. These two bands largely determine the electrical, thermal, and transport properties of the material, and so often they are drawn in isolation, neglecting the lower filled levels and the higher unoccupied levels.

Let E_c be the energy at the bottom of the conduction band and E_v be the energy at the top of the valence band (Fig. 4.8). The **bandgap** is the energy difference between the bands:

$$E_g = E_c - E_v. \tag{4.54}$$

Also define the **Fermi energy**, E_F. At absolute zero $T = 0$, all levels below the Fermi energy are occupied, and above E_F, all levels are empty (Fig. 4.9a). At finite temperature, $T > 0$, thermal fluctuations will excite some of the electrons below E_F to energies above E_F, smoothing the distribution as shown in Fig. 4.9b. So, a

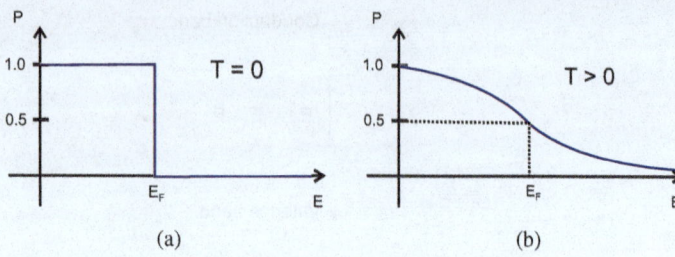

Fig. 4.9 The Fermi energy is the energy at which the occupation probability P of the states passes through $P = \frac{1}{2}$. (a) At zero temperature, all states below E_F are filled, and all states above are unoccupied. (b) For $T > 0$, thermal fluctuations raise the energies of some of the previously lower-energy electrons to create a tail of occupied states above E_F

Table 4.2 Fermi energies and electron densities for a sampling of metals and semiconductors at zero temperature

Element	E_F (eV)	n ($\times 10^{28}$ m^{-3})
Cu	7.00	8.47
Ag	5.49	5.86
Au	5.53	5.90
Ca	4.69	4.61
Ba	3.64	3.15
Fe	11.1	17.0
Zn	9.47	13.2
Hg	7.13	8.65
Al	11.7	18.1
Ga	10.4	15.4
In	8.63	11.5
Sn	10.2	14.8
Pb	9.47	11.0

definition of E_F that holds at both $T = 0$ and $T > 0$ is that the Fermi energy is the energy at which the probability that a state is occupied by an electron is $\frac{1}{2}$.

Nearly every statistical mechanics text derives an expression for the Fermi energy of a material in terms of the electron density $n = N/V$:

$$E_F = \frac{h^2}{8m}\left(\frac{3n}{\pi}\right)^{2/3}. \tag{4.55}$$

Some electron densities and Fermi energies are given in Table 4.2.

Materials may be classified based on the size of the energy gap and the location of the Fermi energy relative to the bands:

- In a **metal**, the conduction band is only partially filled, so the Fermi energy lies inside the band. Every occupied state will have a nearby empty state in the same band which differs from it in energy only by a small amount. So, thermal fluctuations can easily cause electrons to jump from one state to another. In the absence of any electric field, the electron hoppings are all in random directions, with as many moving to the right as to the left. However, even a small electric field will introduce a bias, causing electron motions opposite to the field to predominate. Thus, even small electric fields and voltages lead to currents, and the material is a good conductor.

- In an **insulator**, there is a large gap between the conduction and valence bands, and the conduction band is empty or very nearly so. The Fermi energy then lies in the gap between the bands. Since the valence band is full, there are no unoccupied low-lying energy states for electrons to jump into. The electron can only move if it has sufficient energy to jump across the gap into the conduction band. But recall that at room temperature, $T \approx 300\,\text{K}$, the average thermal energy of an atom is $k_B T \approx 0.025\,\text{eV}$. If the gap between the valence and conduction bands is much larger than this, it is a rare event for an electron to accumulate enough energy to cross the gap. As a result, the electrons are mostly stuck residing in the same localized energy level that they began in, and are unable to travel to other lattice sites, resulting in an insulator. Typical bandgap values for insulators are roughly 4–10 eV.

- In a **semiconductor**, the gap is smaller (typically between about 0.1 eV and $\sim$3.5 eV). As a result, the conductivity of the material is strongly temperature dependent: at low temperatures, few electrons will be able to jump into the conduction band, but as the temperature increases, a growing number of electrons have sufficient energy and so the conductivity increases. If the material very pure, it is called an **intrinsic semiconductor**. For silicon, germanium, and gallium arsenide, the intrinsic electron densities at zero temperature are $1.5 \times 10^{10}\,\text{cm}^{-3}$, $2.4 \times 10^{13}\,\text{cm}^{-3}$, and $1.8 \times 10^{6}\,\text{cm}^{-3}$, respectively. However, often impurity atoms are deliberately added to the material, a process called **doping**, in order to increase the number of charge carriers. The conductive properties of the semiconductor, now called an **extrinsic semiconductor**, can then be altered in a highly controllable manner by appropriate choices of the impurity type and concentration. The impurities can make the material conduct at lower temperature by adding additional charge carriers and by adding additional energy levels in the gap, close to the valence or conduction bands. More will be said about semiconductors in the next section.

Energy gaps for several common semiconductors are given in Table 4.3, along with several insulators for comparison. Notice that when an electron jumps into the conduction band, it leaves a hole (a vacant state) in the valence band. As neighboring valence electrons jump into the hole, they leave a new hole at their previous positions. As a result, the hole can travel through the material just as a particle could. If an electric field turns on, the hole moves in the direction of the field just as a positively charged particle would. For all practical purposes, the hole can be treated as an additional charge carrier. Thus, when an electron is promoted to the conduction band, there are effectively two charge carriers created: a negatively charged conduction electron and a positively charged hole. The fact

Table 4.3 Energy gaps
between valence and
conduction bands for several
insulators and
semiconductors at room
temperature

	Material	E_g
Semiconductors	Silicon (Si)	1.12 eV
	Germanium (Ge)	0.66 eV
	Indium arsenide (InAs)	0.36 eV
	Gallium arsenide (GaAs)	1.42 eV
	Gallium nitride (GaN)	3.4 eV
	Zinc oxide (ZnO)	3.2 eV
Insulators	Carbon (C)	5.47 eV
	Silicon dioxide (SiO_2)	9 eV

that both contribute to the current is important for the operation of devices such as diodes and transistors.

Consider now a conducting material, and consider how the energy bands depend on wavenumber k. We restrict ourselves to one dimension for simplicity. A free electron would have a **dispersion relation** (an energy-momentum or energy-wavenumber relation) given by

$$E = \frac{k^2}{2m}.$$

(4.56)

This describes a parabola if the energy is plotted as a function of kA. At the opposite extreme, an electron tightly bound to a single atom would essentially have zero momentum, so $E(k)$ would be a sharply spiked function such as a Dirac delta function. A real solid is somewhere in between these two extremes: not quite a free particle, but also not completely localized. It moves through a periodic potential created by the regularly-repeating atoms of the crystal lattice. Let a be the **period** of the lattice, the distance between the atoms.

At long wavelengths (low k), for which $\lambda >> a$, the electron only sees the average potential, since the spatial variation of the potential repeats itself many times over each wavelength. As a result, the potential looks constant and the electron behaves like a free particle, with a quadratic dispersion relation, as above. However, as the wavelength decreases, eventually the point comes where the condition $\lambda = 2a$ (or equivalently $k = \frac{\pi}{a}$) is satisfied. Thinking of the electron as a matter wave, at this wavelength, all of the lattice sites could lie either at nodes of the wave, or at antinodes (maxima and minima), as in Fig. 4.10a and b. In either case, standing waves will form between incident and reflected waves. If the incident wave had amplitude ψ_0, then these standing waves will take the form

$$\psi_n(x) = 2\psi_0 \sin \frac{\pi x}{a}$$

(4.57)

$$\psi_a(x) = 2\psi_0 \cos \frac{\pi x}{a},$$

(4.58)

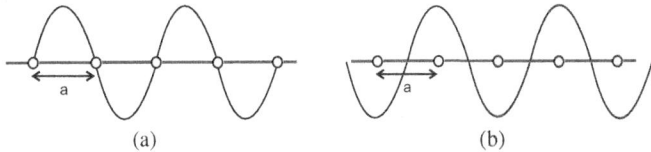

(a)   (b)

Fig. 4.10 Standing waves can occur in a one-dimensional lattice when the (**a**) nodes of the waves are at the lattice sites or when (**b**) the antinodes (or maxima) are at the lattice sites

for nodes and antinodes, respectively. The resulting probability densities have the form

$$|\psi_n(x)|^2 = 4|\psi_0|^2 \sin^2 \frac{\pi x}{a} \tag{4.59}$$

$$|\psi_a(x)|^2 = 4|\psi_0|^2 \cos^2 \frac{\pi x}{a}. \tag{4.60}$$

In a standing wave, the electron amplitude at each x moves up and down over time in a periodic manner, without any horizontal motion. In other words, at $k = \frac{\pi}{a}$ (or more generally at $k = \frac{n\pi}{a}$ for integer n), *the electron does not propagate through the material.* At these momenta, the group velocity of the electron wave vanishes, $v_g = \frac{dE}{dk}$, and so the plot of the dispersion relation $E(k)$ has to have zero slope. The quadratic graph of the free-particle dispersion relation, shown in Fig. 4.11a, is now altered as shown in Fig. 4.11b. Note that there are now gaps in the plot: there are energies that the curve does not cross for any value of k. Thus, destructive interference of the electron waves provides an alternative explanation for the bandgaps that we have already seen exist in the energy spectrum.

Suppose we stay away from the bandgaps, so that k is a wavenumber that will propagate unhindered in the lattice. In this case, we may apply one of the key results of solid state physics, **Bloch's theorem**, which says that in a periodic potential, the wavefunction of a propagating particle will take the form

$$\psi_k(x) = e^{ikx} u(x), \tag{4.61}$$

where $u(x)$ is a periodic function, with the same periodicity as the crystal lattice, $u(x + a) = u(x)$. Note that in addition to being periodic in x, this wavefunction is also periodic in k: $\psi_{k+\frac{2\pi}{a}}(x) = \psi_k(x)$. This implies that the energy must be periodic in k:

$$E(k + \frac{2\pi}{a}) = E(k). \tag{4.62}$$

Thus, the pattern of disconnected curves shown in Fig. 4.11b are really portions of continuous, periodic curves, as shown in Fig. 4.11c. The interval $-\frac{\pi}{a} \leq k \leq \frac{\pi}{a}$ is called the **first Brillouin zone**, and in general the interval $-\frac{(n-1)\pi}{a} \leq k \leq$

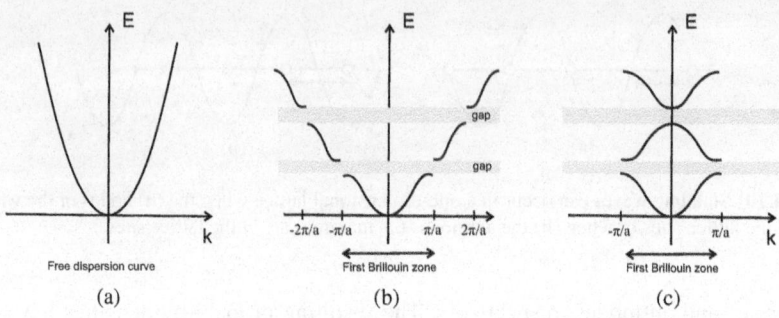

Fig. 4.11 (**a**) The dispersion curve for a free particle. The energy is a quadratic function of momentum. (**b**) Destructive interference alters the dispersion curve, opening up forbidden energy gaps. The regions of k extending from one gap to the next are called Brillouin zones. The first Brillouin zone, from $-\frac{\pi}{a}$ to $\frac{\pi}{a}$, is shown. (**c**) The portions of the dispersion curves lying in the higher Brillouin zone are often translated horizontally to display the all of the curves within the first Brillouin zone. This leads to a set of energy curves or bands separated by energy gaps

$\frac{n\pi}{a}$ is called the **nth Brillouin zone**. Further, since the pattern repeats along the horizontal axis, it is common to show just the first zone, as in Fig. 4.11c; this is called the **reduced-zone representation**. The first three energy bands can be clearly seen arising in the figure.

For a perfect crystal, with perfectly periodic lattice structure, there would be perfect conductivity (zero resistance) away from the Brillouin zone edges, and currents would flow without loss. Backscattering would not exist due to perfect destructive interference of the reflected waves. However, in real life, no such perfect crystal exists. Real crystals have impurities (atoms of other elements embedded in the material) and crystal defects (imperfections in the arrangement of the lattice structure). Even if the crystal was perfectly pure and has a perfectly regular periodicity, at finite temperatures, the atoms at the lattice sites will fluctuate around their equilibrium positions; these time-dependent displacements also act as imperfections in the crystal. So in real crystals, there will always be current losses due to backscattering. In general, these resistive losses will increase with temperature due to the increasing thermal motion of the lattice ions. However, it will be seen in later chapters that in some cases, it is still possible to have currents of electrons (or of light) that travel without backscattering or loss. We will see two mechanisms for this: superconductivity (Chap. 7) and topologically protected states (Chap. 21).

4.3 Semiconductors

Intrinsic semiconductors are generally poor conductors at room temperature. However, by doping (adding impurities), the conductivity can be greatly improved.

Fig. 4.12 Doping with impurities adds additional energy levels inside the gap between conduction and valence bands. *N*-type doping creates new levels just below the conduction band, while p-type doping creates levels above the valence band

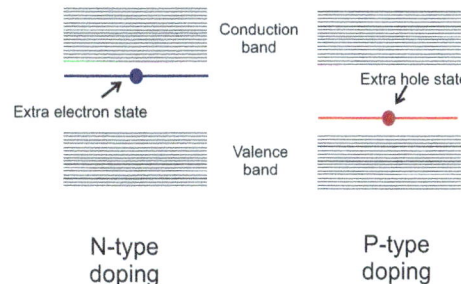

Doping comes in two types. **N-type doping** occurs when the added impurity is a **donor atom**: it has an extra electron compared to the surrounding material. **P-type doping** occurs when the added impurity is an **acceptor atom**: it has fewer electrons and thus contributes extra holes. The extra electrons or holes contribute additional energy levels between the existing bands (Fig. 4.12), altering the conductive properties of the material.

As an example, silicon (Si) is a group IV element on the periodic table, with four valence electrons, i.e., four electrons in its outer shell. Arsenic (As) is a group V element with five valence electrons. When added as an impurity to Si, four of the arsenic electrons bind to neighboring Si atoms, while the fifth remains unattached, providing an extra negative charge carrier that is only weakly bound to nearby atoms, so that it is largely free to move around the material at room temperature. So As is an *N*-type dopant for Si. Nitrogen, phosphorus, and lead, being in the same column of the periodic table as As, will also act as donors in Si. In contrast, aluminum (Al) is an acceptor impurity: it is a group III atom with only three valence electrons. They all bind to neighboring Si atoms, leaving an unoccupied hole in the bond to the fourth Si atom. Once again, this hole can move around, acting as an extra positive charge carrier. Other materials that act as P-type dopants for Si are boron, indium, and gallium.

It is common to view a conductor or a semiconductor as a box containing a weakly interacting electron gas (and a similar gas of holes). Free particles are easy to analyze, but interactions with the crystal lattice means the electrons and holes are not quite free particles, despite their mobility. So a common strategy is to treat the electrons and holes as if they were free particles, but with altered masses. These new momentum-dependent **effective masses** m^* have the effects of interactions swept up inside of them. We can now use the free-particle dispersion relation, with the effective mass replacing the original bare mass:

$$E = \frac{\hbar^2 k^2}{2m^*}. \tag{4.63}$$

This implies that the effective mass is determined by the curvature (second derivative) of the energy function at a given momentum:

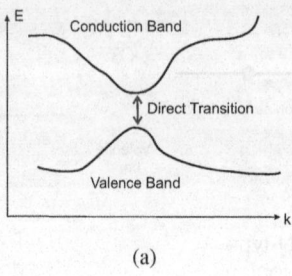
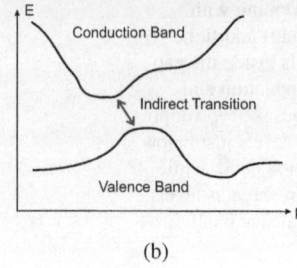

(a)   (b)

Fig. 4.13 (a) A direct transition between extrema of the conduction and valence bands at the same value of k can be accomplished by emission or absorption of a single photon. (b) When the k values of the extrema are different, the transition is indirect and in general will require emission or absorption of both a photon and a phonon

$$m^* = \pm \frac{\hbar}{d^2 E/dk^2},$$ (4.64)

with $+$ for electrons and $-$ for holes. Larger m^* means less mobile charge carriers (lower conductivity), while small m^* corresponds to higher conductivity. Note that electrons and holes can interact differently, so in general, they have different effective masses, $|m_e^*| \neq |m_h^*|$. The minus sign for the effective mass of the hole is due to the fact that exerting a force in some direction on the surrounding electrons will cause the hole to move in the opposite direction.

Near $k = 0$, the electrons and holes should behave like free particles, so that m^* should be similar to the true mass. But near the bandgaps, $d^2 E/dk^2 \to 0$, so that $m^* \to \infty$, consistent with the fact that the electrons and holes should not propagate at these momenta.

The dispersion curves shown earlier are greatly simplified; for real materials, they tend to be more complicated. But in general, the transitions between bands can be divided into two categories. In Fig. 4.13a, the minimum of the conduction band is almost directly above the maximum of the valence band. A transition between them requires a large energy change with a small change in k; such a transition can be accomplished by a single photon and is called a **direct transition**. Direct bandgaps occur materials such as CdTe, InAs, and GaAs. In contrast, the extrema of the conduction and valence bands for silicon or germanium occur at different k values, as in Fig. 4.13b, so both large energy and momentum changes are required for a transition. A photon on its own cannot accomplish this, so it occurs as a two-step process: emission or absorption of a photon has to be accompanied by absorption or emission of a **phonon**, a collective many-electron excitation. (Phonons will be discussed in the next section.) Such a process is called an **indirect transition**. Direct bandgap materials are useful for photovoltaic devices such as solar cells and photodiodes.

Semiconductor devices are formed by creating interfaces between semiconductor regions with different dopings, or interfaces between semiconductors and metals.

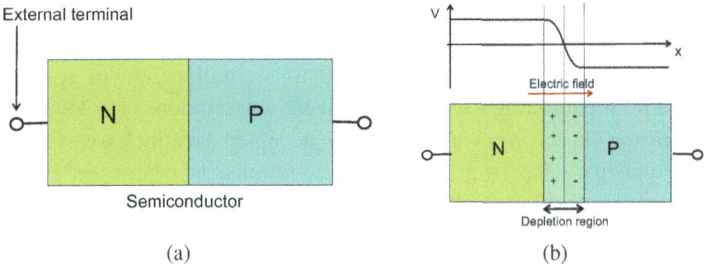

Fig. 4.14 (a) A diode formed by a junction between P- and N-type semiconductors. (b) Diffusion of charges across the junction causes a potential gradient to appear that allows current to flow easily in one direction, but not in the other

Here, we just examine the simplest such device, a **PN junction** or **diode**. Schematically, a PN junction looks like Fig. 4.14a. A block of semiconducting material is doped with excess donors (N-type) on one side of the junction and an excess of acceptors (P-type) on the other side.

Because the electron density is higher in the N-type region, holes tend to diffuse into the P-type region, where the electron density is initially lower. Similarly, holes tend to diffuse toward the N-type region. As a result, charge accumulates around the junction: a positive charge builds up on the left of the junction, due to the diffusing holes, and a negative charge builds up on the right due to the diffusing electrons. This charged region around the junction is called the **depletion zone**. As a result of this charge accumulation, an electric field builds up, pointing from the positive charge toward the negative. Eventually, the field becomes large enough to prevent further diffusion, and the system comes to equilibrium.

The existence of an electric field implies that there is a spatially varying voltage, $E = -\frac{dV}{dx}$ (Fig. 4.14b). The N-type region is of higher voltage, implying that holes cannot easily move up the voltage hill toward the N region unless extra energy is supplied to them from an external source, and electrons similarly can't move up into the P region. Thus, current flows much more easily in one direction than in the other: the PN junction forms a **diode**. A pair of PN junctions are used to form a transistor (see [3] or [4] for more details.)

4.4 Collective Excitations

The electrical and thermal properties of a solid are due to the behavior of the electrons and the atoms in the crystal lattice of the material. However, none of these particles act in isolation: the electrons interact with each other and with the lattice ions, while the lattice ions also interact with each other. So while single-particle excitations can occur (e.g., an electron jumping between energy bands), excitations of the system may also involve *many* particles simultaneously. These multiparticle **collective excitations** are in fact responsible for many condensed matter phenomena. Often these excitations share many of the properties of a well-defined particle: they can be localized to a small region of space, carry a well-defined momentum, and so on. So collective excitations are also often called **quasiparticles**. (There is not a universally accepted terminology: some authors distinguish quasiparticles from collective excitations. But here we will treat the terms as interchangeable.) Probably the best-known quasiparticles are **phonons**, which are essentially quantized acoustic waves in a material, and **Cooper pairs**, which are two-electron states in a superconductor.

Collective excitations can occur in both boson and fermion systems, but the focus here will be quasiparticles in fermion systems. From one point of view, even an electron hole in a conductor or semiconductor can be viewed as a type of collective excitation, since as it moves, its previous locations are filled sequentially by different electrons, and so a hole only exists in a many-body system. Similarly, an electron moving through a material interacts with neighboring particles, altering their positions and momentum. Those particles in turn affect the electron. As a result, the electron has an effective mass and charge that may be different from what it would have moving through vacuum; this so-called *dressed* electron, with the effects of all of the interactions included, can also be viewed as a quasiparticle.

We will look at several types of quasiparticles in the next few subsections; Cooper pairs will be discussed in detail in Chap. 7.

4.4.1   Phonons

Phonons are quantized acoustic vibrations of a crystal lattice. The name was coined in analogy to that of photons: the phonon acts in many ways like a quantum particle propagating through the lattice. For example, the phonon energy is proportional frequency like a photon, $E = \hbar\omega$, where the frequency has a wavenumber dependence to be found below. But unlike photons, which are transverse electromagnetic waves of spin 1, the phonon is a longitudinal sound wave of spin 0. Phonons are important for transport properties (thermal and electrical conduction). They are also important in determining the allowed optical transitions of a material in semiconductor physics, since many semiconductors only have indirect transitions: electrons must absorb or emit both a photon *and* a phonon in order to move between energy bands.

A simplified model of a one-dimensional chain of masses connected by spring-like bonds gives a good qualitative picture of most of the essential features of the phonon. Consider a chain of particles of mass m connected by springs of force constant κ (Fig. 4.15). At equilibrium, the rest position of the nth mass is $x_n = na$, where a is the spacing between masses. Suppose that the particles are now displaced slightly away from equilibrium, $x_n \rightarrow x_n + \epsilon_n$, and then released. The mass feels two restoring forces, due to the springs on either side of it. The change in length of each spring will be depend on the masses at both ends of it. Taking displacements to

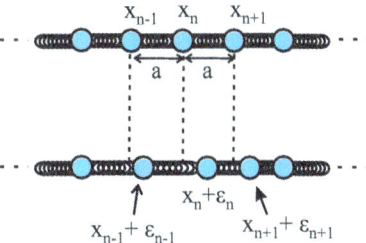

Fig. 4.15 Top: A one-dimensional chain of masses at equilibrium. Each mass is at rest, a distance a from its neighbors. Bottom: The mass at each lattice site x_n is displaced by an amount ϵ_n. As a result, each spring is stretched by an amount $\epsilon_n - \epsilon_{n-1}$ from its equilibrium length of a

be positive if they are to the left and negative to the right, the lengths of the springs on the left and right sides of particle n will change in length by $\epsilon_n - \epsilon_{n-1}$ and $\epsilon_{n+1} - \epsilon_n$, respectively, with the two resulting forces being in opposite directions. So Newton's second law tells us that

$$m\frac{d^2\epsilon_n}{dt^2} = \kappa\left[(\epsilon_{n+1} - \epsilon_n) - (\epsilon_n - \epsilon_{n-1}) \right] \tag{4.65}$$

$$= \kappa\left(x_{n+1} - 2x_n + x_{n-1}\right). \tag{4.66}$$

The solution of this equation can be given in terms of Bessel functions, but it is more useful to find the eigenstates of the system. The first guess is to try a plane wave:

$$\epsilon_n(t) = A\, e^{i(\omega t - nka)}, \tag{4.67}$$

where A is some constant (the wave amplitude). Plugging this solution into Eq. 4.66 and canceling common factors from the two sides gives the result that

$$\omega^2 = \frac{2\kappa}{m}\left(1 - \cos ka\right). \tag{4.68}$$

So, using the trig identity $\sin\frac{\theta}{2} = \sqrt{\frac{1-\cos\theta}{2}}$, the phonon frequency can be written as

$$\omega(k) = \pm\sqrt{\frac{4\kappa}{m}}\sin\frac{ka}{2}. \tag{4.69}$$

This dispersion relation is plotted from $k = 0$ to the edge of the first Brillouin zone in Fig. 4.16. At large k, it approaches a maximum of $\omega_0 = \sqrt{\frac{4\kappa}{m}}$. The group velocity near the maximum is $v = \frac{d\omega}{dk} \approx 0$. So the large k (high frequency) components of the wave are fixed in place: they don't propagate through the chain.

Fig. 4.16 The dispersion curve for a phonon. At small k, $\omega(k)$ is approximately linear, leading to a constant group velocity. At large k, the group velocity drops to zero as $\omega(k)$ approaches a maximum

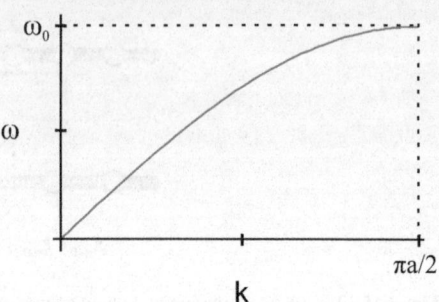

But looking near $k = 0$, we find that the dispersion relation is linear in k:

$$\omega \approx \sqrt{\frac{4\kappa}{m}}\left(\frac{ka}{2}\right). \tag{4.70}$$

This is the dispersion relation expected for acoustic or vibrational waves. So the low-frequency modes behave like quantized sound waves propagating through the chain with group velocity

$$v = \frac{d\omega}{dk} \approx \sqrt{\frac{4\kappa}{m}}\left(\frac{a}{2}\right) = \frac{a}{2}\omega_0. \tag{4.71}$$

The results above can be generalized in various ways. For example, suppose the masses in the chain are not all equal. In particular, suppose that they alternate between two values m_1 and m_2. Then by similar methods as above, the frequencies of the eigenmodes can be shown to be

$$\omega_{\pm}^2 = \frac{\kappa}{m_1 m_2}\left(m_1 + m_2 \pm \sqrt{(m_1 - m_2)^2 + 4m_1 m_2 \cos^2 ka}\right). \tag{4.72}$$

The lower frequency of the two frequencies (with the minus sign) is called an **acoustic mode**, while the higher frequency (+ sign) is called an **optical mode** because in many solids, it is in the right range to couple to infrared radiation. In the acoustic mode, adjacent atoms move in phase with each other (each is moving in the same direction at each instant), while in the optical mode, adjacent atoms are out of phase (moving in opposite directions). There is an energy gap between the bands with minimum size:

$$\Delta E = \hbar\sqrt{2\kappa}\left|\frac{1}{\sqrt{m_1}} - \frac{1}{\sqrt{m_2}}\right|. \tag{4.73}$$

When $m_1 = m_2$, the gap closes and the frequencies reduce back to Eq. 4.69.

In three-dimensional lattices, it is possible for transverse wave modes to arise as well as longitudinal, and so the variety and complexity of phonons will increase beyond the simple 1D model given above. Nevertheless, the one-dimensional picture is useful to keep in mind.

4.4.2 Magnons

In a ferromagnetic material such as iron or nickel, the atom at each lattice site l in the material has its own magnetic dipole moment, μ_l. In the presence of an external magnetic field B, these tend to align with the field. In the absence of external fields, they tend to be distributed randomly at high temperatures, giving a net magnetic moment of zero for the material. At low temperatures, they largely align with each other since each dipole moment creates its own magnetic field that exerts a torque on neighboring moments.

Consider a magnetic moment at a point in the lattice labeled by some vector l. It feels an effective magnetic field from produced by its neighboring spins:

$$B_l = K \sum_{l'} \mu_{l'}, \tag{4.74}$$

where l' labels the positions of the neighboring lattice sites. We don't need to worry here about the exact form of K, which will vary depending on the geometry of the lattice and other factors. In general, the sum is over all $l' \neq l$. Here we will assume that the interaction between the dipole moments drops off rapidly enough with distance so that each atom only interacts with those closest to it. So the sum only needs to be taken over those l', which are nearest neighbors to l. Assume that all sums for the remainder of this section include only nearest neighbors.

Recall that torque gives the rate of change of angular momentum, $\tau = \frac{dL}{dt}$, and that the torque on a magnetic dipole due to a magnetic field is $\tau = \mu \times B$. By Eq. 4.36,

$$\frac{dL}{dt} = \frac{\hbar}{g_L \mu_B} \frac{d\mu}{dt}. \tag{4.75}$$

So the effect on dipole l due to the field from neighboring dipoles will be given by

$$\frac{d\mu_l}{dt} = K' \sum_{l'} \mu_l \times \mu_{l'}, \tag{4.76}$$

where $K' = \frac{g_L \mu_B}{\hbar} K$.

Assume that there is an external field B along the z-axis, so that in the absence of perturbations, the dipole moments will tend to align along the z-axis. Also assume that the temperature is low enough that we don't need to worry much about thermal fluctuations. Now suppose that one dipole moment in the lattice is perturbed away from the z-axis. Through the magnetic torque given above, the perturbation will affect the neighboring moments. These neighboring moments will in turn perturb their neighbors, and the effect of the perturbation will propagate, rippling through the material like a wave. Notice that no material is being displaced in this wave: it is simply the directions of the dipole moments that will be oscillating.

Split $\boldsymbol{\mu}_l$ into Cartesian components $\mu_{lx}, \mu_{ly}, \mu_{lz}$. Assuming that the perturbation is small, the z-component should be roughly constant, but the transverse (x- and y-components) will be oscillating. Recall that electric fields and other oscillating quantities are often taken to be complex, so that the phase of the oscillation can be accounted for by the angle of the complex number in the complex plane. So define a new transverse variable $\tilde{\boldsymbol{\mu}} = \mu_{lx}\hat{\boldsymbol{i}} + i\mu_{ly}\hat{\boldsymbol{j}}$, where both μ_{lx} and μ_{ly} are now allowed to be complex. Substituting a wave solution of the form $\tilde{\boldsymbol{\mu}}(t) = \boldsymbol{\mu}_0 e^{i(\omega t - \boldsymbol{k} \cdot \boldsymbol{l})}$ into Eq. 4.76, it is found that

$$i\omega\mu_0 = K'\left(\boldsymbol{\mu}_z \times \boldsymbol{\mu}_0\right) \sum_{l'} \left(e^{i\boldsymbol{k} \cdot \boldsymbol{l'}} - 1\right). \tag{4.77}$$

If the crystal has inversion symmetry, so that for every lattice vector $\boldsymbol{l'}$ there is a corresponding vector $-\boldsymbol{l'}$, then the sum can be rewritten as

$$\sum_{l'} \left(e^{i\boldsymbol{k} \cdot \boldsymbol{l'}} - 1\right) = \frac{1}{2} \sum_{l'} \left(e^{i\boldsymbol{k} \cdot \boldsymbol{l'}} + e^{-i\boldsymbol{k} \cdot \boldsymbol{l'}} - 2\right) \tag{4.78}$$

$$= \sum_{l'} \left(\cos \boldsymbol{k} \cdot \boldsymbol{l'} - 1\right) \tag{4.79}$$

$$= -2 \sum_{l'} \sin^2 \frac{\boldsymbol{k} \cdot \boldsymbol{l'}}{2}, \tag{4.80}$$

so that

$$i\omega\mu_0 = -2K'\left(\boldsymbol{\mu}_z \times \boldsymbol{\mu}_0\right) \sum_{l'} \sin^2 \frac{\boldsymbol{k} \cdot \boldsymbol{l'}}{2}. \tag{4.81}$$

Going back to the original x- and y-components, it can be checked that they are always $\frac{\pi}{2}$ radians out of phase with each other. In other words, the tip of each $\boldsymbol{\mu}$ vector traces out a circle in the x–y plane. The wave in this case is a wave of rotating dipole moments, called a **spin wave**. The basic quantum of the spin wave is called a **magnon**. Like the phonon, the magnon is a collective excitation involving multiple interacting lattice sites. The dispersion relation again approaches a maximum near the edge of the Brillouin zone. But in contrast to the phonon, $\omega(k)$ is quadratic (rather than linear) at small k,

$$\omega \sim -2 \sum_{l'} \left(\frac{\boldsymbol{k} \cdot \boldsymbol{l'}}{2} \right)^2. \tag{4.82}$$

The group velocity $v_g = \frac{d\omega}{dk} \sim k \sim \sqrt{\omega}$ tends to zero at low frequencies. Low frequencies will tend to remain close to their source. This again is in contrast to phonons, which have a constant, nonzero velocity at low frequencies. High k values will tend toward zero group velocity as the maximum of the sin^2 function is approached. Only intermediate frequencies will propagate away from the perturbation with finite velocity.

Spin wave computing is a large and active field of research, in which information may be encoded into the phase, amplitude, or frequency of the wave. We won't discuss it in detail here, but a review can be found in [5].

4.4.3 Polaritons

Unlike the other collective excitations in this section, which are many-electron excitations, the **polariton** is an excitation made from a photon and an electron-hole pair.

Consider a quantum well sitting inside an optical resonator. As will be discussed in more detail in Chap. 25, an optical resonator is essentially a pair of reflecting surfaces confining photons to a finite spatial region. The finite size of the resonant cavity causes the allowed photon wavelengths to be quantized: the resonant wavelengths λ are those for which the cavity length is an integer multiple of $\frac{\lambda}{2}$. Similarly, a quantum well confines electrons to a finite region, forcing the electron energies to also be quantized.

Suppose that the resonant frequencies of the photon and electron are the same. In this case, a photon passing through the well can excite an electron to a higher-energy band, creating an electron-hole pair. The hole acts like a positively charged particle, so the electron and the hole can orbit each other, forming a bound pair called an **exciton**. But the exciton is unstable, with the electron and hole inevitably recombining after a brief time. During the recombination, a new photon is re-emitted, which can in turn then create a new exciton. We can therefore have excitons and photons constantly converting back and forth into each other. In fact, at any given moment, the system can be in a superposition of a photon and an exciton, forming a hybrid excitation called a **polariton**, or more specifically, an **exciton-polariton**.

The polariton is a bosonic excitation. It has been observed to form Bose-Einstein condensates. Because its mass is small, the excitation can travel at near the speed of light. However, unlike a photon, it can interact with its surroundings, and in fact, it can undergo nonlinear interactions with light. These properties make it a candidate for use in quantum information processing and quantum sensing applications [6].

More generally, polaritons can be defined that consist of electromagnetic waves coupling strongly to any excitation carrying an electric dipole moment. Thus, in addition to exciton polaritons, there are also polaritons formed by coupling of electromagnetic waves to phonons or to surface plasmons (see below).

4.4.4   Plasmons

For many purposes, the electrons in a solid can be treated as a gas of charged particles, in other words, as a plasma. If an electromagnetic wave passes through the plasma, electrons will try to follow the oscillation of the electric field. If the frequency of the wave is slow enough, the electrons have no problem keeping up with the oscillation. In this case, the plasma will absorb the wave and the material will be opaque to the radiation. In contrast, at sufficiently high frequencies, the inertia of the electron will prevent it from accelerating fast enough to keep up with the field oscillation; in this case, the material will be transparent to that frequency of radiation. The dividing line between these two cases is the plasma frequency,

$$\omega_p = \sqrt{\frac{\rho e}{m \epsilon_0}} = \sqrt{\frac{n e^2}{m \epsilon_0}}, \tag{4.83}$$

where ρ and n are the charge and number densities of the electron, related by $\rho = ne$. The plasma frequency is typically in the ultraviolet for metals, so that metals tend to be highly reflective at optical frequencies.

If a wave passes through the electronic plasma in a material, the positive ions in the crystal lattice will also oscillate in response. Since the ions are also moving, the relevant mass in the plasma frequency should really be the reduced mass $m^* = \frac{mM}{m+M}$ where m and M are, respectively, the electron and ion masses. But since $M << m$, it is usually safe to ignore the ion motion and take $m^* \approx m$.

A **plasmon** is a quantum of a collective excitation of the electron plasma-ion lattice system in which the electrons and the ions move out of phase with each other, meaning that when the electrons move left, the ions move right, and vice versa. Again consider one dimension for simplicity. Let ρ_0 be the constant equilibrium charge density of the electron fluid. Following [7], consider an oscillating position-dependent perturbations away from equilibrium:

$$\rho(x) = \rho_0 + \rho_q \cos qx. \tag{4.84}$$

In addition, there will be some charge density from the positive ions ρ_{ion} on the lattice. Since the massive ions move less than the lighter electrons, let's treat ρ_{ion} as constant in a first approximation. Further, in order to make the system electrically neutral, assume that $\rho_0 = -\rho_{ion}$.

Under these circumstances, the Hamiltonian for the system can be put into a harmonic-oscillator-like form (see, e.g., Chap. 1 of [7]). The dispersion relation is of the form $\omega_k = \left(\omega_p^2 + \alpha^2 k^2\right)^{1/2}$, and each plasmon has energy $E = \hbar\omega_k$.

Surface plasmons are plasmon oscillations that appear at the surfaces of conducting materials in contact with nonconducting materials, such as the surface of a metal in air. Surface plasmons can combine with an incident electromagnetic wave to produce a bosonic surface wave called a **surface plasma polariton (SPP)**, which is confined to a sub-wavelength-thick region near the surface.

Surface plasmons have distinct peaks at particular frequencies, which make them useful as nanoscale optical sensors, especially at the surfaces of highly localized conducting objects like metal nanoparticles. They have applications in sub-wavelength microscopy and lithography, as well as in photonic data storage and for detecting the presence of particular proteins and other biomolecules [8]. Plasmonic resonances in metal nanoparticles can act as efficient catalysts for chemical reactions [9], and plasmonic systems have multiple types of applications in solar energy harvesting and energy storage [10].

4.4.5 Other Quasiparticles

A brief and non-comprehensive list of other collective excitations in condensed matter systems would include the following:

- **Excitons:** As mentioned above, these are bound states of electron-positron pairs or of electron-hole pairs. (Positrons are the positively charged antiparticles of electrons.) Excitons are similar to hydrogen atoms, but much lighter, and they are unstable because of the eventual annihilation of the electron and positron or electron and hole. Incident photons of sufficiently high intensity can induce the creation of excitons at the surfaces of solids. These excitations can transport energy along the surface without an accompanying transport of charge.
- **Bogoliubov quasiparticles:** These are electron-pair excitations that appear in superconductors. They will make a brief appearance in Chap. 7.
- **Rotons:** A roton is a quantized vortex appearing in a superfluid. These are bosonic quasiparticles, and Bose-Einstein condensates of rotons have been observed experimentally [16].
- **Majorana fermions:** A Majorana fermion is a fermion that is equal to its own antiparticle. Such particles have been sought for in particle physics for decades, without success. Recently, it has been realized that Majorana-like collective excitations are possible in solids. Majorana fermions are especially interesting because in two-dimensional systems like graphene they can have **anyon** statistics: they act as neither a boson nor a fermion, but as something with statistical properties that are intermediate between the two. Majorana fermions will be discussed in Sect. 21.6.

- **Composite fermions:** Composite fermions can arise in two-dimensional systems in the presence of a strong magnetic field. They appear in fractional quantum Hall systems (Sect. 21.5.5), and in addition to being anyonic, they can also carry fractional charge, meaning that they have a charge that is not an integer multiple of the proton charge e.
- **Solitons:** Solitons are localized, particle-like waves that can propagate through nonlinear systems without dispersing or spreading. Solitons come in a wide variety of types, with an assortment of different properties, and they occur in numerous areas of condensed matter physics, optics, and particle physics, as well as in mechanical and electronic systems. A comprehensive review of solitons and of the related topic of instantons can be found in [17].

4.5 Photonic Crystals and Bragg Gratings

We've seen that in materials with regular, crystalline lattices, the allowed energy levels of the electrons cluster into discrete bands, comprised of many closely spaced discrete levels. The shapes of the bands are determined by the structure of the lattice, and these shapes determine the electrical transport properties of the material. If an electron falls into one of the allowed conduction bands, then it propagates through the material easily, while if it falls into one of the gaps between the bands, it cannot propagate at all. This can be viewed from an electron wave perspective: for electrons inserted into the gaps, destructive interference occurs between forward-propagating electron waves and waves that have been backscattered off of atoms.

By analogy, a similar situation can occur with light: a regular, periodic array of scattering centers can allow photons within some range of energies to propagate while blocking others via destructive interference. Thus, a set of **optical bands** and **optical bandgaps** can appear in a manner similar to those of electrons in conductors. Such a periodic array is called a **photonic crystal** [11–13].

The scattering centers are often taken to be artificially engineered dielectric structures with a dielectric constant and refractive index of different value from the surrounding medium; for example, a periodic arrangement of dielectric micropillars is a common choice in two dimensions. In three dimensions, the scattering centers might be quantum dots, cavities or holes in the materials, or beads suspended in a substrate.

Here we will restrict attention to the case of one-dimensional photonic crystals to see how optical bandgaps arise; the higher-dimensional versions are conceptually similar, although mathematically more complex.

An example of a simple one-dimensional photonic crystal is a **Bragg grating**: a sequence of layers that periodically vary in refractive index. The simplest case is a **step-index grating**, in which the refractive index changes suddenly from one layer to the next, as in Fig. 4.17. To simplify further, assume that all the layers are of the same width, $\Delta x = \frac{d}{2}$, and that the light is travelling horizontally and striking each interface at normal incidence. Label each of the $2N$ layers by an integer, and take

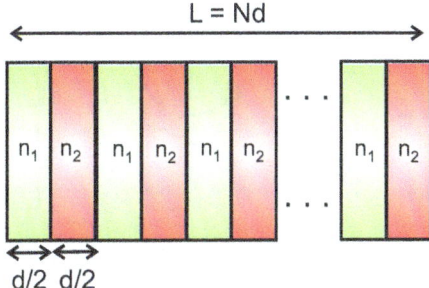

Fig. 4.17 A simple one-dimensional crystal lattice is formed by a Bragg diffraction grating. Layers of equal thickness but alternating refractive indices lead to interference between the beams reflecting off the different interfaces between layers. Whether the interference is constructive or destructive, and thus how much of the light is reflected, is determined by the frequency of the light and the thickness of the layers

the index of refraction to alternate between two values n_1 and n_2 from one layer to the next:

$$n_{2j-1} = n_1, \qquad n_{2j} = n_2, \tag{4.85}$$

for $j = 1, 2, \ldots, N$. Define $\Delta n = n_2 - n_1$. We will assume that $\Delta n \ll n_1, n_2$. As long as $\Delta n \neq 0$, part of the light will reflect each time it strikes the interface between two layers. The total length of the grating is $L = Nd$.

Let the light be propagating in the x-direction. Recall that the electric field in an optical wave can be written in complex form as

$$\boldsymbol{E}(x, t) = \boldsymbol{E}_0 e^{-i(kx - \omega t)} \tag{4.86}$$

The time dependence and polarization direction play no role here, so we will suppress them and focus on the spatial part of the amplitude:

$$E(x) = E_0 e^{-ikx}. \tag{4.87}$$

Let r_1 and t_1 be the amplitude reflection and transmission coefficients at each interface between layers for light coming from the material of index n_1. Similarly, r_2 and t_2 are the coefficients for light incident from the material of index n_2. These can be calculated from the Fresnel equations (Appendix B) if the refractive indices are known, and at normal incidence, they will be real and independent of polarization. Unitarity requires that $|r_k|^2 + |t_k|^2 = 1$ for $k = 1, 2$, so that everything can be expressed solely in terms of the reflection coefficients. From the Fresnel equations, we find that

Fig. 4.18 Light enters from
the left. The reflected light
returning to the left side of
the figure is obtained by
taking a sum over reflections
at each of the interfaces. $\frac{\delta}{2}$ is
the phase gained during a
round trip through one layer
of the material

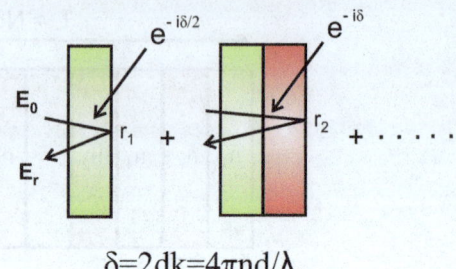

$$\delta = 2dk = 4\pi nd/\lambda_0$$

$$r_1 = -r_2 = \frac{n_1 - n_2}{n_1 + n_2} \approx -\frac{\Delta n}{2n}, \tag{4.88}$$

where the last equality holds if $\Delta n << n_1 \approx n_2 \equiv n$, and the reflection coefficient
is defined as the ratio of the reflected to incident field amplitudes.

For simplicity, we assume that the reflectance at each surface is small, $|r_1| = |r_2| << 1$. Using the binomial expansion, $t_j = \sqrt{1 - r_j^2} \approx 1 - \frac{1}{2}r_j^2 + \ldots$ for
$j = 1, 2$, so that if we keep only terms up to first order in r_j, then $t_1 \approx t_2 \approx 1$.

Let's look at the light reflected back to the left. The incident field coming from
the left at $x = 0$ is E_0 so that before any reflection, the field in the first layer is
$E_1(x) = E_0 e^{-ikx}$. The goal is to find the reflected amplitude E_r. This can be done
in steps. Consider the light reflected from the right surface of the first layer, at $x = \frac{d}{2}$
(first term in Fig. 4.18)). It gains a phase $-i\frac{\delta}{2} = -idk = -\frac{2i\pi nd}{\lambda_0}$, (where λ_0 is the
free-space wavelength) from traveling across the first layer and back, plus a factor
of r_1 from reflection at the first interface, so the first reflection gives a contribution
of $r_1 e^{-i\delta/2} E_0$.

Light reflecting at the second interface travels a total distance of $2d$ (second term
in Fig. 4.18), gaining a phase $e^{-i\delta}$ plus a reflection factor of r_2, thus providing a
contribution $r_2 e^{-i\delta} E_0$. So the first two layers provide a contribution to the reflected
field of

$$E_r(0) = E_0 e^{-i\delta/2} \left[r_1 + r_2 e^{-i\delta/2} \right] = E_0 e^{-i\delta/2} \left[r_1 - r_1 e^{-i\delta/2} \right]. \tag{4.89}$$

Continuing in this manner, one arrives at a finite series, which can readily be
summed:

$$E_r = r_1 E_0 e^{-i\delta/2} \sum_{j=0}^{N-1} \left[e^{-ij\delta} - e^{-i(j+\frac{1}{2})\delta} \right] \tag{4.90}$$

$$= r_1 E_0 e^{-i\delta/2} \left(1 - e^{-i\frac{\delta}{2}} \right) \sum_{j=0}^{N-1} e^{-ij\delta} \tag{4.91}$$

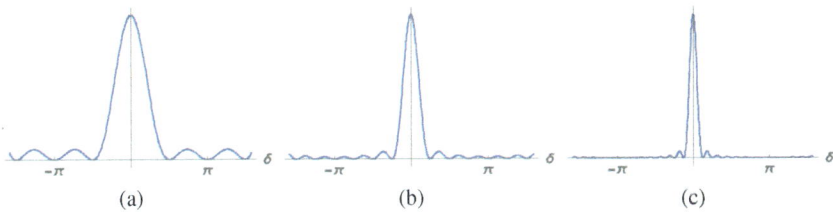

Fig. 4.19 The reflected intensity from a Bragg grating, $I_r = |E_r|^2 = I_0 K \left(\frac{\sin \frac{N\delta}{2}}{\sin \frac{\delta}{2}} \right)^2$, for (**a**) $N = 4$, (**b**) $N = 8$, and (**c**) $N = 16$. The large maximum at the center repeats at $\delta = 2\pi n$ for all integers, n

$$= r_1 E_0 e^{-iN\delta/2} \left(1 - e^{-i\frac{\delta}{2}} \right) \left(\frac{\sin \frac{N\delta}{2}}{\sin \frac{\delta}{2}} \right), \tag{4.92}$$

giving a reflected intensity

$$I_r = |E_r|^2 = I_0 K \left(\frac{\sin \frac{N\delta}{2}}{\sin \frac{\delta}{2}} \right)^2, \tag{4.93}$$

where $K = |r_1 \left(1 - e^{\frac{-i\delta}{2}} \right)|^2$ and I_0 is the incident intensity. This function is plotted in Fig. 4.19.

The maximum reflection occurs when the terms in the sum of Eq. 4.91 constructively interfere, or in other words when the **Bragg condition** holds: $\delta = 2\pi m$, for $m = 1, 2, 3, \ldots$. m is called the **diffraction order**. When the Bragg condition is satisfied, all the terms in the sum are equal (they all equal 1), so the finite sum is just equal to N. The Bragg condition becomes

$$\frac{4\pi nd}{\lambda_0} = 2\pi m, \tag{4.94}$$

so that constructive interference occurs when the free-space wavelength is $\lambda_0 = \frac{\lambda_B}{m}$, where the **Bragg wavelength** is

$$\lambda_B = 2nd. \tag{4.95}$$

The corresponding wavenumbers are $k = mk_B$, where

$$k_B = \frac{2\pi}{\lambda_0/n} = \frac{m\pi}{nd} \tag{4.96}$$

is the Bragg wavenumber. So there is nearly 100% reflection, and no light propagates to the far end of the crystal when the wavenumber of the light is near the values $\frac{\pi}{nd}, \frac{2\pi}{nd}, \ldots$, which should be reminiscent of edges of the Brillouin zones of electrons in a crystal. Because no light propagates at these values, we must have $\frac{d\omega}{dk}$ at these points, indicating vanishing group velocity and the appearance of a photonic bandgap. Restricting ourselves to first-order diffraction ($m = 1$), we find from Eq. 4.91 that

$$E_r = 2r_1 E_0 N e^{-i\delta/2} = -\frac{\Delta n}{n} E_0 N e^{-i\delta/2}, \tag{4.97}$$

so that the intensity reflectance is

$$R = \left| \frac{E_r}{E_0} \right|^2 = \left(\frac{\Delta n}{n} N \right)^2 \tag{4.98}$$

$$= \left(\frac{2\Delta n L}{\lambda_B} \right)^2 \tag{4.99}$$

$$\equiv (\alpha L)^2, \tag{4.100}$$

where $\alpha = \frac{2\Delta n}{\lambda_B}$ is the attenuation coefficient for propagation through the grating, with units of m^{-1}.

Consider a free-space wavelength λ detuned slightly away from the Bragg wavelength, so that $\Delta\lambda = \lambda_0 - \lambda_B$. Then all the terms in the finite sum of Eq. 4.91 will no longer be equal; successive terms will differ in phase by an amount

$$\Delta\phi = -\frac{d\delta}{d\lambda_0}\Delta\lambda = \frac{4\pi nd}{\lambda_0^2}\Delta\lambda. \tag{4.101}$$

From Eq. 4.92, we see that zeros of the intensity will occur when $|\Delta\delta|$ is a multiple of $\frac{2\pi}{n}$. Therefore, the half-width of the Bragg peaks in reflectance is

$$\Delta\lambda = \frac{\lambda_0^2}{2ndN} = \frac{\lambda_B}{N}, \tag{4.102}$$

or in other words, for λ_0 close to λ_B

$$\frac{\Delta\lambda}{\lambda_B} \approx \frac{1}{N}, \tag{4.103}$$

implying that the Bragg peaks become progressively sharper with increasing N.

The width of the spectral bandgaps then becomes

$$\Delta\omega = \left| \frac{d\omega}{d\lambda_0} \right| \Delta\lambda = \left| \frac{d}{d\lambda_0}\left(\frac{2\pi c}{\lambda_0} \right) \right| \Delta\lambda \tag{4.104}$$

$$= \frac{2\pi c}{\lambda_0^2}\Delta\lambda \; = \; \frac{\omega}{\lambda_0}\Delta\lambda, \tag{4.105}$$

so that

$$\frac{\Delta\omega}{\omega} = \frac{\Delta\lambda}{\lambda} = \frac{\Delta n}{n}. \tag{4.106}$$

The photonic bandgap width therefore increases in proportion to the size of the refractive index difference between adjacent layers.

So, we can control the intensity of the reflected light by controlling the thickness of the dielectric layers. Or, conversely, monitoring the reflected intensity allows variations in the thickness of the layers to be measured. More importantly, notice that the reflected intensity depends on the free-space wavelength through the phase shift, δ. As a result, the Bragg grating can be used as a narrowband wavelength filter: only those wavelengths near the λ_B are reflected. The widths of the peaks narrow with increasing N, allowing very narrow passbands to be achieved. The material therefore exhibits narrow allowed optical bands separated by spectral bandgaps.

The above results were for step-index Bragg gratings. However, the qualitative picture remains similar for smoothly and periodically modulated index variations. Two- and three-dimensional optical crystals display similar, though more complicated, optical band structures.

Bragg gratings have a large number of applications, as narrow-bend frequency filters, or as mirrors in lasers, fibers, or integrated photonic systems. Fiber Bragg gratings have become common, in which the refractive index of the core of a fiber is periodically modulated in order to act as a grating. In particular, fiber Bragg gratings are often used as sensors. Small variations in the length of the fiber or in the temperature will cause noticeable changes in the reflection and transmission intensities, which are easily measured. As a result, fiber sensors based on Bragg gratings are used in applications ranging from seismology and undersea sensing to optical interferometry [14, 15].

4.6 Graphene

Carbon is the basis for all organic molecules, and so is one of the basic building blocks of life as we know it. With four valence electrons, it tends to form four covalent bonds with neighboring molecules. Carbon is the fourth most abundant element in the universe (after hydrogen, helium, and oxygen), and it comes in a wide variety of forms. For example, it can form into a crystalline solid with a cubic structure (diamond), into crystals with stacked planar layers (**graphite**), or mesh-like structures called **fullerenes** (such as the famous "buckyball" structure of C_{60}, which resembles a soccer ball), as well as various polymer structures. The physical

properties of these various carbon structures vary widely; for example, diamond is
a very good insulator, while graphite is a good conductor.

The structure of graphite consists of layers a thick single molecule. A single layer
of graphite in isolation is called **graphene**. These thin sheets of carbon were first
isolated in 2004 by Andre Geim and Konstantin Novoselov, who won the Nobel
Prize in 2010 for their work. Their experimental method for the isolation of a single
layer was remarkably simple: they peeled off layers of graphene using ordinary
adhesive tape. Once isolated, these sheets of carbon were quickly verified as having
a number of remarkable properties, and they have found numerous applications in
nanotechnology. In particular, the sheets are essentially two-dimensional (the third
dimension being only one atom thick), and so graphene has become an experimental
platform for implementing theoretical ideas about two-dimensional physics that
have been around for decades. Being two-dimensional, graphene is essentially all
surface, meaning that it readily hosts surface plasmons (Sect. 4.4.4). This makes the
material useful in many sensing applications.

The structure of graphene consists of a honeycomb, with each carbon atom
connected to its three nearest neighbors by covalent bonds. The unit cell of the
two-dimensional lattice is hexagonal in shape, as seen in Fig. 4.20b. The three strong
covalent bonds (called σ bonds) use up three of the four valence atoms of each atom,
while the fourth electron exists in a delocalized state (a π orbital), shared among
the atoms. The π orbitals are what give graphene and graphite their high electrical
conductivity. The crystal lattice divides up into two inequivalent sublattices (the A
and B sublattices), intertwined with each other as shown in Fig. 4.20a. Each A atom
has three B atoms as nearest neighbors and vice versa. The lattice spacing or bond
length is $a = 0.142$ nm. The vectors along the sides of the hexagons, from each A
to its nearest B neighbors, are given by the vectors

$$\delta_1 = \frac{a}{2}\left(\hat{i} + \sqrt{3}\hat{j}\right), \quad \delta_2 = \frac{a}{2}\left(\hat{i} - \sqrt{3}\hat{j}\right), \quad \delta_3 = -a\hat{i}. \qquad (4.107)$$

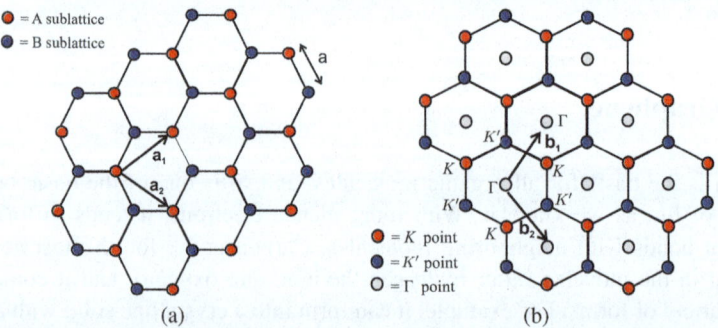

Fig. 4.20 The honeycomb structure of the graphene lattice. (**a**) The interatomic spacing is a. The
A atoms and the B atoms each form a triangular sublattice with primitive lattice vectors $\boldsymbol{a}_1$ and $\boldsymbol{a}_2$.
(**b**) The lattice in reciprocal (momentum) space. γ is the center of the momentum space unit cell,
while K and K' are the inequivalent corners of the cell

Taking one of the A sublattice atoms as the origin, the position of any other A atom can be written as

$$r = m a_1 + n a_2, \qquad (4.108)$$

where m and n are integers, and a_1 and a_2 are the primitive lattice vectors shown in the figure, and given by

$$a_1 = \frac{\sqrt{3}a}{2}\left(\sqrt{3}\hat{i} + \hat{j}\right) \qquad (4.109)$$

$$a_2 = \frac{\sqrt{3}a}{2}\left(\sqrt{3}\hat{i} - \hat{j}\right). \qquad (4.110)$$

Approximate values for the energy levels of the valence and conduction bands are given by the dispersion relation

$$E_{\pm}(k) = E_0 \pm \gamma w(k), \qquad (4.111)$$

where $\gamma \approx 2.8\,\text{eV}$ and k is the two-dimensional Bloch wavevector. The mean energy E_0 can always be set to zero by a choice of origin for the energy scale. The function $w(k)$ was first calculated in the 1940s and is given by

$$w(k) = \left(1 + 4\cos\left(\frac{3k_x a}{2}\right)\cos\left(\frac{\sqrt{3}k_y a}{2}\right) + 4\cos^2\left(\frac{\sqrt{3}k_y a}{2}\right)\right)^{1/2}. \qquad (4.112)$$

The energy gap between valence and conduction bands, $E_g = E_+ - E_- = 2\gamma w(k)$, is on the order of zero to $20\,\text{eV}$ in the first Brillouin zone.

Because the spatial structure is periodic, the structure will also be periodic in momentum space. The reciprocal lattice in momentum space is also hexagonal (Fig. 4.20b). Just as the unit cells in ordinary space can be transformed into each other by translating or displacing them by a vector of the form of Eq. 4.108, the unit cells in reciprocal or momentum space (the Brillouin zones) are related via translation by vectors of the form

$$G = l b_1 + q b_2, \qquad (4.113)$$

where l and q are integers and the momentum-space unit vectors b_1 and b_2 (with dimensions of inverse length) are the primitive lattice vectors of the reciprocal lattice. The real-space and reciprocal-space lattices should be orthogonal to each other:

$$a_i \cdot b_j = 2\pi \delta_{ij}. \qquad (4.114)$$

The general rule for constructing primitive reciprocal lattice vectors in three dimensions is [18]

$$b_i = \frac{2\pi a_j \times a_k}{a_i \cdot (a_j \times a_k)}. \tag{4.115}$$

In two dimensions, the same rule could be used with a third basis vector added that has a nonzero component parallel to the plane of the lattice. Alternately, the reciprocal basis vectors can be found in two dimensions via the rule

$$\begin{pmatrix} b_{1x} \ b_{2x} \\ b_{1y} \ b_{2y} \end{pmatrix} = \frac{2\pi}{a_{1x}a_{2y} - a_{1y}a_{2x}} \begin{pmatrix} a_{2y} & -a_{1y} \\ -a_{2x} & a_{1x} \end{pmatrix}. \tag{4.116}$$

The reader can verify that the primitive vectors are

$$b_1 = \frac{2\pi}{3a}\left(\hat{i} + \sqrt{3}\hat{j}\right), \qquad b_2 = \frac{2\pi}{3a}\left(\hat{i} - \sqrt{3}\hat{j}\right). \tag{4.117}$$

Just as two corners of the real-space unit cell are inequivalent to each other, two of the corners of the reciprocal lattice unit cell are inequivalent and are usually labeled as K and K'. The K and K' points alternate around the corners of the hexagon. The central point of the Brillouin zone is labeled Γ. Taking the central Γ point of one cell as the origin, the locations of the inequivalent corners are at

$$K = \frac{2\pi}{3a}\left(\hat{i} + \frac{1}{\sqrt{3}}\hat{j}\right), \qquad K' = \frac{2\pi}{3a}\left(\hat{i} - \frac{1}{\sqrt{3}}\hat{j}\right). \tag{4.118}$$

At the special points K, K', Γ in momentum space, the gap energies are

$$E_g(\Gamma) \approx 18\,\text{eV}, \qquad E_g(K) = E_g(K') \approx 6\,\text{eV}. \tag{4.119}$$

The K and K' points are called **Dirac points** for the following reason. Taking K as an example, consider a momentum value displaced slightly from the K point, with momentum $q = K + k$, where the displacement k is assumed small, $|k|a \ll 1$. Expanding the dispersion relation $E(k)$ about K, the result is *linear*:

$$E_{\pm} \approx \pm\frac{3}{2}\gamma ak = \pm v_F \hbar k, \tag{4.120}$$

where $k = |k|$. The quantity $v_F = \frac{3\gamma a}{2\hbar} \approx 10^6\,\frac{\text{m}}{\text{s}}$ is the **Fermi velocity**, i.e., the particle velocity at the zero of energy (the Fermi energy). This is remarkable: even though the electron is massive, near the K point, it obeys a linear dispersion relation, which is normally characteristic of massless, relativistic particles. (For a massive nonrelativistic particle, the expectation would be a quadratic dispersion relation,

$E(k) = \frac{\hbar^2 k^2}{2m}$, while in contrast the massless photon has linear dispersion, $E(k) = pc = \hbar ck$.)

In fact, going to momentum space, the Hamiltonian can be put into a form equivalent to the massless Dirac equation. (See [19] for a clear discussion of this.) In a relativistic system, the plus and minus signs in the dispersion relation would correspond to particles and antiparticles with opposite helicities. Here, in the graphene case, the two signs correspond to particles and holes. The situation is similar at the K' point, with the signs reversed.

The fact that excitations appear in nonrelativistic solid-state systems that seem to correspond to a solution to a relativistic quantum field theory is a topic that has received an enormous amount of attention in recent years. It opens up the possibility that exotic particles like Majorana fermions that have eluded detection in high-energy particle physics may have collective excitation analogs in condensed matter systems that are easier to produce and detect. Some of these excitations may play a role in quantum information processing. More will be said on this topic in Sect. 21.6.

Instead of being in a flat plane, the graphene lattice can be rolled up into a cylindrical shape. The resulting structure is known as a **carbon nanotube**. These tubes have diameters on the order of nanometers, but can be up to several centimeters long. They are notable for their extremely high tensile strength: over 100 times greater than that of steel, but at one-sixth of the density. They have thermal conductivity higher than any other known material and are highly stable chemically. Some studies have indicated that carbon nanotubes can be used to make materials 30 times stronger than kevlar, capable of repelling high-velocity artillery. Just as sheets of graphene serve as labs for two-dimensional physics, a carbon nanotube allows exploration of one-dimensional physics.

Applications for carbon nanotubes abound. For example, they can be filled with materials in order to protect those materials from the surrounding environment, which leads to their use as nanomedical drug delivery devices. Electric fields are enhanced near their surfaces, leading to applications as electrical sensors or biosensors. Depending on exactly how the graphene sheets are rolled into a tube, these nanotubes can be either conductors or semiconductors and can be used to construct electronic devices such as transistors, light sources, or batteries at the nanoscale. For a comprehensive review of carbon nanotubes, see [20, 21].

Box 4.2 Semimetals and Weyl Points
Materials are usually classified as insulators or semiconductors if a full valence band is separated from an empty conduction band by a gap, with the size of the gap determining whether the material is an insulator or semi-conductor. In a conductor, there is a partially filled band with a high density of electron states at the Fermi level, which lies within the band. However, in some materials, there is an intermediate situation: the conduction band may

(continued)

Box 4.2 (continued)

overlap the valence band just slightly. In these so-called **semimetals**, there is no bandgap and a low density of states at the Fermi level.

Because of the low density of states at the Fermi level, semimetals have a lower density of charge carriers than metals. Consequently, they have electrical and thermal conductivities that, while greater than a typical insulator, are lower than standard metals. Examples of semimetals include arsenic, antimony, bismuth, boron, and gray tin.

Of special interest are Weyl semimetals. A **Weyl fermion** is a massless chiral fermion, where chiral here means that the spin has a definite, fixed value along the direction of motion. Weyl fermions were first predicted by Hermann Weyl in 1929 and have been sought for in vain by particle physicists ever since. Neutrinos were long believed to be candidates for Weyl fermions; it became clear that this was not the case when it was discovered that neutrinos had a very small (but nonzero) mass. However, it was predicted as early as 1937 by Conyers Herring that Weyl fermions may exist as collective electron excitations in solids. A **Weyl semimetal** is a solid-state crystal in which the low-energy excitations act as Weyl fermions. These excitations have a distinctly topological origin, and Weyl semimetals are closely connected to topological insulators (Chap. 21). The energy bands in the Weyl semimetal touch at isolated points, called **Weyl points** (Fig. 4.21), with a dispersion relation that is linear for small perturbations about the Weyl points. Quasiparticle Weyl fermions were first detected experimentally in the semimetal tantalum arsenide in 2015 [22]. Since then, Weyl fermions have been detected in several other semimetals, as well as in photonic crystals and superfluid Helium-3.

In the vicinity of a Dirac point in two dimensions, the momentum-space Hamiltonian of the form $H_D = v(p_x\sigma_x + p_y\sigma_y)$, where v is the Fermi energy.

(continued)

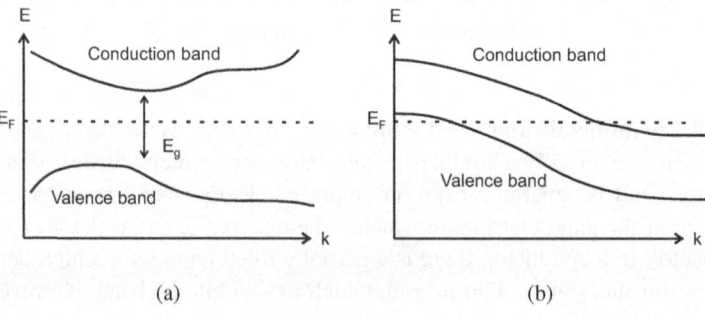

Fig. 4.21 (**a**) In an insulator or a semiconductor, there is an energy gap E_g between the conduction and valence bands. (**b**) In a semimetal, there is no gap: the Fermi energy crosses both bands. However, the overlap is small, with a low density of electrons and holes at E_F

Box 4.2 (continued)

There is no energy gap at $p = 0$, so the two Dirac cones touch and the system is conducting. But perturbations proportional to σ_z would introduce a gap and make the material insulating. The stability of the Dirac points in graphene is due to the existence of time reversal and parity symmetries, which prevent a σ_z term from appearing. But in three dimensions, the Hamiltonian near a Weyl point is of the form $H_W = v(p_x\sigma_x + p_y\sigma_y + p_z\sigma_z)$. Aside from the identity matrix (which would just give an irrelevant shift to the origin of the energy axis), there are no other Hermitian two-by-two matrices that could be added to this Hamiltonian by a perturbation. So perturbations can only move the location of the Weyl point, but cannot open a gap. When a Dirac point becomes gapped, it breaks into two Weyl points of opposite chirality. These Weyl points can only be created or destroyed in pairs and only exist when the discrete symmetries of the system are broken. Interestingly, the Weyl points can be viewed as analogs of magnetic monopoles in momentum space.

Weyl semimetals have a range of potential technological applications. In particular, it has been shown that tantalum arsenide can produce very large photocurrents, with a conversion efficiency from light to electric current that is ten times higher than any other known material [23]. For more on Weyl semimetals, see [24].

Problems

1. A hydrogen atom emits a photon with wavelength 4.59×10^{11} Hz. What were the initial and final principal quantum numbers of the electron state?
2. An electron in a hydrogen atom is in a Bohr orbit of radius 8.464 Å. What are the energy, angular momentum, and orbital speed of the electron?
3. Consider light emitted by the $3^2P_{1/2} \rightarrow 3^2S_{1/2}$ transition of sodium in the sun. (Here the spectroscopic notation $n^{2S+1}L_J$ is used.)

   (a) Assuming that the magnetic field in the sun is approximately $B = 0.5$ T, compute the level splittings ΔE of both the initial and final states.
   (b) Find the splitting $\Delta\lambda$ between the longest and shortest wavelengths emitted in this spectral line.

4. Consider the potential given in Eq. 4.45.

   (a) Find the minimum r_0 of the potential as a function of the parameters m, n, A, and B.
   (b) If a particle is placed inside this potential, but displaced slightly from r_0, it will undergo simple harmonic motion about the minimum. Find the angular

frequency ω of the oscillation. (Hint: expand the potential in a Taylor series about the minimum.)

5. (a) Verify that the primitive vectors of the reciprocal lattice given in Eq. 4.117 satisfy the conditions of Eqs. 4.114 and 4.116.

(b) Show that in two dimensions, Eqs. 4.114 and 4.115 imply 4.116.

6. The mass density of lithium is $\rho = 0.534 \frac{g}{cm^3}$. There is one conduction electron per atom.

(a) Find the number density of conduction electrons.

(b) Compute the Fermi energy and Fermi temperature of lithium. (The Fermi temperature is the temperature at which the mean thermal energy per atom equals the Fermi energy.)

7. Sodium has density of $0.97\,g/cm^3$ and an atomic weight of $23.0\,g/mol$. Given that there is one free electron per atom:

(a) Find the electron density, Fermi energy, and Fermi velocity.

(b) Compute the average spacing between atoms. (Approximate the atoms as being separated by distance $2r$, where r each atom is treated as occupying its own sphere of volume $V = \frac{4}{3}\pi r^3$.)

(c) Find the mean time between electron-atom collisions at the Fermi temperature.

8. Consider an electron in an $n = 2$ state of hydrogen. The atom is placed into a magnetic field of strength $B = 20\,T$, pointing along the z-axis.

(a) Find the Larmor frequency.

(b) What are the allowed l and m_l values? What are the energies of the allowed spin states (relative to the energy in the absence of a magnetic field)?

(c) What is the fractional size of the energy splitting between the highest and lowest levels found in part (b), relative to the Bohr energy E_2 of the state, $\frac{\Delta E}{E_2}$?

9. A Bragg grating is formed inside a silicon fiber by arranging a region where the refractive index varies between two values. The variation Δn is very small compared to the average index of $n = 1.5$. The grating is 0.5 m in length, and a beam of light of free-space wavelength 1550 nm is introduced into the fiber.

(a) The reflectance of the grating is desired to be 5%. What must be the index variation Δn between the two alternating layers?

(b) What is the Bragg wavelength?

(c) Find the spectral widths $\Delta \omega$ and $\Delta \lambda$.

(d) What fraction of the light's initial intensity has been reflected after propagating 0.25 m through the grating?

References

1. W. Thomson, Proc. Roy. Soc. Edinb. **6**, 94 (1867)
2. M. Wertheim, *Physics on the Fringe: Smoke Rings, Circlons, and Alternative Theories of Everything* (Bloomsbury Publishing, 2013)
3. D.A. Neamen, *Semiconductor Physics And Devices: Basic Principles* (McGraw-Hill, New York, 2011)
4. S.M. Sze, M.-K. Lee, *Semiconductor Devices: Physics and Technology*, 3rd edn. (Wiley, New York, 2012)
5. A. Mahmoud et al., J. Appl. Phys. **128**, 161101 (2020)
6. F.I. Moxley, E.O. Ilo-Okeke, S. Mudaliar, T. Byrnes, Emergent Mater. **4**, 971 (2021)
7. P.L. Taylor, O. Heinonen, *A Quantum Approach to Condensed Matter Physics* (Cambridge University Press, Cambridge, 2002)
8. M. Soler, L.M. Lechuga, J. Appl. Phys. **129**, 111102 (2021)
9. U. Aslam et al., Nat. Catal. **1**, 656 (2018)
10. J.A. Dionne, S. Gagli, V.M. Shalaev, Phys. Today **76** (6) 24 (2023)
11. J.D. Joannopoulos, S.G. Johnson, J.N. Winn, R.D. Meade, *Photonic Crystals: Molding the Flow of Light*, 2nd edn. (Princeton University Press, Princeton, 2008)
12. K. Inoue, K. Ohtaka, *Photonic Crystals: Physics, Fabrication and Applications* (Springer, Berlin, 2004)
13. R.S. Quimby, *Photonics and Lasers* (Wiley-Interscience, 2006)
14. A. Othonos, K. Kalli,*Fiber Bragg Gratings: Fundamentals and Applications in Telecommunications and Sensing* (Artech House, 1999)
15. R. Kashyap, *Fiber Bragg Gratings*, 2nd edn. (Academic Press, New York, 2009)
16. L. Chomaz et al., Nat. Phys. **14**, 442 (2018)
17. R. Rajaraman, *Solitons and Instantons: An Introduction to Solitons and Instantons in Quantum Field Theory* (North-Holland, Amsterdam, 1987)
18. N.W. Ashcroft, N.D. Mermin, *Solid State Physics* (Cengage Learning, 1976)
19. E. Witten, La Rivista del Nuovo Cimento **39**, 313 (2016) (arXiv:1510.07698v2 [cond-mat.mes-hall])
20. S. Reich, C. Thomsen, J. Maultzsch, *Carbon Nanotubes: Basic Concepts and Physical Properties* (Wiley-VCH, 2004)
21. S. Nanot et al., in *Handbook of Nanomaterials*, ed. by R. Vajtai (Springer, Berlin, 2013), p. 105
22. S.-Y. Xu et al., Science **349** (6248), 613 (2015)
23. G.B. Osterhoudt et al., Nat. Mater. **18**, 471 (2019)
24. S. Rao, arXiv:1603.02821v2 [cond-mat.mes-hall] 22 (2016)

Chapter 5
Light and Matter: Quantum and Nonlinear Optics

In this chapter, a first look is given at the interactions of light with matter, introducing many of the basic ideas needed for studying real-world quantum systems. We begin with the photoelectric effect and then discuss the quantization of the electromagnetic field and the physics of lasers, before giving a brief introduction to nonlinear optics.

Another useful and more detailed model of light-matter interactions (the Jaynes-Cummings model) will be described in Chap. 24.

5.1 The Photoelectric Effect

Einstein's explanation of the photoelectric effect in terms of absorption of quanta of electromagnetic radiation was one of the first great triumphs of quantum theory. Today, it is still an important topic, forming the basis of solar cells, photodiodes, and many types of photodetectors.

In 1887, Heinrich Hertz was attempting to experimentally verify Maxwell's prediction of electromagnetic waves. He produced sparks between two electrodes (the transmitter), and these sparks produced electromagnetic waves that traveled to another apparatus (the receiver), where they would generate new sparks across a gap between another pair of electrodes. When he placed the receiver in a darkened box to better view the sparks, he found that the maximum gap separation that would allow sparks to cross was decreased. Further investigation by Hertz and others led to a number of findings:

- In a number of metals, the presence of light striking the surface could induce a current (the **photocurrent**) in the material.
- No photocurrent was seen if the frequency of the light was below a certain cutoff frequency ν_c, whose value depended on the type of material. Above the cutoff, the photocurrent is proportional to the intensity of the light.

© The Author(s), under exclusive license to Springer Nature Switzerland AG 2025
D. S. Simon, *Introduction to Quantum Science and Technology*, Undergraduate
Texts in Physics, https://doi.org/10.1007/978-3-031-81315-3_5

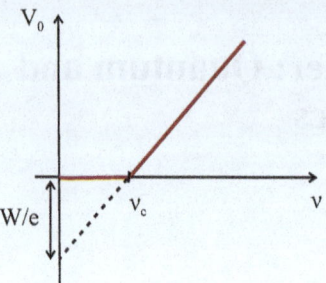

Fig. 5.1 The photoelectric effect. If the stopping potential is plotted versus frequency, it vanishes below the cutoff frequency v_c and is linear above v_c. If the line is extended below the axis (the dotted line), the y-intercept is at $\frac{W}{e}$, where W and e are the work function of the material and the (absolute value of) electron charge

- Applying a voltage opposite to the current, the **stopping potential** V_0 is defined to be the minimum voltage required to cancel the photocurrent. The stopping potential is independent of the intensity of the light, but above v_c, it increases linearly with increasing frequency. The stopping potential versus frequency looks as shown in Fig. 5.1.

A number of attempts were made to explain these effects, but none of the explanations were completely successful until Einstein's paper on the photoelectric effect appeared in 1905 and gave a very simple picture that accounted for all the known facts. He began by assuming that the light was quantized in bundles of energy (now called photons), with each bundle being of size $E = hv = \hbar\omega$. Consider an electron in an atom. It is attached to the atom, and removing it so that it can contribute to a current requires the input of some energy, the **work function** W of the material. If the energy of the photon is smaller than the work function, then either the photon does not get absorbed at all, or else it simply sends the electron into an excited state of the same atom. This immediately implies that there must be a minimum cutoff frequency of

$$v_c = \frac{W}{\hbar} \tag{5.1}$$

for the electron to be ejected from the atom. Since $\lambda v = c$ for each photon, we can also express the cutoff in terms of wavelength: there will be no photocurrent if the wavelength is longer than the cutoff wavelength: $\lambda_c = \frac{c}{v_c} = \frac{hc}{W}$.

If the incoming frequency is above the cutoff, there will be energy left over after overcoming the work function. This excess energy will be the final kinetic energy of the electron:

$$K = hv - W = h(v - v_c). \tag{5.2}$$

The stopping potential is the voltage that must be applied to cancel out this kinetic energy:

$$eV_0 = K \implies V_0 = \frac{h}{e}(v - v_c). \tag{5.3}$$

As expected, this is independent of the light's intensity and linearly dependent on the frequency. However, increasing the intensity of the light means more photons strike the surface per second. This liberates more electrons per unit time, and therefore higher intensities will lead to larger currents.

Before looking in more detail at the interactions between light and matter, we take a detour in the next section to review the raising and lowering operator formalism for quantum harmonic oscillators. This will provide a guide to follow toward treating the incoming light in a fully quantum mechanical fashion.

5.2 Harmonic Oscillators: Raising and Lowering Operators

When the quantum harmonic oscillator is discussed in introductory quantum mechanics texts, it is often described in terms of individual oscillator wavefunctions, which are then given in terms of cumbersome expressions involving Hermite polynomials. But a simpler approach is to introduce raising and lowering operators, which take us up and down the ladder of state vectors describing the oscillator solutions and which form a simple algebraic structure. This operator algebra construction is not only simpler to work with, but it has a generalization to quantum fields, where the raising and lowering operators become formally identical to creation and annihilation operators for particles. In this section, we review the quantum harmonic oscillator using the raising and lowering operator approach. Then later in this chapter, we quantize the electromagnetic field using the analogous creation and annihilation operators for photons.

Recall that classically, a harmonic oscillator of spring constant k with a mass m attached to it is described by the classical energy function:

$$E = \frac{1}{2}mv^2 + \frac{1}{2}kx^2 \tag{5.4}$$

$$= \frac{p^2}{2m} + \frac{1}{2}m\omega x^2, \tag{5.5}$$

where $p = mv$ is the momentum and $\omega = \sqrt{k/m}$ is the resonant angular frequency of the spring. Quantizing, this energy corresponds to the Hamiltonian operator:

$$\hat{H} = \frac{\hat{p}^2}{2m} + \frac{1}{2}m\omega\hat{x}^2, \tag{5.6}$$

where $\hat{p} = -i\hbar\nabla$. The solutions to the corresponding time-independent Schrödinger equation,

$$\hat{H}|n\rangle = E_n|n\rangle, \tag{5.7}$$

form a tower of states labeled by integer $n = 0, 1, 2, 3 \ldots$ with equally spaced energies

$$E_n = \left(n + \frac{1}{2}\right)\hbar\omega. \tag{5.8}$$

Note that even the lowest state (the **vacuum state** or **ground state**) has a nonzero energy $E_0 = \frac{1}{2}\hbar\omega$. The corresponding wavefunctions, $\psi(x) = \langle x|n\rangle$, are given by the standard expressions involving Hermite polynomials [1, 2]. The oscillator states are, of course, orthonormal:

$$\langle m|n\rangle = \delta_{mn}. \tag{5.9}$$

Instead of position and momentum operators, $\hat{x}$ and $\hat{p}$, consider describing the system in terms of an alternate set of operators. To do this, first rescale the position and momentum to get a pair of dimensionless operators:

$$\hat{X} = \sqrt{\frac{m\omega}{2\hbar}}\hat{x} \tag{5.10}$$

$$\hat{Y} = \sqrt{\frac{1}{2m\hbar\omega}}\hat{p}. \tag{5.11}$$

These are known as the **quadrature operators** of the oscillator. Then we define a new pair of (non-Hermitian) operators as linear combinations of the quadratures: the **lowering operator** is

$$\hat{a} = \hat{X} + i\hat{Y}, \tag{5.12}$$

and its Hermitian conjugate,

$$\hat{a}^\dagger = \hat{X} - i\hat{Y}, \tag{5.13}$$

is the **raising operator**. Inverting the previous relations, the quadrature operators can be written as

$$\hat{X} = \frac{1}{2}\left(\hat{a} + \hat{a}^\dagger\right) \tag{5.14}$$

$$\hat{Y} = \frac{i}{2}\left(\hat{a} - \hat{a}^\dagger\right). \tag{5.15}$$

From the commutation relations between $\hat{x}$ and $\hat{p}$, it is readily shown (see Problem 1) that

$$\left[\hat{a}, \hat{a}^{\dagger}\right] = 1. \tag{5.16}$$

From the raising and lowering operators, we construct one further operator, the **number operator**:

$$\hat{n} = \hat{a}^{\dagger}\hat{a}. \tag{5.17}$$

Then the action of the raising and lowering operators is to take us up and down the ladder of states:

$$\hat{a}^{\dagger}|n\rangle = \sqrt{n+1}|n+1\rangle \tag{5.18}$$

$$\hat{a}|n\rangle = \sqrt{n}|n-1\rangle, \tag{5.19}$$

while use of the last two equations leads to the conclusion that the states $|n\rangle$ are eigenstates of the number operator:

$$\hat{n}|n\rangle = n|n\rangle. \tag{5.20}$$

Restricting ourselves to one dimension for simplicity, consider now the operator $\left(\hat{n} + \frac{1}{2}\right)\hbar\omega$:

$$\left(\hat{n} + \frac{1}{2}\right)\hbar\omega = \left(\hat{a}^{\dagger}\hat{a} + \frac{1}{2}\right)\hbar\omega \tag{5.21}$$

$$= \left[\frac{1}{2m\hbar\omega}\left(m\omega\hat{x} - i\hat{p}\right)\left(m\omega\hat{x} + i\hat{p}\right) + \frac{1}{2}\right]\hbar\omega \tag{5.22}$$

$$= \left[\frac{1}{2m\hbar\omega}\left(m^2\omega^2\hat{x}^2 + \hat{p}^2 + im\omega\left[\hat{x}, \hat{p}\right]\right) + \frac{1}{2}\right]\hbar\omega \tag{5.23}$$

$$= \left[\frac{1}{2m\hbar\omega}\left(m^2\omega^2\hat{x}^2 + \hat{p}^2 - m\omega\hbar\right) + \frac{1}{2}\right]\hbar\omega \tag{5.24}$$

$$= \left[\frac{m\omega}{2\hbar}\hat{x}^2 + \frac{\hat{p}^2}{2m\hbar\omega} - \frac{1}{2} + \frac{1}{2}\right]\hbar\omega \tag{5.25}$$

$$= \frac{1}{2}m\omega^2\hat{x}^2 + \frac{\hat{p}^2}{2m} \tag{5.26}$$

$$= \hat{H}. \tag{5.27}$$

Thus, the harmonic oscillator Hamiltonian is given by

$$\hat{H} = \left(\hat{a}^\dagger \hat{a} + \frac{1}{2} \right) \hbar\omega = \left(\hat{n} + \frac{1}{2} \right) \hbar\omega, \qquad (5.28)$$

consistent with the energies given in Eq. 5.8. The Hamiltonian could also be written as $\hat{H} = \left(\hat{X}^2 + \hat{Y}^2 \right)$, where the quadrature operators obey $\left[\hat{X}, \hat{Y} \right] = \frac{i}{2}$. Note that the states $|n\rangle$ are eigenstates of both $\hat{H}$ and $\hat{n}$, consistent with the fact that $\hat{H}$ and $\hat{n}$ commute, $\left[\hat{H}, \hat{n} \right] = 0$.

The ground state is denoted as $|0\rangle$; this state is annihilated by the lowering operator: $\hat{a}|0\rangle = 0$. The various other energy eigenstates can be found by repeatedly acting on the ground state with the raising operator:

$$|n\rangle = \frac{1}{\sqrt{n!}} \left(\hat{a}^\dagger \right)^n |0\rangle. \qquad (5.29)$$

The $|n\rangle$ states are states with definite n value, in the sense that n has zero fluctuations, $\Delta n = 0$ (as shown in Problem 2); this is in contrast, for example, to the coherent and squeezed states to be defined in Chap. 9, which have nonzero number fluctuations.

5.3 The Quantized Electromagnetic Field

Now we consider the quantum version of an electromagnetic wave. First, recall that for a single-mode electromagnetic wave (with a single well-defined k, ω, and polarization), the electric field is of the form

$$E(r, t) = E_0 e^{i(k \cdot r - \omega t)} \hat{e}. \qquad (5.30)$$

Here $\hat{e}$ is a unit vector, giving the polarization direction of the field. In empty space, the field should be transverse to the direction of light propagation, which implies that the polarization vector should satisfy $k \cdot \hat{e} = 0$. Going to multimode systems, we can introduce a sum (or integral) over k vectors and polarization vectors, with $\omega(k)$ being defined for each term in the sum by the system's dispersion relation. For now, we stick to single-mode systems for simplicity.

We can promote the electric field from a classical vector field $E(r, t)$ to a quantum operator, $\hat{E}(r, t)$. We know that this operator represents a physical observable, so it should be Hermitian: $\hat{E}^\dagger = \hat{E}$. Since it should represent propagating waves, it should contain a dependence on position and time of the form $e^{\pm i(k \cdot r - \omega t)}$, for fixed k and ω. We also know that an electromagnetic field passing an atom can stimulate both absorption and emission of photons, so $E(r, t)$ should contain operators that both add and remove photons from the system.

The constraints listed above determine the field uniquely, aside from normalization. The quantized single-mode electric field operator takes the form

$$\hat{E}(r, t) = i \sqrt{\frac{\hbar\omega}{2V\epsilon_0}} \left[\hat{a}(k)e^{i(k\cdot r - \omega t)} - \hat{a}^\dagger(k)e^{-i(k\cdot r - \omega t)} \right] \hat{e}. \tag{5.31}$$

To fix the normalization, the field is assumed to be contained in a box of volume V; this arbitrary volume generally cancels out of physical results. $\hat{a}(k)$ and $\hat{a}^\dagger(k)$ are the photon annihilation and creation operators. The magnetic field can similarly be written as

$$\hat{B}(r, t) = \frac{i}{c} \sqrt{\frac{\hbar\omega}{2V\epsilon_0}} \left[\hat{a}(k)e^{i(k\cdot r - \omega t)} - \hat{a}^\dagger(k)e^{-i(k\cdot r - \omega t)} \right] \hat{k} \times \hat{e}. \tag{5.32}$$

(In these expressions, it is assumed that the polarization vector is real. However, in some cases, such as circular polarization, complex-valued polarization vectors may arise. In such a case, the second term in the expressions for E and B would contain the complex conjugate vector, e^*, instead of e.)

It can be verified that the field operators above satisfy Maxwell's equations. For example, recall that the wavevector operator is $\hat{k} = \frac{\hat{p}}{\hbar} = -i\nabla$, so that the transversality condition $\hat{k} \cdot \hat{e} = 0$ implies Gauss' law in empty space, $\nabla \cdot \hat{E} = 0$.

To see more clearly how the quantized form of the electric field arises, consider a classical electromagnetic field inside a cavity. The electric field of a single-mode standing wave along the z-axis can be written as

$$E(z, t) = E_0 \sin(kz) \sin(\omega t + \phi)$$

$$= E_0 \sin(kz) (\cos\phi \sin\omega t + \sin\phi \cos\omega t)$$

$$= \sqrt{\frac{\hbar\omega}{2\epsilon_0 V}} \sin kz \left[\left(\sqrt{\frac{2\epsilon_0 V}{\hbar\omega}} E_0 \sin\omega t \right) \cos\phi + \left(\sqrt{\frac{2\epsilon_0 V}{\hbar\omega}} E_0 \cos\omega t \right) \sin\phi \right]$$

$$\equiv \sqrt{\frac{\hbar\omega}{2\epsilon_0 V}} \sin kz \left[X \cos\phi + P \sin\phi \right], \tag{5.33}$$

where X and P are quadratures analogous to those of the harmonic oscillator. In the classical case, we see that the quadratures are given by

$$X(t) = \sqrt{\frac{2\epsilon_0 V}{\hbar\omega}} E_0 \sin\omega t \tag{5.34}$$

$$P(t) = \sqrt{\frac{2\epsilon_0 V}{\hbar\omega}} E_0 \cos\omega t. \tag{5.35}$$

In the quantum case, the classical quantities X and P become operators, $\hat{X}$ and $\hat{P}$. These are dimensionless, and they play a role analogous to position and momentum variables for the harmonic oscillator. The two quadrature operators oscillate 90° out of phase from each other and obey the commutation relation

$$\left[\hat{X}, \hat{P}\right] = \frac{i}{2}. \tag{5.36}$$

Consequently, they obey the uncertainty relation

$$\Delta X \Delta P \geq \frac{1}{4}. \tag{5.37}$$

From the quadrature operators, the creation and annihilation operators are then defined analogous to the way that raising and lowering operators were formed from position and momentum:

$$\hat{a}^\dagger = \hat{X} - i\hat{P}, \qquad \hat{a} = \hat{X} + i\hat{P}. \tag{5.38}$$

Inverting, we find that

$$\hat{X} = \frac{1}{2}\left(\hat{a} + \hat{a}^\dagger\right), \qquad \hat{P} = \frac{1}{2}\left(\hat{a} - \hat{a}^\dagger\right). \tag{5.39}$$

These new operators obey

$$\left[\hat{a}, \hat{a}^\dagger\right] = 1. \tag{5.40}$$

They are interpreted as creating and annihilating photons: if $|n\rangle$ represents a state with n photons, then

$$\hat{a}^\dagger |n\rangle = \sqrt{n+1}|n+1\rangle, \quad \text{and} \quad \hat{a}|n\rangle = \sqrt{n}|n-1\rangle. \tag{5.41}$$

In other words, $\hat{a}^\dagger$ adds a photon to the state of the field, while $\hat{a}$ removes one photon. All of the algebraic relations obeyed by the oscillator raising and lowering operators are now also obeyed by the photon creation and annihilation operators: formally, the electromagnetic field can be modeled as a collection of quantum oscillators. The photon number now plays the role previously played by the energy level label or excitation number, n.

Up until now, we have assumed a single-mode field. In general, the field will be multimode: it will have a range of k vectors and can be unpolarized. We therefore have to sum or integrate over all possible values of $\boldsymbol{k}$ and over all polarization directions. The frequency will depend on $k = |\boldsymbol{k}|$ and so will be labeled ω_k. There are two independent polarization states (either linear or circular), and we label them by $s = 0, 1$ and attach an s label on the polarization vectors. Each $\hat{a}$ or $\hat{a}^\dagger$ creates or destroys a photon with a particular value of k and s, so we must attach both k and s labels to them: $\hat{a}(k, s)$ or $\hat{a}^\dagger(k, s)$. So the full multimode field is now of the form

$$\hat{E}(r, t) = i \sum_{k,s} \sqrt{\frac{\hbar \omega_k}{2V\epsilon_0}} \left[\hat{a}(k, s)e^{i(k \cdot r - \omega_k t)} - \hat{a}^\dagger(k, s)e^{-i(k \cdot r - \omega_k t)} \right] \hat{e}_s. \quad (5.42)$$

For the case of continuous momenta, the sum over discrete k becomes an integral:

$$\sum_k \rightarrow \int \frac{d^3 k}{(2\pi)^{3/2}}. \quad (5.43)$$

The field is often split into a positive-frequency or annihilation part $\hat{E}^{(+)}$ and a negative-frequency or creation part $\hat{E}^{(-)}$,

$$\hat{E}(r, t) = \hat{E}^{(+)}(r, t) + \hat{E}^{(-)}(r, t), \quad (5.44)$$

where

$$\hat{E}^{(+)}(r, t) = i \sum_{k,s} \sqrt{\frac{\hbar \omega_k}{2V\epsilon_0}} \hat{a}(k, s)e^{i(k \cdot r - \omega t)} \hat{e}_s \quad (5.45)$$

$$\hat{E}^{(-)}(r, t) = -i \sum_{k,s} \sqrt{\frac{\hbar \omega_k}{2V\epsilon_0}} \hat{a}^\dagger(k, s)e^{-i(k \cdot r - \omega_k t)} \hat{e}_s. \quad (5.46)$$

As implied, $\hat{E}^{(-)}$ and $\hat{E}^{(+)}$ are the parts of the field that, respectively, add or remove a photon from the state. These two parts of the field often enter separately into different situations. For example, detectors implement the positive-frequency part: the detector removes a photon from the field in order to produce a measurable photocurrent, and so the detector is effectively applying the operator $\hat{E}^{(+)}$ to the optical state.

Note the distinction between $\hat{a}(k)$ and $\hat{a}^\dagger(k)$ versus $\hat{E}^{(\pm)}(r)$: $\hat{a}(k)$ and $\hat{a}^\dagger(k)$ destroy or create a photon of fixed momentum $p = \hbar k$, while $\hat{E}^{(\pm)}(r)$ create a photon of unknown momentum at a specific location r.

5.4   Creation and Annihilation Operators for Fermions

Just as creation and annihilation operators $\hat{a}$ and $\hat{a}^\dagger$ are defined for photons, we can define analogous operators $\hat{c}$ and $\hat{c}^\dagger$ for fermions. The difference is that instead of obeying commutation relations, these operators obey *anti-commutation* relations. The **anti-commutator** of any two operators $\hat{A}$ and $\hat{B}$ is defined by

$$\left\{\hat{A}, \hat{B}\right\} = \hat{A}\hat{B} + \hat{B}\hat{A}. \tag{5.47}$$

The fermion creation and annihilation operators then obey the relations:

$$\left\{\hat{c}, \hat{c}^{\dagger}\right\} = 1, \tag{5.48}$$

$$\left\{\hat{c}, \hat{c}\right\} = \left\{\hat{c}^{\dagger}, \hat{c}^{\dagger}\right\} = 0. \tag{5.49}$$

The last line implies that using a creation operator twice will always annihilate a state:

$$\left(\hat{c}^{\dagger}\right)^{2} |\psi\rangle = 0, \tag{5.50}$$

for any $|\psi\rangle$. In particular, applying this fact to the vacuum state means that two fermions cannot be in the same state. The use of anti-commutators instead of commutators therefore encodes the Pauli exclusion principle into the formalism.

Another consequence of the anticommutativity of the operators is that multi-fermion states must be antisymmetric under interchange of identical particles. For example, suppose two identical fermions a and b can each be in one of two locations, labeled 1 and 2. Since there is no way of distinguishing which of the two particles is at which location, we must superpose both possibilities. But the two superposed states must be added with a minus sign:

$$|\Psi\rangle = \frac{1}{\sqrt{2}}\left(|\psi_a(\mathbf{r}_1)\rangle|\psi_b(\mathbf{r}_2)\rangle - |\psi_a(\mathbf{r}_2)\rangle|\psi_b(\mathbf{r}_1)\rangle\right). \tag{5.51}$$

The minus sign makes the wavefunction antisymmetric under $1 \leftrightarrow 2$ (or equivalently under $a \leftrightarrow b$) so that, in accordance with the exclusion principle, the wavefunction vanishes as the two particles approach the same position $\mathbf{r}_1 \to \mathbf{r}_2$.

More generally, for n fermions, the wavefunction has to be antisymmetric under interchange of any two of them and so takes the form of a Slater determinant [2].

5.5   Beam Splitters and Interferometers with Quantum Fields

The action of a beam splitter on a classical field was discussed in Chap. 2. We can now reconsider the action of a beam splitter on quantized light fields. It turns out that the action on a single photon is described by the same matrix that gives the action on a classical field. If a photon is sent into one port of a beam splitter, the output is a photon in a *superposition state*: it is in a superposition of being transmitted and of being reflected.

Let $\hat{a}_0, \hat{a}_1, \hat{a}_2, \hat{a}_3$ be annihilation operators for photons entering or leaving at the four input/output ports of the beam splitter, as in Fig. 2.8. Let 0 and 1 be used as input ports and 2 and 3 for output. Then we can arrange the annihilation operators into two-component operator-valued column vectors representing annihilation of the input and output:

$$\hat{A}_{in} = \begin{pmatrix} \hat{a}_0 \\ \hat{a}_1 \end{pmatrix}, \qquad \hat{A}_{out} = \begin{pmatrix} \hat{a}_2 \\ \hat{a}_3 \end{pmatrix}. \tag{5.52}$$

Then the input and output operators are related by a matrix U_{BS}:

$$\hat{A}_{out} = U_{BS}\hat{A}_{in}. \tag{5.53}$$

This unitary matrix is exactly of the same general form as the classical version in Eq. 2.69:

$$U_{BS} = e^{i\phi_0} \begin{pmatrix} \sin\theta \ e^{i\phi_1} & \cos\theta \ e^{-i\phi_2} \\ \cos\theta \ e^{i\phi_2} & -\sin\theta \ e^{-i\phi_1} \end{pmatrix}. \tag{5.54}$$

For a 50–50 beam splitter, $\theta = \frac{\pi}{4}$, and henceforth we will assume that the phases are given by $\phi_0 = -\phi_1 = \frac{\pi}{2}$ and $\phi_2 = 0$, which leads to the symmetric beam splitter matrix

$$U_{BS} = \frac{1}{\sqrt{2}} \begin{pmatrix} 1 & i \\ i & 1 \end{pmatrix}. \tag{5.55}$$

Then, the input and output annihilation operators are related by

$$\hat{a}_2 = \frac{1}{\sqrt{2}}(\hat{a}_0 + i\hat{a}_1) \tag{5.56}$$

$$\hat{a}_3 = \frac{1}{\sqrt{2}}(i\hat{a}_0 + \hat{a}_1). \tag{5.57}$$

Inverting, we find

$$\hat{a}_0 = \frac{1}{\sqrt{2}}(\hat{a}_2 - i\hat{a}_3) \tag{5.58}$$

$$\hat{a}_1 = \frac{1}{\sqrt{2}}(\hat{a}_3 - i\hat{a}_2). \tag{5.59}$$

Similarly, the creation operators are related by

$$\hat{a}_2^\dagger = \frac{1}{\sqrt{2}} \left(\hat{a}_0^\dagger - i\hat{a}_1^\dagger \right) \tag{5.60}$$

$$\hat{a}_3^\dagger = \frac{1}{\sqrt{2}} \left(-i\hat{a}_0^\dagger + \hat{a}_1^\dagger \right), \tag{5.61}$$

or

$$\hat{a}_0^\dagger = \frac{1}{\sqrt{2}} \left(\hat{a}_2^\dagger + i\hat{a}_3^\dagger \right) \tag{5.62}$$

$$\hat{a}_1^\dagger = \frac{1}{\sqrt{2}} \left(\hat{a}_3^\dagger + i\hat{a}_2^\dagger \right). \tag{5.63}$$

To see the action of the beam splitter on quantum states, first we define some notation to describe the states. A state with n photons in port j will be written as $|n\rangle_j$, where it is understood that the photon is entering the beam splitter as input for $j = 0$ or 1 and is exiting for $j = 2$ or 3. States with photons at two ports simultaneously will be written $|m\rangle_j |n\rangle_k$ for m photons at port j and n photons at port k. Here we assume that the photons are identical and suppress mention of properties such as frequency and polarization that will be irrelevant for the interference effects below.

Example 5.5.1 Send two photons into port 0. The input state is then given by

$$|\psi\rangle_{in} = |2\rangle_0 |0\rangle_1 = \frac{1}{\sqrt{2}} (\hat{a}_0^\dagger)^2 |0\rangle_0 |0\rangle_1. \tag{5.64}$$

The beam splitter then acts on the input creation operator according to

$$(\hat{a}_0^\dagger)^2 \rightarrow \frac{1}{2} \left(\hat{a}_2^\dagger + i\hat{a}_3^\dagger \right)^2 \tag{5.65}$$

$$= \frac{1}{2} \left[(\hat{a}_2^\dagger)^2 + 2i\hat{a}_2^\dagger \hat{a}_3^\dagger - (\hat{a}_3)^2 \right], \tag{5.66}$$

where in the last line, the fact was used that operators at different locations commute, $\left[\hat{a}_2^\dagger, \hat{a}_3^\dagger \right] = 0$.

The output state is then

$$|\psi\rangle_{out} = \frac{1}{2\sqrt{2}} \left[(\hat{a}_2^\dagger)^2 + 2i\hat{a}_2^\dagger \hat{a}_3^\dagger - (\hat{a}_3)^2 \right] |0\rangle_2 |0\rangle_3$$

$$= \frac{1}{2\sqrt{2}} \left[\left(\left(\hat{a}_2^\dagger \right)^2 |0\rangle_2 \right) |0\rangle_3 + 2i \left(\hat{a}_2^\dagger |0\rangle_2 \right) \left(\hat{a}_3^\dagger |0\rangle_3 \right) - |0\rangle_2 \left((\hat{a}_3)^2 |0\rangle_3 \right) \right]$$

$$= \frac{1}{2\sqrt{2}} \left[\sqrt{2} |2\rangle_2 |0\rangle_3 + 2i |1\rangle_2 |1\rangle_3 - \sqrt{2} |0\rangle_2 |2\rangle_3 \right]$$

Fig. 5.2 When two identical photons are sent into port 0, there are three possible outputs: either both exit at port 2 (with probability $\frac{1}{4}$), one exits at each of ports 2 and 3 (with probability $\frac{1}{2}$), or both exit at port 3 (probability $\frac{1}{4}$)

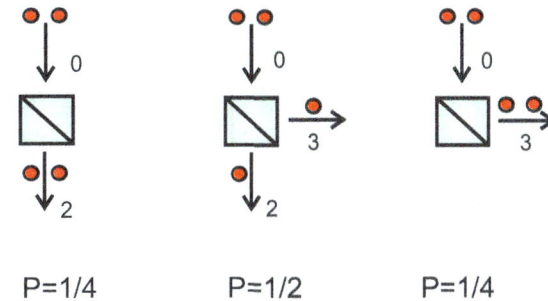

$$= \frac{1}{2}|2\rangle_2|0\rangle_3 + \frac{i}{\sqrt{2}}|1\rangle_2|1\rangle_3 - \frac{1}{2}|0\rangle_2 a|2\rangle_3, \tag{5.67}$$

where the fact that $(a^\dagger)^n|0\rangle = \sqrt{n!}|n\rangle$ was used.

So there are three possible outcomes, both photons at port 2, one at each output port, or both photons at port 3, with respective probabilities $|\frac{1}{2}|^2 = \frac{1}{4}$, $|\frac{i}{\sqrt{2}}|^2 = \frac{1}{2}$, and $|-\frac{1}{2}|^2 = \frac{1}{4}$ (Fig. 5.2).   ∎

Example 5.5.2 Instead of sending two photons into the beam splitter via the same port, consider sending two identical photons into different ports. To be specific, suppose one photon enters port 0 and one enters 1:

$$|\psi\rangle_{in} = |1\rangle_0|1\rangle_1 = \hat{a}_0^\dagger \hat{a}_1^\dagger|0\rangle. \tag{5.68}$$

After the beam splitter, this becomes

$$|\psi\rangle_{out} = \frac{1}{\sqrt{2}}\left(\hat{a}_2^\dagger + i\hat{a}_3^\dagger\right) \cdot \frac{1}{\sqrt{2}}\left(\hat{a}_3^\dagger + i\hat{a}_2^\dagger\right)|0\rangle \tag{5.69}$$

$$= \left(\hat{a}_2^\dagger \hat{a}_3^\dagger + i\left(\hat{a}_2^\dagger\right)^2 + i\left(\hat{a}_3^\dagger\right)^2 - \hat{a}_3^\dagger \hat{a}_2^\dagger\right)|0\rangle \tag{5.70}$$

$$= \frac{i}{2}\left(\left(\hat{a}_2^\dagger\right)^2 + \left(\hat{a}_3^\dagger\right)^2\right)|0\rangle \tag{5.71}$$

$$= \frac{i}{2}\left(|2\rangle_2 + |2\rangle_3\right). \tag{5.72}$$

This is a remarkable result: even though the photons don't interact and even though each of them are choosing an exit port at random, they are somehow still always choosing the **same** exit port each time! Half of the time both photons exit port 2 and half the time they both exit at port 3. They *never* exit at different ports, so there are no coincidence counts. This result is known as the **Hong-Ou-Mandel (HOM) effect**. It is a result of destructive interference: the first and last terms in Eq. 5.70

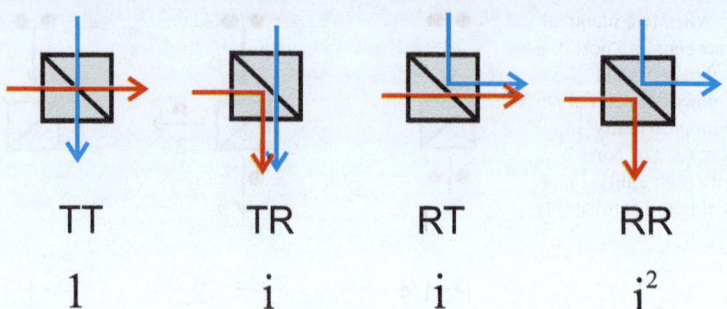

TT	TR	RT	RR
1	i	i	i^2

Fig. 5.3 The HOM effect. Two photons enter a beam splitter. Each of them has an equal amplitude of reflecting and transmitting. Each reflection introduces a phase shift of $e^{i\pi/2} = i$. The amplitudes for both to reflect and that for both to transmit therefore come in with opposite signs, so that the coincidence rate between the two output ports vanishes

give indistinguishable outcomes but enter with opposite signs, and so they cancel, leaving only the terms where the photons are clustered together at the same output port. This is illustrated in Fig. 5.3.

The HOM effect is inherently a two-photon effect: one full two-photon amplitude interferes with another two-photon amplitude. As a result, it is an inherently quantum effect, which cannot be mimicked by classical (single-photon) interference. Note also that although a product state was input to the beam splitter, the output state is entangled. We will return to such two-photon interference effects in later chapters.

5.6   Radiative Transitions in Atoms

Recall that electrons in an atom have a discrete set of energy levels, a ground state E_1, and a set of excited states E_2, E_3, E_4,...Doppler shifts, collisions between atoms, and other processes can broaden these discrete levels into bands of allowed energies. In addition, there is inherent quantum uncertainty, which sets a minimum limit on the widths of these levels. Here, we will ignore these effects and treat each level as precise and well-defined.

Imagine a gas with many atoms of the same type, with the same energy level structure. Consider just the two lowest energy levels, E_1 and E_2. An electron in the ground state can absorb a passing photon and use its energy to jump to the first excited state. This can happen if the frequency ω and energy $E_{ph} = \hbar\omega$ of the photon match the energy difference between the levels:

$$\hbar\omega = E_2 - E_1. \tag{5.73}$$

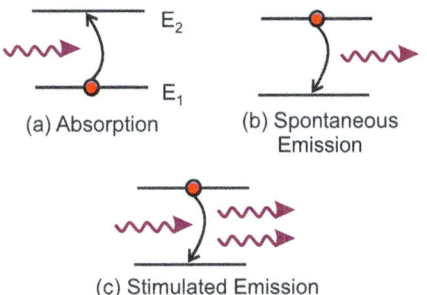

(a) Absorption   (b) Spontaneous
                            Emission

(c) Stimulated Emission

Fig. 5.4 The basic interactions between photons and atoms: (**a**) absorption of a photon, leading to an electron jumping to an excited state; (**b**) spontaneous emission, where an unperturbed excited-state electron drops back to the ground state, emitting a photon in the process; and (**c**) stimulated emission, where the decay to the ground state is caused by perturbation of the atom by a passing photon of the correct frequency, $\omega = (E_1 - E_0)/\hbar$

This photon absorption process is shown in Fig. 5.4a. The absorption process is governed by a term in the Hamiltonian of the form $\kappa \hat{a}\hat{c}_2^\dagger \hat{c}_1$, where κ is some collection of constants. The $\hat{a}$ term annihilates the photon, while the fermion operators $\hat{c}_2^\dagger$ and $\hat{c}_1$ remove an electron from the ground state and create a new one in the excited state.

The inverse process, spontaneous emission, also occurs (Fig. 5.4b), with the electron decaying after some time back into the ground state, emitting a photon in the process to conserve overall energy. The term in the Hamiltonian corresponding to spontaneous emission should be of the form $\kappa^*\hat{a}^\dagger \hat{c}_1^\dagger \hat{c}_2$; note that this is the adjoint of the absorption term, so that when the two terms are added together, they give a Hermitian Hamiltonian:

$$\hat{H} = \kappa \hat{a}\hat{c}_2^\dagger \hat{c}_1 + \kappa^*\hat{a}^\dagger \hat{c}_1^\dagger \hat{c}_2 \tag{5.74}$$

The average time τ before emission occurs is called the **radiative lifetime** of the state.

To look more quantitatively at the absorption and emission rates, let N_1 and N_2 be the numbers of electrons occupying the two energy levels. Then the rate at which emission occurs from the excited state should be proportional to the number of electrons already in that state, so that

$$\frac{dN_2}{dt} = -A_{21}N_2, \tag{5.75}$$

where A_{21} is some constant with units of Hz. The right side is negative because emission *decreases* N_2. The latter equation is easily solved to find the solution

$$N_2(t) = N_2(0)e^{-t/\tau}, \tag{5.76}$$

where the radiative lifetime is given by

$$\tau = 1/A_{21}. \tag{5.77}$$

Absorption can be treated in a similar manner, except that the transition rate should be proportional to the number of photons present, which have frequencies sufficiently close to the transition frequency. The number of such photons should in turn be proportional to the **spectral energy density** of the electromagnetic field, $u(\omega)$, which is defined to be the energy per volume per frequency of the field: $u(\omega) = \frac{du}{d\omega}$. Here u is the energy per volume, so that $u = \frac{U}{V} = \int_0^\infty u(\omega)d\omega$ for a volume V containing total potential energy U. Taking into account the spectral density, then the absorption rate should be of the form

$$\frac{dN_1}{dt} = -B_{12}N_1 u(\omega), \tag{5.78}$$

where $u(\omega)$ has units of Js/m^3 and B_{12} is a constant with units of m^3/Js^2. Assuming that $u(\omega)$ stays roughly constant over time, this also can be integrated to give

$$N_1(t) = N_1(0)e^{-B_{21}u(\omega)t}. \tag{5.79}$$

The existence of absorption and of spontaneous emission should seem fairly obvious, but it was Einstein who first realized that there should be a third process: **stimulated emission**. Just as jiggling an open carton of eggs can make the eggs hop out of the carton and fall to the floor, the oscillating field of a passing photon can cause a periodic perturbation to an electron in the excited state, leading to the electron hopping out of that state and dropping back to the ground state. The presence of one photon of the correct transition frequency therefore stimulates the emission of a second photon of the same frequency (Fig. 5.4c). This process also should be proportional to the number of photons present, or equivalently to the spectral energy density, as well as to the number of excited electrons present. So the stimulated emission transition rate should be of the form

$$\frac{dN_2}{dt} = -B_{21}N_2 u(\omega). \tag{5.80}$$

Combining spontaneous and stimulated emissions, the total rate at which electrons leave the excited state is

$$\frac{dN_2}{dt} = -A_{21}N_2 - B_{21}N_2 u(\omega). \tag{5.81}$$

Collectively, the three constants A_{21}, B_{12}, and B_{21} are called the **Einstein A and B coefficients**. Note that as long as only two levels are accessible, an electron leaving one level must enter the other level, so that the absorption and emission rates must be related by

$$\frac{dN_2}{dt} = -\frac{dN_1}{dt}. \tag{5.82}$$

Therefore, we find that

$$B_{12}N_1 u(\omega) = A_{21}N_2 + B_{21}N_2 u(\omega). \tag{5.83}$$

The three A and B coefficients are in fact not independent of each other. To show this, we need two more ingredients. For the first, consider N atoms in thermal equilibrium at temperature T. If only two levels have nonzero occupancy, then $N_1 + N_2 = N$. Further, let g_1 and g_2 be the degeneracies of each level, or in other words, the number of distinct quantum states (e.g., spin-up and spin-down) of each energy. Then the Boltzmann distribution implies that

$$\frac{N_2}{N_1} = \frac{g_2}{g_1} e^{-\hbar\omega/kT}, \tag{5.84}$$

where k is Boltzmann's constant.

The final fact that we need is the Planck radiation law, which gives us an expression for the spectral density of blackbody radiation:

$$u(\omega) = \frac{\hbar\omega^3}{\pi^2 c^3} \cdot \frac{1}{e^{\hbar\omega/hT} - 1}. \tag{5.85}$$

Rearranging Eq. 5.83 and then using 5.84, we find

$$B_{12}u(\omega) = \frac{N_2}{N_1} (A_{21} + B_{21}u(\omega)) \tag{5.86}$$

$$= \frac{g_2}{g_1} e^{-\hbar\omega/kT} (A_{21} + B_{21}u(\omega)). \tag{5.87}$$

Solving for $u(\omega)$,

$$u(\omega) = \frac{g_2 A_{21}}{g_1 B_{12} e^{\hbar\omega/kT} - g_2 B_{21}}. \tag{5.88}$$

Consistency of the last equation with Eq. 5.85 for all ω and T forces the relations:

$$g_1 B_{12} = g_2 B_{21} \quad \text{and} \quad A_{21} = \frac{\hbar\omega^3}{\pi^2 c^3} B_{21}. \tag{5.89}$$

This provides the relations between the three A and B coefficients. Although we won't derive it here (see, e.g., [3]), it is possible to rewrite these coefficients in terms of a more directly measurable quantity associated with the atom, namely, the matrix element μ of the electric dipole moment between the ground state and the excited state:

$$\mu = -e\langle\psi_2|r|\psi_1\rangle. \tag{5.90}$$

The result is that

$$B_{12} = \frac{\pi}{3\epsilon_0\hbar^2}|\mu|^2 \tag{5.91}$$

$$A_{21} = \frac{\omega^2}{3\pi\epsilon_0\hbar c^3}|\mu|^2. \tag{5.92}$$

The above expressions relating the numbers of electrons occupying different energy levels assumed thermal equilibrium. In equilibrium, the number of electrons in the ground state should be much larger at room temperature than the number in the excited state: $n_1 >> n_2$. However, in the next section, we will see that lasers operate by creating a highly *nonequilibrium* situation: energy is pumped into the system from the outside in order to force $n_2 >> n_1$, a situation known as population inversion.

When photons are emitted or absorbed in an electronic transition, the resulting spectral lines are not infinitely narrow, but span a finite range of frequencies, $\Delta\omega$. Looking at a given emission line, for example, we can see that the maximum brightness of the line is at the resonant frequency, $\omega_0 = \frac{E_1-E_0}{\hbar}$, and the intensity decreases as ω moves away from ω_0. So we define a **spectral line-shape function**, $g(\omega)$, describing how the emission varies with frequency. This function should be peaked at ω_0 and normalized so that $\int_0^\infty g(\omega)\,d\omega = 1$. $g(\omega)$ will have units of inverse angular frequency, $\frac{s}{rad}$. The finite width of this function (the spectral linewidth) is caused by a number of different mechanisms, but the mechanisms can be divided into two categories: homogeneous or inhomogeneous.

Homogeneous line broadening will affect all atoms in the same way. Homogeneous broadening tends to lead to **Lorentzian** line-shape functions or **Cauchy distribution**. Lorentzian functions take the form

$$g(\omega) = \frac{\Gamma}{2\pi}\frac{1}{(\omega - \omega_0)^2 + \left(\frac{\Gamma}{2}\right)^2}, \tag{5.93}$$

where Γ is the full-width at half-maximum (FWHM) of the distribution. Lorentzians decay more slowly for the same Γ than Gaussians (Fig. 5.5).

An example of a Lorentzian or homogeneous broadening mechanism is **lifetime broadening**, also known as **natural or radiative broadening**. This is the line broadening that arises because of the Heisenberg uncertainty in the photon energy or frequency. There is some uncertainty Δt in the emission time of the photon, so the uncertainty relation $\Delta E\Delta t \geq \hbar$ leads to a frequency uncertainty $\Delta\omega = \frac{1}{\Delta t}$. The time uncertainty is often written as $\Delta t = \tau$ and known as the **radiative lifetime**, so

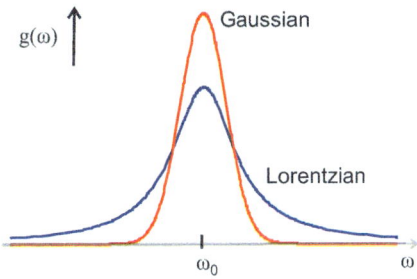

Fig. 5.5 Lorentzian (blue) and Gaussian (red) line-shape functions of the same FWHM

that $\Gamma = \Delta\omega = \frac{1}{\tau}$. For the Lorentzian case, the Fourier transform of $g(\omega)$, which gives the distribution of photon emission times, is of the form

$$g(t) = e^{-i\omega_0 t} e^{-\frac{|t|}{\tau}}. \tag{5.94}$$

Another example of homogeneous broadening mechanism is **collisional** or **pressure broadening**. This also gives Lorentzian broadening with characteristic decay time and width:

$$\tau = \Gamma^{-1} = \frac{1}{\sigma_c P} \sqrt{\frac{\pi m k T}{8}}, \tag{5.95}$$

where P and T are the pressure and temperature (in Pascals and Kelvins), m is the mass of the atom, k is Boltzmann's constant, and σ_c is the collision cross-section between the atoms. (**Cross-sections** are measures of interatomic interaction strengths. They are defined to have units of area; essentially they measure the effective area each atom offers for other atoms to hit and so are proportional to the collision rate.)

The second category of broadening mechanisms is **inhomogeneous broadening**, which is a broadening mechanism that affects different atoms differently. Inhomogeneous broadening leads to a Gaussian line-shape functions. Gaussian or normal line-shapes are given by functions of the form

$$g(\omega) = \frac{1}{\sqrt{2\pi}\,\sigma} e^{-\left(\frac{\omega - \omega_0}{\sigma}\right)^2}. \tag{5.96}$$

σ (not to be confused with the cross-section σ_c) is the standard deviation or the full width at $\frac{1}{e}$ of the maximum. (For a Gaussian, the FWHM and the standard deviation are related by $\Gamma = 2\sqrt{2\ln 2}\,\sigma \approx 2.355\,\sigma$.)

An example of an *inhomogeneous* broadening mechanism is Doppler broadening, in which spectral lines are Doppler shifted by an amount dependent on the atoms' speeds. Since the atomic speeds are randomly distributed, each atom at a given moment will have a different spectral broadening than other atoms, and all of these different widths have to be averaged to give the overall linewidth and spectral line-shape function. The result for Doppler broadening is a standard deviation

$$\sigma = \sqrt{\frac{kT\omega_0}{mc^2}}. \tag{5.97}$$

This corresponds to a FWHM of

$$\Gamma = 2\omega_0\sqrt{\frac{2\ln 2\, kT}{mc^2}}. \tag{5.98}$$

The Fourier transform of the line-shape function is again a Gaussian:

$$g(t) = e^{-i\omega_0 t} e^{-\frac{\pi}{2}\left(\frac{t}{\tau}\right)^2}, \tag{5.99}$$

where the characteristic time for this process is again $\tau = \frac{1}{\Gamma}$.

Note that multiple broadening mechanisms are usually present at the same time. Decay rates add, so that the total characteristic decay time will be found by adding the inverse times of each mechanism separately:

$$\frac{1}{\tau_{tot}} = \frac{1}{\tau_{nat}} + \frac{1}{\tau_{coll}} + \frac{1}{\tau_{dopp}} + \dots \tag{5.100}$$

5.7 Lasers

Lasers have become ubiquitous in everyday life, being used in devices ranging from fiber optics communication systems and welding equipment to grocery store checkout scanners and DVD players. Their uses in science and medicine cover an even broader range of applications. Just to give a few examples, they are used to kill tumors, to make ultraprecise distance measurements, to provide real-time monitoring of rapid chemical reactions, to serve as reference beams for astronomical adaptive optics, and to carry out spectroscopic measurements. Further, their high coherence makes them ideal light sources for interferometry, for quantum optics experiments, and for generating entangled optical states in nonlinear materials.

In this section, we give a brief overview of how lasers work, and in the next section, we give a description of the beam that they generate. Much more extensive coverage of lasers can be found in [4–6].

5.7.1 The Physics of Lasing

The word **laser** is an acronym for *Light Amplification by Stimulated Emission of Radiation*. The action of the laser device when generating a beam is referred to as **lasing**. Building on the theoretical framework laid down around 1916 by Einstein on stimulated emission (see the previous section), masers (the microwave analogs of lasers) were first developed by researchers in the USA and the USSR in the late 1950s, followed by optical lasers in the early 1960s. The chief characteristics of a laser that distinguish it from natural light sources are as follows:

- High coherence: In a thermal light source like the sun, the atoms all emit photons independently of each other, at random times. As a result, the phases of the photons at a given time are unrelated to each other, leading to low coherence. But in the laser, the system is arranged so that many identical transitions occur at the same time, guaranteeing that the photons are all in phase with each other, producing output of much higher coherence. Coherence lengths vary widely depending on the specific type and size of the laser, but they can be anywhere from millimeters to meters, compared to typical coherence lengths of 10^{-6} or less in natural, thermal light sources. The high coherence of lasers means that interference effects can be created with very high visibility.
- The beams are highly collimated, meaning that the light forms a narrow beam. The light rays have only a very small range of propagation directions within this beam. This is compared to natural light sources, where light tends to be emitted in all directions. As a result of this high degree of collimation, laser beams attenuate with distance much more slowly than natural light sources.
- Brightness: Since the light is collimated into a beam of small cross-sectional area, the intensity can be arranged to be very high. Laser beams therefore tend to be very bright.
- Lasers can be made to be highly monochromatic, with a frequency spread that is much smaller than the central frequency of the beam, $\Delta\omega << \omega_0$. (Note that not all lasers have to be quasi-monochromatic: in order to achieve ultrashort pulses, some lasers are engineered to have large $\Delta\omega$.) The spread of frequency is related to coherence: the larger $\Delta\omega$, the lower the coherence.

The photons that are emitted by the laser have one other unusual aspect: large numbers of them tend to cluster into the same quantum state. This is possible due to the bosonic nature of photons: once one photon is in a given state, this increases the probability that others will join the same state. As was seen in the last section, this is the essential idea behind stimulated emission.

The basic structures common to all lasers are shown schematically in Fig. 5.6. A resonant cavity of length L is filled with a material known as a **gain medium**. Energy is input to the system (a process known as **pumping**), in order to excite large numbers of electrons into excited states of the atoms in the gain medium. Suppose that the ground and excited states of the atom are E_1 and E_2. Then any photon passing through the system with frequency close to the resonant frequency

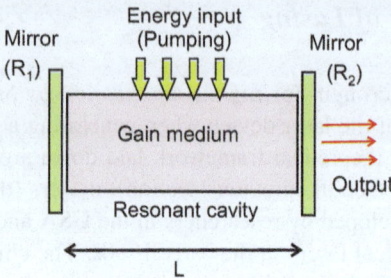

Fig. 5.6 Schematic depiction of a laser. The resonant cavity is filled with a gain medium. Pumping in energy creates a population inversion in the gain medium, so that cascades of stimulated emission occur. The mirrors allow multiple passes of photons through the cavity to allow multiple opportunities to stimulate additional photon emissions. At least one of the mirrors has to have a reflectance $R < 1$ in order for light to escape and form an output beam

of the transition, $\omega \approx (E_2 - E_1)/\hbar$, can stimulate the atoms of the medium to emit more photons. The ends of the cavity are bounded by a pair of mirrors, so each photon may cross the cavity many times, allowing it many opportunities to stimulate the emission of additional photons. Each of the new photons emitted can in turn stimulate emission of additional photons, leading to a rapid cascade of emissions. The photons in the cascade will all be in the same quantum state, having been emitted by identical transitions in identical atoms. At least one of the two mirrors bounding the cavity has to be partially transmitting, so that the photons will eventually escape the cavity, usually after many reflections. Those escaping photons make up the resulting laser beam.

The pumping can come in many forms. For example, in semiconductor lasers and laser diodes, electrical currents are often used to excite the electrons, while in some lasers, the pumping may be the result of exothermic chemical reactions. Another common approach is optical pumping, which uses an external light source (a flash lamp, an arc lamp, or another laser) to provide the necessary energy input. In thermal equilibrium at a fixed temperature, the distribution of electrons between the energy levels is determined by the Boltzmann distribution, which guarantees that the majority of the electrons are in the ground state at normal temperatures. The pumping, however, creates a nonequilibrium situation in which very high numbers of electrons are in the excited state; this situation is called **population inversion**.

The gain medium also comes in many forms. Solids, liquids, and gases are all common gain media. Common gaseous media include a helium-neon (HeNe) mixture, argon, krypton, or nitrogen. Dye lasers use liquid solutions containing organic dyes as the gain material. Solid gain media include crystals doped with rare earths (neodymium, ytterbium, or erbium) or semiconductors (such as gallium arsenide (GaAs), indium gallium arsenide (InGaAs), or gallium nitride (GaN)). Glass such as silicate (silicon dioxide, SiO_2) are also common. Note that SiO_2 is the material that optical fibers are made out of; **fiber lasers**, in which a segment of an optical fiber makes up the resonant cavity and serves as the gain medium, are common.

Box 5.1 From edible lasers to edible electronics

Theodor Hänsch has quoted laser pioneer Arthur Schawlow as saying that "anything will lase if you hit it hard enough" [7]. To prove it, Hänsch and Schawlow tried to make a laser with Jello from the local grocery store as the gain medium. Although their initial experiments failed, they soon achieved an edible laser by using their own recipe of gelatin and dye. They found that they could increase the efficiency of the laser by adding a dash of kitchen detergent to the recipe, although they admit it ruined the edibility of the final device.

In a similar vein, researchers at Rochester created a gin- and tonic-based laser in 1969. The optical power of the laser was very low, and as the researchers admitted: "There are better uses for ethyl alcohol." Nevertheless, their success has inspired many physics students to try their own experiments ever since.

Edible optics is not restricted to lasers. It has been found that optical sensors can be made from sugar, and if the sugar is drawn into long fibers, it can work as a serviceable (but high loss) optical fiber [8]. Sugars and salts can also be used to make simple lenses.

Optics isn't the only area where technology can be made edible. On the electronics front, sodium potassium tartrate (Rochelle salt) has well-known piezoelectric properties and was once used to make microphones and speakers, particularly in old-fashioned gramophones. More recently, researchers at Rice University [9] have found that they can use laser pulses to convert the carbon on the surface of food into graphene. As a result, it is possible to etch edible electronic circuits onto the surfaces of cookies and other foods. These circuits can be used as sensors or to receive or transmit information. The researchers envision the possibility of one day making edible sensors designed to emit a signal if *E. coli* is present or to imprint edible RFID tags onto food to track its source, expiration date, and other information.

Consider light propagating through the gain medium. Normally, as light moves through a material, it is gradually absorbed at a rate proportional to the amount of light present:

$$\frac{dI(z)}{dz} = \alpha I(z). \tag{5.101}$$

This can easily be solved to show that the intensity decays exponentially with distance, a result known as **Beer's law**:

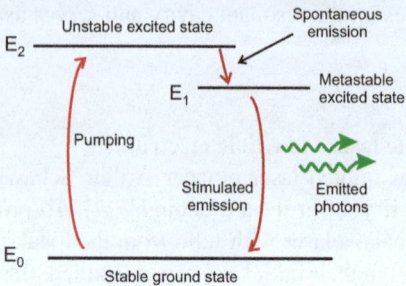

Fig. 5.7 A minimum of three energy levels need to be involved for successful laser generation. Electrons are pumped from E_0 to E_2 and then rapidly decay to E_1. The state of energy E_1 is long-lived enough for the electrons to remain there until emission is stimulated by a passing photon of frequency $\frac{E_1 - E_0}{\hbar}$

$$I(z) = I_0 e^{-\alpha z}, \tag{5.102}$$

where z is the propagation distance and α is the absorption coefficient of the material (at the given frequency of the light). However, in the gain material, it is arranged to have new photons appearing through stimulated emission, which can cause a competing *gain* in intensity. As a result, the intensity becomes

$$I(z) = I_0 e^{(\gamma - \alpha)z}, \tag{5.103}$$

where γ is the **gain coefficient** of the material.

If only two energy levels of the gain medium are involved in the electron transitions, then absorption and emission processes will compete with each other and prevent lasing from occurring. In order to have a practical laser, at least three energy levels (and more commonly four) in the gain medium have to be involved. The simplest three-level scheme is shown in Fig. 5.7. Pumping excites electrons from the ground state energy E_0 to the second excited state, of energy E_2. The state of energy E_2 is unstable, so that the electrons rapidly decay into a slightly lower-energy excited state, of energy E_1. This state is metastable: in the absence of stimulated emission, it has a long lifetime, decaying only very slowly back to the ground state. So population inversion is achieved from E_0 to E_1, and stimulated emission then causes a cascade of laser photons of frequency $\omega = \frac{E_1 - E_0}{\hbar}$.

Lasing will only occur if the gain coefficient exceeds some threshold condition. In order to find this condition, let us assume for generality that the mirrors at both ends of the cavity are partially transmitting. Assume that the proportions of the intensity reflected at the left and right mirrors are, respectively, $\mathcal{R}_1$ and $\mathcal{R}_2$. Intensity can be lost due to both absorption and through transmission through the mirrors, as well as gained via spontaneous emission. So, taking all of these into account, the ratio $\frac{I(2L)}{I_0}$ of final intensity to initial intensity after one round trip is $\mathcal{R}_1 \mathcal{R}_2 e^{(\gamma - \alpha)2L}$. So, in order to have a net gain, rather than loss, we must obey the threshold condition:

$$\mathcal{R}_1 \mathcal{R}_2 e^{(\gamma - \alpha)2L} \geq 1. \tag{5.104}$$

Solving for the gain coefficient, we find the **gain threshold** for lasing to occur:

$$\gamma_{th} = \alpha + \frac{1}{2L} \ln\left(\frac{1}{\mathcal{R}_1 \mathcal{R}_2}\right) = \alpha - \frac{1}{2L} \ln(\mathcal{R}_1 \mathcal{R}_2).$$

Although we won't show it here, the gain coefficient at a given frequency can be written in terms of other material parameters in the form

$$\gamma(\omega) = \frac{\lambda^2}{4n^2(\omega)\tau} \Delta N g(\omega). \tag{5.105}$$

Here, λ is the wavelength of the light. $n(\omega)$ and $g(\omega)$ are the refractive index of the gain material at the corresponding frequency and the spectral linewidth of the emitted photons. $\tau = \frac{1}{A_{21}}$ is the **radiative lifetime** of the metastable state of Fig. 5.7. $\Delta N = E_2 - E_1$ is the **population inversion density**, defined as

$$\Delta N = n_2 - \frac{g_2}{g_1} n_1, \tag{5.106}$$

where n_1, n_2 are the occupation numbers of the two energy levels involved in the transition.

The resonance cavity determines what frequencies will be allowed. As the light reflects back and forth between the mirrors, constructive or destructive interference can occur between the waves propagating in opposite directions. The result of the interference is that the light settles into a standing wave pattern, in which the field amplitudes become larger and smaller over time, but the maxima do not move in space (Fig. 5.8). This standing wave pattern has to have nodes (zeros of amplitude)

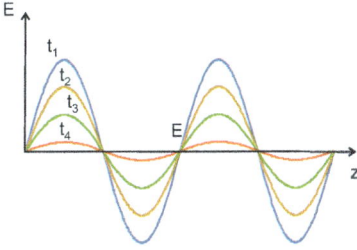

Fig. 5.8 The light in the resonant cavity forms standing waves with nodes at the mirrors. Each standing wave grows and shrinks in amplitude over time, but the crests, troughs, and nodes do not move at all the z-direction. Here, snapshots of the same standing wave are shown at four different times, $t_1, \ldots, t_4$

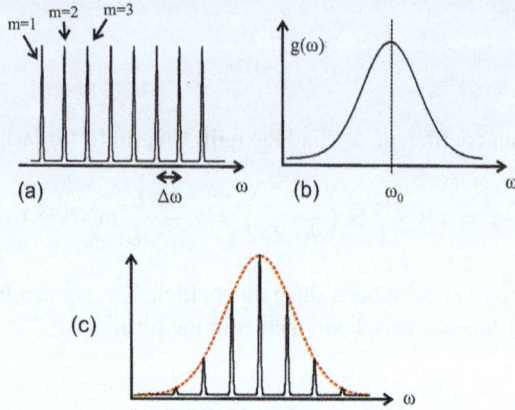

Fig. 5.9 (a) The resonant cavity only allows standing waves near a series of discrete frequencies. These are labeled by an integer mode order m and are separated by $\Delta\omega = \frac{\pi c}{nL}$. (b) The transition between electron energy levels has some spectral width determined by the width of the linewidth function $g(\omega)$. This function is centered at the resonant frequency $\omega_0 = \frac{E_2 - E_1}{\hbar}$ determined by the energy levels of the gain medium atoms. (c) The set of longitudinal modes in (a) are cut off by envelope $g(\omega)$, leaving only a finite number of modes

at the locations of the mirrors, which will restrict the allowed wavelengths and frequencies to a discrete set of values. Requiring nodes to appear at the ends of the cavity is equivalent to requiring that the cavity length L be an integer multiple of half-wavelengths, $L = m\left(\frac{\lambda}{2}\right)$, for $m = 1, 2, 3, \ldots$. m is called the **mode order**. Recalling that the speed of the light in the material is $v = \frac{c}{n}$, where n is the index of refraction, we find that the allowed frequencies (labeled by the integer m) are

$$v_m = \frac{v}{\lambda_m} = \frac{c}{n\lambda_m} = m\left(\frac{c}{2nL}\right). \tag{5.107}$$

So the cavity allows a sequence of **longitudinal modes** separated in frequency by

$$\Delta v = \frac{c}{2nL}, \quad \text{or} \quad \Delta\omega = \frac{\pi c}{nL}, \tag{5.108}$$

as in Fig. 5.9a.

However, recall that the spectral linewidths $g(\omega)$ of the material only allow transitions between energy levels if the photon frequency is within a small range of the resonant frequency (Fig. 5.9b). The spectral line-shape function $g(\omega)$ therefore cuts off the sequence of longitudinal modes as in Fig. 5.9c, leaving only a finite number of frequencies that will survive in the cavity. Depending on the gain material and the cavity length, there may be anything from a single mode to hundreds of thousands of modes that will fit with in the envelope defined by $g(\omega)$. The width of the $g(\omega)$ is called the **laser gain bandwidth**, Δv_{LG}.

Fig. 5.10 Fabry-Perot etalon: as light reflects back and forth multiple times inside a small piece of glass, destructive interference prevents all but a narrow range of frequencies from being transmitted

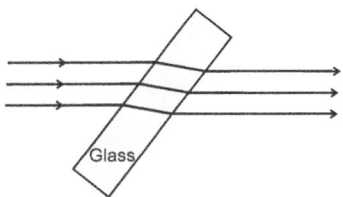

If only one longitudinal mode falls within the envelope defined by $g(\omega)$, of width $\Delta\nu_{LG}$, then the laser is a **single-mode laser**. Otherwise, it is a **multimode laser**.

Even if $\Delta\nu_{LG}$ is broad enough to include multiple modes, single-mode operation can be artificially engineered by adding extra structures to the laser. One possibility would be to add a narrowband frequency filter at the output of the cavity in order to block all but one spectral peak. Another option is to add a **Fabry-Perot etalon** inside the cavity. This is a thin piece of glass tilted at an angle with respect to the beam (Fig. 5.10). Each time light hits one edge of the glass, part of it reflects, so that interference occurs inside the glass between portions of the light that have reflected back and forth different numbers of times. This interference is destructive except for a small range of frequencies, so that the etalon is essentially a narrowband filter. The advantage of using an etalon is that it can be rotated at different angles, which alters the distance the light travels inside the glass between reflections. This in turn alters the frequency at which constructive interference occurs, which means that the laser frequency can be tuned to a desired value (within a small range).

Example 5.7.1 Consider the longitudinal modes in a helium-neon (He-Ne) laser. The central frequency and wavelength are

$$\lambda_0 = 633 \text{ nm}, \qquad \nu_0 = 4.73 \times 10^{14} \text{ Hz}. \tag{5.109}$$

The laser gain bandwidth is typically on the order of $\Delta\nu_{LG} = 1.5 \text{ GHz} = 1.5 \times 10^9 \text{ Hz}$. The corresponding width in the spread of wavelengths is

$$\Delta\lambda_{LG} = \frac{c}{\nu^2}\Delta\nu_{LG} = 0.002 \text{ nm}. \tag{5.110}$$

The spacing between the modes is $\Delta\nu = \frac{c}{2nL}$. $n \approx 1$ for the He-Ne gas, so for a resonator of length $L = 30$ cm, this gives a line spacing of $\Delta\nu = 0.5$ GHz. The number of longitudinal modes emitted by the laser is therefore

$$\frac{\Delta\nu_{LG}}{\Delta\nu} = \frac{1.5 \text{ GHz}}{0.5 \text{ GHz}} = 3. \tag{5.111}$$

In contrast, a titanium-doped sapphire laser has

$$\Delta\nu_{LG} = 128 \text{ THz} = 1.28 \times 10^{11} \text{ Hz}, \tag{5.112}$$

centered at $\lambda_0 = 800$ nm, and supports about 250,000 longitudinal modes.   ∎

Typically, in a multimode laser, each mode oscillates independently, with no fixed phase relation between the modes. Each mode effectively acts as an independent laser, and there is no interference between them. The output intensity is simply the sum of the intensities of the individual modes and is roughly time-independent. In such a case, the laser is referred to as a **continuous-wave (CW) laser**.

However, if there are only a few modes present, it is possible to engineer the laser so that they have a fixed phase relationship. The phases can be arranged to produce a beat pattern, so that interference is alternately constructive and destructive over time. The result is a repeating sequence of short, high-intensity bursts separated by periods of much lower intensity. This called a **pulsed laser**. The separation time between pulses is given by the round-trip time of the light in the cavity: $\tau = \frac{2Ln}{c}$. The inverse of this is the **repetition rate**, $\frac{1}{\tau}$.

Box 5.2 Phonon lasers

Standard lasers and masers emit beams of photons. In principle, a laser can be made that emits beams of any massless boson. These bosons don't need to be true fundamental particles, but could just as easily be collective excitations or quasiparticles. As a result, it has long been speculated that stimulated emission of coherent phonon beams could be possible. Over the past decade or so, several methods have been demonstrated to produce such **phonon lasers**. The phonon laser is also sometimes referred to as a SASER (sound amplification by stimulated emission of radiation). Phonon lasers have been demonstrated in a number of physical platforms including cryogenic atomic systems, solid-state superlattices, and compound microcavities (see [10–12] and references therein).

The system most analogous to a standard laser might be that of the compound microcavity (also called a photonic molecule) described in [10]. Here, a tapered optical fiber coupled evanescently to a pair of optical microcavities provides optical pumping (Fig. 5.11a). The two microcavities (whispering mode galleries) have degenerate resonance frequencies (both equal to ω_0) and are evanescently coupled to each other. Because of the coupling between the two cavities, the energy levels become split: the resulting normal modes of the system have frequencies $\omega_\pm = \omega_0 \pm \delta\omega$, where $\delta\omega$ depends on the coupling. In addition to the optical modes, one of the cavities also has a mechanical vibrational mode of frequency Ω_0. This vibrational mode can gain and lose energy from the surroundings, which amounts to spontaneous absorption and emission of phonons. The optical pumping (if its amplitude is above some threshold value) also leads to *stimulated* emission of phonons. The result is the energy-level diagram in Fig. 5.11b. The Hamiltonian for the system can

(continued)

Fig. 5.11 A phonon laser. (**a**) A fiber is used to optically pump two evanescently coupled cavities. One cavity has mechanical vibrational modes, in addition to the optical mode pumped by the fiber. (**b**) Phonons are absorbed or emitted from the mechanical oscillator when transitions of the oscillator between the normal modes Ψ_+ and Ψ_- occur

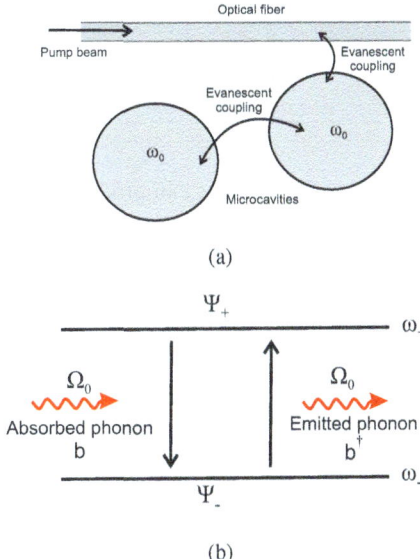

Box 5.2 (continued)
be written in the form

$$\hat{H} = \hbar\omega_+\hat{\Psi}_+^\dagger\hat{\Psi}_+ + \hbar\omega_-\hat{\Psi}_-^\dagger\hat{\Psi}_- + \hbar\Omega_0\hat{b}^\dagger\hat{b} + \frac{\hbar g}{2}\sqrt{\frac{\hbar}{2m\Omega_0}}\left(\hat{b}\hat{\Psi}_+^\dagger\hat{\Psi}_- + \hat{b}^\dagger\hat{\Psi}_-^\dagger\hat{\Psi}_+\right).$$

Here, $\hat{b}$ and $\hat{\Psi}_\pm$ are annihilation operators for phonons and for the normal modes of frequencies $\omega_\pm$. The coupling constant g depends on the resonant frequency ω_0 and the radius of the microcavity with the mechanical vibrational modes. m is the effective mass of the mechanical oscillator. Given this structure, the optical pumping causes population inversion of the normal modes, and the lasing activity works in a very similar manner to optical lasing.

The system described above emits a large numbers of coherent phonons. A system that can produce as few as ten phonons at a time for more pronounced quantum effects uses two coupled ions, which act as a pair of coupled pendula; see [12].

The duration Δt of each pulse is determined by the spread in frequencies. A key result of Fourier theory, the **time-bandwidth theorem**, is essentially a classical analog of the Heisenberg theorem, placing a lower limit on the product of a pulse width in time and the spectral bandwidth $\Delta\nu$ needed to produce it. The exact result depends on the shape of the linewidth. For a Gaussian linewidth containing a set of N modes separated in frequency by spacing $\Delta\nu$, the theorem predicts that the duration of each pulse is

$$\Delta t = \frac{0.441}{N\Delta\nu}. \tag{5.113}$$

Mode-locked lasers can produce ultrashort pulses with durations as small as the picosecond range. For example, for the He-Ne laser in the previous example, this gives $\Delta t \approx 300$ ps, while the titanium-sapphire laser has $\Delta t \approx 3.4$ fs. (Recall that $1\ ps = 10^{-15}$s and $1\ fs = 10^{-12}$s.)

5.7.2   Laser Beams

As has been emphasized before, the light that exits the laser is highly collimated or beam-like. Within the narrow beam that is produced, there can generally be different spatial structures: the intensity within the beam can depend on distance from the axis (r) or on the angle (ϕ) about the axis. Beams with different spatial structures in the transverse plane are referred to as different optical **transverse modes**. The modes are largely determined by the shapes of the cavity and of the mirrors at the ends.

Recall that in free space, propagating light waves have to be transversely polarized, with their electric and magnetic fields perpendicular to the direction of motion. But inside an optical cavity or waveguide, that is not necessarily true. It is possible to have longitudinal field components, polarized along the propagation axis. But the most commonly used modes in laser cavities are the **transverse electric** (TE), **transverse magnetic** (TM), or **transverse electromagnetic** (TEM) modes. These have only transverse electric fields, only transverse magnetic fields, or both fields transverse, respectively.

Restricting ourselves to free space, consider the TEM case [4, 5, 13]. The electric and magnetic fields are denoted E_{mn} and H_{mn}, labeled by a pair of integers m and n.

$$E_{mn}(x, y) \sim H_m\left(\frac{\sqrt{2}x}{w_0}\right) H_n\left(\frac{\sqrt{2}y}{w_0}\right) e^{-\left(\frac{x^2+y^2}{w_0^2}\right)} e^{\phi(z)}, \tag{5.114}$$

where E_0 is the field amplitude, $\phi(z)$ is a phase factor, $H_n(x)$ is the Hermite polynomial), and the parameter w is called the waist size.

We will focus here on the lowest-order case, $m = n = 0$, TEM_{00}. This simplest possible laser mode is called the **Gaussian mode**. This beam has a Gaussian intensity profile in the plane transverse to propagation. The width of the Gaussian varies as it propagates along the z-axis, which is shown in Fig. 5.12. The curves shown consist of points at which the intensity is smaller than the on-axis intensity by a factor of e^2. Let w_0 be the beam radius at its narrowest point (the **beam waist**), and choose coordinates so that $z = 0$ is at the location of the waist. The maximum

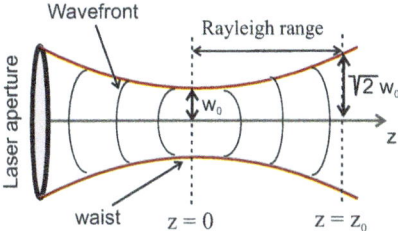

Fig. 5.12 The radius of a Gaussian beam initially becomes smaller after leaving the laser, before slowly growing again. (The rate of spread is greatly exaggerated here to make the behavior clear.) $z = 0$ is taken to be at the waist, or narrowest point. This minimum radius is denoted w_0. The distance at which the beam has area has doubled and the intensity halved (compared to the waist) is the Rayleigh range, z_0. Note that the wavefronts (the surfaces of constant phase) flip their direction of curvature at the waist. Exactly at the waist, the curvature is infinite, so the wavefront is flat. Therefore, near the waist, the laser is a good approximation to a plane wave

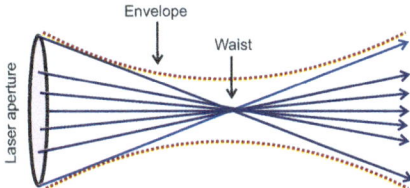

Fig. 5.13 Light rays illustrate how the form of the beam envelope (dashed curve) arises. Initially, most of the light exiting the laser is heading inward, toward the axis. Rays at a range of angles from the z-axis focus over a finite region around the beam waist. The rays then cross and start to move away from each other, causing the beam to expand in radius

intensity I_0 occurs at the origin, on-axis at $z = 0$. The reason that the beam initially contracts and then expands again can easily be seen by drawing a few light rays, as in Fig. 5.13.

After passing the waist, the beam slowly diverges. Since the power is constant, this means that the intensity and the field decrease. The **Rayleigh range** z_0 is the distance at which the beam cross-section is twice as large as at the waist. The on-axis intensity at z_0 is therefore half of its value at the waist: $I(z_0) = \frac{1}{2}I_0$. The Rayleigh range and the waist radius are related by

$$w_0 = \sqrt{\lambda z_0/\pi}, \tag{5.115}$$

implying that a more highly focused beam (smaller w_0) will also diverge more rapidly (smaller z_0). The **depth of focus** (or *confocal parameter*) is also sometimes defined as the distance $b = 2z_0$ over which the intensity varies by a factor of no more than 2.

The amplitude E and intensity I of the Gaussian beam can be expressed in the forms

$$E(r, z) = \sqrt{I_0} \frac{w_0}{w(z)} e^{-\frac{r^2}{w^2(z)}} e^{ikz + \frac{ikr^2}{2R(z)} - i\zeta(z)}, \tag{5.116}$$

$$I(r, z) = |E(r, z)|^2 = I_0 \left(\frac{w_0}{w(z)} \right)^2 e^{-\frac{2r^2}{w^2(z)}}. \tag{5.117}$$

Here, the beam radius at distance z from the waist is

$$w(z) = w_0 \left[1 + \left(\frac{z}{z_0} \right)^2 \right]^{1/2}, \tag{5.118}$$

and the radius of curvature of the wavefront is

$$R(z) = z \left[1 + \left(\frac{z_0}{z} \right)^2 \right]. \tag{5.119}$$

The factor

$$\zeta(z) = \tan^{-1} \left(\frac{z}{z_0} \right) \tag{5.120}$$

is the **Gouy phase**, which varies from $-\frac{\pi}{2}$ to $+\frac{\pi}{2}$ as the beam propagates from $z = -\infty$ to $z = +\infty$. The intensity is not uniformly distributed over the beam area (recall that it is given by a Gaussian distribution), but it can be integrated over the transverse plane to find that the total power carried by the beam is $P = \frac{\pi}{2} I_0 w_0^2$.

Asymptotically, the beam envelope diverges from the axis at an angle

$$\theta_0 \approx \frac{\lambda}{\pi w_0}, \tag{5.121}$$

known as the *beam divergence*. Note that the more focused the beam (the smaller the waist size), the more rapidly the beam diverges (see Eq. 5.115). This has to be the case, due to the Fourier transform relation between transverse position and transverse wavevector (momentum), or equivalently, from the Heisenberg uncertainty relation between position and momentum. In order to better collimate the beam (making it less divergent), the Gaussian profile of the beam has to be made wider. There are no nodes (points of zero intensity) anywhere, which is in contrast to more complicated beam modes, such as the Laguerre-Gauss modes of Sect. 6.3.

Box 5.3 Optical tweezers

One common modern application of lasers is to play the role of **optical tweezers**, which are used to manipulate microscopic particles. Consider a small dielectric particle in the path of the laser beam, and assume that the particle has some electric dipole moment p. Every electromagnetism text shows that the force a nonuniform electric field exerts on an electric dipole is proportional to the gradient of the electric field:

$$F(r) = p \cdot \nabla E(r), \qquad (5.122)$$

where $E(r)$ is the magnitude of the field. At the beam waist, the field strength is at a maximum, so this gradient force vanishes: the waist is an equilibrium position. As the particle moves away from the equilibrium position in either direction along the beam axis, the electric field decreases, so $\nabla E(r)$ points back toward the waist. The gradient force therefore tends to push particles toward the waist and to trap them there. Moving the beam around, the microparticle will follow the waist, and it can therefore be moved to any desired position.

Applications of optical tweezers include the following:

- Constructing nanostructures or sorting nanoparticles of different types
- Isolating individual bacteria, viruses, or tissue cells, rotating them, and tethering them to microscope slides
- Folding or unfolding proteins, in order to study protein structures
- Attaching or removing biological units like amino acids or peptides to specific locations on proteins or other complex biological molecules

Particle traps will be treated in more detail in Chap. 24. For more details on optical tweezers and their uses, see [14].

5.8 Nonlinear Optical Processes

Recall that in a dielectric, an external electric field will polarize the molecules, displacing the positive and negative parts of molecules slightly away from each other. These separated charges will produce a field of their own, which must be added to the external field to find the total field inside the material. As a result, the electric field E, the electric displacement D, and the polarization P inside the dielectric are connected by the constitutive relations [15]:

$$P = \epsilon_0 \chi E \qquad (5.123)$$

$$D = \epsilon_0 E + P = \epsilon_0 \epsilon_r E, \qquad (5.124)$$

where the permittivity of the material is $\epsilon = \epsilon_0 \epsilon_r$ and ϵ_r is the **dielectric constant** or **relative permittivity**. $\chi = \epsilon_r - 1$ is the **electric susceptibility**. Optical properties of the material are determined primarily by the relative permeability, which is related to the refractive index by

$$n = \sqrt{\epsilon_r \mu_r} \approx \sqrt{\epsilon_r}, \tag{5.125}$$

where for a nonmagnetic material, the magnetic susceptibility μ_r can be ignored: $\mu_r \approx 1$.

In introductory electromagnetism and optics courses, it is usually assumed that χ is a constant. At low electric field amplitudes, this is usually a good approximation. But at high amplitudes, such as inside the beam of a laser, it becomes apparent that the susceptibility is actually a function of the field. So suppose that this function can be expanded in a power series:

$$\chi(E) = \chi^{(1)} + \chi^{(2)} E + \chi^{(3)} E^2 + \dots \tag{5.126}$$

$\chi^{(1)}$ is the usual constant part that leads to the linear form of the polarization in Eq. 5.124, while the remaining terms lead to nonlinear corrections:

$$P = \epsilon_0 \left[\chi^{(1)} E + \chi^{(2)} E^2 + \chi^{(3)} E^3 + \dots \right] \tag{5.127}$$

$$= \sum_{n=1}^{\infty} \epsilon_0 \chi^{(n)} E^n. \tag{5.128}$$

(Here for simplicity, we are treating E and P as scalars and the $\chi^{(n)}$ as scalars. In reality, E and P are vectors, and the $\chi^{(n)}$ are tensors. See, e.g., [16] for more precise treatment.) Normally, only the $\chi^{(2)}$ and $\chi^{(3)}$ terms are large enough to be significant in applications.

At normal energies, photons simply pass through each other without interacting. However, inside a dielectric, a pair of photons can interact with each other indirectly, using the crystal lattice as an intermediary: one photon interacts with a charge at one lattice site, which creates a lattice vibration. The vibration then affects charges at neighboring lattice sites, one of which can then interact with a second photon. These nonlinear interactions can effectively cause a single photon to split into several lower-energy photons, or conversely for several photons to merge into one photon of higher energy.

In general, a $\chi^{(n)}$ term leads to an interaction between $n + 1$ photons. For example, consider the $\chi^{(2)}$ term. For a standard dipole interaction between the material and the field, the Hamiltonian corresponding this term will be

$$H^{(2)} = P^{(2)} \cdot E = \left(\epsilon_0 \chi^{(2)} E^2 \right) E = \epsilon_0 \chi^{(2)} E^3. \tag{5.129}$$

Fig. 5.14 In spontaneous parametric downconversion (SPDC), a single incoming photon is split into two outgoing photons via a nonlinear interaction with the crystal. The outgoing photons are entangled in multiple variables (energy, momentum, polarization)

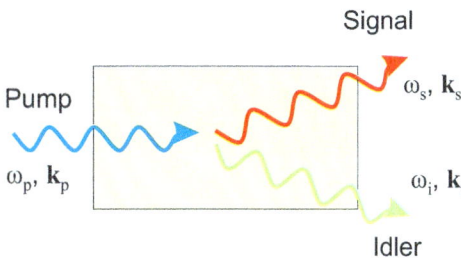

Since each factor of the field (when viewed as a quantum operator) is linear in the photon creation and annihilation operators, each term in this Hamiltonian will involve the creation and/or annihilation of three photons.

One such three-wave process is **spontaneous parametric downconversion (SPDC)**. SPDC is one of the most important nonlinear optical processes, since it not only provides a means of producing entangled photon pairs but also provides a high degree of control over the resulting entangled state. In SPDC, a single high-energy photon called the **pump photon** is split into a pair of lower-energy photons (the **signal** and **idler**), as in Fig. 5.14. (Which output is called the signal and which is called the idler is completely arbitrary.) In order to conserve energy and momentum, the frequencies and wavevectors of the three photons must obey the matching conditions:

$$\omega_p = \omega_1 + \omega_2, \qquad k_p = k_s + k_i. \tag{5.130}$$

The downconversion may be classified as either type I or type II:

- **Type I downconversion:** The signal and pump have the same polarization, with both being perpendicular to the pump polarization.
- **Type II downconversion:** The signal and pump polarizations are orthogonal to each other.

The properties of the outgoing signal and idler can be controlled by varying (i) the properties of the input pump, (ii) the type of crystal used, and (iii) how the crystal is cut. Crystals commonly used for SPDC include beta barium borate (β-BaB_2O_4, known as BBO), potassium dihydrogen phosphate (KH_2PO_4, or KDP), and potassium trihydrogen phosphate (KH_2PO_4, or KTP).

Downconversion can occur at any point in the crystal and the amplitudes for creation at different points will interfere with each other. As a result, the output can be manipulated by cutting the crystal and orienting it relative to the beam so that the interference is constructive only when the signal and idler obey the desired matching conditions. Further control can be obtained by adding polarization or frequency filters after the crystal.

Another three-photon process is **frequency mixing**, in which a pump and idler photon combine to form a higher-frequency signal (Fig. 5.15). A special case of this

Fig. 5.15 (a) Frequency mixing: two photons can mix via a $\chi^{(2)}$ coupling to form photons at the sum and difference frequencies. (b) A special case of frequency mixing is frequency doubling

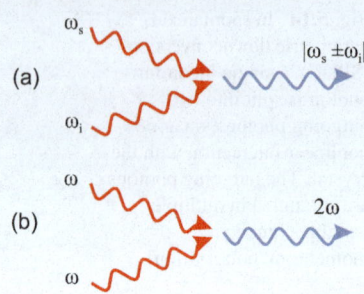

is **frequency doubling**, in which two equal-frequency photons are used to produce a photon of double the original frequency.

Nonlinear four-photon processes occur as well, due to the $\chi^{(3)}$ term in the polarization. One example of such four-wave mixing is when two incoming photons of frequencies ω_1 and ω_2 combine to produce outgoing photons of frequencies $|\omega_1 \pm \omega_2|$. Such a process is useful for converting an undesired input frequency photon to one whose frequency is more useful; for example, in the past, when infrared detectors were less efficient, such an approach would sometimes be used to convert infrared light into more easily detected visible. Four-wave mixing is also used for holographic imaging and for phase conjugation (converting a wave $E(x, t)$ into its complex conjugate $E^*(x, t)$). For an introduction to phase conjugation and its uses, see [17].

5.9 Pockels and Kerr Effects

In the presence of strong electric fields, the index of refraction of a nonlinear optical medium will be altered, opening up new and useful effects. Of particular importance are the Pockels and Kerr effects, alterations of the refractive index which are, respectively, linear and quadratic in the applied fields.

The **Pockels effect** was discovered by Friedrich Carl Alwin Pockels in 1893. It occurs in media with no inversion symmetry (so that $\chi_2 \neq 0$) when light is passed through the material while an external voltage is applied to the medium. Referring to Fig. 5.16, assume a voltage difference V is applied across a slab of material of thickness d, so that the external electric field is $E = \frac{V}{d}$. Assume an optical field is passing through the medium with oscillating field $E_0 \cos \omega t$, where E_0 and ω are the amplitude and frequency of the light. So, taking the two fields to be in the same direction for simplicity, the total electric field strength is $E_{tot} = E + E_0 \cos \omega t$. Keeping only the linear and quadratic terms in the field, the polarization in the material is

$$P = \epsilon_0 \chi_1 E_{tot} + \epsilon_0 \chi_2 E_{tot}^2 + \dots \qquad (5.131)$$

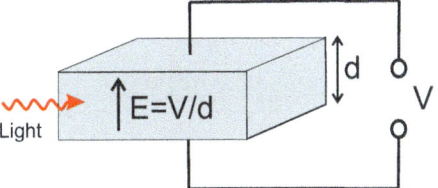

Fig. 5.16 Pockels and Kerr effects. An applied voltage produces a strong electric field in the material, altering the refractive index seen by passing light. The χ_2 and χ_3 contributions, respectively, lead to the Pockels and Kerr effects

$$= \epsilon_0 \chi_1 \left(E + E_0 \cos \omega t \right) + \epsilon_0 \chi_2 \left(E + E_0 \cos \omega t \right)^2 + \dots . \quad (5.132)$$

As seen in the next example, this can be written in the form

$$P = P_0 + P_\omega \cos \omega t + P_{2\omega} \cos 2\omega t + \dots , \quad (5.133)$$

where P_0 is a constant term and for any $n \neq 0$, $P_{n\omega}$ is the coefficient of a term oscillating at frequency $n\omega$.

Example 5.9.1 The expansion coefficients for the polarization can be computed as follows:

$$
\begin{aligned}
P &= \epsilon_0 \chi_1 \left(E + E_0 \cos \omega t \right) + \epsilon_0 \chi_2 \left(E + E_0 \cos \omega t \right)^2 \\
&= \epsilon_0 \left[\left(\chi_1 E + \chi_2 E^2 \right) + \left(\chi_1 E_0 + 2\chi_2 E E_0 \right) \cos \omega t + \chi_2 E_0^2 \cos^2 \omega t \right] \\
&= \epsilon_0 \left[\left(\chi_1 E + \chi_2 E^2 + \frac{1}{2} \chi_2 E_0^2 \right) \right. \\
&\quad \left. + \left(\chi_1 E_0 + 2\chi_2 E E_0 \right) \cos \omega t + \frac{1}{2} \chi_2 E_0^2 \cos^2 \omega t \right],
\end{aligned}
\quad (5.134)
$$

where the identity $\cos^2 \theta = \frac{1}{2}(1 + \cos 2\theta)$ was used in the final line. The coefficients in the expansion can then be read directly off:

$$P_0 = \epsilon_0 \left(\chi_1 E + \chi_2 E^2 + \frac{1}{2} \chi_2 E_0^2 \right) \quad (5.135)$$

$$P_\omega = \epsilon_0 \left(\chi_1 + 2\chi_2 E \right) E_0 \quad (5.136)$$

$$P_{2\omega} = \frac{1}{2} \epsilon_0 \chi_2 E_0^2. \quad \blacksquare \quad (5.137)$$

The P_ω term can be used to define an effective polarizability: $\chi_1' = \chi_1 + 2\chi_2 E$. Using the fact that $n = \sqrt{1 - \chi'}$ and expanding in a binomial series about the leading term, the change in refractive index is then given by

$$\Delta n = \frac{1}{n} \chi_2 E, \tag{5.138}$$

which is more often written as

$$\Delta n = \frac{1}{2} n^3 r E, \tag{5.139}$$

where

$$r = \frac{2}{n^4} \chi_2 \tag{5.140}$$

is called the **Pockels coefficient**, and is measured in meters per volt.

A **Pockels cell** is an optical component made from materials such as lithium niobate ($LiNbO_3$), β-barium borate (BBO), barium titanate (BTO), or monopotassium phosphate (KDP), which exhibit large Pockels effects. Typical values of the Pockels coefficient in such materials may be in the range of 10^{-12}-$10^{-10} \frac{m}{V}$.

A Pockels cell can act as a variable wave plate. In a material in which the index of refraction is polarization-dependent, the electric field will cause each polarization to see a different Δn, leading to a rotation of linear polarization. By varying the external field, the polarization direction of light can therefore be controlled. Further, if the Pockels cell is placed between two crossed polarizers (oriented orthogonal to each other), a controllable optical gate or shutter can be formed. When the external field vanishes, there is no polarization rotation, so any light passing the first filter will be blocked by the second. The gate is therefore closed. But if the electric field is turned on, there will be a value of the field at which the rotation angle is exactly $\frac{\pi}{2}$, allowing the light to pass the second polarizer and opening the gate. Such shutters can switch on and off very rapidly, on nanosecond time scales.

When used as a phase modulator, the phase shift induced by a Pockels cell of thickness L can be written as

$$\Delta\phi = \frac{\pi V}{V_\pi}, \tag{5.141}$$

where the **half-wave voltage** is the voltage needed to shift the phase by π:

$$V_\pi = \frac{d\lambda_0}{r L n^3}. \tag{5.142}$$

Example 5.9.2 The use of a Pockels cell as an amplitude modulator can be illustrated as follows. Suppose the external field is along the y-axis and that an entering optical wave of frequency ω is polarized at $45°$ from the x- and y-axes. It will have components

$$E_x^{(in)}(t) = E_y^{(in)}(t) = \frac{1}{\sqrt{2}} E_0 \cos \omega t. \tag{5.143}$$

The external field E induces a phase shift between the two components, of magnitude

$$\Delta\phi = \frac{2\pi}{\lambda_0}\Delta n L = \frac{2\pi n^3 r L}{\lambda_0}E, \tag{5.144}$$

so that the exiting fields (up to an irrelevant overall phase) are

$$E'_x = \frac{1}{\sqrt{2}}E_0 \cos\omega t \tag{5.145}$$

$$E'_y = \frac{1}{\sqrt{2}}E_0 \cos\left(\omega t + \Delta\phi\right). \tag{5.146}$$

At the output end of the material, there is a polarizer at $90°$ to the input polarization direction. The field that passes this polarizer is the final output field, and is given by

$$E_y^{(out)} = \frac{1}{\sqrt{2}}\left(E'_y - E'_x\right) \tag{5.147}$$

$$= \frac{1}{2}E_0\left(\cos\left(\omega t + \Delta\phi\right) - \cos\omega t\right) \tag{5.148}$$

$$= -E_0 \sin\left(\omega t + \Delta\phi\right)\sin\left(\frac{\Delta\phi}{2}\right). \tag{5.149}$$

This output amplitude can be controlled by altering the external field and thus altering $\Delta\phi$:

$$E_0^{(out)} = E_0^{(in)}\sin\left(\frac{\Delta\phi}{2}\right). \tag{5.150}$$

Because of the minus sign and the $\Delta\phi$ inside the first sine term of Eq. 5.149, the field polarization is also rotated: when the transmission is 100% ($\Delta\phi = \pi$), the polarization has been rotated a full $90°$.   ∎

In addition to the Pockels effect, refractive indices can also be controlled via the Kerr effect, discovered by John Kerr in 1875. Whereas the Pockels effect is linear in the external field, the Kerr effect is quadratic, depending on the nonlinear χ_3 coefficient. The Kerr effect is therefore correspondingly weaker for the same external field; however, unlike the Kerr effect, it exists even in materials in which χ_2 vanishes. There are in fact two distinct Kerr effects, the electro-optic (DC) and optical (AC) effects.

The **DC Kerr effect** or **quadratic electro-optic effect** makes use of a static external field. In the coefficient $P_\omega = \epsilon_0 \chi'_1 E_0$, the field causes an effective χ_1 coefficient of the form

$$\chi'_1 = \chi_1 + 3\chi_3 E^2, \tag{5.151}$$

leading to a refractive index change of

$$\Delta n = \frac{3\chi_3}{2n} E_0^2 \equiv \frac{1}{2} s n^3 E_0^2, \tag{5.152}$$

where the **Kerr coefficient** $s = \frac{3\chi_3}{n^4}$ is typically $10^{-18} - 10^{-14} \frac{m^2}{V^2}$ in crystals and $10^{-22} - 10^{-19} \frac{m^2}{V^2}$ in solids. Kerr cells, made of materials such as nitrotoluene or nitrobenzene, can again cause polarization rotations modulated on nanosecond time scales, but at higher external voltages.

The **AC Kerr effect** or **optical Kerr effect** makes use of the oscillating electric field in the optical wave itself, in place of the static external field. As a result, it was first seen only after the invention of the laser allowed optical intensities large enough for it to become visible. For a wave of amplitude E_0, the effective nonlinear coefficient and the change in refractive index are given by

$$\chi_1' = \chi_1 + \frac{3}{4} \chi_3 E_0^2 \tag{5.153}$$

$$\Delta n = \frac{\Delta \chi}{2n} = \frac{3}{8n} \chi_3 E_0^2. \tag{5.154}$$

The AC Kerr effect has several useful applications. For example, it can lead to beam **self-focusing**, in which a spatially varying intensity can cause an optical beam to focus itself to a point as if a lens was present. The intensity of the light at a given point is given by $I = \frac{1}{2} c n \epsilon_0 E_0^2$, so in terms of the intensity, the shift in refractive index is given by

$$\Delta n = \frac{3\chi_3 I}{4cn^2 \epsilon_0}. \tag{5.155}$$

So, far out from the beam axis, where the intensity is lower, Δn will also be smaller and the light will be moving faster. Near the axis, the light moves slower, due to the larger I and consequently larger Δn. This is similar to a converging lens, which is thinner (producing a smaller phase shift $\Delta \phi$) at the edge and thicker (larger $\Delta \phi$) near the middle. The net result is a spatially varying phase shift that will cause the light to bend inward, toward the region of high-intensity axis, as if it had passed through a converging lens. The degree of focusing is controlled by the intensity of the beam. This self-focusing effect is useful for controlling the shape of a beam inside a material.

A related effect is **self-phase modulation**. The wavenumber can also be modulated by the intensity:

$$k = \frac{n\omega}{c} = \frac{n_0 \omega}{c} + \frac{\Delta n \, \omega}{c} = k_0 + \Delta k, \tag{5.156}$$

where n_0 and k_0 are the vacuum values and $\Delta k = \frac{3\chi_3 I \omega}{4c^2 n^2 \epsilon_0} \frac{\Delta n}{c} \omega$. This leads to a phase shift in the beam:

$$\Delta\phi = \Delta k L = \frac{3\chi_3 I \omega}{4c^2 n^2 \epsilon_0} \frac{\Delta n}{c} \omega L, \tag{5.157}$$

where L is the thickness of the nonlinear material. Thus, the phase of the beam can be modulated by controlling the intensity.

Finally, if two beams of light are propagating in the same nonlinear material, the electric field from one can stimulate a Kerr effect in the other, a phenomenon known as the **cross-Kerr effect**. Needless to say, the cross-Kerr effect is extremely weak at normal intensities, but it has found applications in areas such as quantum nondemolition experiments (Chap. 11) and for quantum information processing. Given two optical fields of frequencies ω_1 and ω_2, the change in refractive index of the ω_1 field is given by

$$\Delta n_1 = \left(\frac{3\chi_3}{4\epsilon_0 c n^2} \right) [I_1 + 2I_2], \tag{5.158}$$

where the two terms represent the self-phase and cross-phase modulation.

5.10 Metamaterials

In addition to natural materials, it is now common to engineer artificial materials designed to have a set of desired properties. One important category of engineered materials are **metamaterials**, in which nanoscale structures are specifically designed to give a material a particular set of optical or electromagnetic properties.

A metamaterial consists of a periodic array of sub-wavelength conducting or dielectric structures, possibly with strategically placed defects. The goal is to arrange for incoming light to diffract off the nanostructures and for the scattered beams to interfere in a desired manner. In this way, the phase, polarization, spectrum, and spatial distribution of the outgoing light can be manipulated.

The earliest metamaterials, in the early 2000s, were simply arrays of metal wires or of gold nanoparticles and operated in the microwave portion of the spectrum. However, conducting materials experience Joule heating and loss, so interest quickly developed in use of dielectric nanostructures. Dielectric structures also have the advantage of large induced electric and magnetic dipole moments, and they can produce highly directional scattering patterns.

Common examples of metamaterials include arrays of metallic wires and dielectric nanoscale rods. Another common structure is an array of split-ring resonators, Fig. 5.17: each ring has a self-inductance, a mutual inductance with the other ring, and capacitances across the gaps in the rings. By controlling all of these parameters,

Fig. 5.17 A split-ring resonator consists of two cylindrical conductors with gaps. The various capacitances and inductances in the unit control the resonant property of oscillations in the system and allow control if its electromagnetic scattering properties

as well as the spatial distribution of the resonators, the relation between incoming and outgoing radiation patterns can be tailored with high precision.

One of the most active areas of research is the development of two-dimensional metamaterials, called **metasurfaces**. These metasurfaces may work in either reflection or transmission modes and can be used, for example, to create flat lenses. Traditional lenses require either the center or the edges to bulge outward, making them bulky. They also tend to be fragile and to have significant optical aberrations. But flat metasurface lenses avoid all of these problems. In particular, being flat, they can be easily incorporated into photonic integrated circuits. Currently, a great deal of effort is being devoted to the development of active metasurfaces, in which the properties of the nanostructures can be controlled in real time, for example, by applying external voltages or mechanical strains. This would allow metasurfaces to function as controllable spatial light modulators.

A wide variety of applications have been proposed for metamaterials and metasurfaces. For example, as mentioned in Chap. 2, a flat metasurface superlens could implement negative refractive index optics and super-resolution imaging. They can also be used to improve the function of solar cells by increasing the range of acceptance angles and by tuning the absorption spectrum to match that of the incoming solar spectrum. One possible application that has received considerable attention is the engineering of invisibility cloaks [18].

Metamaterials can also be used to generate specialized output radiation patterns. In particular, a scheme has been proposed to create entangled light by sending an initial non-entangled optical state through a dielectric-metamaterial interface [19].

They also hold promise as sensors in a variety of context. For example, a negative-index metamaterial can be used to amplify evanescent waves. This not only allows high-sensitivity sensing of evanescent waves at the surfaces of material but also can be used for near-field microscopy. For a review of sensing applications, see [20].

For more detailed reviews of metamaterials, see [21, 22].

Problems

1. (a) Show that the canonical commutation relation for harmonic oscillator operators $\hat{a}$ and $\hat{a}^\dagger$ in Eq. 5.16 is correct.
   (b) Assuming that the photon creation and annihilation operators obey Eq. 5.40, show that Eqs. 5.36 and 5.37 are correct.
2. Consider expectation values of operators in the number state $|n\rangle$.

   (a) Find the expectation value of $\langle \hat{n} \rangle = \langle n|\hat{n}|n\rangle$.
   (b) Find the expectation value of $\langle \hat{n}^2 \rangle = \langle n|\hat{n}^2|n\rangle$.
   (c) Compute the variance $\Delta n^2 = \langle \hat{n}^2 \rangle - \langle \hat{n} \rangle^2$.

3. Consider a $\chi^{(2)}$ nonlinear crystal illuminated by a pump beam of wavenumber k. Let L be the thickness of the crystal in the direction of k.

   (a) Suppose spontaneous parametric downconversion (SPDC) is occurring in the material of the crystal. Approximate the amplitude for production of a downconversion pair by a constant throughout the illuminated portion of the crystal. Then, given that the phase the pump photons gain by traveling a distance x through the crystal is e^{ikx}, show that the intensity of the pump beam at the outgoing side of the crystal is proportional to $L^2 \mathrm{sinc}\left(\frac{kL}{2}\right)$, where the sinc function is defined as $\mathrm{sinc}(x) = \frac{\sin x}{x}$.
   (b) Let the **phase-matching parameter** $\Delta k = k_s + k_i - k_p$ be the wavenumber mismatch between the pump beam and the outgoing signal/idler pair. Then, similar to part (a), the signal and idler will have outgoing intensities proportional to $\mathrm{sinc}^2\left(\frac{\Delta k\, L}{2}\right)$. Clearly, the maximum output intensity I_{max} occurs when the phase-matching parameter is zero, $\Delta k = 0$. Plot $\mathrm{sinc}^2\left(\frac{\Delta k\, L}{2}\right)$ versus Δk, and from this plot, approximate the value of Δk at which the output intensity is $\frac{1}{2}$ of the maximum. What happens to this Δk for thick crystals $L \to \infty$ and for thin crystals $L \to 0$?

4. A parametric amplifier uses a strong pump beam to amplify a weak signal beam. In the process, an idler beam is also created. Let E_p, E_s, and E_i be the pump, signal, and idler field amplitudes in a nonlinear crystal. z is the distance traveled through the crystal. Then the signal and idler amplitudes obey the differential equations [4]:

$$\frac{dE_s}{dz} = i\frac{\omega_s}{nc}\chi^{(2)} E_p E_i^* e^{i\Delta kz}, \qquad \frac{dE_i}{dz} = i\frac{\omega_i}{nc}\chi^{(2)} E_p E_s^* e^{i\Delta kz},$$

where $\Delta k = k_p - k_s - k_i$ is the phase mismatch parameter.

   (a) Assuming that the pump amplitude is approximately constant and that $\Delta k = 0$, find the general solutions for $E_s(z)$ and $E_i(z)$.
   (b) Assume now that there is initially no idler amplitude, $E_i(0) = 0$. Find the amplitudes $E_i(z)$ and $E_s(z)$, and show that as long as $E_s(0) \neq 0$, then both the signal and idler amplitudes grow with increasing z.

5. Verify that the leading correction to the refractive index in the Pockels effect is given by Eq. 5.138.

6. Use the creation and annihilation operator formalism to find the output state of the Mach-Zehnder interferometer as a function of phase ϕ in the upper arm when the input is as follows:

   (a) A single photon entering port 1.
   (b) Two photons entering simultaneously, one each in ports 1 and 2.

7. Consider a He-Ne laser of length $L = 25$ cm and spectral line width $\lambda_{LG} = 0.003$ nm, producing free-space wavelength $\lambda_0 = 663$ nm at room temperature, $T \approx 300$ K. The refractive index of the gain material is $n \approx 1$. Assume the ground and excited states are nondegenerate, $g_1 = g_2 = 1$:

   (a) Find the number of longitudinal modes.
   (b) Find the energy density of the photon gas in the resonant cavity.
   (c) What is the ratio of the Einstein coefficients, $\frac{A_{21}}{B_{21}}$?
   (d) Find the ratio of excited to ground state occupations, $\frac{N_2}{N_1}$.

8. Using Eq. 5.117, compute the power carried by a Gaussian beam as a function of the maximum intensity I_0 and beam waist w_0 at fixed z.

9. The moon orbits at an average distance of about 35,000 km from the earth. The McDonald observatory in Texas does laser-ranging experiments to determine the exact distance over time. They use 1500 mJ pulses of 200 ps width at 532 nm wavelength. The pulses are emitted at a rate of 10 Hz. After reflecting off the moon, they are detected with a telescope of aperture 0.726 m.

   (a) How long does the light take to make the round-trip journey?
   (b) If the laser beam initially has a diameter of 7 mm, what are the diameters of the beam when it reaches the moon and when it returns to the observatory?
   (c) Given the result of part (b), how many photons are detected after each pulse? (Ignore any losses that occur during the trip through the atmosphere or during reflection at the moon's surface.)

10. A ruby laser emits light at 4.32×10^{14} Hz. The ruby gain material has refractive index 1.76:

    (a) If the optical cavity is 20 cm long, find the frequency spacing wavelength spacing between adjacent longitudinal modes.
    (b) Find the wavelengths of the first three emitted modes.

References

1. D.J. Griffiths, D.F. Schroeter, *Introduction to Quantum Mechanics*, 3rd edn. (Cambridge University, Cambridge, 2018)
2. R. Shankar, *Principles of Quantum Mechanics*, 2nd edn. (Plenum Press, New York, 1994)

3. M. Fox, *Quantum Optics: An Introduction* (Oxford University, Oxford 2006)
4. B.E.A. Saleh, M.C. Teich, *Fundamentals of Photonics*, 3rd edn. (Wiley, New York, 2019)
5. A.E. Siegman, *Lasers* (University Science Books, New York,1986)
6. O. Svelto, *Principles of Lasers*, 3rd edn. (Plenum Press, New York, 1989)
7. T. Hänsch, Opt. Photonics News **16**, 14 (2005)
8. S.R. Wilk, Opt. Photonics News **20**, 14 (2009)
9. Y. Chyan et al., ACS Nano **12**, 2176 (2018)
10. I.S. Grudinin, H. Lee, O. Painter, K.J. Vahala, Phys. Rev. Lett. **104**, 083901 (2010)
11. R.P. Beardsley et al., Phys. Rev. Lett. **104**, 085501 (2010)
12. T. Berle et al., Phys. Rev. Lett. 131, 043605 (2023)
13. R.S. Quimby, *Photonics and Lasers* (Wiley-Interscience, New York, 2006)
14. G. Pesce, P.H. Jones, O.M. Marago, G. Volpe, Eur. Phys. J. Plus **135**, 949 (2020)
15. D.J. Griffiths, *Introduction to Electrodynamics*, 5th edn. (Cambridge University, Cambridge, 2023)
16. R.W. Boyd, *Nonlinear Optics*, 4th edn. (Academic Press, New York, 2020)
17. J.-P. Huignard, A. Brignon, *In Phase Conjugate Laser Optics*, ed. by D.R. Vij, A. Brignon, J.-P. Huignard (Wiley, New York, 2003)
18. H. Chen, C.T. Chan, P. Sheng, Nat. Mater. **9**, 387 (2010)
19. M. Siomau, A.A. Kamli, S.A. Moiseev, B.C. Sanders, Phys. Rev. A **85**, 050303(R) (2012)
20. J.J. Yang, M. Huang, H. Tang, J. Zeng, L. Dong, Int. J. Antennas and Propagation, **2013**, Article ID 637270 (2013)
21. L. Solymar, E. Shamonina, *Waves in Metamaterials* (Oxford University, Oxford, 2009)
22. C. Simovski, S. Tretyakov, *An Introduction to Metamaterials and Nanophotonics* (Cambridge University, Cambridge, 2020)

Chapter 6
Angular Momentum, Spin, and Two-State Systems

Spin and orbital angular momentum both play essential roles in applications of quantum mechanics to measurement, communication, and information processing. Most introductory quantum mechanics courses cover the spin of electrons in detail, but photons are a little more subtle: spin appears in disguise, showing up as circular polarization, while orbital angular momentum manages to be present even for photons moving in a straight line. In addition, the gauge invariance of electromagnetism means that only two independent spin states appear for the photon (two polarization states), not the three states that would normally be expected for a spin-1 particle like a photon. In this chapter, the angular momenta of electrons and photons are investigated, and the spin-$\frac{1}{2}$ electron spin is used as a model for more general two-state systems.

6.1 Angular Momentum and Spin

In classical mechanics, the angular momentum of some particle about a fixed point P can be defined by $\boldsymbol{L} = \boldsymbol{r} \times \boldsymbol{p}$, where $\boldsymbol{p}$ is linear momentum of the particle, and $\boldsymbol{r}$ is the vector pointing from P to the particle. By direct analogy, the quantum mechanical orbital angular momentum about some fixed point is given by the operator

$$\hat{\boldsymbol{L}} = \hat{\boldsymbol{r}} \times \hat{\boldsymbol{p}} = -i\hbar \hat{\boldsymbol{r}} \times \nabla. \tag{6.1}$$

Unlike in classical mechanics, all three components of the angular momentum can't be determined simultaneously, since they don't commute with each other. They obey the commutation relations

$$\left[\hat{L}_x, \hat{L}_y\right] = i\hbar\hat{L}_z, \quad \left[\hat{L}_y, \hat{L}_z\right] = i\hbar\hat{L}_x, \quad \left[\hat{L}_z, \hat{L}_x\right] = i\hbar\hat{L}_y, \tag{6.2}$$

© The Author(s), under exclusive license to Springer Nature Switzerland AG 2025
D. S. Simon, *Introduction to Quantum Science and Technology*, Undergraduate
Texts in Physics, https://doi.org/10.1007/978-3-031-81315-3_6

or more compactly,

$$\left[\hat{L}_i, \hat{L}_j\right] = i\hbar \sum_{ijk} \epsilon_{ijk} \hat{L}_k, \tag{6.3}$$

where $i, j, k \in \{x, y, z\}$. ϵ_{ijk} is the completely antisymmetric **Levi-Civita tensor**,, which takes on values

$$\epsilon_{ijk} = \begin{cases} +1, & \text{for } \{ijk\} \text{ a cyclic permutation of } \{xyz\} \\ -1, & \text{for } \{ijk\} \text{ an anticyclic permutation of } \{xyz\} \\ 0, & \text{for any two indices equal to each other.} \end{cases} \tag{6.4}$$

By cyclic permutations, we mean $\{x, y, z\}$, $\{y, z, x\}$, or $\{z, x, y\}$, while anticyclic permutations mean one of $\{z, y, x\}$, $\{y, x, z\}$, or $\{x, z, y\}$. The squared magnitude of the angular momentum,

$$\hat{L}^2 = \hat{L}_x^2 + \hat{L}_y^2 + \hat{L}_z^2, \tag{6.5}$$

commutes with all of the components. So if we take $\hat{L}^2$ and one of the components (traditionally $\hat{L}_z$), they can both be simultaneously measured and simultaneously diagonalized. The eigenvalues of these two operators label a set of eigenstates that can serve as a complete basis of orbital angular momentum states. To define these states, it is easiest to go to spherical polar coordinates (r, θ, ϕ), where θ is the angle from the z axis, and ϕ is the azimuthal angle, i.e., the angle about the z-axis of the projection into the x-y plane. In these coordinates, the two commuting operators take the form

$$\hat{L}_z = -i\hbar \frac{\partial}{\partial \phi} \tag{6.6}$$

$$\hat{L}^2 = -\hbar^2 \left(\frac{1}{\sin\theta} \frac{\partial}{\partial\theta} \right) \sin\theta \frac{\partial}{\partial\theta} + \left(\frac{1}{\sin^2\theta} \right) \frac{\partial^2}{\partial^2\phi}. \tag{6.7}$$

The respective eigenvalues l and m_l are defined by

$$\hat{L}_z|l, m_l\rangle = m_l\hbar \, |l, m_l\rangle \tag{6.8}$$

$$\hat{L}^2|l, m_l\rangle = l(l+1)\hbar^2 \, |l, m_l\rangle. \tag{6.9}$$

The projections of the angular momentum eigenstates into the position-space basis, $Y_{lm_l}(\theta, \phi) = \langle\theta, \phi|l, m_l\rangle$, are given by **spherical harmonics**,

$$Y_{lm_l}(\theta, \phi) = (-1)^m e^{im_l\phi} \sqrt{\frac{(2l+1)}{4\pi} \frac{(l-m_l)!}{(l+m_l)!}} \, P_l^{m_l}(\cos\theta). \tag{6.10}$$

The properties of the **associated Legendre polynomials**, $P_l^{m_l}(x)$, are covered in detail in every standard quantum mechanics textbook (for example, see [1, 2]), so we don't repeat them here, except to mention the orthonormality relation for the spherical harmonics:

$$\int_0^\pi \int_0^{2\pi} Y_l^{m_l} \left(Y_{l'}^{m_l'} \right)^* \sin\theta \, d\phi \, d\theta = \delta_{ll'}\delta_{m_l m_l'}. \tag{6.11}$$

The quantum number l can be any non-negative integer, $l = 0, 1, 2, 3, \ldots$, while the integer m_l obeys $-l \le m_l \le l$.

The fact that l is restricted to integer values implies that the magnitude of the orbital angular momentum can only have a set of discrete, quantized values. The fact that m_l is also required to be an integer means that the z component of angular momentum, or equivalently, the orientation of the angular momentum in space, can also only take on discrete values. Specifically, the angle of L from the measurement axis can only take values satisfying

$$\cos\theta = \frac{m_l}{\sqrt{l(l+1)}}. \tag{6.12}$$

Once the z-component has been measured, the x and y components are completely uncertain, so the tips of the angular momentum could be anywhere on the blue circles in Fig. 6.1

Another distinguishing feature of angular momentum in quantum mechanics is that in addition to orbital angular momentum, a particle can have a built-in spin angular momentum that is the same for all particles of the same species. The components of this spin orbital angular momentum obey the same commutation relations as those of the orbital angular momentum, Eqs. 6.3. They also obey similar

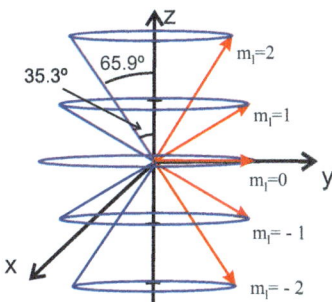

Fig. 6.1 Angular momentum vectors for the case $l = 2$. The magnitude of each vector is $\sqrt{l(l+1)} \, \hbar = \sqrt{6} \, \hbar$. The possible values of L_z are $L_z = m_l \hbar$, with $m_l = -2, -1, 0, 1, 2$. The allowed angles from the z-axis are then given by $\cos\theta = \frac{m_l}{\sqrt{l(l+1)}}$. Since measurement of L_z introduces complete uncertainty into L_x and L_y, the tips of the vectors can be thought of as fluctuating wildly around the blue circles

eigenvalue relations,

$$\hat{S}_z |s, m_s\rangle = m_s \hbar \, |s, m_s\rangle \tag{6.13}$$

$$\hat{S}^2 |l, m_s\rangle = s(s + 1)\hbar^2 \, |s, m_s\rangle, \tag{6.14}$$

with $-s \leq m_s \leq s$. The chief differences from the orbital case are that: (i) the value s is fixed for a given type of particle, and (ii) the spin quantum numbers s and m_s can take on half-integer as well as integer values. Half-odd-integer spin particles ($s = \frac{1}{2}, \frac{3}{2}, \ldots$) are called **fermions**, and they obey the Pauli exclusion principle. Integer spin particles, called **bosons**, do not obey the exclusion principle, and so many of them can condense into the same quantum state. In particular, for a spin-$\frac{1}{2}$ fermion like an electrons, $s = \frac{1}{2}$ and the possible values of m_s are $m_s = -\frac{1}{2}, +\frac{1}{2}$, while for a spin-1 boson like a photon, the values are $s = 1$ and $m_s = -1, 0, +1$.

Recall, in particular, that for spin-$\frac{1}{2}$, the spin operator can be written in terms of the Pauli matrices, $\hat{S} = \frac{\hbar}{2}\sigma$; the Pauli matrices will be further discussed in the next section.

The spin and orbital angular momentum commute with each other,

$$\left[\hat{S}_j, \hat{L}_k\right] = 0, \tag{6.15}$$

for all $j, k \in \{x, y, z\}$. The total angular momentum is the sum of the spin and the angular parts: $\hat{J} = \hat{L} + \hat{S}$, and its components again satisfy relations of the same form as Eq. 6.3. There is again a pair of (integer or half-integer) quantum numbers j, m_j and a complete set of eigenstates:

$$\hat{J}_z |j, m_j\rangle = m_j \hbar \, |j, m_j\rangle \tag{6.16}$$

$$\hat{J}^2 |l, m_j\rangle = j(j + 1)\hbar^2 \, |j, m_j\rangle, \tag{6.17}$$

where now $m_j = m_s + m_l$ and $|l - s| \leq j \leq l + s$.

Just as for the harmonic oscillator states in Sect. 5.2, a set of raising and lowering operators can be defined that take us up and down the ladder of angular momentum states. These are defined by

$$\hat{J}_+ = \hat{J}_x + i\hat{J}_y \tag{6.18}$$

$$\hat{J}_- = \hat{J}_x - i\hat{J}_y. \tag{6.19}$$

From the commutation relations between the x, y, z components, it is readily shown that

$$\left[\hat{J}_z, \hat{J}_\pm\right] = \pm\hbar J_\pm \tag{6.20}$$

$$\left[\hat{J}_+, \hat{J}_-\right] = 2\hbar J_z. \tag{6.21}$$

Then it follows (see Problem 1) that

$$\hat{J}_+|j, m\rangle = \hbar\sqrt{j(j+1) - m(m+1)}\ |j, m+1\rangle \tag{6.22}$$

$$\hat{J}_-|j, m\rangle = \hbar\sqrt{j(j+1) - m(m-1)}\ |j, m-1\rangle. \tag{6.23}$$

6.2   Two-State Systems

Often, it is the case that a physical system has only two accessible states. Common examples of this situation include:

- An electron whose spin is measured along the z-axis. The only two states are spin-up $|\uparrow\rangle$ or spin-down $|\downarrow\rangle$.
- A photon whose linear polarization is measured relative to an orthogonal basis perpendicular to the propagation direction. The two possible states are horizontally polarized along the x-axis $|H\rangle$ or vertically polarized along the y-axis $|V\rangle$.
- An atom bathed in electromagnetic radiation whose energy $\hbar\omega$ is roughly equal to the energy difference between the ground state and the first excited state of the atom. The photons in the radiation have insufficient energy to excite the atom to states beyond the first one, so the two accessible states are the ground state $|g\rangle$ and the first excited state $|e\rangle$.
- In an interferometer with two output ports, each photon could leave at either of the two ports (Fig. 6.2). The two possible output states $|1\rangle$ and $|2\rangle$ are often two different *spatial* modes, with momenta in different spatial directions.
- A photon could be moving either clockwise or counterclockwise inside a resonator ring, a type of optical cavity shaped like a donut.

Fig. 6.2 A Mach–Zehnder interferometer. The output can be thought of as a two-state system, with the output states $|1\rangle$ and $|2\rangle$ being states that exit at different ports, moving in different final directions

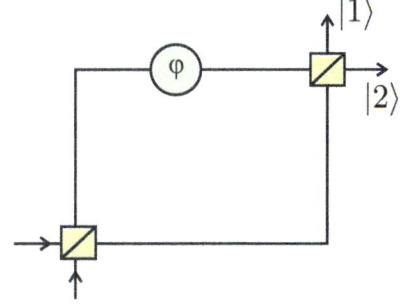

- An atomic nucleus is made of protons and neutrons. A proton and a neutron can be viewed as different states of a single particle by defining a variable called **isospin**, which is mathematically identical to an ordinary spin variable. The protons are then in the isospin-up state, and the neutron is the isospin-down state in some abstract isospin space.

Although all of these systems have different physical origins, they have many properties in common and so can be described in a unified manner by a single mathematical formalism. We will describe the general formalism in this section.

The two states of the system can be used to store one bit of information: one of the two possible states is identified with the logical 0 bit and the other with the logical 1 bit. For this reason, we will often call the two states $|0\rangle$ and $|1\rangle$. It may be helpful to think of the spin states of an electron as an example in the following; in that case, the identifications $|0\rangle = |\uparrow\rangle$ and $|1\rangle = |\downarrow\rangle$ can be made.

We begin by representing our two basis states in matrix form:

$$|0\rangle = \begin{pmatrix} 1 \\ 0 \end{pmatrix}, \quad \text{and} \quad |1\rangle = \begin{pmatrix} 0 \\ 1 \end{pmatrix}. \tag{6.24}$$

These states form a basis for a two-dimensional state space; any other state in the space is a linear combination of these basis states:

$$|\psi\rangle = \alpha|0\rangle + \beta|1\rangle, \tag{6.25}$$

for two complex numbers α and β. Normalization requires that $|\alpha|^2 + |\beta|^2 = 1$. States of this form are often called **qubit** states and will be a topic of central importance in later chapters, and they span a two-dimensional Hilbert space $\mathcal{H}_2$.

Consider transformations between different states in $\mathcal{H}_2$. The transformations should be unitary in order to preserve the normalization of the states, and in order to operate on the two-dimensional states, they should be represented by 2×2 matrices. Examples of such transformations are the three **Pauli matrices**,

$$\sigma_x = \begin{pmatrix} 0 & 1 \\ 1 & 0 \end{pmatrix}, \quad \sigma_y = \begin{pmatrix} 0 & -i \\ i & 0 \end{pmatrix}, \quad \sigma_z = \begin{pmatrix} 1 & 0 \\ 0 & -1 \end{pmatrix}. \tag{6.26}$$

The Pauli matrices are unusual in that they are *both* Hermitian and unitary. In fact, any Hermitian two-by-two matrix can be written as a linear combination of the unit matrix and the three Pauli matrices, which in turn means that any unitary two-by-two matrix can be written as an exponential of the form

$$U = e^{i(a\sigma_0 + \boldsymbol{b} \cdot \boldsymbol{\sigma})}, \tag{6.27}$$

for some vector $\boldsymbol{b} = b_x\hat{\boldsymbol{i}} + b_y\hat{\boldsymbol{j}} + b_z\hat{\boldsymbol{k}}$, where $\sigma_0 = \hat{I}$ is the identity matrix, and $\boldsymbol{\sigma} = \sigma_x\hat{\boldsymbol{i}} + \sigma_y\hat{\boldsymbol{j}} + \sigma_z\hat{\boldsymbol{k}}$ is sometimes called the **Pauli vector**. The fact that unitary transformations on 2×2 (or qubit) systems can be built out of the Pauli matrices is important in quantum computing processes (Chaps. 16 and 17), where quantum logic gates can be formed by implementing Pauli matrix operations.

The most general 2×2 Hermitian matrix can then be written in the forms

$$H = \alpha\sigma_0 + \boldsymbol{\beta} \cdot \boldsymbol{\sigma} = \begin{pmatrix} \alpha + \delta & \beta - i\gamma \\ \beta + i\gamma & \alpha - \delta \end{pmatrix}, \tag{6.28}$$

where $\boldsymbol{\beta} = \alpha\hat{\boldsymbol{i}} + \beta\hat{\boldsymbol{j}} + \gamma\hat{\boldsymbol{k}}$, and α, β, δ, γ are all real.

Example 6.2.1 Consider an electron in a uniform magnetic field $\boldsymbol{B}$. Take the field to be oriented along the z axis, $\boldsymbol{B} = B\hat{\boldsymbol{k}}$. As in Sect. 4.1.2, the magnetic dipole moment of the electron can be written as

$$\boldsymbol{\mu} = -\frac{e}{m}\boldsymbol{S} = -\frac{e\hbar}{2m}\boldsymbol{\sigma}. \tag{6.29}$$

The Hamiltonian or energy operator of the dipole in the magnetic field is

$$\hat{H} = -\boldsymbol{\mu} \cdot \boldsymbol{B} = \frac{e\hbar}{2m}\boldsymbol{\sigma} \cdot \boldsymbol{B}. \tag{6.30}$$

The dot product is a matrix,

$$\boldsymbol{\sigma} \cdot \boldsymbol{B} = B\sigma_z = \begin{pmatrix} B & 0 \\ 0 & -B \end{pmatrix}, \tag{6.31}$$

so the allowed energies of the state are $E = \pm\frac{e\hbar B}{2m}$. Note that the Hamiltonian is of the form of Eq. 6.28, but without the identity (σ_0) term. A term of the form $V(\boldsymbol{r})\sigma_0$ will be added if there is a scalar potential present.

Now define a unit vector

$$\hat{\boldsymbol{n}} = \sin\theta\cos\phi\hat{\boldsymbol{i}} + \sin\theta\sin\phi\hat{\boldsymbol{j}} + \cos\theta\hat{\boldsymbol{k}}. \tag{6.32}$$

The spin operator along this direction is $\frac{\hbar}{2}$ times

$$\boldsymbol{\sigma} \cdot \hat{\boldsymbol{n}} = \begin{pmatrix} \cos\theta & e^{-i\phi}\sin\theta \\ e^{+i\phi}\sin\theta & \cos\theta \end{pmatrix}. \tag{6.33}$$

Suppose that the magnetic field is now along $\hat{\boldsymbol{n}}$ instead of along the positive z-axis: $\boldsymbol{B} = B\hat{\boldsymbol{n}}$. The energies for the spin-up and spin-down (along the z-axis) states can be found:

$$E_\uparrow = \frac{e\hbar}{2m}\langle 0|\boldsymbol{\sigma} \cdot \boldsymbol{B}|0\rangle \tag{6.34}$$

$$= \frac{e\hbar B}{2m} \begin{pmatrix} 1 & 0 \end{pmatrix} \begin{pmatrix} \cos\theta & e^{-i\phi}\sin\theta \\ e^{i\phi}\sin\theta & -\cos\theta \end{pmatrix} \begin{pmatrix} 1 \\ 0 \end{pmatrix} \tag{6.35}$$

$$= \frac{e\hbar B}{2m}\cos\theta \tag{6.36}$$

$$E_\downarrow = \frac{e\hbar}{2m}\langle 1|\boldsymbol{\sigma}\cdot\boldsymbol{B}|1\rangle \tag{6.37}$$

$$= \frac{e\hbar B}{2m} \begin{pmatrix} 0 & 1 \end{pmatrix} \begin{pmatrix} \cos\theta & e^{-i\phi}\sin\theta \\ e^{i\phi}\sin\theta & -\cos\theta \end{pmatrix} \begin{pmatrix} 0 \\ 1 \end{pmatrix} \tag{6.38}$$

$$= -\frac{e\hbar B}{2m}\cos\theta. \tag{6.39}$$

So if the states are initially oriented with spins along an axis at angle θ to the field, then the energies are just reduced by a factor of $\cos\theta$ from the dot product, as would be expected.

However, now a new effect occurs. Consider matrix elements of the Hamiltonian between a spin-up and a spin-down state:

$$\langle 1|\hat{H}|0\rangle = \frac{e\hbar}{2m}\langle 1|\boldsymbol{\sigma}\cdot\boldsymbol{B}|0\rangle \tag{6.40}$$

$$= \frac{e\hbar B}{2m} \begin{pmatrix} 0 & 1 \end{pmatrix} \begin{pmatrix} \cos\theta & e^{-i\phi}\sin\theta \\ e^{i\phi}\sin\theta & -\cos\theta \end{pmatrix} \begin{pmatrix} 1 \\ 0 \end{pmatrix} \tag{6.41}$$

$$= \frac{e\hbar B}{2m}e^{i\phi}\sin\theta. \tag{6.42}$$

Similarly, $\langle 0|\hat{H}|1\rangle = \langle 1|\hat{H}|0\rangle^* = \frac{e\hbar B}{2m}e^{-i\phi}\sin\theta$. The fact that these matrix elements are nonzero implies that the magnetic field induces transitions between the $|0\rangle$ and $|1\rangle$ states, a fact that will be useful for implementing quantum information processing protocols in NMR and atomic systems (Chaps. 23 and 24).   ∎

Particular qubit states that often occur are states that are up or down along the x-axis,

$$|\uparrow\rangle_x = \frac{1}{\sqrt{2}}(|0\rangle + |1\rangle), \qquad |\downarrow\rangle_x = \frac{1}{\sqrt{2}}(|0\rangle - |1\rangle), \tag{6.43}$$

and those that are spin up or down along the y axis,

$$|\uparrow\rangle_y = \frac{1}{\sqrt{2}}(|0\rangle + i|1\rangle), \qquad |\downarrow\rangle_x = \frac{1}{\sqrt{2}}(|0\rangle - i|1\rangle). \tag{6.44}$$

Note that these states are analogous, respectively, to the diagonally polarized and circularly polarized states of light. Just as the polarization states of light can be described by the Poincaré sphere (Fig. 2.17), the spin states of a fermion are

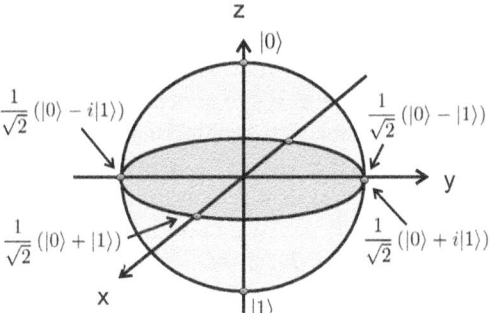

Fig. 6.3 Spin states of a spin-$\frac{1}{2}$ particle are represented by points on the Bloch sphere. The sphere represents the Hilbert space of the spins. Note that the Bloch sphere of spin and the Poincaré sphere (Fig. 2.17) of polarizations are isomorphic, so that the same qubit states can be encoded into either fermion spin or photon polarization

described by the analogous **Bloch sphere** (Fig. 6.3). Moreover, the polarization and spin states are in a one-to-one correspondence, so that the Hilbert spaces of the two systems are isomorphic.

A generic spin state on the Bloch sphere can be described by a unit vector $\hat{n}$. Let θ the angle of $\hat{n}$ from the z axis, and ϕ be the angle from the x-axis of $\hat{n}$'s projection into the xy plane. A state with spin up along $\hat{n}$ is then given by

$$| \uparrow \rangle_n = \cos \frac{\theta}{2} |0\rangle + e^{i\phi} \sin \frac{\theta}{2} |1\rangle, \tag{6.45}$$

with the spin-down state being

$$| \downarrow \rangle_n = \sin \frac{\theta}{2} |0\rangle - e^{i\phi} \cos \frac{\theta}{2} |1\rangle. \tag{6.46}$$

Rotating $\hat{n}$ by some angle $\Delta\Phi$ rotates the states on the sphere by $\frac{\Delta\Phi}{2}$, so that $\hat{n}$ has to rotate by 4π to return to the initial state, as expected for a fermion. As a consequence, states that are orthogonal (at $90°$) to each other in Hilbert space are opposite to each other (at $180°$) on the sphere.

6.3 Photons: Polarization Versus Orbital Angular Momentum

Introductory quantum courses spend a great deal of time on the spin of the electron and on the orbital angular momentum of electrons in atoms. But the spin and orbital angular momentum of photons are also important topics in applications.

Photons are spin-1 particles. For $s = 1$, we would expect to have a triplet of states with different spin orientation along the z-axis: $m_s = -1, 0, +1$, or equivalently, $S_z = \hbar m_s = -\hbar, 0, +\hbar$. However, electromagnetism has a symmetry known as **gauge invariance**. If the vector potential A is shifted by an amount $A \to A - \nabla\lambda(x)$, while simultaneously multiplying the wavefunctions of charged particles by a phase factor, $\psi(x) \to e^{iq\lambda(x)}\psi(x)$, for any function $\lambda(x)$, the behavior of the electromagnetic system remains unchanged. Although we won't prove it here (see, for example, [3, 4]) gauge invariance eliminates the $m_s = 0$ spin components, leaving only two components $m_s = \pm 1$ in propagating light waves. A brief introduction to gauge invariance and gauge fields will be given in Sect. 15.1.

The spin of a photon is usually measured relative to the propagation direction of the light, in which case it is called **helicity**. The helicity is defined to be the projection of the spin onto the direction of motion of the photon,

$$ h = \frac{S \cdot p}{|p|}, \tag{6.47} $$

where p is the photon momentum. Clearly, the helicity can only have two values, $h = \pm 1$, corresponding to the two values of m_s. These two helicity values also correspond to the two possible circular polarizations (Sect. 2.8): $h = -1$ corresponds to left-circular and $h = +1$ to right-circular polarization.

Since photons generally travel in straight lines, the idea that they may carry *orbital* angular momentum is a little more difficult to see, and was in fact, not widely realized until the 1990s. But in fact it is true that an individual photon can carry orbital angular momentum (OAM) $\hat{L}$ about its propagation axis [5], and that this OAM is easy to generate. (Many excellent reviews of optical OAM exist; for more details on the subject see, for example, [10–12].)

We normally picture light as being made from plane waves or spherical waves, but in reality the wavefronts of realistic light fields can have much more complicated spatial structures. Consider, for example, taking a plane wave and passing it through a material that gives it an azimuthally dependent phase shift of the form $e^{il\phi}$, where ϕ is the angle about the propagation axis, z. This amounts to tilting the wavefront, so that it falls farther backward as the propagation axis is circled. The wavefronts then have a corkscrew shape, as in Fig. 6.4. Recall that the Poynting vector, $W = E \times H$, should be perpendicular to the wavefront, which means that it rotates as the wave propagates. The rotation of the Poynting vector implies that the wave has orbital angular momentum about the z-axis.

The OAM is quantized, with $L_z = l\hbar$. Recall that the wavefunction for a state with angular momentum $l\hbar$ contains a phase factor, $e^{il\phi}$ (Eq. 6.10). The phase $\Phi = l\phi$ is increasing linearly as the axis is circled. Notice that (for $l \neq 0$) as you approach the propagation axis from different directions, you seem to get different values for the phase at the $r = 0$. But the phase has to be single-valued; it can't have multiple values at the same point. The only way to reconcile these two facts is if the amplitude of the light vanishes at the axis, so that the phase is unmeasurable there. So for $l \neq 0$, there always has to be a node (a dark spot of zero intensity) on the axis.

Fig. 6.4 Wavefront of an
optical beam with nonzero
orbital angular momentum.
The surfaces of constant
phase are corkscrew-shaped
and rotate as the light
propagates. The Poynting
vector, $\mathbf{W}$, must be
everywhere perpendicular to
the wavefront, so that it also
rotates as the wave propagates
along the z-axis, leading to a
nonzero angular momentum

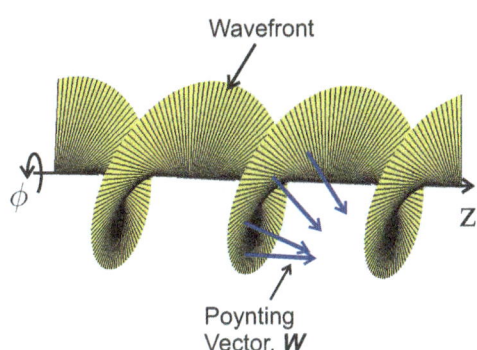

Wavefront

ϕ

z

Poynting
Vector, $\mathbf{W}$

The single-valuedness of the field under $\phi \rightarrow \phi+2\pi$ also forces l to be quantized
with integer values: $e^{il\phi}$ can only equal $e^{il(\phi+2\pi)}$ if l is an integer. We can think of
the $e^{i\Phi(\phi)}$ as a mapping that traces out a curve in the complex plane that encircles
the origin as ϕ varies from 0 to 2π. Different values of l lead to curves that circle the
origin different numbers of times. Since these different paths cannot be continuously
deformed into each other, the quantum number l classifies the states into different
topological classes and is often known as the *topological charge*.

One type of light beam that carries nonzero OAM is the class of the Laguerre–
Gauss (LG) modes. The optical states are labeled by a pair of integers, (l, p). l is
the topological charge defined earlier, while the index p characterizes the radial
structure of the mode, giving the number of nodes (points of zero amplitude) in the
radial direction. The spatial wavefunction of the light is then given by [6]:

$$\psi_{lp}(r, z, \phi) = \langle \mathbf{r} | l, p \rangle \tag{6.48}$$

$$= \frac{C_p^{|l|}}{w(z)} \left(\frac{\sqrt{2}r}{w(z)} \right)^{|l|} e^{-r^2/w^2(r)} L_p^{|l|} \left(\frac{2r^2}{w^2(r)} \right)$$

$$\times e^{-ikr^2 z/(2(z^2+z_R^2))} e^{-i\phi l + i(2p+|l|+1)\arctan(z/z_R)}, \tag{6.49}$$

with normalization $C_p^{|l|} = \sqrt{\frac{2p!}{\pi(p+|l|)!}}$ and beam radius $w(z) = w_0\sqrt{1 + \frac{z}{z_R}}$ at z.

$L_p^\alpha(x)$ are the associated Laguerre polynomials [7], and $z_R = \frac{\pi w_0^2}{\lambda}$ is the Rayleigh
range and the arctangent term is the Gouy phase. The index p characterizes the
radial structure of the mode, giving the number of nodes (points of zero amplitude)
in the radial direction.

In addition to the Laguerre–Gauss beams, a number of other optical beam modes
can carry nonzero OAM, including the higher-order Bessel or Hermite–Gauss
modes.

One way to generate optical OAM states is with a **spiral phase plates**. These
are transparent plates (Fig. 6.5a) whose thickness increases linearly as the angle

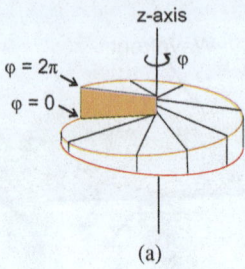

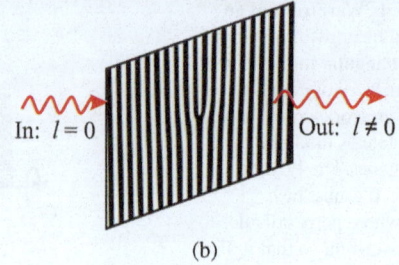

(a)   (b)

Fig. 6.5 Two methods for generating OAM states. (**a**) A spiral wave plate has a thickness that increases linearly as the azimuthal angle ϕ increases around the axis, so that the phase factor accumulated by light passing through it is $e^{il\phi}$. (**b**) A forked hologram: interference leads to a diffraction pattern with several maxima. When illuminated with a plane wave ($l = 0$), the light exiting in the direction of each maximum has OAM determined by the form of the dislocation at the center

Fig. 6.6 Dove prism. Total internal reflection at the horizontal surface causes images to be inverted vertically

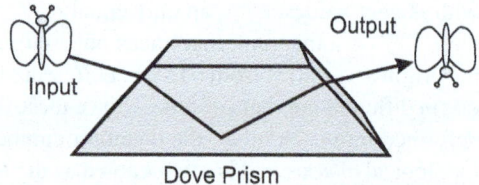

around the axis increases: $\Delta z = \frac{l\phi}{k(n-1)}$ [8], where n is the refractive index, and the surrounding medium is assumed to have index 1.0. This introduces a phase shift $\Delta \Phi = kz$ to the wavefronts, which is proportional to the azimuthal angle ϕ as required.

Another method for generating OAM is by the use of forked diffraction gratings [9] (Fig. 6.5b). Light diffracts from the pattern, and the presence of the discontinuous point in the middle gaurantees that the pattern has a node at the center and an azimuthally varying phase. Increasing the number of prongs in the fork pattern increases the angular momentum change of the outgoing light. These gratings can be in the form of computer generated holograms, or they can be programmed onto the surface of a spatial light modulator (SLM).

Measurement of the OAM of a beam can be accomplished several ways. One common method makes use of an interferometer containing a **Dove prism**. The prism, shown in Fig. 6.6, is designed so that total internal reflection occurs on the bottom surface. Tracing a few rays through the system, it should be easy to see that this results in images that are inverted in the vertical direction, without any corresponding horizontal inversion. Another useful property of these prisms is that they double rotations: if the prism is rotated by an angle θ about the propagation axis, then outgoing images and polarization vectors are rotated by 2θ [13].

The inversion of light reflecting off the bottom of the prism can be used to convert light with a well-defined value of orbital angular momentum, $L = l\hbar$, into light of the opposite OAM: $l \rightarrow -l$.

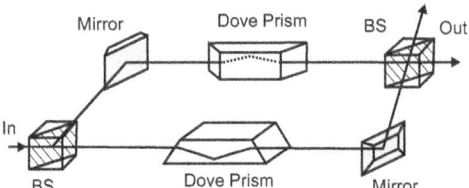

Fig. 6.7 One unit of an interferometer [14] designed to sort OAM values. The Dove prisms rotate the light in the two arms in opposite directions, so that interference at the final beam splitter causes even values of OAM to exit at one port, odd values to exit at the other. Multiple units of this kind can be used to provide additional sorting until the all different values of l present in the system exit at different ports

So one way to measure the OAM of a light beam is to construct an interferometric arrangement with Dove prisms that sorts different l values into different outgoing spatial modes [14–16]. An example of such a sorter is shown in Fig. 6.7 [14]. The basic unit consists of a Mach–Zehnder interferometer with a Dove prism in each arm. The two prisms are oriented at angle $\alpha/2$ relative to each other. When $\alpha = \frac{\pi}{2}$, this unit sorts even and odd values of l into different output ports of the final beam splitter. By concatenating n layers of such units with progressively smaller values of α and introducing appropriate OAM shifts Δl into one outgoing beam of each layer, all values of l up to $l = 2^n$ can be sorted into different output. A similar strategy, with the addition of units containing quarter-wave plates in place of Dove prisms, allows the values of both OAM and spin to be sorted simultaneously [15]. Note that this sorting can be done at the single photon level, since each photon individually has well-defined spin and OAM. Equally important, this sorting is nondestructive: the other properties of the photons are undisturbed by the sorting, so that additional measurements or manipulations can be subsequently done on them.

Beam-like states with $l \neq 0$ always have a singularity along their axis, and the phase $\Phi(\phi)$ wraps around the unit circle l times every time the angle ϕ circles the singular axis once. The phase $\Phi(\phi)$ defines a mapping of the unit circle to itself, $\Phi : S^1 \rightarrow S^1$. Such mappings define topological objects called **homotopy classes**, and the integer l represents the winding number of the given homotopy class. l is the first example we have seen in this book of a topological quantum number: an integer-valued topological invariant that can be used to distinguish different types of spaces or mappings from each other. In Chap. 21, we will look at some topological quantities in more detail and see that they come in a number of physical contexts.

6.4 Q-Plates

As discussed in the previous section, light can carry both intrinsic spin angular momentum **S** and orbital angular momentum **L**. A **q-plate** is a device that can

Fig. 6.8 The molecules of a
liquid crystal have a dipole
moment d, which defines the
director $\hat{n}$. The director
defines a line or axis in space,
giving the orientation of the
spatial molecule

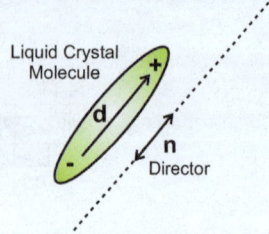

convert these two types of angular momentum into each other while preserving total
angular momentum $J = L + S$.

To discuss q-plates, we must first describe **liquid crystals**. A liquid crystal is a
state of matter that is, to some extent, intermediate between liquid and solid. The
molecules can move around, as in a liquid, but those molecules are very long, which
under some conditions makes it hard for them to move past each other, and they
have strong electric dipole moments, which allows the molecules to be aligned by
an external electric field in such a way that they can exhibit long-range order, similar
to crystalline order. Liquid crystals (first discovered in Austria in 1888) are common
in biological systems such as proteins and tobacco mosaic viruses and in detergents.

Let d be the electric dipole moment of the molecule. The unit vector $\hat{n} = \frac{d}{|d|}$ is
called the **director** (Fig. 6.8). The director and its negative are identified with each
other, $n \sim -n$, so that the director specifies a line or axis, but not a direction (up or
down) along that line. (In group theory terms, unit vectors are normally transformed
into each other via the group $SO(3)$, the group of rotations in three dimensions. But
imposing the equivalence relation $n \sim -n$ means that the director is described by
a **real projective group**, in which directions separated by 180° are identified with
each other. See Appendix E for a brief introduction to group theory.)

Recall that the torque on an electric dipole in the presence of a uniform external
electric field is given by

$$\tau = d \times E, \tag{6.50}$$

which acts to minimize the potential energy

$$U = -d \cdot E. \tag{6.51}$$

So, the director can be rotated by applying an electric field. Depending on the
temperature and the applied electric field, a number of distinct phases can be
defined for liquid crystal. For our purposes, two are important: the nematic ("thread-
like") and smectic phases ("soap-like"). In the **nematic phase**, the dipole moments
or directors are aligned, as shown if Fig. 6.9a. The molecules are free to move
around each other, flowing like the molecules of a liquid, so there is no long-range
positional order for the molecules (no crystal-like structure). But the orientations of
the directors are correlated, so that the dipole moments strongly ordered.

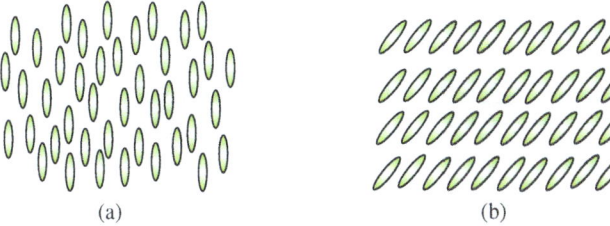

Fig. 6.9 (**a**) In the nematic phase, the molecular axes are aligned so that the molecules can easily move past each other as in a liquid. (**b**) In the smectic phase, the molecules form layers or sheets. This lamellar, or layered sheet, structure prevents motion in one direction, but allows the sheets to slide parallel to each other

At low temperatures, the liquid crystal will often enter a **smectic phase**, in which the molecules form sheets (Fig. 6.9b), with all the directors of the molecules in the same sheet aligned with each order at some angle from the plane of the sheet. Because of the orientation of the molecules, the sheets cannot move through each other: motion is constrained in the direction perpendicular to each layer. But the sheets can easily slide over each other in the parallel direction; this sliding layer behavior is common in soaps, which leads to the name of the phase.

Because of the shape of the molecules, the dipole moment can easily oscillate along the direction of the director, but not perpendicular to it. As a result, there is selective absorption of light in the material: light polarized along the director is much more strongly absorbed than the orthogonal polarization state. The material is also birefringent and optically active, leading to relative phase shifts between different polarizations as the light propagates through the liquid crystal.

Many televisions and computer screens use **liquid crystal displays (LCD's)**. Each pixel in the LCD consists of a liquid crystal wedged between two perpendicular polarizing filters (Fig. 6.10). An electric field is used to control the orientation of the liquid crystal molecules. In the top of the figure, the field rotates gradually as a function of z, so that the molecules rotate from the direction of one polarizer to the direction of the other. Each molecule acts as a polarizer itself, rotating the optical polarization slightly from the direction of the previous molecule. The result is that the light's polarization is rotated 90° as it propagates from the input to the output; thus any light that passed the first filter (on the left) will also pass the second filter on the right. The pixel is then transparent and the input light is transmitted: the pixel is bright.

On the bottom of the figure, the electric field is now oriented along the z-axis, so that all the molecules are oriented parallel to the direction of propagation of the light. There is no rotation of the polarization, so any light that was passed by the first filter will be blocked by the second. The system is therefore opaque, and the pixel is dark. Thus by using the electric field to control the liquid crystal, we can switch the LCD pixel on and off, as desired.

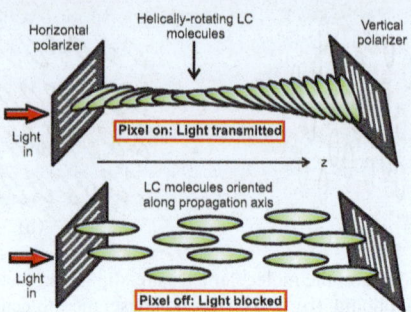

Fig. 6.10 An LCD pixel. Two crossed polarizers have a liquid crystal between them. Top: an electric field is applied to make the molecular orientations rotate as the z-axis is traversed. As the molecules rotate, the transmitted polarization state also rotates, interpolating between the directions of the two polarizing filters. Light can therefore exit at the right, so the pixel is bright. Bottom: the electric field orients the molecules parallel to the z-axis, so there is no rotation of the light's polarization. Therefore, any light entering the first filter will be blocked be the second, and the pixel stays dark

A **q-plate** [17, 18] is then a device made from a thin nematic liquid crystal film wedged between two glass plates. Let ϕ be the angle from some reference axis (the x-axis, say) in the plane of the plate. A electric field is applied in such a way that the director rotates as the propagation axis (z) is circumnavigated; specifically, it is arranged so that the angle α of the director from the reference axis is given by

$$\alpha(\phi) = q\phi + \alpha_0, \tag{6.52}$$

where α_0 is a constant (the value of α on the x-axis). As one complete circling of the z-axis ($\phi \to \phi + 2\pi$) is completed, the director rotates through q turns. If the director was a vector, the requirement that it be single-valued under rotations of α by 2π would force q to be an integer. However, the director is not a vector since it is subject to the equivalence relation $\hat{n} \sim -\hat{n}$; the single-valuedness condition then implies that values of α differing by π also be physically equivalent. This means that q can be either an integer or half-integer. If $q > 0$, then the director rotates counterclockwise with increasing ϕ, and $q < 0$ for clockwise rotation.

ϕ and α are undefined on the z-axis, so the axis is a line of singular points. q is an example of a topological charge or topological index. Defining a unit vector field $\hat{n}(\boldsymbol{r})$, where r is the position vector, q is unchanged under continuous deformations of the field $\hat{n}(\boldsymbol{r})$. The topological nature of q also is apparent in the fact that it measures a global property of the field (the number of rotations of the director around a closed curve), not a local property defined at each point.

Being birefringent, the q-plate also imparts a phase shift δ between orthogonal linear polarization components. δ will depend on not only on the thickness of the liquid crystal film but also on the applied electric field. So, by varying the field, δ can be tuned to a desired value.

So now consider left-handed and right-handed polarization states of incoming light. Using quantum-mechanical notation, we can denote these states as $|L\rangle$ and $|R\rangle$, and write them as column matrices:

$$|R\rangle = \begin{pmatrix} 1 \\ 0 \end{pmatrix}, \qquad |L\rangle = \begin{pmatrix} 0 \\ 1 \end{pmatrix}. \tag{6.53}$$

Then the action of the q-plate on the light is given by

$$|R\rangle \to \cos\frac{\delta}{2}|R\rangle + ie^{-2i\alpha(\phi)}\sin\frac{\delta}{2}|L\rangle \tag{6.54}$$

$$|L\rangle \to \cos\frac{\delta}{2}|L\rangle - ie^{+2i\alpha(\phi)}\sin\frac{\delta}{2}|R\rangle, \tag{6.55}$$

or in matrix form,

$$|\psi_{out}\rangle = U|\psi_{in}\rangle, \tag{6.56}$$

where

$$U = \begin{pmatrix} \cos\frac{\delta}{2} & -ie^{-2i\alpha(\phi)}\sin\frac{\delta}{2} \\ ie^{2i\alpha(\phi)}\sin\frac{\delta}{2} & \cos\frac{\delta}{2} \end{pmatrix}. \tag{6.57}$$

Now consider the case where $\delta = \pi$ and q is a half-integer. Because $\delta = \pi$, Eq. 6.57 tells us that the polarization is reversed,

$$|L\rangle \to ie^{2i\alpha(\phi)}|R\rangle \qquad |R\rangle \to ie^{-2i\alpha(\phi)}|L\rangle, \tag{6.58}$$

which means that the spin of the photon is changed by ± 1 in units of $\hbar$. Because q is half-integer, the exponential factors tell us that the phase changes by an integer multiple of 2π when the z-axis is circled, giving a change in OAM $\pm 2q$ (in units of $\hbar$). In particular, if $q = \frac{1}{2}$, then the total angular momentum of each photon is conserved, with one unit of spin being converted into one unit of OAM (Fig. 6.11).

Q-plates have a number of advantages that make them highly useful in modern optical systems. They have a high conversion efficiency between spin and OAM (typically >95%) and high transmissivity (>85%). They are compact, stable, and easily controlled electrically, and work over a large frequency range.

Q-plates have found a wide variety of uses in quantum information processing and quantum communication applications that involve photon OAM, since they allow easily creation and control of specific OAM values; see, for example, [19–21]. Another application is the creation of vector beams (Fig. 6.12, light beams in which the polarization forms a nontrivial spatial pattern in the transverse plane [22–24]. For a more detailed review of q-plates, see [25].

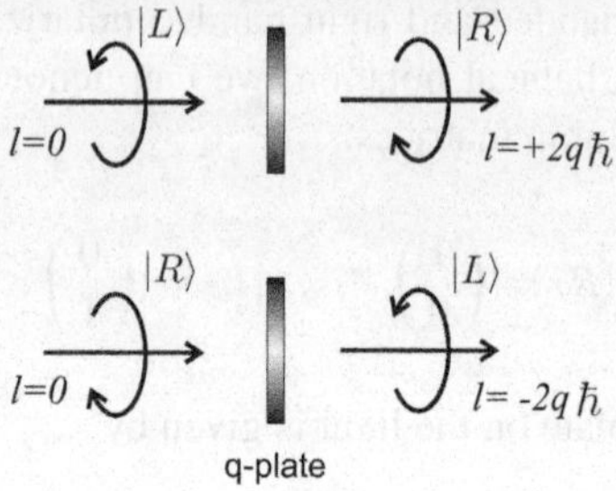

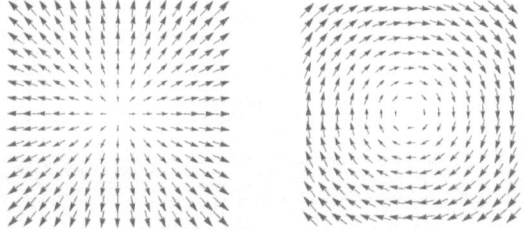

Fig. 6.11 Effect of a q-plate with birefringent phase shift $\delta = \pi$ and topological charge $q = \frac{1}{2}$. The circular polarization of a photon flips, causing a change in spin angular momentum $\Delta s = \pm\hbar$, while the orbital angular momentum l has the opposite shift, $\Delta l = \mp 2q\hbar = \mp\hbar$

Fig. 6.12 An optical vector field is one in which the polarization direction of the field varies from point to point in the field. Here two examples are shown: on the left, the polarization is radially outward from the propagation axis. On the right, the polarization circulates about the axis. In both cases, the light is propagating perpendicular to the page, and the field is singular along the propagation axis: the polarization would not be single-valued at the axis, so the amplitude of the field is forced to vanish there

Problems

1. Prove the relations 6.20 to 6.23 for the total angular momentum raising and lowering operators.
2. For any two vectors a and b, prove that the following identity holds:

$$(a \cdot \sigma)(b \cdot \sigma) = (a \cdot b)\,\hat{I} + i\,(a \times b) \cdot \sigma.$$

3. Verify that the operators $\hat{L}_z$ and $\hat{L}^2$ applied to the spherical harmonics of Eq. 6.10 give back the respective eigenvalues $m\hbar$ and $l(l+1)\hbar^2$.
4. Consider a beam of photons of angular momentum quantum number l.

   (a) If a dielectric particle of mass m absorbs photons from this beam at a rate of $\frac{dN}{dt}$, find the torque exerted on the particle. Because of this torque, OAM beams can be used as optical spanners or optical wrenches.

(b) Suppose that instead of a state of definite angular momentum $|l\rangle$, the photons being absorbed are in the state $|\psi\rangle = \cos\theta|l\rangle + \sin\theta|-l\rangle$. What is the mean torque exerted now?

5. Consider a two-state system. In the absence of interactions, the free Hamiltonian can be written in the form

$$\hat{H}_0 = \epsilon_0|0\rangle\langle0| + \epsilon_1|1\rangle\langle1| = \begin{pmatrix} \epsilon_1 & 0 \\ 0 & \epsilon_2 \end{pmatrix},$$

where ϵ_0 and ϵ_1 are the energies of the two states. Now suppose an interaction potential is turned on:

$$\hat{H} = \hat{H}_0 + \hat{V}, \qquad \text{where} \qquad \hat{V} = \begin{pmatrix} 0 & V_{ab} \\ V_{ba} & 0 \end{pmatrix}.$$

(a) In order for the system to be Hermitian, what constraint(s) must be placed on the entries of $\hat{V}$?

(b) In the presence of the interaction, what are the new energy eigenvalues, $E_\pm$? What are the energy eigenstates?

(c) Now let $V_{ab} = Ve^{-i\phi}$. Define the mean energy $E = \frac{1}{2}(\epsilon_0 + \epsilon_1)$ and the energy gap $\Delta = \frac{1}{2}(\epsilon_0 - \epsilon_1)$. Show that the energies can be written in the forms

$$E_\pm = E \pm \Omega = E \pm \Delta\sec 2\theta,$$

where $\Omega = \sqrt{V^2 + \Delta^2}$ and $\tan 2\theta = \frac{V}{\Delta}$. ($\Omega$ is proportional to the Rabi frequency, which will appear in later chapters.)

(d) Suppose that the system is initially in a state $|0\rangle$ at time $t = 0$. Using the eigenvalues and eigenstates of part (b), write down the state vector $|\psi(t)\rangle$ at later times.

6. Suppose a two-state system has energy levels $E = 0$ and $E = \epsilon$. A collection of N classical particles populate copies of this system.

(a) Find the average occupation numbers N_0 and N_1 of the two energy levels at temperature T.

(b) Find the mean energy of the N-particle system.

7. Consider a two-state system with Hamiltonian $\hat{H} = E_0\sigma_z$. A pair of observables are given by

$$\hat{A} = \hat{\sigma}_y, \qquad \hat{B} = \begin{pmatrix} c & -i\sqrt{c} \\ i\sqrt{c} & 1 \end{pmatrix},$$

for some constant c.

(a) Express $\hat{B}$ as a linear combination of Pauli matrices and the identity matrix.
(b) For $c \neq 1$, show that $\hat{A}$ and $\hat{B}$ are incompatible variables, $\left[\hat{A}, \hat{B}\right] \neq 0$.
(c) Do the two observables commute with the Hamiltonian? Are they conserved?
(d) Find the eigenvectors and eigenvalues of both operators.
(e) Suppose that the qubit is initially in the $\hat{A}$ eigenstate with eigenvalue $+$ 1. Let $p_{A1}(t)$, $p_{A2}(t)$, $p_{B1}(t)$, $p_{B2}(t)$ be the probabilities of measuring each of the possible values of $\hat{A}$ and $\hat{B}$ as functions of time. Assuming free time evolution under the Hamiltonian $\hat{H}$, find each of these probabilities as functions of time.
(f) Again assume the initial state of part (e). Suppose $\hat{B}$ is measured, and then immediately afterward (an infinitesimal time after $t = 0$), $\hat{A}$ is measured. What are the possible combinations of the outcomes for the pair of measurements? What is the probability of each combination?

8. Using Mathematica or any other method, try the following:

(a) Plot both the amplitude and phase (on separate plots) of the Laguerre–Gauss beams $\psi_{lp}(r, z, \phi)$ in the (x, y) plane for fixed z. Do this for $l = 0, 1, 2$ and $p = 0, 1, 2$.
(b) Consider interference patterns formed by equal superpositions of Laguerre–Gauss beam pairs, $I = \left| \frac{1}{\sqrt{2}} \left(\psi_{lp}(r, z, \phi) + \psi_{l'p'}(r, z, \phi) \right) \right|^2$. Plot the intensities of these superpositions for several combinations of l, l', p, p' and see what kinds of patterns can be formed. (z is still held constant.)

References

1. R. Shankar, *Principles of Quantum Mechanics*, 2nd edn. (Plenum Press, 1994)
2. D.J. Griffiths, D.F. Schroeter, *Introduction to Quantum Mechanics*, 3rd edn. (Cambridge University Press, Cambridge, 2018)
3. L.H. Ryder, *Quantum Field Theory* (Cambridge University Press, Cambridge, 1996)
4. A. Zee, *Quantum Field Theory in a Nutshell*, 2nd edn. (Princeton University Press, Princeton, 2010)
5. L. Allen, M.W. Beijersbergen, R.J.C. Spreeuw, J.P. Woerdman, Phys. Rev. A **45**, 8185 (1992)
6. L. Allen, M. Padgett, M. Babiker, Prog. Opt. **39**, 291 (1999)
7. G. Arfken, H. Weber, *Mathematical Methods for Physicists* (Academic Press, London, 2000)
8. M.W. Beijersbergen, R. Coerwinkel, M. Kristensen, J.P. Woerdman. Opt. Commun. **112**, 321 (1994)
9. V. Yu. Bazhenov, M.V. Vasnetsov, M.S. Soskin, JETP Lett. **52**, 429 (1990)
10. A.M. Yao, M.J. Padgett, Adv. Opt. Photon. **3**, 161 (2011)
11. J.P. Torres, L. Torner (eds.), *Twisted Photons: Applications of Light with Orbital Angular Momentum* (Wiley, Hoboken, 2011)
12. S. Franke-Arnold, L. Allen, M. Padgett, Laser Photon. Rev. **2**, 299 (2008)
13. M.J. Padgett, J.P. Lesso, J. Mod. Opt. **46**, 175 (1999)

14. J. Leach, M.J. Padgett, S.M. Barnett, S. Franke-Arnold, J. Courtial, Phys. Rev. Lett. **88**, 257901 (2002)
15. J. Leach, J. Courtial, K. Skeldon, S.M. Barnett, S. Franke-Arnold, M.J. Padgett, Phys. Rev. Lett. **92**, 013601 (2004)
16. C. Gao, X. Qi, Y. Liu, J. Xin, L. Wang, Opt. Commun. **284**, 48 (2011)
17. L. Marrucci, C. Manzo, D. Paparo, Phys. Rev. Lett. **96**, 163905 (2006)
18. L. Marrucci, C. Manzo, D. Paparo, Appl. Phys. Lett. **88**, 221102 (2006)
19. A. Sit et al., Optica **4**, 1006 (2017)
20. I. Nape et al., Opt. Exp. **26**, 26946 (2018)
21. B, Ndagano et al., Nat. Commun. **4**, 2432 (2013)
22. F. Cardano et al., Appl. Opt. **51**, C1 (2012)
23. F. Cardano et al., Opt. Exp. **21**, 8815 (2013)
24. D'Errico et al., Optica **4**, 1350 (2017)
25. A. Rubano, F. Cardano, B. Piccirillo, L. Marrucci, J. Opt. Soc. Am. B **36**, D70 (2019)

Chapter 7
Superconductivity

In this chapter, the basics of superconductivity are covered, along with technological spin-offs, the Josephson function, and the SQUID. Superconducting devices are interesting in their own right, not least because they are an example of a system that can have a coherent wavefunction on a macroscopic size scale. But superconducting materials and devices are also promising systems for implementing quantum computation protocols.

Applications of Josephson junctions in quantum information processing will be discussed in Sect. 26.3. More extensive introductions to superconductivity can be found in Refs. [1–3].

7.1 Superconductors: Basic Properties and Early Models

According to Bloch's theorem, a perfect crystal lattice would transmit electron plane waves without reflection. Such a perfect lattice would have zero resistance. But in real life, no crystal lattice is perfect: there will always be impurities and lattice defects (including the edge of the crystal itself). In addition to these static defects and impurities, at nonzero temperatures, the lattice ions will be vibrating around their equilibrium positions, creating time-dependent defects in the lattice. These lattice vibrations are collective excitations of many ions and can be thought of as quantized sound waves, or **phonons**, propagating through the lattice. The electrons scatter off the phonons, contributing to the resistance. At higher temperatures, more phonons are present leading to increased scattering.

So as a result, most materials at low temperatures have a resistivity with a temperature dependence of the form

$$\rho(T) = \rho_0 + \rho_2 T^2, \tag{7.1}$$

© The Author(s), under exclusive license to Springer Nature Switzerland AG 2025
D. S. Simon, *Introduction to Quantum Science and Technology*, Undergraduate
Texts in Physics, https://doi.org/10.1007/978-3-031-81315-3_7

where ρ_2 is due to phonon scattering and ρ_0 is due to the residual scattering by static defects even in the absence of phonons. In 1911, Heike Kamerlingh Onnes was making measurements of resistivity as a function of temperature. Most of the materials studied had the expected temperature dependence, leveling off to a constant as T approached zero. However, when he examined a sample of solid mercury (mercury becomes solid below $-38.8\,°C$), he found a sudden precipitous drop in resistivity at around 4 K. Further studies verified that the resistivity is exactly zero below the **critical temperature** $T_c = 4.1\ K$. This was the first example of **superconductivity** to be discovered. The temperature dependences of a normal metal and an superconductor are shown in Fig. 7.1.

The critical temperatures of some superconducting materials are given in Table 7.1. The materials in the table are all single elements, but superconducting alloys and compounds also exist.

In 1933, Meissner and Ochsenfeld discovered another property of superconductors: flux exclusion. They placed samples of materials into an external magnetic field above the critical temperature. As the temperature decreased below T_c, they found

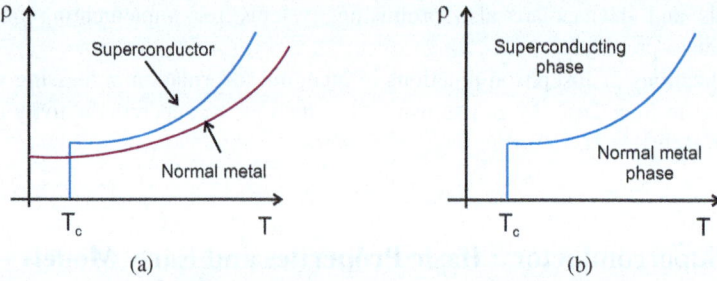

Fig. 7.1 (**a**) The resistivity versus temperature for a normal (non-superconducting) metal and a superconductor. The resistivity of a normal metal approaches a nonzero value as $T \to 0$. But the resistivity of a superconductor drops suddenly to zero when T reaches the critical temperature. (**b**) The superconductor has a phase transition between two phases, with the phase transition along the curve: the superconducting phase is above the curve and the normal phase is below

Table 7.1 Atomic numbers and critical temperatures of some elemental materials that form conventional superconductors at normal pressures. Additional materials superconduct at high pressures. High-temperature superconductors are not included here

Element	Atomic number (Z)	T_c
Aluminum (Al)	13	1.2 K
Titanium (Ti)	22	0.39 K
Vinadium (V)	23	5.3 K
Zinc (Zn)	30	0.88 K
Zirconium (Zr)	40	0.65 K
Niobium (Nb)	41	9.3 K
Tin (Sn)	50	3.7 K
Tungsten (W)	74	0.01 K
Mercury (Hg)	80	4.1 K
Thallium (Tl)	81	2.4 K
Lead (Pb)	82	7.2 K

that the external field was gradually expelled from the sample, eventually becoming completely excluded from the material. Further, if the external field is applied while the sample is already below T_c, the field lines bend to avoid the superconductor, maintaining $\boldsymbol{B} = 0$ inside the material. These flux exclusion effects are known as the **Meissner–Ochsenfeld effects**. In effect, the superconductor acts as a perfect diamagnetic: currents are set up in the superconductor that exactly cancel the applied field. There is a limit to this flux exclusion: if the external field is sufficiently strong, it can destroy the superconductivity.

It is possible for some materials, such as those exhibiting certain types of quantum Hall states, to have vanishing resistivity without displaying any Meissner–Ochsenfeld effect. As a result, the ability to exclude magnetic flux is usually the criterion used to determine whether a material is superconducting or not.

A further effect, known as the **isotope effect**, was discovered in 1950: the critical temperature of a material depends on the isotopic mass of the solid M according to the relation

$$M^{1/2}T_c = constant. \tag{7.2}$$

This implies that T_c should vanish in the absence of lattice vibrations: $T_c \to 0$ is required as $M \to \infty$, in other words, as lattice vibrations are frozen out by the high inertia of the heavy ions. This provided strong evidence that phonons played a role.

Until 1986, all known superconducting materials only became superconductors at very low temperatures. But that year, Bednorz and Müller discovered that the material $La_{2-x}Ba_xCuO_4$ became superconducting at about 35 K, becoming the first known **high-T_c superconductor**. Since then, materials have been found that have critical temperatures up to about 135 K at normal pressures and up to 250 K at extreme high pressure. These high-temperature superconductors work via a different physical mechanism than the so-called conventional (or low-T_c) superconductors. We won't discuss high T_c-superconductors further here; see [4] for more information. More recently, a new class of superconductors called **topological superconductors** have also been the subject of intensive research; see [5, 6].

Consider a superconducting ring, such as that of Fig. 7.2. Because of the vanishing of the resistivity, once a current is created, it can exist indefinitely in the ring without application of an external voltage. These so-called persistent currents have been observed to last for years in rings kept below the critical temperatures. Recall that the electric field $\boldsymbol{E}$ and current density (current per area) $\boldsymbol{j}$ are related by Ohm's law:

Fig. 7.2 A superconducting ring, carrying a supercurrent I. If a magnetic field is applied, it is excluded from the superconductor, but the magnetic flux can be trapped in the hole of the ring

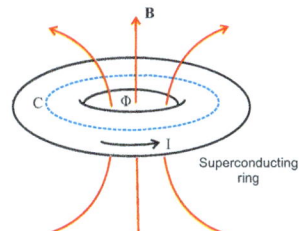

$$E(r, t) = \rho j(r, t). \tag{7.3}$$

For vanishing resistivity, $\rho = 0$, so this implies that the electric field must vanish in the material, $E = 0$. Combined with Faraday's law, it also implies that the flux Φ through the ring should be time-independent:

$$\frac{d\Phi}{dt} = \mathcal{E} = -\int_C E \cdot dl = 0, \tag{7.4}$$

where the integral is over some closed loop C inside the superconducting ring that encloses the central hole, and $\mathcal{E}$ is the induced EMF.

The method of creating a persistent current then consists of applying an external magnetic field B while the material is above the critical temperature, thus creating a flux through the ring. Then lower the temperature below T_c. The magnetic field in the material now vanishes by the Meissner–Ochsenfeld effect, but the time independence of the flux implies that the flux is still nonzero. There must therefore now be a current through the ring to maintain the flux via Ampere's law. Note that the flux is essentially trapped by the superconducting ring: it cannot dissipate because the magnetic field lines cannot move through the superconductor in order to disperse. It will be seen further that the flux is in fact quantized in integer multiples of a basic unit of flux, Φ_0.

Superconductors can be classified into two groups, based on how they behave as the magnetic field H increases. Recall that the magnetic flux density B, the magnetic field H, and the magnetization of the material (the magnetic dipole moment per volume) M are related by

$$B = \mu_0 (H + M). \tag{7.5}$$

Since $B = 0$ in the superconductor, we would expect that $H = -M$. If M is plotted versus H, we should therefore expect a straight line (Fig. 7.3a). The slope of the line is given by the magnetic susceptibility χ, which is

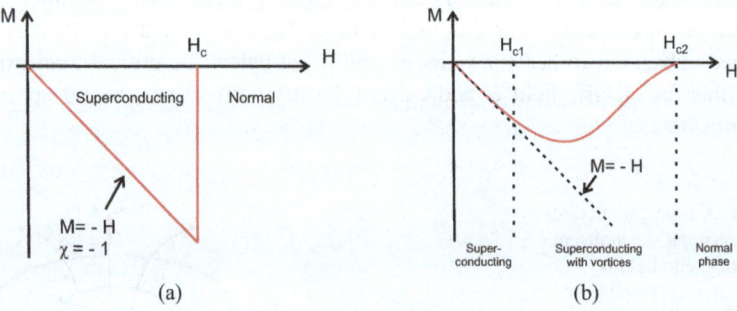

Fig. 7.3 Magnetization versus external magnetic field H for (a) type I and (b) type II superconductors

$$\chi = \frac{dM}{dH} = -1. \tag{7.6}$$

(The fact that $\chi = -1$ is again another way of saying that the superconductor is a perfect diamagnet.) Materials in which relation holds for small H are called **type-I superconductors** (Fig. 7.3b). The linear relation continues until the critical field H_c is reached; once $|H| = H_c$, the superconductivity is destroyed and the magnetization drops suddenly to zero.

For some materials, called **type-II superconductors**, this behavior does not hold. For these substances, there are actually two critical fields, the upper and lower critical fields, denoted respectively as H_{c1} and H_{c2}, with $H_{c1} < H_{c2}$. The plot of M versus H then looks like Fig. 7.3b, with the linear relation breaking down at H_{c1}. The slope then reverses sign, so that M eventually returns to zero in a continuous fashion at H_{c2}. The reason that the magnetization vanishes gradually instead of suddenly is that between H_{c1} and H_{c2}, magnetic fields begin to thread their way through the material, forming tubes of normal phase that are shielded from the surrounding superconducting phase by vortices. Vortices start to form at H_{c1}, and their density in the material increases as the field intensifies, eventually filling the entire volume of the material at $H = H_{c2}$. Plots of the magnetic field versus magnetization and versus temperature for type-I and type-II superconductors are shown in Figs. 7.3 and 7.4.

The vortices aforementioned are called **Abrikosov vortices**. Within each, there is a small core of normal (non-superconducting) material, with a magnetic field B along the vortex axis. Around this core, a supercurrent j circulates; by Ampere's law, j produces its own B field, screening the field of the core from the material outside the vortex. Each vortex has a flux

$$\Phi_0 = \frac{h}{2e}. \tag{7.7}$$

If there are N vortices in a sample of cross-sectional area A, then the total flux in the material is $\Phi = N\Phi_0$ and the average field is $B = \frac{\Phi_0 N}{A} = \frac{Nh}{2eA}$. The density of vortices is therefore given by

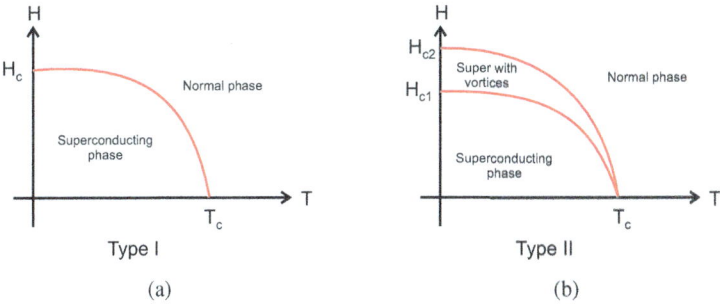

Fig. 7.4 Magnetic field H versus temperature for (**a**) type I and (**b**) type II superconductors

$$n = \frac{N}{A} = \frac{2eB}{h}. \tag{7.8}$$

One of the earliest attempts to construct a theory of superconductivity was due to the brothers Fritz and Heinz London in 1935. They viewed the electron gas in the material as a fluid composed from two distinct components: a superconducting component with number density n_s and a normal component of density n_n, give a total electron density $n = n_n + n_s$. One of the key results of their work is the **London equation**, which relates the superconducting current density j to the vector potential A. First, define a particular gauge for the vector potential, known as the **London gauge**, and defined by the condition that

$$\nabla \cdot A = 0, \tag{7.9}$$

with $A \cdot \hat{n} = 0$ at the superconductor surface, where $\hat{n}$ is a unit vector perpendicular to the surface. Then, in this gauge, the London equation can be written as

$$j = -\frac{n_s e^2}{m} A, \tag{7.10}$$

where m is the electron mass. This is also often written in the form

$$j = -\frac{1}{\mu_0 \lambda^2} A, \tag{7.11}$$

where the **London penetration depth** is defined as

$$\lambda = \sqrt{\frac{m}{\mu_0 n_s e^2}}. \tag{7.12}$$

As the name implies, this measures the depth to which an external magnetic field can penetrate the superconductor; specifically, if the magnetic field is B_0 at the surface, then the field at a depth x below the surface is

$$B(x) = B_0 e^{-x/\lambda}. \tag{7.13}$$

Using the London equation, it can be shown [3] that vortex solutions can appear in the material's interior. Assuming a B field along the z-axis, these solutions have the form

$$B_z(r) = \frac{\Phi_0}{2\pi \lambda^2} K_0 \left(\frac{r}{\lambda}\right), \tag{7.14}$$

where r is the distance from the center of the vortex core and $K_0(r)$ is a modified Bessel function [7]. Near the center of the vortex ($r \ll \lambda$), the material is in a non-superconducting phase and the Bessel function is approximately logarithmic, so that

$$B_z(r) \approx \frac{\Phi_0}{2\pi\lambda^2} \ln\left(\frac{\lambda}{r}\right), \qquad (r \ll \lambda). \qquad (7.15)$$

On the other hand, far from the core center, the Bessel function decays exponentially, giving

$$B_z(r) \approx \frac{\Phi_0}{2\pi\lambda^2}\sqrt{\frac{\pi\lambda}{2r}}e^{-r/\lambda} \qquad (r \ll \lambda) \qquad (7.16)$$

the field therefore vanishes far outside the vortex, as expected from the fact that the surrounding matter is superconducting.

Building on the London model, Ginsburg and Landau constructed a model of superconductivity based on the existence of an **order parameter** ψ such that $\psi = 0$ for $T > T_c$, while $\psi(T)$ is a nonzero function of temperature for $T < T_c$. The **Ginzburg–Landau model (GL)** then proceeds to model the phase transition that occurs at the critical temperature.

Allow the order parameter to be complex-valued and to be position-dependent; in other words, we take $\psi(r)$ to be a complex scalar field. (Here we suppress dependence of ψ on temperature or other parameters.) It will often be useful to write ψ in polar form, in terms of its absolute value and phase,

$$\psi(r) = |\psi(r)|e^{i\phi(r)}. \qquad (7.17)$$

When this field is spatially constant (the spatially varying version will be discussed further), the London super-current density can be written as

$$j_s = -\frac{(2e)^2}{2m^*}|\psi|^2 A = -\rho_s \frac{(2e)^2}{\hbar^2} A, \qquad (7.18)$$

where m^* is the effective mass of the electron and the **superfluid stiffness** ρ_s is defined as

$$\rho_s = \frac{\hbar^2}{2m^*}|\psi|^2. \qquad (7.19)$$

Comparing Eq. 7.18 to the London equation (Eq. 7.10) indicates that we should make the identification $n_s = 2|\psi|^2$.

The free energy per unit volume, $f = \frac{dF}{dV}$, can be expanded in a Taylor series in powers of $|\psi|$ and will take the form

$$f_s(T) = f_n(T) + a(T)|\psi|^2 + \frac{1}{2}b(T)|\psi|^4 + \ldots \qquad (7.20)$$

where f_s is the free energy of superconducting phase, f_n is the free energy in the normal phase (when $\psi = 0$), and near the phase transition point, the higher-order terms should be small. The Taylor expansion coefficients a and b are temperature-dependent, and b is assumed to be positive. a and b can themselves be Taylor expanded in terms of temperature and are assumed of the form:

$$a(T) = c(T - T_c)\ldots, \qquad\qquad b(T) = b_0 + \ldots \qquad (7.21)$$

When $T > T_c$, the coefficient $a(T)$ will then be positive, and the unique minimum of the free energy will occur at $T = 0$, i.e. the normal phase will have lower energy. However, when $T < T_c$, a becomes negative. Minimizing by requiring that $\frac{df_s}{d|\psi|} = 0$ yields $|\psi| = \sqrt{\frac{c}{b_0}}(T - T_c)^{1/2}$, which is nonzero, in agreement that the material is superconducting for $T > T_c$.

Further, the free energy gap per volume between the two phases should be of the form $\Delta f(T) = f_s(T) - f_n(T) = -\frac{1}{2}\mu_0 H_c^2$. Combined with the expressions of f_s and $|\psi|$ earlier, this implies that near the critical temperature, the temperature dependence of the critical field is of the form

$$H_c(T) = \frac{c}{\sqrt{\mu_0 b_0}}(T_c - T). \qquad (7.22)$$

If the order parameter is allowed to be position-dependent, then spatial variations of $|\psi|$ would be expected to increase the energy, so we would expect to add a kinetic energy-type term involving the gradient. The GL free energy density then becomes

$$f_s(T) = f_n(T) + \frac{\hbar^2}{2m^*}|\nabla\psi(r)|^2 + + a(T)|\psi(r)|^2 + \frac{1}{2}b(T)|\psi(r)|^4 + \ldots. \qquad (7.23)$$

When minimized, this gives a Schrödinger-type equation for the order parameter,

$$-\frac{\hbar^2}{2m^*}\nabla\psi(r) + \left(a + b|\psi|^2\right)\psi(r) = 0. \qquad (7.24)$$

This is a *nonlinear* Schrödinger equation (due to the $|\psi|^2$ term) and consequently ψ does not obey a superposition principle. Nonetheless, this bolsters the case for treating the order parameter ψ as a kind of wavefunction for the superconducting system; this will be discussed further.

The effect of a magnetic field on the system can be incorporated using the **minimal coupling procedure**, which is well-known from classical electrodynamics and elementary quantum mechanics. It involves shifting the momentum by a factor proportional to the gauge field,

$$p \to p - qA, \qquad (7.25)$$

where q is the particle charge. Since, as a quantum mechanical operator, the momentum is given by $\boldsymbol{p} = -i\hbar\nabla$, this is equivalent to the substitution

$$\nabla \to \nabla - \frac{iq}{\hbar}\boldsymbol{A}. \tag{7.26}$$

Making this substitution in Eq. 7.24 gives the full equation of motion of the system in the presence of a magnetic field. BCS theory (see further) indicates that the charge in these equations should be taken to be *twice* the electron charge, $q = -2e$.

Box 7.1 Spontaneous Symmetry Breaking and the Higgs Mechanism

Energies of form similar to Eq. 7.23 are common in many areas of physics. At $T = 0$, consider a simplified version with a potential energy of the form

$$V(\phi) = \mu^2\phi^2 + \lambda\phi^4, \tag{7.27}$$

where μ^2 and λ are parameters and ϕ is some scalar (i.e. spinless) field. For simplicity, assume the field is real. This potential has a symmetry under reflection $\phi \to -\phi$. (This generalizes to a rotational invariance in the complex plane, $\phi \to e^{i\theta}\phi$, when the field is complex.) When $\mu^2 > 0$, the potential has a single global minimum at the origin (Fig. 7.5a). But, if μ^2 is allowed to vary, a phase transition occurs when it passes through zero, giving two degenerate minima for $\mu^2 < 0$ (Fig. 7.5b). These minima are often called *vacuum states*. Any quantum state that was initially at the origin ($\phi = 0$) will roll down the hill to one of the new minima. Because of the reflection symmetry of the potential and the degeneracy of the minima, there is no reason for the system to prefer one minimum over the other. But quantum fluctuations will randomly cause it to roll one way or the other, choosing one vacuum state over the other. The result is that, despite the reflection symmetry of the potential, the ground state will be asymmetric. This spontaneous evolution to a state that has less symmetry than the potential is called **spontaneous symmetry breaking (SSB)**.

SSB occurs in many areas of physics, but the most famous example is in particle physics and is known as the **Higgs mechanism** (or sometimes as the Anderson–Englert–Brout–Higgs–Guralnik–Hagen–Kibble mechanism, given its repeated discovery in various contexts). In this case, the field ϕ is complex and represents the Higgs scalar field. When SSB occurs, the field settles into one of the nonzero minima with magnitude $|\phi| = v$, where $v = \pm\sqrt{\frac{-\mu^2}{2\lambda}}$. The Higgs field therefore has a nonzero value even in empty space. In the presence of other fields, such as electron or quark, this nonzero Higgs field interacts with the other fields. The interaction energy is present and of constant

(continued)

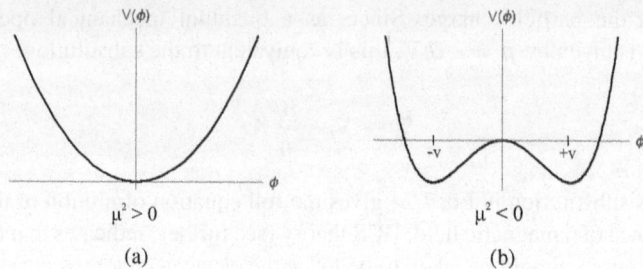

Fig. 7.5 Spontaneous symmetry breaking. When $\mu^2 > 0$, the potential has a minimum at $\phi = 0$. Both the potential and the minimum are symmetric under reflection, $\phi \to -\phi$. (**b**) For $\mu^2 < 0$, the origin becomes unstable, and the system rolls down to one of two lower energy minima at $\phi = \pm v$. The ground state of the system now has less symmetry than the potential.

Box 7.1 (continued)

size everywhere in space, and so it provides the minimum energy a particle represented by the new field can have. This interaction energy provides a nonzero energy gap between the state without the state without the particle and the state with the particle: this Higgs interaction energy has the form of a mass term for the particle. Thus, the Higgs mechanism is responsible for matter particles having nonzero masses.

More information on spontaneous symmetry breaking and the Higgs mechanism can be found in any book on elementary particle physics or quantum field theory. A clear discussion can be found, for example, in [8].

The transition between superconducting and normal phase is an example of spontaneous symmetry breaking. In the high temperature phase, there is a single unique minimum (at $\psi = 0$), and both the free energy and the unique ground state are invariant under phase rotations $\psi \to \psi e^{-i\phi}$. However, although the free energy retains this symmetry in the low temperature phase, the ground state does not. The potential now has a sombrero shape, and the unique ground state bifurcates into a degenerate *ring* of ground states of different ϕ values; transformation $\phi \to \phi + \theta$ rotates the original ground state to a *different* ground state of the same energy. Each element of this degenerate set of ground states now has a trivial symmetry group consisting of just the identity element. This illustrates the hallmark of spontaneous symmetry breaking: in the broken symmetry phase, the ground state has a smaller symmetry group than the free energy or the Hamiltonian (see Box 7.1).

The supercurrent in the presence of phase gradients will now have two terms, one representing the standard quantum-mechanical form for a charged current (of charge $-2e$) and one representing the interaction of that charge with the external magnetic potential,

$$j_s = -i\frac{e\hbar}{2m^*}\left(\psi^*\nabla\psi - \psi\nabla\psi^*\right) - \frac{(2e)^2}{2m^*}|\psi|^2 A \qquad (7.28)$$

$$= -\frac{2e}{\hbar}\rho_s\left(\nabla\phi + \frac{2e}{\hbar}A\right), \qquad (7.29)$$

where the second line follows by using $\psi(r) = |\psi(r)|e^{i\phi(r)}$. Notice that the supercurrent remains invariant under *simultaneous* phase transformations of both the order parameter and the external field:

$$\psi \to \psi e^{i\theta(r)}, \qquad A \to A + \frac{\hbar}{2e}\nabla\theta(r). \qquad (7.30)$$

This is an example of a gauge transformation. Gauge transformations may be familiar to readers from classical electrodynamics, or from particle physics courses; they play a fundamental role in modern physics and are discussed in Chap. 15. More detailed discussions of gauge transformations can also be found in [9] or in nearly any text on particle physics or quantum field theory.

To show that the flux is quantized, look at a region where $j = 0$. Solve Eq. 7.28 for the magnetic potential A,

$$A = -\frac{\hbar}{2e}\nabla\phi(r), \qquad (7.31)$$

so that the flux is (using Stokes' theorem in the second line)

$$\Phi = \int_S B \cdot dS = \int_S (\nabla \times A) \cdot dS \qquad (7.32)$$

$$= \int_C A \cdot dl = -\frac{\hbar}{2e}\int_C \nabla\phi(r) \cdot dl, \qquad (7.33)$$

where C is a closed curve in the superconductor, and S is the surface bounded by C. Any closed curve can be continuously deformed into a circle, so without loss of generality, let C be a circle parameterized by some angle $\alpha \in \{0, 2\pi\}$. Then

$$\Phi = -\frac{\hbar}{2e}\left(\phi(2\pi) - \phi(0)\right) = -\frac{\hbar}{2e}\Delta\phi. \qquad (7.34)$$

But the wavefunction must be single-valued, $e^{i\phi(0)} = e^{i\phi(2\pi)}$, which forces $\Delta\phi$ to be an integer multiple of 2π: $\Delta\phi = 2\pi n$, for $n = 0, \pm 1, \pm 2, \ldots$. Therefore, the flux is quantized,

$$\Phi = n\Phi_o, \qquad (7.35)$$

where $\Phi_0 = \frac{h}{2e} = 2.07 \times 10^{-15} Wb$ is the fundamental **flux quantum**.

7.2 BCS Theory

At the microscopic level, superconductivity is described by **BCS theory** (Bardeen, Cooper, and Schrieffer, 1957), in which the electrons in the metal pair up to effectively form bosons. These bosons then condense into a single macroscopic coherent state for the entire material. Gorkow (1957) demonstrated that the phenomenological GL theory could be derived as a consequence of the microscopic BCS theory, thereby giving a more fundamental understanding of the meaning of the GL order parameter. The BCS theory provides a comprehensive explanation of the behavior of standard (low-temperature) superconductors, but the more recently discovered high-temperature superconductors make use of other mechanisms.

The paired electrons are called **Cooper pairs**. Each electron of momentum k and spin up ($\uparrow$) pairs with another electron of momentum $-k$ and spin $\downarrow$, forming a composite particle of vanishing spin and momentum. In order for this occur, there needs to be an effective attractive potential between the two electrons that is strong enough to overpower their Coulomb repulsion. This occurs as a result of two factors: screening of the electron charge, and phonon exchange with the lattice serving as an intermediary. Screening occurs because as an electron moves through the material, both Coulomb repulsion and fermion statistics cause neighboring electrons to want to move away from it. As the surrounding electron density decreases, it leaves a slight net positive charge from the lattice, reducing the effective electron charge as seen by its neighbors. This screening of the electron charge results in a decrease in the effective interaction potential it exerts on other electrons: instead of the usual Coulomb potential,

$$V(r) = \frac{e^2}{4\pi \epsilon_0 r}, \tag{7.36}$$

the screened potential takes the form

$$V(r) = \frac{e^2}{4\pi \epsilon_0 r} e^{-r/r_0}, \tag{7.37}$$

where r_0 is a constant, called the **Thomas–Fermi screening length** [10].

In addition to a screening that weakens the repulsive force, there can also be an attractive force between electrons, mediated by the crystal lattice of the material. This is due to phonon exchange between electrons (Fig. 7.6). It can be shown that the phonon interaction is of the form

$$V_{ph} \sim \frac{1}{\omega^2 - \omega_D^2}, \tag{7.38}$$

where ω is the phonon frequency and ω_D is the **Debye frequency** [10]. The Debye frequency of a material is given by

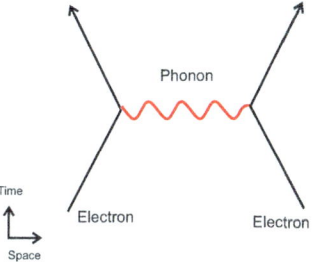

Fig. 7.6 The crystal lattice can mediate interactions between electrons, via phonons (quantized lattice vibrations). The dominant contribution to this interaction is attractive, so if it is strong enough, it can overpower the repulsive Coulomb force and allow Cooper pair formation

$$\omega_D = 2\pi \left(\frac{3n}{4\pi} \right)^{1/3} v_s, \tag{7.39}$$

where v_s is the speed of sound in the material and $n = \frac{dN}{dV}$ is the density of atoms. For $\omega < \omega_D$ (which is the dominant case at normal temperatures), the phonon-mediated potential is negative, and it therefore induces an attractive force between the electrons. As long as this attractive force is stronger than the screened Coulomb repulsion, there will be a net attractive force between nearby electrons. It can be shown that any attractive net interaction, no matter how small, will result in the formation of Cooper pairs.

It should be pointed out that this is a very simplified picture. Since the two electrons in a pair have opposite momentum, they are only close enough to each other to pair up very briefly. So rather than thinking of stable Cooper pairs, a more accurate picture involves a sea of virtual Cooper pairs constantly being created and annihilated, resulting in a many-particle quasi-particle excitation.

The creation operator for the Cooper pair can be written as

$$\hat{C}_k^\dagger = \hat{c}_{k\uparrow}^\dagger \hat{c}_{-k\downarrow}^\dagger, \tag{7.40}$$

where $\hat{c}^\dagger$ is the creation operator for a single electron. If the energy density in the material is u, then the simplest Hamiltonian for a free, non-interacting Cooper pair is

$$\hat{H} = u \left(\sum_k \hat{C}_k^\dagger \right) \left(\sum_{k'} \hat{C}_{k'} \right) \equiv u \hat{\Delta}^\dagger \hat{\Delta}, \tag{7.41}$$

where $\hat{\Delta} = \sum_k \hat{C}_k$. In the ground state, all electron energy levels should be filled up to the Fermi level, k_F, so a reasonable first guess for the ground state of the electrons should be of the form

$$|g\rangle = \prod_{|k| \leq k_F} \hat{C}_k^\dagger |0\rangle. \tag{7.42}$$

But in a superconductor, superpositions of states with no pair and states with one pair should be allowed. So for each fixed k, the state will be of the form

$$|\psi_k\rangle = \left(w_k + v_k \hat{C}_k^\dagger \right) |0\rangle, \tag{7.43}$$

where u_k and v_k are constants satisfying $|u_k|^2 + |v_k|^2 = 1$. The superconducting ground state should therefore be a product of terms like this, one term in the product for each momentum state below the Fermi level:

$$|\Psi\rangle = \prod_{|k| \leq k_F} \left(w_k + v_k \hat{C}_k^\dagger \right) |0\rangle \tag{7.44}$$

Define the expectation value of the operator $\hat{\Delta}$ in the superconducting ground state,

$$\Delta = \langle \hat{\Delta} \rangle = \langle \Psi | \hat{\Delta} | \Psi \rangle \tag{7.45}$$

$$= u \sum_k w_k^* v_k. \tag{7.46}$$

Δ is called the **energy gap** and is a measure of the minimum energy needed to break up a Cooper pair. The energy of the system is then

$$E = 2 \sum_k |v_k|^2 + \frac{1}{u} \Delta^2 \tag{7.47}$$

$$= 2 \sum_k |v_k|^2 + \frac{1}{u} \sum_k w_k^* v_k. \tag{7.48}$$

To get the ground state energy, this needs to be minimized. The details may be found in [11], but the result is that the minimum energy per Cooper pair is

$$E_k = \epsilon_k - \sqrt{\epsilon_k^2 + |\Delta|^2}, \tag{7.49}$$

where the energy gap is

$$\Delta = \frac{|u|}{2} \sum_k \frac{\Delta}{\sqrt{\epsilon_k^2 + |\Delta|^2}} \approx |U| \int \frac{d^3 k}{(2\pi)^3} \frac{\Delta}{\sqrt{\epsilon_k^2 + |\Delta|^2}}. \tag{7.50}$$

$U = uV$ is the total energy in volume V. One of the predictions of the BCS theory is that the gap at zero temperature is given by

$$|\Delta| = 2\hbar\omega_D e^{-1/\lambda} = 1.76 k_B T_c, \tag{7.51}$$

where the electron–phonon coupling constant λ is related to the critical temperature via

$$k_B T_c = 1.13\hbar\omega_D e^{-1/\lambda}. \tag{7.52}$$

The free Fermi gas is gauge invariant. Multiplication of a single-particle fermion wavefunction by a phase $|\psi\rangle \rightarrow e^{i\phi}|\psi\rangle$ (or equivalently, $\hat{c}^\dagger \rightarrow e^{i\phi}\hat{c}^\dagger$) simply multiplies all N-particle wavefunctions by $e^{iN\phi}$, leaving physical observables like particle number, $\hat{n} = \hat{c}^\dagger\hat{c}$, invariant. This is not true in the superconducting phase, where the number of Cooper pairs is not conserved, and where the same gauge transformation takes $\hat{C}^\dagger \rightarrow e^{2i\phi}\hat{C}^\dagger$, leading to non-invariance of physically measurable quantities like the energy gap,

$$\Delta \rightarrow e^{-2i\phi}\Delta. \tag{7.53}$$

So in the superconducting phase, the gauge symmetry is broken.

The superconducting ground state is in fact a macroscopic quantum-coherent state, in which the wavefunction has a well-defined phase. Although the phase at any point is unobservable, phase differences are meaningful (see the next section), as are variation in phase with respect to position or time: the gradient and time derivative of ϕ are respectively responsible for creating the supercurrent and any voltage drop across the material.

To connect the BCS theory to the GL model, consider the creation and annihilation operators $\hat{c}_k^\dagger$ and $\hat{c}_k$. These create or destroy states of definite momentum. But we can Fourier transform to get operators that create and annihilate states of well-defined position instead:

$$\hat{\psi}_\sigma^\dagger(\mathbf{r}) = \frac{1}{\sqrt{V}}\sum_k e^{-i\mathbf{k}\cdot\mathbf{r}} c_{k\sigma}^\dagger \tag{7.54}$$

$$\hat{\psi}_\sigma(\mathbf{r}) = \frac{1}{\sqrt{V}}\sum_k e^{i\mathbf{k}\cdot\mathbf{r}} c_{k\sigma}, \tag{7.55}$$

where σ represents the spin state ($\uparrow$ or $\downarrow$). From these, a pair creation operator can be defined for a spin-singlet state of well-defined position,

$$\hat{\phi}(\mathbf{R}) = \int d^3r \; \phi(\mathbf{r})\hat{\psi}_\uparrow^\dagger(\mathbf{R} + \frac{\mathbf{r}}{2})\hat{\psi}_\downarrow^\dagger(\mathbf{R} - \frac{\mathbf{r}}{2}). \tag{7.56}$$

R is the center of mass of the electron pair, while $\phi(r)$ is some function that describes the spatial profile of the wavefunction. $\phi(r)$ should decay with distance away from the center of the pair, $\phi(r) \rightarrow 0$ as $r \rightarrow \infty$. A two-Cooper pair or four-electron correlation function can then by defined:

$$\rho(R - R') = \langle \hat{\phi}^\dagger(R)\hat{\phi}(R') \rangle \tag{7.57}$$

$$= \int d^3r d^3r' \phi^*(r)\phi(r')\langle \hat{\psi}_\uparrow^\dagger(R + \frac{r}{2})\hat{\psi}_\downarrow^\dagger(R - \frac{r}{2})\hat{\psi}_\downarrow(R' - \frac{r'}{2})\hat{\psi}_\uparrow(R' + \frac{r'}{2})\rangle.$$

At large distances, $R - R' \rightarrow \infty$, the pairs should become uncorrelated with each other so that the correlation function should factor,

$$\rho(R - R') \approx \langle \hat{\phi}^\dagger(R) \rangle \langle \hat{\phi}(R') \rangle \equiv \psi^*(R)\psi(R'), \tag{7.58}$$

where $\psi(R) = \langle \hat{\phi}(R) \rangle$ is the Ginzburg–Landau order parameter. Thus, at distances much larger than the size of a Cooper pair, the BCS theory reproduces the GL model.

7.3 Tunneling in Superconductors: The Josephson Effects

Josephson junctions are essentially a pair of superconducting electrodes separated by an insulator. Such junctions first became important because they can be used as extremely sensitive measuring devices for weak electromagnetic fields. This property is exploited in devices known as SQUID's (superconducting quantum interference devices).

More recently, Josephson junctions have also become useful for constructing qubits. In a linear system like a harmonic oscillator, the energy levels are equally spaced. This can be a problem when trying to construct a qubit system. Qubits are two state-systems. Say the two levels are labeled E_0 and E_1. In a system with many equally spaced levels, the same photon that causes transitions between the two states E_0 and E_1 could also cause a transition between E_1 and a third level E_2. To prevent this, it is desirable to use a nonlinear system, so that the higher level differences like $E_2 - E_1$, $E_3 - E_2$ have different values from the desired transition $E_1 - E_0$. Therefore, by restricting the frequencies of the exciting photons to a narrow range around $(E_1 - E_0)/\hbar$, the higher levels effectively become decoupled, leaving a two-level system. We will see in Sect. 26.3 that the nonlinear inductance of the Josephson junction is capable creating and manipulating several distinct types of qubits.

7.3.1 Josephson Junctions

A Josephson junction is formed by placing a thin insulating gap or barrier between a pair of superconducting electrodes, (Fig. 7.7). The gap width is typically on the order of a nanometer. Assume for simplicity that can treat the system as one-dimensional, with spatial coordinate x. The insulating barrier is at $x = 0$. On each side of the junction there is a macroscopic coherent state wavefunction, ψ_l on the left and ψ_R on the right, and each wavefunction can be written in polar form, in terms of a magnitude and phase,

$$\psi_j(x,t) = |\psi_j(x,t)|\, e^{i\theta_j(x,t)}, \tag{7.59}$$

where $j = L, R$. The insulating barrier between the superconducting terminals is assumed to be thin enough so that there is significant tunneling between the two regions. The tunneling amplitudes in each direction turn out to be determined by the phase difference between the two sides, so that:

$$I(x,t) = I(L \to R) - I(R \to L) \tag{7.60}$$

$$= 2i I_c \left(e^{i(\theta_L - \theta_R)} - e^{i(\theta_R - \theta_L)} \right) \tag{7.61}$$

$$= I_c \sin(\theta_L - \theta_R), \tag{7.62}$$

where the constant I_c is called the **critical current**. Typically, I_c can be anywhere from $\sim nA$ up to μA. We see that a dissipationless current, the **Josephson current**, flows across the junction with vanishing resistance, so that the voltage drop across the junction also vanishes. The existence of this current is called the **DC Josephson effect**.

But if a voltage drop V is applied across the junction, the wavefunction becomes time-dependent, and so the phase difference and current also vary with time. It can

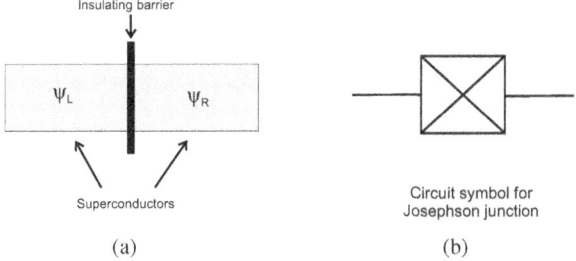

(a)   (b)

Fig. 7.7 Josephson junction: Two superconductors are separated by a thin insulating film. Cooper pairs tunnel between the two superconductors, creating a supercurrent j, whose value depends on the phase difference between the macroscopic wavefunctions on the two sides of the junction. The right figure is the symbolic representation of the junction

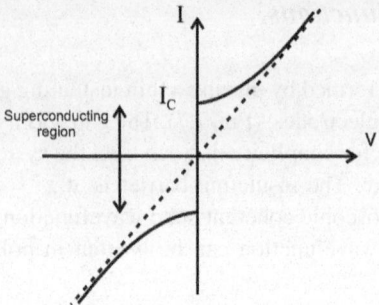

Fig. 7.8 The current–voltage (IV) curve for a superconductor. The material remains supercon-ducting for $I < I_c$. Above I_c, resistance to current flow deveops, leading to a nonzero voltage drop across the ends. For large V, the curve asymptotically approaches a straight line, indicating Ohm's law behavior

be shown that

$$I(t) = I_c \sin\left(\Delta\theta(0) + \frac{2eV}{\hbar}t\right), \tag{7.63}$$

so that the current oscillates at a frequency $v = \frac{2eV}{\hbar}$. In other words, in the presence of a voltage drop, the DC current becomes an alternating current, a phenomenon known as the **AC Josephson effect**. The current–voltage curve is shown in Fig. 7.8

It will be seen in Sect. 26.3 that Josephson junctions can be used to implement qubits in a controllable manner. So, with that discussion in mind, consider the junction's properties in more detail. At this point, let us simplify the notation slightly: denote the phase difference between the two superconductors in Eq. 7.63 simply by Θ. Then Eq. 7.63 is

$$I(t) = I_c \sin\Theta(t), \tag{7.64}$$

where $\Theta(t)$ is a solution to

$$\frac{\partial\Theta}{\partial t} = \frac{2e}{\hbar}V. \tag{7.65}$$

Using the previous two equations, we can write

$$\frac{\partial I}{\partial t} = \frac{\partial I}{\partial \Theta}\frac{\partial \Theta}{\partial t} = \frac{2e}{\hbar}I_c V \cos\Theta, \tag{7.66}$$

or equivalently,

$$V = \frac{\hbar}{2eI_c \cos\Theta}\frac{\partial I}{\partial t}. \tag{7.67}$$

Recall that inductors obey current–voltage relations of the form $V = -L\frac{\partial I}{\partial t}$. So the junction is acting as an inductor, but the inductance is *nonlinear* in the junction phase difference:

$$L = \frac{\hbar}{2eI_c \cos \Theta} \equiv \frac{L_J}{\cos \Theta}, \tag{7.68}$$

where $L_J \equiv \frac{\hbar}{2eI_c}$. The **Josephson inductance**, L_J, is the minimum inductance of the junction. The inductive energy stored in the junction is then $-E_L \cos \Theta$, where $E_L = L_J I_c^2$.

Since the junction has an insulating gap, it should also have a capacitance. Suppose that n Cooper pairs have tunneled through the junction. Then the charge stored on either side of the junction is $\pm Q$, where $Q = 2en$, since each Cooper pair has the charge of two electrons. The capacitive energy is given by $\frac{Q^2}{2C} = 4E_C n^2$, where $E_C \equiv \frac{e^2}{2C}$.

The total Hamiltonian for the junction is then given by

$$H = 4E_C n^2 - E_L \cos \Theta, \tag{7.69}$$

representing a linear capacitor and a nonlinear inductor. n and Θ are conjugate variables. H, n, and Θ should be treated as quantum operators, but we will dispense with the hats over them. The inductive term should be thought of as a potential energy, and the capacitive term as the kinetic energy, with θ and n playing the roles, respectively, of a coordinate and a momentum. Notice that if θ is small, then expanding the cosine to order θ^2 leads to something that looks like a harmonic oscillator potential (shifted by an irrelevant constant). The quartic term in the expansion introduces anharmonicity, which will be essential later. Recall that the energy levels of a simple harmonic oscillator are equally spaced, $E_{j+1} - E_j = \hbar\omega$ for all j. But to form a viable qubit system, the lowest two energy levels (the two states of the qubit) have to be well separated from the higher energy levels, so that the higher states can be ignored. This cannot happen if the energy differences between levels are constant. The anharmonic terms from the cosine introduce the necessary variation in level spacing.

7.3.2 SQUIDs

Now suppose two Josephson junctions are connected by a superconducting ring with input and output terminals at the left and right, as in Fig. 7.9, possibly with some magnetic flux Φ trapped inside the ring. This is the basic architecture of a SQUID. The phase difference across the top junction is $\Delta\theta_1$, and the difference across the bottom junction is $\Delta\theta_2$. Then the total current across the SQUID is

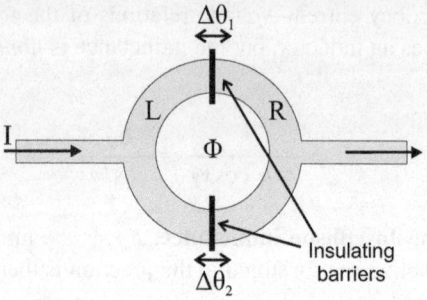

Fig. 7.9 The basic architecture of a SQUID. Two Josephson junctions are formed by placing thin tunnelling barriers at the top and bottom, separating two superconducting regions on the left and right. The superconducting wavefunctions have phase differences $\Delta\theta_1$ and $\Delta\theta_2$ across the two barriers, and the ring may have a magnetic flux Φ trapped in the central region

$$I = I_1 + I_2 = I_{c1}\sin\Delta\theta_1 + I_{c2}\sin\Delta\theta_2. \qquad (7.70)$$

The system can be arranged so that $I_{c1} = I_{c2}$. Suppose that $\Phi = 0$. Then for $I < I_c$, the current will be a constant and the two phase differences will be equal,

$$\Delta\theta_1 = \Delta\theta_2 = \sin^{-1}\frac{I}{2I_c}. \qquad (7.71)$$

Now consider the case where $\Phi \neq 0$. Using Stokes' theorem, we find:

$$\Phi = \int_S \mathbf{B}\cdot d\mathbf{S} = \int_C \mathbf{A}\cdot d\mathbf{l} \qquad (7.72)$$

$$= -\frac{2e}{\hbar}\int_C \nabla\theta\cdot d\mathbf{l} \qquad (7.73)$$

$$= -\frac{2e}{\hbar}(\Delta\theta_1 - \Delta\theta_2) \qquad (7.74)$$

$$= -2\pi\Phi_0(\Delta\theta_1 - \Delta\theta_2) \qquad (7.75)$$

where the line integral is around a closed curve inside the superconductor and enclosing all of the flux, with the surface integral being done over the surface S bounded by C. Let $\Delta\theta = \frac{1}{2}(\Delta\theta_1 + \Delta\theta_2)$ be the average of the two differences. Then we can write

$$\Delta\theta_1 = \Delta\theta + \frac{\pi\Phi}{\Phi_0} \qquad (7.76)$$

$$\Delta\theta_2 = \Delta\theta - \frac{\pi\Phi}{\Phi_0}. \qquad (7.77)$$

Therefore,

$$I = I_c \left[\sin \left(\Delta\theta + \frac{\pi \Phi}{\Phi_0} \right) - \sin \left(\Delta\theta - \frac{\pi \Phi}{\Phi_0} \right) \right] \tag{7.78}$$

$$= 2I_c \sin \Delta\theta \, \cos \left(\frac{\pi \Phi}{\Phi_0} \right). \tag{7.79}$$

So we find that the critical current varies with magnetic flux,

$$I = 2I_c(\Phi) \sin \Delta\theta, \qquad \text{with} \qquad I_c(\Phi) = I_c(0) \cos \left(\frac{\pi \Phi}{\Phi_0} \right). \tag{7.80}$$

Measurement of the critical current therefore gives a means to carry out a highly sensitive measurement of the magnetic field and flux inside the loop; depending on the parameters of the SQUID, it is possible to measure magnetic fields down to the order of $B \sim 10^{-10} T$.

The dynamics of the SQUID can be described by a Hamiltonian. It takes the form

$$H = 4E_C n^2 - 2E_J \cos \left(\frac{\Delta\phi_1 - \Delta\phi_2}{2} \right) \cos \left(\frac{\Delta\phi_1 + \Delta\phi_2}{2} \right) \tag{7.81}$$

$$= 4E_C n^2 - 2E_J(\Phi) \cos \Delta\theta, \tag{7.82}$$

where $E_J(\Phi) = 2E_J \cos \frac{\pi \Phi}{\Phi_0}$, and Eqs. 7.76 and 7.77 have been used. Note that this has the same form as the Hamiltonian for a single junction, but with one advantage: the value of the inductive part of the energy, $E_J(\Phi)$, can be controlled by changing the magnetic flux through the SQUID ring. This will be useful for improving the quality of superconducting qubits in Sect. 26.3.2.

Problems

1. Show that the London equation implies that $\nabla^2 B = \frac{1}{\lambda^2} B$. Consider a magnetic field perpendicular to the surface the magnetic field of the form $B(r) = (0, 0, B_z(x))$. Write down the form of the aforementioned differential equation for this case, and show that the field of Eq. 7.13 is a solution.
2. A thin superconducting ring surrounds a magnetic field $B = 0.02 \, T$ pointing along the ring's axis. The ring has a radius of $r = 1.5 \, cm$ and an inductance of $3 \times 10^{-8} \, H$.

   (a) What is the flux enclosed by the ring? How many fundamental flux quanta Φ_0 are enclosed?
   (b) What is the magnitude of the vector potential A in the ring?

(c) If the field passing through the ring is reduced to zero, find the current induced in the ring.

3. A sample of mercury Hg of atomic mass 199.5 u becomes superconducting at 4.185 K. What are the corresponding critical temperatures for the isotopes ^{202}Hg and ^{204}Hg?

4. For the Cooper pair operators $\hat{C}_k$ and $\hat{C}_k^\dagger$, find the commutation relations $\left[\hat{C}_k, \hat{C}_{k'}\right]$, $\left[\hat{C}_k, \hat{C}_{k'}^\dagger\right]$, and $\left[\hat{C}_k^\dagger, \hat{C}_{k'}^\dagger\right]$.

5. Verify that the gauge transformations of Eq. 7.30 leave the supercurrent and magnetic flux unchanged.

6. Let $\phi(r)$ be the phase of the superconductor wavefunction. Consider a superconductor with a hole in it, i.e., a superconductor with a non-superconducting region R in the middle. Show that for any closed counterclockwise path C around R, we must have $\oint_C \nabla\phi \cdot dl = 2\pi n$ for some integer n, and that n is the same for *any* closed counterclockwise path that encircles R once.

7. Instead of a real field, imagine that the field in Eq. 7.27 is complex. The potential is now $V(\phi) = \mu^2|\phi|^2 + \lambda|\phi|^4$. Let Φ and θ be the magnitude and phase of the field ϕ: $\phi = |\Phi|e^{i\theta}$. This potential will have a continuous rotational symmetry in place of the discrete reflection symmetry of the real case.

(a) Using Mathematica (or by any other method), make a 3D plot of V (vertically) versus the real and imaginary parts of ϕ (horizontal plane) for $\mu^2 > 0$.

(b) Make a similar plot for $\mu^2 < 0$. This is the famous sombrero-shaped potential involved in the Higgs mechanism of particle physics. Verify that the minimum occurs at $\Phi = v \equiv \sqrt{\frac{-\mu^2}{2\lambda}}$.

(c) It should be clear that rotations of the phase, $\phi \rightarrow \phi + \delta\phi$ are symmetries of the potential. Field excitations in this direction can have arbitrarily small energy and therefore are massless. (Mass can be thought of as the minimum energy of a particle at rest.) These massless excitations are called **Goldstone bosons** and appear in any theory with spontaneous breaking of a continuous symmetry. But excitations of the form $\Phi \rightarrow \Phi + \delta\Phi$ do require energy. Given a small perturbation of this form, the field will oscillate up and down the wall of the sombrero. Find the frequency of this excitation for $\mu^2 < 0$.

(d) Will the frequency found in part (c) be changed by small linear perturbations of the potential of the form $V(\phi) \rightarrow V(\phi) + \epsilon\phi$? Will the rotational phase symmetry and the masslessness of the Goldstone excitations be preserved? (Goldstone bosons that gain small masses through the explicit addition of such terms to the potential are called pseudo-Goldstone bosons.)

8. Superfluids are similar to superconductors in many ways. For example, they have a coherent macroscopic wavefunction $\psi(r) = \sqrt{n_s}e^{i\theta(r)}$, where n_s is the number of particles in a superfluid condensate. The particle density can be written as a sum of superfluid and normal densities, $n_0 = n_s + n_n$, and the superfluid component has a particle velocity $v_s = \frac{\hbar}{m}\nabla\theta$ and a particle current $j_s = n_0 v_s$.

Note the similarities to the London theory of superconductors, with θ in place of the gauge field.

(a) Imagine a supercurrent flowing inside a torus-shaped container. Show that the circulation $\kappa \equiv \frac{\hbar}{m} \oint \boldsymbol{v}_s \cdot d\boldsymbol{r}$ around any loop in the fluid can be written as $\frac{\hbar}{m}\Delta\theta$, where $\Delta\theta$ is the change in phase around the integration loop. As a result, show that the circulation must be quantized, $\kappa = \frac{\hbar}{m}p$, for some integer p. What connection does p have to the integration loop and the hole in the torus?

(b) Superfluids must be irrotational, $\nabla \times \boldsymbol{v}_s = 0$. If the superfluid is now in a cylindrical container rotating about its axis, this condition should imply that $\int \boldsymbol{v}_s \cdot d\boldsymbol{r} = 0$ for all curves, which in turn implies no rotation: $\boldsymbol{v}_s$ has a net average component of 0 along the integration curve. But vortices create singular points through which the fluid cannot pass, similar to the hole in the torus in part (a). Show that fluid flowing around a vortex obeys the equation $\frac{1}{r}\frac{\partial}{\partial r}(rv_\phi) = 0$, and that this means the superfluid velocity must be of the form $\boldsymbol{v}_s = \frac{\kappa}{2\pi r}\hat{\phi}$, where $\hat{\phi}$ is a unit vector circulating around the vortex and r is distance from the vortex. The quantized nature of κ then implies that the velocity $\boldsymbol{v}_s$ and the kinetic energy are also quantized.

References

1. M. Tinkham, *Introduction to Superconductivity*, 2nd edn. (Dover, New York, 2004)
2. D.R. Tilley, K. Tilley, *Superfluidity and Superconductivity*, 3rd edn. (Insitute of Physics Publishing, 1990)
3. J.F. Annett, *Superconductivity, Superfluids, and Condensates* (Oxford University Press, Oxford, 2004)
4. A. Bussmann-Holder, H. Keller, Z. Naturforschung **75**, 3 (2020).
5. M. Sato, Y. Ando, Rep. Prog. Phys. **80**, 076501 (2017)
6. Y. Li, Z.-A. Xu, Adv. Quantum Technol. 1800112 (2019)
7. M. Abramowitz, I.A. Stegun, *Handbook of Mathematical Functions with Formulas, Graphs, and Mathematical Tables* (Dover, New York, 1965)
8. F. Halzen, A.D. Martin, *Quarks and Leptons: An Introductory Course in Modern Particle Physics* (Wiley, London, 1984)
9. K. Moriyasu, *An Elementary Primer for Gauge Theory* (World Scientific, Singapore, 1983)
10. C. Kittel, *Introduction to Solid State Physics*, 8th edn. (Wiley, London, 2018)
11. Y.V. Nazarov, J. Danon, *Advanced Quantum Mechanics* (Cambridge University Press, Cambridge, 2013)

Chapter 8
Classical Information and Classical Computation

Computers are a means of processing information. This information needs to be stored in some physical platform, and traditionally, these platforms have been systems that could be treated classically. In contrast, when that platform displays inherently quantum behavior, possibilities open up for new information processing capabilites. In this chapter, we look at classical computing and the quantification of classical information, in order to construct the scaffolding that *quantum* information processing will be built upon in later chapters.

8.1 Binary Arithmetic

Computers use binary numbers to store and process information. Each binary digit (or **bit**) is equal to either 0 or 1. To encode numbers greater than 1 in binary, we need multiple bits. For example, we can encode the base-10 numbers from 0 to 8 as:

$$0 = 0 \tag{8.1}$$
$$1 = 1 \tag{8.2}$$
$$2 = 10 \tag{8.3}$$
$$3 = 11 \tag{8.4}$$
$$4 = 100 \tag{8.5}$$
$$5 = 101 \tag{8.6}$$
$$6 = 110 \tag{8.7}$$
$$7 = 111 \tag{8.8}$$
$$8 = 1000. \tag{8.9}$$

© The Author(s), under exclusive license to Springer Nature Switzerland AG 2025
D. S. Simon, *Introduction to Quantum Science and Technology*, Undergraduate
Texts in Physics, https://doi.org/10.1007/978-3-031-81315-3_8

More generally, in a string of binary digits, a 1 in the $(n + 1)$st position (counting from the *right*, not the left) represents 2^n. So for example:

$$110111 = 2^5 + 2^4 + 2^2 + 2^1 + 2^0 = 32 + 16 + 4 + 2 + 1 = 55.$$

The aforementioned scheme allows any integer to be written as a binary number. Non-integers can be included by using negative exponents:

$$\frac{1}{2} = 2^{-1}, \frac{1}{4} = 2^{-2}, \frac{1}{8} = 2^{-3}, \ldots \tag{8.10}$$

So, for example, to write 0.6875 in binary:

$$0.6875 = \frac{1}{2} + \frac{1}{8} + \frac{1}{16} = 2^{-1} + 2^{-3} + 2^{-4} \longrightarrow 0.1011, \tag{8.11}$$

where the digits to the right of the decimal point represent (from left to right) the $-1, -2, -3$, and -4 powers. Integer and non-integer values can then be added in the obvious way: $3.6875 \longrightarrow 10.1011$.

Recall that **addition mod 2** (denoted $\oplus$) gives a result that always equals 0 or 1: if $x + y$ is even then $x \oplus y = 0$, while if $x + y$ is odd then $x \oplus y = 1$. So for example, $4 \oplus 7 = 1$, but $3 \oplus 5 = 0$. Binary numbers then add mod 2, but with a 1 carried to the left every time two 1's are added. So, for example,

$$01 + 01 = 10$$

$$0010 + 0011 = 0101,$$

$$0110 + 0011 = 1001.$$

Another operation that can be performed is the inversion or **NOT operation**, denoted by a bar. This is simply the interchange of all 0's and 1's. So, for example,

$$A = 0 \longrightarrow \overline{A} = 1$$

$$B = 1 \longrightarrow \overline{B} = 0$$

$$C = 1001 \longrightarrow \overline{C} = 0110.$$

The goal of a computer is to manipulate binary numbers in a controllable manner in order to carry out calculations. These manipulations are implemented physically by devices called logic gates, and the actions of those gates obey algebraic rules known as Boolean logic. These are the subjects of the next sections.

8.2 Logic Gates and Logical Circuits

Logic gates are devices for performing operations on binary numbers. In this section, we discuss classical logic gates in the abstract without worrying about their physical implementation. Classical gates have been mass produced for decades and can be bought at low cost from any electronics company, so their fabrication is well-known and won't be a concern to us. In contrast, when we transition from classical to quantum logic gates in later chapters, the physical implementation of the quantum gates will become a major concern, since how best to implement quantum computing operations in a decoherence-free manner is a major area of current research.

In this section, capital letters $A, B, C, D, \ldots$ will denote a single binary digit, which can take values 0 or 1. We will introduce **single-bit gates** first. These take one binary digit as input and produce a single binary digit as output. For each gate, we define a schematic symbol representing the gate and its operation, and we give a **truth table,**, which is just a tabulation of what outputs are produced for each input.

The simplest single-bit gate is the **identity gate** or **buffer**. This gate just leaves the input unchanged, producing an identical output, $A \to A$. The truth table is:

Input (A)	Output (A)
0	0
1	1

This may not seem very useful, but it serves as a sort of temporary memory, leaving a bit unaltered while waiting for an operation to be performed on some other bit. The schematic symbol for the identity gate is just a straight line (Fig. 8.1a). (In all the logic gate diagrams, it is assumed that input is entering from the left and the output is exiting at the right.)

The most important single-bit gate is the **inverter** or **NOT gate**. Given an input A, the output is the inversion of the input, $A \to \overline{A}$, consisting of interchanging one

Fig. 8.1 Single-bit gates: (**a**) The identity gate or buffer leaves the input unchanged. (**b**) The NOT gate inverts the input, converting 0 into 1 and vice versa

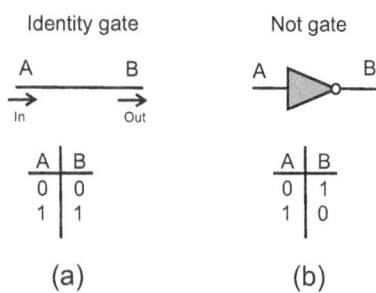

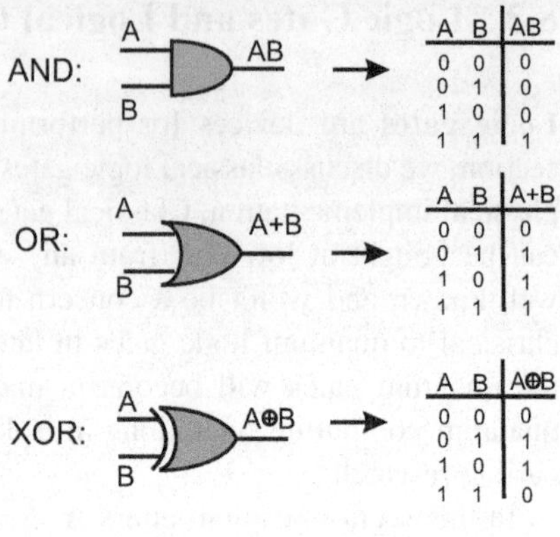

Fig. 8.2 Two-bit gates: The AND gate (top), OR gate (middle), and XOR gate (bottom), with their truth tables

and zero. This can be thought of as addition of 1 mod 2: $NOT(A) = \overline{A} = A \oplus 1$. The truth table is given by

Input (A)	Output $(\overline{A})$
0	1
1	0

.

It might be noted that this is similar to the action of a Pauli sigma matrix: σ_X interchanges spin-up and spin-down states. This will be important when discussing quantum gates in later chapters. Because of this similarity, in quantum contexts, the NOT gate is also sometimes called the σ_X gate, or more simply, the X gate. The symbol for the NOT gate is shown in Fig. 8.1b.

Sometimes, additional one-bit classical gates are defined. For example, a gate can be made that always outputs value 0, regardless of the input. Similarly, a gate can be made that always outputs 1. These constant gates will not have quantum analogs, because, as will be seen later, quantum gates must always be invertible: their input must always be uniquely determinable from the output. The two constant-output gates violate this condition.

Moving on now to two-bit gates, there is more variety. These take two input bits, A and B, and produce either one or two output bits. First is the **AND gate**. The gate is shown schematically at the top of Fig. 8.2 along with its truth table. The output is conventionally written as either AB or $A \wedge B$. The effect of the gate is produced 1 as an output if both A and B are equal to 1; for any other input, the output is 0.

The **OR gate**, shown with its truth table in the middle row of Fig. 8.2, outputs a value of 1 if either A or B equals 1 (or both). If neither of the inputs is 1, then the output is zero. The action of OR is usually written as either $A + B$ or $A \vee B$.

The **exclusive OR**, or **XOR gate** (bottom row, Fig. 8.2), written as $A \oplus B$, is the same as OR, except that the output is zero if *both* A and B equal 1: Notice that the output is just the binary or mod 2 sum of A and B, which justifies the use of the $\oplus$ symbol.

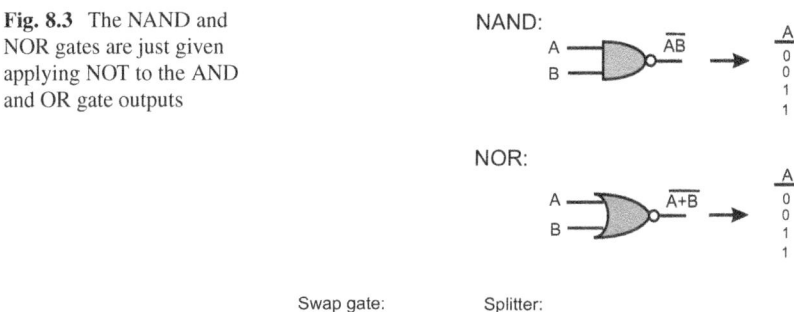

Fig. 8.3 The NAND and
NOR gates are just given
applying NOT to the AND
and OR gate outputs

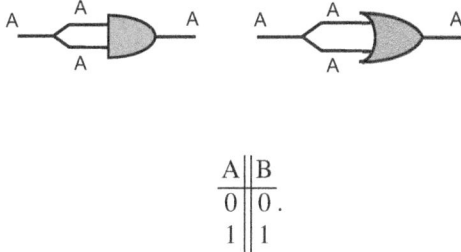

Fig. 8.4 Two "trivial gates": the SWAP gate interchanges the two inputs, while the splitter takes
a single input and produces two outgoing copies of it. The splitter won't be allowed in quantum
computers, due to its lack of reversibility

An inversion can be added after a gate to create a new gate. The two most
common cases (Fig. 8.3) are the **NAND gate** $(\overline{AB})$ and the **NOR gate** $(\overline{A+B})$
of Fig. 8.3.

Finally, two additional "trivial" gates can be defined (Fig. 8.4). The SWAP gate
interchanges the values of the bits on the two incoming lines, while the splitter
makes a second copy of the incoming bit.

Example 8.2.1 Consider following a splitter by an AND gate or an OR gate
(below). It can immediately be checked that they both give the same output:

$$
\begin{array}{c||c}
A & B \\
\hline
0 & 0 \\
1 & 1
\end{array} .
$$

In fact, both are identical to the identity gate.   ■

Now that the basic logic gates have been defined, they can be combined to build
logic circuits, which produce a desired transformation from an arbitrary number of
inputs to a desired set of output states. Some examples are shown further. In each
case, the strategy is to start from the left and to build up the truth table by looking

at the effects of the gates one step at a time, adding a new column to the table each time another gate is encountered.

Example 8.2.2 Consider the following example, with three input bits (A, B, C) and one output. The output can be written as $D = \overline{AB + C}$. To construct a truth table start by listing all possible input combinations. For 3 input bits, there are 2^3 combinations, so there will be 8 rows in the table. Then add a column for the AND gate, listing the values of AB for each line. Finally, add one more column giving the action of the NOR gate on the values of C and AB in each row. The result should look as follows:

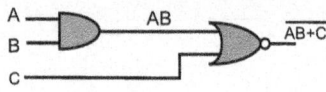

A B C	AB	$\overline{AB + C}$
0 0 0	0	1
0 0 1	0	0
0 1 0	0	1
0 1 1	0	0
1 0 0	0	1
1 0 1	0	0
1 1 0	1	0
1 1 1	1	0

Example 8.2.3 The procedure is similar for the example shown further. There are now four input columns and $2^4 = 16$ rows. Proceeding left to right, apply the AND and OR gates (the fifth and sixth columns) to get the input values of the NAND gate. The final output of the NAND is in the final, seventh, column:

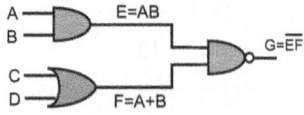

A B C D	E = AB	F = C + D	G = $\overline{EF}$
0 0 0 0	0	0	1
0 0 0 1	0	1	1
0 0 1 0	0	1	1
0 0 1 1	0	1	1
0 1 0 0	0	0	1
0 1 0 1	0	1	1
0 1 1 0	0	1	1
0 1 1 1	0	1	1
1 0 0 0	0	0	1
1 0 0 1	0	1	1
1 0 1 0	0	1	1
1 0 1 1	0	1	1
1 1 0 0	1	0	1
1 1 0 1	1	1	0
1 1 1 0	1	1	0
1 1 1 1	1	1	0

8.3   Boolean Algebra and Universal Gate Sets

In algebra, when given an equation such as $3x^2 + 2y + 5 = 0$, we commonly solve the equation algebraically to find x as a function of y. This is much more efficient than taking individual values of y, plugging them into a calculator to find the numerical value of x for each y, and then constructing a table of (x, y) pairs. The algebraic approach means that the equation only needs to be solved once, rather than many times (infinitely many, in the case of continuous variables).

Similarly for a logical circuit, listing every possible input and finding the corresponding output may not be efficient if the circuit is large. Instead, it is better to take an algebraic approach that solves for the output in terms of an unspecified input variable that is left in symbolic form. The algebra of such logical operations is called **Boolean algebra**, and just like ordinary high school algebra, it has to obey a small set of simple rules:

(i) Order of operations: just as in ordinary algebra, products (AND) are taken before sums (OR) in expressions like $A + BC = A + (BC)$, as indicated by the parentheses.

(ii) AND and OR are commutative:

$$AB = BA, \text{ and } A + B = B + A \tag{8.12}$$

(iii) AND and OR are associative:

$$(AB)C = A(BC) \text{ and } (A + B) + C = A + (B + C) \tag{8.13}$$

(iv) Together the two operations are distributive:

$$A(B + C) = AB + AC \tag{8.14}$$

$$A + (BC) = (A + B)(A + C). \tag{8.15}$$

All of the rules aforementioned are similar to ordinary arithmetic and algebra except for the last one, which may look odd. But it is easy to prove by tabulating the truth tables for the two circuits shown in Fig. 8.5. The left diagram represents $A + BC$ and leads to the table:

A B C	BC	A + BC
0 0 0	0	0
0 0 1	0	0
0 1 0	0	0
0 1 1	1	1
1 0 0	0	1
1 0 1	0	1
1 1 0	0	1
1 1 1	1	1

The right diagram represents $(A + B)(A + C)$ and has the table:

A B C	A + B	A + C	(A + B)(A + C)
0 0 0	0	0	0
0 0 1	0	1	0
0 1 0	1	0	0
0 1 1	1	1	1
1 0 0	1	1	1
1 0 1	1	1	1
1 1 0	1	1	1
1 1 1	1	1	1

The fact that the last two tables have the same output E demonstrates that the identity $A + (BC) = (A + B)(A + C)$ is correct.

Like any associative algebra, there must be additive and multiplicative identity elements. The additive identity element (under the OR operation) is the bit 0, and the multiplicative identity (for the AND operation) is 1. Additional identities include:

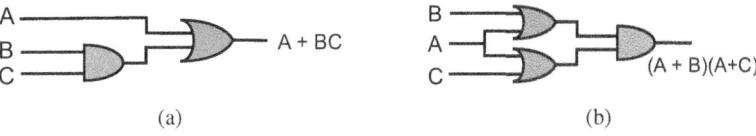

(a)   (b)

Fig. 8.5 Two equivalent circuits that embody the two sides of the identity $A + (BC) = (A + B)(A + C)$

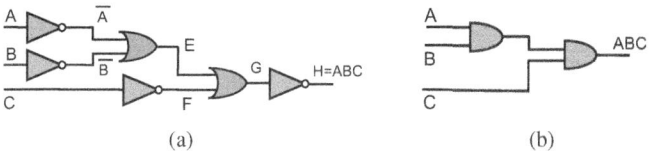

(a)   (b)

Fig. 8.6 Two equivalent circuits, that embody the two sides of the identity $\overline{\overline{A} + \overline{B} + \overline{C}} = ABC$

$$\overline{\overline{A}} = A \qquad AA = A \tag{8.16}$$

$$A \wedge 0 = 0 \qquad A \wedge 1 = A \tag{8.17}$$

$$A + A = A \qquad A + \overline{A} = 1 \tag{8.18}$$

$$1 + A = 1 \qquad 0 + A = A \tag{8.19}$$

$$A + AB = A \qquad A + \overline{A}B = A + B \tag{8.20}$$

Finally, **De Morgan's laws** are:

$$\overline{AB} = \overline{A} + \overline{B} \tag{8.21}$$

$$\overline{A + B} = \overline{A}\,\overline{B}. \tag{8.22}$$

In other words, De Morgan's laws say that under inversion, AND and OR become interchanged. Denoting compositions of operations by $\circ$, another way of saying the same thing is:

$$NOT \circ AND \circ NOT = OR \tag{8.23}$$

$$NOT \circ OR \circ NOT = AND. \tag{8.24}$$

De Morgan's laws are often useful for simplifying logic circuits. For example, consider the circuit in Fig. 8.6a, representing the operation $\overline{\overline{A} + \overline{B} + \overline{C}}$. Using De Morgan's laws, this can be rewritten as

$$\overline{\overline{A} + \overline{B} + \overline{C}} = \overline{\overline{AB} + \overline{C}} = ABC, \tag{8.25}$$

which can be implemented by a simpler circuit, Fig. 8.6b.

It can be shown that the set of gates that have been introduced so far is redundant: some the gates can be built out of the others. This leads to the concept of **universal gate sets**: these are sets of gates from which all possible logic circuits can be built. For example, all possible logic circuits can be built entirely from NOT, AND, and OR gates, so $\{NOT, AND, OR\}$ constitutes a universal set. This set is not unique; many combinations of gates are universal sets. Assuming that splittings are allowed, some possible universal gate sets inlude:

$$\{NOT, AND, OR\}, \quad \{NOT, AND\}, \quad \{NOT, OR\}, \quad \{NOR\}, \quad \{NAND\}. \tag{8.26}$$

Example 8.3.1 By constructing the truth tables or by using Boolean algebra, the reader should be able to easily verify that the NOT, AND, and OR gates can be constructed solely from NAND gates as follows:

8.4   Reversibility and Controlled Logic Gates

So far, the discussion of logic gates has been completely classical. But ultimately, the goal in later chapters is to look at quantum logic gates, and quantum mechanics puts additional constraints on the allowed operations that can be carried out by the gates. The most important is the requirement of **reversibility**.

Quantum mechanics is time-reversible: if a wavefunction $\psi(x, t)$ satisfies the Schrödinger equation, then $\psi(x, -t)$ will satisfy it as well. Effectively, this means that any device that acts on the wavefunction (without making any measurements) must be represented by a unitary matrix. As a consequence, quantum logic gates must be reversible: given the output, it must be possible to run the gate backward to uniquely reconstruct the input. So some of the classical gates we have considered will be off limits for quantum computing.

Consider the AND gate. Multiple input states can lead to the same output: inputs 00, 01, and 10 all lead to the same output value (output $= 1$). So if 1 appears at the output, it is impossible to determine what the original input was. Similarly, NOT, OR, NOR, NAND gates and the splitting operation are all irreversible, and so will not occur in quantum circuits. On the other hand, the NOT and SWAP operations are perfectly reversible.

An easy way to see if a gate is reversible is to check if it satisfies two necessary conditions: (i) It must have equal numbers of input and output bits. This immediately rules out splitting operations, which have one input and two outputs, as well as gates like AND and OR, which have two inputs but only one output. (ii) The truth table must contain all possible combinations of 0's and 1's. For example, if the gate has two output bits, then its truth table should should contain output rows with all four of the combinations 00, 01, 10, and 11. If any of these outputs are missing (or if any occur in two different rows of the table), then the gate is irreversible.

Since most of the one- and two-bit gates we have discussed are irreversible, this seems to leave us with a limited number of gates to work with in quantum systems. But there will be three facts that will allow us to introduce a wider set of logical gate operations: (i) the quantum analogs of classical bits (qubits) are effectively two-dimensional objects, allowing additional operations to be performed on them, such as rotations. (ii) Gates can be formed that involve more than two input and output bits. (iii) We can introduce **controlled gates**, or more generally, gates will ancillary bits. An **ancilla** is a bit that plays a role in an operation but that remains unchanged at the end.

Controlled gates are vital for computing, since they allow the operations performed on one bit (or the interactions between two bits) to be altered depending on the value of a different bit. This allows bits to interact with each other, as well as allowing conditional operations to occur. In one sense, AND and OR gates can already be viewed as conditional gates, since we can think of one input bit as controlling how the gate responds to the other bit. But here we consider some additional controlled gates that also have the advantage that they are reversible.

In a controlled gate, a bit that is potentially changed by the gate is called the **target bit**. The bit that determines what happens to the target is the **control bit**. The control bit is always left unchanged by the action of the gate.

The simplest example of a controlled gate is a **controlled not (CNOT) gate**. A CNOT gate is shown in Fig. 8.7, where the upper line (with the solid dot) represents the control bit, and the bottom line is the target bit. Its action is to apply a NOT gate to B if $A = 1$ and to apply the identity gate if $A = 0$. Either way, A remains unchanged, resulting in the truth table:

Input		Output	
A	B	A	C
0	0	0	0
0	1	0	1
1	0	1	1
1	1	1	0

Fig. 8.7 A controlled not or CNOT gate. If the control A equals 0, B is unchanged. If $A = 1$, then B is inverted. In either case, A does not change

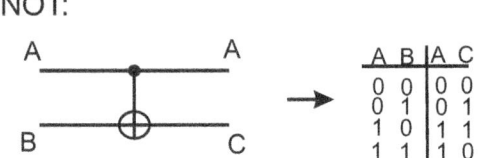

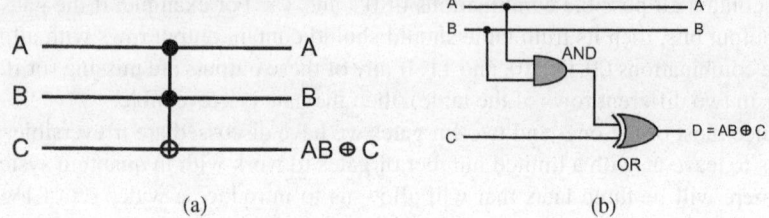

Fig. 8.8 (a) The Toffoli gate is a controlled controlled NOT gate: it acts as a NOT gate on the third only when both of the control bits equal 1. The output of the target bit can be written as $AB \oplus C$. (b) A way of implementing the Toffoli gate in terms of an AND and XOR gate. Notice that this particular arrangement is reversible, despite the fact that the AND and XOR gates by themselves are not

It can be seen that this table satisfies the conditions listed earlier for a reversible operation, so if the output is known, then the input can be uniquely determined.

Sometimes which bit is playing which role is specified by a pair of subscripts, with the control listed first, then the target. In this notation, the gate shown in Fig. 8.7 has A as control and B as target, so it would be called $CNOT_{AB}$; the gate with A as target and B as control would be denoted $CNOT_{BA}$.

A reversible three-bit controlled gate is given by the **Toffoli gate** (Fig. 8.8a). The first two bits, A and B are both control bits, while the third bit C is the target. C is left unchanged if either of the controls are 0, but flips value if both controls equal 1. The controls, of course, are left unchanged, but the output of the target is given by $D = AB \oplus C$. The reader can verify that the Toffoli gate can be implemented in terms of AND and XOR gates as shown in Fig. 8.8b. The truth table is:

A B C	A B D
0 0 0	0 0 0
0 0 1	0 0 1
0 1 0	0 1 0
0 1 1	0 1 1
1 0 0	1 0 0
1 0 1	1 0 1
1 1 0	1 1 1
1 1 1	1 1 0

Another reversible three-bit gate is the **Fredkin gate**, shown in Fig. 8.9. The Fredkin gate is a controlled swap or CSWAP gate; if $A = 0$ all three bits are unchanged, but if $A = 1$, the bits B and C are interchanged. The truth table is:

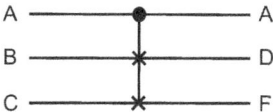

Fig. 8.9 A Fredkin gate acts as a controlled swap (CSWAP) gate, interchanging the last two bits if and only if the control bit A equals 1

A B C	A B D
0 0 0	0 0 0
0 0 1	0 0 1
0 1 0	0 1 0
0 1 1	0 1 1 .
1 0 0	1 0 0
1 0 1	1 1 0
1 1 0	1 0 1
1 1 1	1 1 1

One property reversible gates have that will be of great use when we come to quantum logic circuits is that the action of such gates can be written in the form of square matrices. Consider the CNOT gate, for example. The gate has two inputs (A and B), which have four possible states (00, 01, 10, and 11) and two outputs (A and C) with the same possible states. So CNOT can be written as a four-by-four matrix. We use a basis where the four columns represent the four input states in the order listed earlier, and the four columns represent the same four output states in the same order. The matrix form of the CNOT operation is given by the unitary matrix

$$U_{CNOT} = \begin{pmatrix} 1 & 0 & 0 & 0 \\ 0 & 1 & 0 & 0 \\ 0 & 0 & 0 & 1 \\ 0 & 0 & 1 & 0 \end{pmatrix}. \tag{8.27}$$

So, for example, taking the input to be

$$v_{in} = \begin{pmatrix} 0 \\ 0 \\ 1 \\ 0 \end{pmatrix},$$

representing the state $A = 1$, $B = 0$, the output is

$$v_{out} = U_{CNOT} v_{out} = \begin{pmatrix} 0 \\ 0 \\ 0 \\ 1 \end{pmatrix}, \tag{8.28}$$

which represents the state $A = 1$, $B = 1$.

In a similar manner, the actions of the Toffoli and Fredkin gates can be put into matrix form. There are 3 inputs and output bits, so the number of input and output states will be $2^3 = 8$, requiring 8×8 matrices:

$$U_{Fred} = \begin{pmatrix} 1 0 0 0 0 0 0 0 \\ 0 1 0 0 0 0 0 0 \\ 0 0 1 0 0 0 0 0 \\ 0 0 0 1 0 0 0 0 \\ 0 0 0 0 1 0 0 0 \\ 0 0 0 0 0 0 1 0 \\ 0 0 0 0 0 1 0 0 \\ 0 0 0 0 0 0 0 1 \end{pmatrix} \quad U_{Toff} = \begin{pmatrix} 1 0 0 0 0 0 0 0 \\ 0 1 0 0 0 0 0 0 \\ 0 0 1 0 0 0 0 0 \\ 0 0 0 1 0 0 0 0 \\ 0 0 0 0 1 0 0 0 \\ 0 0 0 0 0 1 0 0 \\ 0 0 0 0 0 0 0 1 \\ 0 0 0 0 0 0 1 0 \end{pmatrix} \tag{8.29}$$

More generally, if a reversible n-bit gate G is represented by the matrix $n \times n$ M, then the controlled M or CM gate is represented by the $2n \times 2n$ matrix

$$CM = \begin{pmatrix} I & 0 \\ 0 & M \end{pmatrix}, \tag{8.30}$$

where I is the $n \times n$ identity matrix.

8.5 Error Detection and Error Correction

In a perfect world, computing devices would always produce accurate results, and communication devices would transmit exact copies of messages from the sender to the receiver. But in real life, there are always errors in these processes. Sources of error are numerous, but a few examples might be:

- Thermal noise in electric circuits and in detectors produces random fluctuations in currents that can lead to computing or communication errors.
- Photon losses in fiber optic circuits or imperfections (aberration or dispersion) in optical elements can lead to errors in optical communication.
- Cosmic rays or charged particles emitted in the decay of radioactive elements in the earth's crust can occasionally alter the bit stored in solid state devices.

Classically, the main worry is bit-flip errors, in which a logical 0 state is accidentally converted to 1, or vice versa. Later, when we discuss quantum systems, a second type of error (phase errors) can become important.

Since errors can occur, it is important to have mechanisms in place for both detecting and correcting these errors. In the following, we assume that information is being transmitted through a system in the form of codewords (strings of bits representing letters of the message), and that the error rate is low enough that the probability of two or more errors occurring in the same codeword is negligible.

Consider error detection first. The simplest **error-detecting code** is a **repetition code**, in which each bit is represented by a codeword that repeats the same digit multiple times. For example, the bit 0 might be represented by the codeword 00, with the bit 1 represented by 11. Errors are then easy to detect. Define the **parity** of the codeword, defined to be the sum (mod 2) of the codeword: 00 and 11 have even parity, while 01 and 10 have odd parity. This is easily implemented by an XOR gate: $0 \oplus 0 = 1 \oplus 1 = 0$ indicates even parity, while $0 \oplus 1 = 1 \oplus 0 = 1$ indicates odd parity. The presence of an odd parity codeword indicates an error. The parity is an example of a **syndrome**, a signal that an error has occurred.

Being able to detect errors is good, but being able to correct those errors is better. A repetition code with three digits per codeword, $0 \rightarrow 000$ and $1 \rightarrow 111$, is the simplest **error-correcting code**. To detect the location of the error, apply an XOR to the first pair of bits, and then apply XOR to the second pair. Letting the codeword be represented by abc, the result is:

abc	$a \oplus b$	$b \oplus c$
000	0	0
001	0	1
010	1	1
011	1	0 .
100	1	0
101	1	1
110	0	1
111	0	0

The values of $a \oplus b$ and $b \oplus c$ are called the **parity bits**. Looking at this table, it is seen that the outcome of the two XOR gates tells us both whether or not an error occurred *and* where it occurred:

$$\text{Parity bits} = 01 \rightarrow \text{error in rightmost codeword digit} \qquad (8.31)$$

$$\text{Parity bits} = 10 \rightarrow \text{error in leftmost codeword digit} \qquad (8.32)$$

$$\text{Parity bits} = 11 \rightarrow \text{error in middle codeword digit} \qquad (8.33)$$

Once the location of the error is note, then a NOT gate can be applied to the affected digit to correct the error.

If two or more errors occur in the same codeword, this method obviously fails, but the hope is that such double errors are sufficiently rare as to be unimportant. Of course, more elaborate error-correcting codes can be devised, but they would inevitably involve more redundancy and be less efficient. This points out an unavoidable trade-off between better error correction and coding efficiency: improving one will damage the other.

8.6   Complexity Classes

An important topic in computation is the quantification of resources needed to carry out a particular computation or algorithm. These resources may include the number of logic gates required, the amount of memory space needed, or the time required to complete the computation. We are concerned with how the required resources scale asymptotically as some parameter becomes large. The scaling rate of the resources serves as a measure of the complexity of the problem being solved by the algorithm.

For example, a typical computational problem might be to ask if a particular integer n is factorable, or what is the shortest route connecting n cities that a salesman wants to visit, assuming that he wishes to visit each city just once and then return to his starting point. The latter is known as the **travelling salesman problem**. In these problems, the scaling parameter is n, and typically, the complexity might be measured by asking how the required computing time (or equivalently, the number of computational steps) scales with increasing n, as n grows large.

The computational time, for instance, might increase no faster than some power of n, say time $\sim n^c$, for some c, as $n \rightarrow \infty$. Then clearly, any polynomial $F(n)$ of order $\leq c$ grows at the same rate or slower. In any of these cases, we write that the computational time scales $\sim O(n^c)$. Any problem that scales in this way (or more slowly) for any c is said to scale polynomially, and the class of such polynomially scaling problems is called complexity class **P**, the class of polynomial-time algorithms. Notice that any problem whose complexity scales logarithmically, or as fractional powers ($n^{1/2}$, $n^{8/3}$) is also of complexity class P.

An algorithm that scales at most polynomially is called an **efficient algorithm**, while one that scales faster than polynomially (exponential scaling like e^n, e^{n^2}, or 2^n) is called **inefficient**. A computational problem that can be solved with an algorithm in class P (or in other words, one for which an efficient algorithm can be found) is often referred to as an **easy** or **tractable**, while those that scale faster than polynomial are called **hard** or **intractable**.

Example 8.6.1 The standard example of a problem in class P is the problem of checking to see if an integer is prime. A brute force method for doing this is simply to compute the ratios $\frac{n}{2}, \frac{n}{3}, \frac{n}{4} \ldots \frac{n}{n-1}$, and see then check to see if any of these ratios is an integer. For any integer $m \leq n$, suppose k computational steps are required to compute the ratio $\frac{n}{m}$ and check to see if it is integer. Then the maximum number of steps required to check the primality of n will be kn. This scales linearly in n, so determining primality is of complexity class P. ■

P is just one of a large number of classical **complexity classes**, with additional classes being defined for quantum algorithms. Here we will describe a few of the more common classical complexity classes.

Class NP contains problems whose solutions can be *checked* in polynomial time. (The letters NP stand for "nondeterministic polynomial".) If a problem is in class NP, then regardless of whether or not there is an efficient algorithm to *solve* the problem, if someone claims to already have a solution then there is an efficient, polynomial time algorithm *to verify or disprove* that proposed solution.

Factoring of integers, n is an NP problem. No efficient algorithm is known for factoring large integers; in fact, some encryption schemes for securely coding sensitive information are based on this fact. However, if someone gives you a collection of integers and claims that these are the prime factors of n, then to check this claim one simply has to multiply them and see if the product equals n. So checking the proposed solution is a class P problem, which makes the original factoring problem NP. Another example of a class-NP problem is the graph isomorphism problem, which asks for a general method of determining whether two networks are equivalent; there is no known polynomial algorithm for finding an isomorphism between two graphs, but once such an isomorphism is proposed, testing it is easy.

A problem in class P is also in class NP: $P \subseteq NP$. It is not known if any problems exist that are in NP, but not in P. For example, the fact that no efficient factoring algorithm is known does not necessarily mean that no such algorithm exists; it could simply be that nobody has been clever enough to find it yet. The open question of whether P is just a subset of NP or if it actually equals NP is called the **P=NP problem** and has been an important problem in computer science for decades.

Another class of problems is the set of **NP-complete problems**. A solution to any one NP-complete problem allows solutions to be found to all NP problems. So the class NP-complete is a clearly subset of class NP: NP-complete $\subset NP$. Examples of NP-complete problems are the traveling salesman problem aforementioned and the solution of $n \times n$ Sudoku problems.

Instead of computational time, the most important resource involved in solving a problem might be the required memory space. The category of **PSPACE** problems is the set of problems whose required memory space scales polynomially in n, with no limits on computational time. A search for a winning strategy on an $n \times n$ checkers board is an example of a PSPACE problem. Not much is known for certain about the relation of PSPACE to the other classes defined earlier. It is believed, based on known examples, that PSPACE is not equal to P or NP, but this has yet to be proven.

8.7 Turing Machines

In 1936, Alan Turing proposed a device that he called an *automatic machine* or *a-machine*. Now known as **Turing machines**, they remain an important topic in computer science, since they provide a model of computation that is simple to work with but that is powerful enough to implement any classical computer algorithm.

A Turing machine is made up of four essential elements:

- **A tape:** This tape consists of a set of cells in a single line along the tape. Each cell can be empty or can contain a single symbol from some finite alphabet $\mathcal{X}$. Unless stated otherwise, we will always assume the binary alphabet $\mathcal{X} = \{0, 1\}$. The tape has the symbol $\triangleright$ in the left-most cell to signal that the end of the tape has been reached, but the tape can extend indefinitely far to the right.
- **A read/write head:** The head moves along the tape, reading the symbols in the cells and overwriting them with new symbols when instructed to by a program. At each time step, after each reading and/or writing, the head can then move one step to the left, one step to the right, or can remain at rest.
- **Register:** The register is a memory device that stores the state of the machine at each moment. Only a finite number of possible states are allowed. This set of states should include a starting state q_s, which signals that the computation should begin, and a halting state q_h, which signals that the computation is complete and no more operations should be carried out.
- **Program:** The program is a list of instructions, telling the machine what it should do at each step, depending on the symbol in the current cell and on the current state of the system. After reading the current cell and the current state, the program tells the head what new symbol to write in the cell, what system state to update to, and which direction to move.

The importance of Turing machines in the theory of computation follows from the **Church–Turing thesis**: every computable problem can be computed with a Turing machine. There is no guarantee that the Turing machine will implement an *efficient* algorithm, but it will, given sufficient time, implement any computable algorithm.

Not all problems are computable. It is possible that attempts to carry out a computation will end up in a closed loop or will otherwise continue forever without reaching an answer or halting. Such problems are called **undecidable**.

The Turing machine can be generalized: the state of the machine can be reset randomly according to some fixed probability distribution at each step in the computation. The result is a **probabilistic Turing machine**. The addition of randomness can often improve the efficiency of algorithms, a fact that ants have long known: by taking frequent random turns, they can efficiently cover large areas in the search for food. This leads to the **Strong Church–Turing thesis**: Any computation can be simulated by a probabilistic Turing machine, using at most polynomial resources.

Both classical and quantum computers obey the regular Church–Turing thesis, but the intense interest in quantum computing is largely due to the widespread belief that quantum computers can violate the strong version of the thesis and perform efficient computations in problems for problems in which classical computers can offer no efficient algorithms.

Although not yet proven, there is an increasing amount of evidence for this **quantum supremacy hypothesis**. For example, as mentioned earlier, there is no known efficient classical algorithm for factoring large numbers. In fact, the common **RSA (Rivest–Shamir–Adleman) encryption** scheme that is used to keep financial transactions and other sensitive information secure is based the fact that factoring large numbers requires exponential resources, so trying to implement decryption algorithms quickly becomes impractical. But the quantum mechanical Shor algorithm (Sect. 17.4) does allow efficient factoring in polynomial time, working exponentially faster than any known classical algorithm. Although this provides evidence in favor of strong Church–Turing violation, it does not provide proof, since it still has not been proven that the factoring problem is not of class P: the fact that no polynomially scaling classical algorithm for factoring has been found so far does not prove that one won't be found at some future time.

A commonly used complexity class for quantum computers is the **BPQ** or **bounded-error quantum polynomial-time class**. As the name implies, it is the class of problems that can be computed in polynomial time on a quantum computer with errors below some fixed bound. At present, little is known for certain about the relation between BPQ and the classical complexity classes. But if a quantum computer is shown to definitively violate strong Church–Turing, that would prove that BPQ is larger than P and would demonstrate the usefulness of quantum computing.

Box 8.1 Looms to Computers: Jacquard, Babbage, and Lovelace

Textiles with complex patterns woven into them were a valuable commodity in the 18th century, but had to be woven by hand and so were time-consuming to produce. By the 1720s, several French inventors had experimented with mechanized looms to speed up production. Over the coming decades, these were gradually improved, and by 1804, Joseph Marie Jacquard had patented the device that came to be known as the Jacquard loom. The Jacquard loom was the first programmable mechanism: cards with different sets of instructions could be fed into the machine, and different punch cards would cause the loom to produce textiles woven into different patterns.

Along with the arithmometer, a gear-driven mechanical calculator built by Thomas de Colmar in 1821, the Jacquard loom inspired Charles Babbage (1791–1871) to try to design programmable computational devices. Babbage was an English mathematician and an avid tinkerer with mechanical devices.

(continued)

Box 8.1 (continued)

At the age of 16, he nearly drowned while testing a pair of shoes that he had designed for walking on water, but despite this, his interest in inventing and building novel mechanical devices continued for the rest of his life. While working with astronomer John Herschel to painstakingly compile a set of mathematical tables by hand, he had the idea of building a device that could automate the process. This led to a device he called the difference engine. He began designing Difference Engine no. 1 in 1822, and an improved design, Difference Engine no. 2 in the 1840's. Neither version was ever completed due to disputes with engineers and a loss of government funding, but a working small-scale model (capable of six-digit calculations) was built. Meanwhile, Babbage's work inspired a Swedish company to design and build a mechanical device that could do 15-digit calculations for use by the British Treasury's General Register Office.

The Difference Engine was essentially a special purpose calculator, designed to be compute the values of polynomial functions. But Babbage soon moved on to more ambitious plans. He began designing the Analytical Engine, a fully programmable mechanical computer. In modern terms, the Analytical Engine would have been capable of simulating a universal Turing machine. Many of Babbage's ideas related to the device sound surprisingly modern today; for example, he designed a printer for the output and discussed its uses in games of strategy such as tic-tac-toe and chess. As with the Jacquard loom, the Analytical Engine could be programmed by inputting a set of punched cards. The punchcard method of programming computers was revived in the early days of electronic computers and was still widely used at least into the 1980s.

In 1833, Babbage met 17 year-old Augusta Ada Byron, the only legitimate child of Lord Byron, at a party. Ada, who later became Augusta Ada King, Countess of Lovelace, already had a strong interest in mathematics (she later studied mathematics under de Morgan and other luminaries), and was fascinated by Babbage's discussion of his computing devices. She translated a treatise on the Analytical Engine, written in French by the Italian engineer Luigi Menebrea, into English. But in addition to carrying out a simple translation, she added copious notes of her own to the manuscript. Among her additions was a detailed description of the steps required to use the engine to compute Bernoulli numbers. As a result of this, she is often credited with creating the first computer algorithm.

Although his engines were never completed in Babbage's lifetime, a working version of Difference Engine no. 2 was finally completed in 1991 (with Babbage's output printer following in 2002), using only gears and other parts that matched the manufacturing tolerances possible in the 19th century, thus finally vindicating Babbage's vision more than a century after his death.

8.8 Classical Information Measures

Information and entropy are key concepts throughout physics, computer science, and communication theory. In this section, we explore these concepts in classical systems. In Chap. 12, they are generalized to quantum systems. For more detail on classical information measures, see [1–3]. First, we use the next subsection to introduce a bit of notation and terminology and to provide some quick reminders of a few basic facts concerning probabilities.

8.8.1 A Brief Review of Probabilities

Let X denote a random variable and x denote a particular value of X. For example, when flipping a coin, X denotes the side that lands face up, and the possible values are $x = $ head or $x = $ tail. The probability distribution $p(x)$ describes the likelihood of the various outcomes. We often think of the possible values of X as forming a set, the **sample space** of X, in which case we will denote the random variable and its sample space by the same symbol. Thus, $x \in X$ means that x is one of the possible values of random variable X.

When more than one random variable is present, we add subscripts to distinguish the probability distribution. For example, if there is a second random variable Y, then we write the probability distributions for X and Y as $p_X(x)$ and $p_Y(y)$. In this case, we write the **joint probability distribution** for the pair of variables taken together as $p(x, y)$. This is the probability that X takes value x **and** Y takes value y.

The **conditional probability** for X to have value x *given that* Y is already known to have value y is denoted $p(x|y)$. Note that conditional probabilities are not symmetric: in general, $p(x|y) \neq p(y|x)$. Recall that joint and conditional probabilities are related by **Bayes' rule**:

$$p(x, y) = p(x|y)p_Y(y). \tag{8.34}$$

Conditional probabilities depend on what prior information the experimenter has available. For example, a medical researcher might ask what is the probability of a randomly chosen person having high cholesterol. Call this probability $p(high)$. Now suppose the researcher is interested in one particular person, and that this person is known to eat four eggs for breakfast every day. Then this knowledge will clearly change the probability of having high cholesterol. Conditioning on the "four eggs condition" will clearly increase the probability: $p(high|4\ eggs) > p(high)$. The probability changes because the extra information changes the sample space over which the probability is being computed: without the condition, the sample space is the general populace, but with the condition, the appropriate sample space is the much smaller subset of people who satisfy the "4 egg" condition.

Table 8.1 The probability distribution for rolling two dice. In the second column, (m, n) means rolling m on the first die and n on the second, to obtain sum $s = m + n$. f, the frequency of the outcome, is the number of entries in the previous column, and the probability is $P_S(s) = \frac{f(s)}{N}$, where $N = \sum f$ is the total number of outcomes

s	Outcomes (die 1,die 2)	f(s)	$P_S(s)$
2	(1,1)	1	1/36
3	(1,2) (2,1)	2	2/36
4	(1,3) (3,1) (2,2)	3	3/36
5	(1,4) (4,1) (3,2) (2,3)	4	4/36
6	(1,5) (5,1) (2,4) (4,2) (3,3)	5	5/36
7	(1,6) (6,1) (2,5) (5,2) (4,3) (3,4)	6	6/36
8	(2,6) (6,2) (3,5) (5,3) (4,4)	5	5/36
9	(3,6) (6,3) (4,5) (5,4)	4	4/36
10	(4,6) (6,4) (5,5)	3	3/36
11	(5,6) (6,5)	2	2/36
12	(6,6)	1	1/36
	Totals	36	1.0

Example 8.8.1 Suppose that two six-sided dice are rolled. Let S be the sum of the two dice; this is a random variable, which can have values $s = \{2, 3, \ldots, 12\}$. Define a second random variable X, which takes the values

$$x = 0 \qquad \text{if the value on the first die is even,} \qquad (8.35)$$

$$x = 1 \qquad \text{if the value on the first die is odd.} \qquad (8.36)$$

The probability for X alone is clearly uniform, $P_X(x) = \frac{1}{2}$ for any $x \in X$. To find the probabilities for S, simply list the possible ways each value s can occur, and count the frequencies f (the number of ways each can occur). The probabilities are then given by the relative frequencies $p = \frac{f}{N}$, where $N = \sum f = 36$ is the total number of possible outcomes. This gives Table 8.1

Suppose we wish to find the probability that $s = 7$ and $x = 0$ (first die even). We see that there are three ways this can happen: $(6, 1), (2, 5), (4, 3)$, so that the joint probability is $p(7, even) = 3/36 = 1/12$. Then Bayes' rule gives the conditional probabilities:

$$p(7|even) = \frac{p(7, even)}{p_X(even)} = \frac{1/12}{1/2} = 1/6 \qquad (8.37)$$

$$p(even|7) = \frac{p(7, even)}{p_S(7)} = \frac{1/12}{1/6} = 1/2. \qquad (8.38)$$

In other words, $1/6$ of the first-die-even events have $s = 7$, but half of the $s = 7$ events have the first die even. ∎

From the example, we see that the conditional probabilities $p(x|y)$ and $p(y|x)$ generally differ from each other and from the unconditioned probabilities. Clearly, the probability of an event is altered by changes in the experimenter's knowledge about events that have already taken place. This will be important later: the

Fig. 8.10 A Venn diagram
for two events

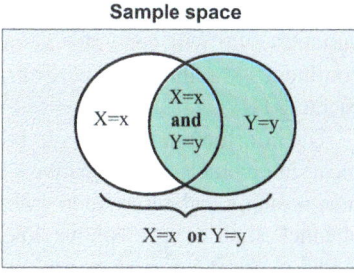

probability of measuring a given quantum event can change depending on what
prior knowledge the experimenter has, or even by what knowledge he or she *could
potentially* obtain from the experimental setup, even if nobody bothers to obtain it.
In a sense, the state is not entirely intrinsic to the system itself, but also depends
on what is or has previously been measured; i.e., what information is available or
potentially available about the system.

Venn diagrams are often useful in conceptually picturing a situation. For two
random variables X and Y, and two events $X = x$ and $Y = y$, the Venn diagram
looks as in Fig. 8.10. The two circles represent the two events of interest. The overlap
region that lies in both circles (the intersection of the two sets) represents the joint
probability of **both** events occuring, $p(x, y)$. The area that lies in *either* one set or
the other, or in both (the union of the sets), represents the probability that either
event (or both) occurs, and is denoted $p(x$ or $y)$. Examination of the Venn diagram
immediately gives the so-called **addition rule** for probabilities:

$$p(x \text{ or } y) = p_X(x) + p_Y(y) - p(x, y), \tag{8.39}$$

where the joint probability has to be subtracted off to prevent double counting of the
overlap region.

Marginal probabilities for one variable (X, say) in a joint distribution are the
probabilities $p_X(x)$ of X having value x, regardless of the value of y. Marginal and
joint probabilities are related by

$$p_X(x) = \sum_{y \in Y} p(x, y) = \sum_{y \in Y} p(x|y) p_Y(y). \tag{8.40}$$

Variables X and Y are **independent** if knowledge of one has no effect on the
probability of the other. For example, X might be the age of your next-door neighbor
and Y is the number of solar neutrinos detected in a week; clearly knowledge
of one of these variables gives you no information about the other. If X and Y
are independent, then the conditional and marginal probabilities are in fact equal,
$p(x|y) = p_X(x)$ and so, by Bayes' rule, the joint probabilities factor:

$$p(x, y) = p_X(x)\, p_Y(y). \quad \text{(independent events)} \tag{8.41}$$

We wish to quantify the information gained by measurements of the random variables. We will generally assume that the variables X and Y are discrete; for continuous variables, the generalizations are usually obvious, with sums replaced by integrals.

The information measures to be defined were developed to characterize communication channels. We consider a situation in which some information is encoded into a signal, which is then transmitted through some classical communication channel, such as a telephone line, an optical fiber, or a radio transmission. The signal at the input is some function $x(t)$, and the goal is to reproduce the original information from the output signal $y(t)$ at the other end. The bits at input and output can be treated as values of random variables X and Y. In a noiseless channel, $y(t)$ will be a deterministic and invertible function of $x(t)$. So $x(t)$ can be determined without error from $y(t)$ in the noiseless case. The Shannon entropy, defined further, characterizes the maximum allowed efficiency of a noiseless communication channel, which simply reduces to the efficiency of encoding the information into the input signal.

Unfortunately, in real-world situations, there will always be noise (thermal fluctuations, background light, electrical static, etc.) that will make the reconstruction difficult. We model the output by a new function $y(t) = x(t) + \mathcal{N}(t)$, where $\mathcal{N}(t)$ describes the noise. $\mathcal{N}(t)$ is some random function, whose value at any given moment is unknown. In most applications, the values of $\mathcal{N}(t)$ will be distributed according to some fixed probability distribution, which may or may not be known. Describing noisy channels will require defining additional information measures, such as the mutual information.

We now proceed to define several information measures and look at what they tell us about noisy and noiseless communication channels. In Chap. 12, some of these will be generalized to produce *quantum* information measures that quantify the amount of information carried by quantum states and the ability of a noisy system to reproduce an initial quantum state from the measurement of a final output state. Such measures will be needed for quantifying things like the maximum data processing rate of a quantum computer or the maximum information an eavesdropper can obtain about the message transmitted between users of a quantum cryptography system.

8.8.2   Shannon Information and Data Compression

Communication and computation are inextricably linked. They are both forms of information processing, and so they share a common set of constraints and underlying principles. In the 1940s, as the telecommunications industry was growing exponentially and computers were being rapidly developed, there was a pressing need to better understand these underlying principles. The result, developed by Shannon, Weiner, and others (building on a mathematical foundation constructed by Nyquist and Hartley in the 1920s), is the modern field of information theory. The basic quantity from which everything else is built is the Shannon entropy $H(X)$, first developed in 1948.

Shannon began with a simple model of communication that can be drawn in the following schematic form:

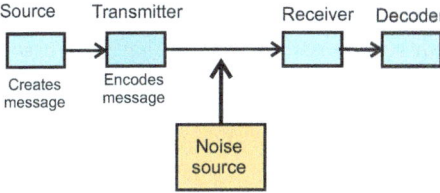

The source (a person typing on a keyboard, output signal of a sensor, etc.) creates the message and feeds it into the transmitter. The transmitter encodes the message into a string of symbols. The set of possible symbols is called the **alphabet**, X of the system. The transmitter then sends the encoded message over a radio link, a telegraph line, or some other communication channel, to a receiver. En route to the receiver, the environment may introduce some noise, which may introduce errors when the message is later read. The receiver then sends the coded message into a decoder, which converts it from the transmitted alphabet back into a form that can be read by the user, such as an English sentence, an image, or a dial reading. The problem is to quantify how fast information can be transmitted through the system and how much effect the noise will have on the transmission. For this section, we focus purely on noiseless channels; noise will be added in later.

The first task is to give a quantitative definition of information. To do this, we think of the message as being a string of symbols chosen randomly according to some probability distribution. We seek a function $I(x)$ that will describe the information gained when a random variable X is measured. The individual measurement outcome is given by the particular value x, which will fluctuate from one measurement to the next. It is often helpful to think of the information gained in a measurement as a "measure of surprise": it should be large when the result is unpredictable, but should vanish when you can reliably predict the result before the measurement is made.

Following Claude Shannon, we require that the information measure satisfies the following conditions:

(i) The information $I(x)$ should only depend on the probability distribution of X: $I(x) = I(p(x))$.
(ii) $I(p(x))$ should be a continuous function of $p(x)$.
(iii) The information carried by pairs of *independent* measurements resulting in values x and y should be additive: $I(x, y) = I(x) + I(y)$.

The solution satisfying these conditions (unique up to a multiplicative constant) is

$$I(x) = -\log_2(p(x)). \tag{8.42}$$

The base two logarithm is used because we measure the information in binary units, or **bits**, which take two possible values, 0 or 1. Suppose we have n binary storage devices (physical binary memory units); for example, these might be n switches that can be on or off, or n spin-$\frac{1}{2}$ particles that can be in spin-up or spin-down states. These devices store a string of n values, each representing logical value 0 or 1. Each possible state of the system represents a length-n string of 0's and 1's. $N = 2^n$ such binary strings of length n are possible, so if we have no reason to believe that any string is more likely than any other, then the probability of any one particular string x is $p(x) = 2^{-n}$. Clearly, by definition 8.42, the number of bits of information that can be stored in this system is n, one per binary digit in the string.

More significant than the information value for a single event x (the outcome of a single measurement), however, is the *mean* information per event, averaged over the set of all possible outcomes. This is the **Shannon entropy**:[1]

$$H(X) = - \sum_x p(x)I(x) = - \sum_x p(x) \log_2 p(x). \tag{8.43}$$

The Shannon entropy represents the uncertainty in the value of X, or the number of bits needed to describe the random variable. Any additional knowledge gained about X should therefore lead to a decrease in the entropy.

Example 8.8.2 Consider a variable X, which can have any one of n possible values, and obeying a uniform distribution:

$$p(x) = \frac{1}{n}, \quad \text{for all } x, \text{ where } x = x_1, x_2, \ldots, x_n. \tag{8.44}$$

Then for any particular value x_j,

$$I(x_j) = -\log_2\left(p\left(x_j\right)\right) = -2\log_2\left(\frac{1}{n}\right), \tag{8.45}$$

so that the Shannon entropy is

$$H = - \sum_x p(x) \log_2 p(x) = - \sum_x \frac{1}{n} \log_2\left(\frac{1}{n}\right) = -\log_2\left(\frac{1}{n}\right) = +\log_2 n. \tag{8.46}$$

In general, for a system with n possible states, $\log_2 n$ will be the maximum possible entropy of the system,

$$0 \le H \le \log_2 n. \tag{8.47}$$

---

[1] A base-10 version of something like the Shannon entropy was actually defined earlier by Alan Turing [4, 5].

The maximum value $H = \log_2 n$ will only be achieved when the outcomes are maximally unpredictable, i.e., when they are given by a uniform distribution. For the binary case, $n = 2$, the maximum entropy occurs when both possible outcomes are equally likely, $p_1 = p_2 = \frac{1}{2}$.

To go to the opposite extreme, suppose that the outcome is completely predictable:

$$p(x_j) = \begin{cases} 1, & \text{for } j = k \\ 0, & \text{for } j \neq k \end{cases}, \tag{8.48}$$

where k is some fixed value. Then,

$$H = -\sum_j p(x_j) \log_2 p(x_j) = -(1) \log_2(1) - (n-1)(0) \log_2(0) = 0, \tag{8.49}$$

where by convention we take $0 \cdot \log_2(0) = 0$. So when the outcome is certain in advance (no surprise is possible), the entropy vanishes. ■

Example 8.8.3 As another simple example, consider the distribution

$$p(x) = \begin{cases} 1/9, & \text{for } x = 0, \\ 2/9, & \text{for } x = 1, \\ 3/9, & \text{for } x = 2, \\ 2/9, & \text{for } x = 3, \\ 1/9, & \text{for } x = 4, \end{cases} \tag{8.50}$$

with $n = 5$ possible outcomes for each measurement. Since this distribution is neither uniform, nor restricted to a single nonzero value, we would expect that the entropy should be somewhere between the possible extremes, $0 \leq H \leq \log_2(5) = 2.32$. It is easy to check that this is the case:

$$H = -\left[\frac{1}{9} \log_2\left(\frac{1}{9}\right) + \frac{2}{9} \log_2\left(\frac{2}{9}\right) + \frac{3}{9} \log_2\left(\frac{3}{9}\right) \right.$$
$$\left. + \frac{2}{9} \log_2\left(\frac{2}{9}\right) + \frac{1}{9} \log_2\left(\frac{1}{9}\right) \right] \tag{8.51}$$
$$= 2.197. \quad ■ \tag{8.52}$$

In the case where there are only two possible outcomes, of probabilities p and $q = 1 - p$, the entropy reduces to the **binary entropy function**,

$$h(p) = -p \log_2 p - (1 - p) \log_2(1 - p). \tag{8.53}$$

Clearly, in the cases where one outcome or the other is a certainty, this function vanishes: $h(0) = h(1) = 0$. The function also takes a maximum when the

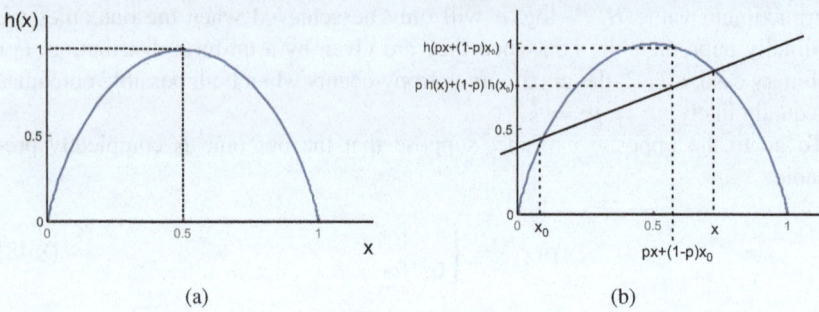

(a)   (b)

Fig. 8.11 (**a**) The binary entropy function describes the case where the random variable can take only two values, or respective probabilities p and $1 - p$. The function is concave down, with a maximum at the point where both outcomes are equally likely, $p = 1 - p = \frac{1}{2}$. (**b**) Concavity: the curve is always above any straight line joining two points on the curve. Here, p is a parameter along the line: $p = 0$ corresponds to the point $(x_0, h(x_0))$ and $p = 1$ is the point $(x, h(x))$. The values of p between 0 and 1 give all the other points on the line

uncertainty is largest, when both outcomes are equally likely: $h(0.5) = 1$. This is consistent with the comments earlier. In addition, this function is symmetric about the maximum and is concave down (Fig. 8.11a). Concavity means that if any two points x_0 and x on the curve are connected by a straight line, then the curve is always above the line (Fig. 8.11b) on the region between x_0 and x. Expressed in equation form, concavity implies that the following inequality always holds:

$$h\,(px_0 + (1 - p)x) \geq p\,h(x_0) + (1 - p)h(x), \tag{8.54}$$

for all $0 \leq p \leq 1$.

The entropy gives a measure of the growth rate in the number of distinguishable sequences that are possible as the length of the sequence increases. Consider a sequence of length n, $\Lambda = \{x_1, x_2, \ldots, x_n\}$, where each x_j is a measured value of some random variable x. Suppose that each x_j can take on N possible values. (So, for a binary sequence, $N = 2$.) Then the number of rearrangements of the elements of Λ will be $n!$, since the first slot can be filled n ways, the second slot $n - 1$ ways, and so on. But not all of these $n!$ rearrangements are distinguishable from each other. Suppose the value j occurs n_j times in the sequence. Then rearranging those n_j elements among themselves doesn't alter the sequence. Which means that to avoid overcounting, we need to divide off the rearrangements that leave the sequence unchanged. The number of rearrangements of the n_j identical elements is $n_j!$, so that the number of distinct, *distinguishable* sequences of length n is

$$N_n = \frac{n!}{n_1!\,n_2!\,\ldots\,n_N!}. \tag{8.55}$$

Assuming that the sequence is long (so that n and all of the n_j are large), Stirling's approximation, $\ln(m!) \approx (m + \frac{1}{2}) \ln m - m$, can be applied:

$$\ln N_n \approx \left(n + \frac{1}{2}\right) \ln n - n - \sum_{j=1}^{N} \left[\left(n_j + \frac{1}{2}\right) \ln n_j - n_j\right] \tag{8.56}$$

$$= \left(n \ln n - \sum_j n_j \ln n_j\right) + \frac{1}{2}\left(\ln n - \sum_j \ln n_j\right), \tag{8.57}$$

where the fact that $n = \sum_j n_j$ has been used. For a long sequence, the term in the second set of parentheses will be negligible compared to the first parentheses, so

$$\ln N_n \approx n \ln n - \sum_j n_j \ln n_j = \sum_j n_j \left(\ln n - \ln n_j\right) = \sum_j n_j \ln \frac{n}{n_j}. \tag{8.58}$$

The probability of choosing a random term in the sequence and finding value j is $p_j = \frac{n_j}{n}$, so that

$$\frac{\ln N_n}{n} \approx \sum_j \frac{n_j}{n} \ln \frac{n}{n_j} \tag{8.59}$$

$$= -\sum_j p_j \ln p_j \tag{8.60}$$

$$= H, \tag{8.61}$$

where H is the Shannon entropy. (Note that we are ignoring the difference between natural logs and base-2 logs here: switching between them just introduces factors on the order of 1, and such factors aren't of much interest to us here.) Rearranging, the number of distinguishable sequences of length n is therefore

$$N_n \sim e^{nH} \tag{8.62}$$

for large n. So, H can be viewed as the logarithmic growth rate of the number of sequences. Since we don't care about overall constant factors, we could just as easily write

$$N_n \sim 2^{nH} \tag{8.63}$$

for large n; this version of the result will be used in the following.

Now suppose we wish to send a message over a noiseless communication channel. Let n be the number of bits of information in the message. We can encode the information into the signal in different ways, and some encodings are more efficient than others, in the sense that some will allow the information to be transmitted with shorter-length strings of transmitted data, at least on average. We have a choice of what alphabet is used to encode the message. Most often we are interested in the binary alphabet $\{0, 1\}$. But other alphabets can be used, such as the English alphabet $\{A, B, C, D, \ldots, Z\}$. Each letter of the alphabet can then be encoded into a block of symbols, called a **codeword**. For example, each letter of the English alphabet might be encoded into a block of five binary digits; each five-digit string representing a single letter is one codeword. For each x in the alphabet, let the codeword be $c(x)$ and the length of the codeword be $l(x)$. If the probability distribution of the letters in the alphabet is given by $p(x)$, then the average codeword length is

$$\bar{l} = \sum_x p(x)l(x). \tag{8.64}$$

As a trivial encoding example, suppose we wish to transmit the answers to a sequence of yes or no questions. Then the alphabet would consist of two elements, $\{yes, no\}$. A simple encoding would be to map $yes \rightarrow c(yes) = 0$ and $no \rightarrow c(no) = 1$, so that $l(yes) = l(no) = \bar{l} = 1$. But this isn't the only possibility. For example, we could use the code $c(yes) = 000$ and $c(no) = 111$, so that each codeword is of length three. Sometimes, such redundant encodings are useful, since they allow error correction. But it is clear that the second encoding scheme is less efficient than the first one, requiring message strings three times as long to transmit the same information.

As a less trivial example, suppose we wish to transmit an arbitrary word in English. We send it as a binary string, and we want the string to be as short as possible. As a first attempt, we could encode each letter as the binary value of its place in the alphabet. So, the first letter in the alphabet, A, would be encoded as 00001, the letter B as 00010, the letter C as 00011, etc. Since there are 26 letters in the alphabet, the blocks have to be at least five binary digits in length to encode them all ($2^5 = 32 > 26$). So the English word "cab" would be given by the string 00010|00001|00011 = 000100000100011 (reading the blocks from right to left), with each five-digit block representing a single letter. But this may not be the most efficient way of encoding letters into binary. Is there a way of defining codewords that would shorten the average length of the binary string needed?

There are indeed more efficient encodings. First, notice that not all letters are equally probable: the letter "e" is about 130 times more likely to occur than the letter "z" in English words (Fig. 8.12). So, a fruitful strategy might be to use shorter codewords for common letters like e, t, a, and o, reserving longer codewords for less frequently used letters like q, x, and z. This might occasionally produce longer than necessary message strings for words like "pizazz" or "pizzapalooza," but should produce shorter messages *on average*.

In fact, if the message is of sufficient length, it may actually be possible to send n bits of information with a message stream containing less than n symbols. This process of **data compression** will entail the risk of some errors in the message

E	12.5%
T	9.3%
A	8.0%
O	7.6%
I	7.6%
N	7.2%
S	6.5%
R	6.3%
H	5.0%
L	4.1%
D	3.8%
C	3.3%
U	2.7%

M	2.5%
F	2.4%
P	2.1%
G	1.9%
W	1.7%
Y	1.7%
B	1.5%
V	1.0%
K	0.5%
X	0.2%
J	0.2%
Q	0.1%
Z	0.1%

Fig. 8.12 An estimate of the relative frequencies of the letters used in the English language. The compiling of such data goes back to at least the 9th century Arab mathematician Al-Kindi, who realized it is useful for breaking alphabetic codes; a similar frequency chart is used for that purpose by a character in the 1843 Edgar Allen Poe story, *The Gold-Bug*

received, but the error probability becomes negligible if the message is sufficiently long. How far the data can be compressed is determined by the Shannon entropy.

Example 8.8.4 Consider a system with a four-letter alphabet $\{A, B, C, D\}$, where the letters occur with the following probability distribution:

$$p(A) = \frac{1}{8} \qquad p(B) = \frac{1}{4} \qquad p(C) = \frac{1}{2} \qquad p(D) = \frac{1}{8} \tag{8.65}$$

The obvious encoding into binary is to map these states into codewords of length 2, for example as: $c(A) = 00, c(B) = 01, c(C) = 10, c(D) = 11$, which gives $\bar{l} = 2$. But a more efficient encoding scheme can be given by attaching longer codewords to the two less probable letters, and a shorter word to the most probable one. For example, consider:

$$c(A) = 110 \qquad c(B) = 10 \qquad c(C) = 0 \qquad c(D) = 111. \tag{8.66}$$

It is readily computed that the average codeword length is

$$\bar{l} = \sum_j l(j)p(j) = 1.75, \tag{8.67}$$

which is shorter than the obvious encoding. We can also check that the Shannon entropy is

$$H = -\sum_j p(j) \log_2 p(j) = 1.75. \tag{8.68}$$

As we will see further (the Shannon noiseless coding theorem), the entropy sets a lower limit on the average codeword length, $H \leq \bar{l}$; so we can be confident that there are no encodings more efficient than this one. ∎

To understand data compression, we must divide strings of transmitted symbols into two categories: **typical** and **atypical**. As the message length increases, atypical sequences are those whose probability of occurrence rapidly decreases toward zero. These are sequences in which the relative frequencies of the alphabet letters are far from the occurrence probabilities of those letters. For example, consider a binary sequence in which $p(0) = 0.1$, while $p(1) = 0.9$. Then in this case, a sequence that has roughly 90% of the digits equal to 1 would be a typical sequence, while a sequence with 90% of the digits equal to 0 would be atypical.

Similarly, if a coin is flipped 2000 times, then you would expect a typical sequence, in which heads comes up very close to 1000 times. A typical sequence consists of half heads and half tails; as the number of flips grows, the relative size of the deviations from this 50–50 distribution should rapidly decrease. On the other hand, if you flip the coin 2000 times and get 1500 heads, then you have reason to suspect something is wrong with the coin; this would be an atypical sequence.

Sequences that differ significantly from the probability distribution of the underlying alphabet's probability distribution become more and more rare as the message length n grows, a result known in statistics as the **law of large numbers**. Think of the entries in a given message sequence as instances of a random variable X. Assume the sequence is **independent and identically distributed (i.d.d.)**: each entry in the sequence is independent of the others, and the probability distribution is the same for each member of the sequence. Then there should be $\sim 2^{nH(X)}$ typical sequences as $n \to \infty$, where H is the Shannon entropy corresponding the random variable X. The probability of each typical sequence is $\sim 2^{-nH(X)}$. Let $|X|$ be the size of the alphabet. The total number of sequences of length n is $|X|^n = 2^{n \log_2 |X|}$. This means that the fraction of typical sequences is

$$f(n) = 2^{n(H(X) - \log_2 |X|)}, \tag{8.69}$$

which obeys $f(n) \to 0$ as $n \to \infty$ for non-uniformly distributed X. At the same time, the probability of a randomly chosen sequence being in the typical set approaches 1 as $n \to \infty$. So for large n, only a vanishingly small fraction of the sequences contribute to the communication. To achieve data compression, the sender need only encode typical strings, and ignore atypical ones. This leads to the possibility of errors, but for n sufficiently large, the probability of such errors tends to zero. The compression rate is then given by

$$\text{compression rate} = \frac{\text{\# of transmitted information bits}}{\text{\# of transmitted source symbols}} = \frac{nH(X)}{n} = H(X). \tag{8.70}$$

Restricting ourselves now to binary alphabets, $|X| = 2$, we have:

$$\text{total number of sequences} = 2^n$$

$$\text{number of typical sequences} \sim 2^{nH(X)}.$$

To be more precise, for any $\epsilon, \delta > 0$, there exists sufficiently large n such that, with probability at of at least $1 - \epsilon$, a sequence $x_1, \ldots, x_n$ occurs satisfying

$$2^{-n(H(X)-\delta)} < p(x_1, x_2, \ldots, x_n) < 2^{-n(H(X)+\delta)}. \tag{8.71}$$

The number of such ϵ-**typical** sequences satisfies

$$2^{n(H(X)+\delta)} \geq N(\epsilon, \delta) \geq (1 - \epsilon)2^{n(H(X)-\delta)}. \tag{8.72}$$

So, using $n\big(H(X) + \delta\big)$ bits, any length-n ϵ-typical sequence can be estimated with error less than ϵ. So the compression rate is

$$\frac{n\big(H(X) + \delta\big)}{n} \to H(X) \text{ as } n \to \infty. \tag{8.73}$$

This is one of basic results of information theory: the Shannon entropy measures the minimum length a message string needs to be to transmit a given number of bits. In other words, it provides a way to tell if a particular encoding scheme is efficient and sets a limit on the maximum possible efficiency (i.e., the maximum data compression). Since there are at most $2^{nH(X)}$ typical sequences of length n, then it takes at most $nH(X)$ bits of information to uniquely identify a given typical sequence. So the most efficient data compression scheme is to simply represent each typical sequence by $nH(X)$ data bits in a typical sequence. We thus have:

Shannon Noiseless Coding Theorem. If the entropy of a random variable is $H(x)$, then to transmit a message with n bits of information over a noiseless channel, a minimum of $nH(X)$ transmitted symbols are needed to convey the information with negligible error as $n \to \infty$. If the message is compressed into a sequence shorter than $nH(X)$, errors will inevitably occur.

In Example 8.8.4, we found an encoding such that the mean codeword length equaled the entropy, $\bar{l} = H$. The noiseless coding theorem then implies that this is a maximum-efficiency encoding: there can be no encoding that leads to shorter message strings for long messages. (The most efficient encoding is not unique: there may be other encoding schemes that also achieve the maximum efficiency.) A long message of n English letters will require a minimum of $nH(X) = 1.75n$ binary digits to convey the full message without error.

The fact that $H(X)$ limits encoding efficiency is one reason why the name *entropy* is appropriate for H: its role in information encoding is similar to the role played by the thermodynamic or Boltzmann entropy, which limits the efficiency of converting heat into work. In fact, in a thermodynamic system for which all available states are equally likely, the Shannon and Boltzmann entropies become the same, up to a constant. For discussion of connections between information entropies and thermodynamic entropies, see [6].

Another way to state the noiseless coding theorem is in terms of the **channel capacity** C of the communication channel, which is a limit on the information transmission rate. Suppose an information source sends out r messages per second with entropy $H(X)$ per message. The information transmission rate is then

$$R = rH(X). \tag{8.74}$$

in bits per second. Then the noiseless coding theorem tells us that it is possible to transmit information across the channel without error if $R < C$, for some maximum value C. An expression for C will be given further after the mutual information is introduced. C is usually measured in units of bits per second (bps).

8.8.3   Joint Entropy

The joint entropy of two variables X and Y is the average information gained when a measurement is made of both:

$$H(X, Y) = -\sum_{x,y} p(x, y) \log_2 p(x, y). \tag{8.75}$$

If the two variables X and Y are **independent** (so that $p(x, y) = p(x)p(y)$ for all x, y), then this simply reduces to

$$H(X, Y) = H(X) + H(Y) \tag{8.76}$$

The joint entropy is simply the obvious generalization of the Shannon entropy from one to two random variables; it measures the combined uncertainty in the pair (X, Y).

8.8.4   Relative Entropy (Kullback-Liebler Distance)

The **relative entropy**, also known as the **Kullback–Liebler distance** is defined by

$$K(X||Y) = -\sum_{x,y} p_X(x) \log_2(p_Y(y)) - H(X) \tag{8.77}$$

$$= \sum_{x,y} p_X(x) \log_2 \frac{p_X(x)}{p_Y(y)}. \tag{8.78}$$

The relative entropy is non-negative, $K(X||Y) \geq 0$, and vanishes if and only if the two distributions are identical, $p_X(X) = p_Y(Y)$. The relative entropy measures the similarity between the distributions and so acts as a distance measure on the space of probability distributions or of random variables. (But is it *not* a metric, since it fails to be symmetric: $K(X||Y) \neq K(Y||X)$.)

8.8.5 Conditional Entropy

The conditional entropy is defined as

$$H(X|Y) = -\sum_{x,y} p(x, y) \log_2 p(x|y) \tag{8.79}$$

where $p(x|y) = \frac{p(x,y)}{p(y)}$ is the conditional probability. If the value of Y is already known, $H(X|Y)$ gives the average additional information gained by also measuring X. Note that $H(X|Y)$ vanishes if the two probability distributions are identical, $p_X(X) = p_Y(Y)$, and it reduces to the Shannon information $H(X)$ if X and Y are independent.

The conditional entropy obeys the relation

$$H(X|Y) \leq H(X). \tag{8.80}$$

In other words, knowledge of Y can never increase the uncertainty in X; it can only leave the uncertainty unchanged (so that $H(X|Y) = H(X)$) if X and Y are independent, or cause a decrease ($H(X|Y) < H(X)$) if they are not independent.

The conditional and joint entropies also related by a type of chain rule:

$$H(X, Y) = H(X) + H(Y|X), \tag{8.81}$$

which follows by taking the logarithm of Bayes' rule, $\log_2 p(x, y) = \log_2 p(y) + \log p(x|y)$ (see Problem 6).

8.8.6 Mutual Information

The mutual information $I(X : Y)$ is defined by

$$I(X : Y) = \sum_{x,y} p(x, y) \log_2 \frac{p(x, y)}{p(x)p(y)} \tag{8.82}$$

is a measure of correlation between the two variables. Notice that it is simply the relative entropy between the joint distribution $p(X, Y)$ and the product distribution $p_X(x)p_Y(y)$, and so it should vanish when X and Y are independent. $I(X : Y)$ can also be thought of how much information the two variables share, as expressed by the following relation:

$$I(X : Y) = H(X) - H(X|Y). \tag{8.83}$$

This can be shown as follows:

$$I(X:Y) = \sum_{x,y} p(x,y) \log_2 \frac{p(x,y)}{p_X(x)p_Y(y)} \tag{8.84}$$

$$= \sum_{x,y} p(x,y) \log_2 \frac{p(x|y)}{p_X(x)} \quad \text{(Bayes's rule)} \tag{8.85}$$

$$= \sum_{x,y} p(x,y) \log_2 p(x|y) - \sum_{x,y} p(x,y) \log_2 p_X(x) \tag{8.86}$$

$$= -\left(-\sum_{x,y} p(x,y) \log_2 p(x|y)\right) - \sum_x p_X(x) \log_2 p_X(x) \tag{8.87}$$

$$= -H(X|Y) + H(X). \tag{8.88}$$

In the next to last line, the fact was used that $\sum_y p(x,y) = p_X(x)$.

Notice from Eq. 8.83 that $I(X:Y) \le H(X)$. This is a purely classical limit; it will turn out that the analog of this relation does not hold in quantum mechanics.

$I(X:Y)$ is symmetric in X and Y, and it is not hard to show that it is equal to the difference in entropy between the two variables taken together as a single system versus the pair of variables taken separately:

$$I(X:Y) = H(X) + H(Y) - H(X,Y). \tag{8.89}$$

$I(X:Y)$ vanishes if X and Y are independent and reduces to $H(X)$ if they share the same probability distribution. Note that this is exactly the opposite of how the conditional entropy behaves.

Example 8.8.5 Suppose that a randomly chosen electron in a system could be moving either to the left or right along the x axis. Because of an external electric field, it has a slight directional bias making it more likely to be right-moving. Similarly, an external magnetic field makes it more likely to be spin up than spin down. The probabilities of each case can be collected in a table:

	Spin up(A)	Spin down(B)
Right-moving (C)	0.4	0.2
Left-moving (D)	0.25	0.15

So the joint probabilities, as read directly from the table, are:

$$p(A,C) = 0.4 \qquad\qquad p(B,C) = 0.2 \tag{8.90}$$

$$p(A,D) = 0.25 \qquad\qquad p(B,D) = 0.15. \tag{8.91}$$

Define two random variables: X is the spin direction, which can take values $x = \{up, down\} = \{A, B\}$, and Y is the direction of motion, with possible values $y = \{right, left\} = \{C, D\}$ The marginal probabilities are obtained by summing the rows and columns. Summing the rows, we find:

$$p(A) = p(A, C) + p(A, D) = 0.65 \tag{8.92}$$

$$p(B) = p(B, C) + p(B, D) = 0.35, \tag{8.93}$$

which gives us the total probability of being spin up or spin down, respectively. These two values completely specify the probability distribution $P(X)$. Similarly, the probabilities of leftward or rightward motion (the elements of the distribution $P(Y)$) are found by adding up the rows, leading to

$$p(C) = p(A, C) + p(B, C) = 0.6 \tag{8.94}$$

$$p(D) = p(A, D) + p(B, D) = 0.4. \tag{8.95}$$

The entropy associated with each of the two random variables is easily found:

$$H(X) = -\big[p(A) \log_2 p(A) + p(B) \log_2 p(B)\big] \tag{8.96}$$
$$= -\big[.65 \log_2(.65) + .35 \log_2(.35)\big]$$
$$= 1.9341$$
$$H(Y) = -\big[p(C) \log_2 p(C) + p(D) \log_2 p(D)\big] \tag{8.97}$$
$$= -\big[.6 \log_2(.6) + .4 \log_2(.4)\big]$$
$$= 1.9710 \ .$$

The joint and conditional entropies are also easily found,

$$H(X, Y) = -\sum_{x,y} p(x, y) \log_2 p(x, y) \tag{8.98}$$
$$= -\big[.4 \log_2 .4 + .2 \log_2 .2 + .25 \log_2 .25 + .15 \log_2 .15\big]$$
$$= 1.9037$$
$$H(X|Y) = H(X, Y) - H(Y) \ = \ 1.9037 - .9710 \ = \ .9327 \tag{8.99}$$
$$H(Y|X) = H(X, Y) - H(X) \ = \ 1.9037 - .9341 \ = \ .9696, \tag{8.100}$$

which leads to the mutual information

$$I(X : Y) = H(X) - H(X|Y) = 1.9341 - .9327 = .0014 \ . \tag{8.101}$$

8.9   Noise

Suppose a message being transmitted through a communication system is encoded into an input signal, $x(t)$, composed of a string of zeros and ones representing information bits. The input bits are viewed as values of a random variable X, described by probability distribution $p_X(x)$. There is a possibility that the noise will cause a **bit flip error**, in which the output bit that is received is a different value than the bit that was sent. In general, the probability p_0 of 0 flipping to 1 and the probability p_1 of 1 flipping to 0 could be different. But we will restrict ourselves to the simplest case of a **binary symmetric channel**, in which the two bit values have equal probability of flipping. So we define $p_0 = p_1 = p$ to be the probability of a flip on an individual bit, with $q = 1 - p$ being the probability of a bit surviving without error.

In addition to signal X, there may be a second random variable N arising in the system, representing the noise. The noise has its own probability distribution $p_N(\eta)$, which is generally independent of $p_X(x)$. If signal X is sent into one end of the system, what comes out the other end is the random variable $Y = X + N$, with values denoted $y = x + \eta$.

The noise means that each measured output value y could correspond to several input values x, and vice versa (Fig. 8.13b). In the presence of noise, there are three relevant Shannon entropies: the source entropy $H(X)$ measures the information input, the receiver entropy $H(Y)$ measures the output information, and the noise entropy $H(N)$ measures the amount of information lost in transmission. The noise increases the uncertainty in each transmitted bit, and its entropy can be viewed as the conditional entropy of the output for a given input:

$$H(Y|X) = H(N). \tag{8.102}$$

The noise therefore reduces the information shared between input and output:

$$I(X : Y) = H(Y) - H(Y|X) \tag{8.103}$$

$$= H(Y) - H(N). \tag{8.104}$$

Recall that for large n, there are $2^{nH(X)}$ typical input sequences of length n. In the absence of noise, the typical outputs would be in one-to-one correspondence with these inputs. But with noise added, there is a source of ambiguity in the relationship between input and output. Let X and Y be the distributions of the input and output, respectively. Each output could now be produced by $2^{nH(X|Y)}$ possible inputs, so the number of non-redundant useful outputs is therefore reduced to $2^{n(H(X)-H(X|Y))} = 2^{nI(X:Y)}$. As mentioned earlier, the channel capacity, C, is defined to be the maximum rate (in bits per second) at which information can be

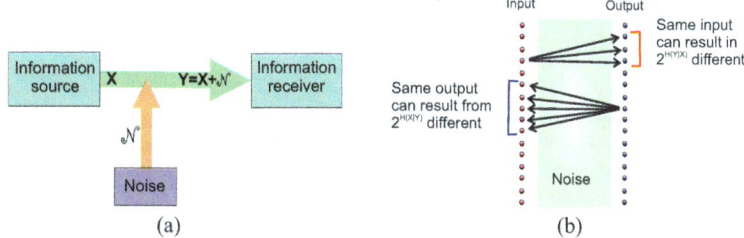

Fig. 8.13 (a) After message X is generated, it becomes partially corrupted by noise N during transmission. As a result, the information recipient will see an altered message Y, and the problem then is to reconstruct X from Y. (b) A noisy channel. For a noiseless channel, each input would correspond to a unique output. But the presence of noise adds uncertainty to the signal so that each output could have resulted from several different inputs. Similarly, each input could lead to several distinct outputs. The conditional entropies $H(Y|X)$ and $H(X|Y)$ measure these added uncertainties

transmitted error-free through a noisy channel. For a discrete, memoryless channel, the number of useful, non-redundant messages of length n is given by

$$2^{n(H(X)-H(X|Y))} = 2^{nI(X:Y)} \leq 2^{nC}, \tag{8.105}$$

so we define the channel capacity to be given by the maximum mutual information:

$$C = \max\left[I(X:Y)\right]. \tag{8.106}$$

The maximum is taken over all possible input probability distributions. More precisely:

Noisy Channel Coding Theorem: Assume a noisy, discrete, memoryless information channel. If information is fed into the channel at rate R (in bits per second), then for any $\epsilon > 0$, there is a coding block length sufficiently long that the information can be transmitted through the channel to produce output with error rate $< \epsilon$, as long as the channel capacity as defined earlier is not exceeded, $R < C = \max\left[I(X:Y)\right]$. If $R > C$, the added uncertainty $H(X|Y)$ caused by noise will satisfy $H(X|Y) > R(X) - C$.

An important case is when the source and the noise both are normally distributed. Then, the channel capacity is given by

$$C = \frac{1}{2}\left(1 + \frac{S}{N}\right), \tag{8.107}$$

where $\frac{S}{N}$ is the signal-to-noise ratio, with S and N being the power of the signal and of the noise, measured in Watts.

8.10   Reversibility Revisited: Thermodynamics and Information

Recall that the thermodynamic entropy S is defined by the heat flow Q in or out of a system,

$$\Delta S = \int \frac{dQ}{T}. \tag{8.108}$$

When all the microstates of a system are equally probable, $p = \frac{1}{\Omega}$, then according to Boltzmann's entropy formula,

$$S = k_B \ln \Omega, \tag{8.109}$$

the entropy can also be written in terms of the number Ω of microscopic configurations that lead to the same macroscopic state. Here k_B is Boltzmann's constant. More generally, if the microstates have differing probabilities, then the Boltzmann formula generalizes to the Gibbs entropy formula,

$$S = -k_B \sum p_i \ln p_i. \tag{8.110}$$

The similarity between the latter formula and the Shannon entropy is obvious.

Hints that thermodynamic entropy is related to information go back at least as far as the introduction of **Maxwell's demon** in 1867. In Maxwell's original discussion, he considered a tiny "finite being" (later dubbed a demon by Lord Kelvin) who exists inside a container of gas (Fig. 8.14). There is a partition dividing the container into

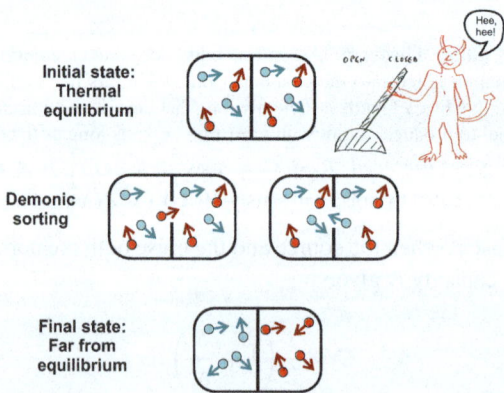

Fig. 8.14 Maxwell's demon: starting with a gas in thermal equilibrium, the demon observes the molecules and opens a gate in the partition to selectively allow hot molecules (colored red here) move to the right side of the container and cold (blue) molecules to the left. The result is that one side of the container becomes hotter, while the other side becomes colder, in seeming contradiction to the second law of thermodynamics

two halves, and a small gate that the demon can open and close in the partition. The gas is initially at equilibrium, with a uniform temperature over both halves of the box. But when the demon sees a hot (i.e. faster than average) molecule on the left side of the container, he opens the gate to allow it to move to the right side of the partition. Similarly, when he sees a cold (slow) molecule on the right side, he opens the gate to allow it to pass to the left side. As a result, the left side cools, while the right side heats up. This is a clear violation of the second law of thermodynamics, which does not allow spontaneous movement away from equilibrium or spontaneous heat flow from the colder side of the partition to the hotter side. The demon can manage to do this because he is observing the molecules and gaining information about them. This collecting of information clearly plays some role in the reduction of entropy that the demon is causing.

Leo Szilard discussed a variation of Maxwell's demon in 1929, allowing a more quantitative discussion. He again imagined a container of gas, but simplified the situation by allowing the gas to consist of a single atom or molecule (Fig. 8.15). The demon now inserts a partition into the box, as before, but this time the partition is free to move without friction to the left or right. Initially, before the demon makes any measurements, the partition is locked in place, in the middle of the container. But then the demon makes an observation of the molecule's location. (In modern terminology, he gains one bit of position information.) If the molecule is on the left side, he attaches a weight of mass m to the left side of the partition and unlocks the partition so that it can slide. Collisions of the molecule with the partition will then lead to a pressure on the wall, forcing it to move to the right. In the process, the moving partition will raise the mass, thus doing work against gravity. Similarly, if the demon sees the molecule on the right side of the box, he attaches the weight on the right, leading again to work. Thus, the demon has again seemingly violated

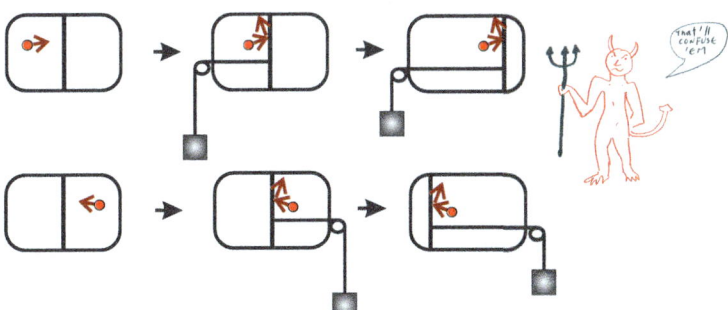

Fig. 8.15 Szilard's demonic engine: starting with a single-molecule gas, the demon observes which side of the partition the molecule is on. If the molecule is to the left, he attaches a weight on the left side, so that when the molecule collides with the partition and pushes it rightward, it will lift the weight (top row). If the molecule is to the right, he attaches a weight on the right side, again causing it to lift the weight (second row). Either way, the demon appears to turns the information he has gained from his observation into work with no heat expended to the environment, again in seeming contradiction to the second law of thermodynamics

the second law by extracting work with no other obvious thermodynamic effect. By randomizing the location of the molecule and then resetting the partition back to the center of the container, the process can be repeated cyclically, leading to what has been called a **Szilard engine**. (Real-world implementations of the Szilard engine have recently been realized experimentally, using single electron systems and colloidal particles [7–9].)

The apparent violation of the second law in these thought experiments was a mystery for decades. Clarification of the situation had to wait until the introduction of the Shannon information, H. The problem was eventually resolved by the work of Landauer, Bennet, Jaynes, and others, in the context of information theory.

In physical systems, thermodynamic entropy S is related to information entropy (Shannon information) H by the relation

$$S = (k_B \ln 2)H. \tag{8.111}$$

(The factor of $\ln 2$ occurs because S is usually defined with a natural logarithm and the Shannon information is defined with base 2 logs.) A corollary of this is **Landauer's principle**: the loss of one bit of information leads to the dissipation an amount of at least $k_B T \ln 2$ of heat into the environment. This can easily be seen for an isothermal process as follows:

$$\Delta S = \frac{\Delta Q}{T} = (k_B \ln 2) \Delta H \quad \longrightarrow \quad \Delta Q = (k_B T \ln 2) \Delta H, \tag{8.112}$$

where it has been assumed that the environment is sufficiently large compared to the system that the temperature does not change appreciably as the heat Q is added to it.

The work generated by the demon in Szilard's engine can then be obtained by looking at the Helmholtz free energy $F = E - TS$ and Shannon information H:

$$W = -\Delta F = -(\Delta E - T\Delta S) < T\Delta S = (k_B T \ln 2) \Delta H. \tag{8.113}$$

So, for each binary left/right measurement made by the demon, the information gain is one bit, $\Delta H = 1$, which then implies that the maximum work done per cycle is $W \leq k_B T \ln 2$. Meanwhile, the information gained has to be stored somewhere. If the demon has finite storage space, then eventually, he will have to start erasing old bits to make space for new ones. Each erasure then leads to heat being dissipated to the environment. Since the heat per erased bit is $Q = k_B T \ln 2$, we see that $Q > W$. In other words, the engine's efficiency is strictly less than 100%, in accord with the predictions of the second law.

The discussion earlier implies that irreversible computation, which involves loss of information, should also entail the generation of heat. This is one reason why all personal computers come with cooling fans built in and why commercial computing centers must spend large sums of money to run enormous cooling systems. Irreversibility enters into computations due to the fact that many of the operations employed, for example, AND and OR gates, take two bits of information as input, but only supply a single bit of output. There is therefore no way to reconstruct the initial state, and one bit of information is lost each time the gate is used.

One of the advantages of quantum computers is that, since they use reversible gates based on unitary operators, quantum computers will generate much less heat and could have much lower energy requirements. There will still be heat generation due to the fact that losses will inevitably occur in any real system: photons will be absorbed in optical fibers and there will be Ohm's law losses in electrical connections. In addition, many quantum computing platforms will require low temperatures to operate, offsetting some of the energy savings. Still, resistance, optical loss, and cooling requirements are practical, rather than fundamental, limitations, and so there is hope of reducing them over time. The lack of irreversibility removes the most important *fundamental* limit on energy dissipation.

For more detailed discussion on the connections between thermodynamics, information, and computational reversibility, see [10–15].

8.11   Parameter Estimation: Fisher Information and the Cramer-Rao Bound

The goal of experiment is often to extract a reliable estimate of the value of some parameter from a set of noisy data. Suppose that the parameter to be estimated is θ, which has some "true" (but unknown) value θ_0. Consider a set of N measurements of some random (θ-dependent) variable x, giving the sample of data points $\chi = \{x_1, x_2, \ldots, x_N\}$. The probability $p_\theta(\chi)$ of measuring the values χ when the parameter has value θ is called the **likelihood function**,

$$L(\chi|\theta) = p_\theta(\chi). \tag{8.114}$$

It is also useful to define the **log-likelihood**,

$$l(\chi|\theta) = \ln L(\chi|\theta). \tag{8.115}$$

If the data values are independent and identically distributed (which we will assume for the remainder of this section), then the likelihood function factors into a product, one for each measurement, so that

$$L(\chi|\theta) = \prod_{i=1}^{N} p_\theta(x_i) = \prod_{i=1}^{N} L(x_i|\theta) \tag{8.116}$$

$$l(\chi|\theta) = \sum_{i-1}^{N} l(x_i|\theta). \tag{8.117}$$

For fixed θ, the likelihood function is the probability of $L(x_i|\theta)$ of obtaining value x_i in a measurement, so that averages of quantities can be obtained by summing (in the discrete case) or integrating (in the continuous case) over the likelihood function,

$$\langle f(x) \rangle = \begin{cases} \sum p_\theta(x) f(x) = \sum f(x) L(x|\theta) & \text{(Discrete case)} \\ \int p_\theta(x) f(x) dx = \int f(x) L(x|\theta) dx & \text{(Continuous case)} \end{cases} \tag{8.118}$$

In particular, the mean value of the log-likelihood is simply the negative of the Shannon entropy up to an overall constant; for the discrete case, for example,

$$\left\langle l(\chi|\theta) \right\rangle = \sum_{x_1,\ldots,x_N} L(\chi|\theta) \ln L(\chi|\theta) = \sum p_\theta(\chi) \ln p_\theta(\chi) = -kH, \tag{8.119}$$

where the constant $k = \ln 2$ simply translates between the natural log in $l(\chi|\theta)$ and the base-2 log occurring inside H: $\ln p_\theta = \ln \left(2^{\log_2 p_\theta}\right) = (\ln 2) \log_2 p_\theta$.

In order to estimate the parameter θ, some function of the data $S(\chi)$ can be proposed as an estimator; this function is called a **statistic**. The **bias** of the estimator is defined by

$$b(S) = \langle S(\chi) \rangle - \theta_0, \tag{8.120}$$

and S is an **unbiased estimator** if $b(S) = 0$.

A fruitful strategy in arriving at the best estimate of θ_0 based on the given data is to start from an unbiased estimator function and then to find the value of θ that maximizes $L(\chi|\theta)$ (or equivalently, that maximizes $l(\chi|\theta)$). From the fact that the mean of $l(\chi|\theta)$ is proportional to the negative of the entropy, it follows that the maximum likelihood estimator minimizes the entropy and maximizes the information gain per measurement.

The variance of the estimator will give a measure of how much the estimate fluctuates around the true value:

$$\sigma_S^2 = \int \left(S(\chi) - \langle S(\chi) \rangle \right)^2 L(\chi|\theta) dx_1 \ldots dx_N, \tag{8.121}$$

or the equivalent expression with sums instead of integrals in the discrete case. The value of σ_S^2 is a measure of the quality of the estimate of θ_0 that is derived from the data: small σ_S^2 implies that the estimates don't fluctuate far from the true value, and so the estimate has high accuracy.

The lower bound on the possible value of the variance is given by the **Cramer–Rao bound**:

$$\sigma_S^2 \geq \frac{1}{F(\theta)}, \tag{8.122}$$

where the **Fisher information** is given (for continuous variables) by

$$F(\theta) = -\left\langle \frac{\partial^2 l(\chi|\theta)}{\partial\theta^2} \right\rangle \tag{8.123}$$

$$= -\int \frac{\partial^2 l(\chi|\theta)}{\partial\theta^2} L(\chi|\theta) dx_1 \ldots dx_N \tag{8.124}$$

$$= -\int \frac{\partial^2 \ln L(\chi|\theta)}{\partial\theta^2} L(\chi|\theta) dx_1 \ldots dx_N. \tag{8.125}$$

The Fisher information measures the amount of information about the parameter that can be extracted from measurements of the random variable x. When F is large, the error in the parameter estimate is expected to be low. Fisher information is widely used in a broad range of fields as diverse as image processing, neuroscience, economics, quantum gravity, and the study of cancer growth. A thorough overview of Fisher information and its applications can be found in [16].

The Cramer–Rao bound should make some intuitive sense: the quantity $\frac{\partial^2 l(\chi|\theta)}{\partial\theta^2}$ measures the curvature of the log-likelihood. Large curvature (i.e., large Fisher information) implies that the likelihood is sharply peaked, which in turn implies that the variance or width of the probability distribution should be small. The derivatives can in fact be pulled out of the expectation value in Eq. 8.123 to prove that the Fisher information is the curvature of the Kullback–Leibler distance (the relative entropy):

$$F(\theta) = \frac{\partial^2}{\partial\theta'^2} K(\theta||\theta')\Big|_{\theta=\theta'}. \tag{8.126}$$

One additional form of the Fisher information is also useful:

$$F(\theta) = \left\langle \left(\frac{\partial l(\chi|\theta)}{\partial\theta} \right)^2 \right\rangle. \tag{8.127}$$

This can be shown as follows.

$$\frac{\partial^2 l}{\partial\theta^2} = \frac{\partial}{\partial\theta} \left(\frac{1}{L} \frac{\partial L}{\partial\theta} \right) \tag{8.128}$$

$$= \left(\frac{\partial}{\partial\theta} \frac{1}{L} \right) \frac{\partial L}{\partial\theta} + \frac{1}{L} \frac{\partial}{\partial\theta} \frac{\partial L}{\partial\theta} \tag{8.129}$$

$$= -\left(\frac{1}{L} \frac{\partial L}{\partial\theta} \right)^2 + \frac{1}{L} \frac{\partial^2 L}{\partial\theta^2} \tag{8.130}$$

$$= -\left(\frac{\partial l}{\partial\theta} \right)^2 + \frac{1}{L} \frac{\partial^2 L}{\partial\theta^2}, \tag{8.131}$$

where the fact that $\frac{\partial l}{\partial \theta} = \frac{\partial \ln L}{\partial \theta} = \frac{1}{L}\frac{\partial L}{\partial \theta}$ has been used. Taking averages of both sides,

$$\left\langle \frac{\partial^2 l}{\partial \theta^2} \right\rangle = -\left\langle \left(\frac{\partial l}{\partial \theta} \right)^2 \right\rangle + \left\langle \frac{1}{L}\frac{\partial^2 L}{\partial \theta^2} \right\rangle. \tag{8.132}$$

But the last term vanishes, since

$$\left\langle \frac{1}{L}\frac{\partial^2 L}{\partial \theta^2} \right\rangle = \int \frac{1}{L}\left(\frac{\partial^2 L}{\partial \theta^2} \right) L\, dx_1 \ldots dx_N \tag{8.133}$$

$$= \int \frac{\partial^2 L}{\partial \theta^2} dx_1 \ldots dx_N \tag{8.134}$$

$$= \frac{\partial^2}{\partial \theta^2} \int p_\theta(\chi)\, dx_1 \ldots dx_N \tag{8.135}$$

$$= \frac{\partial^2}{\partial \theta^2}(1) \tag{8.136}$$

$$= 0, \tag{8.137}$$

which then implies that Eq. 8.127 equals Eq. 8.123.

Example 8.11.1 Consider a normal distribution,

$$p_\mu(\chi) = L(\mu|\chi) = \left(\frac{1}{\sqrt{2\pi}\sigma} \right)^n e^{-\frac{1}{2\sigma^2}\sum_i(x-\mu)^2}, \tag{8.138}$$

for fixed standard deviation σ. The sample $\chi = \{x_1, x_2, \ldots, x_N\}$ is being used to estimate the parameter μ. Taking the logarithm,

$$l(\mu|\chi) = -n \ln \sigma - \frac{n}{2}\ln(2\pi) - \frac{1}{2\sigma^2}\sum_{i=1}^{n}(x_i - \mu)^2, \tag{8.139}$$

so that

$$\frac{\partial l(\chi|\mu)}{\partial \mu} = \frac{1}{\sigma^2}\sum_{i=1}^{n}(x_i - \mu) = \frac{n}{\sigma^2}(\bar{x} - \mu). \tag{8.140}$$

Using Eq. 8.144, this implies that $\langle \bar{x} \rangle - \mu = 0$, so that the sample mean $\bar{x}$ is an unbiased estimator of μ. (The expression $\langle \bar{x} \rangle$ is the average of the sample averages: the sample means $\bar{x}$ are taken from multiple samples, then the average of these multiple $\bar{x}$ values is taken.)

Using $\bar{x}$ as the estimator for μ, the Fisher information can easily be found. Taking the derivative of the score,

$$\frac{\partial^2 l}{\partial \mu^2} = -\frac{n}{\sigma^2}. \tag{8.141}$$

So,

$$F(\mu) = -\left\langle \frac{\partial^2 l}{\partial \mu^2} \right\rangle = +\frac{n}{\sigma^2}. \tag{8.142}$$

The Cramer–Rao bound then tells us that

$$\sigma_{\bar{x}} \geq \frac{\sigma}{\sqrt{n}}. \tag{8.143}$$

This is simply the familiar result that the standard deviation of samples should decrease $\sim \frac{1}{\sqrt{n}}$ as the sample size increases. In statistics texts, the distribution of the sample $\bar{x}$ values around the population or "true" mean is called the **sampling distribution of the mean**. ■

For more on estimation theory, see [17]. The discussion in this section has been entirely classical, but the theory of quantum estimation is also well-developed; see [18, 19].

Problems

1. (a) Find the values of the expressions $AB + (\overline{A} + \overline{B})$ and $AB(\overline{A} + \overline{B})$.
   (b) Construct the truth tables for the expressions in part (a).
   (c) Draw the logic circuit that implements the expressions in part (a).
2. (a) Use Boolean algebra to simplify the expression $A + B(C + \overline{B}) + B\overline{C}$.
   (b) Draw circuit diagrams for the original and simplified expressions in part (a).
3. (a) Construct the truth table and the circuit diagram for the operation $A + BC + \overline{D}$.
   (b) Do the same for $\overline{(A + B)}(C + D)\overline{C}$.
4. Using de Morgan's laws or truth tables, prove each of the identities given in Eqs. 8.16–8.20.
5. Show how a Fredkin gate can be used to implement an AND gate in a reversible manner.
6. Prove the relation of Eq. 8.81, starting from Bayes' theorem.
7. For the binomial distribution, $P(k) = \frac{n!}{k!(n-k)!} p^k (1-p)^{n-k}$, samples of data can be used to estimate the parameter p. Show that the Fisher information is given by $F(p) = \frac{n}{p(1-p)}$, so that the Cramer–Rao bound implies that the variance of the estimate satisfies $\sigma^2 \geq \frac{p(1-p)}{n}$.

8. A random variable X has four possible values, α, β, γ, and δ. Consider the two probability distributions p and q and the encoding $C(x)$ given in the table:

x	$p(x)$	$q(x)$	$C(x)$
α	$\frac{1}{2}$	$\frac{1}{4}$	0
β	$\frac{1}{2}$	$\frac{3}{8}$	1
γ	0	$\frac{1}{8}$	01
δ	0	$\frac{1}{4}$	10

(a) Find the Shannon entropies for $H(p)$ and $H(q)$.
(b) Find the Kullback–Liebler entropies $K(p\|q)$ and $K(q\|p)$.
(c) Find the average codeword length for each distribution.
(d) Find the difference in average codeword length for the two distributions. How is this difference related to the Kullback–Liebler entropies?
(e) What is the minimum possible codeword length for each distribution? Can you find an encoding that minimizes the length for the q distribution?

9. One further quantity often defined in connection with the likelihood function is the derivative $\frac{\partial l(\chi|\theta)}{\partial\theta}$, known as the **score**. Show that the mean score always vanishes,

$$\left\langle \frac{\partial l(\chi|\theta)}{\partial\theta} \right\rangle = 0. \tag{8.144}$$

10. Suppose an information source X produces a signal with alphabet $\{0, 1\}$ and probability distribution $p(0) = p(1) = 0.5$. The bit flip error probability is symmetric (the same for initial bit 0 and for 1) and equal to ϵ. Denote the output variable as Y, with probability distribution $q(X)$.

(a) Write a matrix that takes the input probabilities to the output probabilities. (This is called the **channel matrix**.)
(b) Find the entropies of the source and the output, $H(X)$ and $H(Y)$.
(c) Find the joint distribution $p(X, Y)$ and the joint entropy $H(X, Y)$.
(d) Find the mutual information, $I(X : Y)$. For what values of ϵ will the mutual information be maximized, and what situations do they represent physically?
(e) Find the channel capacity, C. For what value of ϵ does it occur?

References

1. T.M. Cover, J.A. Thomas, *Elements of Information Theory* (Wiley-Blackwell, 2006)
2. C.E. Shannon, W. Weaver, *The Mathematical Theory of Communication* (University of Illinois Press, 1971)

3. M.A. Nielsen, I.L. Chuang, *Quantum Computation and Quantum Information: 10th Anniversary Edition* (Cambridge University Press, Cambridge, 2011)
4. G. Jaeger, *Quantum Information: An Overview* (Springer, Berlin, 2007)
5. I.J. Good Biometrika **66**, 393 (1979)
6. V. Vedral, *Introduction to Quantum Information Science* (Oxford University Press, Oxford, 2006)
7. S. Toyabe, T. Sagawa, M. Ueda, E. Muneyuki, M. Sano, Nat. Phys. **6**, 988 (2010)
8. J. Koski, V. Maisi, T. Sagawa, J.P. Pekola, Phys. Rev. Lett. **113**, 030601 (2014)
9. E. Roldán, I.A. Martínez, J.M.R. Parrondo, D. Petrov, Nat. Phys. **10**, 457 (2014)
10. R. Landauer, Phys. Today **44**, 23 (1991)
11. R. Landauer, Phys. Lett. A **217**, 188 (1996)
12. C.H. Bennet, IBM J. Res. Develop. **32**, 16 (1988)
13. W.H. Zurek, Phys. Today **44**, 36 (1991)
14. E. Lutz, S. Ciliberto, Phys. Today **68**, 30 (2015)
15. J.M.R. Parrondo, J.M. Horowitz, T. Sagawa, Nat. Phys. **11**, 131 (2015)
16. B.R. Friedan, *Science from Fisher Information: A Unification* (Cambridge University Press, Cambridge, 2004)
17. A. van den Bos, *Parameter Estimation for Scientists and Engineers* (Wiley, London, 2007)
18. C.W. Helstrom, *Quantum Detection and Estimation Theory* (Academic Press, New York, 1976)
19. H.M. Wiseman, G.J. Milburn, *Quantum Measurement and Control* (Cambridge University Press, Cambridge, 2010)

Part II
Quantum Mechanics Beyond the Basics

Chapter 9
More on Quantum States

In this chapter, we describe a few additional types of quantum states that appear in applications, such as coherent and squeezed states and $N00N$ states. The application of squeezed states in detecting gravitational waves is discussed as an illustrative example of the use of these states. We also discuss some aspects related to measurement of quantum states. In particular, we examine a subject on which quantum cryptography depends, that of distinguishing between nonorthogonal quantum states.

9.1 Coherent States

An important type of quantum state is the **coherent state**, which can be thought of as a minimal-uncertainty state or the "most classical" state of a quantum system. The coherent state of the harmonic oscillator was first considered by Schrödinger in 1926 and has been generalized in a number of different ways since the work of Klauder and Glauber in the 1960s. Coherent states have found applications in a large array of fields. For example, laser beams can often be modeled as approximations to coherent states, and as discussed in Chap. 7, the wavefunction in a superconductor can be viewed as a macroscopic coherent state formed from collective excitations of Cooper pairs.

Consider a one-dimensional quantum harmonic oscillator. The Hamiltonian operator can be formed from the position and momentum operators $\hat{x}$ and $\hat{p}$ according to

$$\hat{H} = \frac{\hat{p}^2}{2m} + \frac{m\omega^2}{2}. \tag{9.1}$$

As in Chap. 5, we can define a pair of raising and lowering operators,

© The Author(s), under exclusive license to Springer Nature Switzerland AG 2025
D. S. Simon, *Introduction to Quantum Science and Technology*, Undergraduate
Texts in Physics, https://doi.org/10.1007/978-3-031-81315-3_9

$$\hat{a} = \sqrt{\frac{m\omega}{2\hbar}}\hat{x} + i\sqrt{\frac{1}{2m\hbar\omega}}\hat{p} \tag{9.2}$$

$$\hat{a}^\dagger = \sqrt{\frac{m\omega}{2\hbar}}\hat{x} - i\sqrt{\frac{1}{2m\hbar\omega}}\hat{p}, \tag{9.3}$$

which obey the canonical commutation relation,

$$\left[\hat{a}, \hat{a}^\dagger\right] = 1, \tag{9.4}$$

and which act on energy eigenstates states $|n\rangle$ of energy E_n by raising and lowering n:

$$\hat{a}^\dagger|n\rangle = \sqrt{n+1}|n+1\rangle \tag{9.5}$$

$$\hat{a}|n\rangle = \sqrt{n}|n-1.\rangle. \tag{9.6}$$

The number operator $\hat{N} = \hat{a}^\dagger\hat{a}$ acts on these states according to

$$\hat{N}|n\rangle = n|n\rangle. \tag{9.7}$$

Recall also that the excited states can be constructed from the ground state $|0\rangle$ by repeated applications of the raising operator,

$$|n\rangle = \frac{1}{\sqrt{n!}}(\hat{a}^\dagger)^n|0\rangle, \tag{9.8}$$

and that these states are orthonormal,

$$\langle m|n\rangle = \delta_{mn}. \tag{9.9}$$

The spatial wavefunctions $\psi(x) = \langle x|n\rangle$ are given by the standard expression involving Hermite polynomials, given in most quantum texts, such as [1, 2].

Instead of the $|n\rangle$ states, which have a well-defined value of n, let us now imagine states that are in a superposition of different n values. Consider a complex number α, which will be called the **coherent state amplitude**. We define an operator, called the **displacement operator**, defined as

$$\hat{D}(\alpha) = e^{\alpha\hat{a}^\dagger - \alpha^*\hat{a}} = e^{\alpha\hat{a}^\dagger}e^{-\alpha^*\hat{a}}e^{-\frac{1}{2}|\alpha|^2}. \tag{9.10}$$

The second expression follows from the first by means of the Baker–Campbell–Hausdorf (BCH) formula,

$$e^{\hat{A}+\hat{B}} = e^{\hat{A}}e^{\hat{B}}e^{-\frac{1}{2}[\hat{A},\hat{B}]}, \tag{9.11}$$

with $\hat{A} = \alpha \hat{a}^\dagger$ and $\hat{B} = -\alpha^* \hat{a}$. The inverse of this operator is given by $\hat{D}^{-1}(\alpha) = \hat{D}^\dagger(\alpha) = \hat{D}(-\alpha)$. It is readily shown (Problem 3) that this operator shifts the raising and lowering operators by terms proportional to the identity operator according to

$$\hat{D}(-\alpha)\hat{a}\hat{D}(\alpha) = \hat{a} + \alpha \hat{I} \tag{9.12}$$

$$\hat{D}(-\alpha)\hat{a}^\dagger \hat{D}(\alpha) = \hat{a}^\dagger + \alpha^* \hat{I}. \tag{9.13}$$

Successive applications of the displacement operator can be viewed as a single combined displacement, up to an overall phase:

$$\hat{D}(\alpha)\hat{D}(\beta) = e^{(\alpha\beta^* - \beta\alpha^*)/2}\hat{D}(\alpha + \beta). \tag{9.14}$$

Then one way to define the coherent state $|\alpha\rangle$ is as a displaced vacuum state, or in other words, the state that results when the displacement operator acts on the oscillator ground state $|0\rangle$,

$$|\alpha\rangle = \hat{D}(\alpha)|0\rangle. \tag{9.15}$$

By expanding the displacement operator in a power series, it can be shown that the coherent state can be written in terms of the energy eigenstates $|n\rangle$ by the infinite series

$$|\alpha\rangle = e^{-|\alpha|^2/2} \sum_{n=0}^{\infty} \frac{\alpha^n}{\sqrt{n!}} |n\rangle. \tag{9.16}$$

The coherent state is therefore a superposition of an infinite number of energy eigenstates. If $\hat{N}$ is measured, the probability of finding the value n in the coherent state $|\alpha\rangle$ is

$$P_n(\alpha) = |\langle n|\alpha\rangle|^2 = \frac{|\alpha|^{2n}}{n!}e^{-|\alpha|^2}. \tag{9.17}$$

Notice that this is a Poisson distribution. Accordingly, the mean and variance of n should obey the Poisson relation

$$\Delta \hat{N}^2 = \langle \hat{N} \rangle. \tag{9.18}$$

This can be verified by using Eq. 9.16 to calculate (Problem 5) that

$$\langle \hat{N} \rangle = |\alpha|^2, \qquad \langle \hat{N}^2 \rangle = |\alpha|^4 + |\alpha|^2, \tag{9.19}$$

so that

$$\Delta \hat{N}^2 = \langle \hat{N}^2 \rangle - \langle \hat{N} \rangle^2 = |\alpha|^2, \tag{9.20}$$

giving $\Delta \hat{N}^2 = \langle \hat{N} \rangle$ as expected.

The time evolution of a coherent state is easy to determine. For the oscillator, the Hamiltonian has eigenstates $|n\rangle$ with eigenvalues $E_n = (n + \frac{1}{2})\hbar\omega$. So if the initial coherent state is $|\alpha\rangle$, then applying the time evolution operator to the expansion 9.16 gives the state at time t:

$$|\psi(t)\rangle = e^{-i\hat{H}t/\hbar}|\alpha\rangle \tag{9.21}$$

$$= e^{-|\alpha|^2/2} \sum_{n=0}^{\infty} \frac{\alpha^n}{\sqrt{n!}} e^{-i(n+\frac{1}{2})\omega t}|n\rangle \tag{9.22}$$

$$= e^{-\frac{i}{2}\omega t}\left[e^{-|\alpha|^2/2} \sum_{n=0}^{\infty} \frac{(\alpha e^{-i\omega t})^n}{\sqrt{n!}}|n\rangle \right] \tag{9.23}$$

$$= e^{-\frac{i}{2}\omega t}|\alpha(t)\rangle, \tag{9.24}$$

where $\alpha(t) = \alpha e^{-i\omega t}$. In other words, up to the overall phase $e^{-\frac{i}{2}\omega t}$, the state remains a coherent state, with an amplitude of constant magnitude that orbits at rate ω about the origin in phase space.

Using the expansion of Eq. 9.16, it can easily be shown (Problem 6) that the inner product between two coherent states decreases exponentially with the distance $|\alpha - \beta|$ between them in the complex plane,

$$|\langle \alpha | \beta \rangle|^2 = e^{-|\alpha-\beta|^2}. \tag{9.25}$$

This implies that the coherent states are properly normalized, $\langle \alpha | \alpha \rangle = 1$, but it also shows that two coherent states are never completely orthogonal to each other. Recall that the completeness relation for a set of states depending on some continuous, complex parameter λ is of the form $\int |\lambda\rangle\langle\lambda|\, d^2\lambda = 1$, where $\int d^2\lambda$ means integration is done over both the real and imaginary parts of λ. This guarantees that any other state can be constructed as a linear combination of the $|\lambda\rangle$ states. In the case of coherent states, though, the analogous relation is

$$\int |\alpha\rangle\langle\alpha|\, d^2\alpha = \pi. \tag{9.26}$$

The fact that the right side is >1 indicates that the coherent states in fact form an *overcomplete* set. This means that they are not all linearly independent of each other, so that any coherent state can be written as a combination of other coherent states:

$$|\alpha\rangle = \frac{1}{\pi} \int d^2\beta\, |\beta\rangle\langle\beta|\alpha\rangle = \frac{1}{\pi} \int d^2\beta\, e^{-\frac{1}{2}(|\alpha|^2+|\beta|^2)+\alpha\beta^*}|\beta\rangle. \tag{9.27}$$

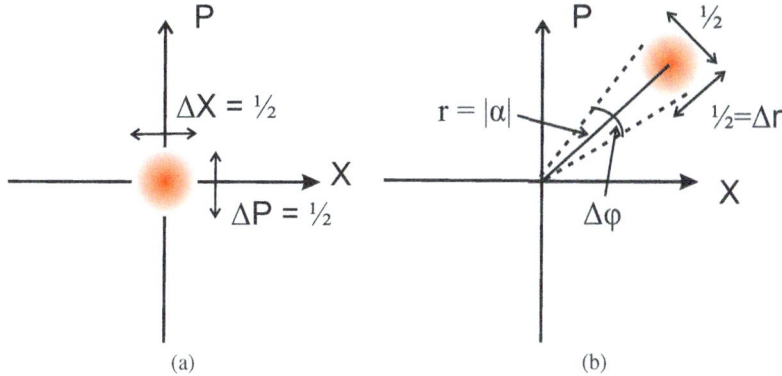

Fig. 9.1 (a) The vacuum state has mean value of zero for both quadratures. But fluctuations about the mean give the state an uncertainty of $\frac{1}{2}$ in each direction. (b) A coherent state is a vacuum state that has been displaced away from the origin. If the amplitude is $\alpha = |\alpha|e^{i\phi}$, then the distance from the origin is $r = |\alpha| = \sqrt{\bar{n}}$, where $\bar{n}$ is the mean photon number. The angle from the X axis is ϕ. The uncertainties in each direction are still bit equal to $\frac{1}{2}$: $\Delta n = |\alpha|\Delta\phi = \frac{1}{2}$

Although the position-space wavefunction for the coherent state is not often useful, it can be found easily enough:

$$\psi_\alpha(x) = \langle x|\alpha\rangle = \psi_0(x - x_0)e^{-\frac{1}{2}(x-x_0)^2 + ip_0 x - \frac{i}{2}p_0 x_0}, \tag{9.28}$$

where ψ_0 is the wavefunction for the vacuum state, and we have defined $\alpha = \frac{1}{\sqrt{2}}(x_0 + ip_0)$.

Referring to Fig. 9.1b, for $|\alpha| \gg 1$, we can approximate $\Delta\theta \approx \tan\Delta\theta \approx \frac{1/2}{|\alpha|} = \frac{\Delta n}{2}$, so the coherent state saturates the lower bound of the phase-photon number uncertainty relation,

$$\Delta\theta\,\Delta n \geq \frac{1}{2}. \tag{9.29}$$

(There is no factor of $\hbar$ on the right because the two variables on the left are both dimensionless.)[1]

Although defined earlier for oscillator systems, the definition of coherent states carries over, more or less unchanged, to optical systems, simply by a reinterpretation of what the $|n\rangle$ states represent. For an optical field, the $\hat{x}$ and $\hat{p}$ operators, rather than representing position and momentum, would represent the quadratures

---

[1] Some complications are being swept under the rug here. Because the phase is not single-valued, $\theta \sim \theta + 2\pi$, there is no straightforward way to define a quantum phase operator $\hat{\theta}$ with a commutation relation of the form $[\hat{N}, \hat{\theta}] = 1$. Nonetheless, the uncertainty relation 9.29 that we would expect from such a commutation relation still holds for $\Delta\theta \ll 2\pi$. See [3] for more details.

of the electric field (Sect. 5.2). $\hat{a}^\dagger$ and $\hat{a}$ then become operators that create and annihilate photons. The $|n\rangle$ states therefore now represent states of the field with a definite number n of photons, and so the probability P_n defined earlier becomes the probability of finding n photons in a given optical pulse or in a fixed-length segment of a continuous beam. The number of photons per pulse is then Poisson distributed for a coherent state; this is exactly the photon-number distribution found in laser beams. Notice that as $n \to \infty$,

$$\frac{\Delta N}{\langle \hat{N} \rangle} \sim \frac{1}{\sqrt{n}}, \tag{9.30}$$

so that fluctuations in photon number become negligible compared with the mean photon number as the classical limit is approached.

Coherent states can be defined for electromagnetic fields in essentially the same manner as for harmonic oscillators, just by replacing the raising and lowering operators by photon creation and annihilation operators. Again define $\hat{a}$ and $\hat{a}^\dagger$ and a pair of dimensionless, Hermitian quadrature operators as in Sect. 5.2,

$$\hat{a} = \hat{X} + i\hat{P} \qquad\qquad \hat{X} = \frac{1}{2}\left(\hat{a}^\dagger + \hat{a}\right) \tag{9.31}$$

$$\hat{a}^\dagger = \hat{X} - i\hat{P} \qquad\qquad \hat{P} = \frac{i}{2}\left(\hat{a}^\dagger - \hat{a}\right). \tag{9.32}$$

The creation and annihilation operators obey $\left[\hat{a}, \hat{a}^\dagger\right] = 1$, from which it follows that

$$\left[\hat{X}, \hat{P}\right] = \frac{i}{2}. \tag{9.33}$$

The amplitude of the coherent state will be related to the electric field amplitude E_0 and the mean photon number $\bar{n}$ by

$$|\alpha| = \sqrt{X^2 + P^2} = \sqrt{\frac{\epsilon_0 V}{4\hbar\omega}} = \sqrt{\bar{n}}. \tag{9.34}$$

The uncertainty relation for $\hat{X}$ and $\hat{P}$ is

$$\Delta X \Delta P \geq \frac{\left[\hat{X}, \hat{P}\right]}{2} = \frac{1}{4}. \tag{9.35}$$

Number states for any finite value of n will clearly exceed this limit,

$$\Delta X \Delta P = \frac{1}{4}(2n + 1), \tag{9.36}$$

while coherent states saturate the bound,

$$\Delta X \Delta P = \frac{1}{4}. \tag{9.37}$$

The vacuum fluctuations of $\hat{X}$ and $\hat{P}$ mean that the vacuum takes up a rotationally invariant region of diameter $\frac{1}{2}$ centered at the origin (Fig. 9.1a); the vacuum state could be found anywhere in this region and cannot be localized further than this. The coherent state can be found in a similar region, except centered at a point away from the origin in the $\hat{X} - \hat{P}$ phases space (Fig. 9.1b).

Coherent optical states can be generated by coupling the field to a classical current. Consider a classical current density $j(r, t)$, assumed for simplicity to be a delta function along the z-axis, $j(r, t) = J(t)\delta(x)\delta(y)$. The interaction potential between the field and the current is

$$V(r, t) = - \int d^3r \; j(r, t) \cdot A(r) = -J(t)AL, \tag{9.38}$$

where L is the length of the current-carrying wire segment, and the field A is assumed constant along the current. Further assuming that the current has the simple time dependence

$$J(t) = J_0 e^{-i\omega t} + J_0^* e^{i\omega t}, \tag{9.39}$$

then in the rotating wave approximation, the interaction becomes [4]

$$V(r, t) \approx -\frac{\mathcal{E}_0}{\omega} \left(J_1 \hat{a}^\dagger + J_1^* \hat{a} \right), \tag{9.40}$$

where $\mathcal{E}_0$ is the electric field amplitude and $J_1 = J_0 \cdot \epsilon^*$. The vacuum state evolves over time into the state $|\Psi(t)\rangle$ given by

$$|\Psi(t)\rangle = e^{-iVt/\hbar}|0\rangle = e^{i\frac{\mathcal{E}_0}{\hbar\omega}(J_1\hat{a}^\dagger + J_1^*\hat{a})t}|0\rangle, \tag{9.41}$$

which is of the form of a coherent state $|\alpha(t)\rangle = D(\alpha(t))|0\rangle$, with $\alpha(t) = -\frac{iVt}{\hbar} = i\frac{J_1\mathcal{E}_0 t}{\hbar\omega}$.

To make measurements on coherent states, homodyne detection (Sect. 22.6) can be used to measure the quadratures, and then amplitude and phase can be obtained from the quadratures,

$$|\alpha| = \sqrt{X^2 + P^2}, \qquad \tan\phi = \frac{P}{X}. \tag{9.42}$$

Example 9.1.1 Consider the action of a beam splitter on a coherent state $|\alpha\rangle$ entering port 1. Writing the coherent state input as

$$|\alpha\rangle_1 = D_1(\alpha)|0\rangle = e^{(\alpha \hat{a}^\dagger - \alpha^* \hat{a})}|0\rangle. \tag{9.43}$$

Then the beam splitter converts this to output state

$$|\psi\rangle_{out} = e^{\frac{1}{\sqrt{2}}\alpha\left(\hat{a}_3^\dagger + i\hat{a}_2^\dagger\right) - \frac{1}{\sqrt{2}}\alpha^*\left(\hat{a}_3 - i\hat{a}_2\right)}|0\rangle \tag{9.44}$$

$$= e^{\frac{1}{\sqrt{2}}\left(\alpha\hat{a}_3^\dagger - \alpha^*\hat{a}_3\right)}e^{\frac{i}{\sqrt{2}}\left(\alpha\hat{a}_2^\dagger - \alpha^*\hat{a}_2\right)}|0\rangle \tag{9.45}$$

$$= D_3\left(\frac{\alpha}{\sqrt{2}}\right)D_3\left(i\frac{\alpha}{\sqrt{2}}\right)|0\rangle \tag{9.46}$$

$$= \left|\frac{\alpha}{\sqrt{2}}\right\rangle_3 \left|\frac{\alpha}{\sqrt{2}}\right\rangle_3 . \tag{9.47}$$

In other words, half of the input intensity exits at each of the two output ports. Each of the two output beams is a coherent state, with the reflected beam gaining a phase shift of $e^{i\pi/2} = i$ in its coherent state amplitude. In other words, the coherent state behaves like a classical light beam would, as should be expected. ∎

9.2 Squeezed States

As we saw in the last section, coherent states are states of minimal uncertainty product, with the uncertainty divided evenly between the two quadratures. Sometimes, however, it is advantageous to look at states of minimal uncertainty product where the uncertainty is not evenly divided: by pushing more of the uncertainty into one of the quadratures, we can reduce measurement uncertainty in the other, allowing one variable to be measured to very high precision. The states that accomplish this are called **squeezed states**.

Define the **squeezing operator**,

$$\hat{S}(z) = e^{\frac{1}{2}\left(z^* \hat{a}^2 - z(\hat{a}^\dagger)^2\right)}, \tag{9.48}$$

where $z = \zeta\, e^{i\phi}$ is a complex number. The magnitude ζ is the **squeezing parameter**. The squeezing operator is unitary, $\hat{S}^\dagger(z) = S^{-1}(z)$. For convenience, define $\hat{A} = -\frac{i}{2}\left(z^* \hat{a}^2 - z(\hat{a}^\dagger)^2\right)$, so that $\hat{S}(z) = e^{i\hat{A}}$.

Notice that the exponent involves squares of the creation and annihilation operators, so nonlinear processes are needed to implement this operation physically. Squeezed light at frequency ω is produced by passing the light through a nonlinear crystal and mixing it with a pump beam of amplitude E_0 and frequency 2ω, with the $\chi^{(3)}$ nonlinear interaction Hamiltonian

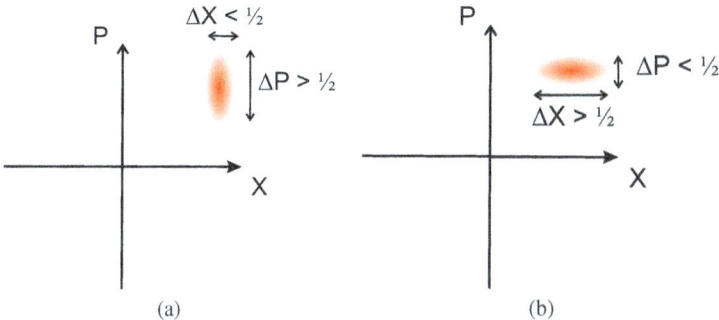

Fig. 9.2 Quadrature-squeezed states. (**a**) Squeezing the $\hat{X}$ uncertainty (reducing ΔX) increases the uncertainty ΔP. (**b**) Similarly, squeezing ΔP increases ΔX. In both cases, the product $\Delta X \Delta P$ remains the same as the coherent state value $\frac{1}{4}$

$$H_{int} = \chi \left(E_0^* \hat{a}^2 - E_0(\hat{a}^\dagger)^2 \right). \tag{9.49}$$

This is an example of degenerate parametric amplification: fluctuations in the light that are in phase with the laser become amplified, while those that aren't will be damped.

The **quadrature-squeezed vacuum state** is defined as $S(z)|0\rangle$. A generic **quadrature-squeezed state** is then obtained by displacing the squeezed vacuum state,

$$|\tau\rangle = D(\alpha)S(z)|0\rangle, \tag{9.50}$$

where the state is specified by a complex number τ to be defined further. Examples of quadrature-squeezed states are shown in Fig. 9.2.

The **squeezed annihilation operator** is

$$\hat{t} = \hat{S}(z)\hat{a}\hat{S}^\dagger(z) \tag{9.51}$$

$$= \hat{a} + \left[i\hat{A}, \hat{a}\right] + \frac{1}{2!}\left[i\hat{A}, \left[i\hat{A}, \hat{a}\right]\right] + \ldots \tag{9.52}$$

$$= \hat{a}\cosh\zeta + \hat{a}^\dagger e^{i\phi}\sinh\zeta. \tag{9.53}$$

The corresponding squeezed creation operator is

$$\hat{t}^\dagger = \hat{a}^\dagger\cosh\zeta + \hat{a}e^{-i\phi}\sinh\zeta. \tag{9.54}$$

The transformation from $\hat{a}, \hat{a}^\dagger \rightarrow \hat{t}, \hat{t}^\dagger$ is called a **Bogoliubov transformation**. Note that these new operators still obey the canonical commutation relation,

$$\left[\hat{t}, \hat{t}^{\dagger}\right] = 1. \tag{9.55}$$

The displacement operator acts on the squeezed annihilation and creation operators by shifts, just as for the unsqueezed case:

$$D(\mu)\hat{t}D^{-1}(\mu) = \hat{t} - \mu, \qquad D(\mu)\hat{t}^{\dagger}D^{-1}(\mu) = \hat{t} - \mu^{*}. \tag{9.56}$$

The squeezed state is an eigenstate of the squeezed annihilation operator,

$$\hat{t}|\tau\rangle = \tau|\tau\rangle, \tag{9.57}$$

where $\tau = \alpha \cosh \zeta + \alpha^{*} e^{i\phi} \sinh \zeta$ is a complex number. It is straightforward to show that the squeezing increases the uncertainty in one quadrature and increases the uncertainty in the other, while keeping the product equal to the minimal, coherent state value:

$$(\Delta X)^2 = \frac{1}{2} e^{-2\zeta}, \qquad (\Delta P)^2 = \frac{1}{2} e^{+2\zeta}, \qquad \Delta X \Delta P = \frac{1}{2}.$$

The position-space wavefunction for the squeezed state can be found. Defining $\alpha = \frac{1}{\sqrt{2}}(x_0 + ip_0)$, the squeezed position-space wavefunction centered at α is

$$\psi(x) = \frac{1}{\pi^{1/4}} e^{\zeta/2} e^{-\frac{1}{2} e^{2\zeta}(x-x_0)^2 + \frac{i}{2} p_0 x - ip_0 x_0}. \tag{9.58}$$

The nonlinear interaction leading to squeezing produces photons in pairs, so the photon number distribution should have vanishing values for odd photon number:

$$p_{2m+1} = 0. \tag{9.59}$$

For even photon numbers,

$$p_n = \binom{2m}{m} \frac{1}{\cosh \zeta} \left(\frac{1}{2} \tanh \zeta\right)^2 m. \tag{9.60}$$

Instead of squeezing the quadratures, or in other words, squeezing along the X or P directions, it is possible to produce states that are squeezed along other phase space directions. For instance, squeezing along the radial direction allows photon number to be measured with higher precision. Or, alternately, the uncertainty in the angular direction can be reduced, allowing improved phase resolution, but at the expense of increasing the uncertainty in photon number. These are shown in Fig. 9.3.

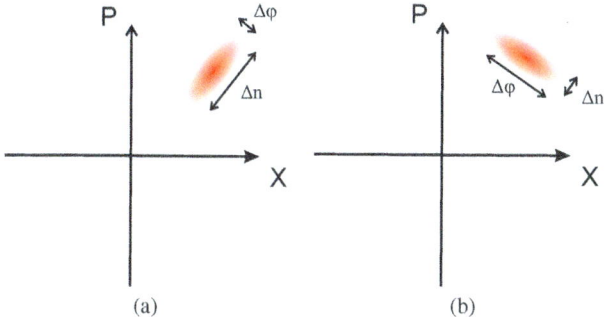

Fig. 9.3 More squeezed states: (**a**) A phase squeezed state and (**b**) a number-squeezed state

9.3 Application: Detecting Gravitational Waves with Squeezed Light

As an application of squeezed states, consider the detection of gravitational waves with a Michelson interferometer. General relativity predicts the emission of gravitational radiation when large masses accelerate. These **gravitational waves** are essentially ripples in spacetime. They are quadrupole excitations, with oscillations in two transverse directions oscillating out of phase with each other. For example, if the wave is propagating in the z-direction, then at a given moment the x-direction may be contracting while the y-direction is expanding.

These oscillations are variations in the distance scale of space-time in each direction. (For those familiar with relativity, it is actually the space-time *metric* that is oscillating.) So objects measured in one direction seem to be getting longer while those in the perpendicular direction seem to be shrinking. Because gravity is a very weak force, these variations are of minuscule size. To quantify the size of the oscillations, define a dimensionless stress variable: if an object is of length L and it changes in length by ΔL, then the stress is defined to be

$$h = \frac{\Delta L}{L}. \tag{9.61}$$

The expected values of stress for large astronomical events are expected to be, at most, of order 10^{-20}–10^{-25} when the resulting gravitational waves reach earth. This explains why, even though gravitational waves were predicted by Einstein at the creation of general relativity around 1916 (and were speculated about by Poincaré and others even earlier), the first experimental detection was not published until a century later, in 2016.

Typical sources of gravitational waves are neutron stars or black holes orbiting around each other or colliding with each other, as well as supernovae and residual signals left over from the Big Bang. The first direct experimental detection [5] was of radiation emitted by two black holes (of 29 and 36 solar masses) 1.3 billion light years from the Earth in a spirally decaying orbit. They produced a peak of radiation when they collided and merged. The maximum strain produced by the wave when it reached the Earth was on the order of 10^{-21}. The waves were of relatively low frequency, $f \leq 250\,\text{Hz}$. The detection of gravitational waves not only provides further confirmation of general relativity but also provides a new means of probing astronomical objects: gravitational telescopes will eventually join optical and radio telescopes as standard tools of astrophysics.

Example 9.3.1 To have an idea of how small the displacements are that need to be detected, consider a simple example. Early gravitational wave searches in the 1960s looked for small gravitational-wave-induced distortions in the dimensions of large metal bars called **Weber bars**. Assume that the wave has angular frequency ω and the bar has mass m. Setting the energy $\hbar\omega$ of the wave equal to the kinetic energy $\frac{p^2}{2m}$ gained by the bar. Then, using the uncertainty relation $\Delta x \Delta p \geq \frac{\hbar}{2}$, we find as a rough order of magnitude estimate:

$$\hbar\omega = \frac{p^2}{2m} \approx \frac{\Delta p^2}{2m} \approx \frac{\left(\frac{\hbar}{2\Delta x}\right)^2}{2m}. \tag{9.62}$$

Solving for Δx, the minimum spatial resolution needed to see a displacement is

$$\Delta x \sim \frac{1}{2}\sqrt{\frac{\hbar}{2m\omega}}. \tag{9.63}$$

Assuming a bar of mass $m = 1000\,\text{kg}$ and wave of frequency $f = 1000\,\text{Hz}$, this gives a result on the order of $\Delta x \sim 10^{-20}\,\text{m}$. Given that this is orders of magnitude smaller than the diameter of an atomic nucleus, such a measurement doesn't seem tenable. So by the 1990s, the focus moved to interferometers with very long baselines (kilometers or more), in conjunction with specialized quantum states to reduce the inherent quantum uncertainty in the measured direction. ∎

The two major experiments currently searching for gravitational waves (as of this writing), LIGO [6] and VIRGO [7], both make use of Michelson interferometers (Sect. 2.5). As the gravitational wave passes, one arm should become longer and the other should become shorter. So the length difference between the two, which is what the Michelson interferometer measures, should oscillate in time. The arms of these interferometers are 3–4 kilometers long in order to achieve high resolution. Nonetheless, it turns out that shot noise in the interferometers will exceed the signals expected from the gravitational waves. So, some method is needed to reduce the noise.

Shot noise is produced by the quantized nature of the light in the interferometer. The photons arrive at random times, producing small random accelerations of the interferometer mirrors. The usual analogy is that shot noise is like the random pattern of sound you hear from raindrops falling on a tin roof. These random

Fig. 9.4 Using squeezed
states to reduce noise in a
Michelson interferometer.
The noise reduction is needed
to detect the gravitational
waves

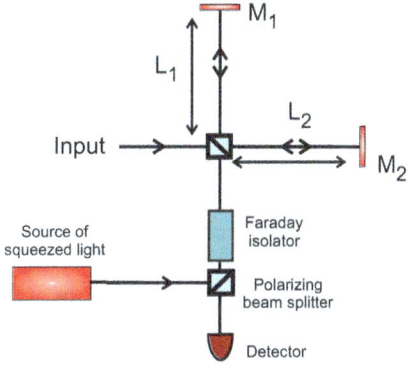

fluctuations introduced into the mirror position and momentum limit the resolution
with which the strain can be measured.

The random fluctuations in photon number have a minimum value given by the
size of the uncertainty region of the vacuum state in Fig. 9.1a. Notice that in the
standard Michelson interferometer, the output port (port B of Fig. 2.10) has light
going out toward the detector, but there is no light being input to that port. The port
is open to the vacuum, and therefore, fluctuations of the vacuum are entering the
system at there.

To reduce the fluctuations, a squeezed state can be fed into the port, as shown in
Fig. 9.4. The introduction of the squeezed state into the system reduces the overall
fluctuations relative to the size of the signal, improving the signal-to-noise ratio. The
Faraday isolator (Sect. 2.8) prevents the squeezed light from passing back into the
detector, but allows the signal to pass. This squeezed state approach has increased
the sensitivity to small signals by a factor of roughly 10 in recent experiments [8],
and further refinements are expected to lead to even better sensitivity in the future.

More detailed information on the use of squeezed states in detection of gravitational waves can be found in [6, 7, 9, 10].

9.4   N00N States

Two-photon entangled states, such as those produced by SPDC are useful in many
applications, some of which will be discussed in later chapters. An obvious question
is to consider if entangled states of more than two particles could produce even
greater benefits. A common type of such state used in many quantum metrology
applications is the **N00N state** [11–13]. Let $|n, m\rangle$ represent a state with n photons in
one spatial mode (for example, in the top branch of a Mach–Zehnder interferometer)
and m photons in a different spatial mode (the bottom branch). Then an N00N state
is defined as

$$\frac{1}{\sqrt{2}} \left(|N, 0\rangle + |0, N\rangle \right). \tag{9.64}$$

This is an entangled N-photon state where all N of the photons will be detected either in one mode or all in the other mode. NOON states up to at least $N = 5$ can be made using parametric down-conversion followed by post-selection [14, 15].

If N independent photons are sent through a phase shifter in one arm of an interferometer, each gains a phase factor of $e^{i\phi}$. These phase shifts are attached to separate photons and do not combine with each other. However, the N photons in a NOON state are not independent of each other, but are part of a single composite state: when sent through the same interferometer, the phase shifts of the photons add to each other, so that the relative phase shift between the two terms of Eq. 9.64 is $e^{iN\phi}$. Effectively, the phase shifts are magnified by a factor of N, allowing the interferometer to measure phase values that are N times smaller than would be possible with classical light.

This phase supersensitivity is the basis for a number of applications. For example, a proposed NOON microscope [16], would allow measurements to be made to biological samples with resolutions that approach the Heisenberg limit, using low illumination to avoid damage to the sample.

A major drawback of these states is that they are extremely fragile in the presence of noise or loss, which would limit the practical implementation of this approach with large N. However, proposals have been made for methods to correct for errors induced by these losses [17].

One of the main motivations for studying NOON states has been their potential use in quantum lithography [12, 18–20]: they can be used to imprint subwavelength structures onto substrates. Optical lithography uses light to etch structures onto wafers of silicon or other materials, in order to make semiconductor chips for computers and other electronic devices. As computer chips get smaller, it is becoming harder and harder to etch structures at progressively finer scales. Since the size of the structures that can be etched is limited by the wavelength of the illuminating light, this can usually be done only by going to shorter wavelengths.

But an alternative that allows etching at smaller scales with the same wavelength is to use multi-photon absorption processes, in which each atom interacts with multiple photons at the same time. In [21], it was demonstrated that use of two-photon absorption could double the lithographic resolution at fixed wavelength. The idea was soon extended from two-photon to N-photon absorption and then to NOON states, again relying on the fact that the N-photon entangled state oscillates N times faster than the single-photon state. As a result, the effective wavelength of the light is reduced by a factor of N. However, practical application of the idea may again be limited by the fragility of the states and by the difficulty of creating high N entangled states on demand. Finding appropriate materials that allow strong N-photon absorption also becomes harder for large N. Prospects for applications can be further limited by low efficiency of the process and the large time required for a sufficient number of entangled photon states to arrive at a given point and interact, making the process very slow.

An analogous classical approach, not needing entanglement, is given by [22, 23]. Using only classical illumination, it has been suggested that improved resolution can ultimately be accomplished using nonlinear optical materials that can mediate N-photon interactions. But these processes again become weaker at high N, requiring high-intensity light that may damage the material. For more on quantum lithography in general, see [24, 25].

Most discussion of $N00N$ states has involved photonic states. However, proposals for generating N-atom $N00N$ states have also been discussed [26–28], with a goal of providing a new testing ground for experiments in the foundations of quantum mechanics.

9.5 Thermal States

Most light sources are thermal sources, like a star or a heated filament. The light emitted is initially in thermal equilibrium with some hot object of temperature T. Since the electrons in the hot material are distributed among states of different energy, the light that is emitted will fluctuate in energy. Even if the light is filtered to make it monochromatic, the number of photons appearing in equal-duration intervals of light will fluctuate. More importantly, the light is emitted at random times by many different atoms, so that the phase will be fluctuating wildly. The result is that the light will have very poor coherence.

Unlike fixed-photon number states, coherent states, and squeezed states, which are *pure* quantum states, thermal light is in a *mixed* state. It is an incoherent sum of states with different photon numbers. Recall from Chap. 3 that a mixed state cannot be described by a state vector, but instead has to be described by a density operator.

Consider thermal light at frequency ω. The density operator can be written as a sum of photon-number states:

$$\hat{\rho} = \sum_{n=0}^{\infty} \rho_n |n\rangle \langle n|. \tag{9.65}$$

The coefficients ρ_n are simply the probabilities of n photons being emitted. In equilibrium, these probabilities will obey the **Boltzmann distribution**,

$$\rho_n = \frac{1}{Z} e^{-\frac{E_n}{kT}}, \tag{9.66}$$

where the energy of the n-photon state is $E_n = n\hbar\omega$, k is Boltzmann's constant, and the **partition function** is the sum over the Boltzmann factors:

$$Z = \sum_{n=0}^{\infty} e^{-\frac{E_n}{kT}}. \tag{9.67}$$

The density operator can also be written in the form

$$\hat{\rho} = \frac{1}{Z} e^{-\frac{\hat{H}}{kT}}, \tag{9.68}$$

with

$$Z = \mathrm{Tr}\left(e^{-\frac{\hat{H}}{kT}}\right). \tag{9.69}$$

Defining $\beta = \frac{\hbar\omega}{kT}$, the partition function can be evaluated, because it is simply a geometric series:

$$Z = \sum_{n=0}^{\infty} e^{-n\hbar\omega/kT} = \sum_{n=0}^{\infty} e^{-n\beta} = \frac{1}{1 - e^{-\beta}}. \tag{9.70}$$

So the density operator becomes

$$\hat{\rho} = \left(1 - e^{-\beta}\right) \sum_{n=0}^{\infty} e^{-n\beta} |n\rangle\langle n|. \tag{9.71}$$

The average photon number is readily found:

$$\bar{n} = \frac{1}{Z} \sum_{n=0}^{\infty} n\, e^{-n\beta} = -\frac{1}{Z} \frac{d}{d\beta} \sum_{n=0}^{\infty} e^{-n\beta} \tag{9.72}$$

$$= \frac{1}{e^{\beta} - 1}. \tag{9.73}$$

Similarly, the spectral energy density (energy per volume per unit of frequency) can be obtained:

$$u(\omega) = \sum_{n} \bar{n}(\omega)\hbar\omega = \left(\frac{\hbar\omega^3}{\pi^2 c^3}\right)\left(\frac{1}{e^{\beta} - 1}\right) = \left(\frac{\hbar\omega^3}{\pi^2 c^3}\right) \frac{1}{e^{\hbar\omega/kT} - 1}. \tag{9.74}$$

9.6 Distinguishing Nonorthogonal Quantum States

Measurement of a state means being able to distinguish between that state and another. Consider just two states for simplicity. If the two states happen to be orthogonal, then they can be distinguished without ambiguity. For example, suppose the two states are horizontal and vertical polarization states of light, $|H\rangle$ and $|V\rangle$. Then if you know the optical state has to be one of these two, then they can be distinguished with polarizing filters: if the light passes a horizontally-oriented filter, then the state is $|H\rangle$, but if it doesn't, it must be $|V\rangle$. Alternatively, a polarizing beam splitter can distinguish between them unambiguously and nondestructively.

However, if the two states are nonorthogonal, $\langle\psi_1|\psi_2\rangle \neq 0$, then they cannot be distinguished unambiguously, at least not in 100% of the measurements. For example, suppose we are trying to distinguish between the horizontal and diagonal states, $|H\rangle$ and $|D\rangle = \frac{1}{\sqrt{2}}(|H\rangle + |V\rangle)$. These are clearly nonorthogonal,

$$\langle H|D\rangle = \frac{1}{\sqrt{2}} = \cos 45°. \tag{9.75}$$

If we try placing a horizontal polarizer in the photon path, the diagonal photon still has a survival probability of

$$P = |\langle H|D\rangle|^2 = \cos^2 45° = \frac{1}{2}. \tag{9.76}$$

Similarly, the $|H\rangle$ state has a 50% probability of passing a diagonal polarizer. So, there is a 50% error rate in distinguishing these states. As we will see in Chap. 18, this fact is essential for the functioning of the BB84 protocol for quantum key distribution.

This is a general characteristic of nonorthogonal states: the minimum error when trying to distinguish between $|\psi_1\rangle$ and $|\psi_2\rangle$ is determined by the angle between them in Hilbert space,

$$\text{Error probability} = |\langle\psi_1|\psi_2\rangle|^2. \tag{9.77}$$

But a better result can be obtained [29–32] if we consider three possible outcomes and make use of POVM's (see Sect. 3.7.2). First, define the angle between the states,

$$\cos\theta = |\langle\psi_1|\psi_2\rangle|, \tag{9.78}$$

and let $|\psi_1^{\perp}\rangle$ and $|\psi_2^{\perp}\rangle$ be the states perpendicular to the states of interest: $\langle\psi_1^{\perp}|\psi_1\rangle = \langle\psi_2^{\perp}|\psi_2\rangle = 0$. Further, let p_1 and p_2 be the probabilities of successfully identifying each state, ψ_1 or ψ_2, when that state was the one initially prepared. Then instead of defining two possible outcomes, suppose we consider the *three* possible outcomes labeled 1, 2, ? and defined by:

$$1: \quad \text{state is } |\psi_1\rangle$$

$$2: \quad \text{state is } |\psi_2\rangle$$

$$?: \quad \text{state is undetermined.}$$

These three outcomes are distinguished by using three nonorthogonal operators forming a POVM:

$$\hat{\Pi}_1 = \frac{p_1}{\sin^2\theta}|\psi_1^{\perp}\rangle\langle\psi_1^{\perp}| \tag{9.79}$$

$$\hat{\Pi}_2 = \frac{p_2}{\sin^2\theta}|\psi_2^{\perp}\rangle\langle\psi_2^{\perp}| \tag{9.80}$$

$$\hat{\Pi}_? = \hat{I} - \hat{\Pi}_1 - \hat{\Pi}_2. \tag{9.81}$$

Clearly, $\hat{\Pi}_1 + \hat{\Pi}_2 + \hat{\Pi}_? = \hat{I}$. If the measurement of $\hat{\Pi}_1$ (respectively, $\hat{\Pi}_2$) registers positive, then we know that the state was $|\psi_2\rangle$ (resp. $|\psi_2\rangle$). In both cases, we have absolute certainty about the state. But if the result is $\hat{\Pi}_?$, then the state remains unknown. So we can have absolute certainty part of the time, but pay for it by having no information at all about the state the remainder of the time.

The probability of failing to identify state j when it is prepared is given by $q_j = 1 - p_j$; the failure probabilities then obey the bound

$$q_1 q_2 \geq |\langle\psi_1|\psi_2\rangle|^2. \tag{9.82}$$

Let η_1 and η_2 be the preparation probabilities of $|\psi_1\rangle$ and $|\psi_2\rangle$, so that the two states are not necessarily equally likely to be present. Then the average failure probability for unambiguously determining the state is

$$\langle q \rangle = \eta_1 q_1 + \eta_2 q_2 = \eta_1 q_1 + \eta_2 \frac{\cos^2\theta}{q_1}, \tag{9.83}$$

where in the last equality we assumed $q_1 q_2$ achieves the minimum of inequality 9.82 in order to write q_2 as a function of q_1. Then minimizing $\langle q \rangle$ with respect to q_1, it is immediately found that the optimal discrimination failure rate is

$$\langle q \rangle \geq q_{min} = 2\sqrt{\eta_1\eta_2}\cos\theta. \tag{9.84}$$

When the production probabilities of the two states are equal, $\eta_1 = \eta_2$, this reduces to the result of Eq. 9.77:

$$q_{min} = 2\cos\theta = |\langle\psi_1|\psi_2\rangle|. \tag{9.85}$$

For more detailed reviews on distinguishing non-orthogonal states, including generalization to more than two states, see [33–35].

9.7   Quantum State Tomography

Suppose a source emits a stream of particles in some unknown state. Possibly we don't even know whether it is a pure or mixed state. The problem is to determine what this unknown state is. This is much harder in quantum mechanics than it would be classically. For example, if you measure the polarization of a photon and find that it passes a horizontal polarizer undisturbed, that doesn't uniquely determine what the polarization actually was: the photon could have been still have been polarized in *any* direction that is not orthogonal to the horizontal measurement axis. Further, if more than one property of the particle needs to be measured to determine the state, then measurement of one property could preclude knowledge of the others. Clearly in both of these examples, a single measurement of one particle is not enough to gain the desired information. To determine complete information about a state, in general it is necessary to make many measurements on multiple particle that have been prepared in the same state.

There are many applications in which these sorts of measurements on identically prepared states may be needed. For example, they could be used to characterize the output of a laser, an ion source, or an entangled-photon source. Or, they may be used to determine the qubit sent in a continuous-variable quantum cryptography protocol.

Here we focus on finding the structure of a quantum state that depends on a pair of continuous quadrature variables, which we will call q and p. The two variables (q, p) provide the phase space on which the system evolves. In the case where q is position, p will be momentum. But recall that the quadrature formalism is more general and can be applied, for example, to systems of oscillators or to electromagnetic fields.

In the case of a pure state, the goal could be to find the wavefunction either in the "position" quadrature space, $\psi(q)$, or in "momentum" quadrature space, $\psi(p)$. More generally, we could allow mixed states, so that the goal is then to characterize the density operator, $\hat{\rho}$. Since we are interested in determining the phase space profile of the state, this really means determining the expectation values $\rho(q) = \langle q|\hat{\rho}|q\rangle$ and $\rho(p) = \langle p|\hat{\rho}|p\rangle$. For an arbitrary unknown state, this leads to **quantum state tomography**, which makes multiple measurements along different slices in phase space and then uses a procedure called the **inverse Radon transform** to reconstruct a particular function (the **Wigner function**) defined on the quantum analog of phase space. Once the Wigner function has been determined, this completely fixes the spatial structure of the density operator. We start in the next subsection by defining the Wigner function and describing its properties.

9.7.1 Wigner Functions and Phase Space Distributions

For many quantum states, such as coherent or squeezed state, the state is determined by a complex amplitude, α. This amplitude can be broken into its real and imaginary parts, $\alpha = q + ip$, where q and p are (up to constants we won't worry about for the moment) the quadratures. q and p act like position and momentum variables, respectively, and more general quadratures can be built up as linear combinations of them.

For squeezed and coherent states, the q and p values fluctuate around their mean values, and these fluctuations are described by a Gaussian distribution. For more general states, we could try to define the state by defining a joint probability density $P(q, p)$ of quadrature values on the quantum analog of phase space. However, this quickly leads to problems. These problems stem from the fact that although we are trying to treat q and p as numbers, they are really Hermitian operators, and in general, they don't commute. Because of this lack of commutativity, a state doesn't occupy a single point in phase space, but rather a cell of finite area (set by the uncertainty principle). More disastrously, any attempt to define a quantum probability distribution $P(q, p)$ will usually lead to the probability density being negative in some regions of phase space. (In fact, the only exceptions to this are the states that are described by Gaussian distributions, such as coherent and squeezed states. All non-Gaussian states will inevitably encounter negative probability densities somewhere.)

Nonetheless, let's plow ahead and try to get as close as we can to a quantum-mechanical phase-space probability distribution. The result will be a **quasiprobability distribution**. These come in several varieties, including the Husimi Q-distribution, the Glauber–Sudarshan P-distribution, and the Wigner distribution. We focus here on the Wigner distribution.

Suppose a state is described by density operator $\hat{\rho}$. The Wigner function of the state is defined to be

$$W(q, p) = \frac{1}{2\pi} \int_{-\infty}^{\infty} e^{ipx} \langle q - \frac{x}{2} | \hat{\rho} | q + \frac{x}{2} \rangle dx. \tag{9.86}$$

Although $W(q, p)$ fails to be a true probability density in general, the marginal distributions obtained by integrating over either q or p,

$$\rho(p) \equiv \langle p | \hat{\rho} | p \rangle = \int_{-\infty}^{\infty} W(q, p) dq \tag{9.87}$$

$$\rho(q) \equiv \langle q | \hat{\rho} | q \rangle = \int_{-\infty}^{\infty} W(q, p) dp, \tag{9.88}$$

are in fact true probability distributions, giving the probability densities for p and q, respectively.

More generally, q and p can be replaced by any pair of quadratures obtained by rotating the phase space by angle θ. Let $\hat{U}(\theta)$ be the operator that rotates $|q\rangle$ and $|p\rangle$ by θ. Then the new state $|\theta\rangle = \hat{U}(\theta)|q\rangle$ will occur with probability density

$$\rho(\theta) = \langle \theta | \hat{\rho} | \theta \rangle = \int_{-\infty}^{\infty} W\left(q', p'\right) dp, \tag{9.89}$$

where $q'(q, p, \theta) = q \cos\theta - p \sin\theta$ and $p'(q, p, \theta) = q \sin\theta + p \cos\theta$ are the rotated quadrature variables. The mappings from $W(q, p)$ to the quadratures given by Eqs. 9.87, 9.88, and 9.89 are examples of **Radon transforms**. Here we define $W(q, p)$ in terms of $\hat{\rho}$, but in practice, the value of $W(q, p)$ is determined first, and then the inverse Radon transform is used to find the density operator, as discussed in the next subsection.

Making one further definition, Eq. 9.86 can be generalized by replacing the density operator by some arbitrary operator $\hat{O}$. So the Wigner function of operator $\hat{O}$ is

$$W_{\hat{O}}(q, p) = \frac{1}{2\pi} \int_{-\infty}^{\infty} e^{ipx} \langle q - \frac{x}{2} | \hat{O} | q + \frac{x}{2} \rangle dx. \tag{9.90}$$

Then, it is fairly easy to prove the following results (see the problems), where all integrals run from $-\infty$ to ∞:

1. The Wigner function is real and normalized:

$$W(q, p) = W^*(q, p), \qquad \int W(q, p) dq\, dp = 1. \tag{9.91}$$

2. The overlap formula:

$$Tr(\hat{O}_1 \hat{O}_2) = 2\pi \int W_{\hat{O}_1}(q, p)\, W_{\hat{O}_2}(q, p) dq\, dp. \tag{9.92}$$

3. Substituting $\hat{O}_1 = \hat{\rho}$ into the previous result, a direct corollary is that expectation values of operators are given by

$$\langle \hat{O} \rangle = Tr(\hat{\rho}\hat{O}) = 2\pi \int W(q, p)\, W_{\hat{O}}(q, p) dq\, dp. \tag{9.93}$$

4. Another immediate corollary of property 3 aforementioned is obtained by letting the two operators be density operators of pure states. For $\hat{O}_1 = |\psi_1\rangle\langle\psi_1|$ and $\hat{O}_2 = |\psi_2\rangle\langle\psi_2|$, we obtain the transition probabilities between pure states:

$$p_{1\to 2} = |\langle\psi_1|\psi_2\rangle|^2 = 2\pi \int W_{|\psi_1\rangle\langle\psi_1|}(q, p)\, W_{|\psi_2\rangle\langle\psi_2|}(q, p) dq\, dp. \tag{9.94}$$

5. Generalizing the previous result from pure states to include potentially mixed states:

$$p_{1\to 2} = Tr(\hat{\rho}_1\hat{\rho}_2) = 2\pi \int W_{\hat{\rho}_1}(q, p)\, W_{\hat{\rho}_2}(q, p) dq\, dp. \tag{9.95}$$

6. Taking $\hat{\rho}_1 = \hat{\rho}_2$ in the previous result, the Wigner function provides a means of finding the purity (Sect. 3.2.5) of a quantum state:

$$\mathcal{P} = Tr(\hat{\rho}^2) = 2\pi \int W^2(q, p) dq\, dp. \tag{9.96}$$

The Wigner function approach therefore allows calculation of a wide range of properties of a quantum system. Unfortunately, in many cases, $W(q, p)$ may be difficult to compute analytically; numerical computations or experimental reconstruction (next subsection) are needed in all but the simplest cases.

However, notice that if the Wigner function is known (from experiment, for example), then all expectation values and transition amplitudes can be computed for the given system. Thus, all physically relevant properties follow directly from $W(q, p)$, without needing explicit knowledge of the wavefunctions or state vectors.

Example 9.7.1 Consider a coherent state with amplitude $\alpha = \frac{1}{\sqrt{2}}(q_0 + ip_0)$. Substituting the coherent state wavefunction

$$\psi(q) = \frac{1}{\sqrt[4]{\pi}} e^{-\frac{1}{2}(q-q_0)^2}, \qquad \psi(p) = \frac{1}{\sqrt[4]{\pi}} e^{-\frac{1}{2}(p-p_0)^2} \tag{9.97}$$

into Wigner's formula (Eq. 9.86) and doing the resulting Gaussian integrals leads to the result (Problem 9)

$$W(q, p) = \frac{1}{\pi} e^{-(q-q_0)^2 - (p-p_0)^2}. \tag{9.98}$$

Setting $q_0 = p_0$ then gives the vacuum state distribution

$$W(q, p) = \frac{1}{\pi} e^{-q^2 - p^2}. \tag{9.99}$$

Generalizing to squeezed states with squeezing parameter ζ leads to

$$W(q, p) = \frac{1}{\pi} e^{-e^{2\zeta} q^2 - e^{-2\zeta} p^2}. \tag{9.100}$$

These are plotted in Fig. 9.5. ∎

The examples aforementioned all have positive-definite $W(q, p)$, so the Wigner function can serve as a true probability distribution in these cases. But this is true only because coherent and squeezed states are described by Gaussian functions. For non-Gaussian states, $W(q, p)$ will no longer be positive definite, and will be a *quasi*-probability distribution, for which only the marginal distributions $\rho(q)$ and $\rho(p)$ will be true probability densities.

An example which is non-positive definite is the Schrodinger cat state formed by a superposition of two coherent states centered at different locations ($+ q_0$ and $- q_0$, with $p_0 = 0$), which will have the Wigner function (see [36]):

$$W(q, p) = C \left[e^{-[(q-q_0)^2 + p^2]} + e^{-[(q+q_0)^2 + p^2]} + 2e^{-(q^2+p^2)} \cos 2pq_0 \right]. \tag{9.101}$$

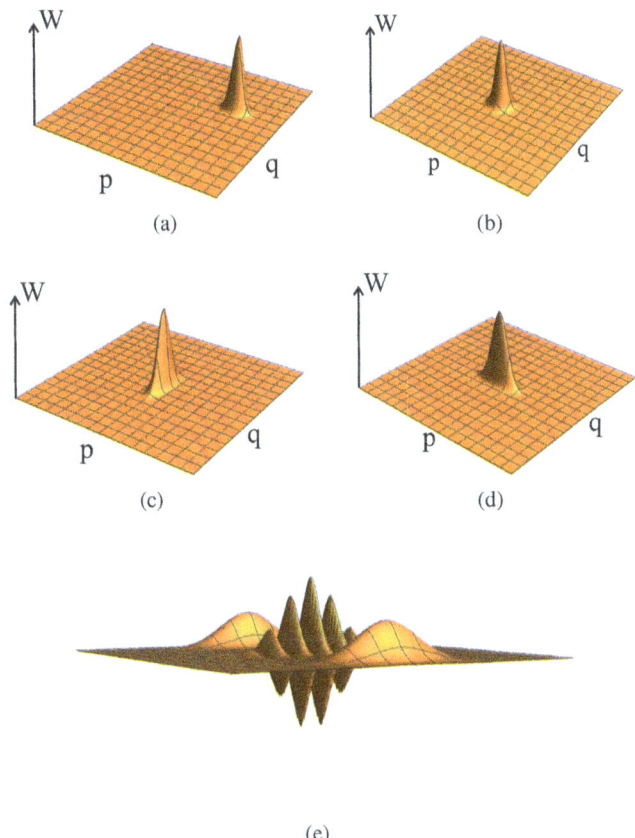

Fig. 9.5 Examples of Wigner functions. (**a**) A coherent state centered at $q_0 = 6$, $p_0 = 9$. (**b**) A vacuum state (a coherent state centered at the origin). (**c**) A squeezed vacuum state with squeezing parameter $s = 0.5$ and (**d**) with $s = -0.5$. (**e**) A superposition of two coherent states centered at $q_0 = \pm 6$. Notice that, unlike in the previous cases, interference between the two terms in the superposition causes the Wigner function to become negative at some points

The last term is an interference term, and because of it the Wigner function can become negative, as can be seen in Fig. 9.5e.

9.7.2   Quantum State Tomography

Tomography (from the Greek word $\tau o \mu o \sigma$, meaning *slice*) consists of measuring the projection of some quantity onto a set of lower-dimensional slices and then piecing together those slices to form a higher-dimensional image. Most often this means taking a series of two-dimensional images and using them to reconstruct a full three-dimensional image. This is a well-developed technique in medical imaging and image recognition analysis. Over the past few decades, a version of it known as quantum state tomography has been developed to reconstruct quantum wavefunctions based on measured probability distributions along spatial slices. This is a highly mathematical subject that requires sophisticated numerical techniques, so here we simply outline the basic idea. See [36, 37] for more detailed treatments. For a tutorial devoted primarily to photonic states, see [38].

Starting from the quadratures q and p, they can be rotated by angle θ to give a new pair of orthogonal quadratures,

$$q'(\theta) = q \cos \theta - p \sin \theta, \qquad p'(\theta) = q \sin \theta + p \cos \theta.$$

For a sequence of θ values, measurements of probabilities can be made in, say, the $q'(\theta)$ plane, holding $p'(\theta)$ fixed, obtaining a distribution $P(q', \theta)$. Then the Wigner function can be computed by taking a two-dimensional Fourier transform,

$$
\begin{aligned}
W(q, p) &= \frac{1}{(2\pi)^2} \int_{-\infty}^{\infty} \int_{0}^{\infty} \int_{-\infty}^{\infty} P(x, \theta) e^{i\xi(q \cos \theta + p \sin \theta - x)} |\xi| \, d\xi \, d\theta \, dx \\
&= \frac{1}{(2\pi)^2} \int_{0}^{\infty} \int_{-\infty}^{\infty} P(x, \theta) K(q \cos \theta + p \sin \theta - x) dx \, d\theta, \quad (9.102)
\end{aligned}
$$

where the integral $K(x)$ is defined by

$$K(x) = \frac{1}{2} \int_{-\infty}^{\infty} e^{ix\xi} |\xi| d\xi. \tag{9.103}$$

After some work (see [36]), the Wigner distribution can then be put into the form of an inverse Radon transform,

$$W(q, p) = -\frac{1}{2\pi} \mathcal{P} \int_{0}^{\pi} \int_{-\infty}^{\infty} \frac{P(x, \theta) dx \, d\theta}{(q \cos \theta + p \sin \theta - x)^2}, \tag{9.104}$$

where $\mathcal{P}$ denotes the Cauchy principle value [39].

Problems

1. Given two operators $\hat{A}$ and $\hat{B}$ such that $[\hat{A}, [\hat{A}, \hat{B}]] = [\hat{B}, [\hat{A}, \hat{B}]] = 0$. By expanding the exponentials, prove the **Baker-Campbell-Hausdorf (BCH) theorem**:

$$e^{\hat{A}+\hat{B}} = e^{\hat{A}} e^{\hat{B}} e^{-\frac{1}{2}[\hat{A}, \hat{B}]}. \tag{9.105}$$

2. Given two operators $\hat{A}$ and $\hat{B}$, show that

$$e^{\hat{A}}\hat{B}e^{-\hat{A}} = \hat{B} + \left[\hat{A}, \hat{B}\right] + \frac{1}{2!}\left[\hat{A}, \left[\hat{A}, \hat{B}\right]\right] + \ldots \tag{9.106}$$

3. Prove Eqs. 9.12 and 9.13.
4. Show that the coherent state $|\alpha\rangle$ is an eigenstate of the annihilation operator, with eigenvalue α: $\hat{a}|\alpha\rangle = \alpha|\alpha\rangle$. Is it also an eigenstate of the creation operator? Why or why not?
5. Show that the coherent state of Eq. 9.16 has the mean values $\langle\hat{N}\rangle$ and $\langle\hat{N}^2\rangle$ given in Eq. 9.19, and so the state is Poisson.
6. Prove the inner product formula for coherent states, Eq. 9.25.
7. Verify that the expansion of Eq. 9.52 leads to the result in Eq. 9.53.
8. Prove that the Wigner function has the properties displayed in Eqs. 9.91–9.94.
9. Show that the coherent state given by the wavefunctions in Eq. 9.97 has the Wigner distribution given in Eq. 9.98.
10. (a) Find the probability that a coherent state pulse has exactly one photon as function of the amplitude $|\alpha|$.
    (b) Find the probability that the pulse has more than one photon.
    (c) Find the probability that the pulse has no photons
    (d) Plot the results of parts (a), (b), and (c) versus $|\alpha|$. How large does $|\alpha|$ have to be to ensure that 95% of the pulses have exactly one photon?
11. For a squeezed state with amplitude $\tau = \alpha\cosh\zeta + \alpha^*e^{i\phi}\sinh\zeta$, write expressions for the classical quadrature variables X and P in terms of $|\alpha|$, ϕ, and ζ.

References

1. D.J. Griffiths, D.F. Schroeter, *Introduction to Quantum Mechanics*, 3rd edn. (Cambridge University Press, Cambridge, 2018)
2. R. Shankar, *Principles of Quantum Mechanics*, 2nd edn. (Plenum Press, 1994)
3. S.M. Barnett, J. Vaccaro, *The Quantum Phase Operator: A Review* (CRC Press, Boca Raton, 2020)
4. S. Haroche, J.M. Raimond, *Exploring the Quantum: Atoms, Cavities, and Photons* (Oxford, 2006)
5. B.P. Abbott et al., Phys. Rev. Lett. **116**, 061102 (2016)
6. M. Tse et al., Phys. Rev. Lett. **123**, 231107 (2019)
7. F. Acernese et al., Phys. Rev. Lett. **123**, 231108 (2019)
8. LIGO Scientific Collaboration, Virgo Collaboration, KAGRA Collaboration et al., Phys. Rev. X, **13**, 041039 (2023)
9. R. Schnabel, Phys. Rep. **684**, 1 (2017)
10. S.E. Dwyer, G.L. Mansell, L. McCuller, Galaxies **10**, 46 (2022)
11. B.C. Sanders, Phys. Rev. A **40**, 2417 (1989)
12. A.N. Boto, P. Kok, D.S. Abrams, S.L. Braunstein, C.P. Williams, J.P. Dowling, Phys. Rev. Lett. **85**, 2733 (2000)
13. H. Lee, P. Kok, J.P. Dowling, J. Mod. Opt. **49**, 2325 (2002)

14. I. Afek, O. Ambar, Y. Silberberg, Science **328**, 879 (2010)
15. Y. Israel, I. Afek, S. Rosen, O. Ambar, Y. Silberberg, Phys. Rev. A **85**, 022115 (2012)
16. Y. Israel, S. Rosen, Y. Silberberg, Phys. Rev. Lett. **112**, 103604 (2014)
17. M. Bergmann, P. van Loock, Phys. Rev. A **94**, 012311 (2016)
18. M. D'Angelo, M. Chekhova, Y. Shih, Phys. Rev. Lett. **87**, 013602 (2001)
19. G. Björk, L.L. Sánchez-Soto, J. Söderholm, Phys. Rev. Lett. **86**, 4516 (2001)
20. G. Björk, L.L. Sánchez-Soto, J. Söderholm, Phys. Rev. A **64**, 013811 (2001)
21. E. Yablonovich, R.B. Vrijen, Opt. Eng. **38**, 334 (1999)
22. S.J. Bentley, R.W. Boyd, Opt. Express **12**, 5735 (2004)
23. R.W. Boyd, S.J. Bentley, J. Mod. Opt. **53**, 713 (2006)
24. P. Kok, S.L. Braunstein, J.P. Dowling, J. Opt. B: Quant. Semiclass. Opt. **6**, S811 (2004)
25. R.W. Boyd, J.P. Dowling, Quant. Inf. Proc. **11**, 891 (2011)
26. F. Soto-Eguibar, H.M. Moya-Cessa, Ann. Phys. **531**, 1900250 (2019)
27. R.J. Lewis-Swan, K.V. Kheruntsyan, Nat. Commun. **5**, 3752 (2014)
28. D.S. Grün, W.K. Wittmann, L.H. Ymai, J. Links, A. Foerster, Commun. Phys. **5**, 36 (2022)
29. I.D. Ivanovic, Phys. Lett. A **123**, 257 (1987)
30. D. Dieks, Phys. Lett. A **126**, 303 (1988)
31. A. Peres, Phys. Lett. A **128**, 19 (1988)
32. G. Jaeger, A. Shimony, Phys. Lett. A **197**, 83 (1995)
33. J.A. Bergou, U. Herzog, M. Hillery, Discrimination of quantum states. Lect. Notes Phys. **649**, 417–465 (2004)
34. A. Chefles, Contemp. Phys. **41**, 401 (2000)
35. S.M. Barnett, Sarah Croke, Adv. Opt. Photon. **1**, 238 (2009)
36. U. Leonhardt, *Measuring the Quantum State of Light* (Cambridge University Press, Cambridge, 1997)
37. A.I. Lvovsky, M.G. Raymer, Rev. Mod. Phys. **81**, 299 (2009)
38. E. Toninelli et al., Adv. in Opt. Photon. **11**, 67 (2019)
39. G. Arfken, H. Weber, *Mathematical Methods for Physicists* (Academic Press, London, 2000)

Chapter 10
Quantum Interference

In Chap. 1, classical optical interferometry was discussed. In this chapter, we look at interference effects more closely. In particular, we wish to consider cases where the interference is inherently quantum, in the sense that it cannot be mimicked by classical optical fields obeying Maxwell's equations. Essentially what we will find is that classical interference arises from **single-photon interference**: each photon interferes only with itself. For example, in the Young two-slit experiment, each photon has an amplitude to go through one slit and an amplitude to pass through the other. The interference arises because those two amplitudes interfere with each other for each photon individually. To have intrinsically quantum interference, it will be necessary to have **two-photon interference** (or more generally, multi-photon interference): a two-photon amplitude for a *pair* of photons will interfere with another amplitude for the *same* two-photon pair to take a different path.

Interference is one of the hallmark effects of quantum mechanics that often distinguishes it from classical physics. We will see that interference between quantum states is an essential ingredient in many quantum information processing or quantum metrology protocols. We focus here primarily on optical interference, but interference of atomic beams and that of neutron beams are also highly developed fields. Most of what is said here can be equally applied to atomic and neutron interferometry. For reviews of atom optics, see [1, 2], and for neutron interferometry, see [3, 4].

10.1 Phase Sensitivity and Entanglement

To see the advantages of using quantum states in interferometry, we can examine the phase sensitivity, $\Delta\phi$, the smallest phase change that can be measured.

Example 10.1.1 Imagine a state of the form $|\phi\rangle = \frac{1}{\sqrt{2}}\left(|0\rangle + e^{i\phi}|1\rangle\right)$. Such a state can be created, for example, using a Mach–Zehnder interferometer with $|0\rangle$ and $|1\rangle$

© The Author(s), under exclusive license to Springer Nature Switzerland AG 2025
D. S. Simon, *Introduction to Quantum Science and Technology*, Undergraduate
Texts in Physics, https://doi.org/10.1007/978-3-031-81315-3_10

identified as the two possible exit points for the photon. The goal is to measure the phase with the minimum possible uncertainty.

Consider [5] a measurement operator $\hat{A} = |0\rangle\langle 1| + |1\rangle\langle 0|$. (The reason for using an operator of this form will be made clear in Sect. 19.1.) Taking the expectation value of $\hat{A}$ in the state $|\phi\rangle$, we find that

$$\langle\phi|\hat{A}|\phi\rangle = \cos\phi. \tag{10.1}$$

Carrying out the measurement on each photon in an N-photon state,

$$_N\langle\phi| \ldots \, _1\langle\phi| \oplus_{k=1}^{N} \hat{A}^{(k)}|\phi\rangle_1 \ldots |\phi\rangle_N = N\cos\phi. \tag{10.2}$$

Noting that $\hat{A}^2 = 1$, it is clear that

$$\Delta\hat{A}^2 = N(1 - \cos^2\phi) = N\sin^2\phi. \tag{10.3}$$

So the minimum phase uncertainty is given by

$$\Delta\phi = \frac{\Delta\hat{A}}{|d\langle\hat{A}\rangle/d\phi|} = \frac{1}{\sqrt{N}}, \tag{10.4}$$

which corresponds to the shot noise or standard quantum limit.

Now consider a similar situation, but with an entangled input state $|\phi_N\rangle = |N0\rangle + e^{iN\phi}|0N\rangle$, where $|NM\rangle$ represents N particles in state $|0\rangle$ and M in state $|1\rangle$. Now the analogous measurement operator is

$$\hat{A}_N = |0N\rangle\langle N0| + |N0\rangle\langle 0N|, \tag{10.5}$$

which is readily seen to have

$$\Delta\hat{A}_N^2 = 1 - \cos^2 N\phi = \sin^2 N\phi. \tag{10.6}$$

Thus, the minimum phase uncertainty is now

$$\Delta\phi = \frac{\Delta\hat{A}_N}{|d\langle\hat{A}_N\rangle/d\phi|} = \frac{1}{N}, \tag{10.7}$$

saturating the Heisenberg limit. This is a clear illustration that adding entanglement to the mix can improve the sensitivities of phase measurements and provides an indication that other types of measurements may show similar enhancements. ∎

More detailed examinations of phase sensitivity and its limits can be found in [10, 11].

10.2 Correlation Functions

To make the differences between classical and quantum interference more precise, we will need the concept of correlation functions. A common situation in physics and engineering is that two mathematical objects (signals, functions, quantum states, vector fields, etc.) need to be compared in order to arrive at a quantitative measure of their similarity. The common approach is to define a **correlation function**. Correlation functions appear throughout science and engineering and, depending on the context, may be known by a variety of names, including *impulse response functions, propagators, coherence functions*, or *Green's functions*.

Start with a pair of ordinary vectors in three-dimensional space. Since we want our measure of similarity to depend only on how close the two vectors are to each other, not on how large they are, it is reasonable to normalize the vectors, or in other words, to restrict attention to unit vectors. That leaves only their directions for comparison, so given two vectors w and v, we form unit vectors $\hat{w} = \frac{w}{|w|}$ and $\hat{v} = \frac{v}{|v|}$, and then the obvious choice for a measure of similarity is their scalar product,

$$\frac{w \cdot v}{|v| \cdot |w|} = \hat{w} \cdot \hat{v} = \cos\theta, \tag{10.8}$$

where θ is the angle between the vectors. This gives a value of 1 if the vectors are identical (perfect similarity) and 0 if they are orthogonal (as dissimilar as possible).

The same idea can be repeated for more sophisticated objects, such as state vectors in Hilbert space or for continuous functions. If an inner product is defined, then we use that inner product as a measure of similarity between the objects. If the objects are not normalized, then we must divide out a factor to account for that. So a reasonable first guess at a definition for correlation between two classical functions $f(t)$ and $g(t)$ might be of the form

$$\frac{\langle f, g \rangle}{\sqrt{\langle f, f \rangle \langle g, g \rangle}}, \tag{10.9}$$

where the inner product represented by the brackets is simply the average taken over some long time interval T:

$$\langle f, g \rangle = \frac{1}{T} \int_{-T/2}^{T/2} f(t)g(t)dt. \tag{10.10}$$

The factors on the bottom are the analog of dividing by length to get unit vectors in Eq. 10.8.

The analogous expression to describe correlations between two (possibly unnormalized) quantum states might be something of the form

$$\frac{\langle \psi_1 | \psi_2 \rangle}{|\psi_1| \cdot |\psi_2|}, \tag{10.11}$$

where the bracket now represents the inner product on the Hilbert space, and $|\psi_j|^2 = \langle \psi_j | \psi_j \rangle$.

This is not quite general enough, however, for many purposes. For example, consider a pair of time-dependent electrical or optical signals. Suppose that the two signals are identical, except that one may have been delayed slightly; for instance, the two signals may have passed through materials with different wave propagation speeds. A definition like the one above would return a low value, indicating lack of correlation, despite the fact that one signal is an identical copy of the other, just shifted in time. To remedy this, we might shift the time parameter in one of the vectors by a delay τ before making the comparison:

$$g(\tau) = \frac{\langle \psi_1(t) | \psi_2(t + \tau) \rangle}{|\psi_1(t)| \cdot |\psi_2(t + \tau)|}, \tag{10.12}$$

This effectively "slides" one signal over the other and checks to see if there is any value of τ at which the correlation spikes. A function such as this is called a **temporal correlation function**.

Note that in the definition of the function $g(\tau)$ aforementioned, there is no t dependence on the left side. This is because we have assumed that all the systems we are interested in are **stationary** in the sense that expectation values of products depend only on the time *difference* between the objects in the product:

$$\langle \psi_1(t) | \psi_2(t + \tau) \rangle = \langle \psi_1(0) | \psi_2(\tau) \rangle, \tag{10.13}$$

for any t. Although nonstationary systems will not arise in the coming chapters, for the sake of completeness we point out that a more general definition of temporal correlation function of the form

$$g(t_1, t_2) = \frac{\langle \psi_1(t_1) | \psi_2(t_2) \rangle}{|\psi_1(t_1)| \cdot |\psi_2(t_2)|} \tag{10.14}$$

will apply equally to stationary and nonstationary cases. In the stationary case, we take $\tau = t_2 - t_1$.

Rather than a correlation between states, another potential object of interest might be the correlation between time-dependent operators. So, given a fixed state $|\psi\rangle$, the temporal correlation between two operators $O_1(t)$ and $O_2(t)$ could be measured by something of the form

$$g(\tau) = \frac{\langle \psi | O_1(t) O_2(t + \tau) | \psi \rangle}{\langle \psi | O_1(t) O_2(t) | \psi \rangle} = \frac{\langle O_1(t) O_2(t + \tau) \rangle}{|\langle O_1(t) O_2(t) \rangle|}, \tag{10.15}$$

where the bracket in the last expression represents the expectation value of the operator product in the state $|\psi\rangle$.

Finally, instead of temporal correlations like those discussed earlier, it is also common to look at spatial correlations, where the states or operators being compared differ by a spatial displacement instead of a time delay; a **spatial correlation function** between two operators, for example, could be defined by:

$$g(\mathbf{y}) = \frac{\langle O_1(\mathbf{x}) O_2(\mathbf{x} + \mathbf{y}) \rangle}{|\langle O_1(\mathbf{x}) O_2(\mathbf{x}) \rangle|}. \tag{10.16}$$

Given these general considerations, we now want to define specific correlation functions that will be relevant to studying classical and quantum correlations in optical systems. This will be the subject of the next two sections.

10.3 First-Order Correlations and Coherence

10.3.1 Classical Coherence

We begin by defining a correlation function for the classical optical field. Consider a situation where light has a choice of taking either of two paths before the paths recombine at some detector. The type of detector is irrelevant at this point: it could be a sophisticated photodetector or simply a screen viewed by eye. The Young two-slit experiment falls into this category. Another example would be the Mach–Zehnder apparatus shown in Fig. 10.1. We assume that there is a time delay τ inserted into one path;. This could be introduced simply by making one path longer than the other. In the figure, this is indicated by the movable or translatable mirror in the lower right corner, which allows the delay to be adjusted by making the path slightly longer or shorter.

We define the **classical first-order correlation function** or **classical first-order coherence function** $g^{(1)}(\tau)$ as

Fig. 10.1 A Mach–Zehnder interferometer with an adjustable time delay in one arm, controlled by moving one of mirrors. This arrangement and the Young two-slit experiment are both prototypical examples of classical or single-photon interference

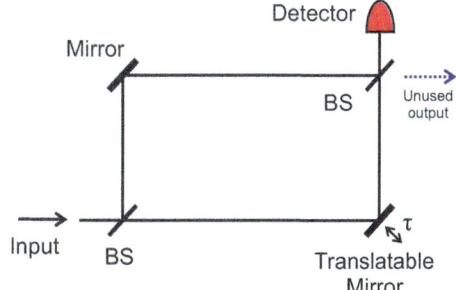

$$g^{(1)}(\tau) = \frac{\langle E^*(t)E(t+\tau)\rangle}{\langle E(t)^*E(t)\rangle}, \tag{10.17}$$

where we have allowed the field to be complex and where

$$\langle E^*(t)E(t+\tau)\rangle = \frac{1}{T}\int_{T/2}^{T/2} E^*(t)E(t+\tau)\,dt. \tag{10.18}$$

T is taken to be very long compared to any other relevant timescales of the system. Such a definition clearly makes sense heuristically. Interference occurs when the classical wave can follow two paths to the same measurement point, one of them possibly taking a longer path and being delayed by time τ. At the measurement point, the two waves are combined by adding the two amplitudes. When this total amplitude is squared to get the intensity, $|E(t) + E(t+\tau)|^2$, there will be terms of the form $|E(t)|^2$ and $|E(t+\tau)|^2$. When averaged over the time T needed to make an intensity measurement, these terms will lead to constant values and simply act as background terms. But the squaring will also lead to cross terms of the form $E^*(t)E(t+\tau)$ and its complex conjugate. When averaged over the measurement interval and normalized, these interference terms are of the form of the expression in Eq. 10.17.

For $\tau = 0$, the numerator and denominator of the correlation function both simply reduce to the intensity, so that

$$g^{(1)}(0) = \frac{\langle I\rangle}{\langle I\rangle} = 1. \tag{10.19}$$

In general, the correlation function should decay as $\tau \to \infty$. There is usually some characteristic value of the time delay at which the correlation begins to noticeably decay; this is called the **coherence time**, τ_c, of the light source (see Fig. 10.2). In general, $g^{(1)}(\tau)$ is a complex function with the limits

$$g^{(1)}(\tau) \to 1 \qquad \text{for} \qquad \tau \ll \tau_c \tag{10.20}$$

$$|g^{(1)}(\tau)| \to 0 \qquad \text{for} \qquad \tau \gg \tau_c. \tag{10.21}$$

The **longitudinal coherence length** can also be defined as $L_c = c\tau_c$, where c is the speed of light. Natural light sources tend to have short coherence times; for example, sunlight has τ_c on the order of femtoseconds ($10^{-15}s$), with a corresponding longitudinal coherence length L_c on the order of hundreds of nanometers. In contrast, single-mode lasers can have coherence lengths on the order of $10^{-4}s$ or more, with L_c as long as tens of kilometers.

A light source with $\tau_c = \infty$ is called **perfectly coherent**. Light sources with $\tau_c = 0$ are **perfectly incoherent**. Perfect coherence and perfect incoherence are often useful idealizations, but realistic light sources always have finite τ_c, making them **partially coherent**.

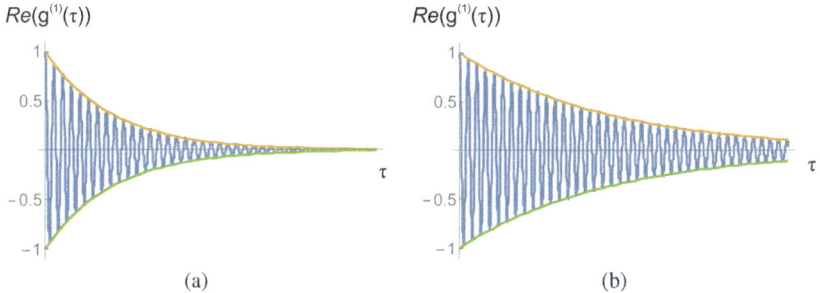

Fig. 10.2 (a) Real part of the correlation function for a quasimonochromatic field of the form given in Eq. 10.23. The function $e^{-i\omega_0\tau}$ in $g^{(1)}(\tau)$ leads to oscillating fringes, which are bounded by a decaying envelope given by $\Phi(\tau) = \langle e^{i[\phi(t+\tau)-\phi(t)]} \rangle$. (b) If the phase function $\phi(\tau)$ varies more slowly, the envelope decays more slowly with increasing delay, and so the coherence time τ_c becomes longer

Example 10.3.1 As an example, consider perfectly monochromatic light of frequency ω_0. We will look at a fixed spatial point, so the spatial dependence of the field will be unimportant and can be ignored. The field can therefore be written in the form $E(t) = E_0 \, e^{-i\omega_0 t}$. Equation 10.17 then gives

$$g^{(1)}(\tau) = e^{-i\omega_0\tau}. \tag{10.22}$$

This doesn't decay at long times as we expect and in fact has an infinite coherence time; but these unphysical properties are due to the fact that a perfectly monochromatic light source itself is unrealistic.

So, as a more realistic example, consider a quasi-monochromatic optical field with a small bandwidth centered at frequency ω_0. The field can be written in the form

$$E(t) = E_0(t) \, e^{-i\omega t} \, e^{i\phi(t)}. \tag{10.23}$$

The width of the frequency spectrum about ω_0 will determine how rapidly $\phi(t)$ varies with time. Inserting this field into Eq. 10.17 leads to

$$g^{(1)}(\tau) = e^{-i\omega_0\tau} \, \Phi(\tau), \tag{10.24}$$

where $\Phi(\tau) = \langle e^{i[\phi(t+\tau)-\phi(t)]} \rangle$. The $e^{-i\omega_0\tau}$ term leads to an oscillation at frequency ω_0 of both the real and imaginary parts of $g^{(1)}(\tau)$, while $\Phi(\tau)$ provides a decaying envelope (Fig. 10.2). If the phase function $\phi(t)$ is rapidly varying, then the factor $e^{i[\phi(t+\tau)-\phi(t)]}$ whizzes rapidly around the unit circle in the complex plane, causing destructive cancelation within the integral that defines the averaging process. The correlation function should then average to a value very close to zero. Thus, the coherence time is determined by the phase function: rapidly varying $\phi(t)$ leads to short τ_c. ∎

Two common situations in applications are Gaussian and Lorentzian correlations:

- A **Lorentzian correlation** function has the form

$$g^{(1)}(\tau) = e^{-i\omega_0 t} e^{-|\tau/\tau_c|}, \tag{10.25}$$

where the coherence time is the inverse of the bandwidth (the spread of frequencies in present in the optical signal):

$$\tau_c = \frac{1}{\Delta\omega}. \tag{10.26}$$

Lorentzian correlation occurs in optical spectra where the dominant broadening mechanisms for the spectral lines are **natural or lifetime broadening** (where the uncertainty principle increases the energy uncertainty) or **collisional or pressure broadening** (in which collisions increase the energy uncertainty of the atoms).

- A **Gaussian correlation** function has the form

$$g^{(1)}(\tau) = e^{-i\omega_0 t} e^{-(\tau/\tau_c)^2}, \tag{10.27}$$

where the correlation time is given by

$$\tau_c = \frac{\sqrt{8\pi \ln 2}}{\Delta\omega}. \tag{10.28}$$

Gaussian correlations often occur, for example, in spectra that are **Doppler broadened**: at high temperatures, spectral lines are broadened by different Doppler shifts from atoms moving toward or away from the measuring device.

Closely related to the first-order correlation function is the interference visibility. Suppose an interference pattern varies between a maximum intensity I_{max} in the bright fringes and a minimum intensity I_{min} in the dark fringes. Then the **interference visibility** measures the contrast between the bright and dark fringes and is defined to be

$$\mathcal{V} = \frac{I_{max} - I_{min}}{I_{max} + I_{min}}. \tag{10.29}$$

Clearly, we must have $0 \leq \mathcal{V} \leq 1$. Examples of interference patterns with different visibilities are shown in Fig. 10.3.

The visibility is an important measure of the quality of an experimental interference pattern. We'll see later that in some situations there are limits on the visibility of classical interference patterns, and that these limits can be violated by quantum interference. The visibility is closely related to the correlation function because it can be shown (see the problems) that:

Fig. 10.3 The visibility measures the contrast between bright and dark fringes in an interference pattern. Examples are shown here with visibilities of 100% and 50%. In both cases, I_0 represents the average intensity

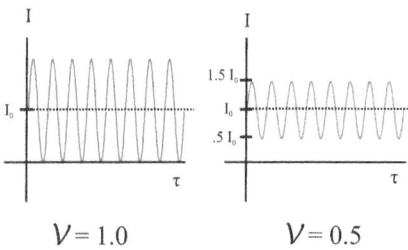

$V = 1.0$ $V = 0.5$

$$V = \frac{2\sqrt{I_1 I_2}}{I_1 + I_2} |g^{(1)}(\tau)|. \tag{10.30}$$

For the case where the two intensities are equal, this simply reduces to

$$V = |g^{(1)}(\tau)|. \tag{10.31}$$

It therefore follows that discernable interference patterns can only be obtained if the photons in the two optical beams were created within a time interval smaller than the coherence time.

Another important related quantity is the **transverse coherence length**. As mentioned earlier, a spatial coherence function can be defined in a similar manner to the temporal coherence function, measuring correlations the transverse plane, perpendicular to the propagation direction of light. The transverse coherence length is the maximum distance Δx_c in this transverse plane over which two optical rays can be combined to form a visible interference pattern. This is why, for example, sunlight (which has a short coherence length) can only produce visible interference after being passed through a pinhole; the pinhole removes portions of the wavefront separated by distances greater than the Δx_c, thereby increasing the overall correlation and visibility.

10.3.2 $g^{(1)}$ for Quantum Fields

So far, we have considered the correlation function for classical fields. In the quantized case, the first obvious change that has to be made is that the electric field must be promoted to an operator. In addition, we need to have an expression for the correlation function that will work even at the level of individual photons.

Recall from Eq. 5.31 that the quantized electromagnetic field can be written as

$$\hat{E}(r, t) = i \sum_k \sqrt{\frac{\hbar \omega_k}{2V\epsilon_0}} \left[\hat{a}(k)e^{i(k\cdot r - \omega_k t)} - \hat{a}^\dagger(k)e^{-i(k\cdot r - \omega_k t)} \right] e_k, \tag{10.32}$$

where e_k is the polarization vector. The electric field is often written in the form

$$\hat{E}(r,t) = \hat{E}^{(+)}(r,t) + \hat{E}^{(-)}(r,t), \tag{10.33}$$

where

$$\hat{E}^{(-)}(r,t) = -i\sum_k \sqrt{\frac{\hbar\omega_k}{2V\epsilon_0}}\left[\hat{a}(k)e^{i(k\cdot r-\omega_k t)}\right]e_k \tag{10.34}$$

$$\hat{E}^{(+)}(r,t) = i\sum_k \sqrt{\frac{\hbar\omega_k}{2V\epsilon_0}}\left[\hat{a}^\dagger(k)e^{-i(k\cdot r-\omega_k t)}\right]e_k. \tag{10.35}$$

Note that these two terms are Hermitian adjoints of each other: $\hat{E}^{(-)} = \left(\hat{E}^{(+)}\right)^\dagger$. $\hat{E}^{(+)}$ is called the **positive frequency part** of the field. It is the part of the field that annihilates photons. $\hat{E}^{(-)}$, the **negative frequency part**, is the piece that contains the creation operator. the positive and negative frequency parts are needed in order to describe photon absorption and emission processes.

In analogy to Eq. 10.17, the quantum version of the first-order correlation function can be written

$$g^{(1)}(\tau) = \frac{\langle\hat{E}^{(+)}(t)\hat{E}^{(-)}(t+\tau)\rangle}{\langle\hat{E}^{(+)}(t)\hat{E}^{(-)}(t)\rangle}, \tag{10.36}$$

where the brackets indicate expectation value in the relevant quantum state of the system.

Consider the case of a single-mode field of fixed polarization, so that the sum in the aforementioned expression for $\hat{E}(r,t)$ reduces to a single term. Further, as in the previous section, focus on the time behavior and consider only a fixed point in space, so that r is a constant. Now $\hat{E}^{(\pm)}$ are both proportional to an oscillating exponential times a creation or annihilation operator. In this case, the expression of Eq. 10.36 reduces to

$$g^{(1)}(\tau) = \frac{\langle\hat{a}^\dagger(t)\hat{a}(t+\tau)\rangle}{\langle\hat{a}^\dagger(t)\hat{a}(t)\rangle}. \tag{10.37}$$

The intensity of the light at the interference point should be proportional to the number of photons arriving there in any fixed period. So at $\tau = 0$, the expectation value appearing in $g^{(1)}(\tau)$ becomes $\langle\hat{a}^\dagger\hat{a}\rangle = \langle\hat{n}\rangle \sim I$. More importantly, consider the rearrangement $\langle\hat{a}^\dagger(t)\hat{a}(t+\tau)\rangle = \langle\psi|\hat{a}^\dagger(t)\hat{a}(t+\tau)|\psi\rangle = \langle\psi^-(t)|\psi^-(t+\tau)\rangle$. Here, $|\psi^-\rangle = \hat{a}|\psi\rangle$ is the state after one photon has been removed from the incoming state $|\psi\rangle$; in other words, it is the state after one photon has been detected (and in the process annihilated) by the detector. The correlation function is therefore measuring the correlations between single photon detections at times separated by delay τ. We will contrast this the case of coincidence detection (detection of photon pairs) in the next section.

10.4 Second-Order Correlations

The first-order correlation function $g^{(1)}$ defined in the previous section measures correlations between *fields* and is essentially a measure of classical correlations. However, there are inherently quantum correlations that are invisible to $g^{(1)}$. We can think of these additional correlations as being correlations between *intensities* (rather than fields) or alternatively as correlations between amplitudes belonging to a *pair* of photons (rather than amplitudes of a single photon). To measure these two-photon correlations, we will define a second-order correlation function $g^{(2)}$. This will detect correlations in coincidence counts (two photons detected simultaneously), rather than in the single-photon detections probed by $g^{(1)}$.

Higher-order correlation functions $g^{(n)}$ with $n > 2$, which measure n-photon correlations, have been extensively studied, both theoretically and experimentally. We will not discuss these here, but numerous studies of them can be found in the literature, including [6–9]

10.4.1 Coincidence Counting: Two-Photon Interference

A common component of optical experiments is a **coincidence counter**. This consists of two detectors, each producing an electric current when a particle is detected. The two outputs from the detector are then connected to a circuit, which records a detection only if both detectors register an event at the same time. (In practice, "at the same time" really means within some small but finite time window determined by the response times of the detectors and the electronics.) If i_1 and i_2 are the currents exiting the two detectors, then the output of the coincidence circuit can be viewed as multiplying them to produce output i proportional to $i_1 i_2$.

Coincidence counting can be used with electrons and other particles, but our main concern here is with photons. The coincidence circuit then counts the detection of *pairs* of photons correlated in time. By building a delay τ into one line of an interferometer, the coincidence circuit measures the occurrence of two photons arriving with a fixed time separation between them; or, from a more classical viewpoint, it measures correlations between intensities at different times

Figure 10.4 shows two interferometers that use coincidence counting. In the figure, solid black lines represent potential light paths, while dashed blue lines are wires carrying electric currents. The coincidence-counting circuit is represented by the circle with crossed lines.

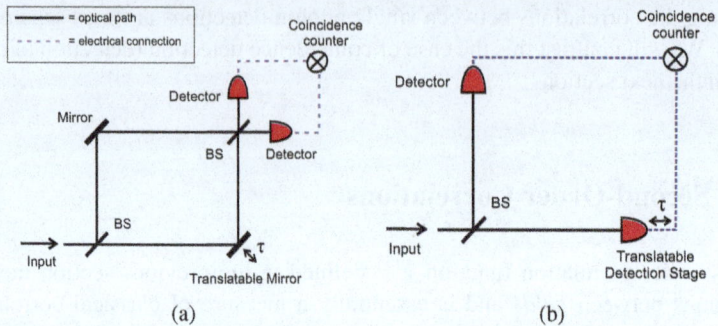

Fig. 10.4 Two interferometers with coincidence counting. (**a**) A Mach–Zehnder interferometer: two photons are input at lower left. (They could enter at either or both of the input ports.) These two photons can either both exit in the same direction (the same output port) at the upper right and reach the same detector, or they can exit in different directions and be detected by different detectors. The coincidence circuit will only register the latter events, not the former. (**b**) A Hong–Ou–Mandel (HOM) interferometer with coincidence counting. The HOM interferometer is discussed in detail in Sect. 10.6

Introducing some notation, the number of coincidence counts in some fixed time interval will be denoted N_c. The **singles counts** (the counts in each detector separately) will be denoted N_1 and N_2. So:

$$i \sim i_1 i_2 \sim N_c \sim N_1 N_2. \tag{10.38}$$

The relations in the last line are proportionalities, not equalities, because the detectors may have non-ideal detection efficiencies and losses may occur in the various optical and electrical lines.

Again denoting the optical intensity in the arms of the interferometer by I_1 and I_2, we can define the **second-order correlation function** as the correlation between the two intensities, rather than between the two fields:

$$g^{(2)}(\tau) = \frac{\langle I_1(t) I_2(t+\tau) \rangle}{\langle I_1(t) \rangle \langle I_2(t) \rangle}. \tag{10.39}$$

Since intensity is essentially the square of the field, these correlators involve products of four fields in the numerator.

As with the first-order function, the definition can be written equivalently in terms of creation operators or photon numbers:

$$g^{(2)}(\tau) = \frac{\langle n_1(t) n_2(t+\tau) \rangle}{\langle n_1(t) \rangle \langle n_2(t) \rangle} \tag{10.40}$$

$$= \frac{\langle \hat{a}_1^\dagger(t) \hat{a}_2^\dagger(t+\tau) a_1(t) a_2(t+\tau) \rangle}{\langle a_1^\dagger(t) a_1(t) \rangle \langle a_2^\dagger(t) a_2(t) \rangle}. \tag{10.41}$$

The brackets in these quantities include both a quantum expectation value over the relevant states and a time average over the detection window determined by the detection response.

Examples of $g^{(2)}(\tau)$ for specific systems will be considered further, but first we consider the general properties of the function. Specifically, let's focus first on *classical* light. There will again be some coherence time τ_c for the light, and we can examine the correlation function at small and large times compared to τ_c. First, define the intensity at a given time to be the average intensity plus some fluctuation term $\Delta I(t)$,

$$I(t) = \langle I \rangle + \Delta I(t), \tag{10.42}$$

where the average is clearly time-independent. Notice that if the average is taken of both sides of the previous equation, then in order for the two sides to be equal, it is necessary that the average of the fluctuations vanish:

$$\langle \Delta I \rangle = 0. \tag{10.43}$$

So consider the average of a product of intensities:

$$\langle I(t)I(t+\tau) \rangle = \left\langle \Big(\langle I \rangle + \Delta I(t) \Big) \Big(\langle I \rangle + \Delta I(t+\tau) \Big) \right\rangle \tag{10.44}$$

$$= \langle I \rangle^2 + \langle I \rangle \Big(\langle \Delta I(t) \rangle + \langle \Delta I(t+\tau) \rangle \Big)$$

$$+ \langle \Delta I(t) \Delta I(t+\tau) \rangle \tag{10.45}$$

$$= \langle I \rangle^2 + \langle \Delta I(t) \Delta I(t+\tau) \rangle. \tag{10.46}$$

In the last line, the first term is a constant, τ-independent background term. The important term is the second, interference term. From it, we see that what is actually measured by $g^{(2)}(\tau)$ is the correlation between intensity *fluctuations* ΔI at different times. If the intensity or (particle number) fluctuations are correlated, then the light must be coherent. For incoherent light, the fluctuations reaching the two detectors should be completely independent, and so their correlation should vanish.

Currents separated by very long delay times, $\tau \gg \tau_c$, should be incoherent with each other, or in other words, they should be statistically independent of each other. This implies that the expectation value of the product in the last equality should factor into separate expectation values of each current independently:

$$\langle \Delta I(t) \Delta I(t+\tau) \rangle = \langle \Delta I(t) \rangle \langle \Delta I(t+\tau) \rangle, \qquad \text{for} \quad \tau \gg \tau_c = 0. \tag{10.47}$$

As a result, we have

$$g^{(2)}(\tau) \to 1, \qquad \text{for } \tau \gg \tau_c. \tag{10.48}$$

Fig. 10.5 Classically, the
second-order correlation
function is expected to decay
monotonically from a
maximum value at $\tau = 0$,
asymptotically approaching a
value of $g^{(2)}(\tau) = 1$ as
$\tau \to \infty$

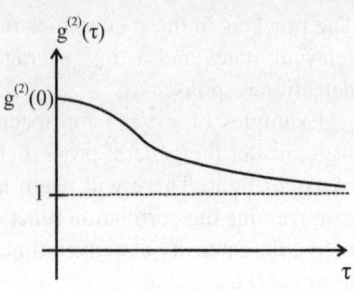

In addition, it is clear from Eq. 10.46 that at $\tau = 0$,

$$\langle I(t)^2 \rangle \geq \langle I(t) \rangle^2. \tag{10.49}$$

So,

$$g^{(2)}(0) \geq 1. \tag{10.50}$$

More generally, it is true that $g^{(2)}(0) \geq g^{(2)}(\tau)$ for all τ. So, classically, we would expect that a generic shape for $g^{(2)}(\tau)$ should be something like that shown in Fig. 10.5.

Some common forms of second-order coherence functions are:

$$g^{(2)}(\tau) = \begin{cases} 1 & \text{(perfect coherence)} \\ 1 + e^{-\pi(\tau/\tau_c)^2} & \text{(Gaussian (Doppler-broadened light))} \\ 1 + e^{-2|\tau|/\tau_c} & \text{(Lorentzian (lifetime-broadened) light)} \end{cases} \tag{10.51}$$

However, the aforementioned considerations apply only to classical light. In quantum mechanics, it turns out that there is a loophole in the argument for the inequality of Eq. 10.50. The inequality of Eq. 10.49 can be viewed as a consequence of the **Cauchy–Schwarz inequality**:

$$\langle X^2 \rangle \langle Y^2 \rangle \geq \langle XY \rangle \tag{10.52}$$

for averages of any random variables X and Y. Although the Cauchy–Schwarz inequality (CSI) is a fundamental result of real analysis and has to hold for all classical probability distributions and classical variables, it can be violated by averages taken over quantum states. In particular, the Cauchy–Schwarz inequality is violated by entangled states, and so CSI violation has even been proposed as a test for entanglement [12]. Therefore, Eq. 10.50 is *not* necessarily true for non-classical states of light: a value $g^{(2)}(0) < 1$ is a clear signal of quantum behavior.

Light can then be classified into three categories based on the value of $g^{(2)}(0)$:

- **Bunched** or **super-Poisson**: $g^{(2)}(0) > 1$ (example: thermal light)
- **Coherent** or **Poisson**: $g^{(2)}(0) = 1$ (example: laser light)
- **Anti-bunched** or **sub-Poisson**: $g^{(2)}(0) < 1$ (example: number states)

The terms bunching and anti-bunching refer to the likelihood of detecting two photons separated by very small delay $\tau \approx 0$. Recall that coherent-state light is Poisson-distributed: the probability of finding n photons in a given time interval is given by the Poisson distribution. But for small time intervals, bunched or super-Poisson light tends to be more clustered: there are more photons on average in small intervals than the Poisson distribution would predict. Similarly, anti-bunched or super-Poisson light has the photons less-clustered, with fewer per small interval than Poisson predicts.

Classical light can be bunched or coherent, but cannot be anti-bunched. There-fore, having anti-bunched light with $g^{(2)}(0) < 1$ is an unambiguous signal of non-classical behavior. (Note that the converse is not true: quantum light can also exhibit Poisson or super-Poisson behavior.) Examples of anti-bunched states are Fock states (states with fixed particle number) or the two-photon states produced by spontaneous parametric down conversion.

Classical light obeys the well-known **Siegert relation**, which relates the first- and second-order coherence functions:

$$g^{(2)}(\tau) = 1 + \left| g^{(1)}(\tau) \right|^2, \tag{10.53}$$

so that $g^{(1)}(\tau)$ completely determines the value of $g^{(2)}(\tau)$. Eq. 10.51 illustrates this relation for the perfectly coherent, Gaussian, and Lorentzian cases.

But, the Siegert relation can be violated by quantum states of light, meaning that in these cases, the second-order coherence measured via coincidence counts can provide new information not available from the first-order coherence. This explains why many quantum experiments often involve coincidence counting. Overviews of the Siegert relation and its uses can be found in [13, 14]; the second of these papers also describes a method for testing the relation in an undergraduate physics lab.

10.4.2 Hanbury Brown and Twiss Interferometer

In the previous subsection, the second-order correlation function was described in the context of quantum coincidence counting experiments, in which the number of photons being detected in any interval was small and so sensitive photodetectors are needed to detect individual photons. However, the coincidence method actually originated in a purely classical astronomical interferometer using higher intensity light. This approach was developed by Hanbury Brown and Twiss in 1956 [15, 16] to provide improved measurements of stellar diameters. Here we discuss the **Hanbury**

Brown Twiss interferometer (HBT), both to provide some historical context and because quantum enhancements of it have been proposed in recent years.

Consider a star of diameter D at a distance of L from the earth (Fig. 10.6). The angular size of the star, as viewed by an earthly astronomer, is

$$\delta\theta_{star} \approx \tan\delta\theta_{star} \approx \frac{D}{L}. \tag{10.54}$$

The goal is to measure the angular size of the star with high precision and to extract the value of D from the measurement.

From the 1920s until the 1950s, the main tool for measuring stellar diameters was the **Michelson stellar interferometer** (not to be confused with the unrelated Michelson interferometer discussed in Sect. 2.5). This consisted of two moveable mirrors, two fixed mirrors, a screen with two slits and a telescope, as shown schematically in Fig. 10.7. The moveable mirrors collected light rays separated by a distance d. The telescope focused the two rays onto a screen or camera, where an interference pattern could form. The arrangement essentially forms a single diffractive system of effective diameter d, so that by the Rayleigh criterion, the minimum resolvable angle should be

$$\delta\theta_{rayl} \approx \frac{1.22\lambda}{d}, \tag{10.55}$$

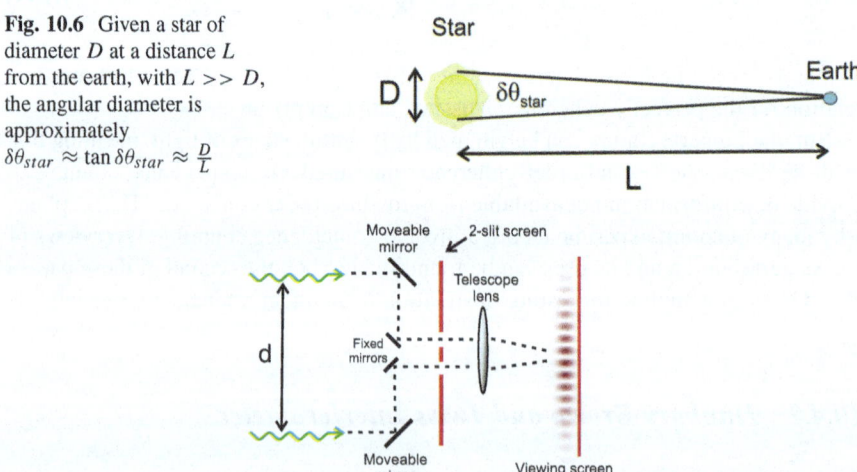

Fig. 10.6 Given a star of diameter D at a distance L from the earth, with $L \gg D$, the angular diameter is approximately $\delta\theta_{star} \approx \tan\delta\theta_{star} \approx \frac{D}{L}$

Fig. 10.7 Michelson stellar interferometer. Mirrors separated by distance d collect light and direct it through a telescope onto a viewing device, where it forms an interference pattern. A screen with two slits blocks all incoming light except that coming from the two collecting mirrors. As d increases, a point is reached at which the interference pattern starts to wash out and the first-order coherence function decays. The value of d at which this occurs is used to measure the diameter of the stellar light source

where it is assumed that the light has been filtered to block all wavelengths except those in a narrow window around λ.

So the star can be clearly resolved if $\delta\theta_{star} > \delta\theta_{rayl}$, in which case a clear, high-visibility interference pattern can form. However, if $\delta\theta_{star} < \delta\theta_{rayl}$, then the star is not resolvable: its image blurs and as a result the interference pattern washes out and becomes a blur as well. Recall that the visibility of the interference pattern is given by the first-order correlation function (in this case, the transverse spatial coherence function), and so the value of d at which the interference pattern decays is equal to the transverse coherence length Δx_c. When the interference is right at the resolution limit, $\delta\theta_{star} = \delta\theta_{rayl}$, it follows from the previous two equations that

$$\frac{D}{L} = \frac{1.22\lambda}{d}. \tag{10.56}$$

So if the star's distance is known (by measuring the Doppler shift of its spectral lines, or by other means), then the diameter of the star can be deduced:

$$D = \frac{1.22\lambda L}{d}. \tag{10.57}$$

By the 1950s, this approach began to reach its limits. To measure stars of smaller size, it was necessary to have larger separations d between the mirrors. However, as d increases, the mechanical stability of the system decreases, and smaller vibrations are capable of causing enough disturbance to disrupt the interference pattern. To measure smaller stars, a fundamentally different method was required.

This was provided by Hanbury Brown and Twiss, whose basic idea (in modern terminology) was to use the second-order correlation function instead of the first-order. Equivalently, they looked at intensity correlations instead of field correlations. Experimentally, this meant using a *pair* of optical detectors, instead of one. The setup is shown schematically in Fig. 10.8. Two moveable photodetectors are now separated by distance d. Each input intensity I_j for $j = 1, 2$ produces an output current i_j from the corresponding detector. The photocurrent is proportional to the optical intensity: $i_j \sim I_j$. These outputs are fed into a multiplier circuit, which in turn produces an output current proportional to the product of the original two photocurrents, averaged over the integration time of the detector:

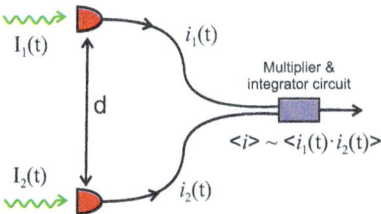

Fig. 10.8 The Hanbury Brown Twiss (HBT) stellar interferometer. Two detectors separated by distance d collect light from a distant star. As d increases, a point is reached where the intensity–intensity correlations drop, allowing the stellar diameter to be determined

$$\langle i \rangle \sim \langle i_1 \cdot i_2 \rangle \sim \langle I_1 \cdot I_2 \rangle. \tag{10.58}$$

When d is small compared to the transverse coherence length, $d << \Delta x_c$, the two detectors should see approximately the same intensity, and so the signal should be large. But when $d > \Delta x_c$, the intensities arriving at the two detectors should be mostly uncorrelated with each other. As a result, the value of $g^{(2)}(d)$ should be close to zero and the output signal should be small. In quantum mechanical terms, the two-photon interference pattern should be washed out when d is too large. So once again, the value of d at which the interference pattern decays should allow the stellar diameter to be measured. This leads to stellar diameter measurements with of higher resolution than previous methods. Because intensities, not fields, are the relevant quantity here, small fluctuations in phase in the detected photons will not affect the coincidence rates.

The HBT method, while remaining useful in astronomy, is now also a staple of quantum optics. The main differences are that the intensities are low (so that the optical inputs are described in terms of photon number, rather than intensity), and that the spatial separation d and spatial correlation $g^{(2)}(d)$ are replaced by a time delay τ and a corresponding temporal correlation function, $g^{(2)}(\tau)$.

Box 10.1 Entanglement-enhanced Stellar Astronomy

The standard HBT interferometer described in the main text is normally a classical device since the detectors used are not of sufficient sensitivity to measure signals at the single-photon or few-photon level and the light sources themselves (stars) are thermal light sources producing purely classical illumination. However, proposals have been made to use entanglement and quantum interference to improve stellar interferometry.

A common form of interferometer for stellar astronomy, shown in Fig. 10.9, is essentially an Hong–Ou–Mandel interferometer (Sect. 10.6) using photons from detectors at two observatories separated by baseline distance d. Suppose two stellar light sources are close together, and the goal is to be able to resolve them as separate objects. In order to make an interferometric measurement, photons from two detectors need to be brought together, but photon loss and phase fluctuations limit the size of d to a few hundred meters, which in turn limits the possible resolution.

To extend the potential baseline and improve the resolution, it has been proposed [17] to add an entangled photon source (EPS) in the middle and to use four detectors instead of two (Fig. 10.10a). The experimenter now looks for four-way coincidences, in which all four detectors fire at once. Two photons come from the EPS and two come from the stellar source. This not only improves the sensitivity and resolution of the measurements but also allows longer baselines due to the high degree of correlation in the entangled quantum states.

(continued)

Box 10.1 (continued)

A further improvement has been suggested using a different four-photon interference scheme [18] that removes the need for an entangled photon source (Fig. 10.10b). Now each of the two interferometers uses one photon from each light source. Coincidences between the detectors attached to the two interferometers are then counted. The presence of prisms to separate frequencies into different detectors allows resolution in frequency as well as angular position. Because there are now no optical connections between the two observatories, stability is greatly increased, and there is essentially no longer any limit on the baseline. It has been estimated that this arrangement can achieve angular resolutions of better than 10^{-5} arcseconds.

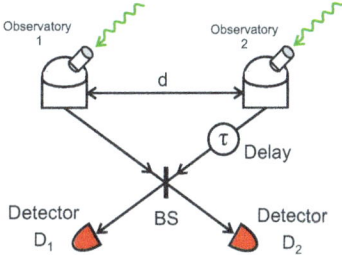

Fig. 10.9 Standard direct-detection interferometer for stellar interferometry. Light from two stellar sources enters a pair of observatories and then is mixed at a beam splitter before reaching a pair of directors. Photon losses and phase fluctuations lead to a limit in how long a baseline d can exist between the two observatories, and so limit the possible resolution

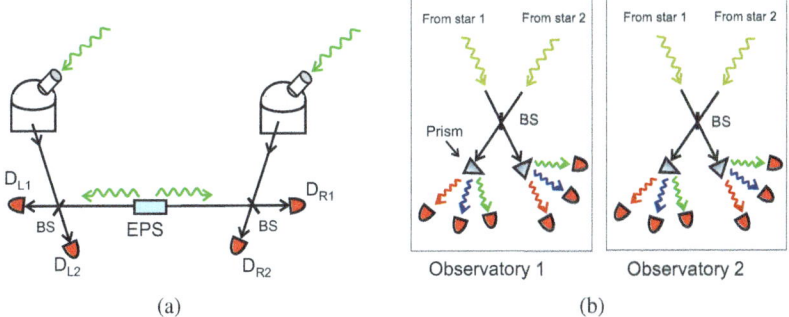

Fig. 10.10 Proposals for entanglement-enhanced stellar interferometry. (**a**) An entangled photon source sends one photon to each observatory, where it is mixed by a beam splitter with a photon entering the telescope. (**b**) By inserting prisms after the beam splitters, different frequencies can be separated, allowing analysis of the spectrum

10.5   Complementarity between First- and Second-Order Interference

We've seen that, roughly speaking, first-order and second-order interference corre-spond to classical and quantum interference, respectively. First-order interference can be seen by monitoring a single detector, but second-order effects require looking at coincidence counts at a pair of detectors. Both types of interference can be viewed in a single two-detector experiment by monitoring singles events and coincidence events simultaneously, but it is seen experimentally that any experiment that has high visibility interference in the singles counts will have low visibility in coincidences, and vice versa. This effect can be quantified to give a complementarity relation between first- and second-order interference effects [19–21].

Consider an experiment in which two photons traverse some apparatus and are detected at the end by a pair of detectors, D_A and D_B. The singles counts at either detector can be measured, giving a singles counting rate $R(j)$ for each detector, with $j = A, B$. Dividing by the rate R_{tot} at which photons enter the system, these give the probabilities per trial of each type of count, $P(j) = R(j)/R_{tot}$. Similarly, the coincidence probability between the detectors can be defined by $P_c(A, B)$. Even in the absence of any correlations, there would be "accidental" coincidences simply by chance. These can be subtracted off from the full coincidence probability to provide a corrected coincidence probability [19], given by

$$P(A, B) = P_c(A, B) - P(A)P(B) + \frac{1}{4} \qquad (10.59)$$

that measures the excess coincidences due to entanglement.

Now allow some parameter (phase, time delay, etc.) to vary, so that the singles and coincidence counts exhibit an interference pattern as the parameter values are scanned over. Denote the first-order visibility (Eq. 10.29) at detector j by $V_1(j)$, $j = A, B$. A **second-order visibility** or **entanglement visibility** can be defined according to

$$V_2(A, B) = \frac{P_{max}(A, B) - P_{min}(A, B)}{P_{max}(A, B) + P_{min}(A, B)}. \qquad (10.60)$$

Then the classical and quantum visibilities obey the **complementarity relations** [19]

$$V_1^2(A) + V_2^2(A, B) = 1 \qquad (10.61)$$

$$V_1^2(B) + V_2^2(A, B) = 1. \qquad (10.62)$$

Thus, when quantum (two-photon) visibility is high, the classical (single-photon) visibility is low, and vice versa.

The second-order visibility can be used to classify the level of entanglement in a system. When $V_2(A, B) < 0.5$, the system only has classical correlations, but if $V_2(A, B) > 0.5$ then entanglement is present. If the visibility is higher still, $V_2(A, B) > \frac{1}{\sqrt{2}} \approx 0.71$ then the system has sufficient entanglement to violate Bell and CHSH inequalities (see Chap. 14).

This complementarity between single- and two-photon visibilities can be seen in spontaneous parametric downconversion (SPDC). The signal and idler photons individually have very low coherence and only produce single-particle interference of low visibility. But the two photons together can produce high-visibility interference patterns in their coincidence counts; in some sense, the coherence of the original pump beam is distributed over the two outgoing photons and only becomes visible when both of them are brought back together in coincidence.

10.6 The Hong–Ou–Mandel Effect

One key difference between classical and quantum systems is how probabilities are handled. In a classical system, if given a set of mutually exclusive outcomes, A, B, C, D,..., then the probability that the outcome will fall inside some subset of those possibilities is just the sum of the probabilities of the events in that subset. So, for example, if it is desired to find the probability that the outcome of an experiment will be either A, B, or C, then the probability of this set of outcomes is simply

$$P(A, \ B, \text{ or } C) = P_A + P_B + P_C. \tag{10.63}$$

However, in quantum systems, what adds is not the *probabilities* themselves, but rather the *probability amplitudes*. So, suppose that the events in the previous paragraph are all indistinguishable. For example, they may correspond to passage through different slits in an opaque screen and then arriving at the same final point in a detector. Then each event has a complex-valued amplitude, $a_A, a_B, \ldots$ and the probability of each event is the absolute square of the amplitude: $P_A = |a_A|^2$, and so on. But the probability of the subset $\{A, B, C\}$ is now

$$P(A, \ B, \text{ or } C) = |a_A + a_B + a_C|^2 \neq P_A + P_B + P_C. \tag{10.64}$$

This differs from the classical expression by the appearance of cross-terms such as $a_A^* a_B$. These cross-terms lead to interference, and they can drastically change the outcome away from what would be expected classically. A dramatic demonstration of this is given by the **Hong–Ou–Mandel (HOM) effect**. This was demonstrated experimentally in [22]; see [23] for a recent review.

Consider a beam splitter with two photons entering it from different ports, as in Fig. 10.11. Each photon can either be reflected or transmitted, with equal probability of each. So for two photons, there are four possible outcomes: (i) both reflect, (ii)

Fig. 10.11 The HOM effect occurs when two photons (the red and green dots) enter a beam splitter simultaneously at different ports

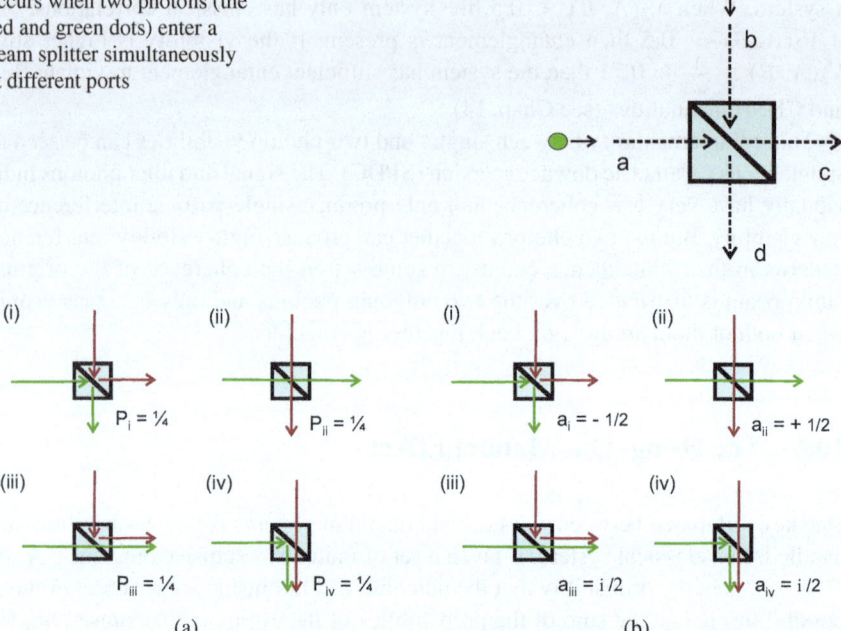

Fig. 10.12 The output from the beam splitter from the previous figure. (**a**) Classically, all four possible outcomes have an equal probability of $\frac{1}{4}$. The coincidence probability is $P_{coinc} = P_i + P_{ii} = \frac{1}{2}$. (**b**) In the quantum case, it is amplitudes, rather than probabilities, that add for indistinguishable outcomes. So now the coincidence probability vanishes: $P_{coinc} = |a_i + a_{ii}|^2 = 0$

both transmit, (iii) the photon from port a transmits and that from port b reflects, or (iv) the photon from port b transmits and that from port a reflects, as shown in Fig. 10.12a. Each of these outcomes has an equal probability of $\frac{1}{4}$. So, classically, the probability of a coincidence event (one photon exiting at c and one exiting at d) should be

$$P_{coinc} = P(\text{outcome (i)}) + P(\text{outcome (ii)}) = \frac{1}{4} + \frac{1}{4} = \frac{1}{2}. \qquad \text{(Classical case)}$$
(10.65)

But in the quantum case (Fig. 10.12b), it is the amplitudes, not the probabilities, of the two indistinguishable outcomes that must be added. Recall (Chap. 2), that each reflection at the beam splitter multiplies the amplitude by a factor of i relative to the amplitude of transmission. So the reflection–reflection amplitude of outcome (ii) is the negative of the transmission-transmission amplitude of outcome (i). The coincidence probability in the quantum case is therefore:

$$P_{coinc} = |a_{(i)} + a_{(ii)}|^2 = \left| \frac{1}{4} + \left(-\frac{1}{4} \right) \right|^2 = 0. \qquad \text{(Quantum case)} \qquad (10.66)$$

The two possible ways that a coincidence can occur destructively interfere, leaving no coincidence events! This is a starkly different outcome from the classical case, where there should be coincidence events on half of the trials.

Notice that there are now just two possible outcomes: both photons exit at c or both exit at d. Even though the photons don't interact with each other and each photon chooses an exit port at random, the two photons always choose the *same* exit. Note also that the output can be viewed as a two-photon $N00N$ state: $|20\rangle + |02\rangle$, where $|mn\rangle$ represents the state with m particles at port c and n is the number at port d.

The setup for the HOM experiment is shown schematically in Fig. 10.13. The two simultaneous input photons are produced via SPDC inside a nonlinear crystal. A time delay is added in the path of one photon. When the delay is zero, both photons enter and exit the beam splitter simultaneously, so that the two photons cannot be distinguished by their detection times. If both photons are also indistinguishable in frequency and polarization then interference between the two coincidence possibilities occurs, and the coincidence count vanishes. However, as the delay increases, it becomes easier to distinguish between the two photons based on their detection times. Since distinguishability destroys interference, the coincidence rate quickly rises back to the classical value. The result is the celebrated HOM dip, in which the

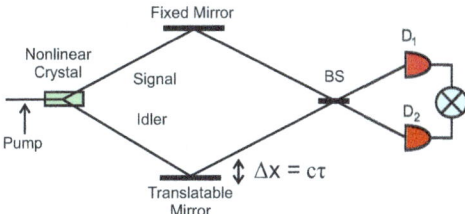

Fig. 10.13 Schematic for the setup used to produce the HOM dip. The input photons are produced by SPDC in the nonlinear crystal. Moving one of the mirrors by distance Δx produces a time delay $\tau = \Delta x / c$ for one of the photons. Interference occurs at the beam splitter, leading to the HOM dip of Fig. 10.14

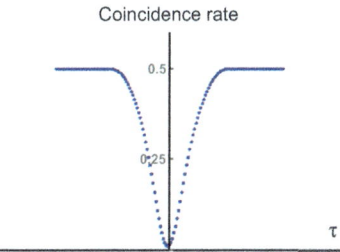

Fig. 10.14 When the delay is zero, the two photons are indistinguishable, and so the amplitudes for the two possible coincidence events exhibit destructive interference, causing the coincidence rate to vanish. As the delay increases, the coincidence rate rapidly rises back to the classical level, as the photons become distinguishable by their arrival times at the detectors

coincidence shows a very sharp decrease as a function of delay near $\tau = 0$. Because of the narrowness of this dip, it can be used to make time measurements with sub-femtosecond resolution. With the use of statistical estimation theory (involving Fisher information and the Cramer–Rao bound), resolution can be reduced to the level of a few attoseconds (10^{-18} s) [24–26].

10.7   The Franson Interferometer and Nonlocal Interference

Classically, two waves have to overlap in space at a given time in order to interfere. However, that is not necessarily the case in quantum mechanics: quantum interference can be nonlocal in the sense that two-particle amplitudes can interfere even if the two particles involved have wavefunctions that never have significant overlap at the same space-time point. The significance of this nonlocal behavior will be discussed in more detail in Chap. 14, in connection with the Bell and CHSH inequalities. Here, we mention two experiments that clearly demonstrate the nonlocal nature of quantum interference effects.

Recall that two-photon interference involves the interference of different amplitudes for the same photon pair. For example, each photon can go in either of two directions at a beam splitter and the amplitudes for these different possibilities interfere with each other. What may not be obvious is that the two photons need not arrive at the beam splitter or at the detectors at the same time in order to produce interference. Only indistinguishability, not temporal overlap, is required, as can be clearly demonstrated by the experiment [27] shown in Fig. 10.15. Type II SPDC is used to create a pair of oppositely polarized photons. Polarizers are used to allow passage only of pairs in which the vertically polarized photon exits the nonlinear crystal on the right, and horizontal on the left. A fixed delay τ_V (greater than the coherence time of the photons) is introduced into the right-hand path prior to reaching the beam splitter. Because of this, the two photons arrive at the beam splitter at different, easily distinguishable times. In order to restore indistinguishability, **postponed compensation** is introduced, in which a selective time delay is added after the beam splitter, arranged so that it only affects the horizontal polarization. The reflection/reflection (RR) and transmission/transmission (TT) cases at the beam splitter are then as in Fig. 10.15. In the RR case, the vertical photon arrives at detector 2, delayed by a time τ_V; the time difference between the two detections is then $\Delta t \equiv t_2 - t_1 = \tau_V$. But in the TT case, the horizontally polarized photon arrives at detector 2 delayed by τ_H, while the vertical photons arrive at detector 1 delayed by τ_V. So, in this case, the detection time difference is $\Delta t \equiv t_2 - t_1 = \tau_H - \tau_V$. If delays are chosen such that $\tau_H = 2\tau_V$, then both cases have the same time difference, $\Delta t = \tau_V$. Therefore, the two cases cannot be distinguished by photon arrival times, and the HOM interference dip appears, even though the two photons never overlap on the beam splitter. The interference exhibits a kind of nonlocality in time, which again demonstrates that the HOM dip is a purely quantum effect.

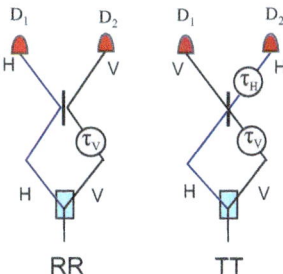

Fig. 10.15 Two photons need not overlap temporally on the beam splitter to interfere and produce an HOM dip. Here, time delay τ_V appears in the right-hand branch before the beam splitter. τ_H is a polarization dependent delay that appears in the right arm after the beam splitter, which causes a delay only for horizontally-polarized photons. If $\tau_H = 2\tau_V$, then both the reflection–reflection (RR) and transmission–transmission (TT) cases lead to the same time delay $\Delta t \equiv t_2 - t_1 = \tau_V$ between detection of photons. The two outcomes are therefore indistinguishable and can interfere

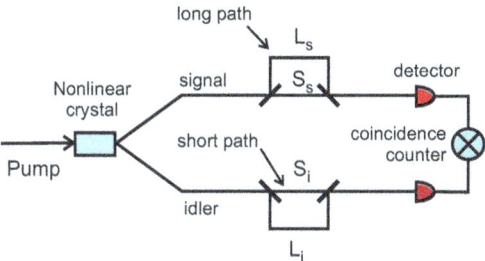

Fig. 10.16 The Franson interferometer. Each photon can travel via a long path or a short path, of respective lengths L_s and S_s for the signal, and similarly of lengths L_i and S_i for the. After post-selection to remove long-short interference, the amplitude for both to follow the long path interferes with the amplitude for both following the short path

The Franson interferometer [28] shown in Fig. 10.16 demonstrates a similar effect. A pump beam at frequency ω_p enters the nonlinear crystal at the left, leading to SPDC. Each photon from a given down conversion pair enters a separate path. Each of those photons then has a choice of two possible subpaths, one long and one short. The difference of path lengths between the long and short paths is much longer than the photon coherence time, so that no single-photon interference occurs between the long and short paths. Any interference that occurs must therefore be two-photon (i.e., quantum) interference, not classical single-photon interference. At the right, the two photons are detected in coincidence.

The time at which each pair is created is unknown, so the cases where both photons take the long path (LL) and where both take the short path (SS) are indistinguishable; these two-photon amplitudes should then interfere with each other. The visibility of this interference can be greatly enhanced by post-selection: arrange the coincidence circuit to have a coincidence window short enough to reject the two cases in which one photon takes the long path and one takes the short path

$(L_1 S_2$ and $S_1 L_2)$. Removing this background, the two-photon visibility is ideally 100%, despite the fact that the two photons are never in the same location at the same time.

Let $\Delta l_s = L_s - S_s$ and $\Delta l_i = L_i - S_i$ be the path difference between the long and short paths for the signal and idler respectively. If $|\omega_s - \omega_i|$ is small compared to the central frequency $\omega_0 = \frac{\omega_{\text{pump}}}{2}$, then the phase difference between the LL path and the SS path is

$$\phi = \frac{\omega_s \Delta l_s}{c} + \frac{\omega_i \Delta l_i}{c} \approx \frac{1}{2c}\omega_{\text{pump}}(\Delta l_i + \Delta l_s). \tag{10.67}$$

The coincidence probability is

$$P_c = |\langle \psi | \psi \rangle|^2 = \frac{1}{4}|1 + e^{i\phi}|^2 \tag{10.68}$$

$$= \frac{1}{2}(1 + \cos\phi) = \frac{1}{2}\left[1 + \cos\left(\frac{\omega_{\text{pump}}}{2c}(\Delta l_i + \Delta l_s)\right)\right], \tag{10.69}$$

producing an interference pattern as the path lengths are varied. This interference was demonstrated experimentally in [29–31].

10.8 The Zou–Wang–Mandel Experiment and Induced-Coherence Imaging

The Franson interferometer works because the time of emission of the photons is unknown, so photons emitted at different times become indistinguishable and can interfere. The **Zou–Wang–Mandel effect** [32, 33] goes one step further: not only the emission time of the photons is unknown, but so is the nonlinear crystal from which the photons are emitted.

Consider the setup in Fig. 10.17. A pump photon hits beam splitter BS_1 and is randomly directed to one of the two identical nonlinear crystals C_1 or C_2. Each

Fig. 10.17 The setup for the Zou-Wang-Mandel effect. C_1 and C_2 are $\chi^{(2)}$ nonlinear crystals, while B is an opaque obstacle that can be moved in or out of beam path e. Paths e and b are well aligned, so that if a photon arrives at detector D_2, there is no way of distinguishing whether it was created by down conversion in crystal C_1 or C_2

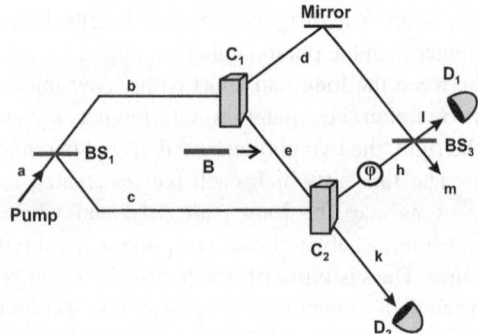

crystal can produce a down conversion pair, emitted along lines e and d for C_1 or lines h and k for C_2. Detectors D_1 and D_2 are placed at outputs l and k. The third output m plays no role, so it can be left unmonitored. Lines e and k are well-aligned, so that when a photon is detected at detector D_2, it can't be determined whether it originated in crystal C_1 or crystal C_2.

So, moving from left to right, we can work out the state of the system. Starting with a pump photon in mode a, we find:

$$|1\rangle_a \to_{BS_1} \frac{1}{\sqrt{2}}\left(|1\rangle_b + i|1\rangle_c\right) \tag{10.70}$$

$$\to_{C_2} \frac{1}{\sqrt{2}}\left[|1\rangle_d|1\rangle_e + i|1\rangle_h|1\rangle_k\right] \tag{10.71}$$

$$\to_{BS_2} \frac{1}{2}\left[|1\rangle_e\left(|1\rangle_m + i|1\rangle_l\right) + ie^{i\phi}|1\rangle_k\left(|1\rangle_l + i|1\rangle_m\right)\right], \tag{10.72}$$

where the action of the phase shifter in line h has also been include in the last line. But, if the crystals are properly aligned, the states $|1\rangle_e$ and $|1\rangle_k$ are indistinguishable, so in the last line $|1\rangle_e$ can by $|1\rangle_k$:

$$\to \frac{1}{2}|1\rangle_k\left[\left(|1\rangle_m + i|1\rangle_l\right) + ie^{i\phi}\left(|1\rangle_l + i|1\rangle_m\right)\right] \tag{10.73}$$

$$= \frac{1}{2}|1\rangle_k\left[i\left(1 + e^{i\phi}\right)|1\rangle_l + \left(1 - e^{i\phi}\right)|1\rangle_m\right] \tag{10.74}$$

$$= -ie^{i\phi/2}|1\rangle_k\left[\cos\frac{\phi}{2}\,|1\rangle_l + \sin\frac{\phi}{2}\,|1\rangle_m\right]. \tag{10.75}$$

First, notice from the first term that if only line l is monitored, then there will be single-photon interference at detector D_1, with detection probability

$$P_1(\phi) = \cos^2\frac{\phi}{2}. \tag{10.76}$$

This arises because any photon detected at D_1 could have originated from C_1 or from C_2; the two crystals essentially act like the two apertures in the Young two-slit experiment. But looking at Eq. 10.75, we see from the $|1\rangle_k|1\rangle_l$ term that there is also interference in the coincidence counts between detectors 1 and 2:

$$P_{12}(\phi) = \cos^2\frac{\phi}{2}. \tag{10.77}$$

Again, this is because the two-photon amplitude could have come from either crystal, and neither detector has any way to distinguish which of the photons it sees.

But now suppose an opaque obstacle (the red line labeled B in the Figure) is placed in the path e. Now that e is blocked, any photon detected at D_2 has to have come from C_2. If a photon is detected at D_1 without an accompanying photon at D_2, then (ignoring the possibility of loss) it must have come from C_1. Thus, we now know which crystal all photons arose from. This path information destroys all interference: P_1 and P_{12} are both now constants, independent of ϕ.

This should be a bit disconcerting (or, in the words of Greenberger, Horne, and Zeilinger, "mind-boggling" [33]): whether or not interference occurs at D_1 depends on the presence or absence of distinguishability of photons e and k, even though neither photon is detected at D_1 or is in the paths joining the pump to D_1. Further, it is not necessary even to monitor D_2 for the blockage at B to destroy the interference at D_1: simply having the *potential* for gaining which-way information is enough to destroy the interference. The experimenter does not actually have to take the trouble to obtain the information (turning on the detector) for its existence to have an effect on interference at D_1. The source of the interference here is sometimes called **induced coherence**, since the coherence between the output of the two crystals is deliberately created by spatially aligning those outputs to make them indistinguishable. For detailed analyses of induced coherence, see [34, 35]

A further twist on this experiment was the realization that it could be used for imaging an object using photons *that never actually interacted with the object* [36, 37]. (For a similar effect using a different mechanism, see Sect. 11.5). To see this, suppose the opaque object B in Fig. 10.17 is replaced by a partially reflecting beam splitter (BS_2), as in Fig. 10.18a. With BS_2 in place, the visibility of the interference patterns decreases as the reflectivity of the beam splitter increases, since the two possibilities become increasingly distinguishable. According to standard understanding of quantum mechanics, the photons reflecting from BS_1 do not need

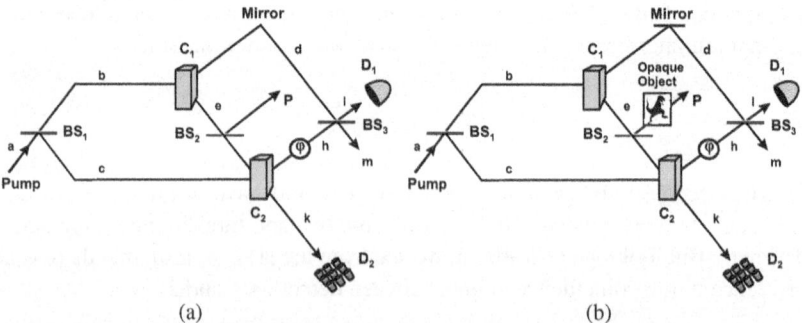

(a)   (b)

Fig. 10.18 (a) The opaque barrier in path e of the previous figure is replaced by a beam splitter, BS_2. The interference visibilities are reduced, because every photon reflected at the beam splitter increases the degree of distinguishability between the amplitudes. (b) Place an object between BS_2 and point P. Points where the object is opaque absorb the photons reflected at the beam splitter. So the visibility of the interference pattern between D_1 and the array of detectors at D_2 varies from point to point in D_2, dependence on the opacity of the corresponding object point. So the image can be reproduced from the spatial dependence of the coincidence rate

to actually be detected in order for the interference to be affected; it suffices that they *could* be detected if someone chose to be there looking for them. For further use, we now replace the single pixel detector D_2 with a CCD camera or an array of single-pixel detectors to allow spatial resolution.

Assume now that BS_2 is a 50/50 beam splitter. A photon emerging at point P means with 100% certainty that the down conversion took place in crystal C_1. If no photon emerges, it could be because it was transmitted (rather than reflected) by BS_2, or it could be because the down conversion took place in C_2 instead. Because there is on average nonzero but imperfect distinguishability between the two possibilities (down conversion in C_1 or in C_2), there will be an interference pattern in the coincidence rate P_{12}, but it will be of reduced visibility. It is important to note, however, that the visibility will be the same for every pixel in D_2.

But now suppose that a small opaque object the size of a single pixel is placed between BS_1 and point P. Then there should be 100% visibility in the interference pattern at the corresponding pixel location in the detectors, since the object blocks the photons from emerging and prevents anyone from using them to determine which crystal the coincidence photons originated in. Therefore, the coincidence rate at this pixel location should increase. Similarly, with a larger object (Fig. 10.18a), the pixels corresponding to opaque object points will have larger interference visibilities than those corresponding to transparent points. Therefore, plotting the coincidence rate interference visibilities versus location in the spatially resolving detector D_2 will provide an image of the object. It will be a negative image (high visibility at opaque points, low visibility at transparent ones).

This goes one step further than traditional ghost imaging (Sect. 20.1.1), in the sense that the image is formed from correlations between two photons detected at D_1 and D_2, *neither* of which ever interacted with the object. So, the method could be referred to as hyperghost imaging or interaction-free ghost imaging. Unlike the case in the usual ghost imaging process setup, it seems to be clear that in this case true entanglement is needed; classical correlations between the photons does not seem to suffice, since the decision to place an object at point P could be taken long after the detected photons have gone on their way to D_1 and D_2.

Problems

1. Suppose that a monochromatic field hits a Young apparatus with slits at r_1 and r_2. The output of the slits is recombined on at point r on a screen to produce a total electric field whose amplitude is of the form

$$E(r, t_1, t_2) = K \left[e^{i\phi_1} E(r_1, t_1) + e^{i\phi_2} E(r_2, t_2) \right]. \tag{10.78}$$

$\phi_1(r_1, t)$ and $\phi_2(r_1, t)$ are the phases gained by the field along the two paths, and t_1, t_2 are the times to reach the observation point on the screen from each of the slits. Further, let ϕ be the phase of the first-order spatial correlation function,

$$g^{(1)}(r_1, r_2) = \frac{\langle E(r_1)E(r_2, t) \rangle}{I_1 I_2} = \left| g^{(1)}(r_1, r_2) \right| e^{i\phi}. \tag{10.79}$$

(a) Show that the total intensity at the screen is given by

$$I = I_1 + I_2 + 2\sqrt{I_1 I_2} \left| g^{(1)}(r_1, r_2) \right| \cos(\phi - \phi_1 + \phi_2).$$

(b) Show that the fringe visibility is given by

$$\mathcal{V} = \frac{2\sqrt{I_1 I_2} \left| g^{(1)} \right|}{I_1 + I_2} \tag{10.80}$$

2. Consider the Young two-slit apparatus illuminated by a quantized state of light. Ignoring polarization, the positive frequency part of the quantized field arriving at the viewing screen after the two slits can be written in the form

$$E^{(+)}(r, t) = f(r)\left(\hat{a}_1 e^{iks_1} + \hat{a}_2 e^{iks_2}\right), \tag{10.81}$$

where $\hat{a}_{1,2}$ are annihilation operators for modes passing through each slit, $s_{1,2}$ the distances from the slits to the viewing screen, and the function $f(r)$ accounts for the spreading of the spherical waves as they move away from the slits. $\hat{\rho}$ is the density operator of the light entering the slits and r is the distance to the viewing screen. The intensity at the screen can be written as $I(r, t) = Tr[\hat{\rho}\hat{E}^{(-)}(r, t)E^{(+)}(r, t)]$.

(a) For a single photon with equal probability of passing through either slit, the input state of the light is $\frac{1}{\sqrt{2}}[|1, 0\rangle + |0, 1\rangle]$. Show that the intensity on the viewing screen is

$$I(r, t) = 2|f(r)|^2 \left(1 + \cos\left(k(s_1 - s_2) + \phi\right)\right), \tag{10.82}$$

where ϕ is the phase of $Tr(\hat{\rho}\hat{a}_1^\dagger \hat{a}_2)$, and the notation $|m, n\rangle$ means the state with m photons passing through the first slit and n through the second.

(b) What is the intensity for the n-photon input state $\frac{1}{\sqrt{2}}[|n, 0\rangle + |0, n\rangle]$, for large n? (For coherent state input, $|\alpha\rangle$, the intensity is of the form $|\alpha|^2 |f(r)|^2 \left(1 + \cos\left(k(s_1 - s_2) + \phi\right)\right)$; see [38, 39].)

3. For the setup of Fig. 10.18a, find the detection probability at detector D_1 as a function of the transmission amplitude t of the beam splitter BS_2.

4. Consider a classical monochromatic wave of oscillating intensity: $I(t) = I_0\big[1 + K \sin(\omega t + \phi)\big]$, for constants K and ϕ.

   (a) Find an analytic expression for $g^{(2)}(0)$.
   (b) Plot $g^{(2)}(\tau)$ versus τ. When computing the time averages, pick arbitrary values of K and ω and some arbitrary averaging time T. Compare the behavior for $\tau \ll T$ and $\tau \gg T$; what does this indicate about the coherence length?

5. (a) Consider a single-photon source, emitting individual photons at a regular sequence of times, $t = nT, n = 0, 1, 2, \ldots$. Find the first- and second-order correlation functions at zero delay, $g^{(1)}(0)$ and $g^{(2)}(0)$.
   (b) Now consider a two-photon source, emitting individual photon *pairs* at a regular sequence of times, $t = nT, n = 0, 1, 2, \ldots$. Again find the first- and second-order correlation functions at zero delay, $g^{(1)}(0)$ and $g^{(2)}(0)$.

6. Consider two photons entering ports a and b of the HOM interferometer. Explicitly calculate the output state, and calculate the first and order correlation functions at zero delay, $g^{(1)}(0)$ and $g^{(2)}(0)$.

7. (a) Send a single monochromatic photon into the Franson interferometer. Writing the input state as $|\psi\rangle = \hat{a}^\dagger|0\rangle$, find the output state (just before the detectors) for phase difference ϕ between the LL and SS paths.

   (b) Find the coincidence probability and verify Eq. 10.69.

8. Imagine a two-beam experiment with two distinguishable photon modes of intensities I_1 and I_2, and annihilation operators $\hat{a}_1^\dagger$ and $\hat{a}_2^\dagger$. In addition to the self-correlations $g_j^{(2)}(\tau) = \frac{\langle I_1(t) I_1(t+\tau)\rangle}{\langle I_1(t)\rangle^2}$ for $j = 1, 2$, also define the cross-correlation

$$g_{12}^{(2)}(\tau) = \frac{\langle I_1(t) I_2(t + \tau)\rangle}{\langle I_1(t)\rangle\langle I_2(t)\rangle}.$$

   (a) Use the Cauchy–Schwarz inequality to show that classically these must obey

$$\left[g_{12}^{(2)}(\tau)\right]^2 \le g_1^{(2)}(0) g_2^{(2)}(0).$$

   (b) Show that if the particles are bosons, then in the quantum case the operators must obey $\langle(\hat{a}^\dagger)^2\hat{a}^2\rangle + \langle\hat{a}^\dagger\hat{a}\rangle - \langle\hat{a}^\dagger\hat{a}\rangle^2 \ge 0$, with a similar relation for the $\hat{b}$ operators.
   (c) For the symmetric case, $\langle(\hat{a}^\dagger\hat{a})^2\rangle = \langle(\hat{b}^\dagger\hat{b})^2\rangle$ and $\langle\hat{a}^\dagger\hat{a}\rangle = \langle\hat{b}^\dagger\hat{b}\rangle$, show that the quantum correlations obey $g_{12}^{(2)}(0) \le g_j^{(2)}(0) + \frac{1}{\bar{n}}$, for $j = 1, 2$. For $\tau = 0$, this is weaker than the classical inequality of part (a). (See [40] for a detailed discussion of classical inequality violations in quantum optics.)

References

1. A.D. Cronin, J. Schmeidmayer, D.E. Pritchard, Rev. Mod. Phys. **81**, 1051 (2009)
2. P. Meystre, *Atom Optics* (Springer, Berlin, 2001)
3. V.F. Sears, *Neutron Optics* (Oxford University Press, Oxford, 1998)
4. H. Rauch, S.A. Werner, *Neutron Interferometry: Lessons in Experimental Quantum Mechanics, Wave-Particle Duality, and Entanglement* (Oxford University Press, Oxford, 2015)
5. H. Lee, P. Kok, J.P. Dowling, J. Mod. Opt. **49**, 2325 (2010)
6. Y. Nieves, A. Muller, Phys. Rev. B **98**, 165432 (2018)
7. D. Elvira et al., Phys. Rev. A **84**, 061802 (2011)
8. M.J. Stevens, S. Glancy, S.W. Nam, R.P. Mirin, Opt. Express **22**, 3244 (2014)
9. A. Rundquist et al., Phys. Rev. A **90**, 023846 (2014)
10. A.S. Lane, S.L. Braunstein, C.M. Caves, Phys. Rev. A **47**, 1667 (1993)
11. Z. Hradil, Quant. Opt. **4**, 93 (1992)
12. T. Wasak, P. Szańkowski, P. Ziń, M. Trippenbach, J. Chwedeńczuk, Phys. Rev. A **90**, 033616 (2014)
13. P. Lasségues et al., Eur. Phys. J. D **76**, 246 (2022)
14. D. Ferreira, R. Bachelard, W. Guerin, R. Kaiser, M. Fouché, Am. J. Phys. **88**, 831 (2020)
15. R. Hanbury Brown, R.Q. Twiss, Nature **127**, 77 (1956)
16. R. Hanbury Brown, R.Q. Twiss, Proc. Roy. Soc. A **242**, 300 (1957)
17. D. Gottesman, T.S. Jennewein, S. Croke, Phys. Rev. Lett. **109**, 070503 (2012)
18. P. Stankus, A. Nomerotski, A. Slosar, S. Vintskevich, *Instrumentation and Methods for Astrophysics*, vol. 5 (2022). https://astro.theoj.org/section/1192-instrumentation-and-methods-for-astrophysics
19. G. Jaeger, M. Horne, A. Shimony, Phys. Rev. A **48**, 1023 (1993)
20. G. Jaeger, A. Shimony, L. Vaidman, Phys. Rev. A **51**, 54 (1995)
21. G. Jaeger, *Quantum Information: An Overview* (Springer, Berlin, 2007)
22. C.K. Hong, Z.Y. Ou, L. Mandel, Phys. Rev. Lett. **59**, 2044 (1987)
23. F. Bouchard et al., Rep. Prog. Phys. **84**, 012402 (2021)
24. A. Lyons, G.C. Knee, E. Bolduc, T. Roger, J. Leach, E.M. Gauger, D Faccio Sci. Adv. **4**, eaap9416 (2018)
25. Y. Chen, M. Fink, F. Steinlechner, J.P. Torres, R Ursin, Quantum Inf. **5**, 1 (2019)
26. H. Scott, D. Branford, N. Westerberg, J. Leach, E. Gauger, Phys. Rev. A **102**, 033714 (2020)
27. T.B. Pittman, D.V. Strekalov, M.H. Migdall, M.H. Rubin, A.V. Sergienko, Y.H. Shih, Phys. Rev. Lett. **77**, 1917 (1996)
28. J.D. Franson, Phys. Rev. Lett. **62**, 7205 (1989)
29. Z.Y. Ou, X.Y. Zou, L.J. Wang, L. Mandel, Phys. Rev. Lett. **65**, 321 (1990)
30. J. Brendel, E. Mohler, W. Martienssen, Euro. Phys. Lett. **20**, 575 (1992)
31. Y.H. Shih, A.V. Sergienko, M.H. Rubin, Phys. Rev. A **47**, 1288 (1993)
32. X.Y. Zou, L.J. Wang, L. Mandel, Phys. Rev. Lett. **67**, 318 (1991)
33. D.M. Greenberger, M.A. Horne, A. Zeilinger, Phys. Today **46** (8), 22 (1993)
34. A. Heuer, R. Menzel, P.W. Milonni, Phys. Rev. Lett. **114**, 053601 (2015)
35. A. Heuer, R. Menzel, P.W. Milonni, Phys. Rev. A **92**, 033834 (2015)
36. G.B. Lemos, V. Borish, G.D. Cole, S. Ramelow, R. Lapkiewicz, A. Zeilinger, Nature **512**, 409 (2014)
37. M. Lahiri, R. Lapkiewicz, G.B. Lemos, A. Zeilinger, Phys. Rev. A **92**, 013832 (2015)
38. C. Gerry, P. Knight, *Introductory Quantum Optics* (Cambridge University Press, Cambridge, 2004)
39. D.F. Walls, G.J. Milburn, *Quantum Optics*, 2nd edn. (Springer, Berlin, 2008)
40. M.D. Reid, D.F. Walls, Phys. Rev. A **34**, 1260 (1986)

Chapter 11
More Interference and Measurement Effects

In this chapter, additional measurement effects are examined that are of an inherently quantum nature. These include weak measurement, quantum nondemolition experiments, delayed choice experiments, the quantum Zeno effect, and interaction-free measurement. These effects all depend in some manner on the existence of superposition and interference effects.

11.1 Strong and Weak Measurements

In Chap. 3, the mathematical formalism for ideal projective measurements and for POVM's was discussed. But in the lab, the final step in a measurement is to couple the quantum system to some classical output device such as a mechanical dial or an electronic digital readout. The idea is that the classical readout device (which we will simply call the meter from now on) should be brought into contact with the system to be measured and allowed to interact with it. The interaction leads to a correlation between the system and the meter, so that quantum observables can be inferred from the state of the meter.

Consider a quantum system S coupled to a classical meter M (Fig. 11.1). The state of the meter is parameterized by some **pointer variable**, q. Examples of pointer variables might be the angle of the pointer on an analog ammeter dial, the size of a photocurrent, or the magnitude of deflection of a charged particle by a magnetic field. The composite system consisting of the classical meter and the quantum system to be measured will have some Hamiltonian of the form

$$\hat{H} = \hat{H}_S + \hat{H}_M + \hat{H}_{int}, \qquad (11.1)$$

where $\hat{H}_S$, $\hat{H}_M$, and $\hat{H}_{int}$ are, respectively, the Hamiltonian of the quantum system, the Hamiltonian of the meter, and the interaction Hamiltonian that describes how

© The Author(s), under exclusive license to Springer Nature Switzerland AG 2025
D. S. Simon, *Introduction to Quantum Science and Technology*, Undergraduate
Texts in Physics, https://doi.org/10.1007/978-3-031-81315-3_11

Fig. 11.1 A quantum system
coupled to a classical
measuring device (the meter).
The pointer variable q of the
meter is arranged to be
correlated with some
quantum observable S of the
system

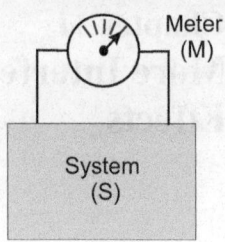

M and S affect each other. Let $\hat{\mathcal{A}}$ be the quantum observable to be measured, with
eigenvalues λ_i and eigenvectors $|\lambda_i\rangle$. Then any pure state of S can be written in the
form

$$|\psi\rangle = \sum_i c_i |\lambda_i\rangle. \tag{11.2}$$

Normally the goal of a measurement is to determine which of the eigenvalues
λ_i occurs when the meter interacts with the system. This is called a **strong
measurement** (Fig. 11.2a). However, a strong measurement of $\hat{A}$ can lead to
quantum back-reaction (Sect. 11.2) that disturbs the measured variable, altering
its value. A different approach [1] is to make **weak measurements** (Fig. 11.2b);
here, the measurement is designed so that the meter only interacts very briefly and
weakly with the system. As a result, the measurement has a high uncertainty and
obtains very minimal information about the state. But conversely, it also creates
minimal disturbance to the system state. A single weak measurement obtains
too little information to pin down the exact eigenvalue. But if sufficiently many
weak measurements are made, then the eigenvalue can be determined to as high a
statistical likelihood as desired (Fig. 11.3).

Let $\hat{q}$ and $\hat{p}$ be any conjugate pair of observables for M, with $\hat{q}$ being the pointer
variable. For example, they could be position and momentum, or angle and angular
momentum. Then, taking the simplest case of a linear interaction, the interaction
Hamiltonian is assumed to be of the form

$$\hat{H}_{int} = \mu g(t) \hat{A} \hat{q}, \tag{11.3}$$

where μ is some coupling constant. The function $g(t)$ is included so that the
interaction between the meter and the system can be turned on only for a short
time T and then turned off again. Consequently, we take $g(t)$ to be nonzero only for
times $0 \leq t \leq T$. The shorter the interaction time or the smaller the coupling μ, the
less the system will be disturbed; however, less information will be gained from the
measurement. We take the function to be normalized, so that $\int_0^T g(t)\, dt = 1$. The
effect of changing the interaction time (i.e. of changing the normalization of g) can
always be absorbed into the value of μ. In the following, the roles of $\hat{q}$ and $\hat{p}$ can
also be interchanged if desired, with similar results.

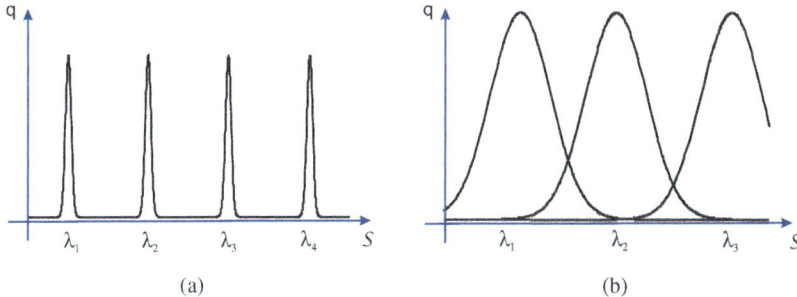

(a)   (b)

Fig. 11.2 (**a**) Strong measurement allows S to be measured with a high degree of certainty: since the distribution of q is tightly localized about the eigenvalues of S, a measurement that falls inside one of the peaks clearly indicates that S equals the corresponding eigenvalue. However, because of the low measurement uncertainty, the system will be strongly perturbed by the measurement. (**b**) In a weak measurement, the uncertainties around each eigenvalue are larger, so that the peaks now overlap. Because of the overlap, a single measurement will have a high uncertainty as to which eigenvalue is present. But, for this high-uncertainty case, the disturbance to the system per measurement will be smaller, allowing multiple measurements to be made with a reasonable probability that the eigenvalue won't change during the sequence of measurements

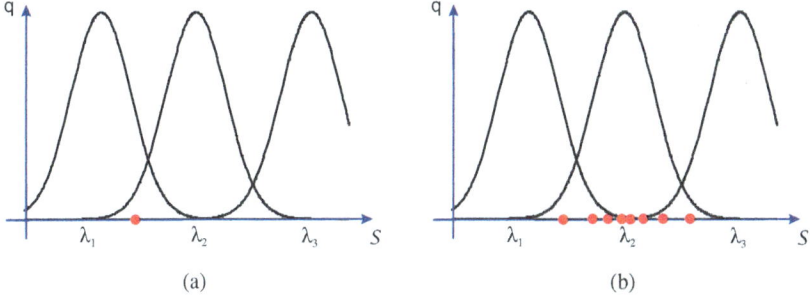

(a)   (b)

Fig. 11.3 (**a**) If a single weak measurement is made, little information is obtained. For example, if the value represented by the red dot is obtained, then we can be fairly sure that $S \neq \lambda_3$, but cannot determine whether the value is λ_1 or λ_2. (**b**) However, if many weak measurements are made, then the distribution of values allows us to determine more information about S: in the case shown, the data represented by the distribution of red dots make it almost certain that $S = \lambda_2$

After the end of the meter–system interaction, at time $t = T$, the additional time evolution introduced by the interaction is given by the unitary operator

$$\hat{U}(T) = e^{-\frac{i}{\hbar} \int H_{int} dt} = e^{-\frac{i}{\hbar} \mu \hat{A} \hat{q} \int g(t) dt} = e^{-\frac{i}{\hbar} \mu \hat{A} \hat{q}}. \tag{11.4}$$

Let $|\psi_i\rangle$ be the initial state of S. Suppose that the initial state of the meter is $|\Phi(p)\rangle$ in the p-representation, so that the full state of the meter and system is

$$|\Psi_i\rangle = |\psi_i\rangle |\Phi(p)\rangle = \sum c_i |\lambda_i\rangle |\Phi(p)\rangle. \tag{11.5}$$

Then the effect of $\hat{U}(T)$ is to take $|\Psi_i\rangle$ to the new state

$$\hat{U}(T)|\Psi_i\rangle = \sum c_i e^{-\frac{i}{\hbar}\mu \hat{A}\hat{q}}|\lambda_i\rangle|\Phi(p)\rangle \tag{11.6}$$

$$= \sum c_i e^{-\frac{i}{\hbar}\mu \lambda_i \hat{q}}|\lambda_i\rangle|\Phi(p)\rangle \tag{11.7}$$

$$= \sum c_i |\lambda_i\rangle|\Phi(p+\mu\lambda_i)\rangle, \tag{11.8}$$

where in the last line, the fact has been used (see Sect. 3.2.3) that if $\hat{q}$ and $\hat{p}$ are conjugate operators, then the q-dependent exponential acts as a translation operator on functions of p: $e^{-\frac{i}{\hbar}bq} f(p) = f(p+b)$. So, the meter state is displaced by an amount proportional to the $\hat{A}$ value, allowing eigenvalue λ_i to be determined from the state of the meter.

A weak measurement then consists of making sure that when the exponential is expanded in powers of μ, the terms higher than linear can be treated as negligible. Taking the product with the final system state $\langle\psi_f|$, we see then that

$$\langle\psi_f|\hat{U}(T)|\Psi_i\rangle = \langle\psi_f|\left(1 - \frac{i}{\hbar}\mu\hat{A}\hat{q}\right)|\psi_i\rangle|\Phi(p)\rangle \tag{11.9}$$

$$= \langle\psi_f|\psi_i\rangle\left(1 - \frac{i}{\hbar}\mu q \frac{\langle\psi_f|\hat{A}|\psi_i\rangle}{\langle\psi_f|\psi_i\rangle}\right)|\Phi(p)\rangle \tag{11.10}$$

$$\equiv \langle\psi_f|\psi_i\rangle\left(1 - \frac{i}{\hbar}\mu q A_w\right)|\Phi(p)\rangle \tag{11.11}$$

$$= \langle\psi_f|\psi_i\rangle e^{-\frac{i}{\hbar}\mu q A_w}|\Phi(p)\rangle \tag{11.12}$$

$$= \langle\psi_f|\psi_i\rangle|\Phi(p+\mu q A_w)\rangle, \tag{11.13}$$

where the **weak value** of $\hat{A}$ is defined to be

$$A_w = \frac{\langle\psi_f|\hat{A}|\psi_i\rangle}{\langle\psi_f|\psi_i\rangle}. \tag{11.14}$$

For small $\langle\psi_f|\psi_i\rangle$, the measurement now results in obtaining the weak value, rather than one of the eigenvalues. Notice that if the overlap between the initial and final S states is small, $\langle\psi_f|\psi_i\rangle << 1$, then the weak value can be large; in fact, the odd situation can occur in which the measured weak value is much larger than any of the eigenvalues and is therefore much larger than any of the possible values that could be obtained in a strong measurement.

The uncertainty in p will be large for the weak measurement, $\Delta p >> A_w$, leading to small disturbance in q. But if such weak measurements are made on an ensemble of N identical particles, then for sufficiently large N, the uncertainty

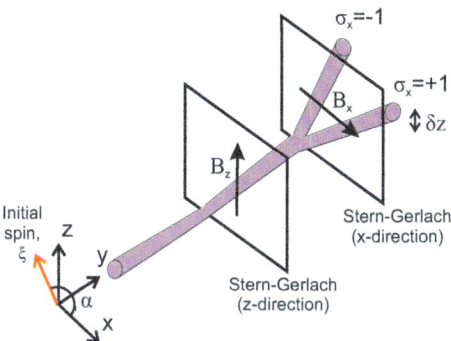

Fig. 11.4 A beam of electrons, initially with spins up along the ξ direction pass through two Stern–Gerlach setups. The first one produces a weak spatial separation of the spins pointing up and down along z, while the second produces a strong separation of spins pointing up and down along x. Measuring the exiting $|\uparrow_x\rangle$ state, there is a displacement of the beam δz that is proportional to the weak value A_w of the observable $\hat{A} = \sigma_z$

in p can be reduced to $\frac{p}{\sqrt{N}} \ll A_w$, allowing the weak value to be obtained with arbitrarily high accuracy.

Example 11.1.1 An example of weak measurement can be provided by the Stern–Gerlach experiment [1]. The interaction Hamiltonian for the particle spin to interact with the external magnetic field is $\hat{H}_{int}(t) = -\mu \boldsymbol{S} \cdot \boldsymbol{B} g(t) = -\frac{\hbar\mu}{2}\boldsymbol{\sigma} \cdot \boldsymbol{B} g(t)$. Assuming that $\boldsymbol{B}$ is along the z-axis, and Taylor expanding $\boldsymbol{B}$ to first order in z,

$$H_{int}(t) = -\frac{\hbar\mu}{2}\sigma_z(B_0 + B_1 z)g(t) = -\hbar\mu'\sigma z(1 + \gamma z)g(t), \qquad (11.15)$$

where $B_0 = B(0)$ and $B_1 = \left.\frac{\partial B(z)}{\partial z}\right|_{z=0}$ are the first two Taylor coefficients, $\mu' = \frac{1}{2}\mu B_0$, and $\gamma = \frac{1}{2}\frac{B_1}{B_0}$. Here the observable to be measured is the particle spin, $\hat{A} = \hat{\sigma}_z$, and the pointer variable is $q = z$.

Let a beam of charged fermions moving along the y axis be sent into a Stern–Gerlach apparatus (Fig. 11.4). Consider two consecutive Stern–Gerlach measurements: a weak measurement of spin along the z-axis, followed by a strong measurement of spin along the x-axis. The weakness of the first measurement is guaranteed by taking the gradient of the field along the z direction small; the two resulting deflected beams are therefore too close together to be distinguished. The second measurement causes the two outgoing beams to be deflected far enough in the x-direction to be discerned unambiguously.

Take the incoming state to be spin up along the ξ-axis, which is at an angle of α from the x-axis, as shown. A spinor at α from x can be written (see Sect. 6.2) as

$$| \uparrow_\xi \rangle = \cos \frac{\alpha}{2} | \uparrow_x \rangle + \cos \frac{\alpha}{2} | \downarrow_x \rangle = \frac{1}{\sqrt{2}} \left(\begin{array}{c} \cos \frac{\alpha}{2} + \sin \frac{\alpha}{2} \\ \cos \frac{\alpha}{2} - \sin \frac{\alpha}{2} \end{array} \right), \qquad (11.16)$$

where in the second equality we have used the fact that in the z-basis the up and down states along the x axis are

$$| \uparrow_x \rangle = \left(\begin{array}{c} 1 \\ 1 \end{array} \right), \quad | \downarrow_x \rangle = \left(\begin{array}{c} 1 \\ -1 \end{array} \right). \qquad (11.17)$$

The weak value of $\hat{\sigma}_z$ is then given by

$$A_w = \frac{\langle \uparrow_x | \hat{\sigma}_z | \uparrow_\xi \rangle}{\langle \uparrow_x | \uparrow_\xi \rangle} = \tan \frac{\alpha}{2}. \qquad (11.18)$$

Given that $\hat{\sigma}_z$ has eigenvalues ± 1, we would normally expect the mean value $\langle \sigma_z \rangle$ to satisfy $-1 \leq \langle \sigma_z \rangle \leq 1$. However, by taking α sufficiently close to π, the tangent in Eq. 11.18 allows for potentially large values, $A_w \gg 1$. (This is the reason for the title of Reference [1], *How the Result of a Measurement of a Component of the Spin of a Spin-1/2 Particle Can Turn Out to be 100*). Suppose that only the final beam with $\sigma_x = +1$ is measured. It can be shown [1] that the displacement of the final detected beam in the z direction, δz, is then proportional to A_w. The displacement δz can therefore be much larger than the original uncertainty Δz, allowing a measurement of A_w from the z deflection.   ∎

11.2 Quantum Nondemolition Measurements

Suppose that we wish to monitor the position of some particle over time. At the first measurement (at time $t = 0$), the position $x(0)$ will have some uncertainty $\Delta x(0)$. The question is, after repeated measurements, how will the uncertainty $\Delta x(t)$ evolve? If any measurements of momentum were made, we know that the uncertainty in x would grow after each one; but here we only make measurements of x, not p, so we might expect that the x uncertainty should be constant, assuming an identical measurement procedure is used each time.

But in fact, Δx will grow with time. This can easily be seen as follows. The first measurement will introduce a momentum uncertainty,

$$\Delta p = \frac{\hbar}{2 \Delta x(0)}, \qquad (11.19)$$

in accordance with the uncertainty principle. This induces an uncertainty in velocity,

$$\Delta v = \frac{\Delta p}{m} = \frac{\hbar}{2m \Delta x(0)}. \tag{11.20}$$

So if a second measurement of x is made at time τ, the uncertainty in x will be larger than it was initially,

$$\Delta x(\tau) = \Delta x(0) + \tau \Delta v = \Delta x(0) + \frac{\hbar \tau}{2m \Delta x(0)}. \tag{11.21}$$

Clearly, the second term is due to the added uncertainty in how far the particle has traveled during the free time evolution between measurements. The first measurement in x induced uncertainty in p, which then induced a new uncertainty in the time evolution of x; this is called **quantum back-action**.

To be more general, suppose we wish to measure some observable $\hat{A}$, and let $\hat{H}$ be the Hamiltonian describing the evolution of $\hat{A}$ over the period from $t = 0$ to $t = \tau$. Then quantum back-action will occur if $\hat{H}$ involves any observable $\hat{B}$, which fails to commute with $\hat{A}$. This is clearly the case in the previous example, where the evolution of x was determined by the momentum-dependent free particle Hamiltonian $H_0 = \frac{\hat{p}^2}{2m}$. In such a case, repeated monitoring of x over time will lead to increasing uncertainty.

If $\hat{H}$ depends on variables that are non-commuting with $\hat{A}$, then $\hat{A}$ *will not commute with itself* at different times. Since

$$\hat{A}(t) = e^{-i\hat{H}t} \hat{A}(0) e^{i\hat{H}t} \tag{11.22}$$

$$= \hat{A}(0) + \frac{it}{\hbar} \left[\hat{H}, \hat{A}(0) \right] + \frac{1}{2!} \left(\frac{it}{\hbar} \right)^2 \left[\hat{H}, \left[\hat{H}, \hat{A}(0) \right] \right] + \dots, \tag{11.23}$$

then the fact that $\hat{H}$ doesn't commute with $\hat{A}(0)$ implies that $\hat{A}(t)$ may not commute with it either. More generally,

$$\left[\hat{A}(t_i), \hat{A}(t_j) \right] \neq 0 \tag{11.24}$$

for any pair of time $t_i \neq t_j$. Given this lack of commutativity, it should be expected from the uncertainty principle that a measurement of $\hat{A}$ at an earlier time will induce uncertainties in its measurements at later times.

However, suppose that the observable $\hat{A}$ in question commutes with the Hamiltonian that governs its time evolution between measurements: $\left[\hat{A}, \hat{H} \right] = 0$. In that case, the back action does not occur and the uncertainty in $\hat{A}$ will not grow with repeated measurements. But notice that the statement that $\left[\hat{A}, \hat{H} \right] = 0$ simply says that $\hat{A}$ is a conserved variable: it doesn't evolve over time at all. Again returning to the case aforementioned of position and momentum, then if $\hat{A} = \hat{p}$ we see that the momentum can be repeatedly measured with no back reaction; in this situation, $\hat{p}$

commutes with $H_0 = \frac{\hat{p}^2}{2m}$ and is simply a constant. The conserved energy also is immune to back reaction under repeated measurements.

The process of measuring only variables that are free of back-action is called a **quantum nondemolition (QND) experiment**. Observables such as $\hat{p}$ and $\hat{H}$ are referred to as **QND observables**. The word "nondemolition" refers to the fact that the state is not destroyed in the measurement. This is in contrast, for example, to the case where back-reaction from measuring x causes a change in the position state or, in more extreme cases, where the particle itself is destroyed by the measurement. An example of the latter is when a photon is absorbed during a measurement by a photodetector.

It can be the case that an observable is not QND at arbitrary times, but is free of back-action only if measured at particular time intervals. In this case, it is referred to as a **stroboscopic variable**. An example of a stroboscopic variable can be seen in the case of a quantum oscillator. Consider an oscillator of mass m and angular frequency ω, with position observable $\hat{q}(t)$. $\hat{q}(t)$ can be written in terms of quadrature operators $\hat{X}$ and $\hat{P}$,

$$\hat{q}(t) = \hat{X}\cos\omega t + \hat{P}\sin\omega t. \tag{11.25}$$

Here, the two quadratures are proportional to the initial position and momentum at $t = 0$: $\hat{X} = \hat{q}(0)$ and $\hat{P} = \frac{\hat{p}(0)}{m\omega}$. The uncertainty principle tells us that

$$\Delta X \Delta P \geq \frac{\hbar}{2m\omega}. \tag{11.26}$$

If we assume that the two quadratures have equal uncertainty (as they would for a coherent state, for example), then they are given by the standard quantum limit,

$$\Delta X_{SQL} = \Delta P_{SQL} = \sqrt{\frac{\hbar}{2m\omega}}. \tag{11.27}$$

The amplitude of the oscillation $A = \sqrt{\Delta X^2 + \Delta P^2}$ has the same value,

$$\Delta A_{SQL} = \sqrt{\frac{\hbar}{2m\omega}}. \tag{11.28}$$

The energy, $E = \hbar\omega\left(n + \frac{1}{2}\right) = \frac{1}{2}m\omega^2 A^2 + \frac{1}{2}\hbar\omega$, then has uncertainty

$$\Delta E_{SQL} = \frac{\partial E}{\partial A}\Delta A_{SQL} = m\omega^2 A \Delta A_{SQL} = \hbar\omega\sqrt{n}. \tag{11.29}$$

Here we have made use of the fact that the amplitude is related to the excitation number n by $A^2 = \frac{2\hbar}{m\omega}n$. Most measurements on the system, without using

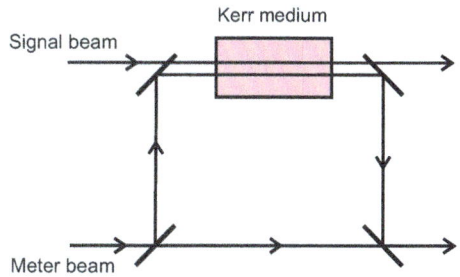

Fig. 11.5 An experiment for QND measurement of optical quadratures. The signal beam interacts with the meter beam through the nonlinear material, allowing information about the signal beam to be extracted from measurements of the meter beam

specialized input states (such as squeezed states) or other specialized techniques, will have uncertainties exceeding the standard quantum limits.

The two quadratures do not commute with themselves at arbitrary times,

$$[\hat{X}(t), \hat{X}(t + \tau)] = [\hat{P}(t), \hat{P}(t + \tau)] = \frac{i\hbar}{m\omega} \sin \omega t, \qquad (11.30)$$

so neither is a QND observable at arbitrary times. But the commutators both vanish when $\tau = \frac{n\pi}{\omega}$ for any integer n. So the variables are both stroboscopic. Either quadrature (but not both, since they don't commute with each other) can be measured repeatedly with accuracy exceeding the standard quantum limit, without back-action, as long as the measurements are made at moments separated by intervals $\tau = \frac{n\pi}{\omega}$.

Example 11.2.1 An example of a QND experiment in optics is shown in Fig. 11.5 [5], consisting of four beam splitters arranged in a Mach–Zehnder-like arrangement, with a nonlinear cross-Kerr medium (Chap. 5) inserted in one arm. The goal is to measure one of the quadratures of a light beam, called the signal beam. The signal beam is inserted into the top branch, as shown; it continues through the Kerr medium and out the other end of the system. Meanwhile, a second beam (the meter beam), which serves as a reference to measure phase against, is inserted into the bottom branch. The meter beam splits, with part of it continuing straight and exiting the other side. The other part enters the Kerr medium, where it affects the signal beam through the nonlinear interaction. The complex amplitudes of the beams can be split into an amplitude quadrature $\hat{X}$ and a phase quadrature $\hat{Y}$. Then the uncertainties of the outgoing signal and meter quadratures are related to those of the incoming quadratures [5] according to

$$\Delta X_{out}^{(s)} = \Delta X_{in}^{(s)} \qquad\qquad \Delta Y_{out}^{(s)} = \Delta Y_{in}^{(s)} - \kappa \Delta X_{in}^{(m)} \qquad (11.31)$$

$$\Delta X_{out}^{(m)} = \Delta X_{in}^{(m)} \qquad\qquad \Delta Y_{out}^{(m)} = \Delta Y_{in}^{(m)} - \kappa \Delta X_{in}^{(s)} \qquad (11.32)$$

where the coupling and other constants are swept into the single constant κ. It is seen that the amplitude quadrature uncertainty of the signal beam is not changed, but that it affects the phase quadrature of the meter beam. So the phase quadrature

$Y^{(m)}$ can be measured to determine the $X^{(s)}$ quadrature with no effect on $X^{(s)}$ itself.
∎

The main problem with QND measurements is that in general only a small amount of information about the variable of interest is obtained with each measurement, and the measurements are often difficult to make; for instance, in the example aforementioned, the Kerr nonlinearity is generally very small so that the measured effect is also small. Detailed reviews of QND measurements may be found in [2–5].

Generalizing the previous example, we define system and probe variables, $\hat{X}_s$ and $\hat{X}_p$, (generalizations of $X^{(s)}$ and $Y^{(m)}$, respectively), and write the Hamiltonian as $\hat{H} = \hat{H}_s + \hat{H}_p + \hat{H}_I$, where the first two terms are the Hamiltonians of the system and probe variables separately, and $\hat{H}_I$ is the interaction between them. Also define the conjugate momenta $\hat{P}_s$ and $\hat{P}_p$. Then a QND process to measure the system via the probe will exist if the following conditions hold [6]:

(i) $\frac{\partial \hat{H}_I}{\partial \hat{X}_s} \neq 0$, so that $\partial \hat{H}_I$ depends nontrivially on $\hat{X}_s$.

(ii) The interaction commutes with the system but not the probe, $\left[\hat{X}_s, \hat{H}_I\right] = 0$ but $\left[\hat{X}_p, \hat{H}_I\right] \neq 0$.

(iii) $\frac{\partial \hat{H}_s}{\partial \hat{P}_s} = 0$, which is needed to prevent back-action on the system.

Box 11.1 Gravitational Redshifts

A simple thought experiment involving a QND measurement can be used to derive the gravitational redshift in general relativity. Following [2], consider an optical cavity in a uniform gravitational field. The cavity contains an optical field of resonant frequency ω. Suppose we try to determine the energy of the field by placing the cavity on a scale and weighing it. Let F_{EM} be the weight of the cavity (after subtracting off the materials making up the cavity itself). If the electromagnetic field in the cavity has energy E, then the effective mass of the field is found from $E = mc^2$. So,

$$F_{EM} = mg = \frac{Eg}{c^2}. \tag{11.33}$$

When the cavity is placed on the tray of the scale, the scale's momentum changes in time τ by

$$\delta p = F_{EM}\tau = \frac{g\tau}{c^2}E, \tag{11.34}$$

so the energy uncertainty in terms of the momentum uncertainty is

$$\Delta E = \frac{c^2}{g\tau}\Delta p. \tag{11.35}$$

(continued)

Box 11.1 (continued)

If there are n photons in the cavity, the energy of the field is $E = (n + \frac{1}{2})\hbar\omega$, and the number-phase uncertainty relation $\Delta n \Delta\phi \geq \frac{1}{2}$ (Sect. 9.1) leads to the inequality

$$\Delta E \Delta\phi \geq \frac{\hbar\omega}{2}. \tag{11.36}$$

If the uncertainty of the scale tray's height during the measurement is Δy, then

$$\Delta\phi = \frac{\hbar\omega}{2\Delta E} = \frac{\hbar}{2\Delta p}\frac{g\omega\tau}{c^2} \approx \frac{g\omega\tau}{c^2}\Delta y, \tag{11.37}$$

assuming that the uncertainty is approximately at its Heisenberg minimum value, $\Delta y \Delta p \approx \frac{\hbar}{2}$.

The phase uncertainty can be attributed to the uncertainty in resonant frequency during the measurement period, which is of duration τ, so that

$$\Delta\omega = \frac{\Delta\phi}{\tau} = \frac{g\omega}{c^2}\Delta y. \tag{11.38}$$

Assuming the uncertainties are all sufficiently small, this can be written in differential form,

$$\frac{d\omega}{\omega} = \frac{g}{c^2}dy. \tag{11.39}$$

Now integrate from $y = 0$ to final height y to find the height dependence of the frequency,

$$\ln\left(\frac{\omega(y)}{\omega_0}\right) = \frac{g}{c^2}y, \tag{11.40}$$

where the integration constant ω_0 is the frequency at $y = 0$. Solving for $\omega(y)$ and then expanding the exponential for small y (keeping only the linear terms), the end result is Einstein's formula for the gravitational redshift in a uniform gravitational field:

$$\omega(y) = \omega_0 e^{\frac{g}{c^2}y} \approx \omega_0\left(1 + \frac{g}{c^2}y\right). \tag{11.41}$$

11.3   Quantum Erasers and Delayed Choice

Interference effects such as the Young two-slit experiment or HOM interference rely on the inability to determine which path each photon took to the interference point. The more information available about which particle took which path, the more the interference pattern is degraded. If the paths are determined, there is no interference, but if the which-way information is then somehow erased, the interference pattern re-emerges. Even more astounding is the fact that this is true even when the decision of whether or not to erase the path information takes place *after* the interference has ostensibly occurred (for example, after the beam splitter in the HOM experiment). Such delayed-choice quantum eraser experiments were were proposed in the 1980s [7] and carried out in the early 1990s. Several such experiments have been conducted; we discuss just one version [8] here.

Consider the HOM effect (Sect. 10.6). Two photons are sent into two different ports of a 50–50 beam splitter. Coincidences can occur two ways, with the paths of the two photons interchanged. As long as there is no way to know which of the two sets of paths occurs, destructive interference causes a sharp dip in the coincidence rate.

Now consider a variation on the HOM experiment. Suppose the two photons initially have the same polarization. This is the case if the photons are produced by Type I SPDC in a nonlinear crystal. To make the discussion simpler, let's assume a particular polarization: take both photons to be horizontally polarized. Label the two paths the photons can take to reach the beam splitter by u and l, for *upper* and *lower*. So the initial state is

$$|\psi_0\rangle = |H\rangle_u |H\rangle_l. \tag{11.42}$$

Now, as in Fig. 11.6a, place a polarization rotator in the lower path. This device rotates the polarization of the bottom photon by angle θ, so the state after the rotator is

$$|\psi(\theta)\rangle = |H\rangle_u \Big(|H\rangle_l \cos\theta + |V\rangle_l \sin\theta\Big). \tag{11.43}$$

After this state encounters the beam splitter, the beam splitter output state is

$$|\psi'(\theta)\rangle = \frac{i}{2}\cos\theta\Big(|2H\rangle_1 + |2H\rangle_2\Big) \tag{11.44}$$

$$+ \frac{1}{2}\sin\theta\Big(|H\rangle_1|V\rangle_2 - |V\rangle_1|H\rangle_2 + i\big(|H\rangle_1|V\rangle_1 + |V\rangle_2|H\rangle_2\big)\Big),$$

where $|2H\rangle_j$ means the state with two photons in arm j. Here, we have used a beam splitter matrix of the form

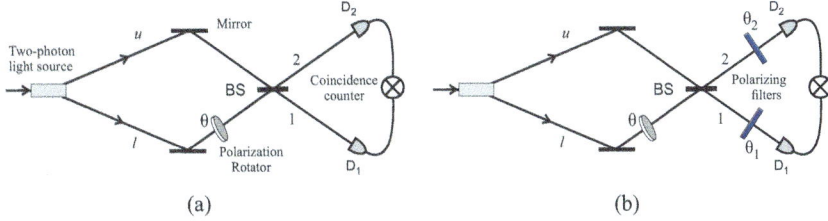

Fig. 11.6 (a) Two photons, initially of the same polarization enter an HOM interferometer. The polarization of one photon is rotated by $\frac{\pi}{2}$ before reaching the beam splitter. All detection events are coincidence events. The polarization rotation makes it possible to determine which photon took which path by measuring the polarization of each detected photon, so no two-photon interference occurs. (b) Inserting polarizing filters before the detectors allows the which path information to be erased, which resuscitates the interference pattern

$$U = \frac{1}{\sqrt{2}} \begin{pmatrix} 1 & i \\ i & 1 \end{pmatrix}. \qquad (11.45)$$

Now set $\theta = \frac{\pi}{2}$, so that the cosine part vanishes. If coincidence counting is done at the output, the last two terms in cosine part also make no contribution. So the output state effectively becomes a Bell state:

$$|\psi'\rangle = \frac{1}{\sqrt{2}}(|H\rangle_1|V\rangle_2 - |V\rangle_1|H\rangle_2) = |\Psi^-\rangle, \qquad (11.46)$$

where $|\Psi^-\rangle$ is one of the maximally entangled Bell states to be discussed further in Chap. 13. Notice two things about this state. First, by measuring the polarizations of the outgoing photons, the path each particle took to reach the beam splitter can be determined. (The polarization measurement can be done, for example, by replacing each single detector with a pair of detectors at the output ports of a polarizing beam splitter.) Second, because of the existence of which-way information, there is no interference between the two-photon amplitudes in the coincidence rate. If a variable time delay or phase shift is introduced, no HOM dip will be seen.

Now make one alteration to this setup. Add polarizers in the two outgoing paths after the beam splitter, as in Fig. 11.6b. Take the polarizers to be oriented at angles θ_1 and θ_2 from the horizontal, respectively, so that they project the photons passing through them onto the states

$$|\theta_1\rangle = |H\rangle_1 \cos\theta_1 + |V\rangle_1 \sin\theta_1 \qquad (11.47)$$
$$|\theta_2\rangle = |H\rangle_2 \cos\theta_2 + |V\rangle_2 \sin\theta_2, \qquad (11.48)$$

so that the coincidence rate will be given by

$$P_{coinc}(\theta_1, \theta_2) = |\langle\theta_1, \theta_2|\psi'\rangle|^2 \qquad (11.49)$$

$$= \frac{1}{2} \left(\cos \theta_1 \sin \theta_2 - \cos \theta_2 \sin \theta_1 \right)^2 \qquad (11.50)$$

$$= \frac{1}{2} \sin^2 (\theta_2 - \theta_1). \qquad (11.51)$$

If $\theta_1 - \theta_2 = \pm \pi$, then this equals 1, indicating that there is no interference. Which path was taken by each photon is still determinable from the outcome. But if $\theta_1 = \theta_2$, both photon polarizations are now being projected onto the same direction, and all which-way information is erased. At this point, the HOM dip reappears.

The two-photon interference occurs when the photons are mixed by the beam splitter, but the decision of which angles to use at the polarizers (i.e., whether or not to erase the path information) can be made after the photons have already passed through the beam splitter. The decision can be made at any point up to the time the photons reach the detectors.

11.4   Quantum Zeno Effect

The Greek philosopher Zeno of Elea (c. 490–430 BC) proposed a series of paradoxes, which claimed to prove that motion was impossible. One of those paradoxes divides time up into infinitesimal points and then states that at each of these frozen instants an object must be motionless at a definite location, and so stringing together many of these motionless instances cannot lead to motion. The conclusion is that motion cannot occur.

Zeno's classical paradoxes were resolved long ago, once Newton and Leibnitz had developed consistent methods of treating infinitesimal quantities and limits. But quantum mechanics leads to a situation that is reminiscent of Zeno's paradox: it asserts that if a system is monitored continuously (or is measured repeatedly at a sufficiently high frequency), then the system won't evolve over time, but will remain in its initial state. Essentially, at the quantum level, watched popcorn won't pop and watched water won't boil! This effect has been noticed in special cases at least as far back as the 1950s, but was first stated in generality in [9].

To describe this **quantum Zeno effect**, consider a system with some observable $\hat{q}$. Assume that $\hat{q}$ has a discrete spectrum of eigenvalues, and that the system starts at $t = 0$ in the nth eigenstate, $\psi_n(0)$, of $\hat{q}$. Continuous monitoring of $\hat{q}$ really means taking a discrete sequence of measurements at times $t = \tau, 2\tau, 3\tau, \ldots$, and then taking the limit as $\tau \to 0$. So if the Hamiltonian is $\hat{H}$, the state just before the first measurement is made at time τ is

$$\psi_n(\tau) = e^{-\frac{i}{\hbar} \hat{H} \tau} \psi_n(0) = \left(1 - \frac{\hat{H} \tau}{i \hbar} + \frac{1}{2!} \left(\frac{\hat{H} \tau}{i \hbar} \right)^2 + \ldots \right) \psi_n(0). \qquad (11.52)$$

The probability of still being in state n is then

$$P(n \rightarrow n, \tau) = |\langle \psi_n(\tau)|\psi_n(0)\rangle|^2 \tag{11.53}$$

$$= 1 - \frac{\tau^2}{\hbar^2}\left[\langle\psi_n(0)|\hat{H}^2|\psi_n(0)\rangle - \langle\psi_n(0)|\hat{H}|\psi_n(0)\rangle^2\right]. \tag{11.54}$$

Note that the linear terms in τ canceled. The terms in the last bracket are simply the variance of the energy

$$\langle\psi_n(0)|\hat{H}^2|\psi_n(0)\rangle - \langle\psi_n(0)|\hat{H}|\psi_n(0)\rangle^2 = \langle E^2\rangle - \langle E\rangle^2 = \Delta E^2, \tag{11.55}$$

so that

$$P(n \rightarrow n, \tau) = 1 - \frac{\tau^2}{\hbar^2}\Delta E^2. \tag{11.56}$$

At time $T = k\tau$, after k measurements, the probability of still being in the initial state is

$$P(n \rightarrow n, T) = [P(n \rightarrow n, \tau)]^k = [P(n \rightarrow n, \tau)]^{T/\tau} \tag{11.57}$$

$$= \left[1 - \frac{\tau^2}{\hbar^2}\Delta E^2\right]^{T/\tau}. \tag{11.58}$$

We now can take $\tau \rightarrow 0$, holding T fixed. The limit of expressions like this can be found in every calculus textbook. The result is

$$P(n \rightarrow n, \tau) \rightarrow e^{-\frac{\tau T}{\hbar^2}\Delta E^2} \rightarrow 1. \tag{11.59}$$

As a result of the repeated measurements, we see that the probability of still being in the initial state approaches 100%: if the measurements are sufficiently rapid relative to the characteristic time scale of the system's evolution, each measurement projects the system back to the initial state.

The quantum Zeno effect has been seen experimentally. The first such experiment (proposed in [10] and carried out in [11]) involved laser-cooled $^9Be^+$ ions in a magnetic trap. The lowest lying states of the system include the three states shown in Fig. 11.7. The lower two levels are close in energy, being due to hyperfine splitting of the ground state. A radio frequency (rf) signal is used to excite all of the electrons from the ground state to E_2. If unmeasured, the electrons should rapidly decay back to the ground state, leaving E_2 largely unoccupied. But a measurement can be made to see whether the electrons are in E_1 or E_2. This measurement consists of sending in short laser pulses at the resonant frequency for the E_1 to E_3 transition, $\nu = \frac{E_3 - E_1}{\hbar}$. If any electrons are in the ground state, they will be excited by the pulse into the level E_3; this level will then rapidly decay back to the ground state, emitting fluorescent photons in the process. If the electrons are instead in state E_2, then the pulse will not excite them into E_3, and the fluorescent signal is absent. Thus, it can

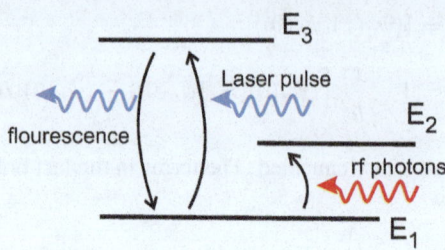

Fig. 11.7 Experimental demonstration of the quantum Zeno effect. Radio frequency (rf) pulses excite the electrons into energy level E_2. Normally, they would rapidly decay back to the ground state E_1. But repeated measurements inhibit the decay, leaving keeping the electrons in the unstable state E_2. Measurement is conducted by sending short laser pulses in at the resonant frequency for the E_1-E_3 transition. Any electrons that have decayed back to E_1 become visible because the pulse excites them up to E_3, and then a fluorescent photon is emitted as it drops back to E_1. Absence of fluorescence indicates that the electrons are still in level E_2

be determined how many electrons are in E_2 versus E_1 by measuring the strength of the fluorescent signal. If the measurement is carried out repeatedly, at intervals τ greater than the characteristic $E_2 \rightarrow E_1$ decay time, τ_c, then strong fluorescence is seen, indicating that the electrons have all decayed to the ground state. But if the measurement is carried out more frequently, the fluorescence decreases and the survival probability of the unstable E_2 state increases. As the time between measurements is lowered to $\tau \ll \tau_c$, the survival probability in E_2 approaches 100%, and the fluorescent signal decays to zero.

11.5   Interaction-Free Measurement

Normally, in order to detect the presence of an object or to measure its properties, it is necessary to interact somehow with the object. For example, to see an object with your eyes, photons need to scatter off it and then enter your pupils. To view an object with an electron microscope, electrons must be scattered from the object and then detected. It is therefore remarkable that according to quantum mechanics, it is possible to view an object or make measurements on it using light that has never interacted with it, a phenomenon that has been called "quantum seeing in the dark" [12].

To understand the basic idea, consider once again the ubiquitous Mach–Zhender interferometer (Fig. 11.8a), with input at only one port. Take the input state to be $\begin{pmatrix} \psi \\ 0 \end{pmatrix}$. Assume that both beam splitters are 50–50. The two branches of the interferometer are labeled O and R, for object and reference arms. With no object present in the object arm, the first beam splitter transforms the input state according to

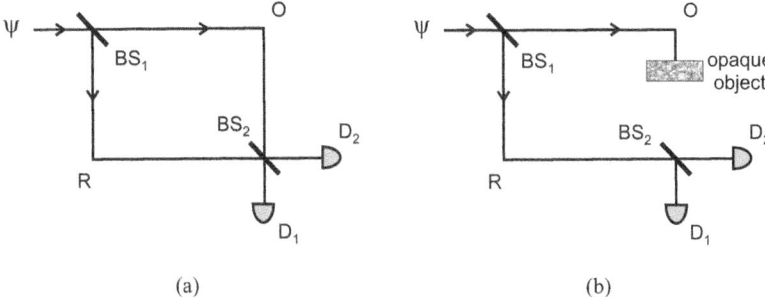

(a)   (b)

Fig. 11.8 (a) A Mach–Zehnder with input at only one port. If both arms are unobstructed, then all light strikes detector D_1, and none reaches D_2. (b) Inserting an opaque object in the object branch O prevents interference from occurring at the second beam splitter. As a result, each detector sees half the photons. The presence of photons striking D_2 is a signal that there is an object present inside the interferometer, so that the object can be detected even though none of the photons that actually struck the object ever reach either detector

$$\begin{pmatrix} \psi \\ 0 \end{pmatrix} \rightarrow \frac{1}{\sqrt{2}} \begin{pmatrix} 1 & i \\ i & 1 \end{pmatrix} \begin{pmatrix} \psi \\ 0 \end{pmatrix} = \frac{1}{\sqrt{2}} \begin{pmatrix} \psi \\ i\psi \end{pmatrix}, \tag{11.60}$$

where the top entry in the vector is the amplitude in the object branch and the bottom is in the reference branch. The second beam splitter then transforms this to

$$\begin{pmatrix} 0 \\ i\psi \end{pmatrix}, \tag{11.61}$$

where the top and bottom entries now represent D_1 and D_2. So all the amplitude arrives at detector D_2; destructive interference occurs so that no photons reach D_1.

But now, imagine an opaque object is placed in the object arm, blocking the beam. The object removes any photons striking it, either by absorbing them or by scattering them out of the interferometer. Then the action of the two beam splitters is now:

$$\begin{pmatrix} \psi \\ 0 \end{pmatrix} \xrightarrow{BS_1} i \begin{pmatrix} 0 \\ \psi \end{pmatrix} \xrightarrow{BS_2} \frac{1}{\sqrt{2}} \begin{pmatrix} -\psi \\ i\psi \end{pmatrix}. \tag{11.62}$$

We see that now, with one arm blocked, each photon is equally likely to hit either detector. Therefore, the detection of any photons at D_1 is a clear signal that an object is present. This is despite the fact that every photon hitting the object is absorbed, which means that *none* of the detected photons at either D_1 or D_2 *ever interacted with the object!*

This effect has been demonstrated experimentally and works as expected [15]. Of course, the efficiency of this protocol is low: 50% of the light is lost in the object branch, and 25% will go to D_2, leaving only 25% to reach D_1. The object detection operation therefore has an efficiency of only 25%. However, by invoking the quantum Zeno effect (see [12]) or by using more complicated interferometric arrangements, the efficiency can be significantly increased. Recently, a variation of interaction-free measurement using microwave pulses in a superconducting transmon circuit achieved detection probability that could exceed 80% in some cases [13].

A more dramatic version of this system was described by Elitzur and Vaidman [14], who proposed the idea of **quantum bomb detection**. The idea is the same, except that the object arm holds a bomb that will explode if it is struck by a photon. By the mechanism described earlier, the bomb can be detected safely 25% of the time by looking for photons arriving at D_2. However, the price is that 50% of the time (when photons enter branch O), the bomb will explode.

The method can be generalized so as to not just detect the object, but to produce an image of it. Imagine that the object is not completely opaque, but allows some light to transmit through it. The transmission probability $T(x)$ is spatially varying, with position variable x. The single detector at D_2 can be replaced by an array of detectors, each measuring photons that arrive at different positions. Then if many photons are sent through the interferometer, many detections will be found at x values where the object is nearly opaque, and few will be detected at x values where the object is nearly transparent. The counts at different locations therefore allow the spatial profile $T(x)$ of the object's transmissivity to be built up over time.

Despite some similarities, notice that this approach is not quite the same as the induced-coherence imaging of Sect. 11.5, which also allows imaging with photons that never reached the object. Induced-coherence imaging depends on two-photon Zou–Wang–Mandel interference, while the method described here is a single-photon interference effect.

Problems

1. Two mutually commuting annihilation operators $\hat{a}_p$ and $\hat{a}_s$ represent a pair of coupled harmonic oscillators and have Hamiltonians $\hat{H}_p = \hbar v_p \left(\hat{a}_p^\dagger \hat{a}_p + \frac{1}{2} \right)$ and $\hat{H}_s = \hbar v_s \left(\hat{a}_s^\dagger \hat{a}_s + \frac{1}{2} \right)$, with an interaction Hamiltonian $\hat{H}_I = \hbar \kappa \hat{a}_s^\dagger \hat{a}_s \hat{a}_p^\dagger \hat{a}_p$. Define system and probe operators [6],

$$\hat{X}_s = \hat{a}_s^\dagger \hat{a}_s, \qquad \hat{X}_p = \frac{1}{2i} \left(\frac{1}{\sqrt{\hat{a}_p^\dagger \hat{a}_p + 1}} \hat{a}_p - \hat{a}_p^\dagger \frac{1}{\sqrt{\hat{a}_p^\dagger \hat{a}_p + 1}} \right).$$

Show that the conditions at the end of Sect. 11.2 are satisfied, so that these define a QND protocol to measure $\hat{A}_s$. ($\hat{X}_P$ is related to the probe phase $\hat{\phi}$, $\hat{a}_P = \sqrt{\hat{a}_p^\dagger \hat{a}_p + 1} \, e^{i\hat{\phi}}$; see Sect. 19.1.)

2. Another example of a QND measurement can be given by an atom in a resonant cavity. The atom is initially in an equal superposition of the ground state $|g\rangle$ and an excited state $|e\rangle$, while the field in the cavity is in the state $|n\rangle$ with n photons. The total initial state of the system is therefore $|\psi(0)\rangle = \frac{1}{\sqrt{2}}\left(|g\rangle + |e\rangle\right)|n\rangle$. After a $\frac{\pi}{2}$-pulse (Sect. 23.1) of electromagnetic radiation, the state of the system at time t will now be [16]

$$|\psi(t)\rangle = \frac{1}{2}\left[\left(e^{-i\chi t(n+1)} - e^{+i\chi tn}\right)|e\rangle + \left(e^{-i\chi t(n+1)} + e^{+i\chi tn}\right)|g\rangle\right]|n\rangle,$$

where χ is a coupling constant. Given this, compute the probabilities $P_g(t)$ and $P_e(t)$ of being in the ground and excited state as a function of time. You should find that the atom now oscillates between the ground and excited state. The rate of the oscillation depends on the number n of photons in the cavity and can be measured without changing the number [17, 18].

3. Imagine a two-level system initially in the ground state, $|g\rangle$. In a particular quantum Zeno experiment, a radiation field is causing oscillations between the excited and ground states. Suppose that the probability of still being in the ground state after time t is $P_g(t) = \cos^2\left(\frac{\Omega t}{2}\right) = \frac{1}{2}\left[1 + \cos\Omega t\right]$, for some frequency Ω.

   (a) What is the probability $P_e(t)$ of being in the excited state $|e\rangle$ after time t? What is the time T_π at which the excited state probability is 100%?
   (b) Set $\Delta t = \frac{T_\pi}{n}$. Again assuming that the system is initially in the ground state, find the probabilities $P_e(\Delta t)$ and $P_g(\Delta t)$ of making a transition or of staying in the ground state during interval Δt.
   (c) Find the same probabilities as in part (b), but now assuming the system was initially in the excited state.
   (d) Suppose two observations are made, at $\frac{T_\pi}{2}$ and at T_π. Given the results aforementioned, find the probability that a system initially in the ground state will be in the excited state after the second measurement. You should find the probability has been cut in half compared to the P_g value of part (b). (Be careful: there are two ways the final state can be reached and both must be counted.)
   (e) Generalize part (e): make n measurements at multiple of $\frac{T_\pi}{n}$ and find the probability of being in the excited state at the end. You should find a result that goes to zero as $n \to \infty$.

4. Consider two states with spin up and spin down along the z-axis, representing $|0\rangle$ and $|1\rangle$ states. A rotation of π about either the x or y axis will convert one state into the other, and so can act as a NOT gate. Instead, suppose we apply a series of rotations by $\frac{\pi}{n}$ about the x axis, for some integer n. A measurement is made of the state after each rotation in the sequence. Assume the initial state was $|0\rangle$. For large n, find the probability that the system will still be found to be in state $|0\rangle$ after every measurement.

5. In an HOM experiment, suppose that the two incoming photons are polarized at an angle θ to each other. Calculate the coincidence rate as a function of

the polarization angle. Show that the rate vanishes at $\theta = \pm\frac{\pi}{2}$ and reaches a maximum at $\theta = 0$. This is a reflection of the fact that interference only occurs only between indistinguishable states (or the *portions* of the states that are indistinguishable), and θ here is serving as a measure of distinguishability.

References

1. Y. Aharanov, D.Z. Albert, L. Vaidman, Phys. Rev. Lett. **60**, 1351 (1988)
2. V.B. Braginsky, F.Ya. Khalili, *Quantum Measurement* (Cambridge University Press, Cambridge, 1992)
3. V.B. Braginsky, F.Ya. Khalili, Rev. Mod. Phys. **68**, 1 (1996)
4. P. Ghose, *Testing Quantum Mechanics on New Ground* (Cambridge University Press, Cambridge, 1999)
5. P. Grangier, J.A. Levenson, J.P. Poizat, Nature **396**, 537 (1998)
6. M.O. Scully, M.S. Zubairy, *Quantum Optics* (Cambridge University Press, Cambridge, 1997)
7. M.O. Scully, K. Druhl, Phys. Rev. A **25**, 2208 (1982)
8. P.G. Kwiat, A.M. Steinberg, R.Y. Chaio, Phys. Rev. A **45**, 7729 (1992)
9. B. Misra, E.C.G. Sudarshan, J. Math. Phys. **18**, 756 (1977)
10. R.J. Cook, Phys. Scr. **T21**, 49 (1988)
11. W.M. Itano, D.J. Heinzen, J.J. Bollinger, D.J. Wineland, Phys. Rev. A **41**, 2295 (1990)
12. P. Kwiat, H. Weinfurter, A. Zeilinger, Sci. Am. 72 (1996)
13. S. Dogra, J.J. McCord, G.S. Paraoanu, Nat. Commun.**13**, 7528 (2022)
14. A.C. Elitzur, L. Vaidman, Founds. Phys. **23**, 987 (1993)
15. P.G. Kwiat, H. Weinfurter, T. Herzog, A. Zeilinger, M.A. Kasevich, Phys. Rev. Lett. **74**, 4763 (1995)
16. C.C. Gerry, P.L. Knight, *Introductory Quantum Optics* (Cambridge University Press, Cambridge, 2005)
17. M. Brune et al., Phys. Rev. Lett. **72**, 3339 (1994)
18. G. Nogues, et al., Nature **400**, 239 (1999)

Chapter 12
Quantum Bits and Quantum Information

Quantum mechanics carries with it extra features, such as superposition, entanglement, and interference, that are not present in classical mechanical systems. As a result, quantum information processing will lead to new features that cannot be mimicked by classical information processing systems. So classical information measures have to be generalized in order to capture these new features. These new quantum information measures should reduce back to the old ones as limiting cases. In a later chapter, it will be seen that quantum measures of information are closely related to quantitative measures of entanglement, and so in addition to limits on communication rates and computation rates, there will also be limits on rates of entanglement sharing.

Quantum information has been a rapidly growing topic since the 1990s, with a quickly mushrooming volume of literature, as evidenced by the many books and papers reviewing the subject. Some good overviews of the subject include [1–6]. A large part of the interest stems from the fact that it provides the underpinnings for quantum computing, to which we will return in Chaps. 16 and 17.

12.1 Classical and Quantum Bits

Classical information is usually represented by a sequence of **bits**. Each logical bit is represented as either a 0 or a 1, with the value stored in some classical two-state system. Many different systems have been devised for storing bits. For example, a capacitor may be charged or uncharged to represent the two possible values in a random access memory (RAM), a transistor can use two charge states in a solid-state drive, standard hard disk drives use small regions with different magnetizations, and on a DVD, the two possible values may signaled by regions of different optical reflectivity.

© The Author(s), under exclusive license to Springer Nature Switzerland AG 2025
D. S. Simon, *Introduction to Quantum Science and Technology*, Undergraduate
Texts in Physics, https://doi.org/10.1007/978-3-031-81315-3_12

In quantum mechanics, the two logical states have to be encoded into two different quantum states of a two-state system, $|0\rangle$ or $|1\rangle$. But, due to the linearity of the Schrödinger equation, the system can also be in a superposition of both states: $|\psi\rangle = \alpha|0\rangle + \beta|1\rangle$, for any complex numbers α and β satisfying the normalization condition $|\alpha|^2 + |\beta|^2 = 1$. This superposition state is called a **qubit**, and since there are an infinite number of possible superpositions, there is now an infinity of distinct states, rather than just the two present for a classical qubit.

The coefficients α and β can be parameterized by a pair of angles, so that the qubit can be written $|\psi\rangle = \cos\frac{\theta}{2}|0\rangle + e^{i\phi}\sin\frac{\theta}{2}|1\rangle$. Clearly, these states are in a one-to-one correspondence with the points on a sphere; the qubit states live on the surface of the sphere. We have seen examples of this in earlier chapters: spin states live on the Bloch sphere, while polarization states live on the Poincaré sphere.

As a qubit is fed through an information processing system, operations can be carried out on both terms of the superposition simultaneously, producing a form of parallel processing in which multiple computations are carried out at once. At the end, when the output of the system is measured, the qubit will collapse into either $|0\rangle$ or $|1\rangle$, so it seems that only one of the parallel processes can lead to anything useful; the other is lost. But, as will be seen in later chapters, a judicious choice of measurements can still often be arranged in order to extract extra information or extra functionality not possible in the classical case. For example, quantum algorithms (Chap. 17) often do not measure the qubit value directly, but rather measure some function that depends on both of the coefficients α and β of the qubit, while superdense coding (Sect. 18.3) allows two classical bits of information to be transmitted in a communication using a just one qubit.

Another important difference between classical bits and quantum qubits is that qubits are much more fragile. Interactions with the computational apparatus and the surrounding environment can lead to decoherence: the relative phase ϕ between the $|0\rangle$ and $|1\rangle$ states can fluctuate. This can wash out the interference effects needed to maintain quantum advantages. As a result, quantum computers often need to be maintained at low temperatures in order to reduce thermal fluctuations, and other safeguards need to be used to further reduce decoherence effects. Quantum error correction mechanisms also need to be put into place to detect and repair errors created by phase fluctuations and other error mechanisms.

Since qubits have properties that are very distinct from those of classical bits, the information measures used to describe them need some adjustments. This is described in the next section.

12.2   Quantum Information Measures

12.2.1   Von Neumann Entropy

Just as the information carried by a classical communication channel can be quantified, so can the information carried by a quantum state. The quantum analog of the Shannon entropy $H(X)$ is the **von Neumann entropy**, $S(\hat{\rho})$. For density operator $\hat{\rho} = \sum_i p_i |\lambda_i\rangle\langle\lambda_i|$, the von Neumann entropy is defined as

$$S(\hat{\rho}) = -\langle \log_2(\hat{\rho})\rangle = -Tr\left(\rho \log_2(\hat{\rho})\right) = -\sum_j p_j \log_2 p_j, \tag{12.1}$$

where the angled brackets denote quantum expectation value and λ_j denote the eigenvalues of $\hat{\rho}_j$. $S(\hat{\rho})$ gives the average information that is carried by the density matrix, or equivalently, it is the minimum number of bits of information needed to fully specify the state.

Example 12.2.1 Consider a pure state $|\psi\rangle = |0\rangle$ of a two-state system. Its density operator is

$$\hat{\rho} = |\psi\rangle\langle\psi| = 1 \cdot |0\rangle\langle0| + 0 \cdot |1\rangle\langle1|. \tag{12.2}$$

In its eigenbasis, the density operator can be written as

$$\hat{\rho} = \begin{pmatrix} 1 & 0 \\ 0 & 0 \end{pmatrix}, \tag{12.3}$$

so

$$S(\hat{\rho}) = -Tr\left[\begin{pmatrix} 1 & 0 \\ 0 & 0 \end{pmatrix}\begin{pmatrix} \log_2(1) & 0 \\ 0 & \log_2(0) \end{pmatrix}\right] = 0, \tag{12.4}$$

where, as usual, we take $0\log_2(0) = \lim_{\epsilon\to 0^+}\left(\epsilon\log_2\epsilon\right) = 0$.

In contrast, suppose the state is maximally mixed, given by a uniform distribution. Then, for a two-state system, the density operator is just proportional to the 2×2 identity matrix, $\hat{\rho} = \frac{1}{2}\hat{I}_2$. So,

$$S(\hat{\rho}) = -\frac{1}{2}Tr\left[\begin{pmatrix} 1 & 0 \\ 0 & 1 \end{pmatrix}\begin{pmatrix} \log_2 \frac{1}{2} & 0 \\ 0 & \log_2 \frac{1}{2} \end{pmatrix}\right] = 1. \tag{12.5}$$

For a maximally mixed system whose Hilbert space dimension is N, this generalized in the obvious manor, with $\hat{\rho} = \frac{1}{2}\hat{I}_N$ and $S(\hat{\rho}) = \log_2(N)$.

As an intermediate case, consider a mixed state with nonuniform probabilities, $p_0 = \frac{1}{3}$, and $p_1 = \frac{2}{3}$. Then in its eigenbasis,

$$\hat{\rho} = \begin{pmatrix} \frac{1}{3} & 0 \\ 0 & \frac{2}{3} \end{pmatrix},$$

(12.6)

leading to entropy

$$S(\hat{\rho}) = -Tr\begin{pmatrix} \frac{1}{3}\log_2\frac{1}{3} & 0 \\ 0 & \frac{2}{3}\log_2\frac{2}{3} \end{pmatrix} = 0.918 \quad \blacksquare$$

(12.7)

The pattern in the previous examples is general: $S(\hat{\rho})$ vanishes for pure states, since the identity of the state is always the same and carries no surprise, but it is nonzero for statistically mixed states. It reaches a maximum for a uniformly mixed state, for which each measurement gives a maximally unpredictable value; in this case, the density operator is proportional to the identity, $\hat{\rho} = \frac{1}{N}\hat{I}$, and the entropy takes on a maximum value of $S(\hat{\rho}) = \log_2 N$ on a Hilbert space of dimension N. So S can be viewed as a measure of the "mixedness" of a quantum state, satisfying

$$0 \le S(\hat{\rho}) \le \log_2 N,$$

(12.8)

with the two extreme values representing completely pure and completely mixed.

The von Neumann entropy is basis-independent, being invariant under unitary transformations:

$$S(\hat{U}\hat{\rho}\hat{U}^\dagger) = S(\hat{\rho}).$$

(12.9)

It is also a convex function on the Hilbert space: if $\hat{\rho} = p\hat{\rho}_1 + (1-p)\hat{\rho}_2$, then

$$S(\hat{\rho}) \ge pS(\hat{\rho}) + (1-p)S(\hat{\rho}).$$

(12.10)

More generally,

$$S\left(\sum_i p_i\hat{\rho}_i\right) \ge \sum_i p_iS(\hat{\rho}_i),$$

(12.11)

for $\sum p_i = 1$.

A useful property of the von Neumann entropy is that it is additive: for a composite system $\hat{\rho} = \hat{\rho}_a \otimes \hat{\rho}_b$, the entropy is the sum of the entropies of the subsystems:

$$S(\hat{\rho}) = S(\hat{\rho}_a) + S(\hat{\rho}_b).$$

(12.12)

In particular, if n (non-interacting) copies of a system are combined, then the entropy of the combined system is just n times the entropy of the original smaller component system,

$$S(\hat{\rho}^{\otimes n}) = n S(\hat{\rho}). \tag{12.13}$$

Just as the Shannon entropy gives the maximum compression rate for a classical communication, the von Neumann entropy gives the maximum compression rate for quantum communication. Suppose Alice is sending information to Bob over a possibly noisy quantum channel. Assume Alice encodes binary logical bits 0 and 1 into physical quantum states according to $0 \rightarrow |0\rangle$ and $1 \rightarrow |1\rangle$. A sequence of n bits is then encoded into a product of n states, $|\psi_1\rangle|\psi_2\rangle \ldots, |\psi_n\rangle$, where each term in the product is either $|0\rangle$ or $|1\rangle$. The source of the message can be thought of as a source of random bits, where 0 and 1 are produced with probabilities p_0 and p_1 equal to the relative frequencies of the bits 0 and 1 in the original message. Each transmitted qubit can then be described by density operator

$$\hat{\rho} = p_0 |0\rangle\langle 0| + p_1 |1\rangle\langle 1|, \tag{12.14}$$

and the full n qubit transmission can be described by n copies of the single qubit density, $\hat{\rho}^{\otimes n}$. Then the quantum analog of the Shannon noiseless coding theorem tells us that this state can be compressed by projecting it onto the space of ϵ-typical sequences, $\hat{\rho}^{\otimes n} \rightarrow \hat{\rho}_n\big|_{\epsilon\ typ}$, and neglecting the atypical sequences. This projection can be made explicit by writing the state as convex linear combination of typical and atypical states:

$$\hat{\rho}^{\otimes n} = (1 - \alpha)\, \hat{\rho}_n\big|_{\epsilon\ typ} + \alpha\, \hat{\rho}_n\big|_{\epsilon\ atyp}. \tag{12.15}$$

For $\alpha \leq \epsilon$, the error coming from neglecting the atypical term $\alpha\, \hat{\rho}_n\big|_{\epsilon\ atyp}$ can be quantified in terms of the fidelity between the original state and the approximate state. For sufficiently large n,

$$\mathcal{F}\big(\hat{\rho}^{\otimes n}, \hat{\rho}_n\big|_{\epsilon\ typ}\big) = 1 - \alpha \geq 1 - \epsilon. \tag{12.16}$$

Following similar reasoning to that of Shannon in the classical case, it then follows that the state $\hat{\rho}^{\otimes n}$ representing the message can (for sufficiently large n) be compressed to a sequence of states of length $n S(\hat{\rho})$ with negligible error and with fidelity arbitrarily close to 1.

12.2.2 Other Quantum Information Measures

Just as in the classical case, it will be useful to define other entropy measures beyond the von Neumann entropy. For example, suppose Alice and Bob have quantum states $\hat{\rho}_A$ and $\hat{\rho}_B$. Then we can define their **quantum joint entropy** as the von Neumann entropy of their combined system, $S(\hat{\rho}_{AB})$. Then in straightforward analogy to the classical case, the **quantum conditional entropy** is defined as

$$S(\hat{\rho}_A|\hat{\rho}_B) = S(\hat{\rho}_{AB}) - S(\hat{\rho}_B), \tag{12.17}$$

describing the information gained from measuring $\hat{\rho}_{AB}$ when $\hat{\rho}_B$ is already known.

The **quantum mutual information**, which measures the correlation between systems A and B, is

$$I(\hat{\rho}_A : \hat{\rho}_B) = S(\hat{\rho}_A) + S(\hat{\rho}_B) - S(\hat{\rho}_{AB}) \tag{12.18}$$

$$= S(\hat{\rho}_A) - S(\hat{\rho}_A|\hat{\rho}_B). \tag{12.19}$$

If the combined state $\hat{\rho}_{AB}$ is a pure state, then the von Neumann entropy vanishes, $S(\hat{\rho}_{AB}) = 0$. In this case, by use of the Schmidt decomposition to be discussed in Sect. 13.4, a basis can always be found such that the state vector and density operator can be written in the forms $|\psi\rangle = \sum_i \lambda_i |\psi_i\rangle_A |\phi_i\rangle_B$ and $\hat{\rho} = \sum_i |\lambda_i|^2 (|\psi_i\rangle\langle\psi_i|)_A (|\phi_i\rangle\langle\phi_i|)_B$. From this it follows that the von Neumann entropies of the partial traces are

$$S(\hat{\rho}_A) = S(\hat{\rho}_B) = \sum_i |\lambda_i|^2 \log_2 |\lambda_i|^2. \tag{12.20}$$

If the state is entangled, then the last expression will be non-zero, indicating that states $\hat{\rho}_A$ and $\hat{\rho}_B$ are *not* pure states. This is a general and important result:

> Partial traces of *pure* entangled states are *mixed* states.

This is the reason that entangling a system with an external environment causes decoherence: even if the total system is in a pure state, the lack of knowledge about the environment leaves the subsystem of interest in a mixed state. Similarly, in SPDC, if the original pump photon is in a pure state, then each of the two daughter photons (the signal and idler), taken alone, is in a mixed state.

Given the result of the previous paragraphs, it then follows from the definitions of the conditional and mutual entropies that for a *pure* entangled composite state $\hat{\rho}_{AB}$:

$$I(\hat{\rho}_A : \hat{\rho}_B) = 2S(\hat{\rho}_A) \tag{12.21}$$

$$S(\hat{\rho}_A|\hat{\rho}_B) = -S(\hat{\rho}_A), \tag{12.22}$$

where the fact that $S(\hat{\rho}) = 0$ for a pure state. Classically, we know that $I(X : Y) \le H(X)$, so we might expect that in the quantum case we should have $I(\hat{\rho}_A : \hat{\rho}_A) \le S(\hat{\rho}_A)$. But the last two equations imply that this is not true. Since the mutual information is a measure of correlation, this implies another important result:

Entangled quantum systems can have stronger correlations than are possible in classical systems.

The **quantum relative entropy** or **quantum Kullback–Liebler distance** is

$$S(\hat{\sigma}||\hat{\rho}) = \text{Tr}\big(\hat{\sigma}\,(\log \hat{\sigma} - \log \hat{\rho})\big). \tag{12.23}$$

Like the classical version, it is not necessarily symmetric, $S(\hat{\sigma}||\hat{\rho}) \neq S(\hat{\rho}||\hat{\sigma})$. It is non-negative,

$$S(\hat{\sigma}||\hat{\rho}) \geq 0, \tag{12.24}$$

with equality only for $\hat{\sigma} = \hat{\rho}$, and has no upper bound. $S(\hat{\sigma}||\hat{\rho})$ acts as a measure of the similarity between two states: it vanishes when $\hat{\rho} = \hat{\sigma}$, and is infinite when the support of $\hat{\rho}$ (the set of points at which $\hat{\rho}$ is nonzero) is perpendicular to the support of $\hat{\sigma}$. Additional properties of the relative entropy include:

- Additivity of product distributions:

$$S\big(\hat{\sigma}_1 \otimes \hat{\sigma}_2||\hat{\rho}_1 \otimes \hat{\rho}_2\big) = S\big(\hat{\sigma}_1||\hat{\rho}_1\big) + S\big(\hat{\sigma}_2||\hat{\rho}_2\big).$$

- Invariance under unitary transformations:

$$S(\hat{U}\hat{\sigma}\hat{U}^\dagger||\hat{U}\hat{\rho}\hat{U}^\dagger) = S(\hat{\sigma}||\hat{\rho}).$$

- The distinguishability of two distributions is reduced by partial traces:

$$S(\text{Tr}_B\hat{\sigma}||\text{Tr}_B\hat{\rho}) \leq S(\hat{\sigma}||\hat{\rho}),$$

  where $\hat{\rho}$ and $\hat{\sigma}$ are bipartite states.
- The probability of mistaking one distribution for the other after n measurement trials is asymptotically

$$P_{err} = 2^{-nS(\hat{\sigma}||\hat{\rho})}, \qquad \text{as } n \to \infty.$$

- Relation to mutual information for composite systems:

$$S\big(\hat{\rho}_{AB}||\hat{\rho}_A \otimes \hat{\rho}_B\big) = I(\hat{\rho}_A : \hat{\rho}_B) \tag{12.25}$$

The quantum relative entropy is often useful for proving properties involving the von Neumann entropy. See, for example, Problem 5 and the proof of subadditivity further. See also [2] for additional applications.

The **subadditivity theorem** states that the joint entropy of two systems satisfies

$$S(\hat{\rho}_{AB}) \leq S(\hat{\rho}_A) + S(\hat{\rho}_B). \tag{12.26}$$

This tells us that the amount of information in the composite $A+B$ system cannot be greater than that already present in the two component systems A and B, regardless of how strongly correlated the two component systems may be with each other. To prove it, start from the non-negativity of $S(\hat{\rho}_{AB}||\hat{\rho}_A \otimes \hat{\rho}_B)$:

$$0 \leq S(\hat{\rho}_{AB}||\hat{\rho}_A \otimes \hat{\rho}_B) \tag{12.27}$$

$$= \mathrm{Tr}\left(\hat{\rho}_{AB}\left(\log_2 \hat{\rho}_{AB} - \log_2 \hat{\rho}_A \otimes \hat{\rho}_B \right) \right) \tag{12.28}$$

$$= \mathrm{Tr}\left(\hat{\rho}_{AB}\left(\log_2 \hat{\rho}_{AB} \right) \right) - \mathrm{Tr}\left(\hat{\rho}_{AB} \log_2 \left(\hat{I}_A \otimes \hat{\rho}_B \right) \right)$$

$$- \mathrm{Tr}\left(\hat{\rho}_{AB} \log_2 \left(\hat{\rho}_A \otimes \hat{I}_B \right) \right) \tag{12.29}$$

$$= -S(\hat{\rho}_{AB}) + S(\hat{\rho}_A) + S(\hat{\rho}_B), \tag{12.30}$$

which completes the proof. Notice that the last line of the proof equals the mutual information, $I(\hat{\rho}_A : \hat{\rho}_B)$, so that this also serves as a proof of Eq. 12.25.

A lower limit is given by the **Araki–Lieb inequality** (see Appendix S of [3] for a proof),

$$S(\hat{\rho}_{AB}) \geq |S(\hat{\rho}_A) - S(\hat{\rho}_B)|. \tag{12.31}$$

So, combining the sub-additivity constraint and the Araki–Lieb inequality, we have

$$|S(\hat{\rho}_A) - S(\hat{\rho}_B)| \leq S(\hat{\rho}_{AB}) \leq S(\hat{\rho}_A) + S(\hat{\rho}_B), \tag{12.32}$$

similar to the triangle inequality on a vector space.

12.3   The Holevo Bound and Accessible Information

Suppose Alice wishes to communicate a message to Bob. She encodes the message into a set of quantum states and sends them to him. Even if Alice's original encoding associated a pure state $|\psi_i\rangle$ with each message, noise in the communication channel means that, in general, Bob will receive a mixed state. So the state $\hat{\rho}_i$ received by Bob might differ from the state that was sent. Since noise will be random, the received state cannot be predicted ahead of time. Instead, there will be a range of possible states with respective probabilities p_i. Bob's goal now is to discriminate between different mixed states $\hat{\rho} = \sum_i p_i \hat{\rho}_i$ in order to extract the maximum amount of information available about the original message. Because of the noise and the mixing, the extraction will be imperfect; the maximum information available will be reduced, and the channel capacity will be lower than it would have been in the pure state case.

Alice's original message can be viewed as a random classical variable X. p_i will be the probability that X will lead to quantum state $\hat{\rho}_i$: $p(X = i) = p_i$. At the other end, Bob will apply a set of measurement operators, arriving at a measurement result $\hat{\rho}_j$ with probability q_j. This leads him to conclude that the original message was $Y = j$. The mutual information, $I(X : Y)$ between the original and reconstructed messages will then be limited by entropy of the corresponding quantum states. In particular, we have the **Holevo bound**:

$$I(X : Y) \leq S(\hat{\rho}) - \sum_i p_i S(\hat{\rho}_i). \tag{12.33}$$

This is often written in the form

$$I(X : Y) \leq \chi(\hat{\rho}), \tag{12.34}$$

where

$$\chi(\hat{\rho}) \equiv S(\hat{\rho}) - \sum_i p_i S(\hat{\rho}_i) = S(\hat{\rho}) - \langle S(\hat{\rho}) \rangle \tag{12.35}$$

is called the **Holevo χ quantity**. Since the channel capacity is defined to be the maximum mutual information, this leads to a result for the quantum channel capacity of a communication channel:

$$C = \chi(\hat{\rho}). \tag{12.36}$$

The Holevo bound limits the amount of accessible information that can be extracted from a transmitted quantum state. A quantum qubit carries more information than a classical bit, since the quantum state is described by a complex number, which can be expressed as two real numbers (two classical bits). So the state of n qubits, being specified by $2^n - 1$ complex numbers, should be able to carry a large amount of information. But in essence, the Holevo bound implies that the information that can be accessed by the recipient at the other end is only that of n classical bits.

12.4   The No-cloning Theorem

The no-cloning theorem [7–10] is a key result underlying many quantum information processing and quantum communication applications. Despite its importance,
it is very simple to demonstrate. Essentially, the theorem says that an unknown
state cannot be copied without making a measurement that will alter or destroy the
original state. More explicitly:

No-cloning Theorem Suppose $|\psi\rangle_A$ and $|\phi\rangle_A$ are two normalized pure
states. Assume that the states are not parallel or orthogonal to each other,
so $(\,|\langle\phi|\psi\rangle|^2$ is not equal to 0 or 1). If you have one of these states, but don't
know which one, then it is impossible to find a unitary operation that will
clone the unknown state. By this, we mean that given some known reference
state $|r\rangle_B$, there is no unitary operator $\hat{U}$ such that

$$\hat{U} : |\psi\rangle_A \otimes |r\rangle_B \to |\psi\rangle_A \otimes |\psi\rangle_B, \qquad \hat{U} : |\phi\rangle_A \otimes |r\rangle_B \to |\phi\rangle_A \otimes |\phi\rangle_B.$$
$$(12.37)$$

Proof Suppose that the desired unitary operation $\hat{U}$ *does* exist. Then take the inner
product of the two states $\hat{U}\left(|\psi\rangle_A \otimes |r\rangle_B\right) = |\psi\rangle_A \otimes |\psi\rangle_B$ and $\hat{U}\left(|\phi\rangle_A \otimes |r\rangle_B\right) =$
$|\phi\rangle_A \otimes |\phi\rangle_B$. We find:

$$\left((\langle\phi| \otimes \langle r|)\hat{U}^\dagger \hat{U} (|\psi\rangle \otimes |r\rangle)\right) = \left((\langle\phi|\psi\rangle)\right)_A \left((\langle\phi|\psi\rangle)\right)_B, \qquad (12.38)$$

or equivalently, since $\langle r|r\rangle = 1$ and $\hat{U}^\dagger \hat{U} = \hat{I}$:

$$\left((\langle\phi|\psi\rangle)\right)^2 = \langle\phi|\psi\rangle. \qquad (12.39)$$

But this is impossible, since by assumption $|\langle\phi|\psi\rangle|$ does not equal 0 or 1. Therefore,
the desired $\hat{U}$ does not exist.   ∎

Another way to see that cloning of quantum states won't work is to consider the
superposition state

$$|\Psi\rangle = \alpha|\psi\rangle + \beta|\phi\rangle, \qquad (12.40)$$

with $|\alpha|^2 + |\beta|^2 = 1$. Assume that neither coefficient is zero, $\alpha, \beta \neq 0$. Then, again
assuming $\hat{U}$ exists, apply it to $|\Psi\rangle_A \otimes |r\rangle_B$. The output will be

$$\hat{U}\left(|\Psi\rangle_A \otimes |r\rangle_B\right) = \alpha|\psi\rangle_A|\psi\rangle_A + \beta|\phi\rangle_B|\phi\rangle_B. \qquad (12.41)$$

But the clone of the original state should be

$$|\Phi\rangle_A \otimes |\Phi\rangle_B = \left(\alpha|\psi\rangle + \beta|\phi\rangle\right)_A \otimes \left(\alpha|\psi\rangle + \beta|\phi\rangle\right)_B \qquad (12.42)$$

$$= \alpha^2 |\psi\rangle_A \otimes |\psi\rangle_B + \beta^2 |\phi\rangle_A \otimes |\phi\rangle_B$$

$$+ \alpha\beta \left(|\psi\rangle_A \otimes |\phi\rangle_B + |\phi\rangle_A \otimes |\psi\rangle_B\right). \qquad (12.43)$$

Clearly, the states of Eqs. 12.41 and 12.43 are not the same, so the cloning process has failed.

The no-cloning theorem makes quantum cryptography possible. If the theorem did not hold, an eavesdropper could make a second copy of an information-bearing state being transmitted between two legitimate users of a system. She could keep one copy for herself and send the other on to its intended recipient. Then, she could make measurements on her copy and obtain all of the information without the two legitimate participants knowing she has been tampering. So the no-cloning theorem allows quantum information to be secure: the legal participants will know if an eavesdropper is stealing information from them because her actions will introduce errors: the states received at one end will differ from those sent at the other end, and a comparison of a small sub-sample of the states will reveal this.

The no-cloning theorem does not preclude making *approximate* copies. Imperfect clones that approximate the original state with fidelities as high as $\frac{5}{6}$ can be made by coupling the state with an ancillary system that evolves unitarily [11].

Problems

1. Consider the Bell states defined by

$$|\Psi^{(+)}\rangle = \frac{1}{\sqrt{2}}\left(|0\rangle|1\rangle + |1\rangle|0\rangle\right) \quad \text{and} \quad |\Phi^{(-)}\rangle = \frac{1}{\sqrt{2}}\left(|0\rangle|0\rangle - |1\rangle|1\rangle\right).$$

   (a) Calculate the von Neumann entropy of $|\Psi^{(+)}\rangle$, and verify that it vanishes.
   (b) Calculate the quantum relative entropy of the states $|\Psi^{(+)}\rangle$ and $|\Phi^{(-)}\rangle$.
   (c) Calculate the quantum relative entropy of $|\Psi^{(+)}\rangle$ and the pure state $|0\rangle_A|0\rangle_B$.

2. (a) Find the quantum mutual information I between the two subsystems in the maximally mixed state $\hat{\rho} = \frac{1}{2}\left(|00\rangle\langle 11| + |11\rangle\langle 11|\right)$.
   (b) Find the quantum mutual information I of the pure entangled state $|\psi\rangle = |01\rangle + |10\rangle$.

3. Consider the mixed state $\hat{\rho} = \frac{1}{3}|0\rangle\langle 0| + \frac{2}{3}|1\rangle\langle 1|$ and the pure state $\hat{\sigma} = |\psi\rangle\langle\psi|$, where $|\psi\rangle = |\phi\rangle = \alpha|0\rangle + \beta|1\rangle$.

   (a) Find the von Neumann entropies $S(\hat{\rho})$ and $S(\hat{\sigma})$.

(b) Let $\rho_{AB} = \frac{1}{3}\Big(|0\rangle\langle 0|\Big)_A \otimes \Big(|0\rangle\langle 0|\Big)_B + \frac{2}{3}\Big(|1\rangle\langle 1|\Big)_A \otimes \Big(|0\rangle\langle 0|\Big)_B$. Verify that the two partial traces are the states $\hat\rho_A\hat\rho$ and $\hat\rho_B = \hat\sigma$ given above.

(c) Find the joint entropy $S(\hat\rho_{AB})$.

(d) Compute the quantum conditional entropy $S(\hat\rho_A|\hat\rho_B)$ and the quantum mutual entropy $I(\hat\rho_A : \hat\rho_B)$. Verify that the relations of Eqs. 12.21 and 12.22 hold.

(e) Compute the quantum relative entropy $S(\hat\rho_A||\hat\rho_B)$. Is it symmetric under interchange, $A \leftrightarrow B$?

4. Consider the two mixed states $\hat\rho = q|0\rangle\langle 0| + (1-q)|1\rangle\langle 1|$ and $\hat\sigma = p|0\rangle\langle 0| + (1-p)|1\rangle\langle 1|$, with $0 \le p, q \le \frac{1}{2}$. Find the Holevo bound for these states.

5. The density operator for the Boltzmann thermal state of Hamiltonian $\hat H$ at inverse temperature $\beta = \frac{1}{k_B T}$ is

$$\hat\rho_\beta = \frac{e^{-\beta\hat H}}{\mathrm{Tr}\, e^{-\beta\hat H}}.$$

It has mean energy $E = \mathrm{Tr}\hat\rho_\beta \hat H$. Show that this state has the largest von Neumann entropy of any quantum state with the same mean energy. (Hint: define another state with the same mean energy, find the von Neumann entropy of both states, then use the non-negativity condition of the relative entropy.)

6. A generalization of the Shannon entropy called the **Renyi entropy** has found a number of applications in areas such as quantum chaos, cryptography, and signal processing. The *quantum* Renyi entropy is defined by $S_\alpha(\hat\rho) = \frac{1}{1-\alpha}\log_2 Tr\, \hat\rho^\alpha$, for real parameter $0 < \alpha < \infty$, $\alpha \ne 1$.

(a) Show that the Renyi entropy is non-negative, and that it equals zero only for pure states.

(b) Show that $S_\alpha(\hat\rho \otimes \hat\sigma) = S_\alpha(\hat\rho) + S_\alpha(\hat\sigma)$, for all α.

(b) Use l'Hôpital's rule to show that $\lim_{\alpha\to 1} S_\alpha(\hat\rho)$ is the von Neumann entropy.

7. One further quantum information measure is the Fano entropy, which is defined in terms of the logarithm of the purity: $F(\hat\rho) = -\log_2\Big(\mathrm{Tr}(\hat\rho^2)\Big)$. Show that $S(\hat\rho) \ge F(\hat\rho)$.

8. Let $\hat\rho = \sum p_i|i\rangle\langle i|$ and $\hat\sigma = \sum q_i|i\rangle\langle i|$ be two density operators expressed in orthonormal basis $|i\rangle$.

(a) Show that the relative entropy can be written as

$$S(\hat\rho||\hat\sigma) = \sum_i \Big(\log_2 p_i - \sum_j P_{ij}\log q_j\Big),$$

where $P_{ij} = \langle i|j\rangle\langle j|i\rangle$.

(b) Prove the Klein identity, $S(\hat\rho||\hat\sigma) \ge \sum_i p_i \log_2 \frac{p_i}{r_i}$, where $r_i = \sum_j P_{ij}q_j$. (Hint: use the fact that the log is a convex function.)

9. Suppose Alice sends Bob a sequence of non-orthogonal pure states,

$$|\psi_1\rangle = \begin{pmatrix} 1 \\ 0 \end{pmatrix}, \quad |\psi_2\rangle = \begin{pmatrix} -\frac{1}{2} \\ \frac{\sqrt{3}}{2} \end{pmatrix}, \quad |\psi_3\rangle = \begin{pmatrix} -\frac{1}{2} \\ -\frac{\sqrt{3}}{2} \end{pmatrix}.$$

Each of these states is sent with equal probability of $\frac{1}{3}$. The non-orthogonality of the states introduces a probability that Bob will identify each state incorrectly.

(a) What is the density matrix for the mixed state obtained by Bob?
(b) What are the error probabilities P_{ij} of misidentifying state ϕ_i as ϕ_j?
(c) What is the channel capacity?

References

1. M.A. Nielsen, I.L. Chuang, *Quantum Computation and Quantum Information: 10th Anniversary Edition* (Cambridge University Press, Cambridge, 2011)
2. V. Vedral, *Introduction to Quantum Information Science* (Oxford University Press, Oxford, 2006)
3. S.M. Barnett, *Quantum Information* (Oxford University Press, Oxford, 2009)
4. G. Jaeger, *Quantum Information: An Overview* (Springer, Berlin, 2007)
5. M.M. Wilde, *Quantum Information Theory* (Cambridge University Press, Cambridge, 2013)
6. J.A. Bergou, M. Hillery, M. Saffman, *Quantum Information Processing*, 2nd edn. (Springer, Berlin, 2021)
7. J. Park, Found. Phys. **1**, 23 (1970)
8. W.K. Wootters, W.H. Zurek, Nature **299**, 5886, 802 (1982)
9. D. Dieks, Phys. Lett. A. **92**, 271 (1982)
10. W.K. Wootters, W.H. Zurek, Phys. Today **62**, 76 (2009)
11. V. Bužek, M. Hillery, Phys. Rev. A. **54**, 1844 (1996)

Chapter 13
Quantifying Entanglement

In recent decades, entanglement has gone from being viewed as a quantum oddity to being seen as a quantifiable resource that be exploited to facilitate applications. In this chapter, entanglement is examined a little more closely, and a range of approaches to detecting and quantifying it are discussed. In particular, Bell states and Schmidt decompositions are defined in order to formalize the description of entangled states, and then measures of entanglement are defined. These measures include concurrence, quantum discord, fidelity, and trace distance, in addition to tests of entanglement such as entanglement witnesses and the PPT criterion.

13.1 Entanglement as a Resource

Recall that entanglement implies that a multiparticle wavefunction cannot be factored into a product of separate single-particle wavefunctions. Consider the simplest case, a bipartite, two-state system. **Bipartite** means two particles (labeled A and B), while two-state mean that each particle can be in state 0 or state 1. Such a system consists of a pair of qubit states. So an entangled state of this system is of the form

$$|\psi\rangle = \alpha|00\rangle + \beta|01\rangle + \gamma|10\rangle + \delta|11\rangle, \qquad (13.1)$$

with the normalization condition $|\alpha|^2 + |\beta|^2 + |\gamma|^2 + |\delta|^2 = 1$. By going to different bases in Hilbert space (i.e., rotating in the $|0\rangle$-$|1\rangle$ plane), the coefficients in this expression will change values. But if the system is truly entangled, then there is no basis in which the right side of Eq. 13.1 reduces to a single term.

More complicated possibilities may occur, in which there are three or more particles entangled, or in which each of those particles can be in three or more

© The Author(s), under exclusive license to Springer Nature Switzerland AG 2025
D. S. Simon, *Introduction to Quantum Science and Technology*, Undergraduate
Texts in Physics, https://doi.org/10.1007/978-3-031-81315-3_13

different basis states. But for now, states such as that of Eq. 13.1 will serve as our model example.

An important concept is the idea of LOCC operations. Consider experimenters Alice and Bob, each working in their own labs, which may be well-separated from each other, possibly by a great distance. Imagine that Alice and Bob share a joint quantum state $\hat{\rho}_{AB}$, which may be either a product state or an entangled state. Each experimenter can only operate on the part of the state in their own labs; in other words, they can only perform **local** operations. Under such local operations, the state transforms according to

$$\hat{\rho}_{AB} \rightarrow \hat{U}_A \otimes \hat{U}_B \, \hat{\rho}_{AB}, \tag{13.2}$$

where the operation $\hat{U}_A$ performed by Alice acts only on the part of the state in her lab, and similarly for $\hat{U}_B$ in Bob's lab. Because the operations in the two labs are independent of each other, they cannot produce any new correlations in the state that were not already present before.

However, if Alice and Bob are allowed to use **classical communication** lines (telephone lines, email, telegrams, smoke signals, tin cans on a string, etc.), then they can coordinate their actions. In this manner, they can introduce additional new *classical* correlations into the joint state. The most general action on the state is then to produce a statistical mixture of classically correlated states:

$$\hat{\rho}_{AB} \rightarrow \sum_n \alpha_n \hat{U}_{A_n} \otimes \hat{U}_{B_n} \hat{\rho}_{AB}, \tag{13.3}$$

with $\sum_n \alpha_n = 1$. The classical correlation appears due to the pairing of each A_n with a corresponding B_n.

So, **local operations and classical communication (LOCC)** can produce classical correlations between widely separated systems. However, they *cannot* produce entanglement, and so cannot increase the values of the quantitative entanglement measures to be introduced in coming sections. Examples of LOCC operations abound: measurements, loss, rotations of particle spin or polarization, and introduction of ancillary particles are all LOCC operations. Generation of entangled states (or increase in the entanglement of states already present) will require some form of purely quantum communication or interaction, and so cannot be LOCC.

Although LOCC operations cannot create or increase entanglement, they *can* be used to manipulate entanglement that is already present in the system, converting one entangled state into another. For example, it will be seen further that, given a large number copies of a mixed, entangled state, LOCC operations can be used to perform entanglement distillation (Sect. 13.9), i.e., to produce a smaller number of copies of the state that have a larger amount of entanglement per copy.

Although entanglement has been studied since the early papers of Schrödinger and of Einstein, Podolsky, and Rosen in the 1930s, a new viewpoint on entanglement has gradually arisen since the 1990s, in which it is seen as a resource for carrying out specific tasks. Entanglement is used as a tool to facilitate a large range of applications, including teleportation of quantum states, implementing dense coding, speeding up computations, detecting eavesdroppers in quantum cryptography, and viewing sub-wavelength objects in microscopes. Some of these tasks simply have improved efficiency in the presence of entanglement, while others are actually impossible without it.

If entanglement is to be viewed as a resource, then it is necessary to have both a means of detecting it in a given quantum system and a means of measuring the amount or quality of the entanglement that is present. In this chapter, we discuss ways of detecting and quantifying entanglement. (A further method, violation of Bell-type inequalities will be discussed at greater length in the next chapter.) Entanglement measures form an enormous field, so here we will just discuss a few of the simplest and most common measures.

Entanglement and information are closely related, so the quantum information measures defined in Chap. 12 will make appearances here. In particular, the von Neumann entropy, previously introduced earlier as a measure of the information carried by a quantum state, will also serve as a simple gauge of the degree of entanglement in a system. Roughly speaking, since entanglement is a form of correlation, it follows that increase or loss of entanglement should lead to an increase or loss of information shared between the two entangled systems.

The work on quantifying entanglement has been extensive; for comprehensive overviews, see [1–3]. Since entanglement and coherence are closely connected, it should not be too surprising that there is also a growing literature on the quantification of coherence as a resource [4–9]

Before discussing entanglement measures, we begin in the next section by introducing a useful set of states that can serve as a convenient basis for bipartite entangled systems.

13.2   Bell States

A bipartite entangled system is spanned by a basis of four two-particle states. These could be taken to be the states $|00\rangle$, $|01\rangle$, $|10\rangle$, $|11\rangle$, as in Eq. 13.1. But often it is more convenient to use a basis in which the entanglement is already built into the basis vectors themselves. The most common such basis is composed of the set of four **Bell states**:

$$|\Psi^\pm\rangle = \frac{1}{\sqrt{2}}\left(|01\rangle \pm |10\rangle\right) \tag{13.4}$$

$$|\Phi^\pm\rangle = \frac{1}{\sqrt{2}}\left(|00\rangle \pm |11\rangle\right). \tag{13.5}$$

0 and 1 could represent the states of any two-state system. For example, if they represent the two spin states of a spin-$\frac{1}{2}$ system, then the Bell states can also be written in the form

$$|\Psi^{\pm}\rangle = \frac{1}{\sqrt{2}}\left(|\uparrow\downarrow\rangle \pm |\downarrow\uparrow\rangle\right) \tag{13.6}$$

$$|\Phi^{\pm}\rangle = \frac{1}{\sqrt{2}}\left(|\uparrow\uparrow\rangle \pm |\downarrow\downarrow\rangle\right). \tag{13.7}$$

The two qubits in these states are *maximally* entangled, in the sense that they produce a maximal violation of the CHSH inequality (Sect. 14.3). Any other bipartite entangled state can be built as a linear combination of the Bell states,

$$|\psi\rangle = \frac{1}{\sqrt{2}}\left(a|\Psi^{+}\rangle + b|\Psi^{-}\rangle + c|\Phi^{+}\rangle + d|\Phi^{-}\rangle\right), \tag{13.8}$$

for some set of complex constants satisfying $|a|^2 + |b|^2 + |c|^2 + |d|^2 = 1$. For example, it takes only a few lines of algebra to show that the state in Eq. 13.1 is of the form of Eq. 13.8 if the coefficients in the two equations are taken to be related by

$$\alpha = c + d, \qquad \beta = a + b, \qquad \gamma = a - b, \qquad \delta = c - d.$$

The Bell states will be useful in a number of applications, such as quantum teleportation (Sect. 18.2) and superdense coding (Sect. 18.3), where manipulation of entangled states is an essential ingredient of the procedure. Some quantum procedures will involve a process called **entanglement distillation**, in which an arbitrary entangled quantum state (possibly mixed, and not necessarily bipartite) is transformed into a set of N Bell states. The number of Bell states that can be distilled will serve as a measure of the degree of entanglement.

13.3   Positive Partial Transpose and Negativity

Many of the following sections in this chapter will define numerical quantities that measure the amount and quality of the entanglement. But another option is to simply apply a test that gives a yes or no answer as to the presence of entanglement. Being all or nothing, these tests are in a sense coarser than numerical entanglement measures, but they are usually easier to implement. In practice, these tests usually give either "yes" or "maybe" as an answer, or else give "no" or "maybe," rather than "yes" and "no."

The most important of these tests is the **positive partial transpose (PPT) condition**, or **Peres-Horodecki criterion**. In order to state this criterion, we must first define the partial trace operation.

Let $|\xi_a\rangle$ and $|\chi_b\rangle$ be sets of basis vectors for subsystems A and B, with the indices a and b running from 1 up to the dimensions of the respective Hilbert spaces. Then a generic joint two-particle density operator for the $A \otimes B$ system is of the form:

$$\hat{\rho} = \sum_{aa'bb'} p_{aa'bb'} |\xi_a\rangle\langle\xi_{a'}| \otimes |\chi_b\rangle\langle\chi_{b'}|. \tag{13.9}$$

The **partial transpose** is defined to be the transpose acting on just one of the subsystems; for example, the partial transpose with respect to system A, T_A, interchanges the bras and kets of the A subsystem only, $|\xi_a\rangle\langle\xi_{a'}| \rightarrow |\xi_{a'}\rangle\langle\xi_a|$, leaving the B subsystem unchanged. The partial transpose with respect to B acts similarly on the B bras and kets, with $|\chi_b\rangle\langle\chi_{b'}| \rightarrow |\chi_{b'}\rangle\langle\chi_b|$.

Apply the partial transpose to the full density matrix. It doesn't matter if T_A or T_B is used, so suppose we use T_A:

$$\hat{\rho}^{T_A} = \sum_{aa'bb'} p_{aa'bb'} |\xi_{a'}\rangle\langle\xi_a| \otimes |\chi_b\rangle\langle\chi_{b'}|. \tag{13.10}$$

Then the **Peres–Horodecki PPT criterion** says:

If $\hat{\rho}^{T_A}$ has a negative eigenvalue, then the system is entangled.

In other words, the system is entangled if $\hat{\rho}^{T_A}$ is *not* a non-negative operator. Note that the converse *does not* hold: the lack of negative eigenvalues does not necessarily imply that the system is separable. For bipartite systems, the positivity of the partial trace is both necessary and sufficient for separability, while for more general systems, it is necessary but may not be sufficient.

Since $\hat{\rho}^{T_A} = \left(\hat{\rho}^{T_B}\right)^T$, both partial transposes have the same eigenvalues. So it doesn't matter whether the partial trace is taken with respect to the A or B subsystem when applying the PPT criterion.

It is easy to see that separable states always lead to $\hat{\rho}^{T_A}$ as non-negative operators. In the separable case, the density operator reduces to the form

$$\hat{\rho} = \sum_j p_j \hat{\rho}_j^A \otimes \hat{\rho}_j^B. \tag{13.11}$$

$\hat{\rho}_j^A$ and $\hat{\rho}_j^B$, the single-particle density matrices, must have non-negative eigenvalues. Taking the transpose of either of them does not change the sign of the eigenvalues. So the partial transpose of $\hat{\rho}$ must have non-negative eigenvalues.

Unfortunately, proving that negative eigenvalues imply entanglement is more difficult and will not be shown here.

Example 13.3.1 Consider an unentangled product state first, for example, $|\psi\rangle = |01\rangle$. The density matrix $\hat{\rho} = |\psi\rangle\langle\psi| = |01\rangle\langle01|$ can be written in matrix form as

$$\hat{\rho} = \begin{pmatrix} 0 & 0 & 0 & 0 \\ 0 & 1 & 0 & 0 \\ 0 & 0 & 0 & 0 \\ 0 & 0 & 0 & 0 \end{pmatrix}, \tag{13.12}$$

where the rows and columns represent the basis $|00\rangle$, $|01\rangle$, $|10\rangle$, $|11\rangle$. Clearly, the eigenvalues are 1 and 0, with 0 being triply degenerate. The partial transpose T_A would interchange the first bit (the A bit) of the ket with the A bit of the bra. It is obvious that this operation will not change the matrix: $\hat{\rho}^{T_A} = \hat{\rho}^{T_B} = \hat{\rho}$, so the eigenvalues remain unchanged. In particular, they remain non-negative, in keeping with the PPT criterion. ∎

Example 13.3.2 Consider an entangled state instead. For example, take the Bell state $|\Psi^{(+)}\rangle = \frac{1}{\sqrt{2}} (|01\rangle + |10\rangle)$. Taking the outer product of this state with itself to get the density operator, we can then express that operator in matrix form using the same basis as above:

$$\hat{\rho} = \frac{1}{2} (|01\rangle + |10\rangle) (\langle01| + \langle10|) \tag{13.13}$$

$$= \frac{1}{2} (|01\rangle\langle01| + |01\rangle\langle10| + |10\rangle\langle01| + |10\rangle\langle10|) \tag{13.14}$$

$$= \frac{1}{2} \begin{pmatrix} 0 & 0 & 0 & 0 \\ 0 & 1 & 1 & 0 \\ 0 & 1 & 1 & 0 \\ 0 & 0 & 0 & 0 \end{pmatrix}. \tag{13.15}$$

The eigenvalues of $\hat{\rho}$ are $(1, 0, 0, 0)$, which are all non-negative, as they must be for a density operator.

So now, apply the partial transpose T_A, interchanging the A bits of each term:

$$\hat{\rho}^{T_A} = \frac{1}{2} (|01\rangle\langle01| + |11\rangle\langle00| + |00\rangle\langle11| + |10\rangle\langle10|) \tag{13.16}$$

$$= \frac{1}{2} \begin{pmatrix} 0 & 0 & 0 & 1 \\ 0 & 1 & 0 & 0 \\ 0 & 0 & 1 & 0 \\ 1 & 0 & 0 & 0 \end{pmatrix} \tag{13.17}$$

The eigenvalues of $\hat{\rho}^{T_A}$ are readily found to be $\left(-\frac{1}{2}, \frac{1}{2}, \frac{1}{2}, \frac{1}{2}\right)$. The presence of the negative eigenvalue then tips us off to the fact that the Bell state is entangled, a fact that we already knew. ■

From the eigenvalues of the partial trace, the **negativity** can also be defined:

$$N(\hat{\rho}) = \left| \sum_j \lambda_j \right|, \tag{13.18}$$

where the sum runs over all negative eigenvalues of $\hat{\rho}^{T_A}$. Clearly, the system is entangled if $N \neq 0$. The concurrence (Sect. 13.7) provides an upper bound on the possible values of the negativity: $N(\hat{\rho}) \leq C(\rho)$. Instead of the negativity, many authors use the logarithmic negativity, $\log_2 N(\hat{\rho})$,, which is greater than 1 for entangled systems. The negativity and its logarithm are both entanglement monotones.

Another relatively simple way to distinguish between entangled and separable states in the bipartite case is to look at reduced density operators. If $\hat{\rho}$ is a product state, its partial traces $\hat{\rho}_A = Tr_A \hat{\rho}$ and $\hat{\rho}_B = Tr_B \hat{\rho}$ are pure states, while the partial trace of an entangled state is mixed (see Problem 2).

13.4   Schmidt Decomposition

Let $|\psi\rangle$ be a pure state of a bipartite composite system, composed of subsystems A and B. Let $|\xi_j\rangle_A$ and $|\chi_j\rangle_B$ be orthonormal bases of A and B respectively. Then it is always possible to find a new basis in which the state takes the form

$$|\psi\rangle = \sum_{j=1}^{N} \lambda_j |j\rangle_A |j\rangle_B, \tag{13.19}$$

where N is the smaller of the Hilbert space dimensions of A and B. $|j\rangle_A$ and $|j\rangle_B$ are new orthonormal bases of A and B. The coefficients λ_j are real, non-negative numbers, called the **Schmidt coefficients**,, and they obey the relation $\sum_j \lambda_j^2 = 1$. This form of the wavefunction is known as the **Schmidt decomposition**. Schmidt decomposition provides a sort of diagonalization, in which the bases of the A and B are aligned with each other.

Assume for simplicity that the Hilbert spaces of A and B are of equal dimension. The Schmidt decomposition can then be proven as follows. In the original basis, the state is of the form

$$|\psi\rangle = \sum_{j,k=1}^{N} A_{jk}|\xi_j\rangle_A|\chi_k\rangle_B. \tag{13.20}$$

The expansion coefficients A_{jk} are elements of a matrix A. By the singular decomposition theorem of linear algebra, the matrix can be diagonalized: there are unitary matrices U and V such that

$$A = UDV, \tag{13.21}$$

where the diagonal matrix D has the eigenvalues of A along its diagonal, so that $D_{lm} = \lambda_l \delta_{lm}$. The state can then be written as

$$|\psi\rangle = \sum_{jklm} U_{jl}\lambda_l\delta_{lm}V_{mk}|\xi_j\rangle_A|\chi_k\rangle_B \tag{13.22}$$

$$= \sum_{jkl} U_{jl}\lambda_l V_{lk}|\xi_j\rangle_A|\chi_k\rangle_B. \tag{13.23}$$

So now define new vectors

$$|j\rangle_A = \sum_{l} U_{lj}|\xi_l\rangle_A, \qquad |j\rangle_B = \sum_{k} V_{jk}|\chi_k\rangle_B, \tag{13.24}$$

leading

$$|\psi\rangle = \sum_{j=1}^{N} \lambda_j |j\rangle_A |j\rangle_B, \tag{13.25}$$

which is the advertised Schmidt decomposition. Unlike Eq. 13.20 in which the matrix formed by the expansion coefficients is in general nondiagonal, the corresponding matrix in Eq. 13.25 has the λ_i on the diagonal and is zero otherwise. Using the orthonormality of the original bases, the reader can easily check that the new bases are orthonormal as well. Notice that the basis used for the Schmidt decomposition is not unique, since simultaneous rotations of both the A and B bases will maintain the Schmidt form of Eq. 13.19 (with different coefficients).

The Schmidt decomposition has a number of applications. For example, using the decomposition, the reduced density matrices of a pure state are immediately seen to be of the forms:

$$\rho_A = \sum_{j} |\lambda_j|^2 |j\rangle_A\langle j|_A \tag{13.26}$$

$$\rho_B = \sum_{j} |\lambda_j|^2 |j\rangle_B\langle j|_B, \tag{13.27}$$

which makes it clear that the reduced density matrices of both component systems have the same eigenvalues, $|\lambda_i|^2$.

The number of nonzero Schmidt coefficients is called the **Schmidt number**, N_s. The Schmidt number is preserved under local unitary transformations. Clearly, in order for the state to be a product state, it is necessary that $N_s = 1$. Any value $N_s > 1$ signals entanglement. The **Schmidt measure** or **Hartley strength** of the state is then defined as

$$E_S = \log_2(N_s). \tag{13.28}$$

E_S measures the entanglement of the state in units of "e-bits," in a similar manner to the way Shannon entropy measures information in terms of qubits.

Sometimes, the Schmidt decomposition is obvious by inspection. For example, consider the state $|\psi\rangle = \frac{1}{2}\left(|00\rangle + |01\rangle + |10\rangle + |11\rangle\right)$. This state factors into the form:

$$|\psi\rangle = \frac{1}{\sqrt{2}}\left(|0\rangle + |1\rangle\right)_A \otimes \frac{1}{\sqrt{2}}\left(|0\rangle + |1\rangle\right)_B, \tag{13.29}$$

so that it can be written in the Schmidt form $|\psi\rangle = 1 \cdot |\lambda_1\rangle_A |\lambda_1\rangle_B + 0 \cdot |\lambda_2\rangle_A |\lambda_2\rangle_B$, where $|\lambda_1\rangle = \frac{1}{\sqrt{2}}\left(|0\rangle + |1\rangle\right)$ and $|\lambda_2\rangle = \frac{1}{\sqrt{2}}\left(|0\rangle - |1\rangle\right)$. The form of $|\lambda_2\rangle$ was chosen to make the set $\{|\lambda_1\rangle, |\lambda_2\rangle\}$ into an orthonormal basis.

Most often, finding the Schmidt basis takes a little more work. A common method for a pure bipartite state $\hat{\rho}$ consists of two steps: first take a partial trace; for example, take the trace over B to find $\hat{\rho}_A = Tr_B \hat{\rho}_{AB}$. Then diagonalize the result. This works because if the Schmidt form of the original state is

$$|\psi\rangle = \sum_j \lambda_j |\lambda_j\rangle_A |\lambda_j\rangle_B, \tag{13.30}$$

then

$$\hat{\rho} = \sum_{ij} \lambda_i \lambda_j |\lambda_i\rangle_A |\lambda_i\rangle_B \otimes {}_A\langle\lambda_j|_B\langle\lambda_j|. \tag{13.31}$$

The partial trace gives $\hat{\rho}_A = \sum_j \lambda_j^2 |\lambda_j\rangle_A {}_A\langle\lambda_j|$, from which the Schmidt form can easily be deduced. Note that the Schmidt basis is simply the eigenbasis of ρ_A (or equivalently, of ρ_B), and that the squared Schmidt coefficients λ_j^2 are simply the eigenvalues.

The Schmidt decomposition is most useful for bipartite states. For composite systems with three or more subsystems, such a decomposition may not exist.

Example 13.4.1 Consider the state $|\psi\rangle = \frac{1}{\sqrt{3}}\left(|00\rangle + |01\rangle + |10\rangle\right)$, which has density operator

$$\hat{\rho} = \frac{1}{3}\begin{pmatrix} 1 & 1 & 1 & 0 \\ 1 & 1 & 1 & 0 \\ 1 & 1 & 1 & 0 \\ 0 & 0 & 0 & 0 \end{pmatrix}. \tag{13.32}$$

So

$$\hat{\rho}_A = Tr_B\hat{\rho} \; = \; \sum_j {}_B\langle\lambda_j|\hat{\rho}|\lambda_j\rangle_B \tag{13.33}$$

$$= {}_B\langle 0|\hat{\rho}|0\rangle_B + {}_B\langle 1|\hat{\rho}|1\rangle_B \tag{13.34}$$

$$= \frac{1}{3}\Big(2|0\rangle\langle 0| + |0\rangle\langle 1| + |1\rangle\langle 0| + |1\rangle\langle 1|\Big)_A \tag{13.35}$$

$$= \frac{1}{3}\begin{pmatrix} 2 & 1 \\ 1 & 1 \end{pmatrix}. \tag{13.36}$$

The eigenvalues λ_i^2 and eigenvectors $|\lambda_i\rangle$ of the previous matrix are easily found. Taking the square roots of the eigenvalues then gives the coefficients in the Schmidt expansion. The result gives Schmidt expansion,

$$|\psi\rangle = \sum_{j=1}^{2} \lambda_j|\lambda_j\rangle_A|\lambda_j\rangle_B,$$

with

$$\lambda_1 = \sqrt{\frac{1}{6}\left(3 + \sqrt{5}\right)} \qquad |\lambda_1\rangle = \sqrt{\frac{2}{5+\sqrt{5}}}\begin{pmatrix} (1+\sqrt{5})/2 \\ 1 \end{pmatrix}$$

$$\lambda_2 = \sqrt{\frac{1}{6}\left(3 - \sqrt{5}\right)} \qquad |\lambda_1\rangle = \sqrt{\frac{2}{5-\sqrt{5}}}\begin{pmatrix} (1-\sqrt{5})/2 \\ 1 \end{pmatrix} \quad \blacksquare$$

Another useful application of the Schmidt decomposition is to show the existence of **purifications**. Given the density matrix, $\hat{\rho}_A$, for a mixed state, recall that a new composite system S can always be found such that $\hat{\rho}_A$ is the reduced density operator for a *pure* state in S. Let the composite system be $S = A \otimes R$, where R is called the reference system. It could be a real physical system or just a fictitious system that is being adjoined to A in order to construct a pure state. Suppose that $\hat{\rho}_A = \sum_j p_j|j\rangle_A {}_A\langle j|$. To construct the purification, define a state of S of the form

$$|\psi_S\rangle = \sum_j \sqrt{p_j}|j\rangle_A|j\rangle_R. \tag{13.37}$$

This is a pure state with density matrix

$$\hat{\rho}_S = \sum_{jk} \left(\sqrt{p_j} |j\rangle_A |j\rangle_R \right) \left(\sqrt{p_k} \langle k|_R \langle k|_A \right) = \sum_{jk} \sqrt{p_j p_k} \left(|j\rangle\langle k| \right)_A \otimes \left(|j\rangle\langle k| \right)_R$$

(13.38)

Taking the partial trace with respect to R:

$$\text{Tr}_R\left(\hat{\rho}_S\right) = \sum_{jk} \sqrt{p_j p_k} \left(|j\rangle\langle k| \right)_A \text{Tr}\left(|j\rangle\langle k| \right)_R \qquad (13.39)$$

$$= \sum_{jk} \sqrt{p_j p_k} \left(|j\rangle\langle k| \right)_A \delta_{jk} \qquad (13.40)$$

$$= \sum_j p_j |j\rangle_A \langle j|_A \qquad (13.41)$$

$$= \hat{\rho}_A. \qquad (13.42)$$

The conclusion is that every state, pure or mixed, can be viewed as a subsystem of a larger pure state. For example, the state of a system A in contact with a heat bath R will be a mixed state; but the system and heat bath are components of a single larger system S, which is in a pure state.

13.5   Fidelity and Trace Distance

Any measure of distance on the Hilbert space, or equivalently any measure of overlap between states, can be put to use as a measure of entanglement. For example, for a state $\hat{\rho}$ one can define the trace distance (see Chap. 3) from the closest un-entangled state:

$$D_{ent}(\hat{\rho}) = \text{Inf}_{\hat{\rho}_{sep}} D(\hat{\rho}) = \text{Inf}_{\hat{\rho}_{sep}} \frac{1}{2} \left| \hat{\rho} - \hat{\rho}_{sep} \right|. \qquad (13.43)$$

Here, the infimum $\text{Inf}_{\hat{\rho}_{sep}}$ means to take the distance minimized over all separable or unentangled states $\hat{\rho}_{sep}$. This will vanish for all separable states, but will be non-zero for entangled states. Likewise, the fidelity of $\hat{\rho}$ with the nearest entangled state forms a similar entanglement measure.

For instance, for the pure entangled bipartite state $|\psi\rangle = \alpha|00\rangle + \beta|11\rangle$, the nearest unentangled state is $\hat{\rho}_{sep} = \alpha^2|00\rangle\langle00| + \beta^2|11\rangle\langle11|$; using this in the definition of D gives $D(\hat{\rho}) = |\alpha\beta|$. Clearly this vanishes if the state is unentangled (either α or β vanishes) and is maximal for a maximally entangled state ($|\alpha|^2 = |\beta|^2 = \frac{1}{2}$).

13.6   Entropy of Entanglement and Entanglement Monotones

Let $\hat{\rho}$ be the density operator for a bipartite system with component subsystems labeled A and B. In the simplest case, A and B might be labels for two entangled particles. Recall (Sect. 3.4) that if we have no knowledge of one of the subsystems (B, say), then the density operator for subsystem A is obtained by summing over all states of B, or in other words, by taking the partial trace over B:

$$\hat{\rho}_A = Tr_B(\hat{\rho}) = \sum_b \langle \chi_b | \hat{\rho} | \chi_b \rangle, \tag{13.44}$$

where the states $|\chi_b\rangle$, $b = 1, 2, 3 \ldots$, are basis states for system B. Similarly, the partial trace with respect to A gives the reduced state of B alone,

$$\hat{\rho}_B = Tr_A(\hat{\rho}) = \sum_a \langle \xi_a | \hat{\rho} | \xi_a \rangle, \tag{13.45}$$

where $|\xi_a\rangle$ are basis states for system A.

The **entropy of entanglement** of a bipartite pure state is then defined as the von Neumann entropy of either of these reduced states:

$$E(\hat{\rho}) = S(\hat{\rho}_A) = S(\hat{\rho}_B). \tag{13.46}$$

Note that it doesn't matter if the partial trace is taken over A or B. The result for $E(\hat{\rho})$ will always be the same, as is clear from Eqs. 13.26 and 13.27.

For a factorable product state, $|\Psi\rangle = |\psi_A\rangle |\phi_B\rangle$ with density operator

$$\hat{\rho} = |\psi_A\rangle |\phi_B\rangle \langle \phi_B | \langle \psi_A |, \tag{13.47}$$

the reduced density operators represent pure states, and so the entropy of entanglement vanishes, since $S = 0$ for pure states. But if the state is maximally entangled, for example $|\psi\rangle = \frac{1}{\sqrt{2}} (|01\rangle - |10\rangle)$, then the reduced densities become maximally mixed, with density operators proportional to the identity matrix. In this case, $S(\hat{\rho}_A) = S(\hat{\rho}_B) = 1$, and so the entropy of entanglement also equals 1, $E(\hat{\rho}) = 1$.

In general, the entanglement of formation satisfies $0 \le E(\hat{\rho}) \le \log_2 N$, where N is the dimension of the Hilbert space, with $E(\hat{\rho}) = 0$ signifying a product state and $E(\hat{\rho}) = \log_2 N$ for maximally entangled states. The entanglement entropy counts the number of entangled bits shared by the two systems, i.e., the number of e-bits.

The entanglement of formation is an example of an **entanglement monotone**. An entanglement monotone $E(\hat{\rho})$ of state $\hat{\rho}$ is a non-negative function that satisfies the conditions:

(i) $E(\hat{\rho}) = 0$ if $\hat{\rho}$ is separable.
(ii) $E(\hat{\rho})$ must be non-increasing under LOCC operations. In particular if a set of local operations acting on $\hat{\rho}$ give a set of potential states $\hat{\rho}_i$, each with

probability p_i, then:

$$E(\hat{\rho}) \geq \sum_i p_i E(\hat{\rho}_i) \geq E\left(\sum_i p_i \hat{\rho}_i\right). \tag{13.48}$$

The entanglement entropy is the unique entanglement monotone that has the additional property of additivity:

$$E(\hat{\rho}^{\otimes n}) = n E(\hat{\rho}). \tag{13.49}$$

In other words, the combined entropy present in multiple copies of a system is simply the sum of the entropies of the individual systems.

13.7 Concurrence

Again consider a two-qubit state. For a pure, entangled state of the form

$$|\psi\rangle = \alpha|00\rangle + \beta|01\rangle + \gamma|10\rangle + \delta|11\rangle, \tag{13.50}$$

the **concurrence** is defined to be

$$C = 2|\alpha\delta - \beta\gamma|. \tag{13.51}$$

In general, the concurrence will satisfy $0 \leq C \leq 1$. If the state is a product state, then $C = 0$. Like the entropy of entanglement, the concurrence is an entanglement monotone.

Unentangled states will always have vanishing concurrence. For example, suppose that $\gamma = \delta = 0$ in the state aforementioned; then the resulting state vector is unentangled, since it can be written as a product: $|\psi\rangle = |0\rangle_A (\alpha|0\rangle_B + \beta|1\rangle_B)$ and $C = 0$. However, if $\beta = \gamma = 0$, then $|\psi\rangle = \alpha|00\rangle + \beta|11\rangle$ is entangled, and the corresponding concurrence is non-vanishing, $C = 2\alpha\delta \neq 0$. For maximally entangled states, such as the Bell states, we find that $C = 1$.

For a mixed state, we can't write a state vector, so a more general definition of C is needed in order to write it in terms of a density operator. To do this, first define a new operator

$$\hat{R} = \sqrt{\sqrt{\hat{\rho}} \tilde{\rho} \sqrt{\hat{\rho}}}, \tag{13.52}$$

called the **spin-flip** operator, where

$$\tilde{\rho} = \left(\sigma_y \otimes \sigma_y\right) \hat{\rho}^* \left(\sigma_y \otimes \sigma_y\right). \tag{13.53}$$

Then diagonalize $\hat{R}$, to find its eigenvalues, λ_j, for $j = 1, 2, 3, 4$. Assume that the eigenvalues are ordered by size, with λ_4 the largest and λ_1 the smallest. Then the more general definition of concurrence is given by

$$C = \begin{cases} \lambda_4 - \lambda_3 - \lambda_2 - \lambda_1, & \text{if } \lambda_4 - \lambda_3 - \lambda_2 - \lambda_1 > 0 \\ 0, & \text{if } \lambda_4 - \lambda_3 - \lambda_2 - \lambda_1 < 0 \end{cases} \qquad (13.54)$$

This reduces to the previous expression for a pure state.

13.8   Entanglement Witnesses

Like the PPT condition, an **entanglement witness** $\mathcal{W}$ [2, 10, 11] gives a yes or no answer for the presence of entanglement. It can unambiguously detect entangled states when it has a value in a particular range, but it gives an indeterminate result in other ranges. To be specific, an *entanglement witness* $\mathcal{W}$ is a scalar function defined on a system's Hilbert space that obeys $\mathcal{W} > 0$ for all separable states.

In general, $\mathcal{W} \leq 0$ is a sufficient condition for entanglement, but it is not a necessary condition: if $\mathcal{W} \leq 0$, then entanglement is present, but if $\mathcal{W} > 0$, then the witness gives no information about entanglement or separability. In some cases, a *strong* entanglement witness may be found, which provides both a necessary *and* sufficient condition for entanglement: a strong witness satisfies $\mathcal{W} \leq 0$ if and only if the state is entangled. Often a witness is strong only for a certain category of states (Gaussian states, for example), but not for others. One may also define an *optimal* entanglement witness. This is a witness $\mathcal{W}$, which detects the largest number of entangled state among all possible witnesses: it detects all entangled states that are detectable by all other witnesses that can be defined on the same category of states.

Scalar entanglement witnesses are usually given as expectation values of operators: $\mathcal{W} = \text{Tr}\left(\hat{\rho}\hat{W}\right)$, such that $\langle \hat{W} \rangle = \text{Tr}\left(\hat{\rho}\hat{W}\right) < 0$ for all separable states $\hat{\rho}$.

Example 13.8.1 States sufficiently close to an entangled state are likely to be entangled, which leads to the idea that an operator of the form

$$\hat{W} = \alpha \hat{I} - |\Psi\rangle\langle\Psi|, \qquad (13.55)$$

where $|\Psi\rangle$ is an entangled pure state may always work as an entanglement witness on bipartite states if α is sufficiently large. It turns out that this works as a witness on bipartite states if $\alpha > \max|\langle\phi|\Psi^{\pm}\rangle|^2$ [12]. This witness can be readily generalized to multipartite states, but it only works on states that are not PPT [13].   ∎

Example 13.8.2 Consider the class of bipartite two-qubit states, $|\psi(\theta)\rangle = \cos\theta|00\rangle + \sin\theta|11\rangle$. Given any pure maximally entangled two-qubit state $|\Psi\rangle$, it can be shown [12, 14] that the operator

$$\hat{W} = \frac{1}{2}\hat{I} - |\Psi\rangle\langle\Psi| \qquad (13.56)$$

serves as an optimal entanglement witness. Thus, for example, the maximally entangled Bell state $|\Psi^-\rangle = \frac{1}{\sqrt{2}}\left(|01\rangle - |10\rangle\right)$ leads to the optimal witness

$$\hat{W} = \frac{1}{2}\hat{I} - |\Psi^-\rangle\langle\Psi^-| \qquad (13.57)$$

$$= \frac{1}{2}\hat{I} - \frac{1}{2}\left(|01\rangle\langle01| + |10\rangle\langle10| - |01\rangle\langle10| - |10\rangle\langle01|\right) \qquad (13.58)$$

$$= \frac{1}{2}\begin{pmatrix} 1 & 0 & 0 & 0 \\ 0 & 0 & 1 & 0 \\ 0 & 1 & 0 & 0 \\ 0 & 0 & 0 & 1 \end{pmatrix}. \qquad (13.59)$$

Similarly, using the Bell state $|\Phi^+\rangle = \frac{1}{\sqrt{2}}\left(|00\rangle + \langle11|\right)$ leads to another optimal entanglement witness,

$$\hat{W} = \frac{1}{2}\left(|01\rangle\langle01| + |10\rangle\langle10| - |00\rangle\langle11| - |11\rangle\langle00|\right) \qquad (13.60)$$

$$= \frac{1}{2}\begin{pmatrix} 0 & 0 & 0 & -1 \\ 0 & 1 & 0 & 0 \\ 0 & 0 & 1 & 0 \\ -1 & 0 & 0 & 0 \end{pmatrix}. \qquad (13.61)$$

Three- and four-qubit generalizations of the witness Eq. 13.56 can be found in [12]. ■

In Chap. 14, the Bell and CHSH inequalities will be shown to place limits on the values of certain quantities for classical systems. Entanglement witnesses can then be formed from these inequalities. For example, the CHSH variable S to be defined in Eq. 14.19 is constrained to obey $|S| \leq 2$ classically, so that $\mathcal{W} = 2 - |S|^2$ will form an entanglement witness.

An entanglement witness divides the Hilbert space of a system into two regions, with all separable states lying on one side of a boundary defined by the condition $\mathcal{W} = 0$ (Fig. 13.1). For a non-strong witness, there can be entangled states with $\mathcal{W} > 0$ that fall on the same side of the plane as all of the separable states. If the surface $\mathcal{W} = 0$ is planar, then $\mathcal{W}$ is a linear entanglement witness; otherwise, it is a nonlinear entanglement witness. The existence of linear entanglement witnesses is guaranteed by the fact that the set of separable states in Hilbert space is always a convex set (any two points in the set can be joined by a straight line that remains

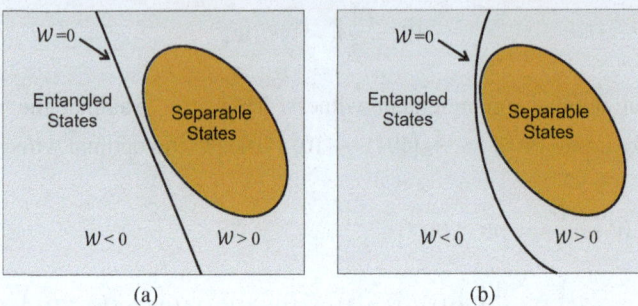

Fig. 13.1 (**a**) A linear entanglement witness: a planar surface divides the full Hilbert space into two regions, with the separable states entirely on one side of the plane. The separable states form a convex set within this region. (**b**) A nonlinear entanglement witness: the boundary surface $\mathcal{W} = 0$ is now a curved surface, which excludes more entangled states from the $\mathcal{W} > 0$ side. This is a finer witness than the one in (**a**)

entirely in the set); for a convex, a plane can always be found such that the set is entirely on one side of the plane.

Given two entanglement witnesses $\mathcal{W}_1$ and $\mathcal{W}_2$ for the same quantum system, $\mathcal{W}_1$, is called *finer* than $\mathcal{W}_2$, if $\mathcal{W}_1$ excludes more of the entangled states from the positive side than $\mathcal{W}_2$. In Fig. 13.1, the nonlinear witness on the right is finer than the linear witness on the left.

Entanglement witnesses are useful for two-qubit systems, but generally become harder to find as the number of qubits increases.

13.9   Entanglement Distillation and Dilution

Suppose an experimenter has a large many-particle entangled quantum state (possibly mixed), or a large number (n) of non-maximally entangled pure states $|\psi\rangle$. Then it turns out that in either case, the system can be distilled down to a smaller number of maximally entangled pure states, in a process called **entanglement distillation**. The number of pure states that can be distilled can serve as a measure of the entanglement of the original system. In the bipartite case, the maximally entangled state can be taken, without loss of generality, to be a Bell state. The number of Bell states obtained in this way is called the number of **e-bits** in the state, and one e-bit is then taken as the basic unit of entanglement.

The distillation down to a smaller number of more highly-entangled states is done using LOCC operations, i.e., Alice and Bob can communicate classically to coordinate their actions, but any transformations one of them applies can act only on the part of the state that is contained locally in their own lab. We don't require that the distilled state $|\phi\rangle$ be identical to the original state $|\psi\rangle^{\otimes n}$, only that the two states have fidelity $\mathcal{F}(|\phi\rangle, |\psi\rangle^{\otimes n})$ that approaches unity as $n \to \infty$. This is done essentially in the same manner that messages are compressed in classical communication theory: thinking of the quantity x defined further (near Eqs. 13.66 and 13.67) as a sequence, we simply ignore all atypical sequences.

The converse operation can also be performed: given a sufficiently large number of Bell states, we can construct any non-maximally entangled state through a process of **entanglement dilution** or **entanglement formation**. For pure states, the measures of entanglement defined using distillation and dilution will always be identical; however, for mixed states they may differ.

First, suppose Alice and Bob share a non-maximally entangled pure state,

$$|\psi_{AB}\rangle = \alpha|00\rangle + \beta|11\rangle, \tag{13.62}$$

where $|\alpha|^2 + |\beta|^2 = 1$, and $|0\rangle$ and $|1\rangle$ are some binary basis. Alice has the first bit in the ket, and Bob has the second. Recall that the state would be maximally entangled only if the two states in the superposition have the same probability, i.e., if $|\alpha|^2 = |\beta|^2 = \frac{1}{2}$. Now suppose multiple copies of this state are produced. (This does not violate the no-cloning theorem: cloning of a state is only prohibited if the state is unknown. The experimenter is free to prepare as many identical copies of a known state as he or she desires simply by repeating the preparation process over and over.) Consider two copies, which gives a four-particle state of the form

$$|\psi_{AB}\rangle^{\otimes 2} = \alpha^2|0_A 0_B 0_A 0_B\rangle + \alpha\beta\Big(|0_A 0_B 1_A 1_B\rangle + |1_A 1_B 0_A 0_B\rangle\Big) + \beta^2|1_A 1_B 1_A 1_B\rangle$$

$$= \alpha^2|00\rangle_A|00\rangle_B + \alpha^2\beta\Big(|01\rangle_A|01\rangle_B + |10\rangle_A|10\rangle_B\Big). \tag{13.63}$$

Similarly, for three copies,

$$|\psi_{AB}\rangle^{\otimes 3} = \alpha^3|0_A 0_B 0_A 0_B 0_A 0_B\rangle + \beta^3|1_A 1_B 1_A 1_B 1_A 1_B\rangle \tag{13.64}$$

$$+\alpha^2\beta\big(|0_A 0_B 0_A 0_B 1_A 1_B\rangle + |0_A 0_B 1_A 1_B 0_A 0_B\rangle + |1_A 1_B 1_A 0_B 0_A 0_B\rangle\big)$$

$$\alpha\beta^2\big(|1_A 1_B 0_A 0_B 1_A 1_B\rangle + |1_A 1_B 1_A 1_B 0_A 0_B\rangle + |0_A 0_B 1_A 1_B 1_A 1_B\rangle\big)$$

$$= \alpha^3|000\rangle_A|000\rangle_B + \beta^3|111\rangle_A|111\rangle_B \tag{13.65}$$

$$+\alpha^2\beta\big(|001\rangle_A|001\rangle_B + |010\rangle_A|010\rangle_B + |100\rangle_A|100\rangle_B\big)$$

$$+\alpha\beta^2\big(|011\rangle_A|011\rangle_B + |110\rangle_A|110\rangle_B + |101\rangle_A|101\rangle_B\big).$$

More generally, for m copies,

$$|\psi_{AB}\rangle^{\otimes m} = \sum_{\{x_1,\ldots,x_m\}\in\{0,1\}} \sum_{n=0}^{m} \alpha^n\beta^{m-n}|x_1, x_2, \ldots x_m\rangle_A|x_1, x_2, \ldots x_m\rangle_B, \tag{13.66}$$

where n of the x_j are equal to 0 and $m - n$ is the number of them that are equal to 1. Notice that this is really a Schmidt decomposition of $|\psi_{AB}\rangle^{\otimes m}$. It be written more compactly as

$$|\psi_{AB}\rangle^{\otimes m} = \sum_x \sqrt{p(x)}|x\rangle_A|x\rangle_B, \tag{13.67}$$

where x is the sequence $x = \{x_1, x_2, \ldots, x_m\}$, and $p(x) = \sum_n |\alpha^n \beta^{m-n}|^2$. We now distill this to a new state by dropping all terms in the sum that contain sequences x that are ϵ-atypical in the sense defined in Sect. 8.8. The resulting state will be non-normalized, so then we renormalize by dividing by the square root of the state's norm. Defining T_ϵ to be the set of ϵ-typical sequences of length m, the un-normalized state is

$$|\xi\rangle = \sum_{x \in T_\epsilon} \sqrt{p(x)}|x\rangle_A|x\rangle_B, \tag{13.68}$$

and the final normalized state is

$$|\phi\rangle = \frac{|\xi\rangle}{\sqrt{\langle\xi|\xi\rangle}}. \tag{13.69}$$

As $m \to \infty$ the number of atypical sequences becomes negligible, so that the fidelity of $|\phi\rangle$ with $|\psi_{AB}\rangle^{\otimes m}$ satisfies $\mathcal{F} \to 1$. Let $\hat{\rho} = \mathrm{Tr}_A|\psi_{AB}\rangle\langle\psi_{AB}| = |\alpha|^2|0\rangle\langle0| + |\beta|^2|1\rangle\langle1|$. Then Alice and Bob both have mixed states,

$$\hat{\rho}_A = \hat{\rho}_B = \sum_{x \in T_\epsilon} p(x)^n(1 - p(x))^{m-n}|x\rangle\langle x| = \hat{\rho}^{\otimes m}. \tag{13.70}$$

The number of ϵ-typical states is at most $2^{m(S(\hat{\rho}_A)+\epsilon)}$, each with probability approaching $2^{-m\,S(\hat{\rho}_A)}$, similar to the classical result. So for sufficiently large m, the distilled state becomes arbitrarily close to

$$|\phi\rangle = \sum_{x \in T_\epsilon} 2^{-\frac{1}{2}mS(\hat{\rho}_A)}|x\rangle_A|x\rangle_B. \tag{13.71}$$

This can be seen as follows: by the definition of typical sequences, it follows that $\langle\psi_{AB}^{\otimes m}|\phi\rangle = \langle\xi|\xi\rangle$, so that

$$\mathcal{F} = |\langle\psi_{AB}^{\otimes m}|\phi\rangle|^2 = \frac{|\langle\psi_{AB}^{\otimes m}|\xi\rangle|^2}{\langle\xi|\xi\rangle|^2} \tag{13.72}$$

$$= \langle\xi|\xi\rangle = \sum_{x \in T_\epsilon} p(x) \tag{13.73}$$

$$= p(x \in T_\epsilon) \tag{13.74}$$

$$\geq 1 - \delta. \tag{13.75}$$

The distilled state is maximally entangled and can be built out of $n(m) = m\,S(\hat{\rho}_A)$ Bell states. The **entanglement of distillation** is then defined to be

$$E_D = \lim_{m \to \infty} \frac{n(m)}{m}, \tag{13.76}$$

the asymptotic number of Bell states that are distilled per copy of the original state. For a collection of pure states, $E_D = S(\hat{\rho}_A)$. Starting from mixed states, there is no known general expression for E_D. It should also be noted that all of the expressions here written in terms of the quantum entropy of state $\hat{\rho}$ can also be written in terms of the classical Shannon entropy of the sequence $x = x_1 x_2 \dots x_m$, since $S(\hat{\rho}) = H(x)$.

Conversely, the **entanglement of dilution** or **entanglement of formation** is

$$E_F(\hat{\rho}_{AB}) = \lim_{m \to \infty} \frac{q(m)}{m} \tag{13.77}$$

where $q(m)$ is the minimum number of Bell states needed to form m copies of a state. The entanglement of formation can also be written in the form

$$E_F(\hat{\rho}_{AB}) = \min \sum_i p_i\, E\big(|\psi\rangle\langle\psi|\big) \tag{13.78}$$

where the minimum is over all possible statistical decompositions of the density operator as mixtures of pure states, $\hat{\rho}_{AB} = \sum_i p_i |\psi_i\rangle$.

For pure states, the entanglements of formation and distillation are equal to each other and to the entanglement entropy:

$$E_D\big(|\psi\rangle_{AB}\big) = E_F\big(|\psi\rangle_{AB}\big) = E\big(|\psi\rangle_{AB}\big). \qquad \text{(pure states)} \tag{13.79}$$

But they may differ for mixed states. In general, E_D and E_F obey the inequality

$$E_D(\hat{\rho}_{AB}) \le E_F(\hat{\rho}_{AB}). \tag{13.80}$$

The difference between the two sides of this inequality is called the **bound entanglement**:

$$B(\hat{\rho}_{AB}) = E_F(\hat{\rho}_{AB}) - E_D(\hat{\rho}_{AB}) \ge 0. \tag{13.81}$$

States with nonzero values of B are called **bound states**. Bound-entangled states are require entanglement to be formed, but from which no Bell states can be distilled. In particular, all bipartite entangled states with non-negative partial transpose are bound states.

The entanglement of formation is related to the concurrence by

$$E_F(\hat{\rho}) = h\big(C(\hat{\rho})\big), \tag{13.82}$$

where $h(y)$ is the binary entropy function, $h(y) = -y \log_2(y) - (1-y) \log_2(1-y)$.

13.10   Quantum Discord

Although entanglement is the most common form of non-classical correlations, it is not the only one. The quantum discord measures quantum correlations in general, including those that may not necessarily be due to entanglement. It measures the gain in information on a composite quantum system when a measurement is made on just one subsystem.

Suppose two quantities I and J are defined for a pair of random variables X and Y:

$$I(X : Y) = H(X) + H(Y) - H(X, Y) \tag{13.83}$$

$$J(X : Y) = H(X) - H(X|Y), \tag{13.84}$$

where $H(X), H(Y)$ are the Shannon entropies and $H(X|Y)$ is the conditional information. Classically, these two quantities are always equal to each other, $I(X : Y) = J(X : Y)$, and they just give the mutual information between X and Y. This follows immediately from the fact that, classically, $H(X|Y) = H(X, Y) - H(X)$.

However, in a quantum system, the latter relation is not necessarily true: it is possible that $H(X|Y) \neq H(X, Y) - H(X)$, since in a quantum-correlated system a measurement on X may affect Y as well [15]. As a result, $I(X : Y)$ and $J(X : Y)$ may not be equal either. The **quantum discord** measures the discrepancy between these two measures of mutual information. To define the discord more precisely, consider a composite system $\hat{\rho} = \hat{\rho}_A \otimes \hat{\rho}_B$. The quantum analog of $I(X : Y)$ is then

$$I(\hat{\rho}) = S(\hat{\rho}_A) + S(\hat{\rho}_B) - S(\hat{\rho}), \tag{13.85}$$

where S is von Neumann entropy. The quantum analog of J is given by

$$J(\hat{\rho}) = S(\hat{\rho}_A) - S(\hat{T}), \tag{13.86}$$

where

$$\hat{T} = \sum_j p_j S(\hat{\rho}_A). \tag{13.87}$$

Here, $\hat{\rho}_A$ is the reduced density matrix of A after a projective measurement has been made on B, and p_j is the probability of the measurement result. j labels the outcomes of the B measurements. The quantum discord is then defined to be

$$\mathcal{D}_A(\hat{\rho}) = I(\hat{\rho}) - \text{Max}\left(J\left(\hat{\rho}\right)\right), \tag{13.88}$$

with the maximum being taken over all possible sets of projective measurements on B.

On pure states, the discord coincides with the entanglement of formation. But the discord can be nonzero even for some separable (i.e. non-entangled) states, so although it measures quantum correlation, it is distinct from entanglement. Note that this definition is asymmetric in terms of A and B; in general $\mathcal{D}_A \neq \mathcal{D}_B$.

Because of the maximization over sets of measurements, the discord can be hard to calculate, although it is possible in some simple cases. For a more detailed review of quantum discord, see [16]. For more on its role in quantum information processing and its relation to other measures of quantum correlation, see [17–20].

Problems

1. Find the Schmidt decomposition and the Schmidt number of the following states.

   (a) $|\psi\rangle = \frac{1}{2}(|00\rangle - |01\rangle - |10\rangle + |11\rangle)$
   (b) $|\psi\rangle = \frac{1}{\sqrt{6}}(|00\rangle + |01\rangle - 2|10\rangle)$
   (c) $|\psi\rangle = \frac{1}{\sqrt{3}}(|00\rangle - |10\rangle + |11\rangle)$

2. Prove that a state $\hat{\rho}$ is a product state if and only if its partial traces $\hat{\rho}_A$ and $\hat{\rho}_B$ are pure states.

3. Show that the three partial traces of the pure tripartite state $|\psi\rangle = \frac{1}{\sqrt{2}}(|000\rangle - |011\rangle)$ do not all have the same eigenvalues; as a result, the Schmidt decomposition does not exist for this state.

4. Consider the density matrices $\hat{\rho}_+ = |\Phi^+\rangle\langle\Phi^+|$ and $\hat{\rho}_- = |\Psi^-\rangle\langle\Psi^-|$, where $|\Phi^+\rangle$ and $|\Psi^-\rangle$ are Bell states. Compute the partial transposes $(\hat{\rho}_\pm)^{T_A}$. Based on the eigenvalues of the resulting matrices, what can be said about the entanglement or separability of these states?

5. Consider the entangled Bell state $|\Psi^+\rangle = \frac{1}{\sqrt{2}}(|01\rangle + |10\rangle)$. Rewrite the state in the diagonal basis $|\pm\rangle = \frac{1}{\sqrt{2}}(|0\rangle \pm |1\rangle)$, and show that the state remains manifestly entangled.

6. (a) Reconsider the entanglement witness $\hat{W} = \alpha\hat{I} - |\Psi\rangle\langle\Psi|$ given in Eq. 13.55. (Recall that $|\Psi\rangle$ is assumed to be entangled.) What is the geometric meaning of such an operator? (Hint: if the answer isn't clear, consult Sect. 17.5.)

   (b) Given a mixed state $\hat{\rho}$, provide an argument that $\hat{W}$ can serve as an entanglement witness for sufficiently large α by looking at the fidelity between $\hat{\rho}$ and $|\Psi\rangle$.

7. The **Werner states** $\rho_\lambda = \lambda|\Psi^-\rangle\langle\Psi^-| + \frac{1-\lambda}{4}\hat{I}$, are mixed states defined in terms of the Bell state $|\Psi^-\rangle$. Calculate the concurrence of these states. What

conclusions can we reach about entanglement or separability for various ranges of the parameter λ?

8. (a) Calculate the entanglement entropy for the Werner state of the previous problem.

   (b) Find $I(\rho)$ for the Werner state. (For a calculation of the alternate quantum mutual information measure $J(\rho)$ and a plot of the resulting quantum discord, see [15].)

9. (a) Let $|\psi\rangle = \alpha|00\rangle + \beta|01\rangle + \gamma|10\rangle + \delta|11\rangle$ be a generic normalized bipartite pure state. Show that the partial trace $\hat{\rho}_A = \mathrm{Tr}_A \hat{\rho}$ is related to the concurrence by $\mathrm{Tr}\hat{\rho}_A^2 = 1 - \frac{C^2}{2}$.

   (b) Verify that Bell states always have $C = 1$.

10. (a) Consider the bipartite state $|\psi\rangle = \frac{1}{\sqrt{2}}(|01\rangle - |10\rangle)$. Compute the entanglement entropy and thus show that this state is maximally entangled.

    (b) Find the entanglement entropy of the state

$$|\psi\rangle = \frac{1}{\sqrt{2 + 2|\alpha|^2}}\Big(|00\rangle + |11\rangle + \alpha\big(|01\rangle + |10\rangle\big)\Big),$$

where α is a complex number. At what values of α will the state be maximally entangled? At what value(s) will it be separable?

References

1. M.B. Plenio, S. Virmani, Quantum Inf. Comput. **7**, 1 (2007)
2. R. Horodecki, P. Horodecki, M. Horodecki, K. Horodecki, Rev. Mod. Phys. **81**, 865 (2009)
3. G. Jaeger, *Quantum Information: An Overview* (Springer, Berlin, 2007)
4. T. Baumgratz, M. Cramer, M.B. Plenio, Phys. Rev. Lett. **113**, 140401 (2014)
5. D. Girolami, Phys. Rev. Lett. **113**, 170401 (2014)
6. T. Chanda, S. Bhattacharya, Ann. Phys. **366**, 1 (2016)
7. Y. Yao, X. Xiao, L. Ge, C. P. Sun, Phys. Rev. A **92**, 022112 (2015)
8. B. de Lima Bernardo, Phys. Scr. **95**, 045104 (2020)
9. A. Streltsov, G. Adesso, M.B. Plenio, Rev. Mod. Phys. **89**, 041003 (2017)
10. M. Horodecki, P. Horodecki, R. Horodecki, Phys. Lett. A **223**, 1 (1996)
11. B.M. Terhal, Phys. Lett. A **271**, 319 (2000)
12. O. Gühne, *Detecting Quantum Entanglement: Entanglement Witnesses and Uncertainty Relations*. University of Hanover Doctoral Thesis (2004). Available at https://d-nb.info/972550216/34
13. M. Bourennane, et al., Phys. Rev. Lett. **92**, 087902 (2004)
14. O. Gühne, G. Tóth, Phys. Rep. **474**, 1 (2009)
15. H. Ollivier, W.H. Zurek, Phys. Rev. Lett. **88**, 017901 (2001)
16. A. Bera, et al., Rep. Prog. Phys. **81**, 024001 (2018)
17. A. Datta, A. Shaji, C.M. Caves, Phys. Rev. Lett. **100**, 050502 (2008)
18. B.P. Lanyon, M. Barbieri, M.P. Almeida, A.G. White, Phys. Rev. Lett. **101**, 200501 (2008)
19. M. Piani, P. Horodecki, R. Horodecki, Phys. Rev. Lett. **100**, 090502 (2008)
20. A. Brodutch, D.R. Terno, Phys. Rev. A **81**, 062103 (2010)

Chapter 14
EPR, Bell Inequalities, and Local Realism

Einstein, Podolsky, and Rosen (EPR) discovered an apparent paradox that seemed to unveil an inconsistency at the heart of quantum mechanics. It wasn't until thirty years later that John Bell pointed out that there was a way to experimentally test the assumptions of the EPR argument and to distinguish between their deterministic version of reality and quantum mechanics. A long succession of experiments done since the 1980s have conclusively come down on the side of quantum mechanics and seem to have shown that the EPR assumption of *local realism* cannot be maintained. In this chapter, we give an introduction to these developments, which provide a theoretical underpinning to applications such as quantum cryptography, quantum secret sharing, and some quantum computing algorithms.

14.1   Local Realism and EPR

Despite the unprecedented success of quantum mechanics in explaining microscopic phenomena and its rapid experimental verification in multiple areas, there was a long-standing resistance to many of its counter-intuitive conceptual underpinnings. In particular, Einstein was a prominent critic of the idea that probabilities were inescapable in quantum mechanics; he insisted that quantum mechanics had to be incomplete and that once the missing ingredient was found, quantum mechanics would be revealed to be a fully deterministic theory along the lines of classical mechanics. In 1935, Einstein, Podolsky, and Rosen (EPR) constructed a thought experiment that seemed to reveal the incompleteness of quantum mechanics. Known as the **EPR paradox**, this result claimed to show that the values of non-commuting variables could be simultaneously determined without any uncertainty in either variable, which would contradict the Heisenberg uncertainty principle.

   Rather than the original version of the EPR paradox [1], here we will sketch David Bohm's reformulation [2, 3]. Bohm's version of the paradox consists of a

© The Author(s), under exclusive license to Springer Nature Switzerland AG 2025
D. S. Simon, *Introduction to Quantum Science and Technology*, Undergraduate
Texts in Physics, https://doi.org/10.1007/978-3-031-81315-3_14

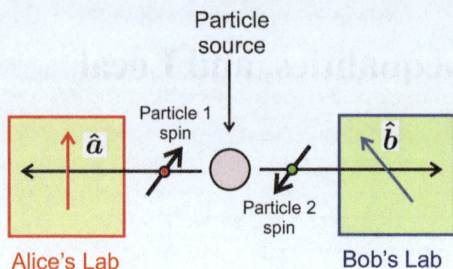

Fig. 14.1 Bohm's version of the EPR thought experiment. A particle source produces two electrons emitted in opposite directions. The spin directions of the electrons are unknown, but spins of two particles are always opposite to each other. Alice and Bob both make measurements of the spin direction for the particle they receive, making their measurements along respective axes $\hat{a}$ and $\hat{b}$

two-particle source, which produces pairs of electrons (Fig. 14.1). These electrons are emitted in opposite directions, and although the directions of the spins are random, the two spins are always perfectly anti-correlated: when measured along any direction, if one spin is measured to be up, then the other must be down. One electron reaches an experimenter named Alice (A), and the other is received by Bob (B).

Recall that the direction (up or down) along direction $\hat{n}$ of a spin can be measured using a **Stern-Gerlach apparatus**. In this apparatus, a non-uniform magnetic field B is present, pointing along direction $\hat{n}$. In the presence of a field gradient, there will be a force on the magnetic dipole moment of the electron, $F = \nabla\left(\mu \cdot B\right)$. For a negatively charged electron, μ is opposite to the spin direction. So, spin-up electrons are deflected in the direction opposite to the field gradient, and spin-down electrons are deflected along the gradient, allowing spin direction to be easily determined.

Alice and Bob will each use a Stern-Gerlach apparatus to measure the spin of the particle they receive along some axis. The axis used by Alice is along unit vector $\hat{a}$, and that used by Bob is along $\hat{b}$. Because of the spin-anticorrelation, if they use the same axis, $\hat{a} = \hat{b}$, then they should always get opposite results, regardless of which way $\hat{a}$ points.

Now consider the case where $\hat{a}$ and $\hat{b}$ are perpendicular. To be specific, take $\hat{a}$ along the z-axis and $\hat{b}$ along the x-axis, so that Alice is measuring the value of $\hat{S}_z$ for one particle and Bob is measuring $\hat{S}_x$ of the other. We know that *for an individual particle* $\hat{S}_z$ and $\hat{S}_x$ are incompatible (non-commuting) variables, $[\hat{S}_z, \hat{S}_z] \neq 0$. So the uncertainty relation tells us that $\Delta S_x \Delta S_z > 0$, and we can't know both values at the same time with precision.

But here, Alice and Bob are making measurements on *different* particles. We can also assume that Alice and Bob are working in widely separated labs (possibly one is on Earth and the other on Mars). The measurements are made very close together in time, so that (assuming special relativity holds) any disturbance caused by measuring one particle would not have time to propagate its effects to the other particle before the second measurement.

So the EPR proposal is this: Alice and Bob synchronize their clocks and agree to make their measurements at a specific time. Alice measures $\hat{S}_z$ on her particle. Suppose that she gets the value $S_z = +1$; then she immediately knows that Bob's particle must have $S_z = -1$. Immediately after Alice makes her measurement, Bob measures $\hat{S}_x$ on his particle. Suppose that he gets the value $S_x = +1$; then he immediately knows that Alice's particle must have $S_x = -1$. Alice then phones up Bob (assuming cell phone reception is good on Mars), and they report their results to each other. The result is that both of them know the values of both S_x and S_z for both particles, in contradiction to the uncertainty principle. To EPR, this demonstrated that quantum mechanics has to be incomplete. They postulated that quantum mechanics must be deterministic and that the appearance of probabilities is due to the fact that there is another underlying variable determining seemingly random variables like the spin direction, and if we measured that underlying deterministic variable as well, then we could determine the values of all observables with certainty.

The EPR argument makes two assumptions:

• It assumes **locality**: the idea that causes can create effects instantly only within their immediate vicinity. To affect distant objects, like the particle in the other lab, time is required for the effect to propagate (at a speed $\leq c$) from one lab to the other. Thus, if the labs are far enough apart in space and the measurements close enough together in time, then the first measurement cannot possibly disturb the second measurement.

• The second assumption is the assumption of **reality**: that observable quantities such as the values of S_x and S_z exist at all times, even if they haven't been measured.

Together, these two assumptions form the concept of **local realism**. A locally real theory of the type envisioned by EPR is known as a **hidden variable theory**.

Quantum mechanics clearly runs afoul of the second assumption, that of realism. As we have seen many times, a particle can be in a superposition state. For example, an electron can be in a superposition of two spins states, or a photon can be in a superposition of two positions (as when passing through two different slits in a screen). So observables like spin, polarization, or position *do not have definite values* until a measurement is made, collapsing the state down to an eigenstate of the measured variable.

As seen in Sect. 10.7, quantum mechanics can also produce nonlocal interference effects. Be careful to note though that despite the violation of locality, quantum mechanics does not conflict with special relativity. *There is no way to use the nonlocality to transmit information faster than the speed of light.* In interferometry experiments, for example, nonlocal effects only become apparent once the results of the two (possibly widely separated measurements) are brought together for comparison, in a coincidence circuit or by some other means. This bringing together of the two results can only be done at subluminal speeds, in accord with relativity.

Example 14.1.1 Following [4], we can give an example of a simple hidden variables theory. Consider Bohm's EPR setup again, and assume that local realism holds so that we can define vectors to specify the unknown but well-defined spin directions of the electrons. Define two unit vectors $\hat{V}_a$ and $\hat{V}_b$ to be the directions of the spins received by Alice and Bob, respectively. Because the source produces perfectly anti-correlated spins, we know that $\hat{V}_b = -\hat{V}_a$. We can then take $\hat{V}_a$ as our unknown, but determinate, hidden variable.

Now assume Alice and Bob are making their measurements along different directions, $\hat{a} \neq \hat{b}$, and define two new observable quantities:

$$\epsilon_a = \text{sign}\left(\hat{a} \cdot \hat{V}_a\right) = \text{sign}\left(\cos\theta\right) \tag{14.1}$$

$$\epsilon_b = \text{sign}\left(\hat{b} \cdot \hat{V}_b\right) = -\text{sign}\left(\hat{b} \cdot \hat{V}_a\right) = -\text{sign}\left(\cos\left(\theta - \phi\right)\right), \tag{14.2}$$

where θ and ϕ are as defined in Fig. 14.2. Taking the product of these new variables,

$$\epsilon_a \epsilon_b = -\text{sign}\left(\cos\theta\right) \cdot \text{sign}\left(\cos\left(\theta - \phi\right)\right) \tag{14.3}$$

$$= -\text{sign}\left(\cos\theta \cos\left(\theta - \phi\right)\right). \tag{14.4}$$

The average of $\epsilon_a \epsilon_b$ over all angles is

$$E_{HV}(\hat{a}, \hat{b}) = \frac{1}{2\pi} \int_0^{2\pi} \epsilon_a \epsilon_b \, d\theta \tag{14.5}$$

$$= -\frac{1}{2\pi} \int_0^{2\pi} \text{sign}\left(\cos\theta \cos\left(\theta - \phi\right)\right) d\theta \tag{14.6}$$

$$= \frac{2}{\pi}\phi - 1, \tag{14.7}$$

where the subscript HV denotes hidden variable theory. In contrast, quantum mechanics predicts

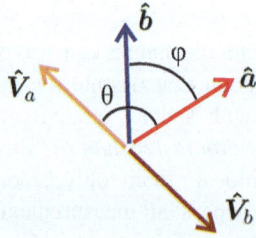

Fig. 14.2 In a local hidden variable theory, electrons have well-defined spin directions, $\hat{V}_a$ and $\hat{V}_b$, even before any measurement is made. Angles θ and ϕ are defined by directions of vectors and of measurement axes, $\hat{a}$ and $\hat{b}$, as shown

$$E_{QM} = -\cos \boldsymbol{a} \cdot \cos \boldsymbol{b} = -\cos \phi. \qquad (14.8)$$

Carrying out multiple trials at different angles and measuring spins on each, the average can be estimated experimentally to determine whether the hidden variable theory or quantum mechanics gives the correct result.

When Alice and Bob measure along the same direction ($\phi = 0$), this expression agrees with quantum mechanics: $E_{HV} = E_{QM} = -1$. But if Alice and Bob orient their measurements along different directions ($\phi \neq 0$), the two theories make different predictions: $E_{HV} \neq E_{QM}$. So the validity of this hidden variable theory can be tested experimentally simply by measuring E for angles other than $\phi = 0$. Needless to say, experiment will agree with quantum mechanics, not with the hidden variable prediction. ∎

However, the results above simply rule out this one particular hidden variable theory. An infinite number of different such theories can be constructed. So how can we be sure that *no* hidden variable theory can reproduce the results of quantum mechanics? We need to have some way to rule out *all possible* hidden variable theories. This is exactly what Bell's theorem accomplishes.

14.2 Bell's Theorem

The question of whether a local hidden variable theory of EPR type was viable was an open question for thirty years, until the mid-1960s, when John Stuart Bell came up with a way to potentially answer it [5–7]. Begin by once again assuming that Alice and Bob measure the spins of their particles along different axes. But now consider **three** directions, $\hat{a}$, $\hat{b}$, and $\hat{c}$, and allow each of the two experimenters to choose any of these directions when making their measurements. Analogous to definitions of ϵ_a and ϵ_c in the hidden variables theory of the last section, define:

$$\epsilon_a = \begin{cases} +1 & \text{if a measurement finds spin up along } \hat{a} \\ -1 & \text{if Alice measures spin up along } \hat{a}, \end{cases} \qquad (14.9)$$

with a similar definition for ϵ_b and ϵ_c. Then the products of these variables can only take the values

$$\epsilon_a \epsilon_b = \pm 1, \quad \epsilon_b \epsilon_c = \pm 1, \quad \epsilon_a \epsilon_c = \pm 1, \qquad (14.10)$$

where we assume Alice is making the first measurement in the product and Bob is making the second one. Let E once again be the average measured value of these products, so that $E(\hat{a}, \hat{b})$ is the average when Alice measures along $\hat{a}$ and Bob measures along $\hat{b}$, $E(\hat{b}, \hat{c})$ is the average when Alice measures along $\hat{b}$ and Bob measures along $\hat{c}$, and so on. Then Bell's theorem says:

Bell's Theorem

With E defined as above, any hidden-variable theory obeying local realism must obey the inequality

$$\left| E_{HV}(\hat{a}, \hat{b}) - E_{HV}(\hat{a}, \hat{c}) \right| \le 1 + E_{HV}(\hat{b}, \hat{c}),   \qquad (14.11)$$

for *any* axes defined by unit vectors $\hat{a}$, $\hat{b}$, and $\hat{c}$.

It can be readily verified that this is obeyed by the hidden-variable theory of the previous section. However, it can easily be seen to be violated by quantum mechanics. If ϕ is the angle between $\hat{a}$ and $\hat{b}$, then quantum mechanics predicts:

$$E_{QM}(\hat{a}, \hat{b}) = \langle \epsilon_a \epsilon_b \rangle = -\hat{a} \cdot \hat{b} = -\cos \phi.   \qquad (14.12)$$

(The minus sign arises because we still assume the spins are created pointing in opposite directions: if Alice and Bob use the same measurement direction (so $\phi = 0$), they have to get opposite results: $\epsilon_a \epsilon_b = -1$.)

Example 14.2.1 Consider unit vectors $\hat{a}$ and $\hat{c}$ that are oriented at angles $\pm 60°$ from $\hat{b}$, as in Fig. 14.3. So,

$$\hat{a} \cdot \hat{b} = \hat{b} \cdot \hat{c} = \cos 60° = \frac{1}{2}   \qquad (14.13)$$

$$\hat{a} \cdot \hat{c} = \cos 120° = -\frac{1}{2}   \qquad (14.14)$$

$$|E_{QM}(\hat{a}, \hat{b}) - E_{QM}(\hat{a}, \hat{c})| = \left| -\left(\frac{1}{2} \right) - \left(+\frac{1}{2} \right) \right| = 1   \qquad (14.15)$$

$$1 + E_{QM}(\hat{b}, \hat{c}) = 1 + \left(-\frac{1}{2} \right) = \frac{1}{2}.   \qquad (14.16)$$

Comparing the last two lines, it is clear that the Bell inequality is violated. ∎

This provides a way to determine experimentally whether an EPR-type local hidden variable theory is lurking beneath what we know of physics or if instead

Fig. 14.3 In Example 14.2.1, the three unit vectors used to determine the Bell inequality are in a single plane, with adjacent vectors separated by 60°

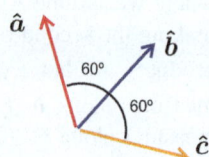

standard quantum mechanics, with its probabilistic nature, is correct: run many trials of an experiment to measure anticorrelated spins along various axes, compute the average E defined above for each set of axes, and check to see if the Bell inequality is satisfied. If the inequality is satisfied in each run of the experiment, then as the number of experiments grows, our confidence that a workable hidden variable theory has been found should also grow. However, if the Bell inequality is *ever* violated in any experiment (and that violation can be reproduced by other experimenters), then we can be confident that the EPR position is incorrect and that the standard version of quantum mechanics is correct, without underlying hidden variables.

Although there have been some experimental tests of the original Bell inequality, most experiments done to date have actually tested a different, but essentially equivalent, relation, called the CHSH inequality. This is described in the next section.

14.3 CHSH Inequalities

The Clauser-Horne-Shimony-Holt (CHSH) inequality [8] is a condition that a locally realistic physical theory must satisfy but that quantum mechanics can violate. As with the Bell inequality, experiment clearly shows that nature violates this inequality.

Once again, assume Alice and Bob share a source of entangled bipartite states with perfectly anticorrelated spins. From each entangled pair, each experimenter measures one particle's spin along a given spatial direction. For measurement along the direction of a unit vector $\hat{n}$, the corresponding result can again take values $\epsilon_{\hat{n}} = \pm 1$, indicating spin up or down along the given direction.

But now, Alice chooses two directions, specified by a pair of unit vectors $\hat{a}$ and $\hat{a}'$, and on each trial, she randomly chooses to measure spin along one of these two directions. Bob does the same along the axes specified by two different vectors $\hat{b}$ and $\hat{b}'$. So on each trial, the pair of experimenters obtains a pair of numbers from among the following set,

$$(\epsilon_a, \epsilon_b), \ (\epsilon_{a'}, \epsilon_b), \ (\epsilon_a, \epsilon_{b'}), \ (\epsilon_{a'}, \epsilon_{b'}). \tag{14.17}$$

Consider the situation according to classical physics first. Even though only two of the four values ϵ_a, ϵ_b, $\epsilon_{a'}$, $\epsilon_{b'}$ can be measured on a given trial, all four of them (according to EPR) have definite values. So it is perfectly acceptable to consider a combination such as $S = (\epsilon_a - \epsilon_{a'})\epsilon_b + (\epsilon_a + \epsilon_{a'})\epsilon_{b'}$. On each trial, it must be true that either $(\epsilon_a - \epsilon_{a'}) = 0$ and $\epsilon_a + \epsilon_{a'} = \pm 2$ or that $(\epsilon_a - \epsilon_{a'}) = \pm 2$ and $\epsilon_a + \epsilon_{a'} = 0$. So it should be easy to convince yourself that the measured values, although random, must always satisfy the condition

$$(\epsilon_a - \epsilon_{a'})\epsilon_b + (\epsilon_a + \epsilon_{a'})\epsilon_{b'} = \pm 2. \tag{14.18}$$

Since the average of a quantity that can only take values $+2$ and -2 must lie between those two values, it follows that if we take an average of the above quantity over many trials,

$$S = E(\hat{a}, \hat{b}) - E(\hat{a}', \hat{b}) + E(\hat{a}, \hat{b}') + E(\hat{a}', \hat{b}'), \tag{14.19}$$

then that average must obey

$$-2 \leq S \leq +2. \qquad \text{(CHSH theorem)} \tag{14.20}$$

To arrive at this result, no assumptions were made about the presence or absence of correlations between the two spins, so this is a completely general result that must be obeyed by any classical system of particles with a two-state spin-like variable. In this context, "classical" means that the measured variables are "real": they exist with definite values at all times, regardless of whether they have been measured or not.

What happens if the requirement of classical realism is jettisoned? Consider now quantum mechanics, in which a variable can be in a superposition of multiple states, settling in to one definite value only when a measurement is made. The assumption of having a single, definite value at each moment is no longer true. Take the unit vectors $\hat{a}$, $\hat{a}'$, $\hat{b}$, $\hat{b}'$ all to be in a single plane, and define a set of angles θ_a, $\theta_{a'}$, θ_b, $\theta_{b'}$ measuring the angles from the unit vectors to some fixed axis in the plane. Define the differences of these angles as $\theta_{ab} = \theta_a - \theta_b$, etc.

Suppose the two particles are entangled in the spin variable. To take a specific case, suppose that the two spins are perfectly correlated: directions of spins fluctuate randomly but are always pointing in the same (random) direction if measured. For example, suppose that spins form a Bell state, $|\Phi^+\rangle = \frac{1}{\sqrt{2}}(|\uparrow\rangle|\uparrow\rangle + |\downarrow\rangle|\downarrow\rangle)$. If Alice measures the spin along axis $\hat{a}$, then Bob's spin must also be in same direction. So if Bob measures along axis $\hat{b}$, then he should get the same value as Alice with probability $|\langle a|b\rangle|^2 = \cos^2 \frac{\theta_{ab}}{2}$ (where the 2 on the bottom comes from the spinor nature of the variable). Similarly, he should get the opposite value from Alice with the complementary probability $1 - |\langle a|b\rangle|^2 = \sin^2 \frac{\theta_{ab}}{2}$.

So, taking averages:

$$\langle \epsilon_a \epsilon_b \rangle = \sum_{\epsilon_a, \epsilon_b} \epsilon_a \epsilon_b P(\epsilon_a, \epsilon_b) \tag{14.21}$$

$$= (+1) \cos^2 \frac{\theta_{ab}}{2} + (-1) \sin^2 \frac{\theta_{ab}}{2} \tag{14.22}$$

$$= \cos \theta_{ab}. \tag{14.23}$$

In the second line above, the first term represents the case where Alice and Bob both measure the same value ($\epsilon_a = \epsilon_b$), and the second term is the case where they measure opposite values ($\epsilon_a = -\epsilon_b$). In the last line, the standard double angle formula from trigonometry was used.

The other three terms in S are computed in a similar manner, leading to the quantum mechanical result that

$$S = \cos\theta_{ab} - \cos\theta_{a'b} + \cos\theta_{ab'} + \cos\theta_{a'b'}. \tag{14.24}$$

To see that this can violate the inequality of Eq. 14.20, we can choose a particular combination of angles. Take, for example, $\theta_{ab} = \theta_{a'b'} = \theta_{ab'} \equiv \theta$ and $\theta_{a'b} = 3\theta$. Then we find

$$S = 3\cos\theta - \cos 3\theta. \tag{14.25}$$

This is plotted in Fig. 14.4. It can be clearly seen that $|S| > 2$ for some angles, violating the CHSH inequality. The maximal violations occur for

$$S_{\max} = 2\sqrt{2} \quad \text{at} \quad \theta = \frac{\pi}{4} \tag{14.26}$$

$$S_{\min} = -2\sqrt{2} \quad \text{at} \quad \theta = \frac{3\pi}{4}. \tag{14.27}$$

This maximum value of S achieved by quantum theory is known as **Tsirelson's bound**.

The results above were derived assuming a spin-correlated Bell state $|\Phi^+\rangle$. But they are much more general. In fact, it is easy to shown that *any* entangled state will lead to CHSH violation [9, 10]. Consider a generic entangled two-particle state, in

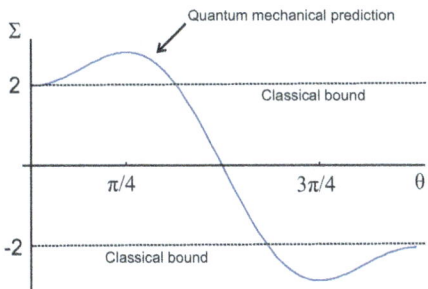

Fig. 14.4 The CHSH inequality states that the spin system described in the text must obey $-2 \leq S < 2$ if classical realism holds. Plotting the quantum mechanical prediction for S as a function of θ (the blue curve) clearly demonstrates that quantum mechanics violates this bound for some ranges of θ. Repeated experimental tests have consistently matched this predicted curve closely, upholding the inherently probabilistic nature of quantum mechanics

which each particle can be in one of two states. By the Schmidt decomposition, a basis can always be found such that the state can be put in the form

$$|\psi\rangle = \lambda_1 |0\rangle_A |0\rangle_B + \lambda_2 |1\rangle_A |1\rangle_B \equiv \lambda_1 |00\rangle + \lambda_2 |11\rangle, \qquad (14.28)$$

where the two states of each particle are labeled $|0\rangle$ and $|1\rangle$ and where $\lambda_1^2 + \lambda^2 = 1$. Again, let θ_a, θ_b, ... be the angles of a set of unit vectors (assumed to be in a single plane) from some fixed axis. The products of the spin operators along the various directions in this state can be found to be of the form

$$E(\hat{a}, \hat{b}) = -\cos\theta_a \cos\theta_b + 2\lambda_1\lambda_2 \sin\theta_a \sin\theta_b, \qquad (14.29)$$

with similar forms for the other products. Again make a specific choice of angles:

$$\theta_a = \pi - \theta_{a'} \equiv \theta, \qquad \theta_b = 0, \theta_{b'} = -\frac{\pi}{2}. \qquad (14.30)$$

It is then readily shown that

$$S = -2 \left(\cos\theta + 2\lambda_1\lambda_2 \sin\theta \right). \qquad (14.31)$$

So, it is clear that $|S|$ will be greater than 2 for some values of θ as soon as $\lambda_1\lambda_2$ is nonzero (no matter how small) or, in other words, as soon as *any* entanglement at all is present, no matter how weak the entanglement is.

The Bell and CHSH inequalities have by now been experimentally verified in a wide variety of different contexts, with potential loopholes in the setup being gradually closed. The majority of these experiments have actually used photon polarization, instead of electron spin; in these experiments, polarizing filters take the place of the Stern-Gerlach apparatus. The first experiments were done using photons produced by annihilation of the electrons and positrons in bound states called positronium or photons emitted in atomic fluorescence [11–13]. Later experiments have generally been carried out using entangled photons from parametric down conversion (notably [14–17]. A review of Bell/CHSH experiments can be found in [18], and a summary of the major experiments, with links to original papers, can be found at https://handwiki.org/wiki/Physics:Belltestexperiments.

The results of all of these experiments give a clear response to the EPR objections. The evidence is now overwhelming that local realism is violated and that quantum mechanics is the best available description of reality. Local hidden variable theories have been definitively ruled out.

More on the EPR and CHSH inequalities can be found in many places, including [4, 7, 10, 19–21].

Box 14.1 BKS Theorem and Contextuality

The uncertainty principle, Bell's theorem, and entanglement are probably the most famous counter-intuitive aspects of quantum mechanics. But there are other, lesser-known aspects that are just as difficult for our classical intuition to grasp. One example is the **Bell-Kochen-Specker (BKS) theorem** [22].

Recall that two variables are incompatible if they don't commute and that incompatible variables obey an uncertainty relation: measuring one variable will increase the uncertainty in the value of the other. So, if all variables in an experiment are compatible with each other (they all commute), is it safe to assume that none of the variables will be affected by measurements of the others? Surprisingly, the answer is no.

Consider two sets of three observable operators, with one observable common to both sets: $\{\hat{A}, \hat{B}, \hat{C}\}$ and $\{\hat{A}, \hat{D}, \hat{E}\}$. Assume all five of these operators commute with each other. Suppose we measure the first triplet of observables on some system and then the second triplet on the same system. Because of the compatibility of all of the operators, none of the measurements should disturb the values of the other variables, and so we would expect that A should have the same value each time. But in fact, this is not the case. The value A still depends on the context of the measurement (what other values are being measured), even when all variables commute. This property, called **contextuality**, is completely independent of the uncertainty relation, and its effects have been seen in experiments. For more detail, see [22, 23].

Problems

1. Consider the Bell state $|\Psi^-\rangle = \frac{1}{\sqrt{2}}(|01\rangle - |10\rangle)$, where $|0\rangle$ and $|1\rangle$ represent spin-up and spin-down states, respectively, of a spin-$\frac{1}{2}$ particle.

   (a) For two unit vectors $\hat{\boldsymbol{v}}$ and $\hat{\boldsymbol{w}}$ at an angle ϕ_{vw} from each other, define $E(\hat{\boldsymbol{v}}, \hat{\boldsymbol{w}}) = \langle|\Psi^-|\hat{\boldsymbol{v}} \cdot \hat{\boldsymbol{\sigma}} \otimes \hat{\boldsymbol{w}}\hat{\boldsymbol{\sigma}}|\Psi^-\rangle$. Calculate $E(\hat{\boldsymbol{v}}, \hat{\boldsymbol{w}})$.

   (b) Now choose four unit vectors $\hat{\boldsymbol{v}}$, $\hat{\boldsymbol{w}}$, $\hat{\boldsymbol{v}}'$, and $\hat{\boldsymbol{w}}'$. Show that the CHSH inequality, Eq. 14.20, can be violated for some choices of these vectors.

   (c) Show that the maximal violation is again given by the angles of Tsirelson's bound, Eqs. 14.26 and 14.27.

2. Show that the hidden variable theory of Example 14.1.1 satisfies the Bell inequality of Eq. 14.11.

3. The CHSH quantity can be generalized. Consider a pair of two-state systems, labeled 1 and 2. Let $\hat{A}$, $\hat{A}'$, $\hat{B}$, and $\hat{B}'$ be spin-based observables attached to the system. Then we can define a generalized S variable:

$$S = \langle \hat{S} \rangle = \left\langle (\hat{A} + \hat{A}')\hat{B} - (\hat{A} - \hat{A})\hat{B}' \right\rangle,$$

where each operator has eigenvalues ± 1. For specificity, let $\hat{A} = Z_1$, $\hat{A}' = X_1$, $\hat{B} = \frac{1}{\sqrt{2}}(Z_2 + X_2)$, and $\hat{B}' = \frac{1}{\sqrt{2}}(Z_2 - X_2)$, where X and Z represent Pauli matrices and the subscripts refer to which particle the operator acts upon.

(a) Write $\hat{S}$ as a matrix in the standard computational basis.
(b) Choose several different separable $|\psi\rangle$ states, and take the mean $S = \langle \psi | \hat{S} | \psi \rangle$ in each of those states. Verify that the inequality $|S| < 2$ is obeyed for each.
(c) Now try a maximally entangled Bell state: find the value of S, and verify that it violates the inequality $|S| < 2$.

4. **Hardy's theorem** (see [24] and Appendix K of [25]): Consider the pure bipartite state $|\psi\rangle = \frac{1}{\sqrt{3}}\left(|00\rangle + |10\rangle + |01\rangle\right)$.

(a) Define the "diagonal" and "antidiagonal" states $|D\rangle = \frac{1}{\sqrt{2}}\left(|0\rangle + |1\rangle\right)$ and $|A\rangle = \frac{1}{\sqrt{2}}\left(|0\rangle - |1\rangle\right)$. These are eigenstates of σ_x: $\sigma_x|D\rangle = +|D\rangle$ and $\sigma_x|A\rangle = -|A\rangle$. Show that $|\psi\rangle$ can be re-expressed in the forms

$$|\psi\rangle = \sqrt{\frac{2}{3}}|D0\rangle + \frac{1}{\sqrt{3}}|01\rangle = \frac{1}{\sqrt{3}}|10\rangle + \sqrt{\frac{2}{3}}|0D\rangle.$$

(b) Now suppose measurements are made of the x and z components of spin of the two particles. Find values of the joint probability $P(\sigma_z^A = -1, \sigma_z^B = -1)$ and of conditional probabilities $P(\sigma_x^B = +1 | \sigma_z^A = +1)$ and $P(\sigma_x^A = +1 | \sigma_z^B = +1)$.
(c) Given the results of part (b), what is the classically predicted (locally realistic) value of $P(\sigma_x^A = -1, \sigma_x^B = -1)$?
(d) Find the value of $P(\sigma_x^A = -1, \sigma_x^B = -1) = |\langle AA|\psi\rangle|^2$ predicted by quantum mechanics. The conflict between this and the result of (c) is an example of a contradiction between entanglement and local realism that does not rely on an inequality.

References

1. A. Einstein, B. Podolsky, N. Rosen, Phys. Rev. **47**, 777 (1935)
2. D. Bohm, *Quantum Theory* (Prentice-Hall, Hoboken, 1951)
3. D. Bohm, Y. Aharonov, Phys. Rev. **108**, 1070 (1957)
4. G. Greenstein, A.G. Zajonc, *The Quantum Challenge: Modern Research on the Foundations of Quantum Mechanics*, 2nd edn. (Jones & Bartlett, Burlington, 2005)
5. J.S. Bell, Physics (N.Y.) **1**, 195 (1964)

6. J.S. Bell, Rev. Mod. Phys. **38**, 447 (1966)
7. J.S. Bell, *Speakable and Unspeakable in Quantum Mechanics* (Cambridge University Press, Cambridge, 1987)
8. J.F. Clauser, M.A. Horne, A. Shimony, R.A. Holt, Phys. Rev. Lett. **23**, 880 (1969)
9. N. Gisin, A. Peres, Phys. Lett. A **162**, 15 (1992)
10. A. Peres, *Quantum Theory: Concepts and Methods* (Springer, Berlin, 1995)
11. A. Aspect, P. Grangier, G. Roger, Phys. Rev. Lett. **47**, 460 (1981)
12. A. Aspect, P. Grangier, G. Roger, Phys. Rev. Lett. **49**, 91 (1982)
13. A. Aspect, J. Dalibard, G. Roger, Phys. Rev. Lett. **49**, 1804 (1982)
14. Z.Y. Ou, L. Mandel, Phys. Rev. Lett. **61**, 50 (1988)
15. J.G. Rarity, P.R. Tapster, Phys. Rev. Lett. **64**, 2495 (1990)
16. P. Kwiat, et al., Phys. Rev. Lett. **75**, 4337 (1985)
17. A.G. White, D.F.V. James, P.R. Englehard, P.E. Kwiat, Phys. Rev. Lett. **83**, 3103 (1999)
18. A. Zeilinger, Rev. Mod. Phys. **71**, 5288 (1999)
19. G. Jaeger, *Quantum Information: An Overview* (Springer, Berlin, 2007)
20. N.D. Mermin, *Boojums All the Way Through: Communicating Science in a Prosaic Age* (Cambridge University Press, Cambridge, 1990)
21. S. Haroche, J.M. Raimond, *Exploring the Quantum: Atoms, Cavities, and Photons* (Oxford University Press, Oxford, 2006)
22. S. Kochen, S.P. Specker, J. Math. Mech. **17**, 59 (1967)
23. D. Mermin, Rev. Mod. Phys. **65**, 803 (1993)
24. L. Hardy, Phys. Rev. Lett. **71**, 1665 (1993)
25. S.M. Barnett, *Quantum Information* (Oxford University Press, Oxford, 2009)

Chapter 15
Gauge Fields and Geometric Phases

In this chapter, non-dynamical Berry phases are introduced. These are phases due to parameter variations and are related to the geometry of the paths traced out by states in Hilbert space. There are strong analogies between Berry connections and Berry curvatures on one hand and field-related quantities in electromagnetism on the other. So to display these more clearly, an introduction to the gauge theory of electromagnetism is also given, before discussing several examples of Berry phases in optical and quantum systems.

Steady-state solutions of energy E to the time-dependent Schrödinger equation generally come with a time-dependent phase factor $e^{\frac{i}{\hbar}Et}$. This is the dynamical phase associated with time evolution of the system. In 1984, Michael Berry [1] realized that if parameters of a system vary slowly over time, an additional phase factor ϕ appears that is completely separate from the dynamical phase. These extra phases tended to have geometrical or topological meanings; for example, for a spin in a magnetic field whose direction varies slowly around a closed loop, the new phase factor will be proportional to the solid angle subtended by the spin state on the Bloch sphere. After Berry's initial work on non-dynamical phases in quantum mechanics, it was soon realized that examples had actually been discovered in classical optical systems, going back as far as the work of Pancharatnam in 1956 [2, 3].

These non-dynamical phases have come to be known as **geometric phases**, **Berry phases**, or **Pancharatnam-Berry phases**. Over the past four decades, they have been found to appear in virtually every area of both classical and quantum physics and to have numerous applications. Applications include some previously known phenomena like the Aharonov-Bohm effect. Even the behavior of the Foucault pendulum involves a classical version of the geometric phase, called the **Hannay angle**.

Mathematically, geometric phases are best described in the setting of fiber bundle theory [4], but here we will stick to a more pedestrian introduction to the subject using only basic quantum mechanics and a minimum of mathematics. A huge

© The Author(s), under exclusive license to Springer Nature Switzerland AG 2025
D. S. Simon, *Introduction to Quantum Science and Technology*, Undergraduate Texts in Physics, https://doi.org/10.1007/978-3-031-81315-3_15

number of variations on the idea of geometric phase have been discussed in the literature. For a more comprehensive discussions of the subject, see [5–7].

As with any type of phase shift, the geometric phase can be detected via interferometry. The only extra complication in measuring it is that it has to be separated from any dynamical phase differences that have accumulated.

It will be seen that the Berry connection and curvature associated with the geometric phase look remarkably similar to the vector potentials and field strength tensors of electromagnetism and other gauge field theories. This is not a coincidence; the electromagnetic field strengths and the Berry curvatures have similar geometric meaning, describing the bending of abstract curved spaces. To make the analogy with electromagnetism clear, we start the chapter with a section reviewing the basics of electromagnetic potentials and gauge transformations.

One note on terminology: geometric phases are also sometimes called *topological phases*; we will avoid this term because it can be confused with the idea of having topological phases of matter (i.e., states of matter with different topological properties), similar to distinct thermodynamic phases. Topological phases in this latter sense will be discussed in Chap. 21.

15.1 Gauge Fields and Gauge Transformations

In classical electrodynamics, the electric and magnetic fields can be viewed as derivable from a scalar potential ϕ and a vector potential A:

$$B(r, t) = \nabla \times A(r, t), \qquad E(r, t) = -\nabla\phi - \frac{\partial}{\partial t}A. \tag{15.1}$$

Defining the four-vector position $r_\mu = (ct, r)$ and four-vector potential $A_\mu = \left(\frac{1}{c}\phi, A\right)$ (see Appendix D for four-vector notation), then the gauge field tensor is defined to be

$$F_{\mu\nu} = \partial_\mu A_\nu - \partial_\nu A_\mu = \begin{pmatrix} 0 & E_x/c & E_y/c & E_z/c \\ -E_x/c & 0 & -B_z & B_y \\ -E_y/c & B_z & 0 & -B_x \\ -E_z/c & -B_y & B_x & 0, \end{pmatrix} \tag{15.2}$$

where $\mu, \nu = 0, 1, 2, 3$ label the components along the time and the three space dimensions and ∂_μ means $\frac{\partial}{\partial x^\mu}$.

Since the days of Maxwell, it has been known that there is a redundancy in the potentials: the scalar and vector potentials ϕ and A can be replaced by new potentials

$$A'(r) = A(r) - \nabla\lambda(r) \tag{15.3}$$

$$\phi'(r) = \phi(r) + \frac{\partial \lambda}{\partial t}, \tag{15.4}$$

where $\lambda(r)$ is any function of position and time; then fields E and B are unchanged. This is seen easily enough:

$$B' = \nabla \times A' \tag{15.5}$$

$$= \nabla \times \left(A(r) - \nabla\lambda(r) \right) \tag{15.6}$$

$$= \nabla \times A(r) - \nabla \times \nabla\lambda(r) \tag{15.7}$$

$$= B, \tag{15.8}$$

since $\nabla \times \nabla f(r) = 0$ for any function f. The invariance in E follows in a similar manner.

In four-vector notation, the gauge transformation would be written as

$$A_\mu(r) \to A'_\mu(r) = A_\mu(r) - \partial_\mu \lambda(r), \tag{15.9}$$

where the four-gradient ∂_μ has components (∂_t, ∇). All physically measurable quantities should be invariant under gauge transformations. So, for example, the tensor $F_{\mu\nu}$, whose components are the physically measurable E and B fields, should be gauge invariant. Applying the gauge transformation to the first equality of Eq. 15.2 immediately shows that $F_{\mu\nu}$ is properly invariant, due to the fact that partial derivatives commute, $\partial_\mu \partial_\nu = \partial_\nu \partial_\mu$.

Classically, electric and magnetic fields are usually treated as fundamental quantities, while scalar and vector fields, with their gauge ambiguity, are treated as auxiliary fields: they are viewed as mathematically convenient but fictitious fields of no physical significance in themselves. In quantum mechanics though, the situation is very different. As shown by the Aharonov-Bohm effect (Sect. 15.4), in a quantum setting, potentials are of more fundamental importance.

The presence of the electromagnetic field will affect the wavefunctions of charged particles. Consider the case of a magnetic field with vector potential A. If a particle of charge q moves along path C in the presence of this potential, Dirac showed that it will gain a phase factor $e^{i\frac{q}{\hbar}\int_C A \cdot dl}$, where dl is the infinitesimal tangent vector field along C. If C is a *closed* curve enclosing area S, then by Stokes' theorem, this phase factor can be written as

$$e^{i\frac{q}{\hbar}\oint_C A \cdot dl} = e^{i\frac{q}{\hbar}\int_S B \cdot dS} = e^{i\frac{q}{\hbar}\Phi} = e^{\frac{2\pi i q}{\hbar}\Phi}, \tag{15.10}$$

where S is the area vector perpendicular to surface S and Φ is the magnetic flux through S. But around a closed curve, the wavefunction must return to its original value; this implies that the phase must be a multiple of 2π. So we see that the flux must be quantized: $\Phi = \frac{nh}{q}$, for integer n.

In the early days of quantum theory, Hermann Weyl attempted to unify gravity and electromagnetism by requiring that physics be invariant under a combined transformation of the matter and the force fields. Weyl's initial attempt involved a change in the magnitude (or "gauge") of the matter field, hence the origin of the name *gauge theory*. His unification model failed, but it was soon found that requiring invariance under transformations of the matter field *phase* (instead of its *magnitude*) was a more fruitful idea.

The full-gauge transformation of electromagnetism in the presence of a charged matter field ψ is therefore

$$A_\mu(r) \to A'_\mu(r) - \partial_\mu \lambda(r), \quad \text{and} \quad \psi(r) \to \psi'(r) = e^{i\lambda(r)}\psi(r) \qquad (15.11)$$

The full theory is required to be invariant under this combined transformation. Notice that the transformation is required to be *local*: the phase of ψ can be rotated independently by an arbitrary amount at each point in space-time, so that $\lambda(r)$ is an arbitrary *function*. In contrast, *global* phase changes, in which λ is a constant, independent of both position and time, are always allowed and have no physical meaning.

Example 15.1.1 Quantum fields are normally described by Lagrangians, rather than Hamiltonians. (For readers who have not seen Lagrangians before, just think of them as functions similar to the Hamiltonian that can be used to derive the equations of motion of a system. No knowledge of Lagrangians beyond this is needed to follow this example.)

So consider the Lagrangian for a complex scalar field $\phi(r)$ of charge $q = -e$ coupled to an electromagnetic field described by gauge field $A_\mu(r)$. In this context, "scalar" simply means spin-0. This system is referred to as **scalar electrodynamics** and is used as a simplified model of real quantum electrodynamics that avoids the complications introduced by spin. The Lagrangian is of the form

$$L = \int \mathcal{L} \, d^3 x, \quad \text{with} \quad \mathcal{L} = (D_\mu \phi)^* (D^\mu \phi) - V(\phi^* \phi) - \frac{1}{4} F_{\mu\nu} F^{\mu\nu}, \qquad (15.12)$$

where the Einstein summation convention is being used (meaning that any index that occurs twice is summed over) and the integral is over all of space-time. D_μ will be defined below. We wish to check that this Lagrangian is invariant under the gauge transformation

$$\phi(r) \to \phi'(r) = \phi(r) e^{i\lambda(r)} \qquad (15.13)$$

$$A_\mu(r) \to A'_\mu(r) = A_\mu(r) + \frac{1}{e} \partial_\mu \lambda(r). \qquad (15.14)$$

(Here, as is traditional, we have rescaled the gauge parameter λ compared to Eq. 15.11 by pulling out an explicit factor of the charge, $-e$.)

The scalar potential V depends only on the product $\phi^*\phi$, not on ϕ or ϕ^* separately. Clearly, this will be invariant under the transformation of Eq. 15.13. We have already seen that $F_{\mu\nu}$ is gauge invariant, so the last term in Eq. 15.13 (the energy of the electromagnetic field by itself) is invariant. The problem therefore reduces to showing the invariance of the first term in the Lagrange density $\mathcal{L}$. This term describes the kinetic energy of ϕ and its coupling with the electromagnetic field.

The **covariant derivative** D_μ is defined by

$$D_\mu = \partial_\mu + iq A_\mu = \partial_\mu - ieA_\mu. \tag{15.15}$$

Using Eqs. 15.13 and 15.14, the covariant derivative acting on the scalar field transforms according to

$$D'_\mu\phi' = \left(\partial_\mu - ieA'_\mu\right)\phi'(r) \tag{15.16}$$

$$= \left[\partial_\mu - ie\left(A_\mu + \frac{1}{e}\partial_\mu\lambda\right)\right]\left(e^{i\lambda(r)}\phi(r)\right) \tag{15.17}$$

$$= e^{i\lambda(r)}\left[i\partial_\mu\lambda + \partial_\mu - ieA_\mu - i\partial_\mu\lambda\right]\phi(r) \tag{15.18}$$

$$= e^{i\lambda(r)} D_\mu\phi(r). \tag{15.19}$$

Notice that $D_\mu\phi$ transforms in the same manner as ϕ itself: both are simply multiplied by the same phase factor,

$$\phi(r) \to e^{i\lambda(r)}\phi(r), \qquad D_\mu\phi(r) \to e^{i\lambda(r)} D_\mu\phi(r). \tag{15.20}$$

This property is *not* shared by the ordinary derivative, which picks up an extra term:

$$\partial_\mu\phi(r) \to \partial_\mu\left(e^{i\lambda(r)}\phi(r)\right) = e^{i\lambda(r)}\left[\partial_\mu + i\left(\partial_\mu\lambda(r)\right)\right]\phi(r). \tag{15.21}$$

The extra $\partial_\mu\lambda(r)$ term will spoil the gauge invariance of any expression containing ordinary derivatives. (Geometrically, the covariant derivative is defined to account for effects due to curvature of the internal space of field configurations.)

So now that we know how $D_\mu\phi$ transforms, it should be immediately apparent that the kinetic term $\left(D_\mu\phi\right)^*\left(D^\mu\phi\right)$ is also invariant. Thus, we conclude that the full Lagrangian density $\mathcal{L}$ (and therefore the Lagrangian L itself) are gauge invariant. ∎

The idea of gauge invariance can be turned around: we can start with a theory that has only charged particles and no electromagnetic field. We then require that the theory be invariant under the transformation $\psi(r) \to \psi'(r) = e^{i\lambda(r)}\psi(r)$ for any function λ. But a theory containing *only* the ψ fields will not be invariant; to restore invariance, the gauge potential $A_\mu(r)$ must be added into the theory. In this view, gauge invariance is a fundamental principle of nature, and maintaining

Table 15.1 Fundamental forces of nature and vector gauge bosons responsible for them. Each of these forces is a consequence requiring gauge invariance. The photon, gluon, and graviton are massless, while gauge bosons of the weak force are massive. Electromagnetism is an Abelian theory (the photon doesn't carry electric charge), and so photons do not interact with other photons. The others are non-Abelian, which means that bosons exchanged between the generalized charges carry the same charge themselves, and so bosons interact with each other

Force	Generalized charge	Gauge bosons	Massive?	Abelian?
Electromagnetic	Electric charge	Photon	No	Yes
Weak nuclear	Weak isospin	$W^{\pm}$ and Z^0	Yes	No
Strong nuclear	Color charge	Gluon	No	No
Gravity	Mass and Energy	Graviton	No	No

it forces the appearance of spin-one bosons, called **vector bosons** (*vector* here simply means spin-one) with appropriate transformation properties. Exchanges of these vector fields are responsible for the forces between the generalized charges of matter fields. In the electromagnetic case, vector bosons are photons that make up the electromagnetic field.

We saw this procedure above, where terms involving ordinary derivatives, such as $\partial_\mu \phi$ or $\partial_\mu \psi$, in a Lagrangian, will always break gauge invariance; to restore invariance, all ordinary derivatives ∂_μ must be promoted to covariant derivatives D_μ. This explicitly shows that gauge invariance of the matter field (e.g., a scalar field ϕ or an electron field ψ) requires the introduction of a gauge field A_μ.

It became clear that this was a useful approach after the 1954 work of Yang and Mills (and independently by Robert Shaw), who showed that nuclear forces could be obtained in the same manner by requiring a generalized version of gauge invariance. It is now known that all forces of nature (including gravity) can be seen as consequences of gauge symmetry, as seen in Table 15.1. As a result, gauge invariance is now taken to be a fundamental principle of nature, on a par with the principle of relativity, as one of the primary building blocks of the standard model of particle physics.

15.2   Berry Phases in Quantum Mechanics

Consider a quantum system that depends on some time-dependent parameter, $\lambda(t)$. For example, λ could be the strength or direction of an external magnetic field, it could be the temperature, or it could be the coupling strength between two optical cavities or between two spin sites on a lattice. Suppose the quantum state occupied by the system depends on the parameter: $|\psi(\lambda)\rangle$. For the moment, assume there is just a single parameter; later in this section, we will generalize to multiple parameters.

First, recall from any introductory quantum mechanics textbook the idea of an **adiabatic process**. Consider a quantum system with a nonzero energy gap ΔE

between its eigenstates; think, for example, of the states of different principal quantum number in the hydrogen atom or of energy levels in a quantum harmonic oscillator. According to the **adiabatic theorem** (first stated by Max Born and Vladimir Fock in 1928), a quantum system remains in its initial eigenstate if a given perturbation acting on it changes the system's parameters slowly enough. For a more precise statement, see, for example, [8]. Essentially, "sufficiently slowly" means that if there is some characteristic time scale T associated with the perturbation, then the rate $\frac{1}{T}$ should be small compared to the transition frequencies of the system, $\frac{1}{T} << \omega = \frac{\Delta E}{\hbar}$. An adiabatic process is any process that obeys this condition.

The basic idea of the adiabatic process should be familiar from classical physics. Consider an egg in an open egg carton. If the carton is moved slowly (i.e., adiabatically), then the egg remains in its initial state: it stays sitting in the carton. This is true even though the state itself is slowly changing (the height of the carton above the floor may be changing). However, if the carton is moved more rapidly (it is given a sudden jerk), the state of the egg changes: it jumps out of the crate and crashes to the floor. Similarly, sudden changes to the parameters in a Hamiltonian create a jolt that causes transitions between states, but sufficiently slow changes leave the particle in the same state, even as the state itself changes.

Box 15.1 Non-Abelian Gauge Fields and Synthetic Gauge Fields

Gauge fields form representations of mathematical groups. A group is said to be **Abelian** if all of its elements commute with each other and **non-Abelian** otherwise. So the electromagnetic field is said to be an Abelian field, because the components of the gauge potentials A_μ are ordinary numbers that commute with each other: $[A_\mu, A_\nu] = 0$. But the nuclear and gravitational forces are all non-Abelian: gauge fields are vector or tensor fields whose spatial components are all non-commuting matrices. These matrices transform a set of equal-energy states into each other: for example, they may rotate different color-charge states of quarks into each other. The fields carry an extra index $a, b, \ldots$ representing the internal generalized charge states. In general, the commutator of two of these fields will be linear combinations of the other fields: $[A_\mu^a, A_\nu^b] = \sum_{c\lambda} C_{c\mu\nu}^{ab\lambda} A_\lambda^c$, where $C_{c\mu\nu}^{ab\lambda}$ are constants and λ is a spatial index. These fields form a mathematical structure called a **Lie algebra** [4, 9].

Non-Abelian fields differ from Abelian fields in a number of important ways. For example, the Abelian electromagnetic field is made up of photons that couple to electric charges of other particles. Because the photon itself is electrically neutral, two photons will not interact with each other. For non-Abelian fields, the gauge field itself carries the same generalized charge that the field couples to: for example, gluons carry color charge, and gravitons carry energy-momentum. As a result, two gluons will interact with each other,

(continued)

Box 15.1 (continued)

as will two gravitons. This makes non-Abelian gauge theories highly non-linear and very difficult to work with.

Non-Abelian fields have long been the province of high-energy physics. But it has become apparent in recent years that quasi-particle excitations in condensed matter systems can exhibit non-Abelian structures. The main requirement for this is a system with a set of degenerate ground states related to each other by a group of symmetry transformations. Generally, in condensed matter contexts, the non-Abelian symmetries couple spatial degrees of freedom to internal degrees of freedom. One example of this would be spin-orbit coupling: spin is an internal degree of freedom, while orbital motion is a spatial motion. Non-Abelian excitations have been found or predicted to exist in a number of systems, including superconducting or superfluid systems, graphene, spin systems, cold atoms [10–12] optics [13], and even electric circuits [14]. Non-Abelian anyons in condensed matter systems will be discussed in Sect. 21.7.

An important concept in many of these systems is the idea of artificial or synthetic gauge fields. Charged particles are easily controlled and manipulated by the use of electric and magnetic fields. But this doesn't work for photons or neutral atoms, which have no charge for electric or magnetic fields to couple to. So an important area of research has been the engineering of artificial or **synthetic gauge fields** that can mimic the effects of a real gauge field.

For example, the Lorentz force from a magnetic field, $F_B = qv \times B$, can be mimicked in many systems by simply rotating the system. This is because the Coriolis force has the same structure as the Lorentz force: $F_C = 2mv \times \omega$. The particles in the system just see the force; they can't tell whether it was produced by a magnetic field B or by a rotation of angular speed $\omega = \frac{qB}{2m}$. Synthetic gauge fields acting on photons have been implemented using, for example, coupled resonant cavities, coupled waveguides, opto-mechanical systems, and metamaterials. For more on synthetic gauge fields in photonic systems, see the references in [15]. For a review of the subject in atomic systems, see [16].

So consider the state $|\psi(\lambda)\rangle$ as λ changes adiabatically from initial value λ_0 to final value λ_f. We will only consider the case where the parameter returns to its initial value, $\lambda_0 = \lambda_f$. So the final state should be the same as the initial state, at least up to an overall phase shift: $|\psi(\lambda_f)\rangle = e^{-i\phi}|\psi(\lambda_0)\rangle$.

First, look at an infinitesimal parameter change, $\lambda + d\lambda$. Expanding in a Taylor series and keeping only terms up to first order in $d\lambda$,

$$|\psi(\lambda + d\lambda)\rangle = |\psi(\lambda)\rangle + \partial_\lambda |\psi(\lambda)\rangle \, d\lambda, \tag{15.22}$$

where ∂_λ is short-hand for $\frac{d}{d\lambda}$. Taking the product of both sides with $\langle\psi(\lambda)|$:

$$\langle\psi(\lambda)|\psi(\lambda+d\lambda)\rangle = 1 + \langle\psi(\lambda)|\partial_\lambda|\psi(\lambda)\rangle \, d\lambda \tag{15.23}$$

Recalling that the Taylor series of the logarithm is

$$\ln(x) = (x-1) + \frac{1}{2}(x-1)^2 + \ldots, \tag{15.24}$$

to leading order in $d\lambda$, we have

$$\ln\langle\psi(\lambda)|\psi(\lambda+d\lambda)\rangle = \langle\psi(\lambda)|\partial_\lambda|\psi(\lambda)\rangle d\lambda. \tag{15.25}$$

Any complex number z can be written in the form $z = e^\Phi$ where the imaginary part of Φ is a phase angle and the real part is $\log|z|$. So set $\ln\langle\psi(\lambda)|\psi(\lambda+d\lambda)\rangle = e^\Phi$ and define the phase change $d\phi = -\text{Im}\Phi$. Then the phase is

$$d\phi = -\text{Im}\left(\langle\psi(\lambda)|\partial_\lambda|\psi(\lambda)\rangle d\lambda\right). \tag{15.26}$$

In fact, the product on the right side is pure imaginary, with vanishing real part. This can be seen using the fact that the state has to remain normalized as λ changes:

$$2\text{Re}\,\langle\psi(\lambda)|\partial_\lambda\psi(\lambda)\rangle = \langle\psi(\lambda)|\partial_\lambda\psi(\lambda)\rangle + \langle\psi(\lambda)|\partial_\lambda\psi(\lambda)\rangle^* \tag{15.27}$$

$$= \langle\psi(\lambda)|\partial_\lambda|\psi(\lambda)\rangle + \langle\partial_\lambda\psi(\lambda)|\psi(\lambda)\rangle \tag{15.28}$$

$$= \partial_\lambda\langle\psi(\lambda)|\psi(\lambda)\rangle \tag{15.29}$$

$$= \partial_\lambda(1) \tag{15.30}$$

$$= 0. \tag{15.31}$$

So the infinitesimal phase change can also be written as

$$d\phi = i\langle\psi(\lambda)|\partial_\lambda|\psi(\lambda)\rangle d\lambda. \tag{15.32}$$

Integrating, the phase change around the full closed loop, $\lambda_0 \to \lambda_f = \lambda_0$, is the adiabatic **Berry phase**:

$$\phi = i\oint \langle\psi(\lambda)|\partial_\lambda|\psi(\lambda)\rangle d\lambda. \tag{15.33}$$

This is often written in the form

$$\phi = \oint A(\lambda)d\lambda, \tag{15.34}$$

where

$$A(\lambda) = i \langle \psi(\lambda)|\partial_\lambda|\psi(\lambda)\rangle \qquad (15.35)$$

is called the **Berry potential** or **Berry connection**.

Generalizing from a single parameter to a set of n parameters, these parameters can be thought of as forming an n-dimensional vector λ in the parameter space. The Berry connection is then also an n-dimensional vector,

$$A(\lambda) = i \langle \psi(\lambda)|\nabla_\lambda|\psi(\lambda)\rangle, \qquad (15.36)$$

where the gradient in parameter space has components $\frac{\partial}{\partial\lambda_1}, \ldots, \frac{\partial}{\partial\lambda_n}$. In the multiparameter case, the **Berry curvature** can also be defined,

$$R(\lambda) = \nabla_\lambda \times A(\lambda). \qquad (15.37)$$

$R(\lambda)$ is a second-rank tensor with components

$$R_{\mu\nu} = \partial_\mu A_\nu - \partial_\nu A_\mu, \qquad (15.38)$$

where μ, ν label components of λ.

As the parameters trace out a closed path C, the Berry phase can then be written in the equivalent forms

$$\phi = \oint_C A(\lambda) \cdot d\lambda = \int_S R(\lambda)dS, \qquad (15.39)$$

where the last equality follows from Stokes' theorem. Here, dS is an area element on the surface S that is bounded by curve C.

Note the similarity to electromagnetism; A and $R_{\mu\nu}$ are analogous to the vector potential and electromagnetic field tensor. This similarity extends to the existence of gauge transformations. A gauge transformation on the state $|\psi(\lambda)\rangle$ consists of multiplying by a parameter-dependent phase factor,

$$|\psi(\lambda)\rangle \rightarrow |\tilde{\psi}(\lambda)\rangle = e^{-i\alpha(\lambda)}|\psi(\lambda)\rangle, \qquad (15.40)$$

where $\alpha(\lambda)$ can be any differentiable function. Under this transformation, it is found that

$$A(\lambda) \rightarrow \tilde{A}(\lambda) = A(\lambda) + \nabla_\lambda \alpha(\lambda). \qquad (15.41)$$

Clearly, the Berry potential is not gauge-invariant. But plugging $\tilde{A}(\lambda)$ into the definition of the curvature immediately shows that the Berry curvature *is* gauge-invariant: $\tilde{R}_{\mu\nu} = R_{\mu\nu}$.

Consider now the Berry phase gained around a closed loop C in parameter space. Since the loop is closed and the state must be single valued, the values of the phase at the initial and final points can only differ by a multiple of 2π: $\phi(\lambda_f) = \phi(\lambda_0)+2\pi\nu$, for some integer ν. In order to maintain this condition, the gauge transformation must obey

$$\Delta\alpha(\lambda) = \int_{\lambda_0}^{\lambda_f} \nabla\alpha(\lambda)\cdot d\lambda = \alpha(\lambda_f) - \alpha(\lambda_0) = 2\pi\nu' \tag{15.42}$$

for some integer ν'. So the effect of the gauge transformation on the geometric phase can only be to shift it by some integer multiple of 2π,

$$\tilde{\phi} = \phi + 2\pi\nu. \tag{15.43}$$

ν is called the **winding number** of the gauge transformation, and it counts how many times $e^{i\tilde{\phi}}$ wraps around the origin in the complex plane for every time $e^{i\phi}$ circles the origin.

Example 15.2.1 Consider a spin-$\frac{1}{2}$ particle in a uniform magnetic field B. Let the direction of B be given by a unit vector $\hat{n} = \sin\theta\cos\phi\,\hat{i} + \sin\theta\sin\phi\,\hat{j} + \cos\theta\,\hat{k}$, so that $B = B\hat{n}$. The Hamiltonian is

$$H = -\gamma B \cdot S = -\frac{1}{2}\gamma\hbar B\,\hat{n}\cdot\sigma = \kappa\begin{pmatrix} \cos\theta & e^{-i\phi}\sin\theta \\ e^{-i\phi}\sin\theta & -\cos\theta \end{pmatrix}, \tag{15.44}$$

where γ is the gyromagnetic ratio, the spin operator is $S = \frac{\hbar}{2}\sigma$, and $\kappa \equiv \frac{1}{2}\gamma\hbar B$. The eigenvectors of H are the vectors pointing up or down along $\hat{n}$,

$$|\uparrow\rangle = \begin{pmatrix} \cos\frac{\theta}{2} \\ e^{i\phi}\sin\frac{\theta}{2} \end{pmatrix}, \qquad |\downarrow\rangle = \begin{pmatrix} \sin\frac{\theta}{2} \\ -e^{-i\phi}\cos\frac{\theta}{2} \end{pmatrix} \tag{15.45}$$

with corresponding energy eigenvalues $E = \pm\kappa$.

Suppose the unit vector $\hat{n}$ now varies adiabatically, slowly tracing out a closed path on the unit sphere. The state vector $|\uparrow\rangle$ will then gain some Berry phase. The Berry connection will be a three-dimensional vector, with components given by

$$A_r = i\left\langle \uparrow\left|\frac{\partial}{\partial r}\right|\uparrow\right\rangle = 0, \tag{15.46}$$

$$A_\theta = i\left\langle \uparrow\left|\frac{\partial}{\partial\theta}\right|\uparrow\right\rangle = 0 \tag{15.47}$$

$$A_\phi = i\left\langle \uparrow\left|\frac{\partial}{\partial\phi}\right|\uparrow\right\rangle = -\left(\sin\frac{\theta}{2}\right)^2 = \frac{1}{2}(\cos\theta - 1). \tag{15.48}$$

The only nonzero components of the curvature are

$$R_{\theta\phi} = -R_{\phi\theta} = \frac{\partial}{\partial\theta}A_\phi = -\frac{1}{2}\sin\theta. \tag{15.49}$$

So, using Stokes' theorem, the geometric phase is given by

$$\phi = \oint_C A \cdot \hat{n} = \int_S R_{\theta\phi}dS = \int_S \frac{1}{2}\sin\theta \, d\theta \, d\phi = -\frac{1}{2}\Omega, \tag{15.50}$$

where Ω is the solid angle enclosed by the curve C on the Bloch sphere.   ∎

15.3   Geometric Phases in Optics

Consider interference between two light beams with (complex) electric fields E_a and E_b. Without loss of generality, we can rescale both fields by a common constant factor in order to normalize them, so that their magnitudes obey

$$|E_a|^2 + |E_b|^2 = 1. \tag{15.51}$$

Doing so allows us to parameterize field amplitudes by an angular variable,

$$|E_a| = \cos\frac{\theta}{2}, \qquad |E_b| = \sin\frac{\theta}{2}. \tag{15.52}$$

Let the phase difference between fields be $\delta\phi = \phi_b - \phi_a$. Up to an overall constant, intensities of these beams are $I_a = |E_a|^2$ and $I_b = |E_b|^2$, so that $I_a + I_b = 1$. Let E_a, E_b, and $E \equiv E_a + E_b$ be represented by points a, b, and c on the Poincaré sphere (Fig. 15.1). Further, let a', b', and c' be the corresponding antipodal points, i.e., points diametrically opposite to a, b, and c on the sphere.

Defining the phase difference between two parallel light components is easy. But cases where the components are at an arbitrary angle to each other is a little more subtle. So, Pancharatnam sought an unambiguous definition of the phase difference between two different, but non-orthogonal, polarization states of light, E_a and E_b. These will be represented by points a and b on the Poincaré sphere (Fig. 2.17). Let $\frac{1}{2}\Theta$ be the angle between the polarization directions in space; the angle between the corresponding optical states on the Poincaré sphere is Θ.

E_b can be broken into components parallel and perpendicular to E_a: $E_b = E_{||} + E_\perp$. Intensities corresponding to these components are $I_{||} = |E_b|^2\cos^2\frac{\Theta}{2}$ and $I_\perp = |E_b|^2\sin^2\frac{\Theta}{2}$. Orthogonal polarizations do not interfere with each other, so the phase difference $\delta\phi$ between the beams is defined to be the phase difference between the two components that are parallel to each other, E_a and $E_{||}$. These interfere to give intensity

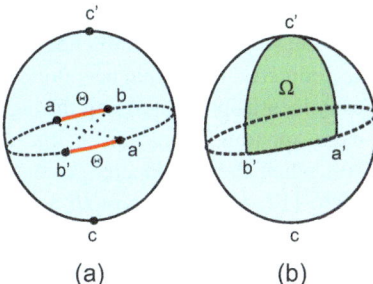

Fig. 15.1 (**a**) Two polarization states of light (*a* and *b*) are located at points separated by angle Θ on the Poincaré sphere, while the combined beam is at point *c*. Points a', b', and c' represent light states orthogonal to *a*, *b*, and *c*. (**b**) The phase shift $\delta\phi$ between beams E_a and E_b is related to the solid angle Ω subtended by geodesic triangle $a'b'c'$ relative to the center of the sphere

$$I' = |E_a + E_{\|}e^{i\delta\phi}|^2 = I_a + I_b \cos^2 \frac{\Theta}{2} + 2\sqrt{I_a I_b} \cos \frac{\Theta}{2} \cos \delta\phi. \tag{15.53}$$

The total intensity is then

$$I = I' + I_b \sin^2 \frac{\Theta}{2} = I_a + I_b + 2\sqrt{I_a I_b} \cos \frac{\Theta}{2} \cos \delta\phi. \tag{15.54}$$

So, experimentally, the phase difference between the two differently polarized optical states can be found by measuring the intensities I, I_a, I_b, and then solving the previous equation for $\delta\phi$. We find that

$$\cos \delta\phi = \frac{I_a + I_b - I}{2\sqrt{I_a I_b} \cos \frac{\Theta}{2}}. \tag{15.55}$$

Suppose that θ_a and θ_b are angles on the sphere from *a* and *b* to *c*, while θ_a' and θ_b' are angles from a' and b' to c'. Pancharatnam noted that when the *a* and *b* beams are added, the intensity at point c' on the Poincaré sphere (orthogonal to the total field *E*) had to vanish, while the b' components of I and I_a must be equal [3]. This leads to the constraints

$$\frac{I_a}{\sin^2 \frac{\theta_a}{2}} = \frac{I_b}{\sin^2 \frac{\theta_b}{2}} = \frac{I}{\sin^2 \frac{\Theta}{2}}. \tag{15.56}$$

As a result, Eq. 15.55 can also be put in the form

$$\cos \delta\phi = \frac{\cos^2 \frac{\Theta}{2} + \cos^2 \frac{\theta_a'}{2} + \cos^2 \frac{\theta_b'}{2} - 1}{2 \cos \frac{\Theta}{2} \cos \frac{\theta_a'}{2} \cos \frac{\theta_b'}{2}}. \tag{15.57}$$

Pancharatnam's key geometric insight was to notice that the expression on the right of the last equation is simply the solid angle Ω subtended by points a', b', and c' on the sphere. $\delta\phi$ is simply the geometric phase that accumulates as a polarization state travels a closed path on the Poincare sphere, passing through points a, b, and c.

Chaio and Wu, who were apparently unaware at the time of Pancharatnam's work, found a similar result when they conducted an experiment to measure the optical Berry's phase in 1986 [17, 18]. They wound a fiber into a helix and sent linearly polarized light into one end. As the light propagates along the fiber, the wavevector k of the light traces out some path in momentum space. Arranging the initial and final ends of the fiber to be parallel, the momentum space path of the light forms a closed loop, C. Since the magnitude of k stays approximately constant (assuming negligible loss), C is a path on the surface of a sphere. It will subtend a solid angle

$$\Omega(C) = 2\pi N \left(1 - \cos\theta\right), \tag{15.58}$$

where N is the number of loops in the fiber optical coil and θ is the pitch angle of the helix formed by the coil, i.e., the angle that the plane of each loop is tilted relative to the plane perpendicular to the central axis of the coil. The geometric phase again turns out to be proportional to the solid angle,

$$\phi(C) = -\sigma \Omega(C), \tag{15.59}$$

where $\sigma = \pm 1$ is the helicity of the light. Furthermore, as the light traverses the fiber coil, the geometric phase also gives the angle $|\phi|$ by which the polarization direction of the outgoing light is rotated relative to its initial orientation (Fig. 15.2).

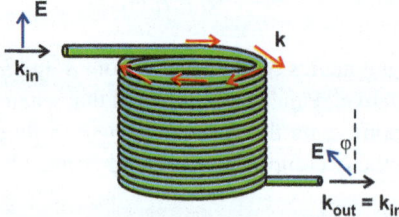

Fig. 15.2 Light traversing a coil of optical fiber. The momentum vector traces out a circle in real space, and the corresponding polarization state traces out a circle on the Poincaré sphere. The angle ϕ through which the polarization vector is rotated is equal to the geometric phase

15.4 Aharonov-Bohm Effect

As mentioned in Sect. 15.1, the fundamental objects in classical electromagnetism are usually taken to be the fields E and B, while the electrostatic and vector potentials $\phi(r)$ and $A(r)$ are treated as just mathematical conveniences that have no intrinsic physical significance. But in quantum mechanics, it turns out that the potentials are the more fundamental objects, with the fields being derived from them. One reason for this shift in viewpoint is the Aharonov-Bohm effect, in which a charged particle gains a phase shift whenever it passes through a spatially dependent vector potential, even if the region through which the particle moves has vanishing magnetic field $B(r)$.

First, recall that the magnetic field and potential are related by $B = \nabla \times A$. There is an ambiguity in the vector potential, since it can be shifted according to $A(r) \rightarrow A(r) + \nabla\alpha$, for any function $\phi(r)$, without affecting B, as a result of the vector identity $\nabla \times \nabla\phi = 0$.

The minimal coupling prescription of electromagnetism tells us that the effect of a magnetic field is to replace $p = -i\hbar\nabla \rightarrow -i\hbar(\nabla - \frac{iq}{\hbar}A)$, so the free-particle Hamiltonian $\hat{H}_0 = \frac{\hat{p}^2}{2m}$ becomes

$$\hat{H} = -\frac{\hbar^2}{2m}\left(\nabla - \frac{iq}{\hbar}A\right)^2. \tag{15.60}$$

Suppose a charged particle has wavefunction $\psi_0(r)$ in the absence of the magnetic field. Then it can be easily verified that in the presence of the potential A, the new wavefunction

$$\psi(r) = e^{\frac{iq}{\hbar}\int_{r_0}^{r} A(r')\cdot dr'}\psi_0(r) \tag{15.61}$$

satisfies the Schrödinger equation with the new Hamiltonian $\hat{H}$. Here, r_0 is some reference point; it is arbitrary but should remain fixed. So the magnetic field introduces an Aharonov-Bohm phase factor $e^{i\phi(r)} = e^{\frac{iq}{\hbar}\int_{r_0}^{r} A(r')\cdot dr'}$. If the particle travels around a closed path C, then the dependence on the reference point drops out, and the phase gain of the wavefunction is

$$\phi = \frac{q}{\hbar}\oint A(r')\cdot dr' = \frac{q}{\hbar}\int_S B\cdot dS = \frac{q}{\hbar}\Phi, \tag{15.62}$$

where S is the area bounded by C, Φ is the magnetic flux, and Stokes' theorem has been used in the next-to-last equality. The point here is that the phase gain of the particle is proportional to the magnetic flux its path encloses.

Now consider the situation shown in Fig. 15.3. A tube of magnetic flux is created, which is confined within a radius R around the z-axis. This flux could, for example, be contained inside a solenoid of radius R.

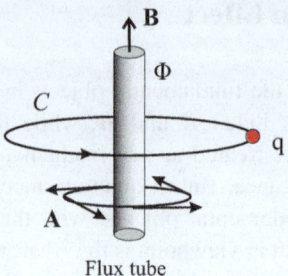

Flux tube

Fig. 15.3 The Aharonov-Bohm effect. A charge q circles a flux tube containing magnetic field **B** (along the z-axis) and flux Φ. Outside the tube, where the charge is present, the field vanishes, but there is a nonzero vector potential circling the z-axis. The particle picks up a phase proportional to the magnetic flux despite being in a region of $B = 0$

Take a particle of charge q traveling around the central singular point in a closed path around the flux, being careful to make sure it never comes closer than a distance R from the z-axis. The magnetic field **B** vanishes along the particle's path. However, the vector potential **A** is nonzero along the path, and this will produce an observable phase shift for the particle's wavefunction. For a path that circles the flux n times, the phase gain is given by $\phi = \frac{nq}{c\hbar}\Phi$.

To make this model look more like a point particle, one may idealize by taking $R \to 0$, so that the magnetic field will be $\boldsymbol{B}(\boldsymbol{r}) = \Phi\delta^{(2)}(\boldsymbol{r})$. Then, by making an appropriate gauge choice, the vector potential can be taken to be

$$\boldsymbol{A}(\boldsymbol{r}) = \frac{\Phi}{2\pi r^2}\left(x\hat{\boldsymbol{j}} - y\hat{\boldsymbol{i}}\right) = \frac{\Phi}{2\pi r}\hat{\boldsymbol{e}}_\phi, \tag{15.63}$$

where $\hat{\boldsymbol{e}}_\phi = \frac{x\hat{\boldsymbol{j}} - y\hat{\boldsymbol{i}}}{r}$ is the unit vector in the azimuthal (ϕ) direction and $r = \sqrt{x^2 + y^2}$. To check that this works, use the fact that $\oint_C \boldsymbol{A} \cdot d\boldsymbol{l} = \int_S \boldsymbol{B} \, d\boldsymbol{S}$. The delta function in **B** makes the right side equal to Φ, while (taking C to be a circle of radius r) the left side becomes

$$\oint_C \boldsymbol{A} \cdot d\boldsymbol{l} = \int_0^{2\pi} \left(\frac{\Phi}{2\pi r}\hat{\boldsymbol{e}}_\phi\right) \cdot \left(rd\theta\hat{\boldsymbol{e}}_\phi\right) = \Phi, \tag{15.64}$$

so the two sides match up.

The Aharonov-Bohm phase can be viewed as a topological effect: the phase is unchanged by smooth deformations of the closed path, as long as the path doesn't cross the singularity in **B** at the origin. It can also be viewed an example of a Berry phase, with the actual physical electromagnetic potential and field, **A** and **B**, playing the roles of the Berry connection and Berry curvature. In Chap. 21, it will be seen that anyons can be modeled as charges with flux tubes attached to them and that the exotic behavior of anyons will follow from the resulting Aharonov-Bohm phase.

15.5 Geometric Phase in Solids: Zac Phase

Consider a solid with gaps between the energy bands. Recall that the wavefunctions must have the Bloch form, which can be written in terms of state vectors as:

$$|\psi_{n,k}\rangle = e^{ikx}|u_{nk}\rangle. \tag{15.65}$$

Here, n labels the band, k is the wave number, and $|u_{n,k}\rangle$ is the Bloch wavefunction, which has the same periodicity as the crystal lattice. Here we assume just one dimension for simplicity. We also set the lattice spacing equal to unity ($a = 1$); this amounts to measuring distances in units of a and wavenumbers in units of a^{-1}.

Recall that when an electric field is applied across a solid, it causes charges in the molecules to move apart from each other and induces a polarization (electric dipole moment per volume) P, which is proportional to the average distance between the charges. (Be careful not to confuse the polarization of the material with the polarization of light, which is completely unrelated.) So, we expect $P = n\rho e\langle\hat{x}\rangle$, where $\hat{x}$ is the position operator, ρ is the particle density (number per volume), and the brackets denote expectation value. But the position operator can be written as a derivative with respect to wavenumber or momentum:

$$P = \rho e\langle\hat{x}\rangle = i\hbar\rho\left\langle\frac{d}{dp}\right\rangle = i\rho e\left\langle\frac{d}{dk}\right\rangle. \tag{15.66}$$

So in the Bloch state for band n:

$$P = i\frac{\rho e}{2\pi}\int_0^{2\pi} dk \; \langle u_{nk}|\partial_k|u_{nk}\rangle, \tag{15.67}$$

where the integral is taken over a full Brillouin zone. But notice that the right side is proportional to a geometric phase:

$$P = \frac{\rho e}{2\pi}\phi. \tag{15.68}$$

This particular example of a geometric phase is called a **Zac phase**. Notice that the parameter that is being taken around a loop here is the momentum (or, equivalently, the wavenumber). The closed loop that is being integrated over is one that encloses an entire Brillouin zone. (Recall that electron states in solids with periodic lattice structures are periodic in k, so that $k = 0$ and $k = 2\pi$ are identified with each other.)

Zac phases come up in a variety of contexts in solid-state physics. For example, they appear in the study of the quantum Hall effect (Chap. 21.5) and appear in the study of topological phases of matter. An example of the latter will be given in Sect. 21.3.

Box 15.2 Geometric Propulsion

There is another, unexpected application of geometric phase in classical physics, having to do with motion in fluids. Recall that one of the key quantities in fluid dynamics is the **Reynolds number**, defined as

$$R_e = \frac{uL}{v}. \tag{15.69}$$

Here, u is the flow speed of the fluid, L is some characteristic size in the problem (e.g., the diameter of an object in the flow), and v is the fluid viscosity. R_e is the ratio of inertial forces to viscous forces. Roughly speaking, in laminar flow (no turbulence), viscous forces are those that keep the fluid layers moving smoothly one over each other, while inertial forces (those due to the momentum of the flow) are those that move particles away from the given layer it resides in. When the inertial forces become too large (high R_e), the fluid layers are no longer well defined, and the flow becomes turbulent.

To get an idea of the sizes of Reynolds numbers in given situations, consider some examples given in [19]: A man swimming in water has $R_e \sim 10^4$, while a goldfish in water has $R_e \sim 10^2$. In both of these cases, pushing against the water creates inertial forces that will propel the man or fish forward. However, for a tiny organism of roughly micron-scale size, the Reynolds number is much lower: $R_e \sim 10^{-5}$ or 10^{-6}. At such a low Reynolds number, the inertial forces on the organism are utterly irrelevant, and the organism should simply be carried along by the laminar flow. How can the organism navigate through the fluid, moving against or perpendicular to the flow?

It turns out that the answer is of geometric origin. Shapere and Wilczek [20] showed that cyclic changes in the shape of an object can lead to net translations or rotations. Thus, a paramecium or amoeba can propel itself through a fluid simply by wriggling around and altering its shape, even though the net force it exerts over the course of the cyclic deformation is zero. This grants mobility to organisms living at low Reynolds number. Similarly, a person trapped in a vat of molasses (high viscosity, low Reynolds number) can't swim effectively by pushing against the molasses but could still propel himself by altering the configurations of his limbs in a cyclic manner. It is interesting to speculate if this knowledge might have saved lives during Boston's Great Molasses Flood of 1919 (a real event; look it up), which led to 21 deaths.

Ref. [21] further showed that quasi-rigid bodies can swim in frictionless curved spacetimes using similar cyclic shape changes. Unfortunately, the effect is too small to ever likely be a practical propulsion method. For example, [21] calculates that in the gravitational field near the surface of the

(continued)

Box 15.2 (continued)
Earth (neglecting friction, air resistance, etc.), meter-scale deformations can only lead to displacements on the order of 10^{-23} m, roughly 10^8 smaller than the radius of an atomic nucleus. But the relevance here is that the motion is essentially the result of a nonzero geometric phase accumulating over the course of each cycle.

Problems

1.(a) In the Pancharatnam setup, when a and b beams are added, the intensity at point c' on the Poincaré sphere must vanish, while the b' components of I and I_a must be equal. Show that this implies Eq. 15.56.

  (b) Show that Eqs. 15.55 and 15.56 imply Eq. 15.57.

2. Apply a gauge transformation $| \uparrow \rangle \rightarrow e^{i\alpha\phi + i\beta\theta} | \uparrow \rangle$ to the state of Example 15.2.1. Show that the Berry connection coefficients are different but that the resulting Berry curvature is unchanged.

3. In Example 15.2.1, show that the correct Berry curvature can also be obtained by using the formula $R_{\theta\phi} = \frac{1}{2}\hat{n} \cdot \left(\partial_\theta \hat{n} \times \partial_\phi \hat{n} \right)$.

4. Many theories of particle physics predict the existence of isolated magnetic charges, called **magnetic monopoles**. In direct analogy to electric charges, a monopole should (up to overall constants) produce a magnetic field of Coulomb's law form: $B = \frac{q_m}{r^2}\hat{r}$, where q_m is the magnetic charge of the monopole.

   (a) Verify that a vector potential of the form $A = \frac{q_m(1-\cos\theta)}{r\sin\theta}\hat{\phi}$ will produce this field. Notice that this potential is singular along the negative z-axis ($\theta = \pi$).

   (b) One way to avoid the singularity is to use two gauge potentials defined on different, but overlapping, regions of space. Define $A_N = \frac{q_m(1-\cos\theta)}{r\sin\theta}\hat{\phi}$, as above, on all of space except the negative z axis, and define $A_S = -\frac{q_m(1+\cos\theta)}{r\sin\theta}\hat{\phi}$, on all of space except the positive z axis. Since they both give the same magnetic field, they must differ only by a gauge transformation on the region where they overlap (everywhere but the z-axis): $A_S - A_N = \nabla\Lambda(r)$ for some function $\Lambda(r)$. Determine the gauge function $\Lambda(r)$.

   (c) Suppose a particle with electric charge e moves through this field. As discussed in the text, the particle's wavefunction must also undergo a gauge transformation when we transform between gauge fields A_S and A_N: $\psi_S = e^{ie\Lambda/\hbar c}\psi_N$. Show that this implies that the product of the electric and magnetic charges must satisfy a quantization condition:

$$\frac{2eq_m}{\hbar c} = \pm n, \tag{15.70}$$

for integer n. (All observed charges in nature are known to be quantized in multiples of the up-quark charge $\frac{1}{3}e$; if a magnetic monopole is found to exist anywhere in nature, it would then explain this fact via the condition stated above.)

5. The Berry phase can also be defined for a discrete sequence instead of a continuous curve. Suppose a system goes through a discrete sequence of N states, $|u_0\rangle$, $|u_1\rangle$, ..., $|u_{N-1}\rangle$, and then back to $|u_0\rangle$. Then the Berry phase is defined $\phi = -Im\left[\ln\left[\langle u_0|u_1\rangle\langle u_1|u_2\rangle \ldots \langle u_{N-1}|u_0\rangle\right]\right]$. Given this, find the Berry phase for the three-state sequence

$$|u_0\rangle = \frac{1}{\sqrt{2}}\begin{pmatrix} 1 \\ 1 \end{pmatrix}, \quad |u_1\rangle = \frac{1}{\sqrt{2}}\begin{pmatrix} 1 \\ e^{2\pi i/3} \end{pmatrix}, \quad |u_2\rangle = \frac{1}{\sqrt{2}}\begin{pmatrix} 1 \\ e^{4\pi i/3} \end{pmatrix}.$$

References

1. M.V. Berry, Proc. Roy. Soc. London A **392**, 45 (1984)
2. S. Pancharatnam, Proc. Indian Acad. Sci. A **44**, 247 (1956)
3. R. Nitayananda, Curr. Sci. **67**, 238 (1994)
4. M. Nakahara, *Geometry, Topology and Physics*, 2nd edn. (CRC Press, Boca Raton, 2003)
5. F. Wilczek, A. Zee, Phys. Rev. Lett. **52**, 2111 (1984)
6. A. Shapere, F. Wilczek, *Geometric Phases in Physics* (World Scientific, Singapore, 1989)
7. D. Vanderbilt, *Berry Phases in Electronic Structure Theory: Electric Polarization, Orbital Magnetization and Topological Insulators* (Cambridge University Press, Cambridge, 2018)
8. J.J. Sakurai, J. J., J. Napolitano, *Modern Quantum Mechanics*, 3 edn. (Cambridge University Press, Cambridge, 2020)
9. R. Gilmore, *Lie Groups, Physics, and Geometry: An Introduction for Physicists* (Cambridge University Press, Cambridge, 2008)
10. Y. Lin, R. Compton, K. Jiménez-García, J. Porto, I. Spielman, Nature **462**, 628 (2009)
11. J. Dalibard, F. Gerbier, G. Juzeliunas, P. Öhberg, Rev. Mod. Phys. **83**, 1523 (2011)
12. N. Goldman, G. Juzeliūnas, P. Öhberg, I.B. Spielman, Rep. Prog. Phys. **77**, 126401 (2014)
13. Y. Chen, R.Y. Zhang, Z. Xiong, Z.H. Hang, J.Li, J.Q. Shen, C.T. Chan, Nat. Comm. **10**, 3125 (2019)
14. J. Wu et al., Nat. Electron. **5**, 635 (2022)
15. M. Hafezi, Int. J. Mod. Phys. B **28**, 1441002 (2014)
16. V. Galitski, G. Juzeliunas, I.B. Spielman, Phys. Today **72**(1), 38 (2019)
17. R.Y. Chaio, Y.S. Wu, Phys. Rev. Lett. **57**, 933 (1986)
18. A. Tomita, R.Y. Chaio, Phys. Rev. Lett. **57**, 937 (1986)
19. E.M. Purcell, Am. J. Phys. **45**, 3 (1977)
20. A. Shapere, F. Wilczek, J. Fluid. Mech. **198**, 557 (1989)
21. J. Wisdom, Science **299**, 1865 (2003)

Part III
Applications

Chapter 16
Quantum Computing: General Considerations

Of all the modern applications that have arisen in recent decades, quantum computing is the one that has attracted the most attention and the most research funding. In this chapter, we discuss what quantum computing is and the advantages it would bring. Focus in this chapter is on general aspects that are independent of the physical platform used to implement the computing hardware. Potential physical platforms are discussed in Part IV, and specific quantum algorithms are discussed in Chap. 17.

16.1 Why Quantum Computing?

Moore's law (first stated by Gordon Moore in 1965) has been widely quoted for decades. It says that the number of components that can fit on a computer chip continues to increase fast enough for the computational speed and available memory of computers to double every 18 months. This statement has continued to hold fairly well for nearly 60 years. But the shrinking size of computer components is approaching an inevitable roadblock that will prevent Moore's law from continuing to hold soon. Sizes of lithographic structures on silicon chips will soon be approaching sizes of individual molecules. Once this quantum-dominated size scale is reached, improved computer performance can no longer be achieved by further miniaturization: some other approach must be taken if computational power is to continue to expand. One such approach is to utilize computers that make use of the laws of quantum physics, rather than classical physics.

Such quantum computers would take advantage of intrinsically quantum phenomena such as superposition, entanglement, and interference to essentially carry out multiple computations simultaneously, an effect known as **quantum parallelism**. Although only one of these many possible outcomes can be measured on a given run of the computer, clever choices of what measurement to make can still

© The Author(s), under exclusive license to Springer Nature Switzerland AG 2025
D. S. Simon, *Introduction to Quantum Science and Technology*, Undergraduate
Texts in Physics, https://doi.org/10.1007/978-3-031-81315-3_16

lead to dramatic improvement in computational time. For some problems, quantum computers are expected to carry out computations *exponentially* faster.

So far, no such dramatic speedup has been documented experimentally, despite enormous efforts and billions of dollars of research funds poured into the effort. The main obstacle to carrying out large-scale computations is the phenomenon of **decoherence**. Interactions with the environment will destroy the coherence of quantum processes and effectively make the system classical again. For this reason, many of the proposed quantum computation platforms require extremely low temperatures and a high level of isolation from the surrounding environment. Those systems that have operated successfully have generally worked with only a very small number of qubits.

Quantum information processing can be carried out in a number of different physical systems, many of which are examined more closely in Part IV of this book. These include nuclear magnetic resonance systems, superconducting materials, cold atoms in optical lattices, photonic systems, and even strands of DNA. Each of these systems has its own advantages and disadvantages. For example, superconducting systems require very low temperatures but allow very stable qubits to be formed. In contrast, photonic systems can operate at room temperature (the photon energy $\hbar\omega$ at optical frequencies is typically about two orders of magnitude than the ambient thermal energy $k_B T$) and have long coherence times; but on the downside, their weak interactions with each other make controlled gates difficult to implement. Photons also come with the difficulty that single photons or entangled photon pairs are generally created probabilistically, at random times; but current work with quantum dots and other systems gives hope of soon producing photons deterministically and on demand in a practical manner.

In this chapter, the focus is mainly on platform-independent aspects of quantum systems and focuses on general physical and computational aspects of quantum information processing. In the next chapter, some specific quantum algorithms that have been proposed or implemented are discussed. For those wishing further detail on quantum computing, there are many books devoted to the topic, including [1–7].

16.2   History and Current Prospects

Before continuing with the physics of quantum computing, we give here a very brief overview of the field's historical development and some idea of current prospects. The idea that quantum information processing might have advantages over classical information processing began germinating in the 1970s and 1980s, when Stephen Weisner, Charles Bennett, Gilles Brassard, and others realized that quantum cryptographic keys could be generated and communicated in a manner that prevented eavesdroppers from gaining access without detection (see Chap. 18). Richard Feynman (as well as Russian mathematician Yuri Manin and others) realized that there were quantum systems that cannot be efficiently simulated using Turing machines that run according to the laws of classical physics. Efficient simulation of these systems would require quantum mechanical computers. This resulted in the idea of **quantum simulators**: single-purpose (nonprogrammable) analog quantum computers designed to simulate a particular phenomenon. This naturally soon led to the idea of programmable quantum computers and the idea of simultaneous processing of quantum superpositions of logical bits (qubits). The theory of quantum Turing machines was developed, beginning with the work of David Deutsch, Baniel Bernstein, Vijay Vazirani, and Andrew Yao.

The field of quantum computing really began to take off with the publication of the several quantum algorithms in the 1990s. The first was Peter Shore's algorithm for factoring integers in polynomial time. This created a ripple, because if a machine could be developed to implement this algorithm, it would compromise the safety of financial and national security information that is currently safeguarded using the RSA encryption procedure. The Shore algorithm was soon followed by the Grover search algorithm, the Deutsch and Deutsch-Josza algorithms, and the quantum Fourier transform.

The first quantum computers to be built used liquid NMR systems. However, the signal produced by NMR systems scales like $\frac{1}{2^n}$ for n qubits, so they are not suitable for scaling up to large-scale computing applications. Current research focuses more on cold atoms in optical lattices, superconducting qubit systems, solid-state qubits, and photonic systems. Photonic systems are very popular for a number of reasons: they don't require low temperatures, photons are easy to produce and detect, they carry multiple useful degrees of freedom (polarization, orbital angular momentum, linear momentum, etc.), and most importantly, they can have very long coherence times. However, long coherence times are a consequence of their weak interactions with their environment, particularly with other photons, and the flip side of this lack of interaction is that it is hard to make controlled gates with them. To use one photon as a control for operations on a second (target) photon, materials with a high degree of nonlinearity are required. Recall that the size of a nonlinear effect decreases as the intensity drops, so it is difficult to achieve the required degree of nonlinearity at the single-photon level. Knill, Laflamme, and Milburn came up with a potential way around this obstacle [8], showing that linear optical systems could achieve controlled operations when combined with the use of measurements. A judicious choice of projective measurement on one photon state can be used to alter the state of the remaining photons, introducing effective interactions. The resulting **KLM formalism** leads to probabilistic computations, which will fail some proportion of the time. But the main problem with the KLM approach is that the required systems scale up in size very rapidly as the number of qubits increases, making them very resource intensive and unsuitable for large-scale computing. A review of optical approaches, including the KLM approach, can be found in [9].

Quantum computers need mechanisms to detect and correct errors, just as classical computers do. In fact, researchers quickly realized that error correction is even more important for quantum computers, since entangled states and coherence are fragile resources that can easily be disturbed by external influences. In fact, the coherence of quantum states can be disrupted by background radiation from the earth's crust or by cosmic rays.

There are two main types of errors: bit flip (or partial bit flip) errors and phase errors. Thinking of the qubit as a point on the Bloch sphere, a partial bit flip is a rotation about either the x or y axis, so that a $|0\rangle$ state is rotated partially toward the $|1\rangle$ state (or vice-versa). A phase error occurs when the relative phase between the two terms ($|0\rangle$ and $|1\rangle$) in a qubit is altered; this amounts to a rotation about the z axis on the Bloch sphere.

So mechanisms needed to be developed to cope with errors. Two main approaches have dominated. The first is quantum error correction codes, in which syndromes are used to detect errors and redundancy in the state is used to correct them; we won't discuss these further here; see [6] or [7] for a detailed discussion at an introductory level. The other approach is to prevent the errors from happening in the first place. One promising way to do that is through topological quantum computing, which takes advantage of conserved topological invariants, such as winding numbers, that do not change under small perturbations of the system. This makes the information in the system resistant to all but the most disruptive environmental effects. Topological quantum computing will be discussed briefly in Sect. 21.7.2; see also [10–12].

16.3   Quantum Versus Classical Computing

As mentioned above, quantum computing requires the physical qubits in the system to remain highly coherent with each other. This is necessary for quantum interference to occur, and it imposes stringent limitations on quantum computing systems. For example, unlike a classical system, the quantum computer's user cannot read out intermediate results and then continue the computation. Any measurement at an intermediate step of a computation will cause a random collapse onto one particular state of any superpositions in the system, effectively rendering the computer back into a classical state. Similarly, intermediate states of the quantum computation cannot be copied without running afoul of the no-cloning theorem.

Another important difference between classical and quantum computing is in the idea of reversibility. In a classical computation, it is not always possible to reconstruct the input from the output. For example, given that the output of the AND gate $A \wedge B$ is 1, there is no way to run back the operation to see if the input was $A = 1$ $B = 0$ or if it was $A = 0$, $B = 1$. Such operations are **irreversible**. Conversely, if the input can be uniquely reconstructed from the output, then the operation is **reversible**. In quantum mechanics, irreversible operations are not allowed. This is because in the quantum case, the logic gates will be unitary operations taking an input state to the output state. For a given gate represented by unitary operator $\hat{U}$, the time-reversed operation will be represented by $\hat{U}^\dagger = \hat{U}^{-1}$. Irreversibility would require that $\hat{U}\hat{U}^\dagger \neq 1$, implying loss of unitarity. Non-unitarity means loss of energy and of information to the surrounding environment. This coupling to the environment is a problem for quantum computing, since it allows the environment to decohere the computer, quickly pushing it back into the classical regime.

The fact that irreversibility forces an energy exchange with the environment has been long known in classical computing. The most obvious example is the fact that erasure of information leads to a heating of the system: one $k_B T$ of heat is generated for each bit of information erased. This is one reason why classical computers always have cooling fans built into them: in addition to the usual Ohm's law heating the circuit elements, the constant erasure of old bits to replace them with new values leads to a large amount of heat generation in the system.

Example 16.3.1 The generation of heat by irreversible processes can be seen heuristically by considering a simple example. Suppose a qubit is initially in state $|0\rangle$, encoding one bit of information. Erasure of this information amounts to replacing the pure state $|0\rangle$ by a maximally mixed state of both bits: $\hat{\rho} = \frac{1}{\sqrt{2}}\left(|0\rangle\langle 0| + |1\rangle\langle 1|\right)$. The initial and final von Neumann entropies are

$$S_0 = -\text{Tr}\left((|0\rangle\langle 0|)\log_2(|0\rangle\langle 0|)\right) = -(1)\log_2(1) = 0, \qquad (16.1)$$

$$S_f = -\text{Tr}\hat{\rho}\log_2\hat{\rho} = -\frac{1}{2}\log_2\frac{1}{2} - \frac{1}{2}\log_2\frac{1}{2} = 1, \qquad (16.2)$$

so that the entropy change is $|\Delta S| = 1$ bit. Equating the von Neumann entropy to the thermodynamic entropy, this leads to a heat of

$$|Q| = |\Delta S|T = T. \qquad (16.3)$$

Going from dimensionless information units to standard thermodynamic units leads to an extra factor of Boltzmann's constant on the right, so we conclude that erasure of one unit of information leads to the creation of $|Q| = k_B T$ Joules of heat. ∎

In order to be useful for quantum computing, a quantum system should satisfy several requirements known as the **Di Vincenzo criteria** [13], which says that a useful physical platform for quantum computing should:

- Be capable of supporting well-defined qubits and scalable upward to support large numbers of qubits
- Have a means to initialize the initial state of the system to a standard reference configuration
- Be robust against noise and environmental perturbations (Effectively, this means that coherence time of qubits should be much longer than the time it takes for each gate to complete its operation, thus allowing many operations before the "quantumness" of the qubit is lost.)
- Have an efficient, low-error means of measuring the final output qubit states

The Di Vincenzo criteria were originally formulated for use with stationary qubits, such as NMR or superconducting qubits. But with the advent of photonic quantum computing platforms, it has become necessary to add two more criteria in order to deal with flying qubits (qubits attached to moving photons):

- The ability to interconvert between stationary and flying qubits
- The ability to deliver the flying qubits to the correct specified locations, without loss of quantum fidelity

16.4   Quantum Gates

As in the case of classical computing, logical circuits can be constructed out of sequences of logical gates, and these gates can act on one, two, or more qubits. As in the classical case, these circuits are feed-forward: the quantum state proceeds step by step in a single direction through the circuit, with no reversals of direction and no loops. The principal extra requirement of quantum circuits is that they have to be reversible.

A quantum gate is a device that performs a unitary transformation on one or more qubits. In general, the transformation on an n-qubit state will be an element of the special unitary group $SU(2n)$ (see Appendix E), meaning that it can be represented by a $2n \times 2n$ unitary matrix of unit determinant. In addition to the gates themselves, the circuit may involve measurements; subsequent gate operations may then depend on the outcomes of those measurements.

16.4.1   Single-Qubit Gates

First, consider gates that act on single qubits. These will implement $SU(2)$ transformations an a qubit represented by a two-component column matrix: $|\psi\rangle = \alpha|0\rangle + \beta|1\rangle = \begin{pmatrix} \alpha \\ \beta \end{pmatrix}$.

The most basic single-qubit gates are those that implement Pauli matrix operations (Fig. 16.1). For example, the Pauli X gate simply operates by applying the Pauli σ_x matrix:

$$X = \hat{\sigma}_x = \begin{pmatrix} 0 & 1 \\ 1 & 0 \end{pmatrix} = NOT, \tag{16.4}$$

which is equivalent to a **NOT gate**. Similarly, Y and Z gates implement the Pauli $\hat{\sigma}_y$ and $\hat{\sigma}_z$ matrices:

$$Y = \hat{\sigma}_y = \begin{pmatrix} 0 & -i \\ i & 0 \end{pmatrix} \tag{16.5}$$

$$Z = \hat{\sigma}_z = \begin{pmatrix} 1 & 0 \\ 0 & -1 \end{pmatrix}. \tag{16.6}$$

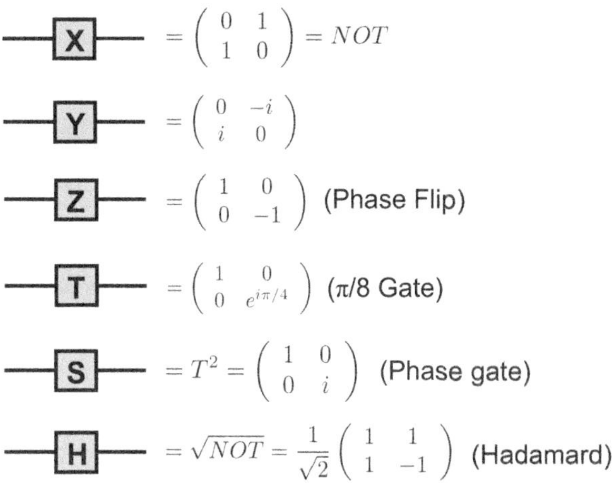

Fig. 16.1 Some common single qubit gates and the unitary transformations they produce on qubits

The Z gate is also called a **phase-flip gate**, since it flips the phase of the $|1\rangle$ state while leaving the $|0\rangle$ state unchanged. Actions of Pauli gates on generic qubits are clearly:

$$X\Big(\alpha|0\rangle + \beta|1\rangle\Big) = \alpha|1\rangle + \beta|0\rangle \tag{16.7}$$

$$Y\Big(\alpha|0\rangle + \beta|1\rangle\Big) = i\Big(-\beta|0\rangle + \alpha|1\rangle\Big) \tag{16.8}$$

$$Z\Big(\alpha|0\rangle + \beta|1\rangle\Big) = \alpha|0\rangle - \beta|1\rangle. \tag{16.9}$$

Another common gate is the $\frac{\pi}{8}$ **gate** or T **gate**, given by the matrix

$$T = \begin{pmatrix} 1 & 0 \\ 0 & e^{i\pi/4} \end{pmatrix} = e^{i\pi/8}\begin{pmatrix} e^{-i\pi/8} & 0 \\ 0 & e^{i\pi/8} \end{pmatrix}, \tag{16.10}$$

with action

$$T\Big(\alpha|0\rangle + \beta|1\rangle\Big) = \alpha|0\rangle + \beta e^{i\pi/4}|1\rangle. \tag{16.11}$$

Up to an irrelevant overall global phase, $|0\rangle$ and $|1\rangle$ states are multiplied by opposite-sign phases. The square of the $\frac{\pi}{8}$ gate then gives the S or **phase gate**:

$$S = T^2 = \begin{pmatrix} 1 & 0 \\ 0 & i \end{pmatrix}, \tag{16.12}$$

with

$$S\Big(\alpha|0\rangle + \beta|1\rangle\Big) = \alpha|0\rangle + i\beta|0\rangle. \tag{16.13}$$

Finally, one more useful single-qubit gate is the **Hadamard gate**, H:

$$H = \sqrt{NOT} = \frac{1}{2}\begin{pmatrix} 1+i & 1-i \\ 1-i & 1+i \end{pmatrix} = \frac{1}{\sqrt{2}}\begin{pmatrix} e^{i\pi/4} & e^{-i\pi/4} \\ e^{-i\pi/4} & e^{i\pi/4} \end{pmatrix}. \tag{16.14}$$

Note that two Hadamard operations give an X or NOT operation,

$$H^2 = X = NOT. \tag{16.15}$$

If the qubit basis is the basis of photon polarizations, it converts horizontal and vertical polarizations into diagonal and anti-diagonal polarizations,

$$H\Big(\alpha|H\rangle + \beta|V\rangle\Big) = \alpha|D\rangle + \beta|A\rangle, \tag{16.16}$$

and vice-versa:

$$H\Big(\alpha|D\rangle + \beta|A\rangle\Big) = \alpha|H\rangle + \beta|V\rangle. \tag{16.17}$$

Similarly, up and down spin states are interchanged by the Hadamard gate with spins at $\pm 45°$.

The quantum states of a qubit can be represented as points on the Bloch sphere. Rotations on the sphere about the three coordinate axes are accomplished by rotation operators generated by three Pauli matrices:

$$R_x(\theta) = e^{-i\theta\sigma_x/2} = \cos\frac{\theta}{2} - i\hat{\sigma}_x \sin\frac{\theta}{2} = \begin{pmatrix} \cos\frac{\theta}{2} & -i\sin\frac{\theta}{2} \\ -i\sin\frac{\theta}{2} & \cos\frac{\theta}{2} \end{pmatrix} \tag{16.18}$$

$$R_y(\theta) = e^{-i\theta\sigma_y/2} = \cos\frac{\theta}{2} - i\hat{\sigma}_y \sin\frac{\theta}{2} = \begin{pmatrix} \cos\frac{\theta}{2} & -\sin\frac{\theta}{2} \\ \sin\frac{\theta}{2} & \cos\frac{\theta}{2} \end{pmatrix} \tag{16.19}$$

$$R_z(\theta) = e^{-i\theta\sigma_z/2} = \cos\frac{\theta}{2} - i\hat{\sigma}_z \sin\frac{\theta}{2} = \begin{pmatrix} e^{-i\theta/2} & 0 \\ 0 & e^{i\theta/2} \end{pmatrix}. \tag{16.20}$$

Recall that products of any two of the Pauli matrices will lead to the third matrix: for $i \neq j$, the identity $\hat{\sigma}_i\hat{\sigma}_j = i\sum_k \epsilon_{ijk}\hat{\sigma}_k$ holds, implying the relation $\hat{\sigma}_x\sigma_y = i\sigma_z$

and its cyclic permutations. So the set of three rotation matrices above is redundant. Any two of the matrices can be used to generate the third. So starting with any initial qubit (any initial point on the Bloch sphere), any two of the rotation operators is sufficient to span the entire sphere. Thus, all single-qubit gates can be built out of any two of the three rotation operators.

16.5 Multiparticle Gates and Universal Sets

In order to do computations, gates are needed that allow two or more qubits to interact with each other. Since these gates need to be reversible, the number of inputs and outputs need to be equal. Running the gate backward needs to be able to reproduce the original input uniquely, so the matrix representation of the gate has to be unitary.

Many multiple-qubit gates are n-qubit controlled gates CU, which leave the input unchanged if the control qubit is $|c\rangle = |0\rangle$ but which apply the unitary operation U to the $n - 1$ target qubits if $|c\rangle = |1\rangle$. These are represented schematically as in Fig. 16.2. They are represented by a $2n \times 2n$ matrix with the identity in the upper left 2×2 block and the unitary matrix filling the lower right $(2n - 2) \times (2n - 2)$ block:

$$CU = \begin{pmatrix} \begin{array}{cc|} 1 & 0 \\ 0 & 1 \\ \hline & \end{array} & \\ & U \end{pmatrix} \tag{16.21}$$

Two useful controlled gates are the controlled NOT (or CNOT) gate discussed in Sect. 8.4 and the **controlled phase** or controlled Z gate, CZ. The action of the CZ gate can be expressed as

$$CZ|ab\rangle = (-1)^{ab}|ab\rangle. \tag{16.22}$$

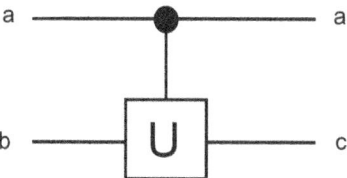

Fig. 16.2 A controlled U gate. The top line is the control, and the bottom is the target. The control is unchanged by the gate. The output c of the target line is $c = b$ if $a = 0$. But if $a = 1$, then the output is $c = Ub$

Fig. 16.3 The CZ gate is symmetric under interchange of control and target qubits

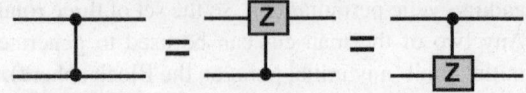

Note that, unlike most controlled gates, the controlled Z gate is symmetric under exchange of the control and target (Fig. 16.3).

The reversible classical gates discussed in Sect. 8.4 (the CNOT, Toffoli, and Fredkin gates) are all suitable for use as quantum gates, as is the SWAP gate. These are the most common two- and three-qubit gates and are more than sufficient for universal quantum computing (see below). An additional quantum gate not mentioned in Sect. 8.4 is the **anti-Toffoli gate**, whose action is given by

$$U_{ATf}|a\rangle|b\rangle|c\rangle = |a\rangle|b\rangle|ab \oplus c\rangle, \quad U_{ATf} = \begin{pmatrix} 1 & & & & & \\ & 1 & & & & \\ & & 1 & & & \\ & & & 1 & & \\ & & & & 0 & 1 \\ & & & & 1 & 0 \end{pmatrix}. \tag{16.23}$$

Just as universal gate sets can be defined that will carry out all classical computations, it is possible to define sets of quantum gates that are universal for quantum computation. These universal sets need to be able to preserve superpositions and entanglement. In order to go beyond the capacity of classical computers, they must be able to implement operations beyond the **Clifford group**. The Clifford group consists of the set of operations that can be generated by the CNOT, H, and S gates. Since the **Gottesman-Knill theorem** says that all Clifford group operations can be efficiently simulated by classical computers, a quantum computer will provide no benefit unless it contains gates that cannot be formed from combinations of the Clifford gates.

Some examples of universal gate sets include the following:

- CNOT, H, T
- CNOT, $R_y\left(\frac{\pi}{8}\right)$
- TOF, H
- $R_x(\theta)$, $R_y(\theta)$, $R_z(\theta)$, S, CNOT
- The controlled Hadamard gate, CH, can serve as a universal set all by itself, since it can simulate both H and the Toffoli gate.

Clearly, quantum analogs of all of the classical logic gates need to be reproducible from elements of a universal quantum gate set. In order to be reversible, there may also need to be ancilla bit. An **ancilla** or **ancillary bit** is a bit that is not of interest in itself, but which is needed for the operation to by carried out; in a sense, it is similar to a catalyst in a chemical reaction. For example, the two-qubit quantum AND and NAND gates can be produced using a three-qubit Toffoli gate,

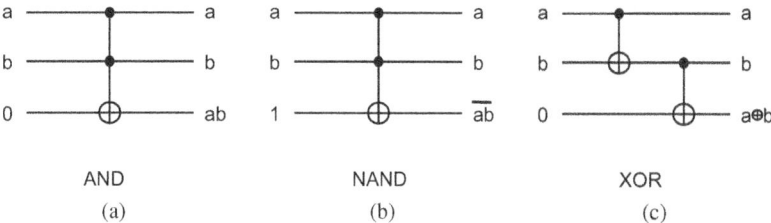

Fig. 16.4 Classical gates such as AND, NAND, and XOR can be implemented reversibly by quantum Toffoli and CNOT gates. To do so requires an extra (ancillary) qubit. Given that the Pauli X gate acts as a NOT gate, de Morgan's laws then allow gates such as OR and NOR to be implemented from these

with one ancilla qubit, as shown in Fig. 16.4a, b. Similarly, a quantum XOR gate can be produced using a pair of CNOT gates, as in Fig. 16.4c.

16.6 Examples of Quantum Logic Circuits

In this section, we give a few examples of quantum circuits built out of simple one- and two-qubit gates. Following common conventions, we take the input to be coming in at the left and output to be exiting at the right of each circuit. In an expression like $|i, j, k, \dots\rangle$, the leftmost entry (i) represents the uppermost line of the circuit diagram, j represents the next one from the top, and so on. (The latter convention is used in most books and papers on quantum computing. One exception is [7], which takes the top line to be at the left of the ket, $|\dots, k, j, i\rangle$.)

In a tensor product operation such as $\hat{A} \otimes \hat{B}$, the first operator, $\hat{A}$, acts on the first qubit (the top line of the diagram), while the second $\hat{B}$ acts on the second qubit (the lower line). For example, consider the two-qubit operations in Fig. 16.5, involving a Hadamard gate acting on one qubit and an identity operation on the other. As usual, the two-qubit basis that will be used is

$$\begin{pmatrix} |00\rangle \\ |01\rangle \\ |10\rangle \\ |11\rangle \end{pmatrix} .$$

In Fig. 16.5a, where the Hadamard gate acts on the first qubit, the resulting operation is

$$H \otimes I = \frac{1}{\sqrt{2}} \begin{pmatrix} 1 & 1 \\ 1 & -1 \end{pmatrix} \otimes \begin{pmatrix} 1 & 0 \\ 0 & 1 \end{pmatrix} \tag{16.24}$$

Fig. 16.5 Product gates $H \otimes I$ and $I \otimes H$

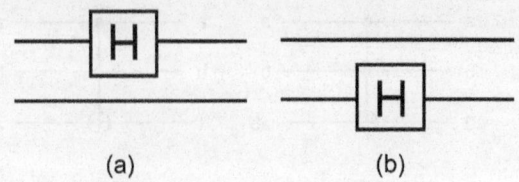

(a)   (b)

$$= \frac{1}{\sqrt{2}} \begin{pmatrix} (1)\begin{pmatrix} 1 & 0 \\ 0 & 1 \end{pmatrix} & (1)\begin{pmatrix} 1 & 0 \\ 0 & 1 \end{pmatrix} \\ (1)\begin{pmatrix} 1 & 0 \\ 0 & 1 \end{pmatrix} & (-1)\begin{pmatrix} 1 & 0 \\ 0 & 1 \end{pmatrix} \end{pmatrix} \tag{16.25}$$

$$= \frac{1}{\sqrt{2}} \begin{pmatrix} 1 & 0 & 1 & 0 \\ 0 & 1 & 0 & 1 \\ 1 & 0 & -1 & 0 \\ 0 & 1 & 0 & -1 \end{pmatrix}. \tag{16.26}$$

Similarly, reversing which qubit is acted on by each gate, as in Fig. 16.5b, we have:

$$I \otimes H = \begin{pmatrix} 1 & 0 \\ 0 & 1 \end{pmatrix} \otimes \frac{1}{\sqrt{2}} \begin{pmatrix} 1 & 1 \\ 1 & -1 \end{pmatrix} \tag{16.27}$$

$$= \frac{1}{\sqrt{2}} \begin{pmatrix} (1)\begin{pmatrix} 1 & 1 \\ 1 & -1 \end{pmatrix} & (0)\begin{pmatrix} 1 & 1 \\ 1 & -1 \end{pmatrix} \\ (0)\begin{pmatrix} 1 & 1 \\ 1 & -1 \end{pmatrix} & (1)\begin{pmatrix} 1 & 1 \\ 1 & -1 \end{pmatrix} \end{pmatrix} \tag{16.28}$$

$$= \frac{1}{\sqrt{2}} \begin{pmatrix} 1 & 1 & 0 & 0 \\ 1 & -1 & 0 & 0 \\ 0 & 0 & 1 & 1 \\ 0 & 0 & 1 & -1 \end{pmatrix}. \tag{16.29}$$

Example 16.6.1 As a first example of a nontrivial quantum circuit, we can make use of the previous results to analyze a circuit formed from a $H \otimes I$ gate followed by $CNOT$ gate, as shown in Fig. 16.6a. To work out what this circuit does, examine it step by step. Take the product of the $H \otimes I$ and CX (= CNOT) matrices; $H \otimes I$ is applied first, so it is on the right in the product. The full operation is then the product of these. Calling the total circuit B, we have:

$$B = CX \cdot (H \otimes I) \tag{16.30}$$

$$= \begin{pmatrix} 1 & 0 & 0 & 0 \\ 0 & 1 & 0 & 0 \\ 0 & 0 & 0 & 1 \\ 0 & 0 & 1 & 0 \end{pmatrix} \cdot \frac{1}{\sqrt{2}} \begin{pmatrix} 1 & 0 & 1 & 0 \\ 0 & 1 & 0 & 1 \\ 1 & 0 & -1 & 0 \\ 0 & 1 & 0 & -1 \end{pmatrix} \tag{16.31}$$

$$= \frac{1}{\sqrt{2}} \begin{pmatrix} 1 & 0 & 1 & 0 \\ 0 & 1 & 0 & 1 \\ 0 & 1 & 0 & -1 \\ 1 & 0 & -1 & 0 \end{pmatrix}. \tag{16.32}$$

To understand the function of this circuit, let it act on each of the basis states in turn:

$$B|00\rangle = \frac{1}{\sqrt{2}} \begin{pmatrix} 1 & 0 & 1 & 0 \\ 0 & 1 & 0 & 1 \\ 0 & 1 & 0 & -1 \\ 1 & 0 & -1 & 0 \end{pmatrix} \begin{pmatrix} 1 \\ 0 \\ 0 \\ 0 \end{pmatrix} = \frac{1}{\sqrt{2}}(|00\rangle + |11\rangle) = |\Phi^+\rangle$$

$$B|01\rangle = \frac{1}{\sqrt{2}} \begin{pmatrix} 1 & 0 & 1 & 0 \\ 0 & 1 & 0 & 1 \\ 0 & 1 & 0 & -1 \\ 1 & 0 & -1 & 0 \end{pmatrix} \begin{pmatrix} 0 \\ 1 \\ 0 \\ 0 \end{pmatrix} = \frac{1}{\sqrt{2}}(|01\rangle + |10\rangle) = |\Psi^+\rangle$$

$$B|01\rangle = \frac{1}{\sqrt{2}} \begin{pmatrix} 1 & 0 & 1 & 0 \\ 0 & 1 & 0 & 1 \\ 0 & 1 & 0 & -1 \\ 1 & 0 & -1 & 0 \end{pmatrix} \begin{pmatrix} 0 \\ 0 \\ 1 \\ 0 \end{pmatrix} = \frac{1}{\sqrt{2}}(|00\rangle - |11\rangle) = |\Phi^-\rangle$$

$$B|11\rangle = \frac{1}{\sqrt{2}} \begin{pmatrix} 1 & 0 & 1 & 0 \\ 0 & 1 & 0 & 1 \\ 0 & 1 & 0 & -1 \\ 1 & 0 & -1 & 0 \end{pmatrix} \begin{pmatrix} 0 \\ 0 \\ 0 \\ 1 \end{pmatrix} = \frac{1}{\sqrt{2}}(|01\rangle - |10\rangle) = |\Psi^-\rangle$$

So this is a **Bell state circuit** or **entangling circuit**: a product state is input and an entangled Bell state is produced. Which Bell state is produced depends on which basis state was fed in. ■

Example 16.6.2 Now consider the circuit in Fig. 16.6b, comprised of three CNOT gates. This implements a quantum SWAP gate. One way to see this is to construct the truth table, working one step at a time from left to right, as was done in classical examples in Chap. 8. A simpler way is use matrix representations of CNOT gates. The first and third CNOT gates are given by Eq. 8.27, while the reader should be able to convince themselves that the middle CNOT is of the form

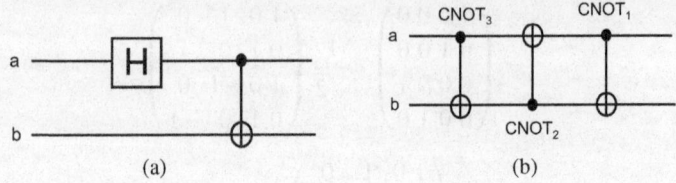

Fig. 16.6 Two simple quantum circuits. (**a**) A circuit to produce Bell states from computational basis states. (**b**) An implementation of a quantum swap gate using three CX gates

$$CNOT_2 = \begin{pmatrix} 1\,0\,0\,0 \\ 0\,0\,0\,1 \\ 0\,0\,1\,0 \\ 0\,1\,0\,0. \end{pmatrix}.$$ (16.33)

So when we multiply the three matrices, the unitary operator for the full circuit is:

$$U = CNOT_1 \cdot CNOT_2 \cdot CNOT_3$$ (16.34)

$$= \begin{pmatrix} 1\,0\,0\,0 \\ 0\,1\,0\,0 \\ 0\,0\,0\,1 \\ 0\,0\,1\,0 \end{pmatrix}_1 \begin{pmatrix} 1\,0\,0\,0 \\ 0\,0\,0\,1 \\ 0\,0\,1\,0 \\ 0\,1\,0\,0 \end{pmatrix}_2 \begin{pmatrix} 1\,0\,0\,0 \\ 0\,1\,0\,0 \\ 0\,0\,0\,1 \\ 0\,0\,1\,0 \end{pmatrix}_3$$ (16.35)

$$= \begin{pmatrix} 1\,0\,0\,0 \\ 0\,0\,1\,0 \\ 0\,1\,0\,0 \\ 0\,0\,0\,1 \end{pmatrix}.$$ (16.36)

Again applying this to the basis vectors, we can verify that it is indeed the SWAP operation:

$$U|00\rangle = |00\rangle$$ (16.37)

$$U|01\rangle = |10\rangle$$ (16.38)

$$U|10\rangle = |01\rangle$$ (16.39)

$$U|11\rangle = |11\rangle.$$ (16.40)

■

Other product identities can be proven in a similar manner. Probably the most important of these involve Hadamard and Pauli gates:

$$HXH = Z$$ (16.41)

$$HYH = -Y \qquad (16.42)$$

$$HZH = X, \qquad (16.43)$$

as can easily be verified by the matrix multiplication method (Problem 4).

Additional examples of quantum circuits will appear over the coming chapters. For example, a circuit that implements quantum teleportation will be studied in Sect. 18.2.

Box 16.1 Cluster State Computing

Although the circuit model of quantum computing is the most common approach, it is not the only one. Another promising approach to quantum computing is **cluster state** or **one-way computing**, which was introduced in [14, 15] and is reviewed in [16–18].

Recall that a measurement on one qubit of a multiqubit state will alter the post-measurement state of the unmeasured qubit. For example, suppose that a three-qubit system starts in the state

$$|\psi\rangle = |0\rangle_a \big(\alpha|0\rangle_b|0\rangle_c + \beta|1\rangle_b|1\rangle_c\big) + |1\rangle_a \big(\alpha|0\rangle_b|1\rangle_c + \beta|1\rangle_b|0\rangle_c\big).$$

If the state of the a subsystem is measured and found to be $|0\rangle_a$, then the bc subsystem collapses to the state $\alpha|0\rangle_b|0\rangle_c + \beta|1\rangle_b|1\rangle_c$. But if the a measurement yields the result $|1\rangle_a$, then bc will collapse to state $\big(\alpha|0\rangle_b|1\rangle_c + \beta|1\rangle_b|0\rangle_c\big)$. Thus, the measurement alters the state of the unmeasured qubits. This is the basis of cluster state computing.

Consider a graph G drawn in a plane, consisting of n vertices connected by a set of edges. The n-qubit cluster state corresponding to this graph is prepared by starting from n qubits (one representing each vertex of the graph). Each qubit is put into reference state $|+\rangle = \frac{1}{\sqrt{2}}\big(|0\rangle + |1\rangle\big)$, and then the system is operated on with a controlled Z gate for each pair of qubits connected by a vertex. For example, the graph of Fig. 16.7 would correspond to the state $CZ_{12}CZ_{23}CZ_{24}|+\rangle_1|+\rangle_2|+\rangle_3|+\rangle_4$. A sequence of measurements is carried out, with later measurements depending on the outcomes of the previous ones. In this way, the system ends up in final state $|\Psi\rangle$. Measurement of this final state then gives the outcome of the computation.

The initial cluster state is highly entangled, but the entanglement decreases over the course of the computation, as measurements are made. Unlike other quantum computing approaches, there is no need to maintain coherence of entangled states over the course of the full computation. It has been shown that the cluster-state approach is equivalent to the circuit model of computation and can perform universal computation. One possible advantage of using cluster states is that it may be useful for reducing the required resources when combined with the KLM approach.

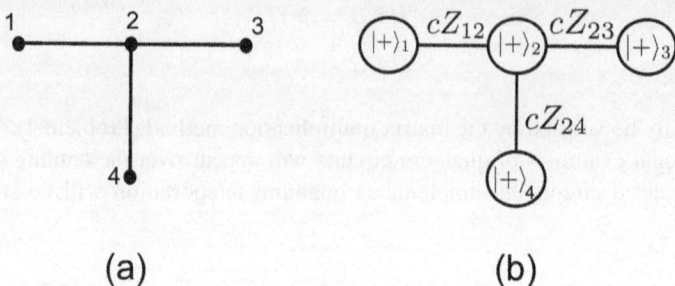

Fig. 16.7 Given a planar graph (**a**), a cluster state can be produced by introducing a qubit $|+\rangle$ at each vertex of the graph and an operator cZ for each edge (**b**). So, the cluster state corresponding to the graph given here is $CZ_{12}CZ_{23}CZ_{24}|+\rangle_1|+\rangle_2|+\rangle_3|+\rangle_4$. (Because the controlled Z gates on different edges commute with each other, the order in which they are applied doesn't matter.) A sequence of measurements on this state then carries out the desired computation

16.7   Adiabatic Quantum Computing

Other approaches to quantum computing exist, beyond the circuit model and cluster state computing. One of the most prominent is the idea of quantum annealing, which has been the basis of the commercially available D-Wave computers. Quantum annealing is well suited to certain types of problems, such as optimization and search problems, but is unsuited to other problems, such as finding solutions to differential equations.

In materials science, annealing consists of heating a material to disrupt its crystal structure. Then, as the material is re-cooled in a controlled manner, the crystal reforms. The new crystal may have different properties from the original, with fewer crystal defects, a different lattice cell orientation, and so on. This method is often used, for example, to reduce the brittleness and increase the ductility of a material. Thermodynamically, what happens in annealing is that the mobility of electrons in the material is increased during the heating, so that during the cooling, those electrons can redistribute themselves to settle into a more stable, lower-energy equilibrium state.

Annealing works by providing the electrons with sufficient energy to roll over barriers, from an initial higher-energy metastable state (e.g., with many lattice defects) to the stable state that forms the global minimum of free energy. Recall that Helmholtz free energy for a system whose states are described by probability distribution $p(x)$ is of the form

$$F[T, V] = \langle E \rangle - TS = \sum_x E(x)p(x) + T\sum_x p(x)\log_2 p(x), \qquad (16.44)$$

where the sum (or integral) is over all classical states x and $\langle E \rangle$ is the mean energy. When the temperature is large, the entropy term dominates in F. Since the entropy, being a convex function, has a single (global) maximum, the free energy will have

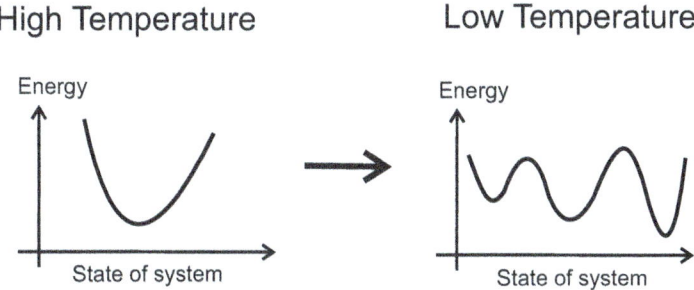

Fig. 16.8 The free energy has a single minimum at high temperatures (left) but can develop several local minima as the temperature is decreased (right). As a result, a system can be trapped in one of the higher-energy local minima, instead of the global minimum

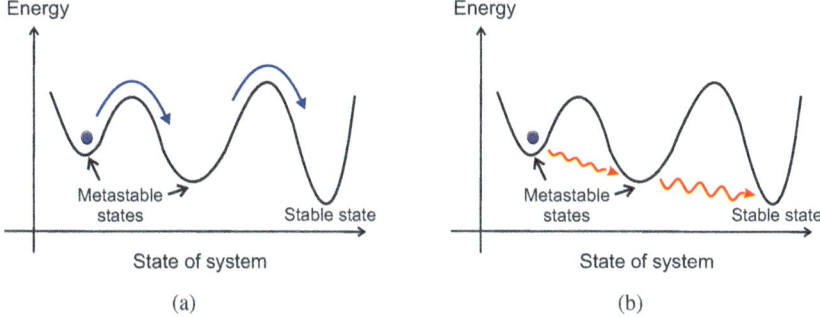

Fig. 16.9 (**a**) Classical annealing. A material is heated above its recrystallization point. The heat provides enough energy for the system to roll over the potential barriers, so that the initial metastable state (e.g., a state with energy-increasing crystal defects) can decay into the stable, lower energy global minimum. (**b**) In quantum annealing, there is tunneling *through* the barriers, in addition to travel *over* the barriers. The rate of decay into the global ground state is much faster

a unique global minimum at high temperatures. As the temperature drops, the potential term begins to dominate. The potential $V(x)$ may have several minima: in addition to the global minimum (the true ground state), there may be several local minima (metastable states) (Fig. 16.8). As the temperature is decreased, the shape of the free energy evolves, and the state of the system moves from the single high-temperature minimum to one of low-energy minima (Fig. 16.9a). Asymptotically (in the long time limit), the probability is high that the system will reach the global ground state.

This is the basis for a classical computing algorithm called **simulated annealing** [19, 20]. In simulated annealing, the programmer starts with a classical Hamiltonian H_0, which has a single known minimum. He or she then slowly varies parameters in the system so that the H_0 evolves into a new Hamiltonian H_1. For example, over time, the form of the Hamiltonian could be $\hat{H}(t) = (1 - \frac{t}{T_0})H_0 + \frac{t}{T_0}H_1$, where T_0 is the time over which the system is allowed to evolve. Starting from the initial state that minimizes H_0, the system evolves into a state that globally minimizes H_1. H_1 is chosen so that its state of minimal energy is the solution to some optimization problem of interest. In this way, the initial state of the system evolves asymptotically to the desired solution state.

Adiabatic quantum computing [21–24] or **quantum annealing** follows the same logic, except that now the solutions can tunnel through barriers (Fig. 16.9b). The initial Hamiltonian, with a unique and known ground state, is evolved slowly to a new Hamiltonian. By the adiabatic theorem, if the parameter evolution is slow enough, the system should remain in its ground state. Given a sufficiently long time, the state of the system will eventually tunnel through any barriers to reach its lowest energy state. It is estimated that the quantum annealer can achieve a speedup of six orders of magnitude or more over simulated annealers.

The company D-Wave Systems has produced several versions of computers based on quantum annealing. These are special purpose computers, rather than programmable, universal computers. The latest version (as of this writing) implements 5000 superconducting qubits. The existence of entangled states within the processers has been demonstrated. Over the years, the company has made a number of claims of quantum supremacy, in other words, of being able to carry out computations or achieve speeds far in excess of any classical computers. These claims have had mixed reviews; some seem to be greatly exaggerated, while others are still being debated or seem to be correct. Part of the confusion has been due to the fact that it is not always clear whether the classical algorithms being used for comparison are indeed the best possible on a classical computer. At the moment, it looks like the most recent D-Wave processor may lead to significant advantages over classical computers. But we will not go into trying to evaluate the competing claims of quantum supremacy here, since new developments are coming rapidly. We simply refer the reader to the literature for the latest updates. In any case, D-Wave processors have demonstrated sufficient promise for a number of large corporations and several defense agencies to have purchased them. They have been fruitfully used in areas that include finance, machine learning, logistics, design of more efficient batteries, and drug discovery. Other companies, including Honeywell, IBM, Google, and Microsoft, are also now building quantum computers based on different architectures, as have a number of university-based start-up companies.

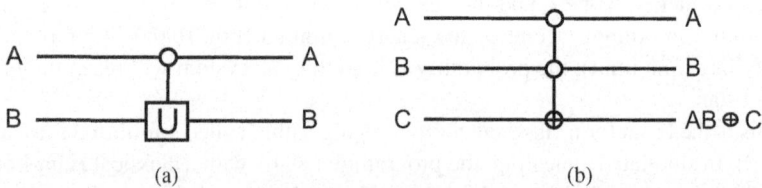

(a)   (b)

Fig. 16.10 (**a**) A controlled gate in which $\hat{U}$ acts on the second qubit if and only if $A = 0$. (**b**) The anti-Toffoli gate

Problems

1. Find the typical wavelength and frequency, λ_{th} and ν_{th}, associated with thermal noise at room temperature. Given the result, compare to typical microwave, infrared, and visible photon frequencies. Which of these spectral ranges could be of practical use for quantum communication or quantum computing at room temperature?

2. Show that the S, T, and H gates can be implemented as combinations of Pauli gates. Draw the corresponding circuit diagrams that implement them.

3. Show that the controlled Hadamard gate, CH, can (depending on the value of the first qubit) simulate both a Toffoli gate and a non-controlled Hadamard gate. As a result, controlled Hadamard gates form a universal set of gates all by themselves.

4. Verify identities 16.41–16.43.

5. Define a new type of controlled gate (Fig. 16.10a) in which the gate operation $\hat{U}$ is carried out if and only if the control qubit on the first line is $|0\rangle$. Then show that the **anti-Toffoli gate** (Fig. 16.10b) acts as an OR gate for $C = 1$ and as a NOR gate for $C = 0$.

6. (a) Write operators $H \otimes X$ and $X \otimes H$ as 4×4 matrices.

   (b) Find results of applying operators of part (a) to the state $|\psi\rangle = \frac{1}{2}(|00\rangle + |01\rangle - |10\rangle + |11\rangle)$.

   (c) Repeat find the result of applying $H \otimes H$ to the state of part (b).

7. Consider a two qubit-system acted upon by a set of quantum gates. Prove the following identities, where subscripts represent which qubit is acted on by each single-qubit gate. For the CNOT gates, 1 is the control qubit.

$$CNOT \cdot X_1 \cdot CNOT = X_1 X_2 \qquad CNOT \cdot X_2 \cdot CNOT = X_2$$
$$CNOT \cdot Y_1 \cdot CNOT = Y_1 X_2 \qquad CNOT \cdot Y_2 \cdot CNOT = Z_1 Y_2$$
$$CNOT \cdot Z_1 \cdot CNOT = Z_1 \qquad CNOT \cdot Z_2 \cdot CNOT = Z_1 Z_2.$$

8. (a) Let CZ_{ij} be a controlled Z gate with i as the control and j as the target. Show that the CZ gate is symmetric: $CZ_{ij} = CZ_{ji}$.

   (b) Show that a CNOT gate can be constructed with a CZ and a pair of Hadamard gates: $CZ = (I \otimes H) \cdot CNOT \cdot (I \otimes H)$. Draw the circuit diagram implementing this.

9. Use CNOT gates to construct a pair of quantum adders that will calculate output states $|a \oplus b\rangle$ and $|a \oplus b \oplus c\rangle$ from input states $|a\rangle |b\rangle$, and $|c\rangle$. Show that the gates are reversible: the input is reconstructable from the output.

10. The circuit below is called a Bell state circuit or entangling circuit. Show that sending untangled product states in on the left can produce all of the Bell states as output on the right. Identify the input state $|ab\rangle$ required to generate each of the four Bell states.

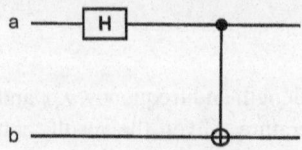

11. Given a fixed Bell state $|\Psi^+\rangle$, construct a sequence of operations that will convert this state into each of the other three Bell states.

References

1. M.A. Nielsen, I.L. Chuang, *Quantum Computation and Quantum Information: 10th Anniversary Edition* (Cambridge University Press, Cambridge, 2011)
2. M. Le Bellac, *A Short Introduction to Quantum Information and Quantum Computation* (Cambridge University Press, Cambridge, 2006)
3. E. Rieffel, W. Polak, *Quantum Computing: A Gentle Introduction* (MIT Press, Cambridge, 2014)
4. J. Stolze, D. Suter, *Quantum Computing: A Short Course from Theory to Experiment*, 2nd edn. (Wiley-VCH, Weinheim, 2008)
5. N.S. Yanofsky, M.A. Mannuci, *Quantum Computing for Computer Scientists* (Cambridge University Press, Cambridge, 2008)
6. N.D. Mermin, *Quantum Computer Science: An Introduction* (Cambridge University Press, Cambridge, 2007)
7. T.G. Wong, *Introduction to Classical and Quantum Computing* (Rooted Grove Press, New York, 2022)
8. E. Knill, R. Laflamme, G.J. Milburn, Nature **409**, 46 (2001)
9. J. O'Brien, Science **318**, 1567 (2008)
10. J.K. Pachos, *Introduciton to Topological Quantum Computation* (Cambridge University Press, Cambridge, 2012)
11. S.H. Simon, Topological Quantum: Lecture Notes and Proto-Book. Available at https://www-thphys.physics.ox.ac.uk/people/SteveSimon/topological2020/TopoBookOct27hyperlink.pdf
12. C. Nayak, S.H. Simon, A. Stern, M. Freedman, S. Das Sarma, Rev. Mod. Phys. **80**, 1083 (2008)
13. D.P. Di Vincenzo, Fortschr. Phys. **48**, 771 (2000)
14. R. Raussendorf, H.J. Briegel, Phys. Rev. Lett. **86**, 5188 (2001)
15. R. Raussendorf, D. Browne, H. Briegel, Phys. Rev. A **68**, 022312 (2003)
16. M.A. Nielsen, Rep. Math. Phys. **57**, 147 (2001)
17. H.J. Briegel, Cluster states, in *Compendium of Quantum Physics Concepts: Experiments, History, and Philosophy*, ed. by D. Greenberger, K. Hentschel, F. Weinert (Springer, Berlin, 2009)
18. P. Kok, B.W. Lovett, *Introduction to Optical Quantum Information Processing* (Cambridge University Press, Cambridge, 2010)
19. S. Kirkpatrick, S.D. Gelett, M.P. Vecchi, Science **220**, 671 (1983)
20. E. Aarts, J. Korst, *Simulated Annealing and Boltzmann Machines: A Stochastic Approach to Combinatorial Optimization and Neural Computing* (Wiley, Hoboken, 1984)
21. M. Ohzeki, H. Nishimori, J. Comp. Theor. Nanosci. **8**, 963 (2011)
22. C. McGeoch, *Adiabatic Quantum Computation and Quantum Annealing: Theory and Practice* (Springer International Publishing, Berlin, 2022)
23. L.P. Yulianti, K. Surendro, IEEE Access **10**, 73156 (2022)
24. A. Rajak, S. Suzuki, A. Dutta, B.K. Chakrabarti, Phil. Trans. A Royal Soc. London **381**, 20210417 (2023)

Chapter 17
Quantum Computing: Algorithms

In the last chapter, physical and structural aspects of potential quantum computers were discussed. In this chapter, we discuss some algorithms that could be carried out on quantum computers and compare their performances to analogous classical algorithms. In each case, we will see that the use of superposition and entanglement leads to increased efficiency.

17.1 Deutsch Algorithm

The Deutsch algorithm was first described by David Deutsch in 1985 [1]. As the first major quantum algorithm proposed, it was one of the key sparks that kindled interest in quantum computing in the following years.

Suppose that a Boolean function is given that maps a classical bit to another classical bit:

$$f : \{0, 1\} \rightarrow \{0, 1\}. \tag{17.1}$$

This function can only have four possible forms: it can be one of two constant functions,

$$f(0) = f(1) = 0 \quad \text{or} \quad f(0) = f(1) = 1,$$

or it can be one of the two balanced functions that output equal numbers of 0s and 1s,

$$f(0) = 0, \ f(1) = 1 \quad \text{or} \quad f(0) = 1, \ f(1) = 0.$$

© The Author(s), under exclusive license to Springer Nature Switzerland AG 2025
D. S. Simon, *Introduction to Quantum Science and Technology*, Undergraduate
Texts in Physics, https://doi.org/10.1007/978-3-031-81315-3_17

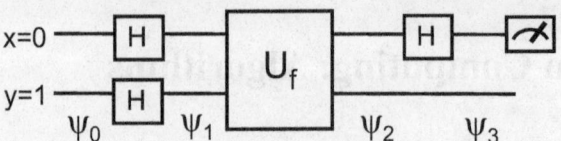

Fig. 17.1 The quantum circuit used to implement the Deutsch algorithm. The gate $\hat{U}_f$ leaves the first bit unchanged, $|x\rangle \rightarrow |x\rangle$, but shifts the second bit according to $|y\rangle \rightarrow |y \oplus f(x)\rangle$

The **Deutsch algorithm** is a quantum algorithm that tests to see whether the function is constant or balanced, without actually determining the function itself. Classically, to determine if a function is balanced or constant, two measurements would have to be made to determine values of $f(0)$ and $f(1)$. The quantum algorithm, however, can make the determination with a single measurement.

A quantum circuit to carry out the algorithm is shown in Fig. 17.1. The input state is $|\psi_0\rangle = |x, y\rangle = |0, 1\rangle$, where the first entry represents the upper line of the figure (the **query register**) and the second entry is the lower line (the **answer register**). The circuit is an example of an **oracle**, a device or quantum circuit designed to determine a specific property of the function, without determining the function itself. In equation form, the quantum circuit implements the operation

$$|\psi_3\rangle = \left(\hat{H} \otimes \hat{I}\right) \circ \hat{U}_f \circ \left(\hat{H} \otimes \hat{H}\right), \tag{17.2}$$

where

$$\hat{H} = \frac{1}{\sqrt{2}} \begin{pmatrix} 1 & 1 \\ 1 & -1 \end{pmatrix} \tag{17.3}$$

is the Hadamard operator and

$$\hat{U}_f |x, y\rangle = |x\rangle |y \oplus f(x)\rangle. \tag{17.4}$$

Working through the operation of the circuit step by step, the left-hand pair of Hadamard gates operate on $|\psi_0\rangle$ according to

$$\begin{aligned}
|\psi_1\rangle &= \hat{H} \otimes \hat{H} |\psi_0\rangle \\
&= \frac{|0\rangle + |1\rangle}{\sqrt{2}} \cdot \frac{|0\rangle - |1\rangle}{\sqrt{2}} \\
&= \frac{1}{2}\Big[|00\rangle - |01\rangle + |10\rangle - |11\rangle\Big].
\end{aligned} \tag{17.5}$$

The $\hat{U}_f$ operation then leads to

$$|\psi_2\rangle = \frac{1}{2}\{|0\rangle|0 \oplus f(0)\rangle - |0\rangle|1 \oplus f(0)\rangle + |1\rangle|0 \oplus f(1)\rangle - |1\rangle|1 \oplus f(1)\rangle\}$$

$$= \frac{1}{2}\left\{|0\rangle|f(0)\rangle - |0\rangle|\overline{f(0)}\rangle + |1\rangle|f(1)\rangle - |1\rangle|\overline{f(1)}\rangle\right\}$$

$$= \frac{1}{2}\left\{|0\rangle\left(|f(0)\rangle - |\overline{f(0)}\rangle\right) + |1\rangle\left(|f(1)\rangle - |\overline{f(1)}\rangle\right)\right\}, \tag{17.6}$$

where $\overline{f(j)} = 1 \oplus f(j)$ is the complement of $f(j)$, for $j = 0, 1$.
In the balanced case,

$$\overline{f(0)} = f(1), \quad \overline{f(1)} = f(0), \tag{17.7}$$

from which we can conclude that

$$|\psi_2\rangle = \pm\frac{1}{2}\left(|0\rangle - |1\rangle\right)\left(|0\rangle - |1\rangle\right). \tag{17.8}$$

In the constant case,

$$f(0) = f(1), \quad \overline{f(0)} = \overline{f(1)}, \tag{17.9}$$

which implies

$$|\psi_2\rangle = \left(|0\rangle + |1\rangle\right)\left(|f(0)\rangle - |\overline{f(0)}\rangle\right) \tag{17.10}$$

$$= \pm\frac{1}{2}\left(|0\rangle + |1\rangle\right)\left(|0\rangle - |1\rangle\right). \tag{17.11}$$

Finally, the last Hadamard gives

$$|\psi_4\rangle = \begin{cases} \pm\frac{1}{\sqrt{2}}|1\rangle\left(|0\rangle - |1\rangle\right) & \text{(balanced)} \\ \pm\frac{1}{\sqrt{2}}|0\rangle\left(|0\rangle - |1\rangle\right) & \text{(constant)} \end{cases} \tag{17.12}$$

$$= \pm\frac{1}{\sqrt{2}}|f(0) \oplus f(1)\rangle\left(|0\rangle - |1\rangle\right) \tag{17.13}$$

So measurement of just the upper query register suffices to determine whether the function is constant or balanced. Again, notice that we cannot determine the actual function from the measurement but that by a judicious choice of measurement, we can extract a specific property of the function (in this case, presence or absence of balance). The required information is obtainable because it is encoded not into the value of one bit or the other but into the two-qubit superposition state.

The **Deutsch-Jozsa algorithm** [2] generalizes the Deutsch algorithm from a one-qubit function to an n-qubit function, $f : \{0, 1\}^{\otimes n} \rightarrow \{0, 1\}$. The algorithm works in essentially the same manner (Problem 1), except that now the input on the query register is the n qubit state $|000\ldots0\rangle$, and the single-qubit Hadamard gates in the upper line of the Deutsch algorithm are replaced by n-qubit Hadamard gates, $\hat{H}^{\otimes n}$. A single measurement is required at the end of the algorithm, whereas classically, n measurements would be needed.

Other oracle-based quantum algorithms that can answer questions with fewer measurements than is possible classically include the Bernstein-Vazirani [3] algorithm, which is discussed in [4], and the Grover search algorithm (Sect. 17.5).

17.2 Quantum Fourier Transform

Given a set of n bits, any integer x in the range $0 \le x \le N - 1$ can be encoded into these bits, where $N = 2^n$. The integer x can be written in binary form as

$$x = x_0 + x_1 2^1 + x_2 2^2 + \ldots x_{n-1} 2^{n-1} \tag{17.14}$$

$$= \left(\frac{x_{n-1}}{2} + \frac{x_{n-2}}{2^2} + \frac{x_{n-3}}{2^3} + \cdots + \frac{x_0}{2^n}\right) 2^n. \tag{17.15}$$

The integer x can then be represented by the binary string $x = x_{n-1} x_{n-2} \ldots x_0$, where each x_j equals either 0 or 1.

Similarly, fractions are often written in binary by listing binary coefficients of the fractional part after a decimal point. Given the integer x above, the fraction $\frac{x}{2^n}$ can be written as

$$\frac{x}{2^n} = \frac{x_{n-1}}{2} + \frac{x_{n-2}}{2^2} + \frac{x_{n-3}}{2^3} + \cdots + \frac{x_0}{2^n} \tag{17.16}$$

$$= 0.x_{n-1} x_{n-2} \ldots x_0 . \tag{17.17}$$

Thus, for example,

$$9\frac{5}{8} = 1 \cdot 2^3 + 0 \cdot 2^2 + 0 \cdot 2^1 + 1 \cdot 2^0 + 1 \cdot 2^{-1} + 0 \cdot 2^2 + 1 \cdot 2^{-3}$$

$$= 1001.101,$$

so that any number, integer, or otherwise can be encoded as a binary string.

Thinking of x as a collection of binary data of size $N = 2^n$, then the classical **discrete Fourier transform** of function $f(x)$ is given by

$$\tilde{f}(y) = \frac{1}{\sqrt{N}} \sum_{x=0}^{N-1} e^{-2\pi i x y / N} f(x), \tag{17.18}$$

where y is also an integer such that $0 \le y \le N - 1$. Since x is integer, the Fourier transformed function is periodic, $\tilde{f}(y + N) = \tilde{f}(y)$. Such transforms are well-known in physics; for example, x might label the position on a one-dimensional crystal lattice, with y the conjugate quasi-momentum.

Given a set of basis states labeled by the integers x $(0 \le x \le 2^{n-1})$ or equivalently by the corresponding binary strings $x_{n-1}x_{n-2}\ldots x_1 x_0$,

$$|x\rangle = |x_{n-1}x_{n-2}\ldots x_0\rangle = |x_{n-1}\rangle|x_{n-2}\rangle\ldots|x_1\rangle|x_0\rangle,$$

we would like to be able to perform a similar operation, the **quantum Fourier transform** (QFT) on the states, $|x\rangle \rightarrow |\tilde{x}\rangle$, implemented by a unitary operator, $\hat{U}_{QFT}$:

$$|\tilde{x}\rangle = \hat{U}_{QFT}|x\rangle = \frac{1}{\sqrt{N}}\sum_{y=0}^{N-1} e^{2\pi i x y / N}|y\rangle. \tag{17.19}$$

The QFT is a building block for many other algorithms and has a range of applications that includes determination of continued fraction representations [5], phase estimation, order finding, and factoring. Some of these will be discussed in later sections.

Expanding y in binary form, each term in the sum of Eq. 17.19 can be written as a product of terms where each term in the product depends only on a single qubit in the y string:

$$e^{2\pi i x y / 2^n} = e^{2\pi i y_{n-1} x / 2} e^{2\pi i y_{n-2} x / 2^2} \ldots e^{2\pi i y_0 x / 2^n} \tag{17.20}$$

$$= \prod_{j=0}^{n-1} e^{2\pi i y_j x / 2^{n-j}}. \tag{17.21}$$

So the Fourier-transformed state can be expanded as well:

$$\hat{U}_{QFT}|x\rangle = \frac{1}{\sqrt{N}} \sum_{y=0}^{N-1} \prod_{j=0}^{n-1} e^{2\pi i x y_j / 2^{n-j}}|y\rangle \tag{17.22}$$

$$= \frac{1}{2^{n/2}} \prod_{j=0}^{n-1} \sum_{y_j=0}^{1} e^{2\pi i x y_j / 2^{n-j}}|y_j\rangle \tag{17.23}$$

$$= \frac{1}{2^{n/2}} \left(\sum_{y_{n-1}=0}^{1} e^{2\pi i x \frac{y_{n-1}}{2}}|y_{n-1}\rangle \right) \otimes \left(\sum_{y_{n-2}=0}^{1} e^{2\pi i x \frac{y_{n-2}}{2^2}}|y_{n-2}\rangle \right) \otimes \ldots$$

$$\ldots \otimes \left(\sum_{y_0=0}^{1} e^{2\pi i x \frac{y_0}{2}}|y_0\rangle \right)$$

$$= \frac{1}{2^{n/2}} \left(|0\rangle + e^{2\pi i x/2}|1\rangle \right) \otimes \left(|0\rangle + e^{2\pi i x/2^2}|1\rangle \right) \otimes \cdots$$

$$\cdots \otimes \left(|0\rangle + e^{2\pi i x/2^n}|1\rangle \right). \tag{17.24}$$

In the last line, the two-term sum over each y_j was expanded out explicitly.

Now use the binary expansion of x to simplify the terms. Consider the exponential in the first factor, for example:

$$e^{2\pi i x/2} = e^{2\pi i \left(\frac{x_0}{2} + x_1 + 2x_2 + \ldots \right)} \tag{17.25}$$

$$= e^{2\pi i \left(\frac{x_0}{2} \right)} e^{2\pi i \times (\text{integer})} \tag{17.26}$$

$$= e^{2\pi i x_0/2}. \tag{17.27}$$

The exponential in the second term can likewise be written

$$e^{2\pi i \frac{x}{2^2}} = e^{2\pi i \left(\frac{x_0}{4} + \frac{x_1}{2} + x_2 + \ldots \right)} \tag{17.28}$$

$$= e^{2\pi i \left(\frac{x_0}{4} + \frac{x_1}{2} \right)} e^{2\pi i \times (\text{integer})} \tag{17.29}$$

$$= e^{2\pi i \left(\frac{x_0}{4} + \frac{x_1}{2} \right)} \tag{17.30}$$

$$= e^{2\pi i x_1 . x_0/2^2}, \tag{17.31}$$

where the binary decimal notation of Eq. 17.17 was used in the last line. In a similar manner, the rest of the exponentials can be rewritten:

$$e^{2\pi i x/2^3} = e^{2\pi i x_2 . x_1 x_0/2^3} \tag{17.32}$$

$$\vdots$$

$$e^{2\pi i x/2^n} = e^{2\pi i x_{n-1} . x_{n-2} x_{n-3} \ldots x_0/2^n} \tag{17.33}$$

So, finally, the transform can be written in the following form:

$$\hat{U}_{QFT}|x\rangle = \frac{1}{2^{n/2}} \left(|0\rangle + e^{2\pi i x_0/2}|1\rangle \right) \otimes \left(|0\rangle + e^{2\pi i x_1 . x_0/2^2}|1\rangle \right) \otimes \cdots$$

$$\cdots \otimes \left(|0\rangle + e^{2\pi i x_{n-1} . x_{n-2} x_{n-3} \ldots x_0/2^n}|1\rangle \right). \tag{17.34}$$

How can this transform be implemented? For a single qubit, the QFT is implemented simply by a Hadamard gate (Fig. 17.2a). Before generalizing to more qubits, we first define a generalized phase gate R_j:

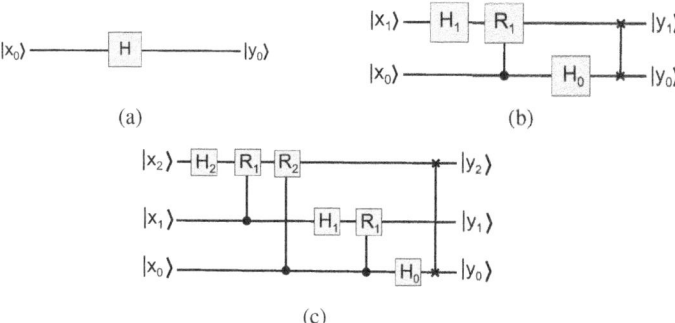

Fig. 17.2 Quantum Fourier transform circuits for (**a**) one-bit, (**b**) two-bit, and (**c**) three-bit states

$$R_j = \begin{pmatrix} 1 & 0 \\ 0 & e^{i\pi/2^j} \end{pmatrix}, \tag{17.35}$$

so that

$$R_j|0\rangle = |0\rangle, \qquad R_j|1\rangle = e^{i\pi/2^j}|1\rangle. \tag{17.36}$$

For small values of j, this just gives us back gates that we are already familiar with:

$$R_0 = \begin{pmatrix} 1 & 0 \\ 0 & -1 \end{pmatrix} = \hat{Z} \tag{17.37}$$

$$R_1 = \begin{pmatrix} 1 & 0 \\ 0 & i \end{pmatrix} = \hat{S} \tag{17.38}$$

$$R_2 = \begin{pmatrix} 1 & 0 \\ 0 & e^{i\pi/4} \end{pmatrix} = \hat{T}, \tag{17.39}$$

where $\hat{Z}$, $\hat{S}$, and $\hat{T}$ are the Pauli $\hat{\sigma}_z$ gate, the $\pi/8$ gate, and the standard-phase gate introduced in Sect. 16.4. Also define the controlled-phase gate CR_j^i, which acts as R_j on qubit x_j if and only if $x_i = 1$.

Then factors in the parentheses in Eq. 17.34 can be written as the result of acting with a Hadamard gate followed by a generalized phase gate. For example, consider the two-qubit case ($n = 2$, $N = 4$). The transform just consists of the first two factors on the right of Eq. 17.34 and can be obtained by the sequence of gates shown in Fig. 17.2b (subscripts on the Hadamard indicate the qubit to which they are applied):

$$H_1|x_1 x_0\rangle = \frac{1}{\sqrt{2}}\Big(|0\rangle + (-1)^{x_1}|1\rangle\Big) \otimes |x_0\rangle$$

$$= \frac{1}{\sqrt{2}}\left(|0\rangle + e^{i\pi x_1}|1\rangle\right) \otimes |x_0\rangle$$

$$CR_1 \cdot H_1|x_1 x_0\rangle = \frac{1}{\sqrt{2}}\left(|0\rangle + e^{i\pi x_1} e^{i\pi x_0/2}|1\rangle\right) \otimes |x_0\rangle$$

$$= \frac{1}{\sqrt{2}}\left(|0\rangle + e^{i\pi x_1 . x_0}|1\rangle\right) \otimes |x_0\rangle$$

$$H_0 \cdot CR_1 \cdot H_1|x_1 x_0\rangle = \frac{1}{2}\left(|0\rangle + e^{i\pi x_1 . x_0}|1\rangle\right) \otimes \left(|0\rangle + e^{i\pi x_0}|1\rangle\right)$$

$$\text{Swap} \cdot H_0 \cdot CR_1 \cdot H_1|x_1 x_0\rangle = \frac{1}{2}\left(|0\rangle + e^{i\pi x_0}|1\rangle\right) \otimes \left(|0\rangle + e^{i\pi x_1 . x_0}|1\rangle\right),$$

verifying that

$$\hat{U}_{QFT} = \text{Swap} \cdot H_0 \cdot cR_1 \cdot H_1$$

for $n = 2$. So the two-qubit QFT is realized by the circuit of Fig. 17.2b. The extension to more than two qubits is then straightforward by repeating similar steps. The three-qubit version is shown in Fig. 17.2c.

The quantum Fourier transform can also be recast in matrix form: defining $\omega = e^{2\pi i/2^n}$, then the matrix representation of the unitary operator can be written in the form

$$\hat{U}_{QFT} = \frac{1}{2^{n/2}}\begin{pmatrix} 1 & 1 & 1 & \cdots & 1 \\ 1 & \omega & \omega^2 & \cdots & \omega^{2^n-1} \\ 1 & \omega^2 & \omega^4 & \cdots & \omega^{2(2^n-1)} \\ \vdots & \vdots & \vdots & \ddots & \vdots \\ 1 & \omega^{2^n-1} & \omega^{2(2^n-1)} & \cdots & \omega^{(2^n-1)(2^n-1)} \end{pmatrix}. \tag{17.40}$$

Example 17.2.1 For a single qubit ($n = 1$), we find that $\omega = e^{2\pi i/2} = -1$, so that the Fourier transform is simply implemented by a Hadamard gate:

$$\hat{U}_{QFT} = \frac{1}{\sqrt{2}}\begin{pmatrix} 1 & 1 \\ 1 & -1 \end{pmatrix} = \hat{H}. \tag{17.41}$$

Thus, under the Fourier transform, the computational basis $\{|0\rangle, |1\rangle\}$ is converted to the diagonal basis:

$$|0\rangle \rightarrow |\tilde{0}\rangle = \frac{1}{\sqrt{2}}\left(|0\rangle + |1\rangle\right) \tag{17.42}$$

$$|1\rangle \rightarrow |\tilde{1}\rangle = \frac{1}{\sqrt{2}}\left(|0\rangle - |1\rangle\right) \tag{17.43}$$

Example 17.2.2 For two qubits, $\omega = e^{2\pi i/4} = i$, and the QFT transformation is given by

$$
\hat{U}_{QFT} = \frac{1}{2}\begin{pmatrix} 1 & 1 & 1 & 1 \\ 1 & \omega & \omega^2 & \omega^3 \\ 1 & \omega^2 & \omega^4 & \omega^6 \\ 1 & \omega^3 & \omega^6 & \omega^9 \end{pmatrix} = \frac{1}{2}\begin{pmatrix} 1 & 1 & 1 & 1 \\ 1 & i & -1 & -i \\ 1 & -1 & 1 & -1 \\ 1 & -i & -1 & i \end{pmatrix}. \tag{17.44}
$$

Suppose that we are given the input state

$$
|\psi_0\rangle = |3\rangle = |11\rangle = \begin{pmatrix} 0 \\ 0 \\ 0 \\ 1 \end{pmatrix}, \tag{17.45}
$$

where we have used that fact that the integer 3 is given by the binary string $(11) = 1 \cdot 2^1 + 1 \cdot 2^0$. Then the Fourier transform of this state is

$$
|\tilde{\psi}_0\rangle = U_{QFT}|\psi_0\rangle = \frac{1}{2}\begin{pmatrix} 1 \\ \omega^3 \\ \omega^6 \\ \omega^9 \end{pmatrix} = \frac{1}{2}\begin{pmatrix} 1 \\ -i \\ -1 \\ i \end{pmatrix}. \tag{17.46}
$$

On the other hand, the input state

$$
|\psi_1\rangle = \frac{1}{\sqrt{2}}\left(|1\rangle + |2\rangle\right) = \frac{1}{\sqrt{2}}\left(|01\rangle + |10\rangle\right) = \frac{1}{\sqrt{2}}\begin{pmatrix} 0 \\ 1 \\ 1 \\ 0 \end{pmatrix} \tag{17.47}
$$

is transformed to:

$$
|\tilde{\psi}_1\rangle = U_{QFT}|\psi_1\rangle = \frac{1}{2\sqrt{2}}\begin{pmatrix} 2 \\ \omega + \omega^2 \\ \omega^2 + \omega^4 \\ \omega^3 + \omega^6 \end{pmatrix} \tag{17.48}
$$

$$
= \frac{1}{2}\begin{pmatrix} \sqrt{2} \\ e^{3\pi i/4} \\ 0 \\ e^{5\pi i/4} \end{pmatrix} \tag{17.49}
$$

$$
= \frac{1}{\sqrt{2}}|00\rangle + \frac{1}{2}e^{3\pi i/4}|01\rangle + \frac{1}{2}e^{5\pi i/4}|11\rangle. \tag{17.50}
$$

■

17.3 Phase Estimation

Suppose that $\hat{U}$ is a unitary operator with an eigenvector $|u\rangle$ of eigenvalue $e^{2\pi i \phi}$. The goal now is to find an estimate of the phase $\phi = 0.\phi_1\phi_2\cdots = \frac{\phi_1}{2} + \frac{\phi_2}{2^2} + \ldots$ that can be made accurate to as many decimal places as desired.

A schematic for a quantum circuit that does this is given in Fig. 17.3. The input consists of $t + n$ qubits; the top t inputs are each prepared in the state $|0\rangle$, while the bottom n are used to input the state $|u\rangle$. So the overall input state of the system is $|0\rangle|0\rangle\ldots|0\rangle|u\rangle$. Boxes acting on the bottom n lines are controlled $\hat{U}^{2^j}$, for $j = 1$ to $j = t$. Since $|u\rangle$ is an eigenstate of $\hat{U}$, the state in the bottom lines remains unchanged, aside from an overall phase, all the way through to the output.

Hadamard gates split each of the first t qubits into $\frac{1}{\sqrt{2}}\Big(|0\rangle + |1\rangle\Big)$, leading to the state

$$\frac{1}{2^{t/2}}\Big(|0\rangle + |1\rangle\Big)\Big(|0\rangle + |1\rangle\Big)\ldots\Big(|0\rangle + |1\rangle\Big)|u\rangle. \tag{17.51}$$

The controlled $\hat{U}^{2^j}$ gates then add phases $e^{2\pi i \cdot 2^j}$ to all factors in the product that contain $|1\rangle$. The final output is then of the form

$$\begin{aligned}
|\psi_{out}\rangle &= \frac{1}{2^{t/2}}\Big(|0\rangle + e^{2\pi i 2^t \phi}|1\rangle\Big)\Big(|0\rangle + e^{2\pi i 2^{t-1}\phi}|1\rangle\Big)\ldots\Big(|0\rangle + e^{2\pi i 2^1 \phi}|1\rangle\Big)|u\rangle \\
&= \frac{1}{2^{t/2}}\Big(|0\rangle + e^{2\pi i 0.\phi_t}|1\rangle\Big)\Big(|0\rangle + e^{2\pi i 0.\phi_{t-1}\phi_t}|1\rangle\Big)\ldots\Big(|0\rangle + e^{2\pi i 0.\phi_1\ldots\phi_t}|1\rangle\Big)|u\rangle \\
&= \frac{1}{2^{t/2}}\sum_{k=0}^{t-1} e^{2\pi i k\phi}|k\rangle|u\rangle \\
&\equiv |\tilde{\phi}\rangle|u\rangle. \tag{17.52}
\end{aligned}$$

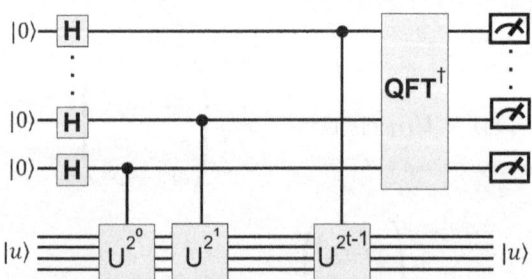

Fig. 17.3 The quantum circuit to implement the phase estimation algorithm. In the first register, n copies of the $|0\rangle$ state are passed through Hadamard gates. In the second register, the eigenstate $|u\rangle$ is input, and t controlled-U^{2^j} gates are used to add phase shifts to the states in the first register. After an inverse quantum Fourier transform, the states in the first register are measured

But compared to Eq. 17.19, $|\tilde{\phi}\rangle$ is the QFT of state $|\phi\rangle$: $|\tilde{\phi}\rangle = \frac{1}{2^{t/2}} \sum_{k=0}^{t-1} e^{2\pi i k \phi} |k\rangle$. So applying the inverse QFT gives the estimator state $|\phi\rangle$. Measurement of the first t qubits of $|\phi\rangle$ will give the values the first t digits in the binary expansion of ϕ.

17.4 Period Finding and Factoring

Let $\mathbb{Z}_2^{\otimes n}$ denote the set of length n binary strings: if $x \in \mathbb{Z}_2^{\otimes n}$, then $x = (x_1, x_2, \ldots, x_n)$, where $x_j \in 0, 1$ for each j. Then consider the following situation. A periodic function f is given on $\mathbb{Z}_2^{\otimes n}$, such that $f(x) = f(y)$ if and only if $y = x \oplus r$ for some fixed $r \in \mathbb{Z}_2^{\otimes n}$, where $\oplus$ is binary addition. The goal is to find the period r of the function f. Classically, the number of operations required to find the period grows exponentially with n. But D. L. Simon [6] found a quantum algorithm that allows the solution to be found and which scales only linearly in n. This algorithm therefore acts as a stark example of quantum speedup, providing a problem that can be solved exponentially faster if a quantum computer is available.

Initialize an n-qubit register in the state $|0, 0, \ldots 0\rangle$. Applying Hadamard gates to each qubit leads to the state

$$|\psi\rangle = \hat{H}^{\otimes n}|0, 0, \ldots 0\rangle = \frac{1}{2^{n/2}} \sum_{x=0}^{N-1} |x\rangle. \tag{17.53}$$

Example 17.4.1 That the Hadamard gates produce an equal superposition of all x values is easy to verify for small N. For example, for the one-qubit case ($N = 2^1$), we find that

$$|0\rangle \rightarrow \hat{H}|0\rangle = \frac{1}{\sqrt{2}}\left(|0\rangle + |1\rangle\right) = \frac{1}{\sqrt{2^1}} \sum_{x=0}^{1} |x\rangle. \tag{17.54}$$

Similarly, for the two-qubit case ($N = 2^2$),

$$|00\rangle \rightarrow \hat{H} \otimes \hat{H}|00\rangle \tag{17.55}$$

$$= \frac{1}{2}\left(|0\rangle + |1\rangle\right) \otimes \left(|0\rangle + |1\rangle\right) \tag{17.56}$$

$$= \frac{1}{2}\left(|00\rangle + |01\rangle + |10\rangle + |11\rangle\right) \tag{17.57}$$

$$= \frac{1}{2}\left(|0\rangle + |1\rangle + |2\rangle + |3\rangle\right) \tag{17.58}$$

$$= \frac{1}{\sqrt{2^2}} \sum_{x=0}^{3} |x\rangle, \tag{17.59}$$

where in the fourth line the binary representation has replaced with the corresponding base-10 value.

For more qubits, the same pattern repeats, giving $|\psi\rangle = \hat{H}^{\otimes n}|0\rangle = \frac{1}{\sqrt{N}}\sum_{x=0}^{N-1}|x\rangle$, for all N. ∎

Now add a second n-qubit register, also initialized in state $|0\rangle = |00\ldots0\rangle$. Define a unitary operator $\hat{U}_f$ that applies the function f by acting on the second register by an amount that depends on the value in the first register:

$$\hat{U}_f|x, y\rangle = |x\rangle|y \oplus f(x)\rangle. \tag{17.60}$$

(This is the same operator that was used in the Deutsch algorithm.) We then have

$$\hat{U}_f|\psi\rangle|0\rangle = \frac{1}{\sqrt{2^{n/2}}}\sum_{x=0}^{N-1}\hat{U}_f|x\rangle|0\rangle = \frac{1}{\sqrt{2^{n/2}}}\sum_{x=0}^{N-1}|x\rangle|f(x)\rangle, \tag{17.61}$$

leaving the first and second registers entangled. Once entangled, a measurement of one register will give us information on the other.

Making a measurement on the second register, suppose that the value obtained is $f(x_0)$. But since $f(x_0) = f(x_0 + r)$, we have no knowledge of whether the value of x is x_0 or $x_0 + r$. By the usual rules of quantum mechanics, the state must therefore be assumed to be in a superposition of both:

$$\frac{1}{\sqrt{2}}\Big(|x_0\rangle + |x_0 + r\rangle\Big)|f(x_0)\rangle. \tag{17.62}$$

From this point on, we will stop writing the second register explicitly; it has played its role and is no longer needed.

Applying $\hat{H}^{\otimes n}$ again to the first register and noting that the action of $\hat{H}$ on bit j can be written $\hat{H}|j\rangle = \frac{1}{\sqrt{2}}\sum_{y=0}^{1}(-1)^{jy}|y\rangle$ leads then to the state

$$|\psi'\rangle = \frac{1}{2^{\frac{n+1}{2}}}\sum_{y}\Big((-1)^{x_0\cdot y} + (-1)^{(x_0\oplus r)\cdot y}\Big)|y\rangle \tag{17.63}$$

$$= \frac{1}{2^{\frac{n+1}{2}}}\sum_{y}(-1)^{x_0\cdot y}\Big(1 + (-1)^{r\cdot y}\Big)|y\rangle \tag{17.64}$$

$$= \frac{1}{2^{\frac{n-1}{2}}}\sum_{y}(-1)^{x_0\cdot y}|y\rangle, \tag{17.65}$$

where the sum is over all y such that $r \cdot y = 0$. (Be careful to note that the dots here are products, not decimal points.) Making a measurement of the state gives us one of the possible y values at random; call it y_1. The obtained value will satisfy the linear equation $r \cdot y_1 = 0$. Repeating the same procedure a second time produces a second value y_2, which will satisfy a second linear equation, $r \cdot y_2 = 0$. By further repetitions of the procedure a sufficient number of times, we obtain enough linear equations to solve for all n binary values making up r.

Since sometimes the measurement will give a value of y that has been obtained before, in general, the procedure will have to be repeated more than n times in order to have enough equations to solve for all n required values. But it turns out that repeat values are of low-enough probability that the number of operations still scales linearly in n for large n. In fact, it can be shown that the probability of having enough information after $n + x$ measurements is given by

$$p = \left(1 - \frac{1}{2^{n+x}}\right)\left(1 - \frac{1}{2^{n+x-1}}\right)\cdots\left(1 - \frac{1}{2^{x+2}}\right) > 1 - \frac{1}{2^{x+1}}. \tag{17.66}$$

(See Appendix G of [5] for a proof.) This probability is already over 99% for $x = 6$ and over 99.9% for $x = 9$.

P. Shor generalized the period (or order) finding algorithm from binary functions periodic under *modulo-2* addition, $f(x) = f(x \oplus r)$, to functions that are periodic under *ordinary* addition, $f(x) = f(x + r)$. Shor's method makes use of quantum Fourier transforms in place of the arrays of Hadamard gates used above. Shor's period finding algorithm is of great practical importance because efficient period finding implies the efficient factoring of large numbers.

Shor considered a function

$$f(x) = b^x \pmod{N}, \tag{17.67}$$

where b is a fixed integer and the possible values of x are $x = 0, 1, 3, \ldots, 2^{n-1}$, with $2^n \geq N$. The goal then was to find the order r of the function, i.e., the smallest nonzero value r such that $f(x) = f(x + mr)$ for any positive integer m. Classically, finding the period requires $\sim \frac{r}{2} \sim \mathcal{O}(2^n)$ operations, which scales exponentially in n, but the quantum algorithm can accomplish the goal in polynomial time, making use of the ability of a quantum system to carry out multiple operations in parallel during the QFT process.

As discussed in Chap. 18, many encryption schemes in current use, such as the RSA scheme, make use of the fact that factoring large numbers efficiently is difficult classically. The best classical factoring algorithm is the general number field sieve, for which the number of operations required grows faster than any polynomial in n. Although we won't go into detail here, it can be shown that the ability to find periods efficiently implies the ability to factor efficiently. So, the Shor order-finding algorithm also allows large numbers to be factored with polynomial scaling. For a detailed discussion of the connection between factoring and periodicity and for details of the Shor factoring algorithm, see [4, 5, 7, 8].

Since efficient factoring algorithms endanger the security of RSA and other encryption schemes that are currently used to protect personal data and financial transactions, the race is on to find new encryption methods that will be immune to hacking by quantum computers. This new field of **post-quantum cryptography** has become a major area of research; introductions and reviews of the field can be found in [9, 10] or in proceedings of the annual PQCrypto conferences [11].

17.5 Grover Search Algorithm

A common category of classical algorithms consists of search algorithms of unstructured data. Unstructured means that there is no ordering or that the ordering is with respect to a different variable than the one being searched (and so provides no help in the search). Examples might be searching a telephone directory to find a desired phone number or identifying a person in an alphabetized data base using their social security number. Let $N = 2^n$ be the number of entries in the database, stored via a list of n bits or qubits, and call the sought-for data point the **marked** value. Classically, the fastest way to carry out the search is simply to compare each data entry to the marked entry, one entry at a time. This is a polynomial-time (in fact linear) algorithm, requiring on average $\frac{N}{2}$ operations. In contrast, a quantum algorithm called the **Grover search algorithm** [12] can find the marked element after roughly $\sqrt{N}$ inquiries, thus leading to a quadratic speedup.

Let the entries in the data base be labelled by coordinate x, with $0 \leq x \leq 2^n - 1$, with x_0 being the coordinate of the marked entry. As usual, let the values of x be used to label a set of quantum states $|x\rangle$. As in the Deutsch algorithm, in addition to the n-qubit data register, we also include an additional one-qubit register $|q\rangle$ (the oracle register), which will be initialized to the state $|q\rangle = \frac{1}{\sqrt{2}}\left(|0\rangle - |1\rangle\right)$. The full $n + 1$-qubit system is then in the state

$$|x\rangle|q\rangle = \frac{1}{\sqrt{2}}|x\rangle \otimes \left(|0\rangle - |1\rangle\right). \tag{17.68}$$

We may define a function

$$f(x) = \delta_{x,x_0} = \begin{cases} 0, & \text{for } x \neq x_0 \\ 1, & \text{for } x = x_0. \end{cases} \tag{17.69}$$

From this function, an oracle $\hat{O}$ may be defined,

$$\hat{O}|x\rangle = (-1)^{f(x)}|x\rangle. \tag{17.70}$$

As mentioned in Sect. 17.1, an **oracle** is an operator that determines some specific property of a qubit without necessarily determining the value of the qubit itself. When $f(x)=0$, this operator has no effect on the state. But if $f(x) = 1$, then when applied to the oracle qubit $|q\rangle$, $\hat{O}$ has the effect in Eq. 17.68 of reversing the sign, $|q\rangle \rightarrow -|q\rangle$, and thus effectively interchanging the $|0\rangle$ and $|1\rangle$ states. Combining both values of f, the effect of the oracle on $|q\rangle$ can be expressed (with $\oplus$ denoting

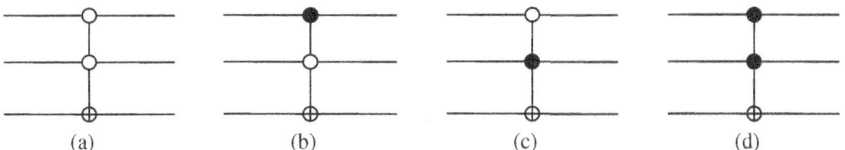

Fig. 17.4 Possible realizations of a two-qubit Grover search oracle operator $\hat{O}$ as quantum gates for cases (**a**) $x_0 = 0$, (**b**) $x_0 = 1$, (**c**) $x_0 = 2$, and (**d**) $x_0 = 3$. The top two lines carry the value of x, while the bottom line takes q as input and gives $q \oplus f(x)$ as output

base-2 addition) as

$$|q\rangle \rightarrow |q \oplus f(x)\rangle = (-1)^{f(x)}|q\rangle, \tag{17.71}$$

and the effect on the full state of the system is

$$|x\rangle \otimes |q\rangle \rightarrow (-1)^{f(x)}|x\rangle \otimes |q\rangle. \tag{17.72}$$

Instead of associating the factor with $|q\rangle$, we can think of it as associated with $|x\rangle$. In other words, we think of the oracle as acting according to $|x\rangle|q\rangle \rightarrow \left((-1)^{f(x)}|x\rangle\right)|q\rangle$, rather than $|x\rangle|q\rangle \rightarrow |x\rangle\left((-1)^{f(x)}|q\rangle\right)$. As a result, the oracle state $|q\rangle$ can be viewed as unchanged, and henceforth we simplify the notation by no longer bothering to write it, although the reader should keep in mind that it will still need to be present in the physical apparatus. The oracle operator can be realized as a quantum Toffoli or anti-Toffoli gate, as in Fig. 17.4 for the two-qubit case.

Now suppose that instead of starting with the data register in one particular state $|x\rangle$, we begin with it in an equal superposition of all the allowed x values. This can be done by initializing the system in the n-qubit state $|0\rangle = |0\rangle_1|0\rangle_2 \ldots |0\rangle_n$ and then applying a Hadamard gate to each qubit as in Sect. 17.4, to get state $|\psi\rangle$:

$$|\psi\rangle = \hat{H}^{\otimes n}|0\rangle = \frac{1}{\sqrt{N}} \sum_{x=0}^{N-1} |x\rangle. \tag{17.73}$$

Now consider the n-qubit operator $\hat{X} = 2|0\rangle\langle 0| - \hat{I}$, where $|0\rangle$ and $\hat{I}$ are the n-qubit zero state and n-qubit identity. This is a conditional phase shift, leaving the $|0\rangle$ state unchanged but flipping the signs of all other states. Referring to Box 17.1, this operator can also be viewed as a reflection; it reflects all states in the Hilbert space about the ray represented by $|0\rangle$. Further, act to the left and right of $\hat{X}$ with Hadamard gates to get the **diffusion gate**

$$\hat{D} = \hat{H}^{\otimes n}\hat{X}\hat{H}^{\otimes n} \tag{17.74}$$

$$= \hat{H}^{\otimes n}\left(2|0\rangle\langle 0| - \hat{I}\right)\hat{H}^{\otimes n} \tag{17.75}$$

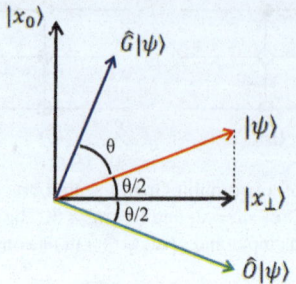

Fig. 17.5 The oracle operator $\hat{O}$ reflects an initial state across the vector $|x_\perp\rangle$. The Grover operator $\hat{G}$ carries out the reflection of $\hat{O}$, followed by a second reflection across the original direction of $|\psi\rangle$. The result is that $|\psi\rangle$ rotates toward the solution $|x_0\rangle$ to the search problem. After applying $\hat{G}$ a number of times $\mathcal{O}(\sqrt{N})$, the original vector converges onto the solution vector

$$= \left(2(\hat{H}^{\otimes n}|0\rangle)(\langle 0|\hat{H}^{\otimes n}) - \hat{I}\right) \tag{17.76}$$

$$= \left(2|\psi\rangle\langle\psi| - \hat{I}\right). \tag{17.77}$$

We see from the last line that $\hat{D}$ is a reflection about the state $|\psi\rangle$.

Finally, the **Grover operator** is defined:

$$\hat{G} = \hat{D}\hat{O} \tag{17.78}$$

$$= \hat{H}^{\otimes n}\hat{X}\hat{H}^{\otimes n}\hat{O} \tag{17.79}$$

$$= \left(2|\psi\rangle\langle\psi| - \hat{I}\right)\hat{O}. \tag{17.80}$$

Then repeated operation of $\hat{G}$ on $|\psi\rangle$ will cause the state to converge to the marked state. To see this, consider Fig. 17.5, in which $|x_0\rangle$ is the state we wish to find, while $|x_\perp\rangle$ is the part of $|\psi\rangle$ perpendicular to $|x_0\rangle$.

$\hat{O}$ also acts as a reflection, as can be seen by decomposing $|\psi\rangle$ into components parallel and perpendicular to $|x_0\rangle$, $|\psi\rangle = \alpha|x_\perp\rangle + \beta|x_0\rangle$. So the action of the oracle is to reverse the sign of the component parallel to $|x_0\rangle$: $\mathcal{O}|\psi\rangle = \alpha|x_\perp\rangle - \beta|x_0\rangle$, consistent with Fig. 17.5.

$\hat{G}$ then is a product of two reflections: $\hat{O}$ reflects $|\psi\rangle$ about $|x_\perp\rangle$, and then $\hat{D}$ reflects about the result about the original direction of $|\psi\rangle$. Since the product of two reflection is a rotation, we should be able to write $\hat{G}$ in the form of a rotation operator. Maintaining proper normalization of all states, $|\psi\rangle$ can be written as

$$|\psi\rangle = \sqrt{1 - \frac{1}{N}}|x_\perp\rangle + \sqrt{\frac{1}{N}}|x_0\rangle = \cos\frac{\theta}{2}|x_\perp\rangle + \sin\frac{\theta}{2}|x_0\rangle, \tag{17.81}$$

where

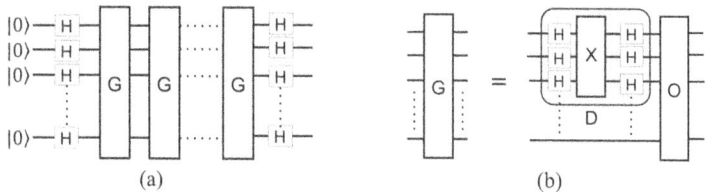

Fig. 17.6 (**a**) The quantum circuit for the Grover search algorithm. There are $n + 1$ lines: the top n represent the data register, the bottom line represents the auxiliary oracle register. The number of repetitions of the Grover gate is given by $k \sim \sqrt{N}$, as in Eq. 17.88. (**b**) The interior makeup of the Grover gate consists of the oracle, a reflection X about the $|0\rangle$ state, and a pair of Hadamard gates on each line

$$\cos \frac{\theta}{2} = \sqrt{1 - \frac{1}{N}}. \tag{17.82}$$

Then, referring to Fig. 17.5, it is easy to see that $\hat{G}$ rotates vectors away from $|x_\perp\rangle$ and toward $|x_0\rangle$ according to:

$$\hat{G}|\psi\rangle = \cos \frac{3\theta}{2} |x_\perp\rangle + \sin \frac{3\theta}{2} |x_0\rangle. \tag{17.83}$$

Notice that $\hat{G}|\psi\rangle$ is closer to the direction of $|x_0\rangle$ than $|\psi\rangle$ was. Further applications bring the state even closer to $|x_0\rangle$. If the Grover operator is applied k times, then

$$\hat{G}^k|\psi\rangle = \cos \frac{(2k + 1)\theta}{2} |x_\perp\rangle + \sin \frac{(2k + 1)\theta}{2} |x_0\rangle. \tag{17.84}$$

The quantum circuit is shown in Fig. 17.6.

How many applications of $\hat{G}$ are needed to converge onto $|x_0\rangle$? In the previous equation, we need $\frac{2k+1}{2}$ to converge to $\frac{\pi}{2}$, or equivalently, we need the cosine term to go to zero:

$$0 = \cos \frac{(2k + 1)\theta}{2} \tag{17.85}$$

$$= \cos k\theta \cos \frac{\theta}{2} - \sin k\theta \cos \frac{\theta}{2} \tag{17.86}$$

$$= \sqrt{1 - \frac{1}{N}} \cos k\theta - \sqrt{\frac{1}{N}} \sin k\theta. \tag{17.87}$$

This will be true when $\tan k\theta = \sqrt{N - 1}$, or equivalently (Fig. 17.7) when $\cos k\theta = \frac{1}{\sqrt{N}}$. Solving for k and then taking N to be large, we find:

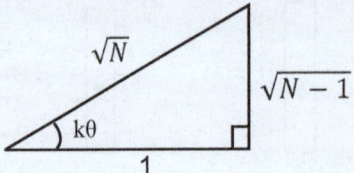

Fig. 17.7 Constructing a right triangle with $\tan k\theta = \sqrt{N-1}$. Choosing the lengths of the two sides to give the correct tangent (and taking one side of length one for convenience), the hypotenuse will equal $\sqrt{N}$, so that $\cos k\theta = \frac{1}{\sqrt{N}}$

Fig. 17.8 Reflecting vector $\boldsymbol{w}$ across vector $\boldsymbol{v}$ results in a new vector $\boldsymbol{w'}$. $\boldsymbol{w'}$ has the same component as $\boldsymbol{w}$ parallel to $\boldsymbol{v}$, but the perpendicular component flips sign

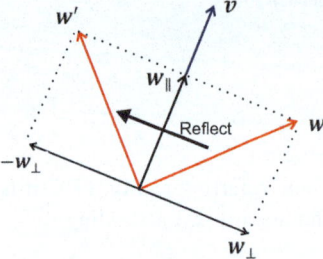

$$k \approx \frac{\sqrt{N}}{2} \left\lfloor \cos^{-1}\sqrt{\frac{1}{N}} \right\rfloor = \left\lfloor \frac{\pi\sqrt{N}}{4} \right\rfloor, \qquad (17.88)$$

where the brackets $\lfloor \ \rfloor$ indicate that the integer part of their content is taken. Here, we have also used the fact that $\frac{\theta}{2} \approx \sin\frac{\theta}{2} = \frac{1}{\sqrt{N}}$ and that the Taylor series for the inverse cosine is $\cos^{-1}x = \frac{\pi}{2} - x + \frac{1}{6}x^3 + \dots$. Thus, the number of operations required for the quantum search algorithm is $\sim \sqrt{N}$, compared to the classical case of $\sim N$. For large N, this can mean a substantial savings in time and resources.

The generalization from a single marked state to M marked states is straightforward; see [5].

Box 17.1 Reflection of Vectors

Consider two vectors $\boldsymbol{v}$ and $\boldsymbol{w}$ in a plane, and imagine that it is desired to find an operator that will reflect $\boldsymbol{w}$ across the direction of $\boldsymbol{v}$. To accomplish this, first break $\boldsymbol{w}$ into parts that are parallel and perpendicular to $\boldsymbol{v}$, $\boldsymbol{w} = \boldsymbol{w}_\parallel + \boldsymbol{w}_\perp$ (Fig. 17.8). The reflected vector then just flips the sign of the perpendicular part: $\boldsymbol{w} \to \boldsymbol{w'} = \boldsymbol{w}_\parallel - \boldsymbol{w}_\perp$.

Suppose we restrict ourselves to the case of interest to us, where the vectors represent quantum states in a Hilbert space. Then the parallel and perpendicular parts can be written in more explicit form. The parallel part is

(continued)

Box 17.1 (continued)
obtained by projecting $|w\rangle$ onto the direction of $|v\rangle$:

$$|w_{\parallel}\rangle = |v\rangle\langle v|w\rangle. \tag{17.89}$$

The perpendicular part is then

$$|w_{\perp}\rangle = |w\rangle - |w_{\parallel}\rangle = \left(\hat{I} - |v\rangle\langle v|\right)|w\rangle, \tag{17.90}$$

where the identity operator can be written in terms of orthonormal basis $|e_j\rangle$ as $\hat{I} = \sum_j |e_j\rangle\langle e_j|$. So the reflected state is

$$|w'\rangle = |w_{\parallel}\rangle - |w_{\perp}\rangle \tag{17.91}$$

$$= \left(|v\rangle\langle v| - \left(\hat{I} - |v\rangle\langle v|\right)\right)|w\rangle \tag{17.92}$$

$$= \left(2|v\rangle\langle v| - \hat{I}\right)|w\rangle \tag{17.93}$$

$$= \hat{R}_v|w\rangle, \tag{17.94}$$

where the reflection operator about $|v\rangle$ is

$$\hat{R}_v = 2|v\rangle\langle v| - \hat{I} = 2|v\rangle\langle v| - \sum_j |e_j\rangle\langle e_j|. \tag{17.95}$$

It should be easy to convince yourself (Problem 6) that two consecutive reflections about two different vectors amount to a rotation.

Problems

1. Generalize the Deutsch algorithm to the n-qubit Deutsch-Jozsa algorithm as outlined in Sect. 17.1. Work out the final state, and show that just measuring the state of the (N-qubit) upper line once is sufficient to distinguish between a balanced function and a constant function, without measuring the bottom qubit or determining the actual function.
2. Draw the quantum gate circuit for the four-qubit QFT.
3. For $n = 2$, find the matrix form of the QFT as given by Eq. 17.40, and show that it is equivalent to the action of the corresponding circuit given in Fig. 17.2c.

4. Compute the quantum Fourier transform of the three-qubit state $\frac{1}{\sqrt{2}}\big(|000\rangle +$ $|111\rangle\big)$.

5. Consider a pair of vectors v and w separated by angle θ in a plane. Show (either algebraically or by geometric reasoning) that the action of reflecting w about v can be represented by the rotation matrix

$$\begin{pmatrix} \cos 2\theta & \sin 2\theta \\ -\sin 2\theta & \cos 2\theta \end{pmatrix}$$

acting on x and y components of w. (The x-axis is taken along the direction of v.)

6. (a) Draw three vectors, u, v, and w, all confined to a single plane. Add to the diagram the images u' and w' of u and w under reflection about v. Now draw a fourth vector y in the same plane. Draw a second diagram with u', w', and y, and add to this diagram images u'' and w'' of u' and w' under reflection about y. It should be easy to convince yourself that the pair of reflections simply acts as a rotation. (b) Prove algebraically that the composition of the two reflections give a rotation. (Hint: give names to the angles between the vectors, and then use the result of the previous problem and some trig identities.)

7. Show that the gates in Fig. 17.4 act to implement the oracle in the Grover search algorithm for $n = 2$.

8. Consider the operator $\hat{U}$ that acts according to $\hat{U}|y\rangle = |xy \pmod{N}\rangle$ for x, y described by n bit binary numbers. Define the states

$$|u_s\rangle = \frac{1}{\sqrt{r}} \sum_{k=0}^{r-1} e^{-\frac{2\pi i s k}{r}} |x^k \pmod{N}\rangle$$

for integers r, s with $0 \le s \le r - 1 \le N - 1$. Suppose that r is the order of x: $x^r = 1 \pmod{N}$.

(a) Show that the $|u_s\rangle$ are eigenstates of $\hat{U}$ with eigenvalues $e^{2\pi i s/r}$.

(b) Show that $\sum_{s=0}^{r-1} |u_s\rangle = 1$. (Hint: Think of exponentials $e^{-2\pi i s k/r}$ as endpoints of vectors in the complex plane. Given that, convince yourself that $\sum_{s=0}^{r-1} e^{-2\pi i s k/r} = r\delta_{k0}$.)

(c) Show then that $\sum_{s=0}^{r-1} e^{-2\pi i s k/r} |u_s\rangle = |x^k \pmod{N}\rangle$

9. Equation 17.88 gives an approximate expression for the number of times that the operator $\hat{G}$ must be applied in the Grover algorithm to converge on the marked value. The exact result for this number is

$$k = \frac{\pi - 2\arctan\left(1/\sqrt{N-1}\right)}{2\arctan\left(2\sqrt{N-1}/(N-2)\right)},$$

where $N = 2^n$ and n is the size of the database.

(a) Verify that the exact expression reduces back to Eq. 17.88 for large N.

(b) Find the percent difference between the exact and approximate expressions for $n = 2, 4, 6, 8$. You should find that the approximation works well even for small values of n.

References

1. D. Deutsch, Proc. Royal Soc. London A. **400**, 97 (1985)
2. D. Deutsch, R. Jozsa, Proc. Royal Soc. London A. **439**, 553 (1992)
3. E. Bernstein, U. Vazirani, SIAM J. Comput. **26**, 1411 (1997)
4. S.M. Barnett, *Quantum Information* (Oxford University Press, Oxford, 2009)
5. M.A. Nielsen, I.L. Chuang, *Quantum Computation and Quantum Information: 10th Anniversary Edition* (Cambridge University Press, Cambridge, 2011)
6. D.R. Simon, SIAM J. Comput. **26**, 1474 (1997)
7. N.D. Mermin, *Quantum Computer Science: An Introduction* (Cambridge University Press, Cambridge, 2007)
8. J. Preskill, Lecture Notes for Physics on Quantum Information and Computation. Available at http://theory.caltech.edu/~preskill/ph219/index.html#lecture
9. D.J. Bernstein, J. Buchmann, E. Dahmen, *Post-Quantum Cryptography* (Springer, Berlin, 2009)
10. L. Chen, et al., Report on Post-Quantum Cryptography, NISTIR 8105 (2016). Available at https://nvlpubs.nist.gov/nistpubs/ir/2016/NIST.IR.8105.pdf
11. *Post-Quantum Cryptography: International Workshop, PQCrypto*. Annual Proceedings, vol. 2–14, various editors (Springer, Berlin, 2008–2023)
12. L.K. Grover, Proceedings of the 28th Annual ACM Symposium on Theory of Computing - STOC '96 (Association for Computing Machinery, Philadelphia, 1996), p. 212

Chapter 18
Quantum Communication and Quantum Cryptography

Quantum communication involves communication via individual quantum states, most often photon states. Quantum communication methods allow opportunities to compress large amounts of data into a small number of photons, as well as to evade disruption by disruptive effects such as turbulence. In addition, quantum cryptography (or more precisely *quantum key distribution (QKD)*) allows greatly enhanced communication security. Although there are a number of practical problems in the way before quantum communication becomes an everyday process, advantages have spurred a large influx of funding into the field from both government and industry. At least one commercially available smartphone already uses weak Poisson pulses to generate single-photon random number generation for QKD, in order to improve data security. This chapter gives an introduction to quantum cryptographic protocols.

18.1 Transmitting Information with Photons

Photons are an ideal platform for communication, for a wide variety of reasons. They can be transmitted over short distances through waveguides or long distances over free space or through optical fibers, with minimal loss. Since photons don't interact significantly with thermal surroundings or with each other at normal intensities, communications sent via photons tend to develop less noise than those sent with electrons. And photons are easy to produce on demand, with a high degree of control over their properties, as well as being extremely easy to detect at most frequency ranges. Finally, photons have several different properties that can be used to encode information: polarization, phase, frequency, arrival time, and orbital angular momentum can all be used to encode information into an individual photon. Intensity modulations or variations in the transverse spatial profile can also be used to encode information into multi-photon beams. Orbital angular momentum, in

© The Author(s), under exclusive license to Springer Nature Switzerland AG 2025
D. S. Simon, *Introduction to Quantum Science and Technology*, Undergraduate
Texts in Physics, https://doi.org/10.1007/978-3-031-81315-3_18

particular, can in principle encode an unlimited amount of information into a single photon since there is no upper limit on the value of the OAM quantum number; of course, in practice, there are practical limitations that make higher values harder to produce. In addition, the higher the angular momentum value, the more fragile it is with respect to disturbances such as atmospheric turbulence.

18.2   Quantum Teleportation

Star Trek fans are familiar with the idea of teleporters: Captain Kirk and crew step into a chamber on their starship and disappear; seconds later, they materialize on the surface of the planet below. So when it was proposed that quantum states could be teleported, it caused a stir in the news media, giving people hope that matter and people could be instantly transported across large distances without passing through the space in between. Unfortunately, the reality of **quantum teleportation** is a little less dramatic than this; it is information, not matter, that is transported, and all the transported information always moves at or below the speed of light. Nevertheless, it is an amazing effect.

Quantum teleportation, first proposed in 1993 [1] and then verified experimentally by several research groups, is a protocol for transmitting an unknown quantum state from one location to another.

Suppose Alice has an initial state $|\psi\rangle = \alpha|0\rangle + \beta|1\rangle$ in her possession. She doesn't know the state, and measuring it would destroy the superposition. She wants to send this unknown state to Bob in another lab. The question is how can this be done without first measuring the state?

In order for Alice to be able to achieve her goal, she and Bob need to already be sharing an entangled two-qubit ancillary state, which could have been readied ahead of time and stored until needed. To prepare this state, they could have started from the initially unentangled state $|00\rangle$. Then they can produce an entangled Bell state, $|\Phi^+\rangle$, using the setup in Fig. 18.1a. The reader should verify that the action of this circuit is given by:

$$|00\rangle \longrightarrow_H \frac{1}{\sqrt{2}} (|00\rangle + |10\rangle) \longrightarrow_{CX} \frac{1}{\sqrt{2}} (|00\rangle + |11\rangle) = |\Phi^+\rangle. \qquad (18.1)$$

At the end, Alice takes the particle from the top line back to her lab, while Bob takes the particle on the bottom line to his.

After this Bell state preparation stage, the entangled state $|\Phi^+\rangle$ is then fed into lines b and c of the main teleportation stage in Fig. 18.1b, while the unknown state $|\psi\rangle$ is fed into line a. Alice has access to the top two lines, a and b, while Bob has access to the bottom line c. Alice allows her half of the entangled state to interact with $|\psi\rangle$ by applying controlled-X and Hadamard gates. The input state on the left is:

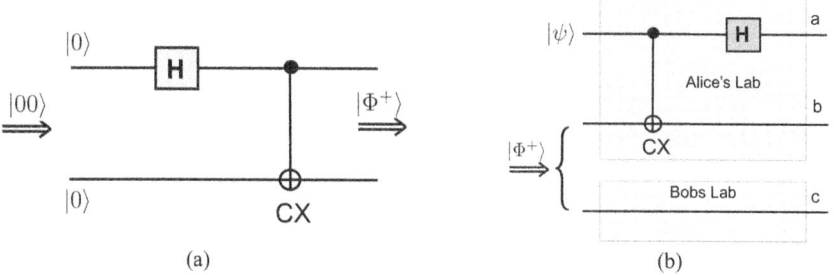

Fig. 18.1 (a) Alice prepares the ancillary qubits in lines b and c. She starts with the initial state $|00\rangle$ and converts it into the Bell state using a Hadamard and a controlled X gate. (b) Another CX and Hadamard are applied, now on lines a and b to teleport the state

$$|\psi\rangle_a|\Phi^+\rangle_{bc} = \frac{1}{\sqrt{2}}\,(\alpha|0\rangle + \beta|1\rangle)_a \otimes (|00\rangle + |11\rangle)_b \tag{18.2}$$

$$= \frac{1}{\sqrt{2}}\Big(\alpha|000\rangle + \alpha|011\rangle + \beta|100\rangle + \beta|111\rangle\Big). \tag{18.3}$$

Applying the CX gate,

$$CX\Big(|\psi\rangle_a|\Phi^+\rangle_{bc}\Big) = \frac{1}{\sqrt{2}}\Big(\alpha|000\rangle + \alpha|011\rangle + \beta|110\rangle + \beta|101\rangle\Big), \tag{18.4}$$

and then the H gate,

$$H \circ CX\Big(|\psi\rangle_a|\Phi^+\rangle_{bc}\Big) = \frac{1}{2}\Big[\alpha|000\rangle + \alpha|100\rangle + \alpha|011\rangle + \alpha|111\rangle \tag{18.5}$$

$$+\beta|010\rangle - \beta|110\rangle + \beta|001\rangle - \beta|101\rangle\Big]$$

$$= \frac{1}{2}\Big[|00\rangle_{ab}\Big(\alpha|0\rangle + \beta|1\rangle\Big)_c + |01\rangle_{ab}\Big(\alpha|1\rangle + \beta|0\rangle\Big)_c \tag{18.6}$$

$$+|10\rangle_{ab}\Big(\alpha|0\rangle - \beta|1\rangle\Big)_c + |11\rangle_{ab}\Big(\alpha|1\rangle - \beta|0\rangle\Big)_c\Big].$$

But by using the explicit forms of the Pauli matrices, it is not hard to see that this output is the same as

$$H \circ CX\,\big(|\psi\rangle_a|\Phi^+\rangle_{bc}\big) = \frac{1}{2}\Big[|00\rangle_{ab}|\psi\rangle_c + |01\rangle_{ab}\Big(X|\psi\rangle_c\Big) \tag{18.7}$$

$$+|10\rangle_{ab}\Big(Z|\psi\rangle_c\Big) + |11\rangle_{ab}\Big(XZ|\psi\rangle_c\Big)\Big].$$

Alice now measures the states of the two qubits she has (a and b), and then, based on what she finds, she communicates classically with Bob and tells him to apply some operation to the state he has (line c). These operations will be the identity operator or some combination of the Pauli X and Z matrices. In particular:

Alice measures:	$\longrightarrow$	Bob applies:
$|ab\rangle = |00\rangle$		No operation
$|ab\rangle = |01\rangle$		X
$|ab\rangle = |10\rangle$		Z
$|ab\rangle = |11\rangle$		XZ

Applying these operators to each of the corresponding terms in Eq. 18.6 or Eq. 18.7 will convert Bob's qubit into the original state $|\psi\rangle$. In fact, the final state in Eq. 18.7 can be written more compactly as

$$|\psi\rangle_{c,out} = \frac{1}{2}\sum_{ab}|ab\rangle \otimes X^a\,Z^b|\psi\rangle. \tag{18.8}$$

Since a and b only take values 0 or 1 and since $X^2 = Z^2 = (XZ)^2 = 1$, Bob's qubit will clearly be converted into $|\psi\rangle$ after the appropriate operator is applied.

So Alice has achieved her goal: by making a joint measurement on the original unknown state and half of the entangled state, and then communicating her results to Bob classically, she has allowed Bob to reconstruct $|\psi\rangle$ in his lab. The state has been teleported from line a in Alice's lab to line c in Bob's lab. Note that the original copy of the state that Alice had has been destroyed, so the no-cloning theorem is obeyed. And since the classical communication that is necessary for the protocol cannot travel faster than light, the process cannot violate special relativity's speed limit. The full procedure, including the initial ancillary state preparation and the final measurements, is summarized in Fig. 18.2.

18.3   Superdense Coding

Dense coding, also called **superdense coding**, can be seen as a sort of inverted version of quantum teleportation. In teleportation, two classical bits are used to transmit one quantum qubit. In dense coding, one qubit is used to send two classical bits. Dense coding was first proposed by Charles Bennett and Stephen Wiesner in 1992 and was achieved experimentally a few years later.

Fig. 18.2 The full teleport procedure, including the state preparation on the left and the measurements on the right

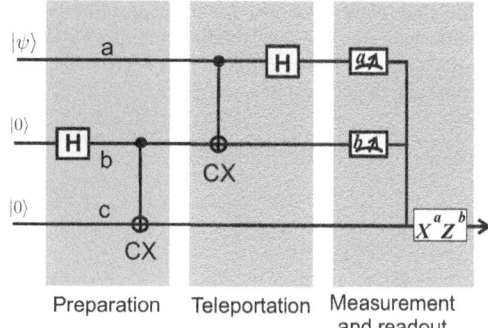

Preparation   Teleportation   Measurement and readout

Table 18.1 The dense coding protocol begins with Alice applying a prescribed set of operations to her half (the first qubit) of the Bell state $|\Phi^+\rangle$. Which operation she applies depends on the values of classical bits (x, y) that she wishes to transmit. After the encoding, she sends her half of the state to Bob

Bits to be transmitted (xy)	Encoding operation	Encoded quantum state
00	I	$(I_A \otimes I_B)|\Phi^+\rangle = |\Phi^+\rangle$
01	X	$(X \otimes I_B)|\Phi^+\rangle = |\Psi^+\rangle$
10	Z	$(Z \otimes I_B)|\Phi^+\rangle = |\Phi^-\rangle$
11	ZX	$(ZX \otimes I_B)|\Phi^+\rangle = -|\Psi^-\rangle$

Alice and Bob initially share one Bell state between them. To be specific, let's assume the state is

$$|\Phi^+\rangle = \frac{1}{\sqrt{2}}\big(|0_A\rangle|0_B\rangle + |1_A\rangle|1_B\rangle\big) = \frac{1}{\sqrt{2}}\big(|00\rangle + |11\rangle\big). \tag{18.9}$$

Alice has two classical bits of information, (x, y), that she wishes to transmit to Bob. She encodes this information by applying a predetermined set of quantum operations to her qubit. Depending on the values she wishes to send, she applies either the identity operator, one of the Pauli operators X or Z, or both of these Pauli operators ZX, according to the scheme shown in Table 18.1. (In the terminology introduced in Chap. 16, the X operation acts as a swap gate, since it swaps the values 0 and 1, and Z is a phase gate.) After doing this, she then sends her newly encoded qubit to Bob.

Bob now possesses both halves of the Bell state and can perform measurements on it. Once he has determined the state he knows which operation Alice performed, and so he knows the values of her bits. One way Bob could determine the entangled state is to first apply the quantum version of a CNOT gate (with the second bit as the control and the first as the target) and then apply the Hadamard gate $H = \frac{1}{2}\begin{pmatrix} 1 & 1 \\ 1 & -1 \end{pmatrix}$, to the second bit. The output of these operations is as shown

Table 18.2 Now that Bob has both halves of the entangled state, he applies a pair of operations to the first qubit (the one sent by Alice) in order to decode the state. He first applies CNOT and then the Hadamard operation, H. The result is given in the last column: the pair of qubits now match the original pair of classical bits, so that Bob can measure them and determine the information sent by Alice

Alice's bits	Bob's received state, $|\psi\rangle$	After CNOT: $\mathrm{CNOT}_{21}|\psi\rangle$	After H: $H_2 \cdot \mathrm{CNOT}_{21}|\psi\rangle$
00	$|\Phi^+\rangle$	$\frac{1}{\sqrt{2}}(|00\rangle + |01\rangle)$	$|00\rangle$
01	$|\Psi^+\rangle$	$\frac{1}{\sqrt{2}}(|11\rangle + |10\rangle)$	$|10\rangle$
10	$|\Phi^-\rangle$	$\frac{1}{\sqrt{2}}(|00\rangle - |01\rangle)$	$|01\rangle$
11	$|\Psi^-\rangle$	$-\frac{1}{\sqrt{2}}(|11\rangle - |10\rangle)$	$|11\rangle$

in Table 18.2: the final qubits match up with the initial values of the classical bits sent by Alice.

Alice has therefore transmitted to Bob two bits of classical information in a single qubit (her half of the Bell state). Of course, this requires that Alice and Bob have both previously shared a Bell state between them and also have previously communicated classically in order to agree upon the set of transformations Alice would apply.

Notice that this communication is also secure: if an eavesdropper (Eve) intercepts the communication between Alice and Bob, she gets only the part sent by Alice. She still has no knowledge of the parts originally in Bob's possession. Without Bob's half of the entangled state, there is no way she can reconstruct Alice's classical bits.

18.4   Entanglement Swapping and Quantum Repeaters

One additional important entanglement-based effect is the idea of **entanglement swapping**. Suppose Alice (A) and Cedric (C) share the qubits of an entangled pair. For the sake of specificity, assume they have the Bell state $|\Psi^-\rangle_{12}$. Separately, Cedric shares a Bell state with Bob: $|\Psi^-\rangle_{34}$. The initial state of the ABC system is therefore

$$|\Psi\rangle_0 = |\Psi^-\rangle_{12}|\Psi^-\rangle_{34}, \tag{18.10}$$

where Alice possesses qubit 1, Bob has qubit 4, and Cedric has possession of qubits 2, 3. By expanding out each of the Bell states into the form $|\Psi^-\rangle = \frac{1}{\sqrt{2}}(|0\rangle|1\rangle - |1\rangle|0\rangle)$ and then rearranging, it is straightforward to show (Problem 2) that $|\Psi\rangle_0$ can be written in the form

Fig. 18.3 Entanglement
swapping. One qubit from
each of the two initially
entangled pairs (1, 2) and
(3, 4) undergoes a Bell state
measurement in Cedric's lab.
As a result, the remaining two
qubits 1 and 4 become
entangled between Alice's
and Bob's labs

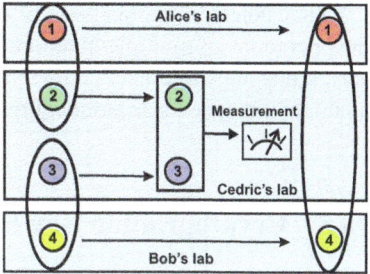

$$|\Psi\rangle_0 = \frac{1}{2}\left(|\Psi^+\rangle_{14}|\Psi^+\rangle_{23} + |\Psi^-\rangle_{14}|\Psi^-\rangle_{23} + |\Phi^+\rangle_{14}|\Phi^+\rangle_{23} + |\Phi^-\rangle_{14}|\Phi^-\rangle_{23}\right).$$

(18.11)

If Cedric now measures which Bell state he has, then the state shared by Alice
and Bob will collapse into one of the four Bell states. Which state their system
collapses to is random and will be determined by which state is revealed by Cedric's
measurement. For example, if Cedric finds that he has the state $|\Phi^+\rangle_{23}$, then the
Alice-Bob system collapses to the entangled state $|\Phi^+\rangle_{14}$. So Alice and Bob, who
have had no communication or interaction with each other, suddenly find themselves
entangled with each other due entirely to the action of Cedric. Cedric's measurement
breaks the two separate entanglements he previously had with Alice and Bob and
swaps them for a single entanglement between the unsuspecting Bob and Alice. This
is illustrated schematically in Fig. 18.3.

This entanglement swapping has several applications. For example, entanglement-
based quantum cryptography and quantum communication systems are limited in
the distance over which they can operate. Entanglement is a fragile property: over
long distances, dispersion, turbulence, and other disruptive effects can lead to the
degradation of the entanglement until the level of entanglement is too small to be
of any use. Classical communications face a similar problem, where dispersion, for
example, may broaden communication pulses and losses may diminish the signal
amplitude to below practically useful levels. But in classical systems, substations
can periodically make copies of the signal, amplifying and cleaning it up before
sending it on to the next station. In this way, the classical signal range can be
extended indefinitely. This option is not allowed for quantum communication
however: the no-cloning theorem prevents the copying of an unknown state. To
make a copy, the state would first have to be measured, which would destroy the
entanglement and may cause unwanted collapses into the wrong measurement basis.
However, the situation can be saved by the use of **quantum repeaters**, which swap
the entanglement of the incoming state and that of an entangled state generated in
the repeater station in order to create a new, more strongly entangled output state
that can then be transmitted to the next repeater or to the final end-user.

Other potential applications of entanglement swapping include using it to link the operations of multiple quantum computers (which is a prerequisite for building, e.g., a quantum Internet) or for potentially linking multiple telescopes to achieve quantum-enhanced astronomical resolution.

18.5   Cryptographic Key Distribution in the Classical World

There is evidence of cryptography being used almost as far back as the invention of writing itself. By the first century BC, Julius Ceasar regularly used a simple substitution cypher to encode messages to his generals. The procedure was just to shift the alphabet by n spaces: each letter was replaced by the letter n places farther back in the alphabet. The recipient was supplied with the value of n ahead of time and so could reverse the shift to decode the message. For example, if $n = 3$, then the word *Caesar* would be encoded as *Fdhvdu*. Although simple to use, this type of substitution code (now call a **Caesar cipher**) is also extremely easy to break.

A slightly more secure approach is to use a substitution cypher with an **encryption key**. The key is a table telling the recipient which letter was switched with which other letter during encryption. For example, the key may tell us to carry out the encryption by making substitutions $a \to x, b \to c, c \to y, d \to q, \ldots$. So the recipient can then undo the swaps and easily reconstruct the original message. Unfortunately, such a code is only slightly harder to break than a Caesar cypher. The key can be made more complicated (using additional symbols from beyond the standard alphabet or using multiple different symbols for the same letter), but these only provide a small layer of extra security. The key has to be agreed upon by the sender and recipient ahead of time, which opens another security problem: an enemy can get a glimpse of the key or even steal a copy, which makes the code completely ineffective.

Modern encryption schemes for financial transactions and other sensitive data are usually based on the difficulty of solving some types of mathematics problems. The idea of **public key encryption** was first publicly proposed in 1976 [2, 3], based on the existence of so-called **one-way functions**: these are functions where $y = f(x)$ is easy to compute but the inverse $x = f^{-1}(y)$ is prohibitively hard. This asymmetry between the difficulties of forward and reverse problems (the encryption and decryption) is the basis for public-key encryption, in which the encryption key can be made publicly available without endangering security, but the decryption key must be kept private.

Elliptic functions are one example of a function that can be used for this purpose. But generalizing from functions to more general one-way operations, the most commonly used version of public key encryption is based on factoring. Factoring a large number is an NP problem: there is no known efficient algorithm for doing the factoring, but if you are given two factors, you can easily verify that they produce the correct product. Given two numbers q and p, finding the product $N = qp$ is easy, but given only N, finding the factors grows rapidly more difficult as N increases. So factoring is a one-way operation in the sense defined above. The most important example of public key encryption is the widely used **Rivest-Shamir-Adleman (RSA) encryption**, described in Box 18.1, which takes advantage of this difficulty of factoring large numbers.[1] The RSA scheme has had great success as a means of keeping private information secure during transactions. But the problem is that if quantum computers are able to violate the strong Church-Turing theorem, then they may be able to break encoding systems based on the difficulty of NP-class problems, including RSA encryption. As a result, there is a great deal of active research in developing new encryption schemes that are resistant to being broken by quantum computers.

Box 18.1 RSA Public Key Encryption

In RSA public key encryption [4], if Bob wishes to receive a private message from Alice, he chooses two prime numbers p and q, with product $N = pq$. The number of binary digits needed to describe N is $n = \log_2 N$. Bob also chooses a large integer c that has no common factors with $r = (p-1)(q-1)$. Since Bob knows c, he can easily compute the mod r inverse, d, such that $cd = 1 \pmod r$. Finding d is, however, impossible for anyone who doesn't know r, and finding r is a difficult problem for anyone who does not know p and q.

So Bob can publicly send the pair of numbers $P = \{c, N\}$ to Alice. This is the public encryption key. Anyone can use it to easily encrypt a message. But Bob keeps the secret decryption key $S = \{d, N\}$ to himself, and without d (or, equivalently, without p and q), the decryption is difficult. Classical factoring algorithms are NP, but are not believed to be P.

Let a be a large number serving as the message that Alice wishes to send. The binary encoding of a will have length $L = \log_2 a$. If $L > n$, she breaks the message into blocks of length $\leq n$ and sends each block separately. To encrypt each block, Alice calculates the number

$$b = a^c \pmod N, \tag{18.12}$$

(continued)

---

[1] The idea of public key encryption had actually been first developed by J. Ellis, C. Cocks, and M. Williamson in 1970, with the first RSA-type cipher developed by Cocks in 1973. However, Ellis, Cocks, and Williamson were working for a British intelligence agency and were not allowed to make their ideas public at the time. Their work was first discussed publicly only in 1997, long after the work of Rivest, Shamir, and Adleman.

Box 18.1 (continued)

which she then sends publicly to Bob. Since Bob knows d, he can then easily decode this message: he calculates b^d (mod N), which by the extension of Fermat's little theorem (Appendix F) equals the original message value: b^d (mod N) $= a$.

The RSA scheme relies on the difficulty of finding prime factors of large numbers, which means that successful practical implementation of the Shor factoring algorithm or of quantum order-finding algorithms is expected to endanger any date encrypted using RSA and similar methods, since these quantum approaches only scale polynomially with key length. For more on RSA encryption, see [5–7].

To discuss encryption more generally, let us start by assuming that the message is encoded in binary, as a string of 0s and 1s. Consider two secret agents Alice and Bob who may be widely separated in space but need to generate a shared secret key for encrypting and decrypting messages. Their problem can be broken into two parts: (i) *the encryption problem*, finding an encryption method that cannot be broken by anyone without access to the encryption key, and (ii) *the distribution problem*, agreeing on such a key over a long distance without an eavesdropper obtaining enough information to reconstruct the key.

The solution to the first part is well-known. The only truly unbreakable code is the **one-time pad** or **Vernam cipher**, named after Gilbert Sandford Vernam, who was working during World War I. In this approach, the secret key k is a random string of binary digits intended to be used just once and then discarded. For example, let m represent the text to be encoded. We assume it has already been converted into the form of a binary string. The encoded message is given by the new binary string $s = m \oplus k$, where $\oplus$ is modulo-two addition. If the encryption key k is truly random, then the bits in the coded message s will be as well. Any patterns that existed in the original message (such as the higher frequency of the letter e compared to the letter z) and that could help an eavesdropper decode the message, are now gone. The key has the maximum possible entropy, and so no useful information can be extracted from it. Note, however, that if the same key is used more than once, correlations between the different messages encoded with the same key *will* allow information about the key to be extracted. This in turn allows information about the encoded messages to be revealed. So it is essential that no key be used more than once.

Because $k \oplus k = 0$, all Bob needs to do to decode the message is to just add the same key to the encoded message: $s \oplus k = m \oplus (k \oplus k) = m$. Therefore, decryption is trivial for anybody with the key but impossible for anyone without it.

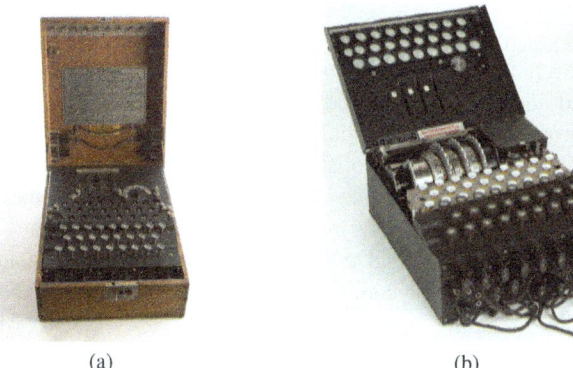

(a)   (b)

Fig. 18.4 Examples of enigma machines. The plaintext message is typed in on the keyboard, and the encoded text appears in sequence of illuminated letters above the keyboard. Front panel folds done to reveal a plugboard used to fix the daily settings (Sources: (**a**) public domain image from the CIA, https://www.cia.gov/legacy/museum/artifact/enigma-machine/. (**b**) Alessandro Nassiri, Museo Scienza e Tecnologia Milano, http://www.museoscienza.org/, under creative commons license.)

Box 18.2 Enigma Machines

Mechanical devices for encoding and decoding text have a long history. Probably the first example was the Hebern rotor machine developed in the early nineteenth century. But the most famous example is the Enigma machine (Fig. 18.4), which was invented by German engineer Arthur Scherbius around 1915. Enigma machines were sold commercially during the 1920s and became widely used by the Axis powers, especially Germany, during World War II.

Enigma machines were portable electromechanical devices, encased in a suitcase. The original message is typed onto a keyboard. A system of three or more rotors are then used to encode the message. The encoded message appears as a sequence of illuminations on a set of 26 lights, which a transcriber records on paper. After each key press, the rotor changes the electromechanical connections, based on a set of machine settings that are changed daily. The daily settings are input by inserting a set of plugs into a plugboard that is revealed when the front panel of the machine case is opened. In order to decipher the message, both the machine and the daily settings are required. Re-entering the ciphertext into another Enigma machine with the same settings converts the message back to the original plaintext message.

German military and intelligence services began using Enigma machines as early as 1926. Operator errors and flaws in the encoding procedures during the encryption of German messages allowed Polish mathematicians

(continued)

Box 18.2 (continued)
to decipher Enigma messages beginning in January 1932. The Polish Cipher Bureau passed their results on to Britain and their Allies. As Germans gradually improved their cryptographic procedures, further work was needed to break the new codes. The Polish code-breakers built replica Enigma machines, which had to be destroyed when Germany invaded Poland. Some of the people involved eventually made it to England, where they worked with British mathematicians and cryptologists at Bletchley Park to finally break the improved German codes, allowing the Allies to intercept Nazi communications.

A quantum analog of Enigma machines was proposed in [8] and demonstrated in [9].

The Vernam cypher with a one-time pad completely solves the problem of unbreakable encryption, but it still leaves the problem of safely *distributing* the key among the legitimate users. Because Alice and Bob must use the *same* key, an eavesdropper, traditionally known as Eve, may be able to intercept the passing of the key from one to the other, enabling her to decode encrypted messages exchanged between Alice and Bob.

In the classical world, there is no foolproof means for completely secure key distribution. But quantum mechanics allows for such means via an approach called **quantum key distribution (QKD)**. Indeed, the goal of QKD is to generate one-time encryption keys shared between two widely separated legitimate users (Alice and Bob) in such a way that laws of quantum mechanics prevent ill-intentioned eavesdroppers (Eve) from obtaining such keys without revealing her interference. The eavesdropping itself is not preventable, but if it occurs, then legitimate agents can unambiguously detect it. Then, when Eve's tampering is detected, Alice and Bob will always know their communication line has been compromised and so they switch to another, more secure, key distribution channel. Thus, Alice and Bob can generate as many one-time pads as needed without worry that the messages they encode with them will be decrypted by enemy agents.

One remaining problem is that Alice and Bob each need to be sure that they are actually distributing key material with each other, not with impostors. So, it is usually assumed that they already share a small amount of secret information between them in order to authenticate each other's identities, before proceeding with the key generation and message encryption. The key distribution procedure can thus be thought of as a **key expansion protocol**, in which an initially shared, short authentication key grows into a much longer encryption key.

For more in-depth discussion of classical cryptology (including public key encryption and the enigma code), see [10].

18.6   Quantum Cryptographic Key Distribution

Quantum key distribution (QKD) is based on the inability of an experimenter to distinguish unambiguously between two non-orthogonal states (Sect. 9.6). The basic idea is most easily demonstrated in terms of a two-state system, such as the spin states of an electron or the polarization states of a photon.

Consider photon polarizations to illustrate. Suppose that Alice sends a sequence of photons to Bob and that she randomly chooses one of two orthogonal polarization states for each one. For example, she may be choosing randomly between horizontal and vertical polarizations. She prepares the random polarization states by having a polarization filter randomly rotate back and forth between the two orthogonal directions. The states $|H\rangle$ and $|V\rangle$ can be equated, respectively, with the logical 0 and 1 states, so that the random string of polarizations will represent a binary string. This string is to be used as a secret, one-time encryption key.

When Bob receives the photons, he can easily measure the polarization state. For example, he could use a polarizing beam splitter (PBS) that reflects vertically polarized photons and transmits those that are horizontally polarized. If H photons are directed to one detector and V photons are sent to a different detector, then the sequence of detector firings allows Bob to reproduce the sequence of polarizations sent by Alice. The secret encryption key is thus shared between the two agents.

Notice, though, that this process depends on Bob knowing the basis that Alice used when preparing the polarizations. Suppose that Alice forgot to tell Bob that she was using the horizontal-vertical (HV) basis. Bob may mistakenly think that she is using the diagonal-antidiagonal (DA) basis, and so he may have his PBS oriented at $45°$ from the orientation of Alice's polarizers. In this case, one of Bob's detectors will be detecting diagonally polarized photons (at $+45°$ from the positive x-axis), and one will be detecting antidiagonal ($-45°$) photons. Assume that Bob assigns values 0 to diagonal and 1 to antidiagonal. Recall that

$$|D\rangle = \frac{1}{\sqrt{2}}(|H\rangle + |V\rangle) \qquad\qquad |H\rangle = \frac{1}{\sqrt{2}}(|D\rangle + |A\rangle) \qquad (18.13)$$

$$|A\rangle = \frac{1}{\sqrt{2}}(|H\rangle - |V\rangle) \qquad\qquad |V\rangle = \frac{1}{\sqrt{2}}(|D\rangle - |A\rangle). \qquad (18.14)$$

So if Alice sends a horizontal photon (representing 0), what is the probability that Bob measures the state $|D\rangle$ and correctly gets the value 0 that Alice intended him to receive? Clearly,

$$P(0 \to 0) = P(H \to D) = |\langle D|H\rangle|^2 = \frac{1}{2}. \qquad (18.15)$$

He only has a 50% change of getting the correct value. Similarly, the probability of Bob receiving the wrong value is

$$P(0 \rightarrow 1) = P(H \rightarrow A) = |\langle A|H \rangle|^2 = \frac{1}{2}. \tag{18.16}$$

Bob's error rate is 50%: his results are completely uncorrelated with the signal Alice sent.

So, if Alice and Bob are using the same basis, they can communicate without error, and their bit strings are perfectly correlated. But if they use mismatched bases, then their results are uncorrelated and Bob simply receives random noise. Thus, in addition to the quantum communication channel (the photons), some classical communication channel is needed in order for them to make sure they are using the same basis.

Return now to the case where Alice and Bob are correctly using the same basis, but suppose that the eavesdropper, Eve, has placed a polarizing beam splitter in the path of the photon in order to measure its polarization. *If Eve knows the basis* that Alice and Bob are using, then she can measure the polarization and determine the corresponding bit of the encryption key. She can then create a new photon with the same polarization as the one she measured and send it on to Bob. As long as Eve doesn't introduce a noticeable delay as she performs her shady manipulations, Bob has no way to know that this is not the same photon Alice sent. He can reconstruct the key perfectly and can see no indication that Eve has been tampering with the light. Eve can repeat the same trick for each photon sent and learn the entire encryption key perfectly, with no evidence left to tip off Alice and Bob that she has even been there.

However, if Eve *doesn't* know the basis Alice is using, then she has a problem. If Alice is flipping randomly between the HV and DA bases between each transmitted photon, then the best that Eve can do is guess randomly in which basis to measure. She will guess correctly 50% of the time, and each time she guesses the wrong basis, she has a 50% probability of measuring the wrong bit value and of consequently sending Bob the wrong value. Thus, she introduces an error rate e into Bob's results, and this error rate is at least 25%:

$$e = \text{(fraction of wrong guesses)} \cdot \text{(probability of wrong bit)} \geq \frac{1}{2} \cdot \frac{1}{2} = \frac{1}{4}.$$

By comparing part of their data, Alice and Bob can easily detect these errors and determine that Eve has been engaging in nefarious activities.

So, the strategy for Alice and Bob is the following:

(i) Alice randomly chooses one of the two bases (HV or DA) before creating each photon. Once the basis is chosen, she randomly chooses one of the two polarization states in that basis and sends Bob a photon with that polarization.

(ii) Bob randomly chooses one of the two bases to make his measurement in. Half the time, he will measure in the same basis Alice used to prepare the state and have no errors. Half the time, he will choose the wrong basis, and his measurements in these cases will have a 50% error rate.

(iii) Alice and Bob now communicate over a classical communication line. They tell each other which basis they used for each trial. But Alice **does not** tell Bob what state was sent on each trial, and similarly, Bob **does not** tell Alice what result he got on his measurement.

(iv) Alice and Bob throw out the trials on which they used different bases. This means that, on average, half of the data is wasted. But on the remaining trials (the ones where Alice's preparation basis and Bob's measurement basis agree), they should (in the absence of eavesdropping) have a string of bit values that match perfectly, with zero error rate. Let the set of trials that they kept be denoted S. This step is called **sifting**.

(v) To test for eavesdropping, they randomly divide S into two disjoint subsets, $S = C \cup K$. Usually, the set C is a small fraction of S, with K being much larger.

(vi) The set C is used to check security: Alice and Bob now tell each other their actual bit values for each photon in set C. If there was no eavesdropping, then their values should agree. However, if Eve was active, then they should see a discrepancy of about 25% between their values. If the latter is the case, they shut down the communication channel and seek an alternate means of communicating.

(vii) If there was no substantial error rate in the check set C, then the communication is secure. So the values in the set K should agree perfectly between the two users and can be used to form the secure encryption key. Notice that Alice and Bob did not tell each other the values in set K, only the bases used to create and measure each one.

Since the actual bit values in K were never communicated over the classical channel between the two legitimate users, Eve can't evade this procedure by tapping into the classical line to steal them. In fact, the entire classical communication could be done in public: it could be carried out over a public radio station or posted on billboards by the side of the highway, and Eve could gain no advantage from it.

Of course, the procedure outlined here is greatly idealized. In real life, there will be sources of error in C that have nothing to do with eavesdropping. Bob's detectors will always have some built-in error rate, stray photons from other sources may accidentally enter the communication stream, static in the classical communication channel may cause Alice and Bob to mistake each other's classical information, and so on. So there will always be some error rate found even in Eve's absence, which means that the maximum natural error rate ϵ expected in the system has to be estimated beforehand. As long as $\epsilon \ll 25\%$, then Alice and Bob can assume the communication is secure when $e \leq \epsilon$.

Note that the same procedure can be carried out with electrons, using spin along some axis in place of polarization. Instead of two polarization bases, two spin-measurement axes would be used, with spin either up or down along that axis at each measurement. The polarizing beam splitter in this case would be replaced by a Stern-Gerlach apparatus. Otherwise, essentials of the protocol remain the same.

The procedure outlined above is called the **BB84 protocol**, first proposed by Bennett and Brassard in 1984. Most other QKD protocols are some variation or generalization of the same idea.

The two bases used in the BB84 protocol are rotated at 45° to each other. This choice of angle introduces the maximum uncertainty in the state when measured in the wrong basis. The two bases are said to form a set of **mutually unbiased bases**. Consider two bases of dimension d, where the unit vectors in the two bases are denoted $\{\hat{e}_1, \hat{e}_2, \ldots \hat{e}_d\}$ and $\{\hat{f}_1, \hat{f}_2, \ldots \hat{f}_d\}$. Then the bases are mutually unbiased if

$$\left| \hat{e}_i \cdot \hat{f}_j \right|^2 = \frac{1}{d} \tag{18.17}$$

for all $i, j = 1, 2, \ldots d$. For two dimensions, this gives $\cos^2 \theta = \frac{1}{2}$, or $\theta = \pm 45°$. The mutually unbiased condition guarantees that a vector in the preparation basis (say, the $\hat{e}$ basis) will have equal probabilities of any possible result when measured in the wrong measurement basis (the $\hat{f}$ basis), so that the measurement gives absolutely no information about the prepared state. As the bias increases (in other words, as the inner products $\left| \hat{e}_i \cdot \hat{e}_j \right|^2$ begin to differ from each other as i and j vary), the eavesdropper can gain more information from her measurement and over many trials may be able to decipher part of the key.

In the BB84 case, two linear polarization bases were used (HV and AD). But bases from left-handed and right-handed circular polarizations are also mutually unbiased with respect either of those linear polarization bases. Thus, we have three possible mutually unbiased bases in two dimensions:

$$HV: \qquad |0\rangle \text{ and } |1\rangle \tag{18.18}$$

$$DA: \qquad \frac{1}{\sqrt{2}} (|0\rangle + |1\rangle) \text{ and } \frac{1}{\sqrt{2}} (|0\rangle - |1\rangle) \tag{18.19}$$

$$\text{Circular}: \qquad \frac{1}{\sqrt{2}} (|0\rangle + i|1\rangle) \text{ and } \frac{1}{\sqrt{2}} (|0\rangle - i|1\rangle) \tag{18.20}$$

where $|0\rangle$ and $|1\rangle$ represent unit vectors along the x and y axes. (Note that if all three of these basis are simultaneously rotated by any fixed angle, they remain mutually unbiased.) Any two of these three bases can be used for the BB84 protocol. If the dimension d is prime, it is known that $d + 1$ mutually unbiased bases exist. The question of how many exist in an arbitrary dimension d remains unsolved.

In the BB84 protocol, one of the legitimate users (Alice) of the system created a single-particle state and sent it to the other user (Bob). An alternative approach is to use a pair of entangled particles, with each of the two users having possession of the two particles. Assuming that the polarizations or spins are perfectly correlated (or perfectly anticorrelated) in the entangled states, the two can make their measurements in random bases and then, after classical communication, keep only those trials in which they used the same basis. The rest of the protocol proceeds largely in the same manner as before. This is known as the **Ekert, E91**, or **EPR protocol** [11]. In this entanglement-based protocol, security can be maintained by testing the rate of Bell inequality violation, since any tampering by Eve on one of the photons will degrade the level of entanglement. One interesting feature of the E91 protocol is that the source of the entangled photons doesn't need to reside in either Alice's lab or Bob's. It could be under the control of a third party. In fact, it could be operated by the eavesdropper herself, and she would be unable to gain any advantage from controlling it.

18.7   Other Quantum Cryptography Protocols

Over the past few decades, a wide variety of QKD protocols have been developed, each with their own distinct features. Here, we briefly list a few of the more common protocols. See the references for additional details.

18.7.1   B92 Two-State Protocol

As seen in the last section, the BB84 protocol requires two bases for state preparation and measurement. In 1992, Bennett found a variation that requires preparing states in only a single basis. Known as the B92 [12] protocol, it uses a single two-dimensional basis in which the two basis vectors are nonorthogonal to each other. For example, the basis formed by the two vectors

$$|0\rangle = |\rightarrow\rangle = \begin{pmatrix} 1 \\ 0 \end{pmatrix}, \qquad |1\rangle = |\nearrow\rangle = \frac{1}{\sqrt{2}} \begin{pmatrix} 1 \\ 1 \end{pmatrix} \tag{18.21}$$

may be used. Since the $|0\rangle$ and $|1\rangle$ states are not orthogonal, they cannot be unambiguously distinguished.

Each trial consists of Alice randomly preparing one of these two states ($|\rightarrow\rangle$ or $|\nearrow\rangle$) and sending it to Bob. Bob still makes measurements in two bases, switching randomly between them, in exactly the same manner as in BB84. As before, he uses the HV basis and the diagonal DA ($|\nearrow\rangle$ and $|\searrow\rangle$) bases.

Depending on the state sent and the measurement basis used, the outcomes are listed in Table 18.3. Suppose Bob measures in the HV basis: then if he detects $|\uparrow\rangle$, he knows with certainty that $|\nearrow\rangle$ was sent, since the other possible state is orthogonal to the detected state. But if he detects $|\rightarrow\rangle$, he can't tell which state was sent, since his result has equal overlap with both of the potential transmitted states. Similarly, suppose Bob measures in the DA basis: then he knows the state unambiguously if he detects $|\searrow\rangle$ (indicating that $|\rightarrow\rangle$ was sent) but has gained no knowledge if he detects $|\nearrow\rangle$. As a result, he discards any trials in which he detects either $|\rightarrow\rangle$ or $|\nearrow\rangle$, and he tells Alice over a classical (possibly public) channel

Table 18.3 Bob measures the state he received from Alice in one of two randomly chosen bases. If the result of his measurement is either $| \uparrow \rangle$ or $| \searrow \rangle$, then he can determine which state was sent, because each of these is orthogonal to one of Alice's possible states. If he measures $| \rightarrow \rangle$ or $| \nearrow \rangle$, then he can't determine the transmitted state because his measured state is not orthogonal to either and so shares an overlap with each

Bob's basis	Bob's result	Bob's conclusion
HV	$| \uparrow \rangle$	A sent $| \nearrow \rangle = | 1 \rangle$
HV	$| \rightarrow \rangle$	Indeterminate: A sent either $| \rightarrow \rangle$ or $| \nearrow \rangle$
DA	$| \searrow \rangle$	A sent $| \rightarrow \rangle = | 0 \rangle$
DA	$| \nearrow \rangle$	Indeterminate: A sent either or $| \rightarrow \rangle$ or $| \nearrow \rangle$

which trials were discarded. Alice and Bob then check security on a subset $\mathcal{C}$ of trials, as in BB84. If the security subset survives the check, then the remaining trials $\mathcal{K}$ form the key.

At the end of this process, Alice and Bob share a common, secure key. The inability to unambiguously distinguish between two vectors of a single *nonorthogonal* basis now plays the same role that inability to unambiguously determine vectors created and detected in two *different*, mutually unbiased, bases did in the BB84 protocol.

One flaw of this procedure is that it is insecure in the presence of loss. This is because Eve can intercept the signal, making the same type of measurements as Bob. Any inconclusive measurements, she then discards. When she gets a definite result, she sends to Bob a new photon in the same state, and then she covers her tampering by sending additional photons of her own to replace those that she discarded. Alice and Bob can restore security by using *pairs* of pulses, one dim and one bright, with a phase difference between them [12]. Bob monitors the interference between the two pulses. Eve can't remove the strong pulse without making the drop in intensity obvious, and if she blocks the weak pulse, then the expected interference between the two pulses does not appear. Fabricating false pulses to replace those she took introduces noticeable errors due to trials on which her result will be inconclusive: she won't be able to transmit replacement photons with the correct phase.

18.7.2 Six-State Protocol

Another common approach is to use a six-state protocol [13, 14], with three mutually unbiased bases. States are prepared and measured in one of three randomly chosen bases, the HV, DA, and circular polarization bases. The protocol then uses the states

$$|0\rangle, \quad |1\rangle, \quad | \nearrow \rangle = \frac{1}{\sqrt{2}} (|0\rangle + |1\rangle), \quad | \nearrow \rangle = \frac{1}{\sqrt{2}} (|0\rangle + |1\rangle), \quad (18.22)$$

$$|r\rangle = \frac{1}{\sqrt{2}}\left(|0\rangle + i|1\rangle\right), \quad |l\rangle = \frac{1}{\sqrt{2}}\left(|0\rangle - i|1\rangle\right) \tag{18.23}$$

Since Alice and Bob each choose from among three bases, more trials are lost in reconciliation. But the greater symmetry of the six states on the Poincaré sphere simplifies the security analysis, and the increase in number of bases decreases Eve's probability of guessing the correct basis, lowering from $\frac{1}{2}$ to $\frac{1}{3}$, which means that she introduces a larger error rate when she tries to intercept bits while gaining lower-average information per trial.

18.7.3 Decoy State and SARG04 Protocols

The QKD protocols described above depend on the ability of one party to prepare and send either a single photon or a pair of entangled photons on demand. The problem, though, is that in reality, pure single-photon states are hard to produce. What is really done is that a coherent state from a laser is attenuated so that most pulses will contain a single photon. But coherent state pulses have Poisson distributed photon numbers: there is a small but finite probability that two or more photons will be present in the "single-photon" transmission. Similarly, if an entangled photon pair is needed, a laser is sent into a nonlinear crystal, and then a tiny fraction of the input photons will produce an output entangled pair. Most of the time, only one pair will be produced in a small time window, but the probability of producing two or more pairs remains nonzero.

This possibility of multiple-photon productions opens an opportunity for Eve, who can exploit it to evade security procedures. In a *photon-number-splitting (PNS) attack*, [15–20] Eve steals photons from multiphoton pulses to gain partial information about the key. By sending on the remainder of the pulse to Bob without tampering with it, she introduces minimal error into Bob's measurements. But she is free to make measurements on the excess photon that she kept, from which she can gain information for free: since Bob isn't expecting this excess photon, he makes no attempt to measure it and so does not see its loss as an error.

PNS attacks can be mitigated by using decoy states. The idea is to intersperse two sequences of pulses, with different mean photon numbers. One set of pulses is used to generate the key. The other is a set of decoy pulses used to enhance security. Eve doesn't know which pulses are decoys and which are key-generating. By comparing the pulse detection probability versus photon number, Eve's action appears as a change in the probability distribution of photon number in the pulses. By using decoy states, stronger pulses may be used, allowing the same security level to be maintained over longer distances.

The *SARG04 protocol* [21–23] is another QKD protocol, which is automatically resistant to PNS attacks. In this protocol, the classical bits are not encoded into a single state but rather are encoded into *pairs* of nonorthogonal states. One pair represents a bit value of 0, and a different pair represents 1. Instead of communicating a basis choice to Bob, Alice tells Bob which nonorthogonal pair the state was chosen from. But she does not tell Bob which state within the pair was used. Then, similar to the B92 protocol, Bob can deduce the correct bit if he detects a state perpendicular to one of the two states in the corresponding pair but can reach no conclusion if he detects a state that has any overlap with either member of the pair. The four states used could even be the same states as in the BB84 protocol, so that the only change necessary from BB84 is the classical communication. For example, one pair could be $| \uparrow \rangle$ and $| \nearrow \rangle$, while the other pair could be $| \rightarrow \rangle$ and $| \searrow \rangle$.

The SARG04 provides greater security against PNS attacks for the following reason. If Eve steals a photon from a two-photon pulse, she can only learn that it is in one of two non-orthogonal states, which does not determine the bit. This is in contrast to BB84, in which Eve gains complete information about the state if she can store it until after finding out the correct measurement basis from Alice and Bob's classical exchange. The SARG04 therefore provides greater security against PNS attacks and can be shown to have high security against one- and two-photon attacks. In order to obtain a definite bit value, Eve must steal two photons and perform unambiguous state discrimination (Sect. 9.6) on them. Thus, she needs to attack three-photon pulses, which will be very rare. As a result, Eve can obtain much less information per pulse, and she must work much harder to gain it.

18.7.4   Continuous Variables and Other Protocols

All of the QKD protocols discussed so far have involved discrete sets of states defined in terms of two-state variables such as polarization or spin. But continuous families of states labeled by variables such as position and momentum can also be used. Since single photons or entangled-photon pairs are not needed, these continuous-variable protocols have the advantage that they can be used with high-intensity beams, allowing a more rapid key-generation rate. For the same reason, cheaper and less sensitive detectors can be used. These continuous variable protocols use strong coherent state or squeezed state beams, with the information encoded into the X and P quadratures, as measured via homodyne detection (Sect. 22.6).

Eve can steal a portion of the beam and make measurements without noticeably diminishing the intensity. But security is maintained by the fact that if Eve measures one quadrature of the beam, she will increase the uncertainty in the other quadrature. This of course can be seen when Bob makes measurements of that second quadrature, and so again, Eve's interference will be revealed. Unfortunately, the security of continuous-variable protocols still tends in general to be lower than for discrete variable protocols. For more detailed discussions of these protocols, see the reviews [24, 25].

Another avenue for generalization is to discrete Hilbert spaces with dimensions higher than two, which allows a greater coding capacity and a higher degree of security. So, for example, instead of switching between two polarizations, the switching can be done between three or more different orbital angular momentum values. OAM values are used in many higher-dimensional QKD approaches; see, for example, [26–37], among others. Another possibility is time-bin QKD: the time of arrival of a particle or a pulse is used to encode the binary values [38]

18.8 Other Communication-Related Protocols

18.8.1 Quantum Bit Commitment

In the real world, multiple parties who do not necessarily trust each other still need to find a way to work together. It turns out that quantum mechanics can help with such situations. One type of protocol that does this is called **quantum bit commitment**.

Suppose Alice needs to send Bob some information. Focus on one bit of this information. She wants to make sure that Bob can't read the bit until she decides to allow it. But Bob wants to make sure that the bit is fully committed and that she can't change its value after sending it. Metaphorically, Alice is sending the bit inside a locked safe; once the safe is sent, the bit inside is committed, and she cannot change it. But Bob cannot open the safe to learn the bit value until Alice sends him the key to the safe.

One possible way to achieve this is the following. Suppose Alice wants to give an answer to a yes or no question. Bob doesn't want her to change her answer after the deadline for committing, but Alice doesn't want Bob to be able to access the answer until some later time that she controls. So Alice prepares a set of N qubits; for specificity, let's assume these are N photon polarizations. If the answer is yes, then she randomly prepares photons with equal probability in either $| \uparrow \rangle$ or $| \rightarrow \rangle$ states. If the answer is no, they are randomly prepared in $| \nearrow \rangle$ or $| \searrow \rangle$ states. She makes a note of the state prepared for each photon but takes care to make sure that the sequence of polarizations remains perfectly random within the given basis. In each basis, we can think of one state as representing 0 and one as representing 1. She sends the qubits to Bob, who puts them into storage until Eve chooses to reveal her answer. Currently, quantum memories to store qubits are not practical outside the lab, but some progress has been made in that direction (see references in Sect. 27.1), so it is realistic to believe that this procedure could be carried out in the foreseeable future.

If Bob measures the states early to try to prematurely read the answer, he doesn't know what basis to measure in. He will have to choose an arbitrary orthogonal basis for measurement. Suppose he chooses two orthogonal basis vectors $|\psi\rangle$ and $|\phi\rangle$. Regardless of what pair he picks or which answer Alice chose, he will find that his probabilities of measuring the 0 and 1 states are always equal,

$$p(0) = p(1) = \frac{1}{2}. \tag{18.24}$$

So he just gets random noise. This is true even if by chance Bob happens to choose the same basis that Alice used, because she is generating the two states in that basis randomly, with equal probabilities. So when Alice sends the qubits to Bob, she can also safely let him know that horizontal-vertical basis corresponds to yes and diagonal-antidiagonal corresponds to no without giving away the answer.

Once Alice is ready to unveil her answer, she sends that answer to Bob, along with her record of the sequence of states she sent. If Bob now measures in her chosen basis, he should get identical results to what Alice recorded. If she changed her mind (e.g., telling Bob "no" after committing the qubits to "yes"), then she will be telling Bob to measure in the wrong basis, and he will find a 50% error rate.

Early attempts at quantum bit commitment procedures, such as the one described above, had loopholes in them and have been proven to be insecure. In order to ensure security, it is necessary to take relativity into account and require that information cannot be communicated faster than light speed. Here, we briefly describe one procedure [39] that accounts for relativistic constraints.

The process begins with Bob sending Alice a set of N photons for each bit to be committed. These photons are randomly polarized in one of four directions, taken to be $| \uparrow \rangle$, $| \rightarrow \rangle$, $| \nearrow \rangle$, or $| \searrow \rangle$. Note that together these make up a pair of mutually unbiased bases. Bob does not reveal the polarizations to Alice. When Alice receives these photons, she makes a polarization measurement. The basis in which she measures is determined by whether the bit b that she wishes to commit is equal to 0 or 1. If her bit is $b = 0$, she measures in the $\{| \uparrow \rangle, | \rightarrow \rangle\}$ basis, while for a bit value of $b = 1$, she measures in the $\{| \nearrow \rangle, | \searrow \rangle\}$ basis.

Two distant labs are located at space-time points Q_0 and Q_1, where Alice has a pair of trusted agents (Fig. 18.5). She communicates her measurement results to both of them. Alice's agents then pass the information on to a corresponding pair of Bob's agents at the next desk from them in the same labs. Bob's agents compare the two sets of results to make sure they are identical and that they arrived at the expected, predetermined times. They then pass the data on to Bob. Bob compares

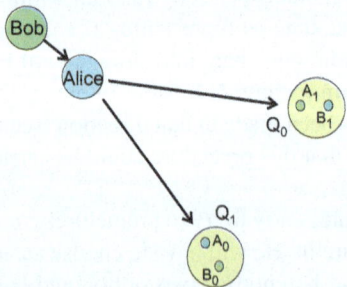

Fig. 18.5 The participants in a quantum bit commitment protocol. Bob sends a polarized photon to Alice, who imprints her bit onto it by making a measurement. Alice sends copies of her results to her agents A_0 and A_1, located at space-time points Q_0 and Q_1. Bob's agents B_0 and B_1 verify the arrival of the information at the appointed time, and when Alice's agents reveal the data to them, they then send it on to Bob who decodes the bit that was committed by Alice

the results to make sure they are consistent with the polarizations he sent, and if so, he can extract Alice's bits from the data.

When Alice measures in the same basis that a photon was prepared, she should obtain Bob's state correctly, but when she measures in the wrong basis, she should get a 50% error rate. So when results are communicated back to Bob, as long as N is large enough, he can tell what basis she used by the error rate, and thus he can determine her bit. But from her measurement results, Alice cannot determine what states Bob sent, since she doesn't know if she is using the correct basis or not. So if she later wants to change her bit value after she committed, she cannot do anything to alter the measurement data in a consistent manner.

This achieves the goal: Alice can't change her results after the appointed time; otherwise, communications would arrive at the agents' labs later than expected, but by making distances between Alice and the agents sufficiently large, Alice can guarantee that Bob can't obtain the information early.

This scheme was implemented experimentally in [40], with the agents separated by 20 km. This introduced a 30 μs delay, which was sufficient to ensure that the probability of cheating was below 5%. A variation on the same procedure was carried out in [41], with agents in labs in Singapore and Geneva, which are separated by a 9354 km straight-line distance through the earth, introducing a 15 ms time delay, sufficient to reduce the cheating probability to about 1 in 18 million.

18.8.2 Quantum Secret Sharing

Now suppose that there are three or more participants in the problem. Alice wants to communicate some secret information to the other parties, but she needs them to cooperate with each other and doesn't trust them to do so voluntarily. Or she may suspect that one of them is dishonest, but may not be certain which one. Or, possibly, she may want to make sure none of them can gain any information before the others, so that nobody will have an advantage in some competition. For example, the information might be about a financial stock, and she wants to make sure that nobody has a head start on the trading floor. **Quantum secret sharing** is about making sure that several participants can only act on information simultaneously by requiring them to cooperate in order for the information to be unlocked and to do so in such a way that other, unauthorized parties cannot intercept the information first.

Consider the case where Alice wants to share the secret information with two parties, Bob and Charlie. Then Alice can generate a random key k, consisting of a string of random binary digits, and then she can use it to encode the binary message m through binary (mod 2) addition. The result is an encrypted message $c = m \oplus k$. If the key k is truly random, then c is also completely random, so no information can be extracted from either k or c alone. So all Alice has to do is send k to Bob and c to Charlie. Neither of them alone can extract the message from the string of random digits that they receive, but if they cooperate they can reconstruct the message simply by adding their random strings together:

$$c \oplus k = (m \oplus k) \oplus k = m. \qquad (18.25)$$

The procedure above can be implemented classically. But just, as in any key distribution procedure, the problem is to prevent an eavesdropper from obtaining any information. This is where quantum mechanics again comes in.

One approach [42] makes use of the entangled three-particle **Greenberger-Horne-Zeilinger (GHZ) state** defined by

$$|\psi\rangle = \frac{1}{\sqrt{2}}\Big(|000\rangle + |111\rangle \Big), \qquad (18.26)$$

where, for example, $|0\rangle$ might represent horizontal polarization and $|1\rangle$ represents vertical. Each of the three participants has one particle from this state. Each of them randomly chooses to make a measurement in either the diagonal-antidiagonal basis or the circular basis, and then they announce publicly the basis they used but not the result of the measurement.

The GHZ state can be written in each of the other bases. So, for example, suppose that Alice and Bob both measured in the diagonal-antidiagonal basis:

$$|\psi\rangle = \frac{1}{2}\Big[\Big(|D\rangle_A |D\rangle_B + |A\rangle_A |A\rangle_B \Big)|D\rangle_C + \Big(|D\rangle_A |A\rangle_B + |A\rangle_A |D\rangle_B \Big)|A\rangle_C \Big]. \qquad (18.27)$$

So, from this expression, it is clear that if results of Alice's and Bob's measurements are known, then results of Charlie's measurement will be known, assuming he also used the same basis. Similar rewritings of the states can be made for other combinations of Alice and Bob measurement bases. It turns out that if Bob and Charlie both measured in the same basis, then if each of them knows Alice's outcomes (regardless of Alice's basis), they can reconstruct the other's results. Thus, once Alice reveals her results, Bob and Charlie will gain simultaneous knowledge of the full state. This will work 50% of the time, in the cases where Bob and Charlie are using the same basis. Since they have publicly announced their basis choices, all three participants know which trials are useful and which should be discarded.

A full table of possible outcomes can be constructed (Fig. 18.6). If Alice's and Bob's measurements are listed on the outside of the table, then Charlie's will be in the middle. If Charlie's results are listed on the side of the table instead of Bob's, then the same table will give Bob's results in the middle.

Fig. 18.6 The table of possible outcomes for a quantum secret sharing protocol. Alice and Bob's measurement results are on the top left, while corresponding results for Charlie are in the interior region of the table

		Alice			
		D	A	R	L
	D	D	A	L	R
Bob	A	A	D	R	L
	R	L	R	A	D
	L	R	L	D	A

Charlie

See [42] for more details on this protocol and for a discussion of security. The advantage of this approach is that it can be generalized for any number of participants. For example, if Alice wants to share a secret with three participants, then everyone would need to share a four-particle entangled state $|\psi\rangle = \frac{1}{\sqrt{2}}\Big(|0000\rangle + |1111\rangle\Big)$.

Problems

1. With the aid of the frequency chart in Chap. 8, decipher the following message written using a substitution cipher: κβθ θμθπαξλκ ξλτ κβθ δτηγθ τξλζθτ ξλ δυληθκ νλ κβθ μνεακ τφ κβθ μτψθμχ ρηξλκηδ δττλ.
2. Show that Eq. 18.10 can be written in the form of Eq. 18.11.
3. In Sect. 18.2, the quantum teleportation procedure was described assuming that Alice and Bob initially shared entangled qubits in the Bell state $|\Phi^+\rangle$.

    (a) Work out the teleportation procedure for the case when the two participants share an initial $|\Psi^+\rangle$ state, instead of $|\Phi^+\rangle$.
    (b) Design a state preparation stage analogous to Fig. 18.1a to prepare the $|\Psi^+\rangle$ from a $|00\rangle$ state.

4. Again imagine Alice wants to teleport the state $|\psi\rangle = \alpha|0\rangle + \beta|1\rangle$ to Bob, but now suppose they share the entangled state

$$|\Psi\rangle = \frac{1}{\sqrt{2}}\big(|000\rangle + |111\rangle\big),$$

where Alice is in possession of the first two qubits and Bob has the third. (This state is called a Greenberger-Horne-Zeilinger (GHZ) state).

   (a) Work out the state obtained by sending the state $|\psi\rangle|\Psi\rangle$ through the circuit below.

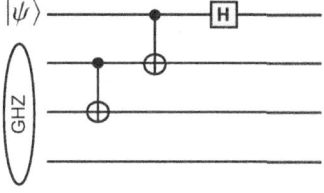

   (b) Now Alice makes a measurement on her qubits. Make a list of her possible results, and for each of them, determine what operation she must instruct Bob to carry out in order to reproduce the original state.
5. Consider the entangled three-qubit pure GHZ state, $|\Psi\rangle = \frac{1}{\sqrt{2}}(|000\rangle + |111\rangle)$.

(a) Take the partial trace of the density matrix with respect to any one of the qubits. What do you get? Verify that the result is a mixed, but non-entangled state. (This is one odd property of the GHZ state: the complete set of three qubits is entangled, but there is no pairwise entanglement between any two of them.)

(b) However, measurements on one qubit can cause the remaining qubits to become entangled. Suppose that the third qubit is measured. Suppose further that the measurement is done in the diagonal basis $\frac{1}{\sqrt{2}}(|0\rangle \pm |1\rangle)$, rather than in the computational $|0\rangle, |1\rangle$ basis. Now the result will be maximally entangled: show that the remaining two qubits will form Bell states. Which Bell state will appear as a result of which measurement result?

6. Imagine Alice and Bob attempting to generate a key using the BB84 protocol, but with a pair of bases that are not mutually unbiased:

Basis 1: $|0\rangle$ and $|1\rangle$

Basis 2: $|\alpha\rangle = N\Big((1+\epsilon)|0\rangle + (1-\epsilon)|1\rangle\Big),$

$|\beta\rangle = N\Big((1-\epsilon)|0\rangle - (1+\epsilon)|1\rangle\Big),$ where $(N = 1/\sqrt{1+\epsilon^2})$.

(a) Find the mean error rate in the absence of eavesdropping when Alice and Bob use the same basis. What is the mean error rate when they are using different bases?

(b) Alice and Bob discard the trials on which they used different bases. On the remaining trials, describe a strategy that Eve can use to eavesdrop that will reduce the error rate she causes below the 25% rate that would occur when the bases are mutually unbiased. What is the minimum error rate she introduces now?

7. Text on computer systems is usually coded using the ASCII (American Standard Code for Information Interchange) scheme, in which each symbol (letters, numbers, punctuation, etc.) is represented by a codeword containing seven binary symbols. (Note that blank spaces count as symbols.)

(a) Alice wants to privately send Bob the message "Party at 8, don't tell Eve." How many bits long is this message when encoded into ASCII?

(b) Suppose Alice sends this message using the BB84 protocol. Half of the transmitted bits are used for the security check, and if no eavesdropper is found, then the remaining half are used for the message. On average, what is the total number of bits that Alice will need to send Bob?

8. In four dimensions, it is possible to find up to five mutually unbiased bases. Consider the basis consisting of the unit vectors $(1, 0, 0, 0)$, $(0, 1, 0, 0)$, $(0, 0, 1, 0)$, and $(0, 0, 0, 1)$. Find three other (possibly complex) sets of basis vectors that are mutually unbiased with respect to this basis and to each other.

9. Draw the circuit diagram that will implement the dense coding protocol of Sect. 18.3.

10. Verify the statement that Bob will always get the two equal probabilities of Eq. 18.24 regardless of the choice of orthogonal basis vectors $|\psi\rangle$ and $|\phi\rangle$ he chooses in the non-relativistic bit commitment protocol that was described in Sect. 18.8.1.

11. (RSA example) From a given verbal message, we can construct a corresponding numerical message by replacing $a \rightarrow 1$, $b \rightarrow 2$, $c \rightarrow 3$, etc. So the word "qubit" is now represented by the numerical string $x = x_1 x_2 x_3 x_4 x_5 = 17, 21, 2, 9, 20$. We wish to encode and decode this message using an RSA encryption with values $p = 5$, $q = 7$, and $c = 5$.

(a) What are N, r, and d?

(b) Encrypt $x_1, x_2, \ldots, x_5$ using $y_j = x_j^c (mod\ N)$.

(c) Now decrypt using $x_j = y_j^d (mod\ N)$, and see that you get back the original message.

References

1. C.H. Bennet, G. Brassard, C. Crèpeau, R. Jozsa, W. Wooters, Phys. Rev. Lett. **70**, 1895 (1993)
2. W. Diffie, M. Hellman, IEEE Trans. Inf. Theory **IT-22**, 644 (1976)
3. R. Merkle, Commun. ACM **21**, 294 (1978)
4. R.L. Rivest, A. Shamir, L.M. Adleman, Commun. ACM **21**, 120 (1978)
5. M.A. Nielsen, I.L. Chuang, *Quantum Computation and Quantum Information: 10th Anniversary Edition* (Cambridge University Press, Cambridge, 2011)
6. N.D. Mermin, *Quantum Computer Science: An Introduction* (Cambridge University Press, Cambridge, 2007)
7. N. Koblitz, *A Course in Number Theory and Cryptography*, 2nd edn. (Springer, Berlin, 1994)
8. S. LLoyd, arXiv:1307.0380 [quant-ph] (2013)
9. D.J. Lum, M.S. Allman, T. Gerrits, C. Lupo, V.B. Verma, S. Lloyd, S. Wo. Nam, J.C. Howell, Phys. Rev. A **94**, 022315 (2016)
10. R.E. Klima, N.P. Sigmon, *Cryptology: Classical and Modern, with Maplets* (CRC Press, Boca Raton, 2012)
11. A.K. Ekert, Phys. Rev. Lett. **67**, 661 (1991)
12. C.H. Bennett, Phys. Rev. Lett. **68**, 3121 (1992)
13. D. Bruß, Phys. Rev. Lett. **81**, 3018 (1998)
14. D.G. Enzer, P.G. Hadley, R.J. Hughes, C.G. Peterson, P.G. Kwiat, New J. Phys. **4**, 45 (2002)
15. W.Y. Hwang, Phys. Rev. Lett. **91**, 057901 (2003)
16. X.B. Wang, quant-ph/0410075 (2004)
17. X.B. Wang, quant-ph/0411047 (2004)
18. H.K. Lo, X. Ma, K. Chen, Phys. Rev. Lett. **94**, 230504 (2005)
19. X. Ma, quant-ph/0503057 (2004)
20. D. Gottesman, H.-K. Lo, N. Lütkenhaus, J. Preskill, Quant. Inform. Comp. **4**, 325 (2004)
21. V. Scarani, A. Acín, G. Ribordy, N. Gisin, Phys. Rev. Lett. **92**, 057901 (2004)
22. C. Brianciard, N. Gisin, B. Kraus, V. Scarani, Phys. Rev. A **72**, 032301 (2005)
23. N. Lütkenhaus, Phys. Rev. A **61**, 052304 (2000)
24. S.L. Braunstein, P. van Loock, Rev. Mod. Phys. **77**, 513 (2005)
25. C. Weedbrook, S. Pirandola, R. Garcia-Petrón, N.J. Cerf, T.C. Ralph, J.H. Shapiro, S. Lloyd, Rev. Mod. Phys. **84**, 621 (2012)
26. A. Vaziri, G. Weihs, A. Zeilinger, Phys. Rev. Lett. **89**, 240401 (2002)
27. S. Groblacher, T. Jennewein, A. Vaziri, G. Weihs, A. Zeilinger, New J. Phys. **8**, 1 (2006)

28. D.S. Simon, N. Lawrence, J. Trevino, L. dal Negro, A.V. Sergienko, Phys. Rev. A **87** 032312 (2013)
29. D.S. Simon, A.V. Sergienko, New J. Phys. **16**, 063052 (2014)
30. M. Krenn, R. Fickler, M. Fink, J. Handsteiner, M. Malik, T. Scheidl, R. Ursin, A. Zeilinger, New J. Phys. **16**, 113028 (2014)
31. M. Mirhosseini, O.S. Magaña-Loaiza, M.N. O'Sullivan, B. Rodenburg, M. Malik, M.P.J. Lavery, M.J. Padgett, D.J. Gauthier, R.W. Boyd, New J. Phys. **17**, 033033 (2015)
32. F.M. Spedalieri, Opt. Commun. **260**, 340 (2006)
33. L. Aolita, S.P. Walborn, Phys. Rev. Lett. **98** 100501 (2007)
34. M. Stutz, S. Groblacher, T. Jennewein, A. Zeilinger, Appl. Phys. Lett. **90**, 261114 (2007)
35. C.E.R. Souza, C.V.S. Borges, A.Z. Khoury, J.A.O. Huguenin, L. Aolita, S.P. Walborn, Phys. Rev. A **77**, 032345 (2008)
36. J.T. Barreiro, T.-C. Wei, P.G. Kwiat, Nat. Phys. **4**, 282 (2008)
37. W. Boyd, A. Jha, M. Malik, C. O'Sullivan, B. Rodenburg, D.J. Gauthier, in *Advances in Photonics of Quantum Computing, Memory, and Communication IV*, ed. by Z. Ul Hasan, P.R. Hemmer, H. Lee, C. M. Santori, Proceedings of the SPIE, vol. 7948 (2011), p. 79480L
38. I. Marcikic, H. De Riedmatten, W. Tittel, V. Scarani, H. Zbinden, N. Gisin, Phys. Rev. A **66** 062308 (2002)
39. A. Kent, Phys. Rev. Lett. **109**, 130501 (2012)
40. Y. Liu, et al., arXiv:quant-ph/1306.4413 (2013)
41. T. Lunghi, et al., Phys. Rev. Lett. **111**, 180504 (2013)
42. M. Hillery, V. Bužek, A. Berthiaumme, arXiv:quant-ph/9806063 (1998)

Chapter 19
Quantum-Enhanced Metrology and Sensing

One area in which quantum-based technology is already having a significant impact is in the area of metrology and sensing. **Metrology** is the science of high-precision measurement. As discussed in the following sections, quantum superposition, entanglement, and squeezing allow unprecedented resolution and sensitivity in a broad range of measurements. These improvements are in part due to the reduction of disruptive effects such as turbulence and dispersion and in part due to the more rapid phase oscillations of entangled states. Applications that are discussed include sensing of gravitational fields, quantum gyroscopes, magnetoencephalography, and atomic clocks.

Some of the applications described in this chapter are still in development and may be difficult to translate into real-world, commercially available devices. Often this is because of the fragility of the quantum states involved. However, other quantum-based applications, such as the quantum gravimeters described in Sect. 19.4, are already available as commercial devices. In addition, even some of the quantum applications that may never move out of the lab have inspired classical applications that probably never would have been discovered if the quantum-based version had not been found first; an example of such a quantum-inspired application is ghost imaging (see Sect. 20.1).

In the first few sections of this chapter, we discuss general considerations related to quantum sensing. Much of this discussion will revolve around interferometric phase measurements, since many sensing technologies ultimately rely on phase measurements. Then, in the last few sections, we discuss several examples of measurement and sensing devices that are already in scientific or medical use. Additional examples, such as quantum radar, quantum-enhanced microscopes, and superconducting sensors, are discussed in later chapters.

© The Author(s), under exclusive license to Springer Nature Switzerland AG 2025
D. S. Simon, *Introduction to Quantum Science and Technology*, Undergraduate
Texts in Physics, https://doi.org/10.1007/978-3-031-81315-3_19

19.1   Phase Measurement: Heisenberg and Standard Quantum Limits

Phase measurement, the determination of phase differences between particles following different trajectories, is fundamental to many areas of physics. This is the object of interferometry, and an enormous range of applications follow from interferometric measurements. To name just a few applications, measurements of star sizes (Chap. 10), structure of complex molecules or the lattice spacing in crystals, or magnetic flux passing through a superconducting ring (Chap. 7 and Sect. 19.5) all depend on precision phase measurements. So an important question to ask is what are the fundamental limits imposed on phase measurements by quantum mechanics? We focus here on phase measurement limits in optical interferometry [1, 2] and the ensuing fundamental bounds on the **phase sensitivity**, i.e., the minimum uncertainty $\Delta\phi$ in the measured phase value.

Quantum mechanically, we should expect the photon number operator $\hat{N}$ and the phase operator $\hat{\phi}$ to be a canonically conjugate operator pair, with their commutator leading to a Heisenberg uncertainty relation that would provide an ultimate physical limit on phase measurements. But a complication in studying quantum phase is that, unlike most physically measurable quantities, phase is not represented by a Hermitian operator $\hat{\phi}$ that gives the correct canonical commutation relations. The root of the trouble is that ϕ is indistinguishable from $\phi+2\pi$; in other words, it is only $e^{i\phi}$ that can be measured, not ϕ. This problem has been widely discussed [3–15], and there are several possible resolutions to it. (See [16] for a review of the history of the search for a quantum phase operator up to the early 1990s.) Here we discuss just one method, which makes use of the non-Hermitian **Susskind-Glogower (SG)** operator [3]

$$\hat{G} = \widehat{e^{i\phi}} = \sum_{n=0}^{\infty} |n\rangle\langle n+1| = (\hat{N}+1)^{-1/2}\hat{a}, \tag{19.1}$$

where, $\hat{a}$, $\hat{N}$, and $|n\rangle$ are, respectively, the photon annihilation operator, the photon number operator, and the Fock state of photon number n. The eigenstates of $\hat{G}$ are

$$|\phi\rangle = \sum_{n=0}^{\infty} e^{in\phi}|n\rangle. \tag{19.2}$$

The corresponding eigenvalue relation is given by

$$\hat{G}|\phi\rangle = e^{i\phi}|\phi\rangle, \tag{19.3}$$

where ϕ is the phase of the state. To get a physical observable, we need to somehow construct a phase-dependent Hermitian operator from $\hat{G}$. One way to do that is to add $\hat{G}$ to its Hermitian conjugate to form the Hermitian operator

$$\hat{C} = \frac{1}{2}\left(\hat{G} + \hat{G}^{\dagger}\right) = \sum_{n=0}^{\infty}\left(|n\rangle\langle n+1| + |n+1\rangle\langle n|\right). \tag{19.4}$$

(A special case of this operator was used in Sect. 10.1; the example given below is just a slight generalization of the argument in that chapter.) The expectation value $\langle\hat{C}\rangle$ measures the rate of single-photon transitions between adjacent energy levels. Now this phase-dependent observable can be measured while acting on some quantum state and from the measured value the state's phase can be extracted:

$$\langle\psi|\hat{C}|\psi\rangle = \frac{1}{2}\left(e^{i\phi} + e^{-i\phi}\right) = \cos\phi. \tag{19.5}$$

Example 19.1.1 Consider an interferometer, with optical input at one single port. Various types of N-photon input could be considered, such as a Fock state of fixed photon number N or N separate measurements on a stream of single-photon states. In each of these cases, the phase sensitivity $\Delta\phi$ scales for large N as $\frac{1}{\sqrt{N}}$.

To see this, consider N measurements of consecutive single-photon inputs. Let the photons enter port A of the Mach-Zehnder interferometer in Fig. 19.1a. Each photon can traverse the lower path, reflecting off mirror M_2, or it can traverse along the upper path, reflecting off M_1 and gaining an extra phase shift of ϕ. Any phase shifts due to reflection at the mirrors can be absorbed into the definition of ϕ, and so we don't need to consider them explicitly. Denote the state leaving the first beam splitter along the lower path by $|0\rangle$ and the state leaving along the upper path by $|1\rangle$. Then the state arriving at the second beam splitter is the superposition state

$$|\psi\rangle = \frac{1}{\sqrt{2}}\left(|0\rangle + e^{i\phi}|1\rangle\right). \tag{19.6}$$

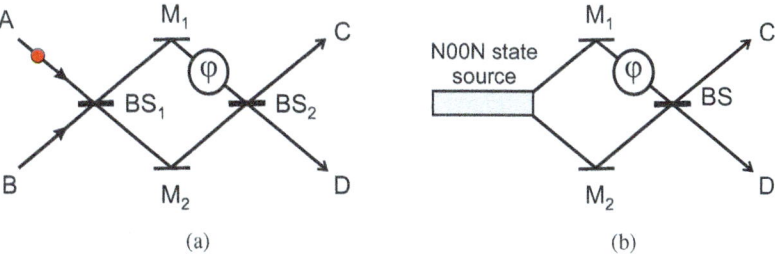

(a)   (b)

Fig. 19.1 (a) Input port A of a Mach-Zehnder interferometer has two possible ways of reaching the final beam splitter, BS_2. The amplitude for traversing the upper path gains an extra phase shift of ϕ relative to the lower path. The second beam splitter mixes the two amplitudes, making it impossible to determine which path was followed. The erasure of the which-path information leads to interference at the output ports, from which the phase ϕ can be extracted. (b) Using an entangled NOON state source, the phase sensitivity can be enhanced by a factor of $\frac{1}{\sqrt{n}}$

Since only one photon at a time is entering the system, the photon number n never exceeds 1. So, $\hat{C}$ can be truncated to $\hat{C} = |0\rangle\langle 1| + |1\rangle\langle 0|$. When wedged between two superposition states, $\hat{C}$ picks out the interference or cross terms; it links the $|0\rangle$ component of one state to the $|1\rangle$ component of the other. In fact, using a computational basis of the form

$$|0\rangle = \begin{pmatrix} 1 \\ 0 \end{pmatrix}, \qquad |1\rangle = \begin{pmatrix} 0 \\ 1 \end{pmatrix}, \tag{19.7}$$

$\hat{C}$ becomes the Pauli spin matrix: $\hat{C} = \sigma_x = \begin{pmatrix} 0 & 1 \\ 1 & 0 \end{pmatrix}$, and the state of Eq. 19.6 becomes $|\psi\rangle = \frac{1}{\sqrt{2}} \begin{pmatrix} 1 \\ e^{i\phi} \end{pmatrix}$.

Taking the expectation value in the state of Eq. 19.6, we find

$$\langle \hat{C} \rangle = \langle \psi | \hat{C} | \psi \rangle = \cos\phi, \tag{19.8}$$

$$\langle \hat{C}^2 \rangle = \langle \psi | \hat{C}^2 | \psi \rangle = 1. \tag{19.9}$$

Therefore, the uncertainty in the measurement of $\hat{C}$ is

$$\Delta C = \sqrt{\langle \hat{C}^2 \rangle - \langle \hat{C} \rangle^2} = \sqrt{(1 - \cos^2\phi)} = |\sin\phi|. \tag{19.10}$$

Repeating the experiment N times on an identical ensemble of states, standard statistical theory tells us that the uncertainty should be reduced by a factor of $\sqrt{N}$, so that $\Delta C = \frac{|\sin\phi|}{\sqrt{N}}$. Our best estimate of ϕ is then the value $\bar{\phi} = \cos^{-1}\langle \hat{C} \rangle$, and the uncertainty in this phase estimate is given by

$$\Delta\phi = \frac{\Delta C}{|\frac{d\langle \hat{C} \rangle}{d\phi}|} = \frac{|\sin\phi|}{\sqrt{N}|\sin\phi|} = \frac{1}{\sqrt{N}}. \qquad \blacksquare \tag{19.11}$$

The last equation is the **standard quantum limit** or **shot noise limit**. In the case of single-photon states being measured N times, this uncertainty can be viewed as being due to photonic shot noise or "sorting noise," the Poisson-distributed random fluctuations of the photon number in each arm of the apparatus due to the random choice made by the beam splitter as to which way to send each photon. If the experimenter instead uses a single coherent state pulse with mean photon number $\langle \hat{N} \rangle = N$, the result for the uncertainty is the same, since random fluctuations of photon number in the coherent state beam have the same Poisson statistics as the sorting noise.

Example 19.1.2 Instead of a sequence of N independent photons, the input state could be an entangled state, so that the entire N-photon set is described by a single joint quantum state. Imagine an entangled N-photon state, with the photons fed into two branches of an apparatus. At any horizontal position in apparatus, denote by $|n_1\rangle|n_2\rangle = |n_1, n_2\rangle$ the state with n_1 photons in the top branch and n_2 photons in the lower branch. Consider, for example, a $N00N$ state (Sect. 9.4), $\frac{1}{\sqrt{2}}(|N, 0\rangle + |0, N\rangle)$ as input [17]. The state reaching the final beam splitter is $|\psi_N\rangle = \frac{1}{\sqrt{2}}\left(|0, N\rangle + e^{iN\phi}|N, 0\rangle\right)$. Because of the entanglement, the phase shifts from the N photons act collectively, giving a total phase shift $N\phi$ to one component of the entangled state. In place of $\hat{C}$ above, an appropriate measurement operator to extract this total phase shift might now be $\hat{B}_N = |0, N\rangle\langle N, 0| + |N, 0\rangle\langle 0, N|$. It is readily checked that the expectation values and uncertainties are given by

$$\langle\hat{B}_N\rangle = \langle\psi_N|\hat{B}_N|\psi_N\rangle = \cos N\phi, \tag{19.12}$$

$$\langle\hat{B}_N^2\rangle = \langle\psi_N|\hat{B}_N^2|\psi_N\rangle = 1 \tag{19.13}$$

$$\Delta B_N = \sin N\phi \tag{19.14}$$

$$\Delta\phi = \frac{1}{N}. \tag{19.15}$$

This explicitly demonstrates that this entangled system beats the standard quantum limit. It in fact saturates the **Heisenberg limit**, the fundamental physical bound imposed by quantum mechanics [18].   ■

This example clearly shows that entanglement can give a $\frac{1}{\sqrt{N}}$ advantage in phase sensitivity over unentangled N-photon states. The increase in oscillation frequency, signaled by the factor of N inside the cosine in Eq. 19.12, indicates super-resolution as well. Experimental demonstration of NOON states to carry out phase microscopy near the Heisenberg limit was carried out in [19]. Although the improved phase measurement ability with $N00N$ states has been demonstrated in the lab, $N00N$ states are difficult to create for large N and so to a large extent are not yet practical in real-life settings outside the lab.

In some situations, such as imaging, the optimal phase uncertainty may be obtained via information theoretic means through the Cramer-Rao bound [20, 21], by maximizing the mutual information [22, 23] (Sect. 12.2.2) or by using Bayesian analysis [24, 25] to provide strategies for optimizing the estimation strategy.

Other entangled N-photon states [26, 27] may be used instead of $N00N$ states. Some of these states give slightly better sensitivity or are slightly more robust to noise, but overall they give qualitatively similar results to those derived from NOON states.

Although entangled states can be used to overcome the standard quantum limit, similar results can also be achieved with non-entangled states. For example, squeezed states have been shown to be capable of achieving $\frac{1}{N^{3/4}}$ phase sensitivity [28]. However, up to this point, only entangled states seem to be capable of fully reaching the Heisenberg bound.

19.2   Dispersion Cancelation and Related Effects

Suppose you have a microscope, a telescope, or some other optical device whose purpose is to transfer light from an object to a viewing device and then to form an accurate image. Along the way from the object location to the image location, there are many possible sources of distortion that can affect the propagating light, resulting in a distortion of the image and a loss of resolution. Similar effects can also reduce the visibility of interference-based optical devices. Examples of such distortions include turbulence from motion of the fluid medium that light is propagating through and aberration from the lenses in the optical system itself. One practical application of photon entanglement is the mitigation or even the total elimination of many of these disturbing effects, leading to enhanced resolution and visibility. Removing or reducing phase distortions is essential not only to high-resolution quantum measurements but also to transmission of high-quality quantum states for communication and information processing.

Appropriately chosen entangled states have been shown to lead to turbulence mitigation (Sect. 20.1.3), aberration cancelation [29–35], and reduction of chromatic dispersion and polarization mode dispersion. Here we will focus on chromatic dispersion [36–43].

19.2.1   Group and Phase Velocities and Dispersion

In introductory courses on waves or electromagnetism, the focus is usually on monochromatic plane waves. These are the simplest possible waves in two senses: their wavefronts are flat, and the wavefronts are equally spaced. Recall that a wavefront is a surface of constant phase. For example, a wavefront can be a surface made up of points that are all at the crest of a wave. For a plane wave, the wavefronts are all flat planes, extending off to infinity in the two transverse directions (perpendicular to the wave propagation). For a monochromatic (single frequency) wave, these wavefronts are also spaced equal distances apart, and they repeat forever along the propagation direction. For a monochromatic plane wave, we can assign a unique speed to the wave. This is the speed that all of the wavefronts are moving through space. For an electromagnetic wave, this unique speed will be c in vacuum or $v = \frac{c}{n}$ in a medium of refractive index n.

As convenient as they are, monochromatic plane waves don't exist in real life, except as approximations over small regions of space. The plane wave requirement that the wavefronts extend off to infinity in the transverse directions would require an infinite amount of energy to be carried by the wave. The monochromaticity requires that the wave repeats forever in both directions along the propagation axis, which is also unrealistic, because it requires the wave to have existed for all time, into the infinite past and future. In real life, electromagnetic waves may have life spans that could be billions of years long, as they travel through interstellar space, but those life spans do not extend into the infinite past: the photons making up the wave have to have been emitted from a light source at some point in time. Similarly, most or all of those photons will at some point in the future be absorbed by matter. So real light waves don't exist as eternally unchanging plane waves. Rather, they are **wave pulses** or **wave packets** of finite duration. Fourier analysis tells us that any wave packet that has a finite life span must also contain a finite spread of frequencies: it cannot be monochromatic.

Here, we will continue to approximate light as a sequence of planar wavefronts for simplicity. By focusing attention on light sufficiently close to some optical axis, this will be a sufficiently good approximation for our current purposes. But we will remove the monochromatic assumption and allow the light to be formed from a range of frequencies $\Delta\omega$ spread about some central frequency ω_0.

Once we start considering wave packets, it becomes necessary to distinguish between two different definitions of wave velocity. The **phase velocity** is the ratio of frequency to wavenumber,

$$v_p = \frac{\omega}{k}. \tag{19.16}$$

v_p is the speed at which individual wavefronts propagate. On the other hand, **group velocity**,

$$v_g = \frac{d\omega}{dk}, \tag{19.17}$$

provides the speed at which the wavepacket as a whole moves through space. For a plane wave in vacuum these two velocities are equal and independent of the optical frequency. But inside a transparent material, the refractive index can depend on frequency, causing some of the frequency components to travel faster than others. As a result, group and phase velocities can differ, and they are generally frequency dependent. This phenomenon is known as **dispersion**. Notice that if the phase velocity is frequency dependent, then some parts of the wave packet will move faster than others, and as a result, the *shape* of the wavepacket will be changing as the light pulse propagates. In general, the pulse *duration* Δt will also increase or broaden over time.

So now, imagine that the light traverses a dispersive material of thickness L. The relation between photon wavenumber k and frequency ω is known as a **dispersion relation**. In a nondispersive material such as vacuum, the dispersion relation is linear,

$$k(\omega) = \frac{n\omega}{c} = \frac{\omega}{v}, \tag{19.18}$$

where the refractive index n is a frequency-independent constant. But in a dispersive medium, the refractive index is frequency dependent, $n(\omega)$, so that k becomes a nonlinear function of ω.

To analyze a nonlinear dispersion relation, the first step is usually to expand $k(\omega)$ as a Taylor series about the mean frequency of the wavepacket,

$$k(\omega) = \frac{n(\omega)\omega}{c} = k_0 + \alpha(\omega - \omega_0) + \beta(\omega - \omega_0)^2 + \dots, \qquad (19.19)$$

where $k_0 = \frac{n(\omega_0)}{c}\omega_0$ is the wavenumber at the mean frequency. In most cases of interest, the frequency range $\Delta\omega$ is small enough that the terms in this expansion are negligible beyond the first few.

The phase shift $\phi(\omega)$ acquired by the wavepacket as it traverses the medium may be written as

$$\phi(\omega) = k(\omega)L = \phi_0 + c_1(\omega - \omega_0) + c_2(\omega - \omega_0)^2 + c_3(\omega - \omega_0)^3 + \cdots \qquad (19.20)$$

Here, $\phi_0 = k_0 L$ is the phase shift acquired by photons at mean frequency ω_0. The **group delay** term $c_1 = \alpha L$ represents a time delay for the pulse as a whole, causing an overall translation of the pulse. $c_2 = \beta L$ represents the **group velocity dispersion (GVD)**, which causes broadening of wave packets and of any interference peaks derived from them (Fig. 19.2). Third-order and higher-order terms introduce skewing and other types of distortions; we will not consider these higher-order terms here.

The group velocity can now be written as a function of frequency,

$$v_g(\omega) = \frac{d\omega}{dk} = [\alpha + 2\beta(\omega - \omega_0) + \dots]^{-1} \qquad (19.21)$$

$$\approx \frac{1}{\alpha} - \frac{2\beta(\omega - \omega_0)}{\alpha^2}, \qquad (19.22)$$

where a binomial approximation has been applied in the last line. The velocity difference between frequencies separated by the full bandwidth $\Delta\omega$ of the pulse is given by

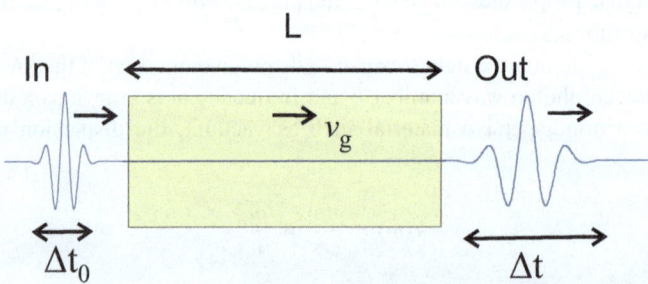

Fig. 19.2 A wave packet passing through a dispersive material is broadened by the $c_2 = \beta L$ group velocity dispersion term. In a communication system, information is carried by a sequence of such pulses. If the broadening is sufficiently large, consecutive pulses will begin to blend into each other, limiting the rate of information transfer

$$\Delta v_g = |v_g(\omega + \Delta\omega) - v_g(\omega)| = \frac{2\beta}{\alpha^2}\Delta\omega. \tag{19.23}$$

The time to traverse distance L is $t = \frac{L}{v_g}$, so if Δt_0 and $\Delta t = \Delta t_0 + \tau$ are the initial and final widths of the pulses, then the temporal broadening of the pulse is given by

$$\tau = |\Delta t - \Delta t_0| = \left| \frac{L}{v_g} - \frac{L}{v_g + \Delta v_g} \right| = \frac{L}{v_g}\left| 1 - \frac{1}{1 + \frac{\Delta v_g}{v_g}} \right| \tag{19.24}$$

$$\approx \frac{L\Delta v_g}{v_g^2} \tag{19.25}$$

$$\approx 2\beta L\Delta\omega, \tag{19.26}$$

which shows that the amount of broadening is proportional to the second-order dispersion coefficient β.

This temporal broadening of pulses limits how many information bearing pulses can be sent through a system per time. If the separation between pulses is T, then the pulses begin to overlap when $\tau \sim T$. This limits the maximum information transfer rate through a communication system or the maximum computational rate in an information processing system.

19.2.2 Dispersion Cancelation

Dispersion cancelation was first demonstrated in quantum optics in the early 1990s using frequency-anticorrelation of entangled photon pairs produced in SPDC to cancel even-order dispersion terms, such as the GVD term β. Even-order cancelation is observed [38] when dispersive material is added inside the Hong-Ou-Mandel (HOM) interferometer (see Sect. 10.6). As the time delay in one arm of the interferometer is varied, an interference pattern occurs in the measured coincidence rate; this pattern is contained inside a triangularly shaped envelope (the HOM dip). The resulting interferogram depends only on the odd-order dispersion coefficients of the material; the effects of the even-order terms cancel out. In particular, the HOM dip is not broadened by the dispersion, demonstrating insensitivity to the second-order group coefficient, c_2 or β. A separate, nonlocal type of dispersion effect occurs when entangled photons are sent through opposite-sign dispersive materials [36].

Dispersion cancelation has found a number applications, including high-resolution timing measurements [44], improved clock synchronization [45], and enhanced polarization mode dispersion measurements [46–48]. The effect appears naturally in quantum optical coherence tomography (QOCT) [49–51], which has applications in biological imaging; QOCT will be discussed in Sect. 20.3.

It has also been shown [52] that a quantum interferometer can be constructed, which separates the effects of even-order and odd-order dispersion terms into different portions of an interferogram in a way that allows the different terms to be measured or manipulated separately. One part of the interferogram exhibits even-order cancelation, while other parts exhibit odd-order cancelation.

The original forms of dispersion cancelation made use of entangled photons and so were inherently quantum effects. It has been shown more recently [53–57] that many aspects of the effect may be mimicked using classically correlated light.

19.2.3 Steinberg-Kwiat-Chiao Dispersion Cancelation

The first demonstration of dispersion cancelation used an apparatus such as that shown in Fig. 19.3. An SPDC source sends entangled photon pairs into an HOM interferometer, and the coincidence counts at the output are measured. One branch of the interferometer contains a dispersive element of length L before the beam-splitter, while dispersion is negligible in the other branch. The dispersion relation is taken to be of the form given in Eq. 19.19. In the lower branch, the mirror is moveable. We denote by δl the distance the mirror has been moved away from the balanced position (the position in which the that two branches have equal path lengths).

The state of the light emitted from the crystal is given by

$$|\Psi\rangle = \int d\omega \; \Phi(\omega)|\omega_0 + \omega, \omega_0 - \omega\rangle, \tag{19.27}$$

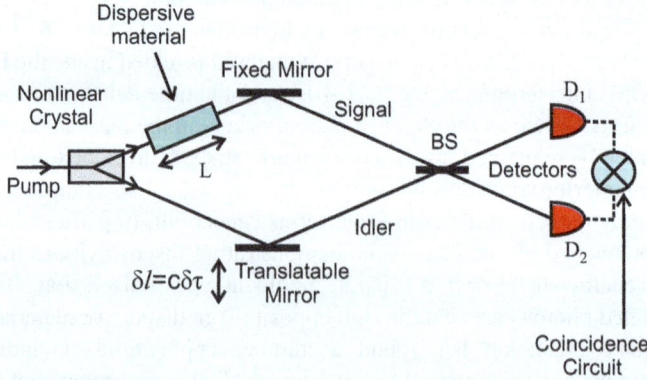

Fig. 19.3 HOM interferometer with dispersive material in one arm. The coincidence rate in the interferogram constructed as the lower mirror moves will exhibit even-order dispersion cancelation. Solid lines represent light paths, while dashed lines are electrical connections

where the first and second arguments of the ket on the right correspond to signal and idler modes, respectively. (Note that ω here is not the frequency but is the frequency *difference* from the central value ω_0.) Equation 19.27 describes the finite spread of the wave pulse around the central frequency. The spectral function $\Phi(\omega)$ gives the amplitude for the SPDC source to produce each frequency. It has a maximum at $\omega = 0$ and then decays as the frequency difference $|\omega|$ increases. We don't need the precise form of $\Phi(\omega)$ here. We only need to know that it is symmetric about the central frequency, $\Phi(-\omega) = \Phi(\omega)$.

If the signal and idler annihilation operators are denoted by $\hat{a}_s(\omega)$ and $\hat{a}_i(\omega)$, then after the beam splitter and propagation phase shifts, annihilation operators at detectors 1 and 2 are given by

$$\hat{a}_1(\omega_1) = \frac{1}{\sqrt{2}} \left(i\hat{a}_s(\omega_1)e^{ik(\omega_1)L} + \hat{a}_i(\omega_1)e^{i\omega_1\delta l/c} \right) \tag{19.28}$$

$$\hat{a}_2(\omega_2) = \frac{1}{\sqrt{2}} \left(i\hat{a}_i(\omega_2)e^{i\omega_2\delta l/c} + \hat{a}_s(\omega_2)e^{ik(\omega_2)L} \right). \tag{19.29}$$

Terms involving $e^{ik(\omega)L}$ account for the extra phase difference one photon gains inside the dispersive material, while terms involving $e^{i\omega\delta l/c}$ involve the extra path length traveled by the other photon. (Any common phase differences that are the same for both photons have been dropped, since they have no effect on the outcome.)

The positive- and negative-frequency components of the field (Sect. 5.3) at the detector j ($j = 1, 2$) may also be written in terms of the creation and annihilation operators,

$$\hat{E}_j^+(t_j) = \int d\omega_j \hat{a}_j(\omega_j)e^{-i\omega_j t_j} \tag{19.30}$$

$$\hat{E}_j^-(t_j) = \int d\omega_j \hat{a}_j^\dagger(\omega_j)e^{i\omega_j t_j} = \left(\hat{E}_j^+(t_j) \right)^\dagger. \tag{19.31}$$

The coincidence rate can be computed from

$$R_c = \int_{-T/2}^{T/2} dt_1 dt_2 \langle\Psi|\hat{E}_1^-(t_1)\hat{E}_2^-(t_2)\hat{E}_1^+(t_1)\hat{E}_2^+(t_2)|\Psi\rangle, \tag{19.32}$$

where T is the integration time of the detector. The integrand is of the form $\langle\phi(t_1, t_2)|\phi(t_1, t_2)\rangle$, where $|\phi(t_1, t_2)\rangle = \hat{E}_1^+(t_1)\hat{E}_2^+(t_2)|\Psi\rangle$ is the unnormalized state obtained after the detectors remove two photons from the electromagnetic field during the coincidence detection.

The detectors are slow compared to the rate at which down conversion occurs, so one can take $T \to \infty$, in which case, time integrals that occur when Eqs. 19.30 and 19.31 are substituted into Eq. 19.32 become proportional to delta functions:

$$\frac{1}{2\pi} \int_{-\infty}^{\infty} e^{-i\Omega t} dt = \delta(\Omega), \tag{19.33}$$

Substitution of Eqs. 19.28–19.31 into R_c and use of the commutation relations

$$\left[\hat{a}_j(\omega), \hat{a}_k^\dagger(\omega')\right] = \delta_{jk} \, \delta(\omega - \omega') \tag{19.34}$$

$$\left[\hat{a}_j(\omega), \hat{a}_k(\omega')\right] = \left[\hat{a}_j^\dagger(\omega), \hat{a}_k^\dagger(\omega')\right] = 0, \tag{19.35}$$

lead (after a bit of algebra, see Prob. 2) to the result

$$R_c = \int d\omega_1 d\omega_2 d\omega d\omega' \, \Phi(\omega)\Phi^*(\omega') \, |\psi|^2, \tag{19.36}$$

where the coincidence amplitude is given by

$$\psi = \langle 0|\hat{a}_1(\omega_1)\hat{a}_2(\omega_2)\hat{a}_i^\dagger(\omega_0 - \omega)\hat{a}_s^\dagger(\omega_0 + \omega)|0\rangle \tag{19.37}$$

$$= \frac{1}{2}\left\{-\delta(\omega_1 - \omega_0 - \omega)\delta(\omega_2 - \omega_0 + \omega)e^{i[k(\omega_1)L + \omega_1\delta l/c]}\right. \tag{19.38}$$

$$\left. + \delta(\omega_1 - \omega_0 + \omega)\delta(\omega_2 - \omega_0 - \omega)e^{i[k(\omega_2)L + \omega_2\delta l/c]}\right\}.$$

The exponentials can be expanded in power series in the frequencies. Keeping only terms up to order ω^2, then using the delta functions to do three of the frequency integrations in R_c, the coincidence rate reduces to

$$R_c = \frac{1}{8\pi^2} \int d\omega \, |\Phi(\omega)|^2 \left\{1 - \cos\left[[k(\omega_0 - \omega) - k(\omega_0 + \omega)]L + \frac{2\delta l}{c}\omega\right]\right\}$$

$$= \frac{1}{8\pi^2} \int d\omega \, |\Phi(\omega)|^2 \left\{1 - \cos\left[2\left(\alpha L - \frac{\delta l}{c}\right)\omega\right]\right\}, \tag{19.39}$$

where the fact that $\Phi(-\omega) = \Phi(\omega)$ was used. Note that β no longer appears in the coincidence rate, and so there is no broadening of wave packets, demonstrating the cancelation of the GVD term in this setup. More generally, *all* even-order dispersion coefficients cancel, leaving the coincidence rate dependent only on the odd-order dispersion.

19.2.4 Franson Dispersion Cancelation

In the Steinberg-Kwiat-Chiao method of dispersion cancelation, the two photons in each pair are mixed at the beam splitter. An apparatus with a different type of dispersion cancelation, in which no mixing of the two photons occurs, is shown in

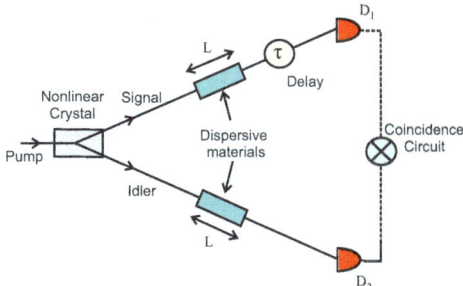

Fig. 19.4 Franson-type dispersion cancelation. A parametric down conversion source illuminates a pair of dispersive materials. The coincidence rate exhibits dispersion cancelation if the dispersion coefficients in the two arms are opposite in sign and equal in magnitude. Solid lines represent light paths, while dashed lines are electrical connections

Fig. 19.4 [36, 37]. There is now dispersion in both branches, and the goal is to make the dispersive effects of the two branches cancel each other. Because the photons are never brought together at a beam splitter, this is referred to as *non-local* dispersion cancelation.

Assume for simplicity that the input light has a Gaussian spectral distribution

$$\Phi(\Omega_0 + \omega) = \Phi(\Omega_0)e^{-\omega^2/2\eta^2} \tag{19.40}$$

of width η. Due to dispersion, the wave vectors in the two arms can be written as

$$k_1 = k_0 + \alpha_1\omega + \beta_1\omega^2 \tag{19.41}$$

$$k_2 = k_0 + \alpha_2\omega + \beta_2\omega^2, \tag{19.42}$$

where α_1, β_1 are the dispersion coefficients in one arm and α_2, β_2 are the coefficients in the other arm. k_0 is the wave number at the central frequency, $k_0 = k(\omega_0)$. In the absence of dispersion, a single beam pulse arriving at one of the detectors would be measured to have a duration $\Delta t_0 = \frac{1}{\eta}$. With the dispersive material included, the second-order coefficient β_i in each arm will broaden the incoming light pulse, increasing the width of the temporal distribution at each detector. But if we look at coincidence counts instead, we see a different result.

It can be shown that the variance in the distribution of coincidence times in Fig. 19.4 is

$$\Delta t_Q^2 = \frac{1/\eta^4 + (\beta_1 + \beta_2)^2 L^2}{1/\eta^2} = \frac{\Delta t_0^4 + (\beta_1 + \beta_2)^2 L^2}{\Delta t_0^2}. \tag{19.43}$$

So it is seen that if the second-order dispersion coefficients are equal and opposite in the two branches, $\beta_2 = -\beta_1$, then the dispersive broadenings in the two beams cancel, leaving a coincidence distribution equal in width to that which would occur in the absence of any dispersion at all:

$$\Delta t_Q = \Delta t_0. \tag{19.44}$$

There is no therefore dispersive broadening in the coincidence rate.

This can be contrasted with the corresponding classical case, where the same apparatus is illuminated by a pair of uncorrelated but statistically identical pulses of light:

$$\Delta t_C^2 = \frac{2/\eta^4 + (\beta_1^2 + \beta_2^2)L^2}{1/\eta^2}. \tag{19.45}$$

In the classical case, β_1 and β_2 are squared *before* adding (instead of *after* as in the quantum case), so that there is no possibility of cancelation. This is the result of adding probabilities classically, rather than adding amplitudes quantum mechanically.

In this section, we have looked at the cancelation of dispersive effects in interference experiments, where the analysis is easier. But dispersive effects also occur in imaging experiments, meaning that dispersion cancelation can lead to improved resolution in microscopy and other areas.

19.3   Quantum Gyroscopes

The standard method of measuring rotation classically is by means of a **Sagnac interferometer** (Fig. 19.5). At the entrance of the interferometer, a beam splitter sends each half of the beam circulating around the loop in opposite directions. If the interferometer is rotating, then the beam splitter will be moving as the light is propagating around the loop, so that light moving in one direction will have to travel farther to return to the beam splitter than light travelling the other direction. Since it will take longer for one portion of the beam to return to the beam splitter than the other portion, the two counter-propagating beams will be out of phase by an amount proportional to the rotation rate and the size of the loop. By looking at the resulting interference pattern, the rotation rate can easily be determined.

To obtain an expression for the phase shift, imagine that the interferometer is formed by a circular loop of radius r. (It is drawn as a rectangle in the figure, but in fact the result will hold for an interferometer of *any* shape; so we derive it for a circular loop for simplicity.) Let $t_\pm$ be the times for the light to travel clockwise $(-)$ and counterclockwise $(+)$ around the loop. In the absence of any rotation, we would have

$$t_\pm = \frac{2\pi r}{c}, \tag{19.46}$$

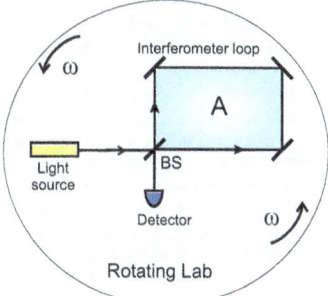

Fig. 19.5 A Sagnac interferometer, used to measure rotation rates. A beam splitter sends the light into a superposition of two beams: one propagating clockwise around the loop, the other counter-clockwise. The beams are recombined at the beam splitter so that their relative phase can be measured. The area of the loop is A

where c is the speed of light. However, if the interferometer is rotating counterclockwise at angular speed ω, then the beam splitter moves a distance $vt = r\omega t$ between the entrance time and the exit time of the light. So the light's counterclockwise path becomes longer by a distance $r\omega t_+$, and the clockwise path becomes shorter by $r\omega t_-$, giving:

$$t_\pm = \frac{2\pi r \pm r\omega t_\pm}{c}. \tag{19.47}$$

Solving for the times,

$$t_\pm = \frac{2\pi r}{c \mp r\omega}. \tag{19.48}$$

The time difference between the two paths is

$$\Delta t = |t_+ - t_-| \tag{19.49}$$

$$= \frac{2\pi r}{c} \left| \frac{1}{1 + \frac{r\omega}{c}} - \frac{1}{1 - \frac{r\omega}{c}} \right| \tag{19.50}$$

$$= \frac{2\pi r}{c} \left(\frac{2r\omega/c}{1 - \left(\frac{r\omega}{c}\right)^2} \right). \tag{19.51}$$

Since the rotational speeds are usually small enough that $\frac{r\omega}{c} \ll 1$, it is safe to approximate this as

$$\Delta t = \frac{4\pi r^2 \omega}{c^2} = \frac{4A\omega}{c^2}. \tag{19.52}$$

Here, we have used the fact that the area of the circular loop is $A = \pi r^2$. The difference between the arc lengths traversed by the two paths is $\Delta s = c\Delta t$, so the phase difference is

$$\Delta\phi = k\Delta s = \left(\frac{2\pi}{\lambda}\right) \cdot (c\Delta t) = \frac{8\pi A\omega}{\lambda c}. \tag{19.53}$$

Although the derivation here is nonrelativistic, the same result still holds when relativistic effects are accounted for, at least up to first order in $\frac{v}{c}$ [58].

By making the area large (e.g., by replacing the single loop by a coil with a large number N of loops), the Sagnac interferometer can measure very small values of ω. But for many applications, for example, in gravitational physics, a signal-to-noise ratio is needed that is beyond that obtainable by the classical Sagnac interferometer. So, a number of quantum-enhanced interferometers, known as **quantum gyroscopes**, have been proposed or developed in recent years.

Several quantum gyroscopes based on entangled N00N states have been proposed; see [59, 60]. Squeezed states have also been used to enhance sensitivity to small rotational speeds, for example, in [61–64].

A superconducting quantum gyroscope [65] is shown schematically in Fig. 19.6. Superfluid He^4, cooled to about $2\ K$, is contained in an annular vessel. A weak partition spans the right side of the ring; this is a thin membrane with thousands of tiny apertures. In the superfluid state, the helium experiences no viscosity. As a result, if the ring is rotating, the fluid stays at rest while the partition moves through it. The surprising thing is that the superfluid not only passes through the apertures but in fact oscillates, creating an acoustic signal in the fluid, known as **quantum whistling** [66]. This oscillation can be detected, and in fact a recording of the sound can be heard at http://www.jpl.nasa.gov/heliumwhistle/. Quantum whistling in superfluids is closely analogous to the electrical Josephson effect in superconductors (Sect. 7.3), and the structure of the superconducting quantum gyroscope is closely analogous to that of a SQUID (Sect. 7.3.2). When the ring is rotating, the excess pressure on one side of the membrane plays the same role that the applied voltage plays in the SQUID. The oscillations produced by each hole are very weak, but since the fluid is in a single macroscopic quantum state, the oscillations from the collection of small holes will interfere constructively, producing a signal strong enough to be measured.

Fig. 19.6 A superconducting quantum gyroscope. The partition has thousands of tiny holes in it. When the annular vessel rotates, superfluid is forced through the holes, creating an oscillation called quantum whistling via the superfluid Josephson effect

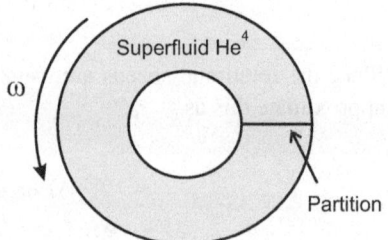

Another quantum-based option for measurement of rotations is an atom interferometer. This has been used to give a high-precision measurement of the Earth's rotation [67]. The starting point is a cloud of cold atoms. It is hit by a sequence of laser pulses, which act as beam splitters and mirrors. Using an approach similar to that used in NMR (to be discussed in Chap. 23), a so-called $\frac{\pi}{2}$ pulse serves as beam splitter, putting the atoms into a superposition of two states: a ground state deflected to the left is superposed with an excited state deflected to the right. A pair of π pulse acts as mirrors, with a final $\frac{\pi}{2}$ pulse undoing the effect of the first one. The result is a cold atom analog of our old friend, the Mach-Zhender interferometer (Sect. 2.5). If the Earth is rotating, then one beam will take longer than the other to reach the final beam splitter, introducing a phase shift that can be measured from the resulting interference pattern. The initial experiment demonstrating this effect achieved a resolution of 26 nrad/s and measured the Earth's rotation with an error of under 1.2% [67].

19.4 Quantum Gravity Sensing

An example of a quantum sensor that has already become commercially available and proven itself in the field is the **quantum gravimeter** [68–72]. Gravimeters are used to measure the local gravitational field strength. An *absolute gravimeter* measures the value of g at a point, while *differential gravimeters* or *gravitational gradiometers* measure the field at two slightly displaced points in order to measure spatial variations in the local field. Gravimeters have a wide variety of uses. For example, they can be used to detect the presence of underground pipes, tunnels, and abandoned mine shafts without having to excavate the area. In archeology, they can be used to detect the remains of ancient buildings that have become buried or to detect undiscovered chambers in pyramids and crypts. In marine navigation, gravimeters can be used to produce accurate maps of the seabed, allowing ships to navigate and to avoid shallow water even when GPS and other external communications go out.

Traditional gravimeters use masses hanging from springs. One end of the spring is hung from the top of the device, and then the mass at the other end is displaced downward by an amount proportional to the gravitational field, thus giving a measure of the local value of g. Although such devices can be very accurate, they have several drawbacks. If the apparatus is moved, then it takes time for transient motions of the spring to die out, making measurements slow. If there are small seismic tremors, or vibrations from nearby car traffic or construction, then noise in the measured signal becomes a problem. Furthermore, springs get stretched out over time, so that traditional gravimeters need periodic recalibration.

A new generation of *quantum* gravimeters evade these problems using atom interferometry [73–77]. These contain a gas of cold rubidium or other atoms in a chamber. The gas is released and allowed to fall under the influence of gravity, while laser pulses near a resonant frequency are used to put the atoms into a superposition of two atomic states (see Chap. 24). Further pulses are used to split the falling atomic cloud into multiple distinct paths. The pulses are **chirped** (their wavelength varies over time), so that when interacting with the laser, each atom gains a phase shift proportional to the position at which the interaction occurred. At the bottom of the fall, the two beams are recombined in order to measure the phase difference between them. This allows high-precision reconstruction of the acceleration and therefore of the gravitational field. The atoms are then recollected via a magneto-optical trap (Chap. 24) and allowed to fall again, about once every half second. For a more detailed description of the operation, see the supplementary material for [69].

Moreover, any external vibrational noise, such as nearby highway traffic, will affect all parts of the apparatus equally and will therefore cancel out of the interferometric measurements. Noise is thus greatly reduced. In addition, the devices are fast and easily portable, and they require no recalibrations. Quantum gravimeters have begun to find use in seismology: one has been used for several years on Mount Etna, an active volcano in Sicily, collecting data on the density changes of rock, magma, and underground gas. Other scientific applications that have been proposed include precision measurements of Newton's gravitational constant, G, and the search for cosmic dark matter.

19.5   Magnetoencephalography

Modern neurology attempts to understand the workings of the brain by mapping the activity of different brain cells during various activities. Knowledge of the functioning of the network of neurons in the brain is essential to obtaining an understanding of disorders like epilepsy and dementia, as well as diagnosing these problems in individual patients. To obtain a useful map of brain cell functioning, it is necessary to measure very small currents arising in those cells. Ideally, this should be done in a non-invasive manner, without having to insert probes into the brain itself. **Magnetoencephalography (MEG)** attempts to do this by measuring the tiny magnetic fields these currents create outside the skull and then using mathematical modeling methods to reconstruct the currents in the neurons. The difficulty arises from the fact that these fields, typically on the order of 10^{-13}T, are nearly a billion times smaller than the field produced by the Earth itself (which ranges between 20 and 60 μT), in addition to being minuscule compared to other ambient magnetic fields from nearby electrical devices.

SQUIDs (Sect. 7.3.2) are capable of the necessary sensitivity to small fields, but they come with some drawbacks. For example, they need to be cooled to cryogenic temperatures, about 4 K. The subject's skull obviously needs to remain near room temperature, so SQUID sensors have to be separated from the skull by a thermally insulating gap; this increases the distance of the sensor from the source of the currents, reducing sensitivity. In addition, the cryogenic dewar needed to keep the SQUID cold can weigh hundreds of pounds. As a result, the whole apparatus is fixed in place. Meanwhile, the patient must be kept immobile, since any movements relative to the fixed apparatus will blur the resulting images. The equipment itself costs several million dollars, in addition to the operating costs.

So it would be a tremendous improvement if a similar level of magnetic field sensitivity could be obtained from a smaller device that operates at room temperature. If the sensors are sufficiently small, they can be fitted onto helmets; the sensors would then be capable of moving along with the patient's head, removing the need for the subject to remain unmoving inside a large chamber. This also means that helmets of different sizes can be produced, accommodating the mapping of brain function for adults, children, and even babies by having an array of different-size helmets that can be plugged into the same recording devices.

Such a possibility is provided by the use of **optically pumped magnetometers (OPMs)**. These achieve similar sensitivity to small fields as a SQUID does, but manage to do it at room temperature. They can be made as small as ~ 1 cm (or even millimeter-sized, though with reduced sensitivity). Unlike the other sensors in this chapter, the OPM is not an interferometer, but it *is* a quantum device in the sense that it relies on the interactions of individual photons with individual atoms in a vapor.

The basic structure of the OPM is shown in Fig. 19.7a. The vapor cell contains a gas of rubidium (Rb^{87}) or another alkali atom. The circularly polarized laser pumps the vapor into a state in which all the atomic spins are aligned with the angular momentum direction of the beam. The resulting magnetization makes the vapor sensitive to any external magnetic fields. The laser continues flipping spins until the all the spins are aligned. After this no additional photons can be absorbed by the input beam, because there is no accessible state the atom can transition to with the given photon energy. At this point magnetization of the vapor reaches a maximum, and since no light is being absorbed, the output intensity reaches a maximum as well. This steady state will be maintained as long as there is no magnetic field present.

If a magnetic field is turned on, oriented in a direction that differs from the spin direction of the magnetized vapor, it will cause spins of some of the vapor atoms to flip. This means that some of the atoms are now able to absorb photons from the laser again. Absorption of these photons from the beam causes the output intensity to drop by an amount that depends on the field magnitude (Fig. 19.7b), allowing the field to be measured from the dip in light intensity.

The fact that the Earth's field and other environmental fields are much larger than the magnetic fields of interest means that unwanted fields have to be screened out somehow. This is done with a two-prong strategy. First, *passive* screening is carried out by performing the testing in a room whose walls are lined with conducting materials, to block out any unwanted external fields. Second, *active* screening is carried out by having additional sensors near the location of the head. This measures any remaining fields in the vicinity, and then mathematical modeling is used to calculate their effect on the measurements and to subtract off those effects. By means of these screening operations, the minuscule fields produced by the neuronal currents can then be clearly discerned. After screening, quantum correlations between spins in the vapor can be maintained with coherence times on the order of a minute or more.

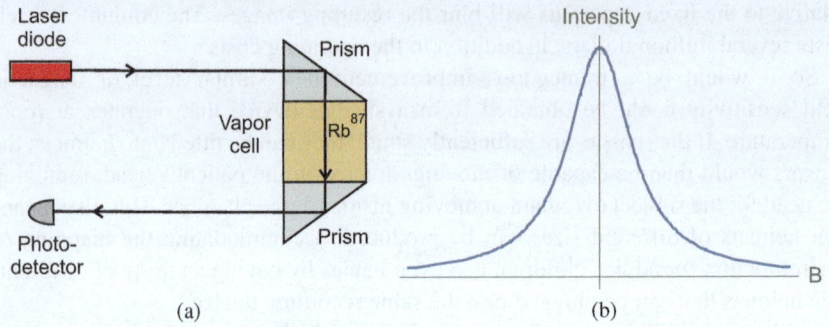

Fig. 19.7 (**a**) Schematic of setup for an OPM. Light is passed through a cell containing rubidium vapor. The intensity of the surviving light after passage is measured by a photodetector. (**b**) As the magnitude of B increases, the measured intensity drops, providing a means of measuring very small fields. The curve is Lorentzian

A similar method can be used for **magnetocardiographs (MCGs)** to measure currents in the heart muscle. OPM-based MCGs and MEGs are both commercially available already. For more information on the operation of these devices, see [78] for a discussion of the MEG and [79] for the MCG.

19.6   Atomic Clocks

Keeping accurate time is essential for many everyday activities. But for some applications, such as implementing GPS navigational systems or making precision measurements of fundamental constants or general relativistic effects, clocks need to have accuracies far beyond what is needed for everyday purposes. Since the 1940s, clocks based on measurements of atomic or molecular energy-level transitions have been developed that have increased the accuracy of time measurements to unprecedented levels. In fact, since 1967, the current International System of Units (SI) definition of the standard second has been based on an atomic clock: one second is defined to be 9,192,631,770 times the transition period of electrons between the hyperfine levels of the ground state of the Cesium-133 atom at rest, at $T = 0$ K. Since this definition was agreed upon, atomic clocks have continued to improve in accuracy and stability. Currently, the best quantum clocks can make measurements that are accurate to within one second every 100 million years. Reviews of atomic clocks can be found in [80–82].

Traditional mechanical or electrical clocks all have two main elements in common: (i) an oscillation with a stable reference frequency to provide "ticks" of a standard size and (ii) a means of counting these ticks in order to determine the lengths of time intervals. In a grandfather clock, for example, the stable oscillation is provided by an oscillating pendulum. Each oscillation then turns a set of gears, which in turn move hands on the dial in order to count the ticks and read off the time. In many wrist watches, a quartz crystal is used as the oscillator: mechanical vibrations of the quartz occur at a precise frequency determined by the size and cut of the crystal. Quartz is piezoelectric, meaning that it can convert mechanical oscillations into electrical oscillations and vice versa. Applying a current to the quartz causes it to oscillate at a fixed frequency like the pendulum of a grandfather clock.

In a typical atomic clock, the stable tick is provided by a local oscillator field (a laser), and the counting of the ticks is accomplished by interference with femtosecond frequency combs (see Box 19.1). Atomic clocks then add a third component: a feedback mechanism to prevent the local oscillator frequency from drifting; this is provided by the atomic energy-level transitions.

The idea of building a clock using the vibrational frequency of an electromagnetic wave goes back to James Clerk Maxwell, while Lord Kelvin suggested the use of transitions in gases of sodium and hydrogen in 1879. I. I. Rabi and coworkers developed most of the ideas necessary for atomic clocks during the 1940s and 1950s, leading to the demonstration of an ammonia clock by researchers led by Harold Lyons at the US National Bureau of Standards in 1949. This was based on the transitions between the two configurations of the NH_3 molecule shown in Fig. 19.8, which occur at a frequency of 23.87 GHz. Placed in a quartz oscillator circuit, the electronic oscillation was converted into a radio wave beamed into the ammonia gas. Time was read by interrogating the gas with a long microwave pulse and measuring the electromagnetic power that exits the other side of the gas chamber. When the quartz circuit is oscillating at the ammonia resonance frequency, most of the energy in the wave will be absorbed. Thus when the frequency is correct, the transmitted power on the other side of the chamber drops; it grows when the oscillation frequency moves away from resonance. This provides feedback used to keep the oscillation of the quartz circuit stable. The result was a clock with errors of less than one part in a billion.

The long radio pulses used in the ammonia clock introduced uncertainties from Doppler shifts that prevented the accuracy from being substantially improved. Norman Ramsey introduced a major improvement by using a pair of shorter microwave pulses, reducing the Doppler effect and narrowing the line widths. This

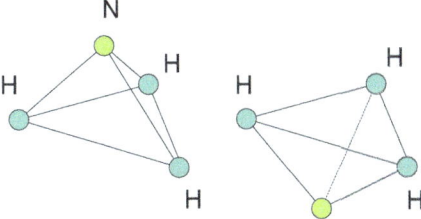

Fig. 19.8 The ammonia molecule, NH_3. There are two degenerate ground-state configurations, with the N atom either above or below the plane of the three H atoms. The fixed-frequency 23,870 MHz oscillations between these two ground states were used as the basis of the NH_3 clock

Ramsey interferometer approach (see Sect. 23.1) has been the standard mode of operation for atomic clocks ever since. Essentially, the idea is to perform a Hadamard operation by illuminating with a microwave $\frac{\pi}{2}$ pulse (see Sect. 23.1). This puts the atomic ground state into a superposition of the ground and excited states. The superposition state oscillates until the second $\frac{\pi}{2}$ pulse puts the atom back into the ground state. The phase gained during the free propagation of the superposition state can be measured, which then allows the time to be determined, since for fixed frequency, the phase is $\phi = \omega T$.

The first cesium clock was demonstrated in 1955 by L. Essen at the National Physics Lab. Using the Ramsey approach, oscillations involving the hyperfine splittings of the ground states in cesium, rubidium, hydrogen and other atoms have since led to improved stability and accuracy, culminating in the adoption of the cesium-133 standard for the SI second. Until the 1990s, all cesium clocks were cesium *beam* clocks, in which the atoms were sent as a beam through an evacuated chamber. Subsequently, however, atomic *fountain* clocks have been found to be more accurate. In a cesium fountain clock, six lasers along the three axes are used to trap and cool atoms in the chamber to sub-milli-Kelvin temperatures. Additional lasers then kick the atoms upward about 1 meter, after which they fall back under the influence of gravity. The round trip up and down takes about a second, allowing long measurement times. As a result, very narrow linewidths and very high time resolution can be achieved, stable to about one part in 10^{15}.

The drawback of cesium fountain clocks is that to get such high accuracy, many measurements have to be averaged over a long time. Reducing the length of this averaging period is possible by replacing the microwave pulses by optical pulses. These optical clocks, which are currently under development, operate at frequencies 10,000 times higher than the microwave frequencies of current cesium fountain clocks. Since clock accuracy and stability are generally proportional to the oscillator frequency, optical fountain clocks are expected to achieve the same stability as microwave clocks by averaging measurements over just a few seconds, instead of days. With longer averaging times, they may eventually reach stabilities of 10^{17} or better.

Finally, most atomic clocks use charged ions. These ions exhibit interactions with each other, leading to one further source of error to be eliminated. The elimination is achieved by using neutral atoms, which can be manipulated while trapped in optical lattices (see Sect. 24.4). Thousands of atoms can be trapped in the same lattice site, allowing very-high-accuracy measurements, with high stability and high signal-to-noise ratios. These optical lattice clocks hold the potential for stabilities on the order of one part in 10^{18} or more. Such clocks will be able to measure gravitational redshift effects at the millimeter scale, approaching the level of resolution needed to test some theories of quantum gravity and dark matter. See [83] for a detailed review.

As with many other topics covered in this book, it has been proposed that atomic clocks can be improved with the use of entanglement. In [84] it is suggested that entangling the atoms in an optical lattice clock will allow Heisenberg-limited time measurements.

Box 19.1 Frequency Combs

Unlike microwave and radio-wave frequencies common in electronics, waves in the optical portion of the spectrum oscillate too fast for the number of oscillations in a given time interval to be counted directly. This had long limited atomic clocks to operation at microwave frequencies and had provided a barrier to the optical-frequency operation that would allow more precise timing resolution. The development of the frequency comb changed this, opening up new possibilities for precision optical metrology and leading to J. L. Hall and T. W. Hänsch to share in the 2005 Nobel prize.

An optical **frequency comb** [85–87] is a frequency spectrum that has discrete delta function peaks (or approximations to them) at equally spaced frequencies. Frequency combs have applications across a broad spectrum of applications, including precision distance measurements, spectroscopy, medical diagnostics, and dark matter searches. Frequency combs can be created several ways, including through the use of mode-locked lasers or through four-wave mixing in nonlinear crystals.

Consider a mode-locked laser, for example. The resonant cavity allows multiple longitudinal modes of equally spaced frequencies,

$$f_m = f_0 + mf_r, \tag{19.54}$$

where f_0 is the carrier frequency offset and f_r is the frequency spacing of the comb (or the repetition rate of the pulsed laser), with $f_0 < f_r$. The result is a sequence of ultrashort laser pulses, equally spaced in time and of identical shape, with an optical field of the form

$$E(t) = \sum_m C_m e^{i(f_0 + mf_r)t + \phi_m} \equiv \sum_m C_m e^{if_m t + \phi_m}. \tag{19.55}$$

The frequency spacing is determined by the refractive index n_g of the laser gain medium and the laser cavity length L according to

$$f_r = \frac{2\pi c}{2n_g L}. \tag{19.56}$$

This field is periodic in time, at least up to a phase: $E(t + \tau) = E(t)e^{2\pi i \tau f_0 / f_r}$.

(continued)

Box 19.1 (continued)

In mechanical clocks, systems of different-sized gears allow the conversion of rotational frequencies to higher or lower frequencies (Fig. 19.9). The frequency comb acts in a similar manner, allowing vastly different frequencies to be meshed and interconverted between each other. For example, an optical frequency oscillation can be converted to a much slower radio-frequency oscillation, allowing the number of oscillations (e.g., the ticks of a clock) to be counted. This is because an optical frequency $f_{opt} \gg f_0$ can be compared to one of the teeth of the optical comb (a high harmonic of f_0),

$$f_{opt} = f_m \quad \Longrightarrow \quad f_r = (f_{opt} - f_0)/m. \qquad (19.57)$$

Keeping the atomic transition rate steady implies that the faster f_{opt} is stabilized, which in turn keeps the much lower f_0 stable. Counting the ticks n_0 occurring at rate f_0 is easy and allows the number of ticks $n_{opt} = mn_0$ occurring at the higher frequency f_{opt} to be determined. By this means, input microwave frequencies can be multiplied by a factor $\sim 10^5$ to produce output optical frequencies.

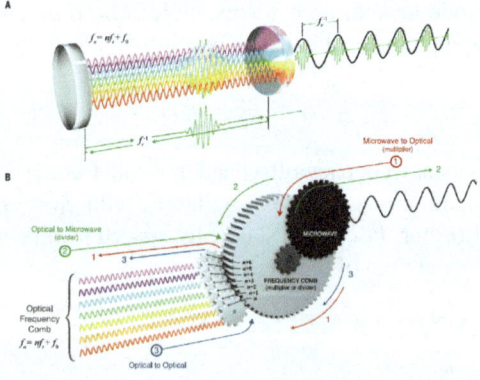

Fig. 19.9 (a) The cavity in a mode-locked laser produces waves of equally spaced frequencies, forming an optical frequency comb. The laser output is a periodic sequence of pulses, with faster oscillations inside the pulse envelope. (b) As described in Box 19.1, the frequency comb acts as a wave analog of a mechanical gear chain, allowing vastly different frequencies to be converted into each other. In the Figure, the black sinusoidal curve represents the slower microwave or radio-wave frequency, while faster optical oscillations are represented by the green curve inside the sinusoidal envelope (Figure reproduced with permission from [86])

Problems

1. In addition to the operator $\hat{C}$ defined in Eq. 19.4, we could also define an operator $\hat{S} = \frac{1}{2i}(\hat{G} - \hat{G}^\dagger)$.

   (a) Show that $\hat{G}^\dagger \hat{G} = 1 - |0\rangle\langle 0|$, which implies that $\hat{G}$ is not unitary.
   (b) Show that $[\hat{C}, \hat{N}] = i\hat{S}$ and $[\hat{S}, \hat{N}] = -i\hat{C}$.
   (c) Show that the commutator $[\hat{C}, \hat{S}]$ is proportional to the projection onto the vacuum state, $|0\rangle\langle 0|$.
   (d) For the fixed-number state $|n\rangle$ with $n \geq 1$, show that $\Delta C = \Delta S = \frac{1}{\sqrt{2}}$, independent of n.

2. (a) Show that the coincidence amplitude defined in Eq. 19.37 is given by Eq. 19.38.
   (b) Show that the coincidence rate in Eq. 19.32 can be written in the form of Eq. 19.36.

3. Suppose that a wave pulse is made from a superposition of a *finite* set of frequencies. Then the discrete version of the group velocity will be $v_g = \frac{\Delta \omega}{\Delta k}$. Consider a superposition of two frequencies ω_1 and ω_2 moving through a material with phase velocities $v_1 = \frac{\omega_1}{k_1}$ and $v_2 = \frac{\omega_2}{k_2}$.

   (a) Show then that the group velocity v_g satisfies $(v_1 - v_g)(v_2 - v_g) > 0$. From this it follows that either both phase velocities are less than v_g (normal dispersion) or both are greater than v_g (anomalous dispersion).
   (b) Suppose that the two frequency components in part (a) come from the signal and idler produced in collinear SPDC. Let ω_p and v_p be the frequency and phase velocity of the pump. Show that $\frac{\omega_1}{v_1} + \frac{\omega_2}{v_2} = \frac{\omega_p}{v_p}$.
   (c) Also show that $\frac{\omega_1}{v_1} - \frac{\omega_2}{v_2} = \frac{\omega_1 - \omega_2}{v_g}$, where v_g is the group velocity of part (a). Hence, show that the phase velocities and frequencies of the signal, pump, and idler are related to the group velocity by

   $$\left(\frac{\omega_1}{v_1}\right)^2 - \left(\frac{\omega_2}{v_2}\right)^2 = \frac{\omega_p(\omega_1 - \omega_2)}{v_g v_p}.$$

4. A Gaussian pulse passing through a dispersive material has (ignoring the x and y dependence) an electric field of the form $E(0, t) = A_0 e^{-\alpha t^2 + i\omega_0 t}$ at $z = 0$, where α is a constant and ω_0 is the central frequency of the pulse. The pulse disperses, so that at distance z, it has the form

   $$E(z, t) = A(z, t)e^{i(\omega_0 t - k_z z)} \exp\left(\frac{4iza(t - z/va)^2}{1/\alpha^2 + 16a^2 z^2}\right).$$

Here, $a = \frac{1}{2}\frac{d^2 k_z}{d^2\omega}$ measures the group velocity dispersion. If the initial time dura-

tion of the pulse is τ_0, then the duration at distance z is $\tau(z) = \sqrt{1 + \left(\frac{8az\ln 2}{\tau_0^2}\right)}$.

(a) Find an expression giving the distance the pulse must travel in order for the pulse duration to double. (Note that the answer depends on τ_0: shorter pulses will double in duration faster.)

(b) Find the frequency of the wave, $\omega(z, t) = \frac{\partial\phi}{\partial t}$ at distance z, where $\phi(z, t)$ is the phase of the wave. From this, you should see that not only the pulse duration but also the frequency will change as the wave propagates. A time-dependent frequency is said to be **chirped**.

5. (a) Suppose a Sagnac interferometer consisting of a single loop of optical fiber is being used with light of $\lambda = 1546$ nm. If each loop encloses an area 715 m², how large of a phase shift is produced by the Earth's rotation?

(b) A recent experiment [64] (with the same parameters as in part (a)) measured the Earth's rotation using entangled $N00N$ states, $\frac{1}{\sqrt{2}}\left(|0N\rangle + |N0\rangle\right)$, in a Sagnac interferometer, with a superposition of N photons moving clockwise and N counter-clockwise. This leads to an N-fold increase in the size of the phase shift. In the experiment, a phase shift of about 5.5 mrad was found. What value of N was used?

(c) How large a value of N would be needed to produce a similar size phase shift if the interferometer was used to measure the rotation rate of the moon?

6. (EEG versus MEG) The electric potential at a distance r from a nerve axon of radius a can be written [88] approximately in the form of a dipole potential,

$$V = \frac{\mathbf{p} \cdot \mathbf{r}}{4\pi\sigma_o r^3} = \frac{\Delta v\sigma_i a^2}{r\sigma_o r^2}\cos\theta,$$

where Δv is the amplitude of the voltage signal in the axon and the magnitude of the dipole moment is of the form $p = \pi a^2\sigma_i\Delta v$. r and θ are the magnitude and direction of the vector pointing from the center of the axon to the observation point where the field is measured. The ratio of conductivities inside and outside of the nerve cell is usually on the order of $\frac{\sigma_i}{\sigma_o} \sim 10$. Assume radius $a = 1.0$ µm and action potential $\Delta v = 70$ mV.

(a) Estimate the electric potential at a distance of $r = 2$ mm directly above the center of the axon ($\theta = 0$).

(b) Estimate the electric field strength and the magnitude of the force exerted on a 1 mC charge.

(c) The magnetic field at the same point takes a similar form, $\mathbf{B} = \mu_0\frac{\mathbf{p}\times\mathbf{r}}{4\pi r^3}$. Find the force it exerts on a 1 mA current passing through a 1 cm length of wire. (In general, MEG has the advantages over EEG of higher spatial resolution and shorter response time but tends to be more expensive.)

7. (Gravimetry) The force between the Earth (radius R_e and mass M_e) and a mass m at height h above the earth's surface is given by Newton's law of gravity, with $r = R_e + h$.

   (a) For $\Delta h << R_e$, find an approximate formula for the difference in acceleration, Δa, between two equal size masses that differ in height by Δh. Estimate Δa for $\Delta h = 10$ m.

   (b) Suppose that an object of volume $V = 1$ m^3 and density $\rho_0 = 8$ g/m^3 is buried 3 m underground. What is the change in gravitational acceleration at a height 0.1 m above ground, compared to a point where there is no such object? Assume the density of the dirt near the Earth's surface is about 8 g/m^3 (Hint: What is the change in gravitational force compared to when the same volume is filled with dirt? The mass of the rest of the Earth can be ignored since it is the same in both cases.)

8. Atomic clocks are currently in use in satellite- and spacecraft-based experiments to detect dark matter in the universe. Dark matter is expected to create local variations in the fine structure constant α, which will in turn create slight shifts in the resonant frequencies ω associated with atomic energy level transitions. Many theoretical models predict that the shifts will obey $\frac{\delta\alpha}{\alpha} \approx \frac{\delta\omega}{\omega}$.

   (a) Using the precision estimate given in this chapter for cesium atomic clocks, estimate the size of fine-structure variation $\delta\alpha$ that can be detected with a cesium fountain clock.

   (b) Do the same for an optical clock.

9. (Frequency comb) The gas in a mode-locked helium neon laser has refractive index $n_g \approx 1$. Suppose a laser of cavity length $L = 30$ cm is producing light of frequency 4.5×10^{14} Hz.

   (a) What is the frequency spacing in the output beam?

   (b) Suppose the frequency comb is to be used to convert the optical frequency to a radio frequency of 10^9 Hz. Estimate how high a harmonic needs to be used.

References

1. K.T. Kapale, L.D. Didomenico, H. Lee, P. Kok, J.P. Dowling, Concepts Phys. **2**, 225 (2005)
2. Z. Hradil, Quantum Opt. **4**, 93 (1992)
3. L. Susskind, J. Glogower, Physics **1**, 49 (1964)
4. P. Carruthers, M.M. Nieto, Rev. Modern Phys., **40**, 411 (1968)
5. J.-M. Levy-Leblond, Ann. Phys. (N.Y.) **101**, 319 (1976)
6. D.T. Pegg, S.M. Barnett, Phys. Rev. A **39**, 1665 (1989)
7. J.A. Vaccaro, D.T. Pegg, Opt. Commun. **70**, 529 (1989)
8. S.M. Barnett, S. Stenholm, D.T. Pegg, Opt. Commun. **73**, 314 (1989)
9. J.A. Vaccaro, D.T. Pegg, J. Modern Opt. **37**, 17 (1990)
10. S.M. Barnett, D.T. Pegg, Phys. Rev. A **41**, 3427 (1990)

11. J.A. Vaccaro, D.T. Pegg, Phys. Rev. A **41**, 5156 (1990)
12. G.S. Summy, D.T. Pegg, Opt. Commun. **77**, 75 (1990)
13. S.M. Barnett, D.T. Pegg, Phys. Rev. A **42**, 6713 (1990)
14. D.T. Pegg, J.A. Vaccaro, S.M. Barnett, J. Mod. Opt. **37**, 1703 (1990)
15. J. Bergou, B.-G. Englert, Ann. Phys. (N.Y.) **209**, 479 (1991)
16. M.M. Nieto, Phys. Scripta **T48**, 15 (1993)
17. A.N. Boto, P. Kok, D.S. Abrams, S.L. Braunstein, C.P. Williams, J.P. Dowling, Phys. Rev. Lett. **85**, 2733 (2000)
18. A.S. Lane, S.L. Braunstein, C.M. Caves, Phys. Rev. A **47**, 1667 (1993)
19. Y. Israel, S. Rosen, Y. Silberberg, Phys. Rev. Lett. **112**, 103604 (2014)
20. H. Cramer, *Mathematical Methods of Statistics* (Princeton University Press, Princeton, 1946)
21. C.R. Rao, Bull. Calcutta Math. Soc. **37**, 81 (1945)
22. D.S. Simon, A.V. Sergienko, T.B. Bahder, Phys. Rev. A **78**, 053829 (2008)
23. T.B. Bahder, Phys. Rev. A **83**, 053601 (2011)
24. Z. Hradil, R. Myška, J. Peřina, M. Zawisky, Y. Hasegawa, H. Rauch, Phys. Rev. Lett. **76**, 4295 (1996)
25. L. Pezzé, A. Smerzi, G. Khoury, J.F. Hodelin, D. Bouwmeester, Phys. Rev. Lett. **99**, 223602 (2007)
26. B. Yurke, S.L. McCall, J.R. Klauder, Phys. Rev. A **33**, 4033 (1986)
27. S.D. Huver, C.F. Wildfeur, J.P. Dowling, Phys. Rev. A **78**, 063828 (2008)
28. C.M. Caves, Phys. Rev. D **23**, 1693 (1981)
29. C. Bonato, A.V. Sergienko, B.E.A. Saleh, S. Bonora, P. Villoresi, Phys. Rev. Lett. **101**, 233603 (2008)
30. C. Bonato, D.S. Simon, P. Villoresi, A.V. Sergienko, Phys. Rev. A **79**, 062304 (2009)
31. D.S. Simon, A.V. Sergienko, Phys. Rev. A **80**, 053813 (2009)
32. L.A.P. Filpi, M.V. da Cunha Pereira, C.H. Monken, Opt. Exp. **23**, 3841 (2015)
33. A.N. Black, E. Giese, B. Braverman, N. Zollo, S.M. Barnett, R.W. Boyd, Phys. Rev. Lett. **123**, 143603 (2019)
34. M.D. Mazurek, K.M. Schreiter, R. Prevedel, R. Kaltenbaek, K.J. Resch, Sci. Rep. **3**, 1582 (2013)
35. M. Jensen, N. Misraelsen, M. Maria, T. Feuchter, A. Podoleanu, O. Bang, Sci. Rep. **8**, 9170 (2018)
36. J.D. Franson, Phys. Rev. A **45**, 3126 (1992)
37. J.D. Franson, Phys. Rev. A **80**, 032119 (2009)
38. A.M. Steinberg, P.G. Kwiat, R.Y. Chiao, Phys. Rev. Lett. **68**, 2421 (1992)
39. J. Jeffers, S.M. Barnett, Phys. Rev. A **47**, 3291 (1993)
40. S.-Y. Baek, Y.-W. Cho, and Y.-H. Kim, Opt. Express **17**, 19241 (2009)
41. T. Zhong, F.N.C. Wong, Phys. Rev. A **88**, 020103 (2013)
42. J. M. Lukens, A. Dezfooliyan, C. Langrock, M.M. Fejer, D.E. Leaird, A.M. Weiner, Phys. Rev. Lett. **112**, 133602 (2014)
43. J. Ryu, K. Cho, C.-H. Oh, H. Kang, Opt. Express **25**, 1360 (2017)
44. A.M. Steinberg, P.G. Kwiat, R.Y. Chiao, Phys. Rev. A **45**, 6659 (1992)
45. V. Giovannetti, S. Lloyd, L. Maccone, F.N.C. Wong, Phys. Rev. Lett. **87**, 117902 (2001)
46. E. Dauler, G. Jaeger, A. Muller, A. Migdall, J. Res. Natl. Inst. Stand. Technol. **104**, 1 (1999)
47. D. Branning, A. L. Migdall, and A. V. Sergienko, Phys. Rev. A **62**, 063808 (2000)
48. A. Fraine, D.S. Simon, O. Minaeva, R. Egorov, A.V. Sergienko, Opt. Exp. **19**, 22820 (2011)
49. A. Abouraddy, M.B. Nasr, B.E.A. Saleh, A.V. Sergienko, M.C. Teich, Phys. Rev. A **65**, 053817 (2002)
50. M.B. Nasr, B.E.A. Saleh, A.V. Sergienko, M.C. Teich, Phys. Rev. Lett. **91**, 083601 (2003)
51. M.B. Nasr, B.E.A. Saleh, A.V. Sergienko, M.C. Teich, Opt. Express **12**, 1353-1362 (2004)
52. O. Minaeva, C. Bonato, B.E.A. Saleh, D.S. Simon, A.V. Sergienko, Phys. Rev. Lett. **102**, 100504 (2009)
53. K.J. Resch, P. Puvanathasan, J.S. Lunden, M.W. Mitchell, K. Bizheva, Opt. Express **15**, 8797 (2007)

54. P. Kaltenbaek, J. Lavoie, D.N. Biggerstaff, K.J. Resch, Nat. Phys. **4**, 864 (2008)
55. J. Lavoie, P. Kaltenbaek, K.J. Resch, Opt. Express **17**, 3818 (2009)
56. P. Kaltenbaek, J. Lavoie, K.J. Resch, Phys. Rev. Lett. **102**, 243601 (2009)
57. V. Torres-Company, H. Lajunen, A.T. Friberg, New J. Phys. **11**, 063041 (2009)
58. W. Englelhardt, Annales de la Fondation Louis de Broglie **40**, 149 (2015)
59. M. Fink, et al., New J. Phys. **21**, 053010 (2019)
60. F. de Leonardis, R. Soref, M. de Carlo, V.M.N. Passaro, Sensors **20**, 3476 (2020)
61. K. Liu, C. Cai, J. Li, L. Ma, H. Sun, J. Gao, Appl. Phys. Lett. **113**, 261103 (2018)
62. M. Mehmet, et al., Opt. Lett. **35**, 1665 (2010)
63. M.R. Grace, C.N. Gagatsos, Q. Zhuang, S. Guha, Phys. Rev. Appl. **14**, 034065 (2020)
64. R. Silvestri, et al., Sci. Adv. **10**, eado0215 (2024)
65. R.E. Packard, S. Vitale, Phys. Rev. B **46**, 3540 (1992)
66. E. Hoskinson, R.E. Packard, T.M. Haard, Nature **433**, 376 (2005)
67. P. Berg et al., Phys. Rev. Lett. **114**, 063002 (2015)
68. R. Rummel, W. Yi, C. Stummer, J. Geod. **85**, 777 (2011)
69. V. Ménoret, P. Vermeulen, N. Le Moigne, S. Bonvalot, P. Bouyer, A. Landragin, B. Desruelle, Sci. Rep. **8**, 12300 (2018)
70. R. Caldani, K.X. Weng, S. Merlet, F. Pereira Dos Santos, Phys. Rev. A **99**, 033601 (2019)
71. A.-K. Cooke, C. Champollion, N. Le Moigne, Geosci. Instrum. Method. Data Syst. **10**, 65 (2021)
72. C. Janvier, V. Ménoret, B. Desruelle, S. Merlet, A. Landragin, F. Pereira dos Santos, Phys. Rev. A **105**, 022801 (2022)
73. A. Peters, K.Y. Chung, S. Chu, Metrologia **38**, 25 (2001)
74. J.B. Fixler, G.T. Foster, J.M. McGuirk, M.A. Kasevich, Science **315**, 74 (2007)
75. G. Rosi, F. Sorrentino, L. Cacciapuoti, M. Prevedelli, G.M. Tino, Nature (London) **510**, 518 (2014)
76. D.-K. Mao, X.-B. Deng, H.-Q. Luo, Y.-Y. Xu, M.-K. Zhou, X.-C. Duan, Z.-K. Hu, Rev. Sci. Instrum. **92**, 053202 (2021)
77. P. Asenbaum, C. Overstreet, T. Kovachy, D.D. Brown, J.M. Hogan, M.A. Kasevich, Phys. Rev. Lett. **118**, 183602 (2017)
78. E. Boto et al., Nature **555**, 657 (2018)
79. K. Jensen et al., Sci. Rep. **8**, 16218 (2018)
80. H. Lyons, Sci. Am. **196**(2), 71 (February 1957)
81. A.D. Ludlow et al., Rev. Mod. Phys. **87**, 637 (2015)
82. M.A. Lombardi, T.P. Heavner, S.R. Jefferts, Measure **2**, 74 (2007)
83. A. Derevianko, H. Katori, Rev. Mod. Phys. **83**, 331 (2011)
84. J.D. Weinstein, K. Beloy, A. Derevianko, Phys. Rev. A **81**, 030302(R) (2010)
85. Th. Udem, R. Holzwarth, T.W. Hänsch, Nature **416**, 233 (2002)
86. S.A. Diddams, K. Vehala, T. Udem, Science **369**, eaay3676 (2020)
87. T. Fortier, E. Baumann, Comm. Phys. **2**, 153 (2019)
88. R.K. Hobbie, B.J. Roth, *Intermediate Physics for Medicine and Biology, 5th ed.* (Springer, New Year, 2015)

Chapter 20
Quantum Imaging and Related Topics

One prominent application of quantum mechanics is in the development of new imaging methods. In quantum ghost imaging, a photon that did not interact with the object is used to produce an image using its entanglement with a photon that did, while in quantum illumination low numbers of entangled photons are used to detect objects in situations where classical illumination would be overpowered by the surrounding thermal background. Other entanglement-based imaging methods such as entangled photon microscopy and quantum optical coherence tomography allow advantages such as greater subsurface imaging depth and higher resolution that make them promising tools for biomedical imaging applications. In this chapter, we discuss some of these methods.

20.1 Ghost Imaging

An optical detector is called a **bucket detector** or **single-pixel detector** if it has no spatial resolution: it can detect the presence or absence of light, but cannot give any information about how the intensity of the light varies over the surface of the detector. It would seem obvious that such a detector could never produce an image of any object that has spatial structure. Similarly, it seems clear that photons that never interacted with the object could also never produce an image of it. Although it is impossible for either of these two ingredients (photons that never saw the object and photons that are observed only by a bucket detector) to form an image separately, it turns out that taken together, the two of them can in fact form images. This effect seemed so counterintuitive and spooky when it was discovered that it came to be known as **ghost imaging**. Also often called **quantum imaging**, the initial experiments in which it was observed required entangled photons. Since then, it has become apparent that ghost imaging can also be observed with classical light beams, albeit with lower visibility and lower signal-to-noise ratio. The key ingredient in

© The Author(s), under exclusive license to Springer Nature Switzerland AG 2025
D. S. Simon, *Introduction to Quantum Science and Technology*, Undergraduate Texts in Physics, https://doi.org/10.1007/978-3-031-81315-3_20

ghost imaging turns out simply to be strong spatial *correlation* (either classical or quantum) between the illumination in the two beams. As a result, ghost imaging is also called **correlated imaging**.

Beyond being an interesting physical phenomenon, ghost imaging has a number of useful features. For example, the two light beams need not be at the same frequency. So one beam can probe the object in the infrared, ultraviolet, or terahertz portions of the spectrum, while the other beam produces an image with visible light; this allows images to be produced in portions of the spectrum where detectors may be expensive, of low resolution, or inefficient, using only cheap, high-resolution, visible-light detectors. The quantum version of ghost imaging also allows high-resolution images to be produced at very low light levels in situations (e.g., in biological or medical applications) where high-intensity light may damage the object. On the down side, quantum imaging is slower than classical imaging. It has recently been shown [1] that when combined with neural network processing, quantum ghost imaging can produce subwavelength super-resolution imaging. Ghost imaging also shows a measure of resistance to the distorting effects of turbulence in the medium through which the light is propagating and can be arranged to cancel some dispersive effects in the medium (see Sect. 20.1.3).

More detailed reviews of ghost and quantum imaging include [2–8].

20.1.1   *Quantum Ghost Imaging*

Recall (Sects. 3.6 and 5.8) that the photons produced in SPDC are individually of low coherence. So if a double slit is placed in the path of the signal beam from down conversion, no interference pattern is will be formed (Fig. 20.1a). However, suppose that a small detector is placed after the idler beam as well (Fig. 20.1b), and then the coincidence count between the two detectors is measured as the small detector is scanned over different locations. The diffraction pattern expected from the double slit now reappears [9] in the coincidence rate! Although the signal and idler are both of low coherence and therefore unable to produce discernable interference fringes, the high coherence of the pump beam is still lurking, hidden in the correlation between the two outgoing beams. It only becomes visible again when the information from the two beams is brought together at the coincidence counter. The resulting two-photon interference became known as **ghost interference**.

Once ghost interference was observed, it was natural to see if a similar ghost *imaging* effect could be produced, and such an effect was quickly verified [9, 10]. Down-conversion produces signal and idler beams of low intensity. Without ever interacting with the object, idler photons strike either a large detector with spatial resolution (a CCD camera, e.g., or an array of avalanche photodiodes) or a small detector that can be scanned over the beam to map out the spatial distribution of the arriving light. The spatially resolving idler detector will be called D_2. Meanwhile, signal photons reflect off or transmit through some object (the transmission case is shown) and then proceed to a bucket detector D_1. A coincidence circuit then records

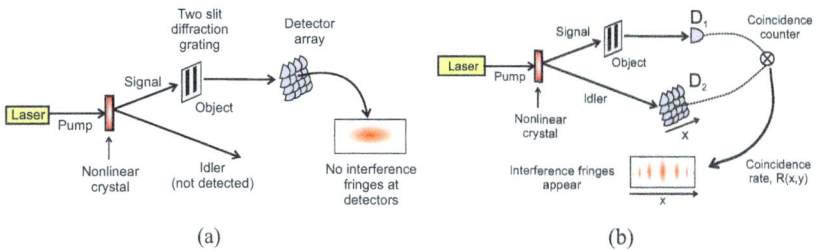

Fig. 20.1 (a) Due to their low coherence, signal photons from SPDC are unable to produce interference fringes on their own. The same is true for the idlers. (b) Suppose that the signal photons are passed through a diffraction grating, and then the signal and idler photons are detected in coincidence. Then interference fringes reemerge as the coincidence rate is plotted versus detected idler position. The high coherence of the original pump beam is still present but is distributed in a nonlocal manner between the signal and the idler beams

events in which detection occurs in both detectors within a short time window. Let $T(x, y)$ be the transmission profile of the object, where x and y are spatial coordinates in the object plane: the image should then produce an image of $T(x, y)$, possibly rescaled or inverted. $E_p(x, y)$ is the spatial profile of the laser field in the output plane of the thin crystal.

A schematic diagram of one potential quantum ghost imaging apparatus is shown in Fig. 20.2 (a second variation can be found in the Problems). In the version shown, a lens of focal length f is placed in branch 2 and the object in branch 1. d_1 and d_2 are distances from the nonlinear crystal to, respectively, the object and the lens. s_1 and s_2 are then the respective distances from the object to D_1 and from the lens to D_2. The distance from the object through the crystal and on to the lens is $s_0 = d_1 + d_2$. This distance is then arranged to obey the imaging condition,

$$\frac{1}{s_0} + \frac{1}{s_2} = \frac{1}{f}. \tag{20.1}$$

Let x_2 and y_2 be the position coordinates in the spatially resolving detector D_2. When the coincidence rate is then plotted versus position in D_2, the image of the original object reappears in the plot. The imaging process is therefore a highly nonlocal process and in fact can be used to investigate the apparently nonlocal causal structure of quantum mechanics to be discussed in Chap. 14.

The form of the imaging condition makes sense if the system is viewed using the **Klyshko "backward wave" picture**. Think of the signal as moving backward in time. So the bucket detector D_1 acts like a light source instead of a detector; the light travels through the object, reflecting off the nonlinear crystal as if it was a mirror, and then passes through the lens to reach detector D_2.

Let $r_2 = x_2\hat{i} + y_2\hat{j}$ and $r_0 = x_0\hat{i} + y_0\hat{j}$ be the positions in detector 2 and in the object, respectively. Position r_1 of detector 1 is a fixed constant. Although we won't derive it here, the coincidence rate can be shown [11] to be of the form

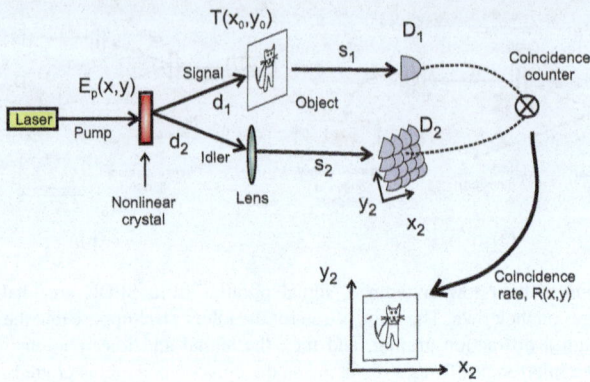

Fig. 20.2 Quantum ghost imaging. Signal photons from SPDC interact with an object but reach a detector with no spatial resolution. Idler photons don't interact with the object but reach a spatially resolving detector (either a detector with multiple pixels or a single-pixel detector that be moved to scan over the object). Neither detector by itself has sufficient information to reconstruct an image, but the image will reappear in the coincidence rate

$$R_c(r_2) = \int d^2r_0 \left| h_{s_1}(r_1, r_0) h_0(r_2, r_0) T(r_0) \right|^2, \tag{20.2}$$

where

$$h_0(r_2, r_0) = \int d^2r \, E_p(r) h_{d_2+s_2}(r_2, r) h_{d_1}(r_0, r). \tag{20.3}$$

Factors of the form $h_d(r, r')$ are the propagation or impulse response functions taking the field a distance d between points r and r'. Here, in the Klyshko picture, the pump field E_p, acting as a mirror, converts the backward-propagating signal photon entering the crystal into an idler photon going out. Clearly, the image of the object is imprinted onto the coincidence rate, with a resolution determined by the function $h_0(r_2, r)$ and independent of the resolution (or lack thereof) in D_1.

Conceptually, it is not too hard to see why ghost imaging works. Propagation directions of signal and idler directions are anticorrelated: measurement of the location of one will give the position of the other. Every time a signal photon hits the bucket detector, D_1, it opens a coincidence window; the corresponding idler then hits the spatially resolving detector, determining the location of both photons. This will happen more often when the object is transparent (or reflective), allowing more signal photons to reach D_2, and less often when the object is opaque (or nonreflective), so the coincidence rate will be proportional to the transmissivity (or reflectivity) of the object.

20.1.2 Classical Ghost Imaging

The early ghost imaging experiments used entangled pairs of photons from an SPDC source. But eventually it was realized that entanglement wasn't needed; the only requirement was spatial correlation between the light hitting the two detectors, and this could be a purely classical correlation. The drawback of using classical correlation is that visibility and signal-to-noise ratios are smaller than with entangled light sources. Ghost imaging is one of a number of examples of **quantum-inspired technologies**, where a useful effect was first discovered in an entangled system but then was found to be reproducible in classical systems. These quantum-inspired effects are often of lower quality (e.g., lower visibility or lower resolution) in the classical systems but are often of more practical use outside the lab due to the fragility of entangled quantum states in real-world environments. In addition, by using classical light sources, the intensity can be higher, leading to shorter data-collection times.

The first successful classical ghost imaging method [12] used a beam splitter to split a classical beam into two copies. One copy was directed to the image and the bucket detector, the other passed directly to the spatially resolving detector, without interacting with the object. Recall that the beam splitter splits each photon into two amplitudes, one for each path, and it is these two potential amplitudes that interfere to form the coincidence rate. An essential point is that not only average properties of photons but also fluctuations (in photon number and phase, or in quadratures) are identical in the amplitudes arriving at each detector.

Thermal light from a classical light source can also produce ghost images [13, 14]. In experiments, the thermal light is often simulated by sending the light through a ground-glass diffuser. Irregularities in size, shape, and distribution of bits of ground glass simulate the spatial fluctuations in a thermal light source. The diffuser is then rotated about its central axis to simulate the temporal fluctuations.

The terms in the correlation functions for a ghost imaging experiment can be divided into two types: phase sensitive and phase insensitive. Phase-sensitive correlations measure expectation values of the form $\langle E(x, t) E(y, t') \rangle$, while phase-insensitive correlations measure those of the form $\langle E^\dagger(x, t) E(y, t') \rangle$. Thermal ghost imaging turns out to depend only on phase-insensitive correlations, while quantum ghost imaging includes phase-sensitive correlations [15]. Because of this, visibility of classical interference patterns and imaging never exceeds 71%, whereas in principle, visibility can reach 100% with quantum (i.e., entangled) interferometry and imaging.

20.1.3 Ghost Imaging and Turbulence

One further advantage of ghost imaging is that it can, to some extent, mitigate the distorting effects of turbulence. Turbulence occurs in a fluid when thermal fluctuations cause the density of the fluid to vary in a random manner. Fluctuations are random in both space and time but follow a well-defined probability density. As density varies, this leads to fluctuations in the refractive index of the fluid, so that light passing through the medium gains randomly varying phase shifts. Interference between nearby light rays that gained slightly different phases leads to time-varying fluctuations in brightness (e.g., think of the twinkling of stars when viewed through the atmosphere). When a static image is formed by averaging these twinkling images over some data collection time, the result will be blurring and loss of resolution as a result of the turbulence. This is a problem in many applications, including ground-based optical astronomy and underwater imaging, as well as many biomedical imaging applications, where images may be taken through turbulent blood flow in capillaries or turbulent air flow in lungs. So methods that can remove effects of turbulence in imaging have wide ranges of application.

Consider a ghost imaging setup that is allowed to have a turbulent fluid in the light path. There are a number of possible variations, since the turbulent fluid could be between the light source and object, between the object and bucket detector, between the source and lens, between lens and resolving detector, or any combination of these. In addition, the light source could be coherent or incoherent. Numerical results in [16] indicated that the use of ghost imaging led to a reduction of distortion in the image, compared to standard imaging methods. Further theoretical analysis [17–19] and experimental work [20] confirmed that under many circumstances, a significant reduction in turbulent distortion occurred. The amount of reduction depended on the strength and location of the turbulence and the level of coherence of the illumination.

Given that effects of turbulence in imaging can be reduced by ghost imaging, it was natural to investigate if other distorting effect could be reduced or eliminated by using ghost imaging. It was soon shown in a variety of theoretical and experimental investigations that dispersive effects and some effects from aberration can be mitigated or completely canceled by use of correlated imaging methods [21–25].

20.2   Quantum Illumination and Quantum Radar

Quantum illumination, which uses pairs of entangled photons, was first proposed in [26]. It was further elaborated in [27–32] and has been experimentally demonstrated in [33, 34] for optical photons. It is a method for determining the presence or absence of a low-reflectivity object through the noise provided by the extraneous photons produced by the environment. As with other methods described in this chapter, the essential idea is to use coincidence counting to measure intensity-intensity correlations between two photons, as in the Hanbury Brown-Twiss interferometer. The noise is uncorrelated between the paths to the two detectors, so that it does not contribute to the correlation function and coincidence rate.

Standard illumination consists of shining a sequence of N uncorrelated photons (Fig. 20.3a) onto an object. Any reflected photons are recorded by a detector. The

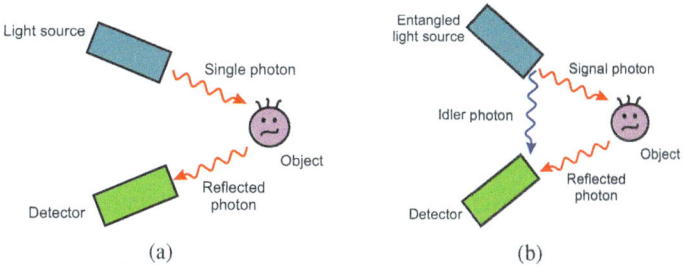

Fig. 20.3 (**a**) With standard illumination, a single photon is reflected off an object and then detected. (**b**) In quantum illumination, an entangled photon source is used. One photon is detected after reflection off the object; the other travels straight to the detector without interacting with the object. The detector only registers a count when a pair of matching photons is detected

goal is to detect the object with this light, but it is desirable to use very dim light (low N) to avoid the illumination being noticed by others. For example, the illumination might be aimed at detecting an enemy fighter jet without tipping off the pilot that the vehicle had been detected. The problem is that at low N, signal photons may be outnumbered by thermally produced photons from the environment, and there is no way of sorting the small number of signal photons from the much larger number of noise photons.

So the alternative is **quantum illumination** (Fig. 20.3b), in which the sequence of N single-photon pulses is replaced with a sequence of N entangled-photon pairs. One photon, the ancilla or idler, travels straight to the detector (possibly after a controlled delay to make the two photon arrival times coincident). The other photon, the signal, is sent to the target region. If a photon is detected coming from that region, then it must be decided whether it is the original signal photon that has been reflected back by the object or if it is a thermal noise photon. A measurement is made on the pair formed by the ancilla and the returned photon, in which the state of this pair is compared to the initial state of the ancilla-signal pair. If the original signal photon has been lost, then the ancilla goes from its initial pure state into a mixed state, as the signal degrees of freedom get traced out. Consequently, the probability of the new two-photon state being able to deceive the experimenters decreases by a factor of d, where $d = 2^N$ is the number of signal and detector modes available and N is the number of entangled pairs or the number of e-bits of entanglement. (Recall that 1 e-bit is the amount of entanglement possessed by a Bell state.) This means a corresponding increase in the signal-to-noise ratio (SNR) by a factor of d. The source of this increase can be traced back to an increase in the mutual information between the incident and reflected beams [35].

One thing that is especially notable about this effect is that as long as the initial state was entangled, the SNR improvement persists even after the entanglement has been lost through coupling to the environment. This means that the effect, though quantum in origin, may be robust enough to be used in real-world settings even after the entanglement has decayed away. The initial phase-sensitive cross-correlation between the signal and idler is stronger than allowed by classical physics, and it remains in excess of any correlations with surrounding thermal photons even after the entanglement has decayed away. The fact that few photons are needed and that it can be used at room temperature has led to suggestions that it might be useful for noninvasive biological imaging in situations where higher-intensity light could cause tissue damage.

The work of [36, 37] extends the idea from the visible to the microwave region, making it more suitable for use in remote sensing of distant objects and opening up the possibility of what the authors referred to as **quantum radar**, which is reviewed at length in Ref. [38]. A further application of quantum illumination occurs in quantum secure communication. Suppose that an eavesdropper (Eve) tries to listen in on a private communication between two participants, Alice and Bob. Such eavesdropping introduces errors in the communication. In most quantum cryptography scenarios, these eavesdropper-induced errors are beneficial because they help enforce security: detection of the excess errors signals to Alice and Bob that their communication has been compromised (see Chap. 18). Another possibility, though, is that some mechanism can be introduced that prevents errors between Alice and Bob, while decreasing the amount of information Eve can extract about the message by increasing her error rate. A method for doing this has been presented [30, 32, 39, 40] in which the use of quantum illumination can lower the error rate for Alice and Bob by orders of magnitude relative to Eve's error rate. Unlike the quantum key distribution protocols discussed in Chap. 18, this method involves direct communication between Alice and Bob, with no need to generate a secret key. It has potential to allow secure communication at gigabyte/second rates over tens of kilometers even in the presence of photon losses.

Further analysis [41] shows that in realistic situations, the advantages of quantum illumination are likely to be much more modest than the original analyses suggested. But the fact that entangled photon probes can improve performance even after the entanglement has been destroyed is an important development, which strongly corroborates the importance of entanglement for applications and shows it is likely to have unexpected benefits in other systems.

20.3 Optical Coherence Tomography

Optical coherence tomography (OCT) is an interferometric method that determines how much light is reflected as a function of distance from the detector. This can be used as a range-finding method, to detect the distance of a discrete object. But more often, it is used as an imaging method: by scanning a sample at different locations and measuring the reflectivity as a function of depth below the surface, a 3D image of the object's subsurface structure can be built up. Further, OCT is noninvasive and provides no exposure to ionizing radiation. As a result, OCT has become a common tool in many areas of biology and medicine. Here, we first look at the basic working of classical OCT. Then we look at how entangled light can

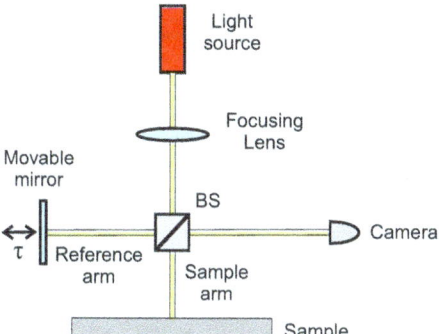

Fig. 20.4 Schematic depiction of classical optical coherence tomography (OCT). The beam splitter sends half the light into the reference arm containing the moveable mirror and the other half into the sample. The sample and reference beams then both reflect back to the beam splitter, which mixes them and sends the resulting interference pattern to the camera. Moving the mirror introduces a delay τ, and so the image can be focused to view different depths $z = v\tau$ within the sample

implement a quantum version by means of the HOM effect. This quantum OCT has advantages over the classical version in terms of signal-to-noise ratio and resolution.

20.3.1 Classical OCT

OCT [42–44] is based on the use of a Michelson interferometer. The wavefront from a classical broadband, low-coherence light source is split by a beam splitter (Fig. 20.4) into a reference arm (R) and a signal arm (S). The broadband light is described by a complex-valued spectral function $\Phi(\omega)$, giving the relative amplitude of frequency $\omega_0 + \omega$ in the light. Here, ω_0 is the central frequency, ω is the distance from that center, and Φ is assumed to be symmetric about the center, $\Phi(-\omega) = \Phi(\omega)$. After encountering the adjustable mirror and the sample, respectively, the reference and sample beams are recombined by the beam splitter before they arrive at the detector.

Consider a single frequency first. If the mirror is moved distance $\delta l = c\tau$, the extra path length in that arm gives the reference beam an extra phase factor $e^{i(\omega_0+\omega)\tau}$. Meanwhile, suppose that the beam in the sample arm penetrates to a depth z into the sample before reflecting. Let $\phi(z, \omega)$ be the phase gained during the distance-z propagation in the material, and let $r(z, \omega)$ be the complex amplitude reflectivity at depth $z = \delta l$. Then that frequency component is multiplied by a factor $r(z, \omega)e^{2i\phi(z,\omega)}$ after reflection in the medium. The 2 in the exponent accounts for the fact that it is a round trip (down to depth z and then back). Since the reflection can occur at any depth within the sample, we must therefore integrate over z, leading to the **transfer function**

$$H(\omega) = \int dz \, r(z, \omega) e^{2i\phi(z,\omega)}. \tag{20.4}$$

Thus, given an input amplitude $\Phi(\omega)$, the output amplitude at fixed frequency after recombining the two beams will be of the form

$$\left[e^{i(\omega_0+\omega)\tau} + H(\omega)\right]\Phi(\omega). \tag{20.5}$$

Now we need to add up the contributions of all the frequency components. These sum incoherently, meaning that we must add the intensities rather than the amplitudes. Since intensities are absolute squares of field amplitudes (up to an overall constant), we end up with an intensity of the form

$$I \sim \int \left|\left[e^{i(\omega_0+\omega)\tau} + H(\omega)\right]\Phi(\omega)\right|^2 d\omega \tag{20.6}$$

$$= \int \left\{1 + |H(\omega)|^2 + 2\mathrm{Re}\left[H(\omega)e^{-i(\omega_0+\omega)\tau}\right]\right\} |\Phi(\omega)|^2 \, d\omega \tag{20.7}$$

$$= \Gamma_0 + 2\mathrm{Re}\left[\Gamma_1(\tau)e^{-i\omega_0\tau}\right], \tag{20.8}$$

where

$$\Gamma_0 = \int d\omega \left[1 + |H(\omega)|^2\right] S(\omega) \tag{20.9}$$

$$\Gamma_1 = \int d\omega \, H(\omega) e^{-i\omega\tau} S(\omega) \tag{20.10}$$

$$S(\omega) = |\Phi(\omega)|^2 \tag{20.11}$$

$S(\omega)$ is called the **spectral power function** and describes the relative power at each frequency in the source beam. The self-interference term, Γ_0, is a constant (delay-independent) background term, while the cross-interference term Γ_1 leads to an interference pattern as delay τ is varied. The height of the peaks and dips at delay τ give a measure of the amount of light reflected from depth $z = v\tau$, where v is the speed of light in the material, allowing a one-dimensional profile of reflectance versus z to be constructed. By scanning the apparatus over the two-dimensional surface of the sample, this then allows a three-dimensional image to be built up of the internal subsurface structure of the material. This image can extend to a depth of a couple of millimeters into the samples interior in biological material before the signal becomes too attenuated to provide useful information, and it has resolutions of a few μm. High data acquisition rates allow a sequence of static images to be compiled to form video images of time-varying samples.

OCT has become common in a diverse range of applications. For example, it is widely used in optometry to image the corneas, retinas, and optical nerves of patients. It can be used to guide the placement of stents in cardiology or to image

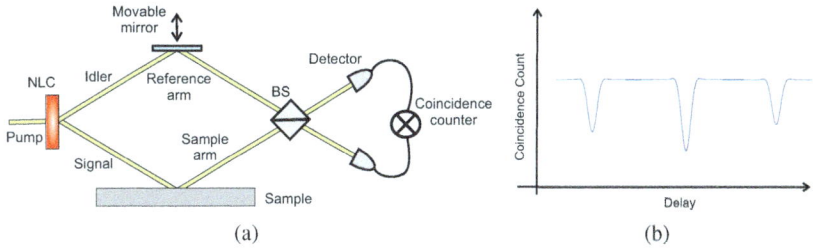

Fig. 20.5 (a) Schematic depiction of quantum optical coherence tomography (QOCT). A nonlinear crystal (NLC) splits a pump beam into signal and idler beams that enter the reference and sample arms of an interferometer. Adjusting the position of the movable mirror alters the depth within the sample that can be viewed in focus. A BS mixes the signal and idler before sending them to a pair of detectors counting in coincidence. (b) Moving the mirror introduces a delay between the two beams. When an interface occurs between two regions with different indices of refraction, an HOM dip occurs when the delay is adjusted to make the two path lengths equal. Shown is the case when the material contains three such discrete interfaces. For a continuous variation of the index of refraction, the interferogram is more complex, and image reconstruction will require computer analysis

subsurface layers of the skin in dermatology. It has also become a tool for analyzing subsurface layers of paint by art historians and art conservation experts. In recent years, OCT systems have been developed that can be integrated into smart phones [45].

20.3.2 Quantum OCT

The quantum version of optical coherence tomography [46–49] provides subsurface imaging of biological tissues and other samples, with the advantages of higher signal-to-noise ratio, reduced distortion (due to even-order dispersion cancellation, see Sect. 19.2), and with resolution that is improved by a factor of two over classical OCT. QOCT consists of placing the adjustable mirror and the sample into the two branches of an HOM experiment (Fig. 20.5), so that one photon from each down conversion pair goes through each of them. The two photons are measured in coincidence after the beam splitter. When the phase shifts in the two arms are equal, we know that the HOM effect should cause a sharp dip in the coincidence rates.

To be more quantitative, consider an entangled two-photon SPDC state of the form used in Sect. 19.2.2,

$$|\Psi\rangle = \int d\omega \ \Phi(\omega)|\omega_0 + \omega, \omega_0 - \omega\rangle, \tag{20.12}$$

where the first and second entries of the state vector are signal and idler states. Let τ be the delay due to the moveable mirror. By an analysis similar to the classical case, but with terms for two photons propagating simultaneously in opposite arms, the coincidence rate can be found:

$$R(\tau) \sim \Lambda_0 - Re[\Lambda_1(2\tau)], \tag{20.13}$$

where

$$\Lambda_0 = \int d\omega \, |H(\omega)|^2 \, S(\omega) \tag{20.14}$$

$$\Lambda_1(\tau) = \int d\omega H(\omega) H^*(-\omega) S(\omega) e^{-i\omega t}. \tag{20.15}$$

Comparing to the classical case, several advantages of QOCT are clear from these equations. For example, notice that the factor of 1 is missing in the self-interference term, resulting in a smaller background term and a higher interference visibility. Since the coincidence rate involves two photons, rather than one, the cross-interference term varies twice as fast with changing τ (as seen by the factor of 2 inside Λ in Eq. 20.13); this means that the resolution in the axial direction (along z) is doubled. Furthermore, since the sample is being probed simultaneously by two anticorrelated frequencies, $\omega_0 \pm \omega$, even-order dispersion cancellation occurs, leading to further image improvement.

When the object consists of several reflecting surfaces, the coincidence interferogram will exhibit an HOM dip for each surface (Fig. 20.5b). The amplitude of the dip will depend on the reflectance of the surface. In the more general case, where there the reflection occurs continuously as a function of z, the interferogram can give a map of the density or refractive index variations in the material.

20.4   Quantum-Enhanced Microscopy

As discussed in Chap. 2, diffraction limits the resolution that is possible with classical optical systems. However, classical resolution bounds such as the Rayleigh limit have a number of loopholes built into their assumptions, and the use of quantum mechanics allows experimenters to take advantage of these loopholes in order to achieve super-resolution, i.e., resolution beyond the classical limits. As a result, a large number of quantum-based approaches have been developed to enhance both the resolution and the signal-to-noise ratio (SNR) in microscopy [50–65]. Because the field has become large, here we give just a brief overview of a few methods and refer the reader to the references for more detail on these and other approaches. In addition to improved SNR and resolution, these quantum approaches often have additional advantages: (i) They use low photon number, and so they don't cause the tissue damage that may occur at the high intensities often required to achieve high SNR in classical imaging methods. (ii) Stray light from the surroundings can be more easily suppressed. (iii) In approaches that involve entangled photon pairs, the detected photon and the photon used to interact with the biological tissue may have very different frequencies. This is important because some frequencies may cause much more tissue damage than others (e.g., x-ray and UV are much more damaging than visible and IR) or particular frequencies may be need to stimulate a particular transition in a particular biological molecule. In addition, on the detection side, there are some frequency ranges in which detectors may be of lower resolution, more expensive, require brighter illumination, or otherwise be unfit for the particular application. So being able to have different frequencies on the detection and sample sides of the apparatus can be very useful.

In general, there are at least three approaches [50] to quantum-enhanced super-resolution microscopy: (i) Model-based approaches, based on the use of the quantum Fisher information and quantum Cramer-Rao bound, take a computational approach to estimating the image that provides the optimal match to the data. (ii) Photon anti-bunching of quantum light can be used to alter the photon-number statistics, which in turn can enhance the SNR. (iii) Entanglement can be used gather more information per detection than is possible for classical illumination.

Here we focus on the last of these possibilities, entanglement-based super-resolution microscopy.

20.4.1 Entangled Two-Photon Fluorescence Microscopy

Most high-resolution biological microscopes are based on fluorescence. A laser pulse excites the ground state of a molecule (usually a fluorescent dye) to an excited state, which then decays to lower energy again (Fig. 20.6a). The excited state usually loses some energy to non-radiative phonon scattering before decaying back to the ground state, so the photon emitted in the decay will have a lower frequency than that of the excitation pulse. This allows the decay photons from the target molecule to be easily separated from the excitation photons, which then allows the locations of the target molecules to be mapped, building up an image. The sample is usually scanned sequentially, with only a small area being viewed at a given moment.

The possibility of two-photon microscopy was first raised by Maria Goeppert Mayer in 1931 but only became practical after the work of W. Denk, J. H. Strickler, and W. W. Webb around 1990. (See [66] for a detailed review.) Here, instead of using a single excitation photon, each excitation requires *two* photons whose energies sum to the required energy (Fig. 20.6b). In general, there is no intermediate electronic state present to act as a stepping stone between the ground and excited states, although it is often useful to think of a "virtual state" through which the electron passes.

The advantages of two-photon microscopy include reduced cell damage due to the use of excitation photons with lower frequency and lower energy. In addition, there is improved sectioning ability, or in other words, thinner slices or cross-

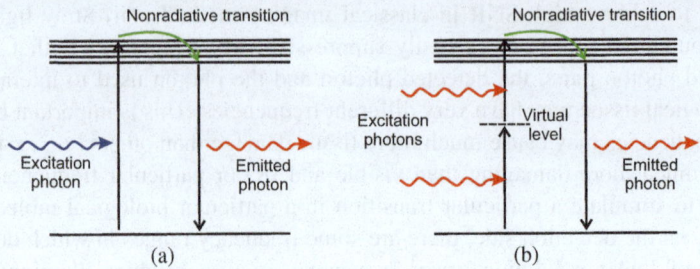

Fig. 20.6 (**a**) Single-photon fluorescence. An input photon creates an electron excitation to a higher energy level. After phonon emission leads to a non-radiative transition to a lower level, emission of a new photon of lower frequency takes the electron back to the ground state. (**b**) Two-photon fluorescence. Now two input photons are required to create the initial excitation. There does not need to be a real energy level for the electron to reside in between the two absorptions; rather, one can imagine the existence of a "virtual level" that only exists during the excitation process

sections of the sample can be viewed, leading to reduced blurring due to extraneous light scattering from nearby layers. Infrared light also scatters less than visible light in matter, allowing viewing of cross sections lying deeper inside the sample. On the other hand, resolution is often lower than similar confocal microscopes, and it takes a long time to build up images. The latter is true because the probability of two photons of appropriate energies being absorbed by the same molecule at the same time is low.

This is where the addition of entanglement provides an advantage. The experimenter can ensure that two frequency-entangled photons will be in the same spatial mode (i.e., leaving the same creation point with the same momentum direction), greatly enhancing the probability of two-photon transitions in a molecule. The two-photon cross-section (a measure of the reaction rate) can be amplified by a factor of 10^8 or more [67], allowing much faster image formation. The entangled-photon approach also enhances the SNR and allows focusing to a smaller focal region. Several approaches to two-photon entangled fluorescence microscopy can be found in Refs. [68–70], and the subject is reviewed in detail in [50].

20.4.2   Other Entangled Microscopes

Proposals for using entanglement to improve resolution in lithography go back at least to [71]. **Lithography** is the writing of patterns onto semiconductor structures using light. Lithography and microscopy (writing microscopic structures and reading structures, respectively) are clearly very closely related, so it was an obvious leap from subresolution quantum lithography to subresolution quantum microscopy. Two quantum microscopy methods that could evade the diffraction limit using coincidence counting and entanglement were proposed in [72], both involving two-photo cascades emitted by atoms. These are shown schematically in

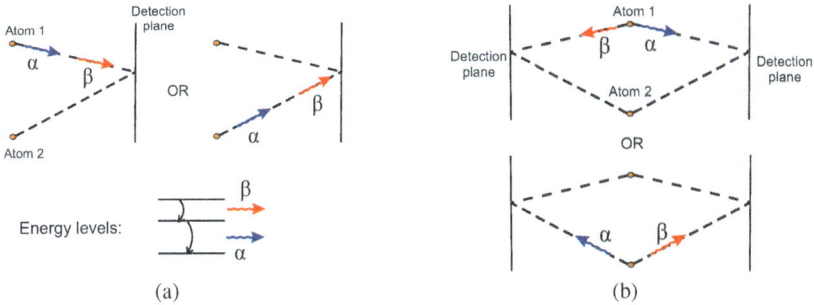

Fig. 20.7 (**a**) Two-photon entangled spatial microscopy: Two identical atoms with the energy levels shown at the bottom can each undergo a two-step decay. If a pair of decay photons arrive in the observation or detection plane, there is no way to know if the photons originated with atom 1 or atom 2. If the detector is moved parallel to the plane, an interference pattern will form. The two-photon interference pattern provides higher spatial resolution than a comparable classical, single-photon pattern. (**b**) Two-photon entangled spatial microscopy: Now each atom emits one photon to the left and one to the right. If two photons are detected in one of the detection planes, we again have no information about which atom emitted them. Because different frequency photons are interfering in each plane, the spectral resolution is improved, allowing the atomic energy levels to be determined with higher precision

Fig. 20.7. Both involve two-step atomic decays, in which an excited electron first decays to a short-lived unstable state and then to the ground state, emitting two photons in the process. The two photons are often path- and frequency-entangled. Such two-photon atomic decays were a common source of entanglement before the widespread use of SPDC.

In Fig. 20.7a, two atoms are located a small distance apart. Each has an amplitude to emit a two-photon pair, with the choice of decays arranged so that both photons will be emitted in the same direction. When a pair of photons are detected on the screen, it is unknown which of the two atoms emitted the pair, and so the state arriving at the detector is path-entangled. Labelling the two photons α and β and the paths from the two atoms 1 and 2, the state will be

$$|\Psi\rangle = \frac{1}{2}\Big(|\alpha, \beta\rangle_1 \otimes |0\rangle_2 + |0\rangle_1 \otimes |\alpha, \beta\rangle_2\Big), \tag{20.16}$$

where $|0\rangle$ is the state with no photons. Here, $|\alpha, \beta\rangle$ represent the *spatial* modes of the two photons, as specified by their path. As discussed in Chaps. 9 and 10, two-photon interference and imaging have resolution that is improved over single-photon approaches by a factor of two. Essentially, two-photon interference patterns with light of wavelength λ oscillate at the same rate as single-photon patterns at wavelength $\lambda/2$. Similarly, an N photon generalization of this system should lead to an N-fold resolution improvement.

A variation on the same arrangement again uses atoms emitting a two-photon cascade, but now the decay is chosen so that the two photons move in opposite directions to *different* observation screens Fig. 20.6b. Path lengths on the two sides may be different, leading to different arrival times, but are still not known which atom emitted the pair. α and β have different frequencies. So, there is now *frequency*-entanglement, since it is not known which of the two different-frequency photons arrived at which detector. This is a form of **spectral microscopy**, in which the energy levels of the emitting atom can be measured with line-widths that are half as large as those of standard single-photon microscopy.

Another form of entangled microscopy was experimentally demonstrated in [73]. This was an entangled-photon version of **differential interference contrast (DIC) microscopy** or **Nomarski microscopy**. DIC uses polarized light, sending two spatially displaced, mutually coherent, and orthogonally polarized beams into the sample. After passing through the sample, the polarization information is erased using diagonal polarizers, and then the beams are recombined. Interference provides information about the composition of the material at each point in the sample, since changes in index of refraction (which is highly sensitive to polarization) will lead to shifts of the interference pattern. There is thus prominent changes in the interference at the edges between different materials, allowing high-contrast images to form. In a proof of principle experiment, Ref. [73] showed that a variation on the DIC microscope that used two-photon entanglement had a SNR of 1.35 times better than a comparable classical phase contrast microscope, also allowing resolution at greater depths within the sample.

The entangled phase microscope of Ref. [73] is also a confocal microscope [74–76]. In confocal microscopy, two lenses are used to narrow the field of view to a small region formed by the overlap of images of two pinholes in opaque screens, as in Fig. 20.8. Only light that was in the overlap of *both* the illumination lens focal spot and the detector lens focal spot will contribute to the image. This weeds out stray light from other points in the sample that would otherwise reduce the visibility of the material point being viewed. The microscope then scans over the sample, viewing it one point at a time to construct a full image. Confocal microscopes are well-known to improve resolution in both the imaging plane and in the transverse

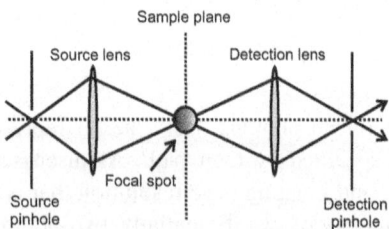

Fig. 20.8 Schematic depiction of a confocal microscope. Two pinholes exist in opaque screens lying between the sample and light source and between the sample and detector. The images of these pinholes, via a pair of lenses, overlap in the sample, and only light that originates in this overlap region will be detected. This removes stray like from elsewhere in the sample, narrowing the point spread function and improving the resolution in all three spatial dimensions

direction (perpendicular to the image plane). Several proposals have been made to use entangled photon pairs to further shrink the focal spot in confocal microscopes [77, 78], which would lead to greatly enhanced resolution.

Several other entangled microscopy techniques involving $N00N$ states, structured light, and other approaches have been either proposed or demonstrated, and some of these can be found in [79–83].

Problems

Several of the problems below refer to the ghost imaging setup in Fig. 20.9, in which the signal and idler both pass through lenses of focal length f. Up to overall constants, the coincidence rate $R(r_1, r_2)$ and the two-photon amplitude $A(r_1', r_2')$ at the output of the crystal are

$$R(r_1, r_2) = \left| \int h_s(r_1, r_1') h_i(r_2, r_2') A(r_1', r_2') \, d^2r_1' d^2r_2' \right|^2 \tag{20.17}$$

$$A(r_1', r_2') = \int \frac{d^2q}{(2\pi)^2} e^{iq \cdot (r_1' - r_2')} U(q) V(-q), \tag{20.18}$$

where U and V are momentum-space functions determined by the down conversion process. The impulse response function in the signal branch is given by

$$h_s(r_1, r_1') = -\frac{i}{\lambda f} e^{-\frac{2\pi i}{\lambda f} r_1 \cdot r_1'} T(r_1').$$

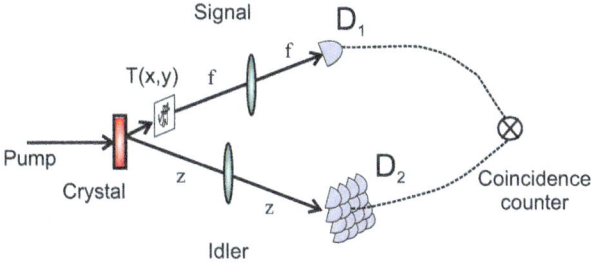

Fig. 20.9 Ghost imaging setup for Problems 1 and 2. The single-pixel detector and the target with transmission amplitude $T(r)$ are both one focal length from the signal lens. (The distance from the nonlinear crystal to the target is negligible.) The idler lens is distance z from both the crystal and the spatially resolving detector

1. Referring to the setup of Fig. 20.9, consider the case where $z = f$ in the bottom (idler) branch. In this case, the idler impulse response function is

$$h_i(r_2, r_2') = -\frac{i}{\lambda f} e^{-\frac{2\pi i}{\lambda f} r_2 \cdot r_2'}.$$

Show that the coincidence rate is proportional to the squared Fourier transform of the target:

$$R(r_1, r_2) \sim \left| U\left(-\frac{2\pi}{\lambda f} r_2\right) V\left(\frac{2\pi}{\lambda f} r_2\right) \tilde{T}\left(\frac{2\pi}{\lambda f}(r_1 + r_2)\right) \right|^2.$$

This is a ghost *diffraction* pattern.

2. In contrast to the previous problem, assume that $z = 2f$ in the bottom branch. Now,

$$h_i(r_2, r_2') = \delta(r_2 + r_2') e^{-\frac{\pi i}{\lambda f}|r_2|^2}.$$

Show that this leads to a ghost *imaging* output, with a coincidence proportional to the squared target amplitude itself (not the Fourier transform):

$$R(r_1, r_2) \sim \left| U\left(\frac{2\pi}{\lambda f} r_1\right) V\left(-\frac{2\pi}{\lambda f} r_1\right) \right|^2 |T(-r_2)|^2.$$

3. Look up Ref. [72], and fill in the detailed steps going from Eq. (18) to Eq. (2) of that paper.

4. An optical wave on OCT is of the form $E(z, t) = A \sin \omega\left(\frac{x}{v} - t\right)$, where v is the speed of the wave in the material, and we are ignoring the transverse directions. The reference beam travels a distance x_r, while the sample beam travels a distance $x_2 = x_1 + \delta z$. The two beams are recombined, and the intensity $\langle I \rangle(x) = \langle E^2 \rangle$ of the interference pattern is measured, where the average is taken over a time much longer than the period of the optical wave. (Assume both beams travel through material of roughly the same refractive index.)

    (a) Calculate the average intensity $\langle I \rangle$.

    (b) Show that $\langle I \rangle \to 0$ when $\delta z \gg v\tau_{coh}$, where τ_{coh} is the coherence time.

    (c) If 600 nm light with $\tau_{coh} \approx 5 \times 10^{-11}$ s is used, estimate the maximum depth that can be viewed inside the tissue. Assume refractive index $n = 1.4$.

5. In an OCT setup, assume a transfer function $H(\omega) = r_1 + r_2 e^{2in\omega L/c}$, where r_1 and r_2 are reflectances at the front and back surfaces of a sample of thickness L and n is the refractive index. Calculate Λ_0 and $\Lambda_1(\tau)$ of Eqs. 20.14 and 20.15 for a beam of spectral power $S(\omega)$. (The answer should depend on the Fourier transform $\tilde{S}(\tau)$ of $S(\omega)$. Assume the spectrum is normalized, $\int d\omega\, S(\omega) = 1$, and that it varies slowly enough that integrals such as $\int d\omega\, S(\omega) \sin \omega t$ or $\int d\omega\, S(\omega) \cos \omega t$ will average to zero.)

6. Using the propagation functions from Sect. 2.2, explicitly construct the function h_0 of Eq. 20.3. For simplicity, assume the pump field is a plane wave, so that E_p is constant, and assume pump and idler frequencies are equal.

References

1. C. Moodley, A. Forbes, Sci. Rep. **12**, 10346 (2022)
2. Z.Y.J. Ou, *Multi-Photon Quantum Interference* (Springer, Berlin, 2007)
3. M.I. Kholobov (ed.) *Quantum Imaging* (Springer, Berlin, 2007)
4. Y. Shih, *An Intro. to Quantum Optics: Photon and Biphoton Physics* (CRC Press, Boca Raton, 2011)
5. L. Basano, P. Ottonello, Am. J. Phys. **75**, 343 (2007)
6. A. Gatti, E. Brambilla, M. Bache, L.A. Lugiato, Ghost imaging, in *ghostkolobov*
7. Y. Shih, arXiv:0805.1166 [quant-ph] (2008)
8. B.I. Erkmen, J.H. Shapiro, Phys. Rev. A **77**, 043809 (2008)
9. D.V. Strekalov, A.V. Sergienko, D.N. Klyshko, Y.H. Shih, Phys. Rev. Lett. **74**, 3600 (1995)
10. T.B. Pittman, Y.H. Shih, D.V. Strekalov, A.V. Sergienko, Phys. Rev. A **52**, R3429 (1995)
11. A.F. Abouraddy, B.E.A. Saleh, A.V. Sergienko, M.C. Teich, J. Opt. Soc. Am. B **19**, 1174 (2002)
12. R.S. Bennink, J. Bentley, R.W. Boyd, J.C. Howell, Phys. Rev. Lett. **92**, 033601 (2004)
13. A. Valencia, G. Scarcelli, M. D'Angelo, Y.H. Shih, Phys. Rev. Lett. **94**, 063601 (2005)
14. F. Ferri, D. Magatti, A. Gatti, M. Bache, E. Brambilla, L.A. Lugiato, Phys. Rev. Lett. **94**, 183602 (2005)
15. B.I. Erkmen, J.H. Shapiro, Phys. Rev. A **77** 043809 (2008)
16. J. Cheng, Opt. Exp. **17**, 7916 (2009)
17. P. Zhang, W. Gong, X. Shen, S. Han, arXiv:1005.5011 (2010)
18. C. Li, T. Wang, J. Pu, W. Zhu, R. Rao, Appl. Phys. B **99**, 599 (2010)
19. K.W.C. Chan, D.S. Simon, A.V. Sergienko, N.C. Hardy, J.H. Shapiro, P. Ben, Dixon, G. Howland, J.C. Howell, J.H. Eberly, M.N. O'Sullivan, B. Rodenburg, R.W. Boyd, Phys. Rev. A **84**, 043807 (2011)
20. P.B. Dixon, G. Howland, K.W.C. Chan, C. O'Sullivan-Hale, B. Rodenburg, N.C. Hardy, D.S. Simon, J.H. Shapiro, A.V. Sergienko, R.W. Boyd, J.C. Howell, Phys. Rev. A **83**, 051803(R) (2011)
21. C. Bonato, D.S. Simon, P. Villoresi, A.V. Sergienko, Phys. Rev. A **79**, 062304 (2009)
22. D.S. Simon, A.V. Sergienko, Phys. Rev. A **80**, 053813 (2009)
23. D.S. Simon, A.V. Sergienko, Phys. Rev. A **82**, 023819 (2010)
24. M.V. da Cunha Pereira, L.A.P. Filpi, C.H. Monken, Phys. Rev. A **88**, 053836 (2013)
25. L.A.P. Filpi, M.V. da Cunha Pereira, C.H. Monken, Opt. Exp. **23**, 3841 (2015)
26. S. Loyd, Science **321**, 1463 (2008)
27. S.H. Tan, B.I. Erkmen, V. Giovannetti, S. Guha, S. Lloyd, L. Maccone, S. Pirandola, J.H. Shapiro, Phys. Rev. Lett. **101**, 253601 (2008)
28. S. Guha, B.I. Erkmen, Phys. Rev. A **80**, 052310 (2009)
29. C. Weedbrook, S. Pirandola, J. Thompson, V. Vedral, M. Gu, arXiv:1312.3332 (2013)
30. J.H. Shapiro, Phys. Rev. A **80**, 022320 (2009)
31. J.H. Shapiro, S. Lloyd, New. J. Phys. **11**, 063045 (2009)
32. Z. Zhang, M. Tengner, T. Zhong, F.N.C. Wong, J.H. Shapiro, Phys. Rev. Lett. **111**, 010501 (2013)
33. E.D. Lopaeva, I. Ruo Berchera, I.P. Degiovanni, S. Olivares, G. Brida, M. Genovese, Phys. Rev. Lett. **110**, 153603 (2013)
34. E.D. Lopaeva, I. Ruo Berchera, S. Olivares, G. Brida, I.P. Degiovanni, M. Genovese, Phys. Scr. **T160**, 014026 (2014)

35. S. Ragy, I. Ruo Berchera, I.P. Degiovanni, S. Olivares, M.G.A. Paris, G. Adesso, M. Genovese, J. Opt. Soc. Am. B **31**, 2045 (2014)
36. S. Barzanjeh, S. Guha, C. Weedbrook, D. Vitali, J.H. Shapiro, S. Pirandola, Phys. Rev. Lett. **114**, 080503 (2015)
37. S. Barzanjeh, S. Pirandola, D. Vitali, J.M. Fink, Sci. Adv. **6**, eabb0451 (2020)
38. R.G. Torromé, N. Ben Bekhti-Winkel, P. Knott, arXiv:2006.14238v3 [quantum-ph] (2021)
39. J.H. Shapiro, Z. Zhang, F.N.C. Wong, Quant. Inf. Process. **13**, 2171–2193 (2014)
40. Q. Zhuang, Z. Zhang, J. Dove, F.N.C. Wong, J.H. Shapiro, arXiv:1508.01471 [quant-ph] (2015)
41. J.H. Shapiro, IEEE Aerosp. Electron. Syst. Mag. **35**, 8 (2020)
42. J.M. Schmitt, IEEE J. Sel. Top. Quant. Electron. **5**, 1205 (1999)
43. B.E. Bouma, et al., Nat. Rev. Methods Primers **2**, Article number 79 (2022)
44. M.E. Brezinski, *Optical Coherence Tomography: Principles and Applications* (Academic Press, Cambridge, 2006)
45. J.D. Malone, I. Hussain, A.K. Bowden, Biomed. Opt. Exp. **14**, 3138 (2023)
46. A.F. Abouraddy, M.B. Nasr, B.E.A. Saleh, A.V. Sergienko, M.C. Teich, Phys. Rev. A **65**, 053817 (2002)
47. M.B. Nasr, B.E.A. Saleh, A.V. Sergienko, M.C. Teich, Phys. Rev. Lett. **91**, 083601 (2003)
48. M.B. Nasr, B.E.A. Saleh, A.V. Sergienko, M.C. Teich, Opt. Exp. **12**, 1353 (2004)
49. M.C. Teich, B.E.A. Saleh, F.N.C. Wong, J.H. Shapiro, Quant. Inf Proces. **11**, 903 (2012)
50. W.P. Bowen, et al., Contemp. Phys. 1–25. https://doi.org/10.1080/00107514.2023.2292380
51. C.A. Casacio, L.S. Madsen, A. Terrasson, et al., Nature **594**, 201 (2021)
52. M.A. Taylor, W.P. Bowen, Phys. Rep. **615**, 1 (2016)
53. Y. Zhang, Z. He, X. Tong, et al., Sci. Adv. **10**, eadk1495 (2024)
54. T. Li, F. Li, X. Liu X, et al., Optica **9**, 959 (2022)
55. Z. Xu, K. Oguchi, Y. Taguchi, et al., Opt. Lett. **47**, 5829 (2022)
56. K.E. Dorfman, F. Schlawin, S. Mukamel, Rev. Mod. Phys. **88**, 045008 (2016)
57. M.G. Raymer, T. Landes, M. Allgaier, et al., Optica **8**, 757 (2021)
58. G.B. Lemos, V. Borish, G.D. Cole, et al., Nature **512**, 409 (2014)
59. E.A. Santos, T. Pertsch, F. Setzpfandt, et al., Phys Rev Lett. **128**, 173601 (2022)
60. J.M. Cui, F.W. Sun, X.D. Chen, et al., Phys. Rev. Lett. **110**, 153901 (2013)
61. Y. Israel, R. Tenne, D. Oron, et al., Nat Commun. **8**, 14786 (2017)
62. M. Unternährer, B. Bessire, L. Gasparini, et al., Optica **5**, 1150 (2018)
63. R. Tenne, U. Rossman, B. Rephael, et al., Nat. Photonics **13**, 116 (2019)
64. H. Defienne, P. Cameron, B. Ndagano, et al., Nat. Commun. **13**, 3566 (2022)
65. N.P. Mauranyapin, A. Terrasson, W.P. Bowen, Adv. Quantum. Technol. **5**, 2100139 (2022)
66. P.T.C. So, C.Y. Dong, B.R. Masters, K.M. Berland, Ann. Rev. Biomed. Eng. **2**, 399 (2000)
67. D. Tabakaev, A. Djorović, L. LaVolpe, et al., Phys. Rev. Lett. **129**, 183601 (2022)
68. M.C. Teich, B.E. Saleh, Česk. Čas. Fyz. **47**, 3 (1997)
69. G. Scarcelli, S.H. Yun, Opt. Exp. **16**, 16189 (2008)
70. O. Varnavski, T. Goodson III, Am. Chem. Soc. **142**(30), 12966 (2020)
71. A.N. Boto, P. Kok, D.S. Adams, S.L. Braunstein, C.P. Williams, J.P. Dowling, Phys. Rev. Lett. **85**, 2733 (2000)
72. A. Muthukrishnan, M.O. Scully, M.S. Zubairy, J. Opt. B: Quant. Semiclass. Opt. **6**, S575 (2004)
73. T. Ono, R. Okamoto, S. Takeuchi, Nat. Commun. **4**, 2426 (2013)
74. M. Minsky, Scanning **10**, 128 (1988)
75. R.H. Webb, Rep. Prog. Phys. **59**, 427 (1996)
76. J.C. Mertz, *Introduction to Optical Microscopy* (Roberts and Company, Greenwood Village, 2009)
77. D.S. Simon, A.V. Sergienko, Opt. Exp. **18**, 9765 (2010)
78. D.S. Simon, A.V. Sergienko, Opt. Exp. **21**, 22147 (2010)
79. S. Karmakar, R.E. Meyers, Y. Shih, J. Biomed. Opt. **20**, 016008 (2015)
80. A. Classen, J. von Zanthier, M. Scully, G.S. Agarwal, Optica **4**, 580 (2017)

81. O. Schwartz, J.M. Levitt, R. Tenne, S. Itzhakov, Z. Deutsch, D. Oron, Nano. Lett. **13**, 5832 (2013)
82. M. Li, C.-L. Zou, D. Liu, G.-P Guo, G.-C. Guo, X.-F. Ren, Phys. Rev. A: At., Mol., Opt. Phys. **98**, 012121 (2018)
83. G.R. Jin, W. Yang, C.P. Sun, Phys. Rev. A **95**, 013835 (2017)

Chapter 21
Topological Materials

Since the discovery of the quantum Hall effect in 1980, topological states of matter have become a subject of intense interest in condensed matter physics, a trend that has only accelerated since the discovery of topological insulators and the realization that it may be possible to create many-electron excitations that can mimic exotic states such as Majorana fermions and anyons. Many of these topologically based phenomena have now found analogs in other areas such as photonic and acoustic systems. The combination of topological ideas with advances in quantum mechanics has potential to produce a broad range of new effects and technological advances, such as topological quantum computing. In this chapter, we give an introduction to topological ideas in physics, along with brief discussions of a few potential applications.

21.1 Why Topology?

Geometry studies the relations between angles and lengths of rigid objects. **Topology** arises when the requirement of rigidity is dropped: any two objects that can be bent or stretched continuously into each other, without the objects being ripped, are topologically equivalent. In other words, topology is the study of properties that are invariant under *continuous* deformations. For example, a pancake is topologically equivalent to a loaf of bread, since one can be stretched and compressed into the shape of the other, but neither is equivalent to a pretzel, since holes would have to be ripped into them to produce a pretzel shape. For decades, topology has been an important tool in theoretical physics, especially quantum field theory and condensed matter physics, and it has now become essential in a number of more applied areas such as data analysis and artificial intelligence. A surge of new interest in applied topology has risen with the discovery of topological excitations in solids that led to unusual states of matter, such as the quantum Hall effect, topological insulators,

© The Author(s), under exclusive license to Springer Nature Switzerland AG 2025
D. S. Simon, *Introduction to Quantum Science and Technology*, Undergraduate
Texts in Physics, https://doi.org/10.1007/978-3-031-81315-3_21

Majorana fermions, and anyons, as well as with the realization that some of these could be used to implement topological quantum computing, a form of quantum computing in which the states of qubits are protected by topological considerations, reducing the need for error correction. Many of the considerations in this chapter first arose in condensed matter systems, but they have since given birth to analog effects in photonic, acoustic, electronic, and even mechanical systems.

Most properties of interest in physics can be determined by local measurements: measurements of electric field, energy, charge, and so on determine the amount of these quantities *at a specific point*, or at least within a small volume that is covered by the detection system. These are measurements of *local* properties. Topological properties, however, are *global*, and cannot be determined by local measurements. For example, looking at an infinitesimal region of a plane or of a torus, they seem identical. To distinguish between them, measurements have to made over large regions; for example, a series of closed curves can be traced out on the surface and then compared, to see if they can all be continuously deformed into each other. In the planar case, they can. But in the case of the torus, closed loops that circle the holes different numbers of times are not equivalent and cannot be stretched or twisted into each other without leaving the surface of the torus. So global, large-scale measurements are capable of detecting the difference, while local measurements cannot.

Topological properties are preserved under local perturbations and can be altered only by making changes to the global structure. As a result, when topological properties of a system are conserved, the system displays a high degree of robustness against many types of disturbances. We will see that this leads, for example, to immunity to scattering and loss in some systems, even in the presence of high levels of impurities and disorder.

In systems discussed in this chapter, topology will enter when the wavefunctions wrap in a non-trivial way around the Brillouin zone. Topologically distinct states will exist as long as an energy gap remains between the allowed energy bands. Transitions between different topologies can only occur when the energy gap closes. At those closing points, topological invariants become undefined, and the global structure of the wavefunctions changes. The wavefunction topology will turn out to be closely related to the Berry phase discussed in Chap. 15.

21.2   Topological Invariants

To a large extent, topological properties of a space are determined by the set of holes in the space. (The word "space" in this chapter will always mean "topological space," a space which has structure that allows continuity and limits to be well defined; see [1] for a more precise definition.) So, for example, the real plane is simply connected (it has no holes), and so it is not topologically equivalent to the torus, despite the fact that they are both two dimensional. As discussed above, distinguishing between the two spaces is possible because every closed path in

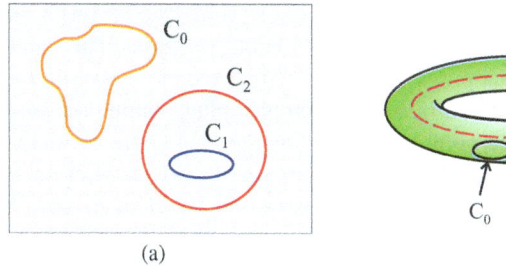

Fig. 21.1 (**a**) In the plane, all closed curves can be continuously deformed into each other. All the curves shown are topologically equivalent to each other, and, in particular, each can be continuously contracted to a single point. (**b**) Topologically inequivalent curves on a torus. Curves like C_0 can be continuously contracted to a single point, but curves like C_1 and C_2 cannot because the holes of the torus obstruct the contractions. None of these three curves can be contracted into each other, so they represent three distinct topological classes

the plane can be continuously contracted to a point (Fig. 21.1a) or to any other closed curve. But this is not possible for the torus. There, two sets of closed curves exist that cannot be deformed to a point or to each other (see Fig. 21.1b). By determining the set of inequivalent classes of curves that cannot be deformed to each other, the topological structure of the space can be determined. This is the subject of **homotopy theory**, which we will only discuss very sketchily here; see the references for more detail.

The closed curves in Fig. 21.1 can be viewed as images of circles under some mapping. Recall that a mapping is just a rule taking points in one space to points in another space. So, for example, the exponential function $f_n(x) = e^{inx}$ for fixed n and for real-valued variable x maps the real number line $\mathbb{R}$ onto the unit circle S^1 in the complex plane, $\mathbb{C}$. So f can be viewed either as a map $f_n : \mathbb{R} \to S^1$ or as $f_n : \mathbb{R} \to \mathbb{C}$. Similarly, the map $g_{mn}(x, y) = e^{imx+iny}$ for real-valued variables x and y and fixed values of m and n can be seen as a map from the real plane to the two-dimensional torus, $g : \mathbb{R}^2 \to T^2$.

Keep in mind that the word "torus" always means the hollow shell forming the surface of the donut shape, not the solid interior of the donut. Similarly, the circle S^1 and sphere S^2 also always mean the circumference and the hollow spherical shell, not the solid disk or ball formed by their interiors.

We often view a space (called the **target space**) as the image of some standard reference space (e.g., a circle or sphere) under a mapping, and then we categorize the target space based on the set of equivalence classes of images under continuous deformation of the map. For example, consider the plane. The reference space can be taken to be the unit circle, and then we can look at mappings from the circle to the plane and ask if these images are all equivalent. In fact, they are: every image of the circle is a closed curve that can be continuously collapsed to a point, and every closed curve can be continuously deformed into every other closed curve. So there is only one equivalence class (or **homotopy class**) of curves: the identity class consisting of curves contractible to a point. The plane is topologically trivial.

Now however, consider the plane with a single point (the origin) removed. The introduction of this hole changes the allowed homotopy classes. This can be seen by looking at the family of mappings $f_n(\theta) = e^{in\theta}$ indexed by integer n. f_n wraps the unit circle n times around the origin in the complex plane, taking the point at angle θ on the initial circle to the point at angle $n\theta$ on the final circle. So mappings with different n values produce images that wrap around the origin different numbers of times. These images can't be deformed into each other without crossing the origin, but since the origin is not part of the space, this is not allowed. The curves that circle the hole different numbers of times are now topologically inequivalent to each other.

The number of times the curve circles the origin as θ runs from zero to 2π is called the **winding number** of the map f_n and is usually denoted ν. Thus, in this example, $\nu = n$. More generally, consider a map γ that takes the unit circle to a closed curve in the complex plane, $\gamma(\theta) \to z(\theta) = r(\theta)e^{i\phi(\theta)}$. $r(\theta)$, which is required to be nonzero, plays no essential role, so it can simply be set to a constant value. The winding number about the origin is then given by

$$\nu = \frac{1}{2\pi i} \int \frac{dz}{z} = \frac{1}{2\pi i} \int d(\ln z) = \frac{\Delta\theta}{2\pi}, \tag{21.1}$$

where the integral runs from $z(\theta = 0)$ to $z(\theta = 2\pi)$. Bending and stretching the image curve continuously will not change the winding number, so ν is an example of a **topological invariant**, a number which is invariant under continuous deformations of some curve or space. Two spaces cannot be topologically equivalent if they have different sets of topological invariants. The winding number characterizes the **first homotopy class** of the target space. The set of first homotopy classes of a space M forms a mathematical group called the **first homotopy group**, or the **fundamental group**, denoted by $\pi_1(M)$ [2–4]. (See Appendix E for a brief introduction to groups.)

So, the plane, which has a fundamental group consisting of a single equivalence class of curves, can be written as $\pi(M) = 1$, where 1 represents the identity element (in this case, the *only* element) of the group. The punctured plane has a homotopy group whose elements are characterized by a single integer, so it has the fundamental group $\pi_1(M) = \mathbb{Z}$, where $\mathbb{Z}$ represents the set of integers. (This set is a group under ordinary addition, with the number 0 playing the identity element.)

There may be multiple winding numbers. So, for example, consider again the map $g_{mn}(x, y) = e^{imx+iny}$. Assume that m and n are integers. This is a map from the xy-plane to a product of a pair of circles. Moving around these two circles will sweep out a two-dimensional torus as in Fig. 21.1b, T^2, so g can be viewed as a map from the plane $\mathbb{R}^2$ to the two-torus, T^2, $g_{mn} : \mathbb{R}^2 \to T^2$. The torus can also be thought of as a square with opposite sides identified: in Fig. 21.2, if you glue the two vertical sides together, you get a cylinder. Then gluing the ends of the cylinder together (identifying the top with the bottom), you obtain a torus. This is similar to a video game in which a spaceship flying off one side of the screen reappears on the opposite side: the space on which the spaceship lives is a torus. In the case of the torus, two winding numbers can be defined (call them μ and ν). μ counts how many

Fig. 21.2 The torus can be viewed as a rectangle with opposite sides identified. Gluing the two sides labeled *A* together gives a cylinder; gluing the *B* sides together yields a torus

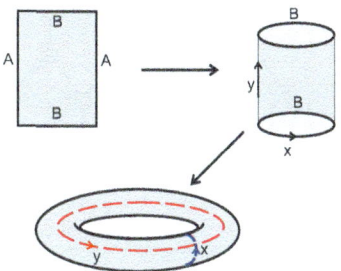

times the curves circles the "big" direction of the torus, and ν counts the number of times it circles the "small" direction. Two curves with different values of the pair (μ, ν) cannot be continuously deformed into each other without leaving the surface of the torus and traversing one of the holes. The fundamental group of the torus is therefore made up of two copies of the integer group: $\pi(M) = \mathbb{Z} \times \mathbb{Z}$.

This can be generalized to more dimensions. Instead of using the unit circle S^1 as the reference space, higher-dimensional spheres S^m can also be used. Then one can ask how many times the initial sphere wraps around a unit sphere in the target space and whether these differently wrapped image spheres can be deformed into each other. This gives new integers, which characterize the **higher homotopy classes**. The homotopy class using the dimension-m sphere S^m is called the mth homotopy class. For a topological space of dimension n, all of the mth homotopy classes up to the nth are independent of each other and may or may not be trivial. All of the mth homotopy classes for $m > n$ are trivial. The set of distinct homotopy classes of images of the n-sphere form the **nth homotopy group**, $\pi_n(M)$.

For oriented spaces of dimension $2n$, an important invariant will be the nth **Chern number**. The first Chern number, which is defined in two dimensions and characterizes $\pi_{2n}(M)$, will be encountered below.

21.3 The SSH Model: An Introduction to Topological Matter

A simple one-dimensional model that illustrates the existence of topological states of matter is the **Su-Schrieffer-Heeger (SSH) model**. It first arose as a model of polyacetylene, but it has since been shown that a number of other systems can be mapped to it. The basic SSH model demonstrates the existence of nonzero winding numbers, topological defects, solitons, and scattering-free propagation. The model has been generalized in many ways, for example, adding non-Hermitian couplings (loss and gain), next-nearest-neighbor couplings, and other embellishments, which give rise to additional effects. Here we describe just the basic model.

Polyacetylene is a polymer formed by a linear chain of hydrogen and carbon atoms. Recall that carbon has four valence electrons, and so it tends to form four bonds to neighboring atoms. Each carbon bonds to two neighboring carbons

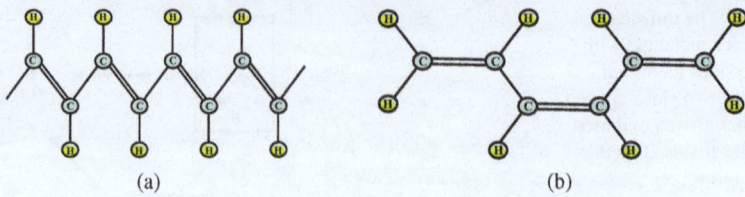

Fig. 21.3 The two forms of polyacetylene: (**a**) trans form and (**b**) cis form

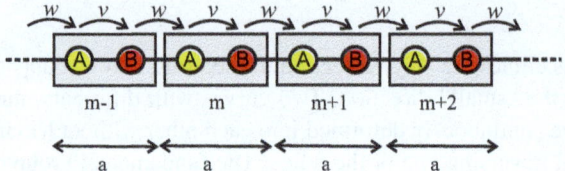

Fig. 21.4 The SSH model. Unit cells are labeled by a discrete index $m \in \{1, 2, 3, \ldots, N\}$, and within each unit cell, there are two inequivalent lattice sites, labeled A and B. The intracellular hopping amplitude from A to B within the same unit cell is v, while w is the intercellular hopping amplitude between A of one cell and B of the adjacent cell. The ratio $\frac{v}{w}$ determines the topological state of the system. The lattice spacing between each unit cell is a

and one hydrogen. To use up all the valence electrons, each carbon therefore has to participate in one double bond. There are two ways to do this, called cis-polyacetylene and trans-polyacetylene, as shown in Fig. 21.3. The cis form has a non-degenerate ground state and is thermodynamically unstable at room temperature, which makes it uninteresting for our purposes. However, the trans form has two degenerate electronic ground states: focusing on the carbon chain, the single and double bonds can be at the locations drawn in the figure, or they can be interchanged. These two states have equal energies and the extra electron can tunnel back and forth, causing the position of the double bond to fluctuate back and forth.

The trans-polyacetylene system can be drawn in a more schematic form as in Fig. 21.4. The chain of molecules is written as the collection of unit cells (the rectangular boxes), with two sublattice sites inside each cell. These sub-sites are called A and B, and they represent the two inequivalent positions of the excess electron or, equivalently, the location of the double bond. The lines connecting the A and B sites represent the hopping amplitudes between the sites. The *intracell* hopping amplitudes between A and B sites within the same unit cell are denoted v, and the *intercell* hopping amplitudes between A and B sites in different unit cells are called w. The lattice spacing between unit cells will be denoted a. We assume that v and w are independent of position along the chain, and without loss of generality, we assume that they are real-valued. We label the unit cells by an integer m, where m runs from 1 to N and N is the number of unit cells. Notice also that a particle in the A state can only hop to a B site and vice versa. There are no hoppings within the same sublattice, i.e., no A-to-A or B-to-B hoppings. This is important because it ensures that the chiral symmetry discussed below is not broken.

Now, this more abstract SSH model, divorced from the specific example of the polyacetylene model, can be thought of as a model of a particle hopping between different sites (the generalization of the double bond hopping around). The excess particle in the more abstract setting could be any particle (electron, photon, etc.). In the limit of infinite chain length, $N \to \infty$, the choice of which sites to label A and which to call B is arbitrary (meaning that the choice of which bonds correspond to v and which to w is also arbitrary). But when N is finite, so that the chain has edges, the ambiguity is removed: the last link on each end of the finite chain will correspond to amplitude v. For the moment, we assume that either $N \to \infty$ or that there are periodic boundary conditions in which the $n = N$ site is connected back to the $n = 1$ site, allowing us for now to ignore edge effects.

Now consider a single excess particle that can hop between different locations over time. We can denote the states of the system by a set of vectors $|m\alpha\rangle = |m\rangle \otimes |\alpha\rangle$, where $m = 1, 2, \ldots N$ labels the unit cell and $\alpha = A, B$ labels the sub-site within the unit cell. We also think of time as a series of discrete time steps Δt, so that v and w are, strictly speaking, the transition amplitudes per unit time Δt.

Imagine a particle hopping from state $|m A\rangle$ to state $|m B\rangle$. The amplitude for this is v, which (using the fact that the states are normalized, $\langle m A|m A\rangle = \langle m B|m B\rangle = 1$) can be written as

$$v = \langle m B|m B\rangle \cdot v \cdot \langle m A|m A\rangle = \langle m B| \cdot \left(|m B\rangle v \langle m A| \right) \cdot |m A\rangle. \qquad (21.2)$$

The last expression on the right is written with the initial state on the right and the final state on the left. The expression in the parentheses between them is therefore the operator that turns the initial state into the final state during time Δt. But the operator that governs transitions between different states over time is (up to an overall constant that can be absorbed into the constants v and w) the Hamiltonian:

$$\langle \psi(t)|U(t)|\phi(0)\rangle = \langle \psi(t)|e^{-\frac{it}{\hbar}\hat{H}}|\phi(0)\rangle \qquad (21.3)$$

$$= \langle \psi(t)|\left(\hat{I} - \frac{it}{\hbar}\hat{H} + \ldots \right)|\phi(0)\rangle \qquad (21.4)$$

$$= \text{constants} \cdot \langle \psi(t)|\hat{H}|\phi(0)\rangle, \qquad (21.5)$$

where $U(t)$ is the time evolution operator between initial and final states $|\phi(0)\rangle$ and $|\psi(t)\rangle$. Here we have assumed that the initial and final states have no overlap, $\langle \psi(t)|\phi(0)\rangle = 0$, so that the identity term does not contribute. The hopping amplitude therefore represents a matrix element of the Hamiltonian.

Now that we know that transition amplitudes between consecutive times are equal to terms in the Hamiltonian, knowledge of the allowed discrete-time hopping amplitudes allows us to immediately construct the Hamiltonian from them. Effectively, an operator of the form $|\phi\rangle\langle \psi|$ removes a particle from state $|\psi\rangle$ and replaces it with a particle in state $|\phi\rangle$. For the SSH model, it should therefore be easy to see that the Hamiltonian is of the form

$$\hat{H} = \sum_{m=1}^{N} \left\{ v|m, B\rangle\langle m, A| + w|m + 1, A\rangle\langle m, B| \right. \tag{21.6}$$

$$\left. + v^*|m, A\rangle\langle m, B| + w^*|m + 1, B\rangle\langle m, A| \right\},$$

or more briefly,

$$\hat{H} = \sum_{m=1}^{N} \left\{ v|m, B\rangle\langle m, A| + w|m + 1, A\rangle\langle m, B| + H.C. \right\}, \tag{21.7}$$

where $H.C.$ represents the Hermitian conjugate of the terms that are written explicitly. By redefining the phase of the wavefunctions if necessary, the values of v and w can always be chosen to be real, so we will take them to be real henceforth.

In order to see how topological effects arise, carry out a Fourier transform to the (discrete) momentum space. For the rest of this section, we will assume that units are chosen so that $\hbar = 1$, so there is no need to distinguish between momentum and wavenumber: $k = p$. The position and momentum space states are related by

$$|m\rangle = \frac{1}{\sqrt{N}} \sum_{k=1}^{N} e^{imka}|k\rangle, \tag{21.8}$$

where the momentum is quantized, with allowed values $k = \delta, 2\delta, 3\delta, \ldots, N\delta$, where $\delta = \frac{2\pi}{N}$. Using the series representation of the Kronecker delta,

$$\frac{1}{N} \sum_{m=1}^{N} e^{-ima(k-l)} = \delta_{k,l} = \begin{cases} 1, & \text{if } k = l \\ 0, & \text{if } k \neq l \end{cases}, \tag{21.9}$$

the Hamiltonian becomes (for real v, w):

$$\hat{H} = \frac{1}{N} \sum_{klm} \left\{ v e^{ilma}|lA\rangle\langle kB|e^{-ikma} + w e^{il(m+1)a}|lA\rangle\langle kB|e^{-ikma} + H.C. \right\}$$

$$= \sum_{kl} \left[v\delta_{lk}|lA\rangle\langle kB| + w\delta_{lk}|lA\rangle\langle kB|e^{ila} + H.C. \right]$$

$$= \sum_{k} \left[v|kA\rangle\langle kB| + w|kA\rangle\langle kB|e^{ika} + H.C. \right]$$

$$= \sum_k |k\rangle\langle k| \left[\left(v + we^{ika}\right)|A\rangle\langle B| + \left(v + we^{-ika}\right)|B\rangle\langle A| \right]. \qquad (21.10)$$

The position-space and momentum-space versions of the Hamiltonian can both be written in matrix form. In position-space, the basis vectors can be ordered this way: $|1, A\rangle, |1, B\rangle, |2, A\rangle, |2, B\rangle, \ldots, |N, A\rangle, |N, B\rangle$. Using this ordering to label the rows and columns of the matrix, the position-space Hamiltonian is given by the block-diagonal matrix:

$$\hat{H} = \begin{pmatrix} 0 & v & 0 & 0 & 0 & 0 & \ldots \\ v & 0 & w & 0 & 0 & 0 \\ 0 & w & 0 & v & 0 & 0 \\ 0 & 0 & v & 0 & w & 0 \\ 0 & 0 & 0 & w & 0 & v \\ 0 & 0 & 0 & 0 & v & 0 \\ & & & \vdots & & \end{pmatrix}. \qquad (21.11)$$

In momentum space, the Hamiltonian is also block diagonal,

$$\hat{h}(k) = \begin{pmatrix} h(k_1) & 0 & 0 & \ldots \\ 0 & h(k_2) & 0 \\ 0 & 0 & h(k_3) \\ & \vdots & & \ddots \end{pmatrix}, \qquad (21.12)$$

where $k_n = n\delta$ for $n = 1, 2, \ldots N$. The two-by-two blocks are all of identical form

$$h(k) = \begin{pmatrix} 0 & v + we^{ika} \\ v + we^{-ika} & 0 \end{pmatrix} = \begin{pmatrix} 0 & z \\ z^* & 0 \end{pmatrix}. \qquad (21.13)$$

The rows and columns of $h(k)$ represent the A and B states at fixed k, and the "complex energy" value z is given by $z = v + we^{ika}$.

We take the zero of energy to be the Fermi energy of the system, midway between the two energy levels. The energy levels themselves become the conduction and valence bands in more complicated models. Notice that the Hamiltonian is periodic in momentum space as a consequence of the lattice periodicity: $\hat{H}(k + \frac{2\pi}{a}) = \hat{H}(k)$. The range $-\pi \leq ka \leq \pi$ constitutes the first Brillouin zone.

The time-independent Schrödinger equation for $h(k)$ gives the energy eigenvalues:

$$h(k)|\psi(k)\rangle = E|\psi(k)\rangle. \qquad (21.14)$$

Writing

$$|\psi(k)\rangle = \begin{pmatrix} a(k) \\ b(k) \end{pmatrix}, \tag{21.15}$$

where the upper component corresponds to the A substate and the lower to the B substate, this becomes

$$\begin{pmatrix} 0 & v + we^{ika} \\ v + we^{-ika} & 0 \end{pmatrix} \begin{pmatrix} a(k) \\ b(k) \end{pmatrix} = E(k) \begin{pmatrix} a(k) \\ b(k) \end{pmatrix}. \tag{21.16}$$

Energy eigenvalues are easy to find:

$$E_{\pm}(k) = \pm\sqrt{(v + we^{ika})(v + we^{-ika})} \tag{21.17}$$

$$= \pm\sqrt{v^2 + w^2 + 2vw\cos ka}. \tag{21.18}$$

It is then readily verified that eigenvectors are (up to an arbitrary overall phase):

$$|E_{\pm}\rangle = \frac{1}{\sqrt{2}} \begin{pmatrix} \pm e^{-i(\theta(k))} \\ 1 \end{pmatrix}. \tag{21.19}$$

Note that $E_{\pm}(k) = \pm|z|$. In fact, $z = E_k e^{i\theta(k)}$, where

$$\theta(k) = \tan^{-1}\left(\frac{\text{Im}(z)}{\text{Re}(z)}\right) = \tan^{-1}\left(\frac{w\sin ka}{v + w\cos ka}\right), \tag{21.20}$$

so $h(k)$ can be written in polar coordinates (E_k, θ) in the complex energy plane:

$$h(k) = E_k \begin{pmatrix} 0 & e^{-i\theta(k)} \\ e^{i\theta(k)} & 0 \end{pmatrix}. \tag{21.21}$$

There is a gap between the energies for most values of k, given by

$$\Delta E = \text{Min}(E_+ - E_-) = |v - w| \equiv 2\Delta. \tag{21.22}$$

When $\Delta \neq 0$, it takes a finite amount of energy to jump between bands, and so the material acts as an insulator. However, for $v = w$, the gap closes at $k = \pi$: $E_-(\pi) = E_+(\pi)$ (Fig. 21.5). The presence or absence of energy gaps will be important below. Notice that $h(k)$ can also be written as $h(k) = \boldsymbol{d}(k) \cdot \boldsymbol{\sigma}$, where the vector $\boldsymbol{d}(k)$ has components

$$d_x(k) = v + w\cos ka, \qquad d_y(k) = w\sin ka, \qquad d_z(k) = 0. \tag{21.23}$$

As k varies from 0 to 2π, $\boldsymbol{d}(k)$ traces out a closed curve (in fact, a circle) in the $\{d_x, d_y\}$ plane. The two cases where $v < w$ and $w < v$ are shown in Fig. 21.6.

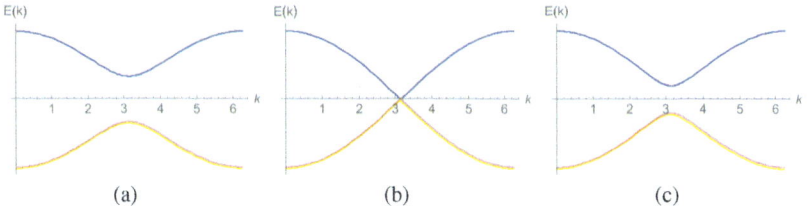

Fig. 21.5 The energy bands for the SSH model for the three possible cases: (a) $v < w$, (b) $v = w$, (c) $v > w$. At $v = w$, the energy gap closes, and the winding number becomes undefined. Here, we have set $a = 1$

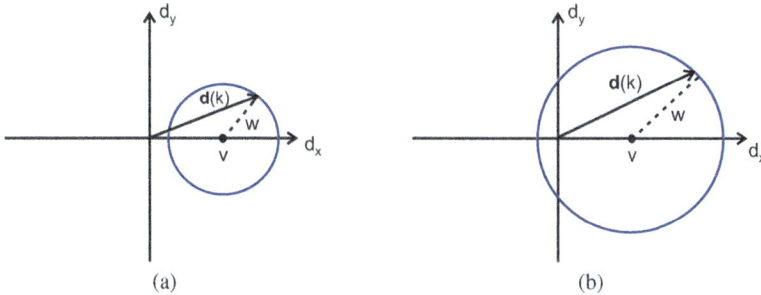

Fig. 21.6 As k varies, the vector $\boldsymbol{d}(k)$ traces out a circle centered at $d_x = v$. The radius of the circle is w, so that the origin is not enclosed for $v > w$, but for $v < w$, the circle encloses the origin once. These two cases have winding numbers 0 and 1, respectively

In both cases, the circle is centered at on the x-axis at $x = v$ and has radius w. But crucially, the circle encloses the origin when $v < w$ and does not when $v > w$.

Recall (Chap. 15) that the Berry connections for the two energy bands are given by

$$A_\pm(k) = i\langle E_\pm(k)|\frac{d}{dk}|E_\pm(k)\rangle = \frac{1}{2}\frac{d\theta}{dk}. \tag{21.24}$$

The Zac or Berry phase is then the integral of the Berry connection over the Brillouin zone:

$$\phi_\pm = \int_{-\frac{\pi}{a}}^{\frac{\pi}{a}} A_\pm(k)\,dk = \frac{1}{2}\Delta\theta, \tag{21.25}$$

where $\Delta\theta$ is the change in $\theta(k)$ across the Brillouin zone. By looking at Fig. 21.6 for the two cases $v < w$ and $w > v$, it can be seen that

$$\phi_\pm = \begin{cases} \pi & \text{if } |v| < |w| \\ 0 & \text{if } |v| > |w|, \end{cases} \tag{21.26}$$

The winding number of these two cases is

$$\nu_\pm = \frac{\phi_\pm}{\pi} = \begin{cases} 1 \text{ if } |v| < |w| \\ 0 \text{ if } |v| > |w|. \end{cases} \tag{21.27}$$

If $v = w$, both ν and ϕ are undefined.

So, there are two topological phases: in one of them, the Zac phase is $\phi = \pi$ for both energy bands, and the winding number of each is $\nu = 1$. In the other phase, $\phi = \nu = 0$. A system can only go from one of these two insulating phases by passing through the conducting phase with $v = w$ and the gap closed: $\Delta E = 0$. This is a general property of topological systems: different topological phases exist when there is an energy gap, and transitions can only occur when the gap closes. At the gap-closing point, the winding number (and topological invariants in general) becomes undefined.

So far, we have only considered the interior of the SSH chain, i.e., the bulk of the material, and ignored the edges of the chain. So now allow the SSH chain to be of finite length N, with ends at $n = 1$ and $n = N$. In Sect. 21.5, it will be seen that the quantum Hall system is an insulator in the bulk, but it can conduct along the surface. There are localized edge states that cannot propagate into the bulk of the material. A similar situation holds in the SSH model. If the SSH chain is in the topologically nontrivial $\nu = 1$ state, then at the edge, there is a boundary between different phases: the nontrivial phase inside the chain and the trivial phase of the adjacent vacuum. There is a localized edge state: it decays exponentially with distance from the edge, and cannot propagate into the bulk. This will occur at all boundaries between different topological phases, so we will refer to them as **edge states** or **boundary states** interchangeably.

The SSH system, being one-dimensional, has a zero-dimensional boundary, consisting of a single point at each end. In the case of a two- or three-dimensional solid, the boundary will be, respectively, one or two dimensional. In these cases, boundary states can propagate along the boundary surface, which means that the material becomes a conductor at the surface, even though it remains an insulator in the interior bulk region. On the energy band diagram, the presence of the zero-energy boundary or edge states connects the conduction and valence bands, as in Fig. 21.7. Recall that the group velocity of a state is given by the slope of the dispersion relation graph, $v_g = \frac{1}{\hbar}\frac{dE}{dk}$. It is clear from the plots that the edge states have slopes that are either always positive or always negative. So the edge states are *chiral*: each can move only in one direction. If a single chiral edge state exists, then the state is topologically protected: regardless of impurities in the system or irregularities of the boundary shape, the propagation is free from back-scattering, since there is no accessible state in the opposite direction for the particles to scatter into.

There is a relation between edge or boundary states and the topological quantum numbers of the bulk regions on the two sides of the boundary. If N is the winding number or some other topological number in the bulk and ΔN is the change in N

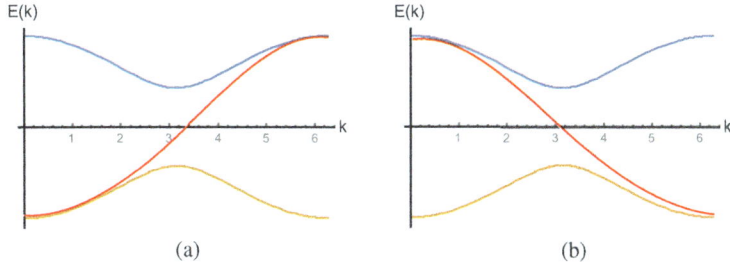

Fig. 21.7 When the system has a boundary (or, more generally, an interface between regions of different topological state), there can be states connecting the two bands. These states are localized on the edge or boundary and are unidirectional or chiral: the state in (**a**) is right-moving (positive group velocity), while the state in (**b**) is left-moving (negative group velocity)

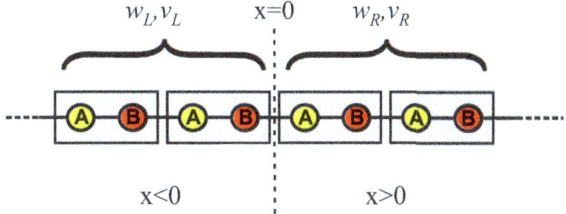

Fig. 21.8 A boundary between different regions of the SSH system. The hopping amplitudes are v_L and w_L to the left of the origin and are v_R and w_R to the right. If $v_L < w_L$ and $v_R > w_R$ (or vice versa), then the winding numbers on the two sides of the boundary will be different, leading to a bound state of zero energy exponentially localized near the origin

across the boundary, then the number of topologically protected edge states is given by $n = \Delta N$, a result known as the **bulk-boundary correspondence**.

Example 21.3.1 An explicit form of the zero energy state at the boundary between two topological regions can be found. When convenient, we will write the hopping amplitudes in terms of the energy gap Δ and the average hopping amplitude $t = \frac{v+w}{2}$:

$$v = t + \Delta, \qquad w = t - \Delta. \tag{21.28}$$

Suppose the boundary is at $x = 0$, with amplitudes v_L, w_L, t_L, Δ_L on the left ($x < 0$) and v_R, w_R, t_R, Δ_R on the right ($x > 0$) (Fig. 21.8). We will take the ends of the chain to be at $\pm\infty$.

First, look at a point away from the $x = 0$ boundary on either side. Dropping the L, R subscripts for the moment in order to treat both sides simultaneously, the momentum space Hamiltonian is given in Eq. 21.13. The minimum gap occurs at $k = \frac{\pi}{a}$, so expand about this minimum by defining $k = \frac{\pi}{a} + q$, assuming q small ($q \ll \frac{1}{a}$). Then to leading order in q,

$$h(q) = \begin{pmatrix} 0 & \Delta + iqwa \\ \Delta - iqwa & 0 \end{pmatrix}, \tag{21.29}$$

where the fact has been used that $e^{i\pi} = -1$. Since we might expect the solution to be either a plane wave or a decaying exponential, try a solution of the form

$$\psi(x) = \begin{pmatrix} \phi_1 \\ \phi_2 \end{pmatrix} e^{i\kappa x}, \tag{21.30}$$

where κ could be either real or imaginary. Here ϕ_1 and ϕ_2 are assumed to be independent of position and represent the amplitude of the A and B components in each lattice cell. (The values of κ, ϕ_1, and ϕ_2 may be different on the two sides of the boundary, as we will see below.) The momentum q in the position space representation is $q = -i\frac{\partial}{\partial x}$, which simply pulls down a factor of κ from the exponent. So we can write

$$h(q) = \begin{pmatrix} 0 & \Delta + i\kappa wa \\ \Delta - i\kappa wa & 0 \end{pmatrix}. \tag{21.31}$$

Diagonalizing (see Problem 2), the eigenvalues are

$$E = \pm\sqrt{\Delta^2 + w^2\kappa^2 a^2}, \tag{21.32}$$

which can be solved for κ:

$$\kappa = \pm\frac{1}{wa}\sqrt{E^2 - \Delta^2}. \tag{21.33}$$

Eigenvectors on each side of the boundary are then of the form

$$\psi(x) = N\begin{pmatrix} \Delta + iw\kappa a \\ E \end{pmatrix} e^{i\kappa x}, \tag{21.34}$$

where N is a normalization constant.

Now consider the boundary between the two regions. We need two boundary conditions. The first one should be obvious:

$$\psi_L(0) = \psi_R(0), \tag{21.35}$$

where $\psi_{L,R}(x)$ are solutions of form 21.34 to the left and right of the origin. To get the second boundary condition, note that the probability current density $j(x)$ has to be conserved at the origin. In every introductory quantum mechanics class, the current density is defined as

$$j(x) = \frac{\hbar}{2mi}\left(\psi^* \frac{\partial \psi}{\partial x} - \psi \frac{\partial \psi^*}{\partial x}\right). \tag{21.36}$$

Here, the role of mass is played by the energy gap Δ. So, by using Eq. 21.35, the second boundary condition $j_L(0) = j_R(0)$ becomes:

$$\frac{1}{\Delta_L}\frac{\partial \psi_L(0)}{\partial x} = \frac{1}{\Delta_R}\frac{\partial \psi_R(0)}{\partial x}, \tag{21.37}$$

which can be put into the form (Problem 2)

$$\Delta_L^2 E^2 = \Delta_R^2 E^2. \tag{21.38}$$

This can only be satisfied either by

$$\Delta_L = \pm \Delta_R \tag{21.39}$$

or by

$$E = 0. \tag{21.40}$$

So, zero energy solutions exist, and for these we find from Eq. 21.33 that $\kappa_{L,R} = \pm i \frac{\Delta_{L,R}}{w_{L,R}a}$. The proper choice of sign is determined by the fact that the solutions should decay as $x \to \pm \infty$. So the result is that

$$\kappa_L = -i\frac{\Delta_L}{w_L a}, \qquad \kappa_R = +i\frac{\Delta_R}{w_R a}. \tag{21.41}$$

Plugging everything back into Eq. 21.34, we arrive at the final expression for the localized, zero-energy boundary state:

$$\psi(x) = \begin{cases} \psi_L(x), & \text{for } x < 0 \\ \psi_R(x), & \text{for } x > 0 \end{cases}, \tag{21.42}$$

where

$$\psi_L(x) = N\left(\frac{\Delta_L + iw_L\kappa_L a}{E}\right)e^{\Delta_L x/w_L a}, \tag{21.43}$$

$$\psi_R(x) = N\left(\frac{\Delta_R + iw_L\kappa_R a}{E}\right)e^{-\Delta_R x/w_R a}. \tag{21.44}$$

If the parameters on each side of the boundary correspond to different winding numbers, then this boundary state cannot propagate in either direction, and so it is localized at $x = 0$, decaying exponentially in both directions. Notice that there will always be a boundary at the edge of a finite chain, so if the chain has winding number $\nu = 1$, then the empty space beyond the chain's end will be a trivial state of $\nu = 0$, and we would expect a similar exponentially localized boundary or edge state at the end of the chain.   ∎

21.4 Symmetries

A **unitary symmetry** exists when there is a unitary transformation $\hat{U}$, which commutes with the Hamiltonian:

$$\hat{U}\hat{H}\hat{U}^{\dagger} = \hat{H}, \tag{21.45}$$

or equivalently,

$$[\hat{U}, \hat{H}] = 0. \tag{21.46}$$

Continuous unitary symmetries are an important topic in quantum mechanics, due to **Noether's theorem**: the existence of a continuous unitary symmetry implies the existence of a corresponding conservation law. For example, translation and rotational symmetries imply linear and angular momentum conservation, respectively, while time-translation symmetry implies energy conservation. Discrete unitary symmetries such as parity and time reversal invariance are also important (see below), but do not lead to conservation laws.

In the study of topological phases, a second type of symmetry often plays an essential role: a **chiral symmetry** is an operator that *anticommutes* with the Hamiltonian:

$$\hat{\Gamma}\hat{H}\hat{\Gamma}^{\dagger} = -\hat{H}, \tag{21.47}$$

or equivalently,

$$\hat{\Gamma}\hat{H} + \hat{H}\hat{\Gamma} = 0. \tag{21.48}$$

The chiral operator has the property $\hat{\Gamma}^{\dagger}\hat{\Gamma} = \hat{\Gamma}^2 = 1$.

The SSH system has a chiral symmetry, also known as **sublattice symmetry** or **inversion symmetry**. It consists of flipping the sign of the states on one sublattice, which we take to be the B sublattice:

$$\hat{\Gamma}|mA\rangle = +|mA\rangle, \qquad \hat{\Gamma}|mB\rangle = -|mB\rangle. \tag{21.49}$$

This will clearly reverse the sign of the Hamiltonian. In the internal two-dimensional AB space of each unit cell with basis $|A\rangle = \begin{pmatrix} 1 \\ 0 \end{pmatrix}$ and $|B\rangle = \begin{pmatrix} 0 \\ 1 \end{pmatrix}$, the sublattice symmetry operator can be taken to simply be $\hat{\Gamma} = \sigma_z$. In terms of this chiral symmetry operator, we can define a pair of projection operators,

$$\hat{P}_A = \frac{1}{2}(1 + \hat{\Gamma}), \qquad \hat{P}_B = \frac{1}{2}(1 - \hat{\Gamma}), \tag{21.50}$$

so that $\hat{P}_A + \hat{P}_B = \hat{I}$ and $\hat{P}_A \hat{P}_B = \hat{P}_B \hat{P}_A = 0$. $\hat{P}_A$ projects states onto the A sublattice, removing any portions of the state residing on the B sublattice, with $\hat{P}_B$ similarly projecting onto the B sublattice.

The existence of the chiral symmetry has a number of consequences. For example, notice that

$$\hat{H} = \hat{P}_A \hat{H} \hat{P}_B + \hat{P}_B \hat{H} \hat{P}_A, \tag{21.51}$$

which implies that

$$\hat{P}_A \hat{H} \hat{P}_A = \hat{P}_B \hat{H} \hat{P}_B = 0. \tag{21.52}$$

This guarantees that there is no hopping within the same sublattice, i.e., no $A \rightarrow A$ or $B \rightarrow B$ transitions, in chirally symmetric models.

Notice also that if $|\psi_n\rangle$ is an energy eigenstate of energy E_n, $\hat{H}|\psi_n\rangle = E_n|\psi_n\rangle$, then the state $\hat{\Gamma}|\psi_n\rangle$ is an eigenstate with energy $-E_n$:

$$\hat{H}\left(\hat{\Gamma}|\psi_n\rangle\right) = -\hat{\Gamma}\left(\hat{H}|\psi_n\rangle\right) = -E_n\hat{\Gamma}|\psi_n\rangle. \tag{21.53}$$

So in the presence of chiral symmetry, the eigenstates come in pairs with opposite energies $\pm E_n$.

There is one exception to the last statement: a state of zero energy can be its own chiral partner, $\hat{\Gamma}|\psi_0\rangle = |\psi_0\rangle$, and thus be unpaired. This makes zero-energy states exceptionally stable. A perturbation to the system can only cause the energy of this state to become nonzero if it causes the single zero-energy state to split into *two* states, of opposite energy. Therefore, chiral symmetry protects the topological phase of the system: unless a perturbation breaks chiral symmetry or unless it is severe enough to alter the energy level structure, either by creating new states or by destroying some of those already present, the system cannot cross zero energy to change topological phase.

In addition to the chiral sublattice symmetry, a system may have other discrete symmetries. Most important among these are time-reversal symmetry and charge conjugation (or particle-hole) symmetry.

Time-reversal symmetry is an anti-unitary symmetry implemented by an operator $\hat{\mathcal{T}}$ such that $\left[\hat{\mathcal{T}}, \hat{H}\right] = 0$, which reverses the flow of time, $t \rightarrow -t$. In particular, the position operator is left unchanged, $\hat{x} \rightarrow \hat{x}$, but linear and angular momentum are reversed, $\hat{p} \rightarrow -\hat{p}$ and $\hat{L} \rightarrow -\hat{L}$. Recall that for a unitary operator $\hat{U}$,

$$\langle \hat{U}\psi | \hat{U}\phi \rangle = \langle \psi | \phi \rangle. \tag{21.54}$$

In contrast, an **antiunitary** operator satisfies the relation

$$\langle \hat{\mathcal{T}}\psi | \hat{\mathcal{T}}\phi \rangle = \langle \phi | \psi \rangle = \langle \psi | \phi \rangle^*. \tag{21.55}$$

If the $\hat{\mathcal{T}}$ operator is applied to the canonical commutation relation $[\hat{x}, \hat{p}] = i\hbar$, the left side gives us:

$$\hat{\mathcal{T}}[\hat{x}, \hat{p}]\hat{\mathcal{T}}^{-1} = \left(\hat{\mathcal{T}}\hat{x}\hat{\mathcal{T}}^{-1}\right)\left(\hat{\mathcal{T}}\hat{p}\hat{\mathcal{T}}^{-1}\right) - \left(\hat{\mathcal{T}}\hat{p}\hat{\mathcal{T}}^{-1}\right)\left(\hat{\mathcal{T}}\hat{x}\hat{\mathcal{T}}^{-1}\right) \tag{21.56}$$

$$= \hat{x}\left(-\hat{p}\right) - \left(-\hat{p}\right)\hat{x} \tag{21.57}$$

$$= -[\hat{x}, \hat{p}] \tag{21.58}$$

$$= -i\hbar. \tag{21.59}$$

So, if the canonical commutation relations are to be preserved, we are forced to conclude that $\hat{\mathcal{T}}i\hat{\mathcal{T}}^{-1} = -i$. In other words, the time-reversal operator must act to carry out complex conjugation on complex quantities. Suppose the complex conjugation operator is denoted by $\hat{K}$: $\hat{K}i\hat{K}^{-1} = -i$ with $\hat{K}^{-1} = \hat{K}$. So scalar-valued wavefunctions obey $\hat{K}\psi(x)\hat{K}^{-1} = \psi^*(x)$. This property can also be shown directly from the Schrödinger equation (see Problem 5).

For a spinless particle, we can simply take $\hat{\mathcal{T}} = \hat{K}$. But matrix-valued wavefunctions, such as those that carry spin or other internal variables, may be more complicated. For example, time-reversal should flip spins. Spin-up along the z-axis can be flipped to spin-down by rotating the spin by π about one of the other two axes; traditionally the rotation axis is taken to be along y. Since such a rotation is generated by the operator

$$e^{-\frac{i\pi}{\hbar}\hat{S}_y} = e^{-\frac{i\pi}{2}\hat{\sigma}_y} = -i\sigma_y,$$

the time-reversal operator on spin-$\frac{1}{2}$ wavefunctions can be taken to be

$$\hat{\mathcal{T}} = e^{-\frac{i\pi}{\hbar}\hat{S}_y}\hat{K} = -i\sigma_y\hat{K}. \tag{21.60}$$

It can be verified then that the spin operator obeys

$$S \to \hat{T}S\hat{T}^{-1} = -S. \tag{21.61}$$

For spin-0 wavefunctions, the operator $\hat{T} = \hat{K}$ clearly implies that $\hat{T}^2 = +1$. However, for a spin-$\frac{1}{2}$ wavefunction, the operator $\hat{T} = i\sigma_y\hat{K}$ obeys $\hat{T}^2 = -1$. This is true for fermions and bosons in general: $\hat{T}^2 = +1$ for bosons and $\hat{T}^2 = -1$ for fermions.

If the Hamiltonian is time-reversal invariant, one important consequence of this is **Kramer's theorem**: For a particle or system that obeys $\hat{T}^2 = -1$, every energy level is doubly degenerate. Specifically, if state $|\psi\rangle$ has energy E, then the state $\hat{T}|\psi\rangle$ is a distinct state ($\hat{T}|\psi\rangle \neq |\psi\rangle$) with the same energy.

Note that the presence of magnetic fields will break time reversal symmetry: the source of a magnetic field is an electric current. But currents will flip direction under time-reversal, and so the magnetic field will reverse as well. The reversal of $\boldsymbol{B}$ will then change the sign of magnetic dipole couplings $\boldsymbol{\mu} \cdot \boldsymbol{B}$ and possibly other terms in a Hamiltonian.

Finally, an additional discrete, anti-unitary, chiral symmetry $\hat{C}$ may be present in a system. In particle physics, this symmetry is usually called **charge conjugation**, since it interchanges particles with their opposite-charge antiparticles. In condensed matter systems, it is usually called **particle-hole symmetry**, since it interchanges electrons and holes. If a Hamiltonian is charge-conjugation symmetric, then it satisfies

$$\hat{C}\hat{H}\hat{C}^\dagger = -\hat{H}, \tag{21.62}$$

The charge conjugation operator is conventionally taken to be $\hat{C} = \sigma_x\hat{K}$ on fermions. Like time-reversal, $\hat{C}$ can square to either $+1$ or -1.

Notice that the product of time-reversal and charge-conjugation on a fermion system gives a chiral inversion symmetry:

$$\hat{T}\hat{C} = (-i\sigma_y\hat{K})(\sigma_x\hat{K}) = -i\sigma_y\sigma_x = -\sigma_z = -\hat{\Gamma}.$$

So time-reversal and charge conjugation symmetries together imply inversion symmetry. (The reverse does not necessarily hold.)

The integer quantum Hall effect will only occur in the presence of a time-reversal breaking magnetic field, but other effects such as the anomalous quantum Hall effect require time-reversal symmetry to be intact. The presence or absence of topological states will always depend only on global properties of the system: specifically, it depends on the dimension and the types of discrete symmetries possessed by the Hamiltonian. As a result, tables can be constructed of distinct topological classes of physical systems based on their dimension and symmetry structure. This leads to the so-called **periodic table of topological insulators**. To display this table, first define a set of discrete symmetry classes, as shown in Table 21.1

Table 21.1 Definitions of the general symmetry classes, based on whether the system has chiral, time-reversal, or charge-conjugation symmetry

Symmetry Class	Chiral Symm. $\hat{\Gamma}$	Time-reversal $\hat{\mathcal{T}}$	$\hat{\mathcal{T}}^2$	Charge Conj., $\hat{C}$	$\hat{C}^2$
A	No	No	–	No	–
AIII	Yes	No	–	No	–
AI	No	Yes	$+1$	No	–
BDI	Yes	Yes	$+1$	Yes	$+1$
D	No	No	–	Yes	$+1$
DIII	Yes	Yes	-1	Yes	$+1$
AII	No	Yes	-1	No	-
CII	Yes	Yes	-1	Yes	-1
C	No	No	–	Yes	-1
CI	Yes	Yes	$+1$	Yes	-1

Table 21.2 The sets of values for the topological invariants, for each symmetry class. $\mathbb{Z}$ and $2\mathbb{Z}$ represent the sets of integers and of even integers, respectively. $\mathbb{Z}_2$ is the set of integers mod 2: $\mathbb{Z}_2 = \{0, 1\}$

Symmetry Class	$d = 0$	$d = 1$	$d = 2$	$d = 3$
A	$\mathbb{Z}$	0	$\mathbb{Z}$	0
AIII	0	$\mathbb{Z}$	0	$\mathbb{Z}$
AI	$\mathbb{Z}$	0	0	0
BDI	$\mathbb{Z}_2$	$\mathbb{Z}$	0	0
D	$\mathbb{Z}_2$	$\mathbb{Z}_2$	$\mathbb{Z}$	0
DIII	0	$\mathbb{Z}_2$	$\mathbb{Z}_2$	$\mathbb{Z}$
AII	$2\mathbb{Z}$	0	$\mathbb{Z}_2$	$\mathbb{Z}_2$
CII	0	$2\mathbb{Z}$	0	$\mathbb{Z}_2$
C	0	0	$2\mathbb{Z}$	0
CI	0	0	0	$2\mathbb{Z}$

The group structure of the topological invariant associated with the system can then be determined based on the symmetry class and the dimensionality, as shown in Table 21.2.

21.5 Quantum Hall Effects and Topological Insulators

The classical Hall effect, known since 1879, has long formed the basis for numerous types of high-precision measurements and sensing devices. It can be used, for example, to determine the densities of charge carriers in a material or to conduct high-precision measurements of magnetic fields. The various *quantum* Hall effects that have been discovered since the 1980s have revealed an array of underlying topological structures. In particular, the integer quantum Hall system serves as the archetype of a category of materials known as topological insulators, which act as insulators in their interior (the *bulk*) but allow scattering-free conduction on their surfaces. These exhibit two-dimensional or three-dimensional analogs of the topological phases seen in the SSH model. Detailed reviews of quantum Hall effects may be found in [5–8]. Reviews of topological insulators and the related topic of topological superconductors include [9–12].

21.5.1   Classical Hall Effect

In the **classical Hall effect**, a static magnetic field across a material causes the appearance of an electric potential difference in the direction perpendicular to the field. Consider a slab of conducting material with a current traveling along the x-direction due to a voltage difference V_0 between the ends of the material. Let L_x, L_y, and L_z be the widths in the x, y, and z directions, respectively, as in Fig. 21.9. Turning on a uniform magnetic field pointing in the z-direction, the electrons in the material will initially be moving opposite to the current direction, and so they will feel a Lorentz force $\boldsymbol{F}_B = -e\boldsymbol{v} \times \boldsymbol{B}$ in the y-direction. As a result, electrons accumulate on one side of the material, leaving a deficit of electrons and a resulting positive charge on the other side. An electrostatic voltage difference V_H and an electric field $\boldsymbol{E}_y$ therefore begin to grow in the y-direction as a result of the charge separation on the opposite surfaces.

As the charge builds up on the edges of the material, electric and magnetic forces will eventually become equal, and so an equilibrium is reached. At equilibrium, the electric and magnetic fields will obey

$$eE_y = e|\boldsymbol{v} \times \boldsymbol{B}| \quad \rightarrow \quad E_y = vB. \tag{21.63}$$

So now, in addition to the potential difference V_0 along the longitudinal x-direction, there will also be a voltage difference across the transverse y-direction,

$$V_H = E_y L_y = vBL_y. \tag{21.64}$$

This is called the **Hall voltage**.

Fig. 21.9 The classical Hall effect. A voltage V_0 causes a current to flow along the x-direction. But a uniform magnetic field in the z-direction causes the charges to deflect along the y-direction. The resulting charge buildup creates a transverse Hall voltage V_H along the y-direction

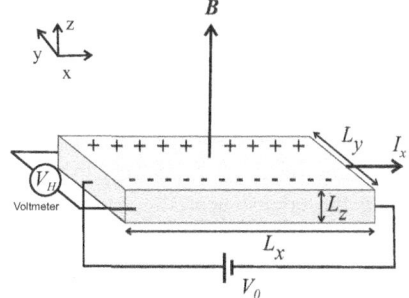

21.5.2   Two Dimensions

Suppose now that $L_z \ll L_x, L_y$, so that the material can be treated roughly as two dimensional. In two dimensions, the current density J is defined as current per *length* (in the y-direction), rather than per area. Define n_e to be the two-dimensional electron number density (number per area). Then $J = n_e e v$ and the total current is $I = J L_y = n_e e v L_y$. The Hall voltage can then be written in terms of the current as

$$V_H = \frac{BI}{n_e e}. \tag{21.65}$$

Recall that a particle in a box of size L has quantized energy levels $E = \frac{n^2 h^2}{8mL^2}$, where $n = 1, 2, 3 \ldots$ So as the material becomes thinner in the z-direction, excitations beyond the ground state ($n > 1$) become energetically inaccessible at low temperatures, and the degree of freedom in the z direction is essentially frozen out. A similar effect occurs in high-energy physics: in superstring theory, extra space-time dimensions beyond the usual four are needed for mathematical consistency of the theory. To explain why we don't see those extra dimensions, they are assumed to be *compactified* or curled up into tiny circles. If the circles are of sufficiently small radius, then the universe we observe would behave as if it was four dimensional in most ways, since there would be insufficient energy to excite transitions to the excited states of the compactified dimensions.

The **longitudinal resistance** is the usual resistance defined by the Ohm's law ratio of voltage across the x-direction to current in the x-direction: $R_{xx} = \frac{V_0}{I}$. But with the presence of the magnetic field there is also movement of charge in the transverse direction, so one can define a **transverse resistance** as the ratio of y voltage to x current:

$$R_{xy} = \frac{V_H}{I} = \frac{B}{ne}. \tag{21.66}$$

Now assume that B (perpendicular to the two-dimensional system) is strong and consider electrons in the interior of the material, far from the edges. The magnetic field causes the electrons to undergo uniform circular motion (Fig. 21.10) in the xy-plane at the cyclotron frequency,

$$\omega = \frac{eB}{m} \tag{21.67}$$

But if the electrons are moving in circles, how do currents arise? The current travels along the edges: near the edge, the circular motion can't be completed without hitting the edge and reflecting back elastically. This leads to the skipping motion shown on the top and bottom surfaces in Fig. 21.10, which carries current along the boundary of the material. The quantum Hall material is therefore a **topological insulator**: the interior is insulating, while the surface conducts. Further, note that

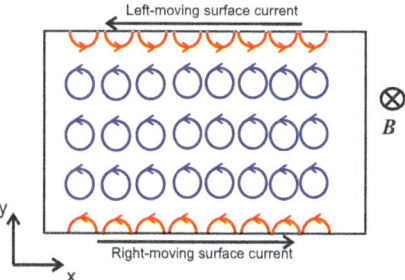

Fig. 21.10 Motions of electrons in a material in the presence of a uniform magnetic field pointing into the page. The electrons in the interior undergo cyclotron motion due to the field, with the energy levels (the Landau levels) of the circular motions quantized. Since there is no net motion along the x-axis, these circular motions lead to no net longitudinal current. But at the surfaces of the material, electrons are forced to perform a skipping motion, leading to unidirectional currents at each surface

along each edge, the electrons can only travel in one direction and that the directions are opposite on the two sides: the surface currents are unidirectional or **chiral**.

On each surface, there are no nearby "wrong direction" energy states that the electrons of the surface currents can scatter into, and so surface currents are highly stable against backscattering. This makes surface currents highly robust against defects in the material or irregularities in shape of the surface.

21.5.3   *Conductivity and Resistivity*

Since the voltage in one direction can lead to current flow in a different direction, the conductivity of the material is given by a matrix representing a second-rank tensor: the matrix element σ_{ij} is the ratio of current density in direction i to the electric field in direction j, where $i, j = x, y$. So Ohm's law is

$$J = \sigma E, \tag{21.68}$$

(or, in component form $J_i = \sum_j \sigma_{ij} E_j$) where the conductivity is:

$$\sigma = \begin{pmatrix} \sigma_{xx} & \sigma_{xy} \\ -\sigma_{xy} & \sigma_{xx} \end{pmatrix}. \tag{21.69}$$

(The equalities $\sigma_{yy} = \sigma_{xx}$ and $\sigma_{yx} = -\sigma_{xy}$ follow by assuming the material is isotropic.) The resistivity is given by the inverse of the conductivity:

$$\rho = \sigma^{-1} = \begin{pmatrix} \rho_{xx} & \rho_{xy} \\ -\rho_{xy} & \rho_{xx} \end{pmatrix}, \tag{21.70}$$

which leads to

$$\sigma_{xx} = \frac{\rho_{xx}}{\rho_{xx}^2 + \rho_{xy}^2}, \qquad \sigma_{xy} = -\frac{\rho_{xy}}{\rho_{xx}^2 + \rho_{xy}^2}. \tag{21.71}$$

Thinking of the material as a box containing a 2D electron gas at low temperature, the **Drude** or **free-electron model** [13–15] of solid state physics tells us that the current density in the x-direction in the absence of the B field can be written as

$$J_x = \sigma_0 E_x, \qquad \text{where} \qquad \sigma_0 = \frac{n_e e^2 \tau}{m}. \tag{21.72}$$

Here, τ is the mean time between electron scatterings. (The mass on the bottom is actually the *effective* mass m^* of the electron, but we will continue to simply write it as m.) With the magnetic field turned on, the Lorentz force on this current density is of the form $F_B = -e\boldsymbol{v} \times \boldsymbol{B} = \frac{J_x \times B}{n_e}$, and so the induced electric field $E_y = \frac{F_B}{-e}$ in the y-direction can be thought of as a correction to the external field E and to the total current density J, with each component of J affected by the other components:

$$\boldsymbol{J} = \sigma_0 \left(\boldsymbol{E} - \frac{\boldsymbol{J} \times \boldsymbol{B}}{n_e e} \right). \tag{21.73}$$

The electron number density can also be found, at least up to a constant. In the next subsection, it will be shown that the maximum number of electrons per area in the material that will fill all the available states is $\left(\frac{N}{A} \right)_{max} = \frac{eB}{h}$, so if a fraction ν of the levels is filled, then

$$n_e = \nu \left(\frac{N}{A} \right)_{max} = \frac{\nu e B}{h}. \tag{21.74}$$

So, substituting the components of J into Eq. 21.73 and using Eq. 21.68, we find after a bit of algebra (see Problem 7) that classically:

$$\sigma_{xx} = \frac{n_e e^2 \tau / m}{1 + (e\tau B / m)^2} = \frac{\sigma_0}{1 + \omega^2 \tau^2} \tag{21.75}$$

$$\sigma_{xy} = \nu \frac{e^2}{h} \left(1 - \frac{1}{1 + (e\tau B / m)^2} \right) = \frac{\sigma_0 \omega \tau}{1 + \omega^2 \tau^2}. \tag{21.76}$$

Taking $\omega \tau = \frac{eB\tau}{m} \to \infty$, we find

$$\sigma_{xx} = 0, \qquad \sigma_{xy} = \nu \frac{e^2}{h}, \tag{21.77}$$

which in turn leads to

$$\rho_{xx} = \frac{m}{n_e e^2 \tau}, \qquad \rho_{xy} = \frac{B}{n_e e} = \frac{h}{e^2 \nu}. \tag{21.78}$$

The **Hall coefficient** (in m^3/C) is then defined as

$$R_H = -\frac{E_y}{J_x B} = \frac{\rho_{xy}}{B}, \tag{21.79}$$

which gives

$$R_H = \frac{1}{n_e e}. \tag{21.80}$$

The classical behavior of the conductivity is given by Eqs. 21.75 and 21.76, with nonzero σ_{xx} and with σ_{xy} being a linear function of B^{-1}. ν in Eq. 21.76 is allowed to be a continuous variable. In contrast, it will be seen in the next section that at low temperatures and strong magnetic fields, Eqs. 21.77 hold, with the added feature that ν turns out to be quantized to integer values, giving a clear signal of quantum effects.

21.5.4   Landau Levels and the Integer Quantum Hall Effect

By the 1970s, it was possible to produce very thin films (thicknesses of $< 10\,\text{nm}$) that acted like true two-dimensional materials to a high degree of accuracy, with $L_z << L_x, L_y$. When placed in magnetic fields and cooled to temperatures below $1\,K$, where thermal noise is insufficient to disrupt the circular electron motions, it was found that the linear dependence of R_H on B predicted classically breaks down. Instead, as B is increased, the value of R_H exhibits a sequence of flat, well-defined plateaus (Fig. 21.11). This is the **integer quantum Hall effect**, first seen in 1980 [16]. It generally occurs in impure or disordered material samples, and it can only occur in systems in which time reversal invariance is broken, usually by the presence of a magnetic field.

The plateaus can be labeled by an integer $\nu = 1, 2, 3 \ldots$ Plateau number ν is centered at

$$B_\nu = \frac{2\pi \hbar n_e}{\nu e} \equiv \frac{n_e}{\nu} \Phi_0, \tag{21.81}$$

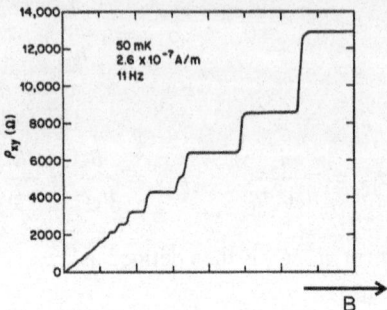

Fig. 21.11 The integer quantum Hall effect. As the magnetic field strength B increases, ρ_{xy} increases in a series of discrete steps. This contradicts the classical prediction that ρ_{xy} should increase as a smooth, linear function of B. The longitudinal resistance ρ_{xx} vanishes on each of the plateaus (Figure reproduced with permission from [17], ©1982 American Physical Society)

where n_e is the electron density and $\Phi_0 = \frac{2\pi\hbar}{e} = \frac{h}{e}$ is the fundamental flux quantum. The fact that the plateaus are labeled by an integer is a hint that the integer quantum Hall effect is of topological origin. On each plateau, the longitudinal resistivity ρ_{xx} vanishes, while $\rho_{xy} \neq 0$.

To understand the plateaus, consider the circular electron motions. The energies of the cyclotron orbits are quantized, leading to a series of discrete energy levels $E_n = \left(n + \frac{1}{2}\right)\hbar\omega$, called **Landau levels** (with $n = 1, 2, \ldots$), where ω is the cyclotron frequency. The plateaus appear at the values

$$\rho_{xy} = \frac{h}{\nu e^2} \approx \frac{25,812.80745\ \Omega}{\nu}, \tag{21.82}$$

where $\nu = 2, 3, 4, \ldots$. While on these plateaus, the ordinary (non-Hall) resistance of the material vanishes, while the transverse conductivity is found to be quantized:

$$\sigma_{xy} = \frac{e^2}{h}\nu. \tag{21.83}$$

The plateaus occur because increasing B cannot affect ρ_{xy} until it contributes enough energy to promote an electron to the next higher Landau level. ν, called the **filling fraction**, measures the number of electrons per flux quantum in the level.

More quantitatively, we can consider the wavefunction and energy levels in the system. In order to define the vector potential, a gauge choice has to be made. Two possibilities are the symmetric gauge

$$A(x, y) = \frac{1}{2}B\left(x\hat{j} - y\hat{i}\right) = -\frac{1}{2}Br\phi, \tag{21.84}$$

(where $r = \sqrt{x^2 + y^2}$ and ϕ is the unit vector in the angular direction) and the Landau gauge

$$A(x, y) = Bx \hat{j}. \tag{21.85}$$

The reader can check that these both give the same magnetic field, $B = \nabla \times A = B\hat{k}$.

Working in the Landau gauge, separation of variables can be attempted by factoring the wavefunction, $\psi(x, y, z) = \Phi(x)\Theta(y)\Omega(z)$. The choice of gauge has broken the translational symmetry in the x-direction, but translation symmetry still holds in the y and z directions (at least approximately; the finite size of the material prevents it from being exact). So the y and z parts of the solutions are just plane wave,

$$\Theta(y) = e^{ik_y y} \quad \text{and} \quad \Omega(z) = e^{ik_z z} \tag{21.86}$$

where k_y and k_z are quantized by the finite size of the material: $k_y = \left(\frac{2\pi}{L_y}\right) n_y$ and $k_z = \left(\frac{2\pi}{L_z}\right) n_z$, where n_y, n_z are integers. Note for use below that the spacing between adjacent k_y values is $\Delta k_y = \frac{2\pi}{L_y}$. The full wavefunction should then be of the form

$$\psi(r) = \Phi(x)e^{ik_y y + ik_z z}. \tag{21.87}$$

In the Landau gauge, the Hamiltonian is

$$\hat{H} = -\frac{\hbar^2}{2m}\left(\nabla - \frac{iq}{\hbar}A\right)^2 = -\frac{\hbar^2}{2m}\left[\partial_x^2 + \partial_z^2 + \left(\partial_y + \frac{ieB}{\hbar}x\right)^2\right]. \tag{21.88}$$

Using the wavefunction 21.87, a few lines of algebra show that the time-independent Schrödinger equation $\hat{H}\psi = E\psi$ becomes

$$-\frac{\hbar^2}{2m}\frac{d^2\Phi(c)}{dx^2} + \frac{\hbar^2}{2m}\left(k_y + \frac{eB}{\hbar}x\right)^2\Phi(x) = E_{eff}\Phi(x), \tag{21.89}$$

where $E_{eff} = E - \frac{\hbar^2 k_z^2}{2m}$. But comparing the second term of 21.89 to the standard oscillator potential, this is just the Schrödinger equation for a harmonic oscillator centered at location

$$x_0 = -\frac{\hbar k_y^2}{2m} \tag{21.90}$$

with effective spring constant $K_{eff} = \frac{e^2 B^2}{m}$ and with frequency given by the cyclotron frequency

$$\omega = \sqrt{\frac{K_{eff}}{m}} = \frac{eB}{m}. \tag{21.91}$$

Right away, we can write down the energies:

$$E_{n,n_z} = E_{HO} + \frac{\hbar^2 k_z^2}{2m} = \left(n + \frac{1}{2}\right)\hbar\omega + \frac{\hbar^2}{2m}\left(\frac{2\pi}{L_z}\right)^2 n_z^2. \tag{21.92}$$

If L_z is sufficiently small, the last term becomes very large for $n_z = 0$, and so at low temperatures, only the ground state $n_z = 0$ will contribute. Thus, we can ignore this term and simply set $E = \left(n + \frac{1}{2}\right)\hbar\omega$, as stated above. Here, the oscillator quantum number n will label the various Landau levels.

The x direction is of width L_x, so $-\frac{L_y}{2} < x_0 < \frac{L_y}{2}$, which implies by Eq. 21.90 that $|k_y| < \frac{eB}{2\hbar}L_x \equiv k_{y,max}$. So there is a span of width $2k_{y,max} = \frac{eBL_x}{\hbar}$ in momentum space that the states can fill. If $\mathcal{A} = L_x L_y$, the number of allowed states is therefore

$$\frac{2k_{y,max}}{\Delta k_y} = \frac{eBL_x/\hbar}{2\pi/L_y} = \frac{eB\mathcal{A}}{h}, \tag{21.93}$$

which is commonly written in terms of the **magnetic length** $l_B = \sqrt{\hbar/eB}$ as

$$N = \frac{\mathcal{A}}{2\pi l_B^2}. \tag{21.94}$$

In this latter form, N has a simple interpretation: $2\pi l_B^2$ is the area taken up by each cyclotron orbit, and so the degeneracy N of the Landau level is simply the area $\mathcal{A}$ of the available material divided by the area taken up by each orbit.

If the density of electrons (number per area) is n_e, then the filling factor, or fraction of available states filled, is $\nu = 2\pi l_B^2 n_e$. When the Landau level is full ($\nu = 1$), resistivity should vanish: electrons can't scatter off impurities because there are no empty states in the same level to scatter into. As a result, the scattering time diverges, $\tau \to \infty$: currents exhibit no backscattering and will flow around obstacles and defects in their path with current loss. If the magnetic field is increased, electrons will remain on the same plateau until the field supplies sufficient energy to go to the next Landau level.

At each point in the two-dimensional Brillouin zone, a Berry curvature can be defined for the mth energy band,

$$F_m = \partial_{k_x} A_{my} - \partial_{k_y} A_{mx}, \tag{21.95}$$

corresponding to a Berry potential A_m. The Brillouin zone is a torus T^2 in momentum space. The transverse conductivity can then be shown to be the value of F_m averaged over the zone,

$$\sigma_{xy} = \frac{1}{(2\pi)^2} \int_0^{2\pi} \int_0^{2\pi} \sigma_{xy}(k)dk_x dk_y = \frac{e^2}{(2\pi)h} \int_{T^2} F\, d^2k = \frac{e^2}{h} C_m, \qquad (21.96)$$

where C_m is the first Chern number of the band. The total Chern number or **TKNN invariant** [18] is then the sum of this expression over all occupied bands, $C = \sum_m C_m$. The quantization of the transverse conductivity is due to the integer nature of the topological invariant C.

21.5.5   More Quantum Hall Effects

As the amount of disorder or level of impurities in a material decreases, the integer plateaus of the quantum Hall effect begin to shrink and eventually disappear. In the process, less prominent plateaus at fractional values of ν start becoming noticeable. In this **fractional quantum Hall effect**, first discovered in 1982 [19, 20], ν now takes on simple fractional values such as $\nu = \frac{1}{3}, \frac{1}{5}, \frac{1}{7}, \frac{2}{3}, \frac{2}{5}, \ldots$. Whereas the integer effect is independent of the presence or absence of interactions between the electrons, the presence of the fractional effect is strongly dependent on the presence of electron-electron interactions, and it is only visible in highly pure samples of material.

In the fractional quantum Hall effect, electrons form multiparticle quasiparticle excitations in which charge screening leads to each excitation seeming to have a fractional charge. The total charge present is still an integer multiple of the e, but different parts of the each electron's charge seem to exist in widely separated parts of the material: the charge becomes fractionalized. Charge fractionalization can be seen in other topological systems, including the SSH model. It is accompanied by a fractionalization of the quasiparticle exchange statistics, leading to anyonic behavior (see Sect. 21.7). The N-particle ground state wavefunction of a system of filling fraction $\nu = \frac{1}{p}$ was first postulated by Laughlin [20] to be of the form

$$\psi_m(z_1, z_2, \ldots, z_N) = \mathcal{N} \prod_{1 \le i \le j \le N} (z_i - z_j)^n \prod_{k=1}^{N} e^{-|z_k|^2}, \qquad (21.97)$$

where the jth particle has complex coordinate $z_j = \frac{1}{2l_B}(x_j + iy_j)$ and $\mathcal{N}$ is a normalization constant. This quasiparticle state takes into account the repulsion between the electrons (it vanishes as $z_i \to z_j$ for any i and j). Excitations above the ground state, located at position z, are then given by

$$\psi(z) = \prod_{i=1}^{N}(z_i - z)\psi_m. \tag{21.98}$$

Such an excitation can be shown to carry a fraction $\frac{N}{m}$ of the electron charge and to behave as a particle of spin $\frac{1}{2m}$, with anyon exchange angle $\frac{\pi}{m}$.

Other variations on the quantum Hall effect have been uncovered since the 1980s. For example, Haldane [21] showed that a Hall-like effect can occur in periodic systems even in the absence of Landau levels or a net external magnetic flux. This **anomalous quantum Hall effect** was seen experimentally in 2013 [22] and is based on the concept of a synthetic gauge field: a Berry connection that is modulated periodically in space, so that the net flux through each unit cell of the material is zero. Because the net flux vanishes, the system is time-reversal invariant.

Similarly, in a system with time reversal symmetry and no external $\boldsymbol{B}$, but with strong spin-orbit couplings, an effect known as the **spin quantum Hall effect** [23–25] can occur. This looks like two copies of the integer quantum Hall effect, with opposite magnetic fields, superimposed on top of each other. Spin-up electrons travel in one direction, while spin-down travels the other way. This leads to a transport of spin, with no accompanying transport of charge. While the standard quantum Hall effect is effectively a unidirectional charge pump, pushing charge from one side of the material to the other, the spin quantum Hall effect acts as a one-way spin pump.

Box 21.1 Topological Lasers

Normally in quantum mechanics, the Hamiltonian is assumed to be Hermitian, $\hat{H}^{\dagger} = \hat{H}$, in order to obtain real energy eigenvalues. However, in the real world, most physical systems are open: they exhibit gain and loss as they exchange particles with their environment. These open systems have non-Hermitian Hamiltonians $\hat{H}^{\dagger} \neq \hat{H}$ and therefore have non-unitary time evolution. Energy eigenvalues become complex, with imaginary parts representing the rate of energy gain or loss. Non-Hermiticity is especially natural in optical systems, where thermal photons are constantly being generated, signal photons are being absorbed by detectors, and so on. In particular, lasers are inherently non-Hermitian systems, since the laser cavity must be pumped with energy from the outside.

So it is natural to ask if lasers can be formed using systems with topologically protected edge or boundary states. Lasing by localized photonic modes with topological protection should be robust against mechanical and thermal disturbances, impurities, and fabrication errors. Because the lasing is localized to the boundaries, which have small volumes compared to the bulk,

(continued)

Box 21.1 (continued)

the resulting light produced should be more intense for the same pumping rate.

The question is how to implement topological lattices in a medium with sufficient optical gain. One successful approach to doing this [26] Is using arrays of micropillars with microcavity polaritons confined to quantum wells. These polaritons are quasiparticles formed by coupling photons to quantum well excitons (bound electron-hole pairs) confined inside a small cavity. Micropillars consist of two distributed Bragg gratings that act like mirrors, confining light is a small gallium arsenide quantum well wedged between them. Details of excitations are not needed here, except for the fact that they can have nonzero angular momentum, labelled by a quantum number $l = 0, 1, 2, \ldots$. The s-state ($l = 0$) is spherically symmetric and is of no use to us. But the p-state ($l = 1$) has distinct axes parallel and perpendicular to a pair of high-amplitude lobes (Fig. 21.12). If a set of micropillars are arranged in a zigzag pattern as shown in Fig. 21.12a, then hopping of states between pillars will have a different amplitude for transitions along the parallel and perpendicular axes (Fig. 21.12b). The result is that the system can be described by an SSH model. However, the material in the quantum wells also exhibits optical gain and so can act as a laser when pumped with energy. The spectrum of one chain of such pillars exhibits a gap between energy bands. But when two such chains are brought into contact, a pair of zero-energy eigenstates appear in the middle of the gap, representing the topologically protected boundary states. These boundary excitations, when pumped, will brightly lase, and the output intensity and spectrum will be highly insensitive to defects or irregularities in the system.

In addition to the micropillar approach described above, topological lasers have been implemented in many other physical systems, including microring resonators, photonic crystals, metamaterials, nonlinear optical waveguides, and periodically driven Floquet systems. For a review, see [27].

21.6 Majorana Fermions

Every student of quantum mechanics learns about the **Dirac equation**, which describes relativistic spin-$\frac{1}{2}$ fermions such as the electron. The Dirac equation is a matrix-valued differential equation. One version of it is

DBG
DBG
Quantum well
(a)   (b)

Fig. 21.12 (**a**) A series of micropillars arranged in a zigzag pattern. Each pillar has a GaAs quantum well wedged between two distributed Bragg gratings (DBG) that act as mirrors. (**b**) The pillars viewed from above. The excitations in each pillar have two distinct axes, parallel and perpendicular to a pair of amplitude lobes. Alternate pairs of pillars are oriented along different axes relative to these, leading to alternating hopping amplitudes between the pillars

$$\left(i\hbar \sum_\mu \gamma^\mu \frac{\partial}{\partial x_\mu} - mc \right) \psi(x) = 0, \tag{21.99}$$

or, more compactly,

$$\left(i\hbar\gamma^\mu \frac{\partial}{\partial x_\mu} - mc \right) \psi(x) = 0, \tag{21.100}$$

where the index μ runs from 0 to 3, with $\mu = 0$ representing the time component and $\mu = 1, 2, 3$ being the three time components. In the second version, the Einstein summation convention is used: any index that occurs twice, once up and once down, is summed over. The γ_μ are 4×4 matrices, acting on the 4-component column vector ψ representing the fermion state. Recalling that space and time components come into four-dimensional products with opposite signs, the Dirac equation can also be written a bit more explicitly as

$$i\left(\gamma^0 \frac{\partial}{\partial t} - \boldsymbol{\gamma} \cdot \nabla - \frac{mc}{\hbar} \right) \psi(\boldsymbol{x}, t) = 0. \tag{21.101}$$

The **Dirac gamma matrices**, γ_μ, are required to obey an anti-commutation relation:

$$\{\gamma_\mu, \gamma_\nu\} = 2\eta_{\mu\nu}, \tag{21.102}$$

where the anti-commutator, $\{\hat{A}, \hat{B}\}$, of operators $\hat{A}$ and $\hat{B}$ is defined by

$$\{\hat{A}, \hat{B}\} = \hat{A}\hat{B} + \hat{B}\hat{A}, \tag{21.103}$$

and η is the flat-space metric of special relativity (see Appendix D). The Dirac equation implies the Klein-Gordon equation: applying the operator $\left(i\gamma^\mu \frac{\partial}{\partial x_\mu} + \frac{mc}{\hbar} \right)$

to the Dirac equation, we find:

$$0 = \left(i\gamma^\mu \frac{\partial}{\partial x_\mu} + \frac{mc}{\hbar}\right)\left(i\gamma^\nu \frac{\partial}{\partial x_\nu} - \frac{mc}{\hbar}\right)\psi(x) \tag{21.104}$$

$$= -\left(\gamma^\mu \gamma^\nu \partial_\mu \partial_\nu + \frac{m^2 c^2}{\hbar^2}\right)\psi(x) \tag{21.105}$$

$$= -\left(\frac{1}{2}\{\gamma^\mu, \gamma^\nu\}\partial_\mu \partial_\nu + \frac{m^2 c^2}{\hbar^2}\right)\psi(x) \tag{21.106}$$

$$= -\left(\eta^{\mu\nu}\partial_\mu \partial_\nu + \frac{m^2 c^2}{\hbar^2}\right)\psi(x), \tag{21.107}$$

with ∂_μ being shorthand for $\frac{\partial}{\partial x_\mu}$.

The four components of column matrix $\psi(x)$ represent (i) a spin-up particle state, (ii) a spin-down particle state, (iii) a spin-up antiparticle state, and (iv) a spin-down antiparticle state. A massive Dirac particle requires all four of these states to exist. As for the gamma matrices, there are multiple different representations, each useful in different situations; we will not need explicit representations, so we will not display them here. (Explicit expressions for them can be found in any standard quantum mechanics text with a chapter on relativistic wave equations.)

The Dirac equation and the four-component Dirac spinors were first introduced in 1928. Dirac spinors were found to be inherently complex-valued; this was a requirement for distinct particles and anti-particles to exist. In 1937, the Italian physicist Ettore Majorana discovered that the Dirac equation could be recast in a different form with real solutions. These real solutions, called **Majorana states**, represented particles that were equal to their own antiparticles. Since particles and antiparticles must have opposite charges, Majorana fermions must be electrically neutral.

For Majorana excitations, the four-component Dirac equation reduces to a pair of two-component Majorana equations:

$$(i\hbar\partial_t - \boldsymbol{p} \cdot \boldsymbol{\sigma})\psi_R - mc\psi_L = 0 \tag{21.108}$$

$$(i\hbar\partial_t + \boldsymbol{p} \cdot \boldsymbol{\sigma})\psi_L - mc\psi_R = 0, \tag{21.109}$$

where $\boldsymbol{p}$ is the spatial momentum and the two-component **Majorana spinors** ψ_L and ψ_R are the left- and right-handed components of the Dirac spinor:

$$\psi = \begin{pmatrix} \psi_R \\ \psi_L \end{pmatrix}. \tag{21.110}$$

Particle physicists have been searching unsuccessfully for Majorana fermions in nature for decades. One possible candidate is the neutrino: since it is an electrically uncharged fermion, it is a possible candidate for a Majorana particle. But since the neutrino is exceedingly weakly interacting, confirming its Majorana nature has been difficult, and it remains unclear today whether it is described by a real Majorana wavefunction or a complex Dirac wavefunction.

However, in recent years, it has been realized that it may be possible for a collective excitation in a condensed matter system to form a Majorana quasiparticle. This has led to a surge of interest in Majorana states, especially in light of the fact that they could be used to implement topological quantum computing (see the next section). Henceforth, we will use the phrases *Majorana state* or *Majorana particle* to refer to both collective Majorana excitations and fundamental single-particle Majorana fermions.

Box 21.2 The Mystery of Ettore Majorana

Ettore Majorana was born in Catania, on the Sicilian coast, in 1906, and quickly gained renown as a child prodigy. After beginning studies in civil engineering, he switched to theoretical physics and found a place in Enrico Fermi's research group in Rome. In a career that only lasted 10 years, he tackled and made progress on some of the most difficult problems in the physics of the day. His published work was mostly in various areas of atomic and nuclear physics, but he also wrote papers on geophysics, engineering, and mathematics; most of these papers were never published during his (presumed) lifetime. His most famous work, on what we now call Majorana fermions, was his final published paper.

Fermi, normally highly critical of the work of other physicists, made no secret of his high regard for Majorana. As quoted by Giuseppe Cocconi, Fermi is said to have declared of Majorana: " There are also those of high standing, who come to discoveries of great importance, fundamental for the development of science.... But then there are geniuses like Galileo and Newton. Well, Ettore was one of them. Majorana had what no one else in the world had ... " (quote taken from *Ettore Majorana: Unpublished Research Notes on Theoretical Physics*, ed. by S. Esposito, et al, Springer, 2009). Among other things, he was the first to correctly interpret an experiment by Iréne Joliot-Curie and Frédéric Joliot as requiring the existence of the particle now known as the neutron; he never published his results, and the Nobel prize for the discovery of the neutron later went to Chadwick in 1932.

But what catapulted Majorana into legend was his disappearance. He had been depressed for years and had written multiple suicide notes over time. On March 25, 1938, he mailed a suicide note to Antonio Carrelli, Director of the Naples Physics Institute, before boarding a ship to Palermo. At some point, he bought another ticket from Palermo to Naples and then sent a telegram and another letter to Carrelli, rescinding his previous suicide note and reassuring Carrelli that he was alright. Then he was never seen or heard from again.

(continued)

Box 21.2 (continued)

It is widely believed that he changed his mind yet again and committed suicide after all. But no body was ever found, leaving speculation to run rampant. Such speculation has been intensified by the claim that he took the equivalent of up to $70,000 US with him on his final journey. For years afterward, there were Elvis-like sightings of Majorana on multiple continents. Various theories were advanced, including claims that he had changed his name and moved to the USA, Argentina, or Venezuela or that he had entered a monastery. As recently as 2015, an investigation by the Public Prosecutor's Office in Rome concluded that Majorana had been alive and living in Valencia, Venezuela, until at least 1959, based largely on a photograph that is alleged to show Majorana in the company of a local mechanic. But this conclusion still remains contested by many.

Recall that the creation and annihilation operators, $\hat{c}_j^\dagger$ and $\hat{c}_j$, for a set of electrons obey the anticommutation relations

$$\{\hat{c}_i, \hat{c}_j\} = \{\hat{c}_i^\dagger, \hat{c}_j^\dagger\} = 0, \qquad \{\hat{c}_i^\dagger, \hat{c}_j\} = \delta_{ij}, \tag{21.111}$$

where the subscripts could label, for example, different positions on a crystal lattice. Consider a new set of operators $\hat{\gamma}_{j1}$ and $\hat{\gamma}_{j2}$, which we will denote generically as $\hat{\gamma}_{j\alpha}$, for $\alpha = 1, 2$. (Do not confuse these new operators with the Dirac gamma matrices. Unfortunately, the notation $\hat{\gamma}_{j\alpha}$ is widely used.) The $\hat{\gamma}_{j\alpha}$ are new annihilation operators, constructed as linear combinations of the old creation and annihilation operators:

$$\hat{\gamma}_{j1} = \hat{c}_j^\dagger + \hat{c}_j, \qquad \hat{\gamma}_{j2} = i(\hat{c}_j^\dagger - \hat{c}_j). \tag{21.112}$$

Inverting these relations, we find that the original operators can be written as

$$\hat{c}_j = \frac{1}{2}(\hat{\gamma}_{j1} + i\hat{\gamma}_{j2}), \qquad \hat{c}_j^\dagger = \frac{1}{2}(\hat{\gamma}_{j1} - i\hat{\gamma}_{j2}). \tag{21.113}$$

So, in a sense, $\hat{\gamma}_{j1}$ and $\hat{\gamma}_{j2}$ are like the real and imaginary parts of $\hat{c}_j$. But comparison of the two relations in Eq. 21.113 imply that the $\hat{\gamma}_{j\alpha}$ must be self-adjoint:

$$\hat{\gamma}_{j1}^\dagger = \hat{\gamma}_{j1}, \qquad \hat{\gamma}_{j2}^\dagger = \hat{\gamma}_{j2}; \tag{21.114}$$

this can also be seen from Eqs. 21.112. This self-adjoint property means that the corresponding states created by these operators will be real (as opposed to the original Dirac electron fields, which are complex in general), and more to the point, these new states will represent particles that are equal to their own antiparticles.

These are the Majorana particles. The reader can verify that the new Majorana creation/annihilation operators obey

$$\{\hat{\gamma}_{i\alpha}, \hat{\gamma}_{j\beta}\} = 2\delta_{ij}\delta_{\alpha\beta}. \tag{21.115}$$

Note the strangeness of these new Majorana particles: Eq. 21.114 implies that the same operator that creates the excitation also serves to annihilate them. Also, note that any fermion can be split into a pair of Majoranas in this manner. However, usually the resulting Majoranas are strongly overlapping in position and momentum, leaving them inextricably mixed together; thus there is no purpose in treating them as separate objects. It is only when the two Majorana components $\hat{\gamma}_{j1}$ and $\hat{\gamma}_{j2}$ can be sufficiently separated from each other in space that there is any benefit to treating them as separate particles. So the question is whether there are circumstances in which they can be well separated.

Remember that in condensed matter physics, it is holes that play the role of antiparticles to electrons. So in a solid, each Majorana quasiparticle can be viewed as a coherent superposition of an electron and a hole. But we saw in Chap. 7 that superconductors support exactly this sort of electron-hole superposition and that these superposition states can remain coherent over macroscopic distances. So superconductors should be a prime candidate for hosting Majorana excitations. In fact, in superconductors, the well-known **Boguliobov-de Gennes (BdG) equation**, which is used to describe quasiparticle excitations in the BCS theory, is closely related to the Majorana equations.

A number of proposals have now been made for generating Majorana excitations in superconductors, many of them centering on the edge or boundary modes that occur between trivial and nontrivial topological phases or on vortices bound to defects. These states are in fact **Majorana zero modes**, Majorana states that are also topologically protected zero-energy states. What is particularly important about these Majorana zero modes are that they are *anyonic* (see the next section), which makes them ideal for use in topological quantum computing. In particular, Kitaev [28] proposed that Majorana zero modes at the ends of supeconducting nanowires could provide a basis for the construction of topologically-protected qubits.

21.7   Anyons, Braiding, and Topological Computing

Error correction is vital in all computational systems, and especially so in quantum computation. The standard approach to error correction is to use error-correcting codes with a level of built-in redundancy to allow for the detection of errors and manipulations to correct them. An alternative approach is to try to prevent the errors from happening in the first place. One way to do this is by means of **topological quantum computing**. The idea is to encode information into topological invariants such as the winding number of a wavefunction. These topological invariants are discrete numbers that are dependent on the global, rather than the local, properties of a system. As a result, they are highly resistant to being affected by small perturbations of the system. The most common approach to building topological computing protocols is by means of braiding of anyons. Braiding and anyons are the topics of the following subsections. For additional reviews of anyons and their uses in topological quantum computing, see [29–31].

21.7.1 Exchange Statistics and Anyons

In three dimensions, it is well-known that particles come in two generic types: fermions and bosons. In addition to having integer spin, the other characteristic property of bosons is that their multiparticle wavefunctions are symmetric under interchange: when two identical bosons are interchanged, the wavefunction remains the same. Spin-$\frac{1}{2}$ fermions, on the other hand, have antisymmetric wavefunctions: interchanging two of them introduces a minus sign into multiparticle wavefunctions.

Why are fermions and bosons the only possibilities in three space dimensions? To see the reason, consider two identical particles at locations x_0 and x_1 (Fig. 21.13). Let $\hat{A}$ be the operator that interchanges the two particles,

$$\hat{A}\psi(x_1, x_2) = \psi(x_2, x_1).$$

Because the two particles are identical, interchanging them once should give a wavefunction parallel to the original in Hilbert space:

$$\hat{A}\psi(x_1, x_2) = \psi(x_2, x_1) = C\psi(x_1, x_2). \tag{21.116}$$

Maintaining normalization requires $|C| = 1$, so that it should be of the form $C = e^{i\theta}$ for some real number θ.

Now interchange the particles a second time, so that

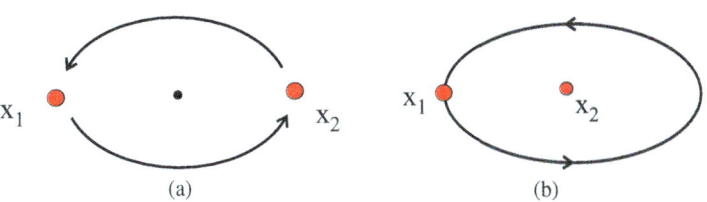

(a)   (b)

Fig. 21.13 (a) Exchange of two identical particles, located at positions x_1 and x_2. The particles are not allowed to pass through each other, so the configuration $x_1 = x_2$ (represented by the black dot) acts as a hole in the allowed configuration space. Two consecutive exchanges return the particles to their initial positions. (b) The double interchange is continuously deformable to leaving one particle fixed and taking the other particle around it. The hole in the configuration space can now be taken to be the position of the stationary particle

$$\hat{A}^2 : \ \psi(x_1, x_2) \to e^{2i\theta} \psi((x_1, x_2)).$$

But any path through space that exchanges the two particles twice can be continuously deformed into a path in which one particle stays at rest and the other circles around it (Fig. 21.13). And in three dimensions, any such path can then be deformed into the trivial path, in which neither particle moves at all. Thus, we should expect interchanging the particles twice to be equivalent to doing nothing, which implies that $C^2 = e^{2i\theta} = 1$. So, in three dimensions, we must have either $\theta = 0$ under interchange (bosons) or $\theta = \pi$ (fermions). Note that this argument hinges on the fact that in three dimensions, any path in which one particle circles the other can be "slipped over" the stationary particle and deformed to the trivial path.

The "statistical angle" θ can also be written as $\theta = 2\pi s$. For the two possible three-dimensional cases, we find $\theta = 0 \to s = 0$ and $\theta = \pi \to s = \frac{1}{2}$. This is exactly what we would expect for bosons and fermions, respectively, if we identify s with the spin of the particle.

What about two dimensions? Can the same argument be applied? Clearly not: we can still hold the particle at x_2 stationary and take the particle at x_1 in a circle around it. But this path can no longer be deformed continuously to the trivial path. The particles can't pass through each other, so the one at x_2 acts as an obstruction, or a hole in the space of allowed positions of the other particle. The path can no longer be deformed to the trivial one without lifting the moving particle out of the two-dimensional space to avoid the obstruction. The condition that $C^2 = 1$ no longer necessarily holds. The most we can say is that under interchange, the wavefunction must pick up a phase factor $e^{i\theta}$, where θ can now be anything between 0 and π. The spin is then also somewhere intermediate between 0 and $\frac{1}{2}$.

Such particles have been named **anyons** [32–34]. Depending on the value of θ, it can be a boson, a fermion, or it can have statistical properties anywhere in between the two. Such particles can only exist in two dimensions. Since we live in a three-dimensional world, it may seem that they can have no physical relevance. But very thin films or single-atom-thick graphene sheets behave to a very high approximation as two-dimensional systems, since particles have very little room to move in one of the three dimensions. (Recall that for a potential well or particle in a box, the energy gap between the ground state and excited states increases as the box becomes smaller. So for a sufficiently thin object, despite the fact that the thin direction still has finite thickness, motions in that direction become energetically inaccessible at low temperatures.)

As a result, collective excitations that behave like anyons are expected to exist in condensed matter systems. In fact, in 2020, research groups in Paris and at Purdue presented strong experimental evidence of anyons in two very different solid-state systems [35, 36]. It is believed that anyons also play a role in the fractional quantum Hall effect and can exist at the surfaces of topological insulators.

An anyon can be usefully modeled in the following manner. Consider a particle of charge q that has attached to it a magnetic dipole carrying magnetic flux Φ, as in Fig. 21.14. If one such particle (call it P_2) is stationary at point x_2, an identical

Fig. 21.14 Anyons can be modeled as composite particles with both electric charge and an attached magnetic flux tube. The exchange statistics and spin are then determined by the charge and flux

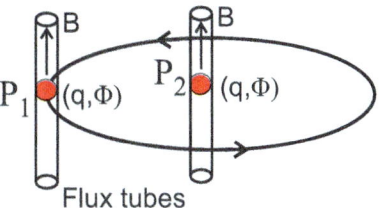

Flux tubes

such composite particle (P_1) starting at x_1 is taken in a closed path around P_2. Because of the flux attached to P_2, P_1 will gain an Aharanov-Bohm phase $e^{\frac{iq\Phi}{c\hbar}}$. But equivalently, we can think of P_2 as circling around P_1 in this process, so the wavefunction of P_2 also gains a phase $e^{\frac{iq\Phi}{c\hbar}}$. The total phase gain of the two-particle wavefunction is therefore $e^{2\frac{iq\Phi}{c\hbar}}$. The spin each particle is then given by $s = \frac{\phi}{2\pi} = \frac{q\Phi}{2\pi c\hbar}$. The spin is therefore determined by the charge and flux carried by the particle.

21.7.2 Non-Abelian Anyons and Topological Quantum Computing

The anyons discussed above are called **Abelian anyons**. Abelian simply means commutative: if multiple paths are traversed by one of the particles, the resulting phase factors can be multiplied in any order: they commute with each other, i.e., $e^{i\theta_1}e^{i\theta_2} = e^{i\theta_2}e^{i\theta_1}$. However, that is not necessarily the case: it is possible that particles have internal quantum numbers (such as the color charge carried by quarks), so that different *overall* configurations lead to the same *spatial* configuration. In this case, wavefunctions have some additional index a, $\psi_a(x_1, x_2)$, so that under interchange, the wavefunction can be a unitary *matrix*, $\hat{U}$:

$$\psi_a(x_1, x_2) \rightarrow \psi_b(x_2, x_1) = U_{ba}\psi_a(x_1, x_2),$$

where U_{ba} the matrix element, $U_{ba} = \langle\psi_b|\hat{U}|\psi_a\rangle$. In this case, excitations are noncommuting: if $\hat{U}^{(\alpha)}$, $\hat{U}^{(\beta)}$ are unitary matrices accrued along (non-topologically equivalent) closed paths α and β, then $\hat{U}^{(\alpha)}\hat{U}^{(\beta)} \neq \hat{U}^{(\beta)}\hat{U}^{(\alpha)}$.

The color charges of particle physics provide a set of degenerate states of the quarks. In this case, the degeneracy of the ground state arises because of a local symmetry: the Hamiltonian is invariant under a group of local transformations (called the *color SU(3) group*), which forces different color states to have the same energy. Although electrons in a solid do not have an analogous symmetry or non-Abelian internal degrees of freedom, if they are strongly interacting, then they can form quasi-particle excitations that behave very differently than the original electrons. In particular, the ground state may have degeneracies that do not arise as a result of local symmetries, and permutations of these degenerate ground states may be non-Abelian transformations.

Consider N objects located at positions $x_1, \ldots, x_N$. If you rearrange the objects among the possible locations, this set of rearrangements forms an Abelian group S_N, called the **permutation group** or the **symmetric group**. This is a discrete, finite group, of order $N!$. Notice that repeating any interchange twice in a row leads to the identity operation (in other words $g_i^2 = I$ for all i).

Now imagine that each of the objects is attached to a string, so that as two of the objects are interchanged, there are two different ways that strings can cross (the left strand going either over or under the right strand). Repetition of the same operation twice in a row no longer equals the identity, since two successive twists of the strands in the same direction do not cancel. The resulting group is therefore no longer S_N. It is the **braid group** on N strands, B_N. For $N \geq 2$, this is a group of infinite order, since no matter how many times you twist a pair of strands in the same direction, they never untangle (in other words, no powers of any of the elements give back the identity element). The inverse operation of braiding a pair of strands clockwise consists of braiding the same pair counterclockwise.

Consider a set of N lines or strands initially parallel to each other, as in Fig. 21.15 Then the braid operation b_n consists of interchanging the upper ends of the nth and $(n + 1)$st line, with line $n + 1$ passing *over* line n (as shown in Fig. 21.16 for a set of three strands). Equivalently, we can think of the two strands as being rotated *clockwise* (as viewed from the top of the page) about the point halfway between them. The inverse operation b_n^{-1} consists of rotating the two strands *counterclockwise* (the right strand going *under* the left strand).

In addition to the requirements for forming a group, a braid group must satisfy two other conditions:

Fig. 21.15 A set of N strands before braiding

$$x_1 \quad x_2 \quad \cdots\cdots \quad x_{n-1} \quad x_n \quad \cdots\cdots \quad x_N$$

Fig. 21.16 Braiding operations and their inverses on a set of three lines

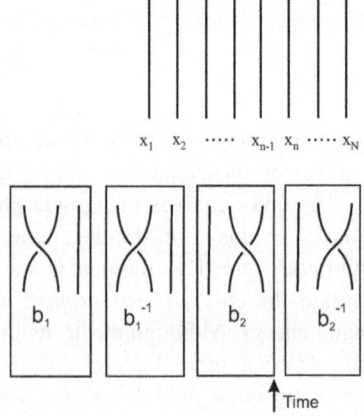

$b_1 \qquad b_1^{-1} \qquad b_2 \qquad b_2^{-1}$

↑ Time

Fig. 21.17 The meanings of
the conditions $b_i b_j = b_j b_i$
for $|i - j| \geq 2$ (left) and
$b_n b_{n+1} b_n = b_{n+1} b_n b_{n+1}$
(right). Time is increasing
from the bottom of the figure
toward the top. In each case,
it is clear that the diagrams on
the two sides of the equality
can be smoothly deformed
into each other

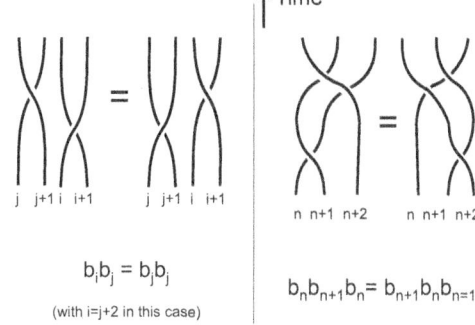

$$b_i b_j = b_j b_i, \text{ if } |i - j| \geq 2 \tag{21.117}$$

$$b_n b_{n+1} b_n = b_{n+1} b_n b_{n+1}, \text{ for all } n \tag{21.118}$$

The second condition is called the **Yang-Baxter equation**. The meanings of these conditions are illustrated in Fig. 21.17.

For $N = 1$, group B_1 is trivial, with the group containing of only one element, the identity. For $N = 2$, group B_2 is isomorphic (in a one-to-one correspondence) to the group of integers, with ordinary addition as the group operation. All braid groups are non-Abelian for $N \geq 3$ and of infinite order for $N \geq 2$. The infinite braid group B_N is generated by the finite set of braiding operations, $b_1, b_2, \ldots, b_N$, and their inverses.

Braid groups come up in many places in physics, for example, braided vortex lines play a role in the study of superconductors and other fluids. But for us, the relevance has to do with anyons and particle statistics. Suppose that the lines in the diagrams of the previous section represent the trajectories of a set of **identical** particles through spacetime. Then the braiding of two lines represents the interchange of two identical particles. In three space dimensions, the result can only be a multiplication of the initial state vector by a factor of ± 1. But in two spatial dimensions, the particles could be anyons, so that the result of each interchange is multiplication by some phase factor $e^{i\phi}$, where ϕ could now be anything and the phases can be different for different types of particles. The output state is related to the input state by some unitary matrix built from these phases; the size of this matrix equals the dimension of the representation of the braid group. *This matrix represents a unitary group operation that can be viewed as describing the action of a quantum gate*, making anyons and braids relevant to quantum computing.

Because the unitary matrix is determined by the nonlocal braiding process, the outcome is topologically stable. Assume that the start and end positions are held fixed and only the trajectories joining them can change. Then small deviations in trajectories won't alter the result of the braiding operation as long as the lines don't cross each other (i.e., as long as particles that are following the lines don't collide). The goal of topological quantum computing is to make the computational hardware (the braided anyons) error-resistant, instead of correcting for errors at the software level with redundant coding and error correcting codes.

Problems

1. Consider a closed path γ in the punctured complex plane, the complex plane with the origin removed. The winding number in the complex plane is defined by Eq. 21.1, $v = \frac{1}{2\pi i} \oint_\gamma \frac{dz}{z} = \frac{1}{2\pi i} \oint_\gamma d(\ln z)$.

   (a) If x and y are the real and imaginary parts of z, show that v can also be written as $v = \frac{1}{2\pi i} \oint_\gamma \left(\frac{x}{r^2} dy - \frac{y}{r^2} dx \right)$, where $r = \sqrt{x^2 + y^2}$.

   (b) A more general definition for the winding number of a map $f : t \to f(t)$ is $v = \frac{1}{2\pi i} \oint_\gamma \frac{d}{dt} \log f(t) \, dt$, where t is a parameter along curve γ. Show that this reduces to the definitions above for $z(\theta)$ playing the role of $f(t)$.

2. (a) Diagonalize $h(k)$ for the SSH model to verify that the energy and the eigenstates are of the forms given in Eqs. 21.32 and 21.34.

   (b) Find the normalization constant in Eq. 21.44.

   (c) Fill in the missing steps to arrive at the SSH boundary conditions of the forms of Eqs. 21.37 and 21.38.

3. Consider a one-dimensional lattice four sites long, with Hamiltonian [37]

$$
H = \begin{pmatrix} 0 & v+g & 0 & 0 \\ v-g & 0 & w+g & 0 \\ 0 & w-g & 0 & v+g \\ 0 & 0 & v-g & 0 \end{pmatrix}, \tag{21.119}
$$

   where g is a real constant and v, w are the hopping amplitudes.

   (a) Solve for the energy eigenvalues and show that they can be written in the form $E = \pm\sqrt{A \pm B}$, where $A = -\frac{3}{2}g^2 + v^2 + \frac{1}{2}w^2$ and $B = \sqrt{A^2 - (g^2 - v^2)}$.

   (b) Find the values of g at which the energy eigenvalues transition from real to complex values. At these points, pairs of eigenvectors coalesce into a single eigenvector. Such parameter values are called **exceptional points (EPs)** and have become an active research area because of the possibility that sensors can be made super-sensitive near these points. See [38, 39] for an introduction to EPs and their uses in sensing.

4. Derive the orbital frequency for an electron in a uniform magnetic field, Eq. 21.67.

5. Starting from the Schrödinger equation, show that $\hat{\mathcal{T}}$ must act on scalar wavefunctions by complex conjugation, $\hat{\mathcal{T}} = \hat{K}$, and that therefore it must be anti-unitary.

6. Derive Eqs. 21.71 relating the resistivity and conductivity in a two-dimensional quantum Hall system.

7. Using Ohm's law and Eq. 21.73, show that conductivities σ_{xx} and σ_{xy} have the forms given in Eqs. 21.71 and 21.76.

8. Verify that Eq. 21.61 holds.
9. (a) Show that for a spin-$\frac{1}{2}$ particle, $\hat{\mathcal{T}}^2 = -1$.
   (b) Using the result of part (a), prove Kramer's theorem.
10. The Dirac gamma matrices can be written in the Majorana representation as

$$\gamma^0 = \begin{pmatrix} 0 & \sigma_y \\ \sigma_y & 0 \end{pmatrix}, \quad \gamma^1 = \begin{pmatrix} i\sigma_z & \\ & \sigma_z \end{pmatrix},$$

$$\gamma^2 = \begin{pmatrix} 0 & -\sigma_y \\ \sigma_y & 0 \end{pmatrix}, \quad \gamma^3 = \begin{pmatrix} -i\sigma_x & \\ & -i\sigma_x \end{pmatrix}.$$

Using these matrices, show that the Dirac equation splits into the two Majorana equations (Eqs. 21.108 and 21.109).

References

1. J. Munkres, *Topology*, 2nd edn. (Pearson, New York, 2021)
2. C. Nash, S. Sen, *Topology and Geometry for Physicists* (Dover Publications, New York, 2011)
3. M. Nakahara, *Geometry, Topology, and Physics*, 2nd edn. (CRC Press, New York, 2003)
4. D.S. Simon, *Topology in Optics: Tying Light in Knots*, 2nd edn. (IOP Publishing, New York, 2022)
5. S. M. Girvin, in *Topological Aspects of Low Dimensional Systems*, ed. by A. Comtet, T. Jolicoeur, S. Ouvry, F. David (Springer, Berlin, 2000)
6. P.L. Taylor, O. Heinonen, *A Quantum Approach to Condensed Matter Physics* (Cambridge University, Cambridge, 2002)
7. M.O. Goerbig, arXiv:0909.1998v2 [cond-mat.mes-hall] (2009)
8. D. Tong, arXiv:1606.06687 [hep-th] (2016)
9. T.D. Stanescu, *Introduction to Topological Quantum Matter and Quantum Computation* (CRC Press, New York, 2017)
10. B.A. Bernevig (with Hughes T L), *Topological Insulators and Topological Superconductors* (Princeton University, Princeton, 2013)
11. M.Z. Hasan, C.L. Kane, Rev. Mod. Phys. **82**, 3045 (2010)
12. J.K. Asbóth, L. Oroszlány, A.P. Pályi, *A Short Course on Topological Insulators: Band Structure and Edge States in One and Two Dimensions* (Springer, Berlin, 2017)
13. N. Ashcroft, D. Mermin D, *Solid State Physics* (Brooks Cole, California, 1976)
14. M.A. Omar, *Elementary Solid State Physics* (Addison-Wesley, New York, 1975)
15. D.J. Griffiths, *Introduction to Electrodynamics*, 4th edn. (Cambridge University, Cambridge, 2017)
16. K. von Klitzing, G. Dorda, M. Pepper, Phys. Rev. Lett. **45**, 494 (1980)
17. M.A. Paalanen, D.C. Tsui, A.C. Gossard Phys. Rev. B **25**, 5566 (1982)
18. D. Thouless, M. Kohomoto, M. Nightingale, M. den Nijs, Phys. Rev. Lett. **49**, 405 (1982)
19. D.C. Tsui, H.L. Stormer, A.C. Gossard, Phys. Rev. Lett. **48**, 1559 (1982)
20. R.B. Laughlin, Phys. Rev. Lett. **50**, 1395 (1983)
21. F.D.M. Haldane, Phys. Rev. Lett. **61**, 2015 (1988)
22. C.Z. Chang et al. Science **340**, 167 (2013)
23. C.L. Kane, E.J. Mele, Phys. Rev. Lett. **95**, 226081 (2005)
24. J. Maciejko, T.L. Hughes, S.C. Zhang, Ann. Rev. Cond. Mat. Phys. **2**, 31 (2011)
25. X.L. Qi, S.C. Zhang, Phys. Today **63**, 33 (2010)

26. P. St-Jean et al., Nat. Photonics **11**, 651 (2017)
27. Y. Ota et al., Nanophotonics **9**, 547 (2020)
28. A. Yu. Kitaev, Phys.-Usp. **44**, 131 (2001)
29. J.K. Pachos, *Introduction to Topological Quantum Computation* (Cambridge University, Cambridge, 2012)
30. S.H. Simon, *Topological Quantum: Lecture Notes and Proto-Book*, available at https://www-thphys.physics.ox.ac.uk/people/SteveSimon/topological2020/TopoBookOct27hyperlink.pdf
31. C. Nayak, S.H. Simon, A. Stern, M. Freedman, S. Das Sarma, Rev. Mod. Phys. **80**, 1083 (2008)
32. J. Leinaas, J. Myrheim, Nuovo Cim. Soc. Ital. Fis. B **37**, 1 (1977)
33. F. Wilczek, Phys. Rev. Lett. **48**, 1144 (1982)
34. F. Wilczek, Phys. Rev. Lett. **49**, 957 (1982)
35. H. Bartolomei et al., Science **368**, 173 (2020)
36. J. Nakamura, S. Liang, G.C. Gardner, M.J. Manfra, Nat. Phys. **16**, 931 (2020)
37. D.S. Simon, C.R. Schwarze, A. Ndao, A.V. Sergienko, J. Opt. Soc. Am. B **xx**, xx (2024)
38. W.D. Heiss, J. Phys. A: Math. Theor. **4**, 444016 (2012)
39. J. Wiersig, Photonics Res. **8**, 1457 (2020)

Part IV
Physical Implementations

Chapter 22
Optical Sources and Detectors

Many of the imaging, sensing, and information processing applications discussed in this book are based on optical systems. In addition, even in atomic, ionic, and solid-state systems, electromagnetic waves are used to excite the systems to higher energy levels, to manipulate superpositions of states, to move atoms around, and to measure the states via fluorescent readouts. As a result, optical sources and detectors are vital ingredients in applications, regardless of the primary physical platform being used. In this chapter, we look at some common optical sources and optical detectors and discuss some of their basic properties.

22.1 Optical Detectors

Many of the applications in this book involve the creation and detection of photons, either singly or in pairs. So in this chapter, we cover some of the basic background related to optical sources and detectors. We begin in the next few sections with detectors.

In general, optical detectors destroy the measured photon. They absorb the photon at some surface, removing it from the electromagnetic field and converting the absorbed energy into an electrical current that can be measured. The photon detection process is therefore not an ideal von Neumann measurement, but rather would be described by a generalized POVM measurement.

Detectors can be classified in a number of different ways. For example, we can distinguish them based on their ability to form images. **Bucket detectors** or **single-pixel detectors** simply detect whether a photon was detected or not, but gives no information on precisely *where* the photon hit the detector, so that it provides no spatial resolution. A **spatially resolving** or **multi-pixel detector** provides information on the location where the photon was absorbed, so that the spatial variations in light intensity can be resolved. This allows images or spatial

© The Author(s), under exclusive license to Springer Nature Switzerland AG 2025
D. S. Simon, *Introduction to Quantum Science and Technology*, Undergraduate
Texts in Physics, https://doi.org/10.1007/978-3-031-81315-3_22

interference patterns to be obtained. The detectors can also be categorized based on their low-intensity behavior; in particular, by their ability or inability to resolve the number of photons absorbed.

Detectors can also be grouped based on the means used to generate a signal from the absorbed photon. **Thermal detectors** convert the absorbed energy into heat, which is then measured by some means, while **photodetectors** use the photoelectric effect to convert the absorbed energy directly into an electric current.

A number of useful figures of merit can be defined for optical detectors [1]. A few useful quantities include the following:

- For photoelectric detectors, the **quantum efficiency** η is the fraction of absorbed photons that produce photoelectrons. Let P_{opt} be the power in the optical signal. Then, since

$$\frac{\text{charge}}{\text{time}} = \frac{\text{energy/time}}{\text{energy/photon}} \cdot \frac{\text{charge}}{\text{electron}} \cdot \frac{\text{electrons}}{\text{photon}}, \quad (22.1)$$

  we find that the electron current generated is

$$i_{el} = \frac{P_{opt}}{h\nu} \cdot e\eta. \quad (22.2)$$

- For photoelectric detectors, the **detector gain** G is then defined by

$$G = \frac{Q}{e}, \quad (22.3)$$

  where Q is the total current flowing into the detector for each electron that absorbs an incoming photon. The gain is a useful measure of how much the detector amplifies the signal before readout; high gain allows lower intensity light to be detected. In the presence of gain, the photocurrent detected will be greater than the electron current generated: $i_{det} = G i_{el}$.
- Another measure of the detector quality is the **responsivity** or current out per unit optical power in

$$\mathcal{R} = \frac{di_{det}}{dP_{opt}} = \frac{Ge\eta}{h\nu}, \quad (22.4)$$

  (measured in amps per watt). From the equations above, we see that the photocurrent can be related to the responsivity as

$$i_{det} = \mathcal{R} P_{opt}. \quad (22.5)$$

- The speed of a detector can be measured using its **response time**, τ. This is the time it takes the transient electric current $i(t)$ due to photon absorption to reach its maximum value, i_0: $i(t) = i_0(1 - e^{-t/\tau})$.

We briefly look at various types of detectors in the following sections, beginning with an overview of thermal detectors. General references for most of this chapter include [2–4].

22.2   Thermal Detectors

Thermal detectors are designed to detect light by measuring the heat deposited in some material as the light is absorbed. Thermal detectors tend to have several disadvantages: they are slow, they often are not very sensitive to low light levels, and they usually provide very little ability to discriminate between photons of different frequencies. On the plus side, they are often more sensitive to infrared and other parts of the spectrum than other detector types.

In general, thermal detectors have three main components:

• A **sensor**: this is the material that actually absorbs the light. It should have a large surface area to absorb the maximum number of photons. But it should also be of small volume so that the energy of the absorbed photons can create a measurable temperature change in the material. This means that the sensor should therefore be in the form of some thin sheet with a large area A in one plane and small thickness d in the perpendicular direction. The surface of area A is called the **active surface** (Fig. 22.1).

• A **heat sink**: In order for heat to flow from the sensor, it has to be in contact with some material of lower temperature, known as the heat sink. The heat sink should stay at constant temperature as heat flows into it from the sensor, which means that it should be much larger in mass and volume than the sensor.

• A **measuring device**: some means to measure the heat flow and convert it into an electrical signal that can be read or stored by accompanying electronics.

Schematically, thermal detectors look as shown in Fig. 22.1. The sensor and heat sink are connected by a linking material, which we assume for simplicity has length l and constant cross-sectional area A'. Initially (in the absence of light), the sensor and heat sink are both assumed to be at some temperature T_0. When light is absorbed,

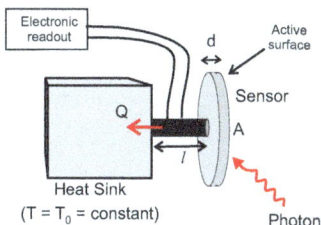

Fig. 22.1 Schematic of a thermal light detector. Light is absorbed by the active area A of the sensor, causing the sensor to rise in temperature. This leads to heat Q flowing from the sensor to the heat sink. The heat flow is measured and converted to an electronic readout

the sensor increases in temperature by ΔT, while the heat sink temperature remains approximately constant. Let an amount Q of heat flow from the sensor to the heat sink after the absorption. The goal then is to measure Q and to relate it to the photon energy absorbed during the measurement interval.

Looking at the material linking the sensor to the sink, recall that the heat flow per time through the link should be of the form

$$\frac{dQ}{dt} = \frac{kA'}{l} \Delta T, \tag{22.6}$$

where k is the thermal conductivity of the material. This is often written more simply as

$$\frac{dQ}{dt} = C\Delta T, \tag{22.7}$$

where $C = \frac{kA'}{l}$ is the **thermal conductance**.

Let P_{opt} be the optical power absorbed by the sensor (in Watts), and let m and c be the mass and specific heat (in $\frac{J}{kg\,°C}$) of the sensor. Then energy conservation tell us that

(net energy change of the sensor per time)

= (energy per time in from light) - (heat per time out to sink),

or

$$mc\frac{d(\Delta T)}{dt} = P_{opt} - C\Delta T. \tag{22.8}$$

This equation is readily solved for $\Delta T(t)$ (it is mathematically identical to the equation for a charging or discharging capacitor). After a photon is absorbed, the temperature change first rises from 0 to ΔT_{max},

$$\Delta T(t) = \Delta T_{max}\left(1 - e^{-t/\tau}\right), \tag{22.9}$$

then decays back to zero

$$\Delta T(t) = \Delta T_{max}e^{-t/\tau}, \tag{22.10}$$

where the time constant for the decay is $\tau = \frac{mc}{C}$.

We see that once a photon is absorbed, it takes time for the heat to dissipate. Until that happens, another photon cannot be detected. So the time constant determines the maximum response rate of the detector: small τ means rapid response. We see that τ can be made small by taking m small. However, if ρ is the density of the

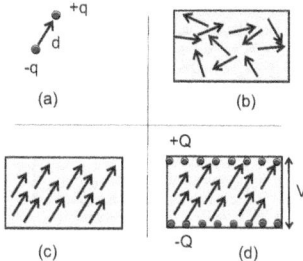

Fig. 22.2 (a) In a pyroelectric material, the molecules have nonzero dipole moments, $p = qd$. (b) At high temperatures, the orientation of the dipoles is random due to thermal agitation, leaving an average polarization of zero. The positive ends of the dipoles all cancel with neighboring negative ends, leaving a net charge of zero throughout the material. (c) At lower temperatures, the dipoles can align, leading to nonzero polarization. Neighboring charges still cancel to leave net charge of zero in the bulk of the material, but at the surfaces there are no opposite sign charges to provide cancellation. The result is that each surface gains a total charge $\pm Q$ from the sum of all the uncancelled dipole charges on it. (d) The separation of the charges $\pm Q$ at the surface leads to a potential difference V across the material

material in the detector, then $m = \rho d A$. So there is a trade-off: making m small by reducing the active area A will also reduce the sensitivity to dim light.

Thermal detectors fall into two main categories, thermoelectric and pyroelectric. **Pyroelectric detectors** are made from materials whose molecules have electric dipole moments $p = qd$ (even in the absence of external electric fields) whose magnitudes are temperature dependent (Fig. 22.2). p maintains a fixed nonzero value at low temperatures, which decay with increasing T, dropping to zero for T greater than some critical temperature. This **pyroelectric effect** had been first observed by Johann Georg Schmidt in 1707, who noticed that hot coal ash could attract nearby iron. Examples of pyroelectric materials include tourmaline (a mineral composed mainly of boron and silicates), lithium niobate ($LiNbO_3$), and lithium tantalate ($LiTaO_3$). Pyroelectric effects have also been seen in semiconductors such as gallium nitride and in noncrystalline materials, including the bone and tendon.

The use of pyroelectric materials in optical detectors makes use of the fact that, in a slab of such material, as the dipole moment grows, more positive charge will collect on one side of the slab and more negative charge will collect on the opposite surface. The resulting charge separation leads to the growth of a voltage V across the surface, called the **pyroelectric potential**, V. A circuit to measure temperature changes using the material is shown in Fig. 22.3. The material is placed inside a capacitor, so that the pyroelectric potential forces charges onto the plates of the capacitor. As the temperature changes, the charges on the plate also have to change, forcing some of the charge to flow from one plate to the other through the load resistor, R_L. This current produces a signal voltage V_s across R_L, which can be measured. Note that the signal is only nonzero when the temperature is changing.

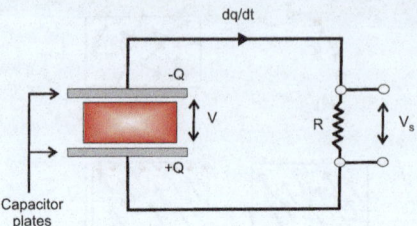

Fig. 22.3 The positively charged side of the pyroelectric material attracts negative charges onto the neighboring capacitor plate (and vice versa at the other plate). When the pyroelectric slab changes temperature, the resulting change in pyroelectric potential V causes the charges on the capacitor plates to change. This forces a current $i = \frac{dq}{dt}$ through load resistor R. Measuring the voltage V_s across the resistor then allows the temperature change of the material to be determined

The insensitivity to constant background temperatures is one advantage of this type of detector. But it creates a corresponding problem: how to measure the intensity of a steady light beam, in which photons are constantly bombarding the material and temperature should be roughly constant? The answer is to add *modulation* (temporal variation) to the light. One common means of doing this is to have a disk with a slot in it rotating in front of the optical sensor. Light strikes the material when the slot is in the beam path, raising the temperature. Then the slot rotates out of the beam path, resulting in a temperature drop. The thermoelectric material then senses the size of the drop and converts it into a measurable signal voltage.

The second major category of thermal detector consists of **thermoelectric detectors**. These rely on the **thermoelectric effect**, first discovered by Volta in 1794 and rediscovered by Seebeck in 1821. In a thermoelectric material, the energy bands vary strongly with temperature. If two such materials at different temperatures are placed in contact, a voltage difference appears between the energy bands on the two sides, and so a current will flow across the junction between them. The junction itself is called a thermocouple, and the size of the resulting current will be proportional to the temperature difference between the two sides of the junction. At room temperatures, typical voltage differences across the junction may be on the order of 10–50 μV per °C.

In a thermoelectric optical detector, the sensor is again attached to a heat sink. A pair of thermocouples then link the heat sink and the sensor to wires leading to a voltage detector as shown in Fig. 22.4. If the two thermocouples are at the same temperature, the currents i_1 and i_2 will be equal, and so will cancel, leading to a zero net current and voltage across the resistor. However if the sensor absorbs a photon and heats up, then $i_2 > i_1$ and a nonzero voltage will appear.

Fig. 22.4 Schematic of a thermoelectric detector. Thermocouples on the sensor and heat sink produce currents that depend on the temperatures of the materials. If the temperatures are equal, the currents will cancel. If not, then a net current will flow in the resisistor, producing a signal voltage

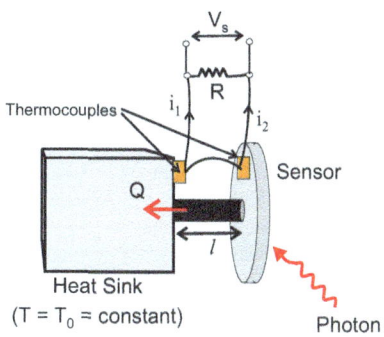

22.3 Photodetectors

Most modern detectors are non-thermal. Rather than measuring the heat from absorbing a photon, the absorbed photon energy is instead used to generate an electric current, via the photoelectric effect. These photodetectors come in a number of varieties. In each case, a photon stimulates the emission of an electron, and the problem is to amplify the signal from that one electron into a current large enough to detect easily.

22.3.1 Vacuum Photodiodes

The vacuum photodiode is a fairly old and relatively simple type of photodetector that has rapid response, but it is not very sensitive to low light levels. It also has the disadvantage of requiring high voltages. It is shown schematically in Fig. 22.5.

The vacuum tube contains two terminals, a positively charged metal plate called the **anode** and a negatively charged photoemissive material called the **cathode**. Typical materials for use in the cathode might be GaAS or InGaAs, which have high efficiencies in the visible and infrared ranges. Other materials can be used for sensitivity in the UV range. A high-voltage power supply provides a large operating voltage, V_0, between the two terminals. V_0 creates an electric field pointing from the anode toward the cathode. When light falls on the cathode, electrons are ejected from the material. They are accelerated toward the anode by the electric field. Since there is vacuum between the terminals, there is no scattering of electrons with air molecules to impede their progress between the terminals. The electrons absorbed at the anode then flow around the circuit, creating a clockwise current around the circuit. The resulting current pulse through the load resistor can then be measured.

Let d be the anode-to-cathode distance and τ to be the transit time to cross this distance. To find the response time, we assume a simple model in which the electrons start from rest at the cathode and the electric field in the terminal gap is uniform, $E = \frac{V_0}{d}$. In that case, the electrons have constant acceleration in the gap,

Fig. 22.5 Vacuum photodiode. Incident light causes electrons to be emitted by the cathode. They are accelerated by an electric field toward the anode and then flow around the circuit back to the cathode. The resulting current through the load resistor signals the presence of the light

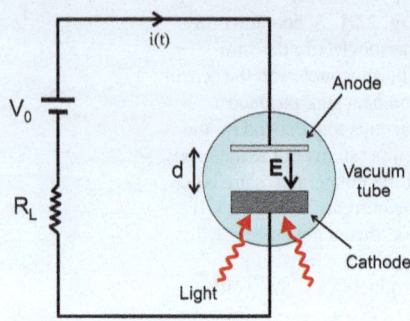

allowing their speeds and the resulting current to be easily found. The acceleration is $a = \frac{eE}{m}$ = constant, so it follows immediately that for $0 \le t \le \tau$,

$$v(t) = at = \frac{eE}{m}t \tag{22.11}$$

$$x(t) = \frac{1}{2}at^2 = \frac{eE}{2m}t^2. \tag{22.12}$$

So, setting $x(\tau) = d$ and solving for τ

$$\tau = \sqrt{\frac{2md}{eE}} = \sqrt{\frac{2md^2}{eV_0}}. \tag{22.13}$$

So by making the operating voltage sufficiently large, τ, the response time can be made very short. For example, if the V_0 is on the order of kilovolts, τ can be on the order of a few picoseconds.

To find the current, note that the power P_L lost in the load resistor must equal the energy gained per time $\frac{dW}{dt}$ by the electrons while accelerating from the cathode to the anode. The energy gain during the acceleration is given by the work done by the field:

$$dW = F\,dx = eE \cdot v\,dt = \frac{eV_0 v}{d}dt. \tag{22.14}$$

So,

$$P_L = \frac{dW}{dt} \implies V_0 i(t) = \frac{eV_0}{d}v(t). \tag{22.15}$$

Therefore,

$$i(t) = \frac{e}{d}v(t) = \frac{eE}{V_0}v(t). \tag{22.16}$$

Fig. 22.6 A photocell. The structure is similar to the vacuum photodiode, except that the light-sensitive material fills the gap between the anode and the cathode, leading to higher sensitivity

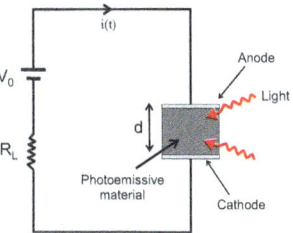

So, $v(t)$ and $i(t)$ both have the same temporal shape: each photoelectron creates a current that rises linearly until time τ and then drops rapidly to zero. Each pulse exists for time τ, justifying the name response time and limiting how close together in time two electrons can be and still be distinguishable as separate events.

The total charge transfer between the terminals during the current pulse is just

$$Q = \int_0^\tau i(t)dt = \frac{eE}{V_0} \int_0^\tau v(t)\, dt \qquad (22.17)$$

$$= \frac{eE}{V_0} \frac{eE}{m} \int_0^\tau t\, dt = \frac{e^2 E^2}{2m V_0}\tau^2 \qquad (22.18)$$

$$= e, \qquad (22.19)$$

giving a detector gain of $G = 1$.

22.3.2 Photocells

A **photocell** or photoconductive detector is similar to a vacuum photodiode, except that both terminals are metal and the space between them is filled with a photoemissive material that serves as the active region for absorbing light (Fig. 22.6). Because of the larger active region, photocells are much more sensitive to low light levels. They also tend to have much wider frequency response. The disadvantage, however, is that they tend to be much slower, with longer response rates τ.

To find the detector gain, first we must define the electron and hole mobilities μ_e and μ_h in the material. These are measures of how effective an applied electric field is in making electrons and holes move in a material. They are simply defined as the ratios of the respective velocities in the material to the field strength

$$v_e = \mu_e E, \qquad v_h = \mu_h E, \qquad (22.20)$$

where the mobilities have units of $\frac{m^2}{V \cdot s}$. Although both electrons and holes contribute to the photocurrent, the electrons usually have much higher mobility and provide the main contribution to the current. So for simplicity, we ignore the holes and consider just the electrons.

We saw before (Eq. 22.16) that $i(t) = \frac{eE}{V_0} v(t)$. However, now $v(t)$ will be the drift velocity of the electron in the material, which will be roughly constant. So,

$$i(t) = \frac{eE}{V_0} (\mu_e E) = \frac{e\mu_e E}{d}, \tag{22.21}$$

which is constant, rather than linearly increasing, over the response time. The total charge transfer between the terminals is therefore

$$Q = \int_0^\tau i(t)\, dt = \left(\frac{e\mu_e V_0}{d^2} \right) \tau, \tag{22.22}$$

so the gain $G = \frac{Q}{e}$ is

$$G = \left(\frac{\mu_e V_0}{d^2} \right) \tau. \tag{22.23}$$

The gain can therefore be made very large by making d small; however doing this comes at the expense of making the volume of active region smaller, reducing the sensitivity to dim light.

22.3.3   Photomultipliers

Photomultipliers have been used since the early days of electronic light detection. Due to their high sensitivity to small numbers of photons, they are still in widespread use in areas ranging from astrophysics to elementary particle physics. The basic idea is to have each emitted electron collide with a sequence of terminals, with each collision causing the emission of additional electrons, resulting in a large cascade of particles.

The photomultiplier starts with a setup similar to a vacuum photodiode, except that between the anode and the cathode, we place a sequence of secondary metal electrodes called **dynodes**, as shown in Fig. 22.7. Suppose there are a total of N dynodes. If V_0 is the voltage between anode and cathode, then the voltage at the nth dynode (relative to the cathode) is $\frac{n}{N+1} V_0$, for $n = 1, 2, \ldots, N$. The process starts with a photon causing the emission of an electron from the cathode. The electric field $E = \frac{V_0}{d}$ accelerates the electron to high energy, so that when it collides with the first dynode it is energetic enough to knock several additional electrons out of the metal. Let δ denote the number of secondary electrons produced in each collision. These new electrons are now accelerated by the electric field, so that when they strike the second dynode, each of them produces δ outgoing electrons. After N collisions, the final number of electrons reaching the anode (per initial electron) is

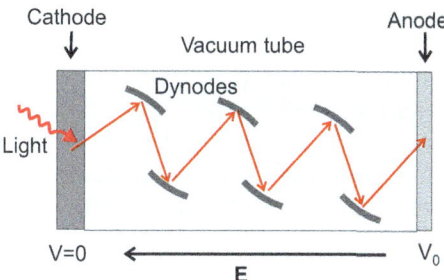

Fig. 22.7 Photomultiplier. An initial electron liberated from the cathode by a photon hits a sequence of dynodes. The electric field in the tube accelerates the electron so it has enough energy to liberate several more electrons at each collision. The result is the single initial electron initiates a large cascade of electrons, leading to very high gain in the signal at the anode

δ^N, so that the detector gain is

$$G = \delta^N. \tag{22.24}$$

Increasing the operating voltage V_0 will increase the violence of each electron-dynode collision and will therefore increase the value of δ. Typical values might be $V_0 \sim 1$ kV, $\delta = 5$–10, $N \approx 10$, giving a gain of at least $G \sim 5^{10} \sim 10^7$. This high gain is the chief advantage of the photomultiplier, making it sensitive to very low-intensity light. However, the photomultiplier is slow (requiring time for multiple electron-dynode collisions before detection) and require high voltages.

22.3.4 Charge-Coupled Devices (CCDs)

A **charge-coupled device** or **CCD** is an array of solid-state detectors, each collecting a single pixel of an image. These can be used as cameras, with high resolution and very short exposure times. They have very high quantum efficiencies: in comparison to the human eye that has an efficiency of about 20% and photographic emulsions of around 10% or less, CCD cameras can have quantum efficiencies of over 90%. Each CCD pixel is very sensitive at low photon numbers, so that it can detect very dim light. In addition, it can detect dim objects that are close to much brighter objects, without losing the dim object in the glare of the other one. The dynamic range of a typical CCD (the intensity ratio of brightest to dimmest signals detectable) is typically on the order of 10^5, about a thousand times better than photographic films and plates.

A schematic of a CCD camera is shown in Fig. 22.8a. A p-doped silicon substrate is covered with a thin insulating layer of silicate, SiO_2. An array of metal gates is placed on top of the silicate, with electrodes attached to separate voltage sources, so that each gate can be controlled independently. This is a **metal-oxide-semiconductor (MOS)** structure, similar to that used in diodes and other devices. Applying a positive voltage V_G to one of the gates attracts electrons to the vicinity of the silicate layer, holding them there in a potential well. A layer of positive charges forms on the bottom of the silicon substrate in compensation, as shown. As in Fig. 22.8b, suppose that the left-most gate is held at positive V_G, while the gates to its right are all at zero voltage. If a photon strikes the semiconductor near the first gate, it will produce additional electron-hole pairs. The electrons will remain confined near the gate by the potential well. The number of excess electrons will be proportional to the number of photons absorbed. Suppose that the CCD is briefly exposed to light. If now the positive voltages are cycled toward the right, the electrons will be carried along with the voltage, as the potential well itself moves toward the right-hand edge. The charge can then be collected from the edge, and the intensity of the light hitting the first pixel can then be determined from the size of the measured charge. If it takes time T to move the control voltage one unit the right, then electrons collected at time NT after the exposure correspond to photons striking the device N pixels to the left of the charge-collection region. This provides spatial resolution along the x-axis. If there are rows of such pixels perpendicular to the page, then resolution in the full two-dimensional plane of the surface will be obtained. Since the pixels can be made very small, very high spatial resolution is possible.

CCD's have the further advantage of requiring low power consumption, in contrast to other detectors such as photomultipliers, which require high voltages. As a result of the desirable properties of large dynamic range, high speed and efficiency, high resolution, and low power requirements, CCD cameras have being

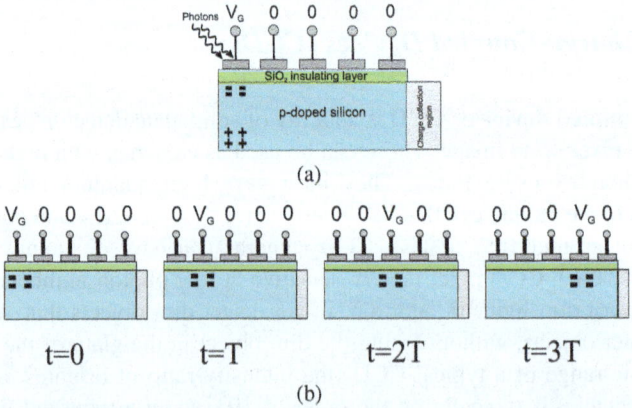

Fig. 22.8 (a) Schematic of a CCD camera. The positive gate voltage holds any electrons in the silicon layer in a localized potential well near that gate. Incoming photons create electron-hole pairs that increase the electron density near the gate where they hit. (b) Cycling the positive voltage to the right, the electrons will follow. The amount of charge reaching the charge collection area on the right will be proportional to the number of photons striking the gate

come standard in a wide variety of applications, ranging from consumer video cameras to astronomical and biomedical imaging.

22.4 Superconducting Detectors

When working in the quantum regime, simply detecting light is often not enough. Often it is also necessary to know how many photons have been absorbed by the detector. Such photon-counting detectors are possible, and their function-ing usually involves superconducting systems. Superconducting photon-number counting detectors generally fall into two categories, transition-edge sensors or superconducting nanowire detectors.

Transition-edge sensors (TES) work by keeping the active region (the material that will be absorbing the photons) right at the critical temperature marking the boundary between superconductor and ordinary phases. It needs to be supercon-ducting, but teetering right on the edge of a transition into the non-superconducting phase. This is achieved by placing the active material in contact with a much larger thermal bath that will keep it at a fixed temperature. A small current is passed through the superconducting material, which can be continuously monitored. The detector is shown schematically in Fig. 22.9a.

If a photon is absorbed by the active region, then its incident energy will go into heating the material, tipping it over into the non-superconducting phase. As a result, the resistance of the material will become nonzero, and the current will drop, signaling the presence of the photon.

The curve of resistance (shown in Fig. 22.9b) is roughly linear as a function of temperature (at least for small temperature changes), while the temperature is in

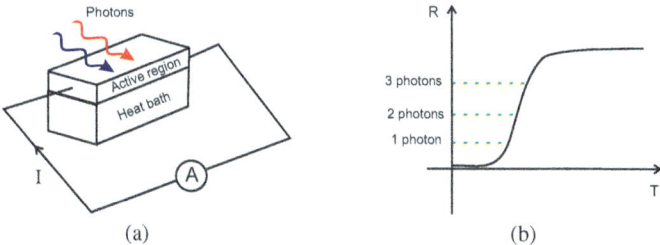

(a) (b)

Fig. 22.9 (**a**) Transition-edge sensor. Photons absorbed in the active region heat the material and cause it to transition from the superconducting to non-superconducting phase. This causes a resistance to appear, which will be apparent from a sudden drop in the current, I. (**b**) For small temperature increases, the change in resistance of the active region is roughly linear with respect to temperature. So (for photons of the same frequency) the resistance will be proportional to the number of photons absorbed, making photon counting possible. Note that if the number of photons is too large, the resistance saturates at a maximum va, so the device works best at small photon numbers

turn linear in the energy of the absorbed photon. As a result, the number of photons can be read off from the size of the resistance, as shown. This works as long as the number of photons is small, but once it becomes too large, the detector saturates: the curve bends to become nearly horizontal, so that the number of detected photons can no longer be determined.

TES detectors can be highly efficient at photon detection, with efficiencies of over 90%. However, after each detection there is a recovery time required for the active material to return to the superconducting phase. Until this happens, further detections cannot occur. There is thus a trade-off to be made: making the active region larger increases the efficiency, since there is a larger volume available for the photons to interact with and be detected in. But increasing the volume also means that it takes longer to cool the material back down, and so recovery times become longer.

In a **superconducting nanowire detector (SND)**, it is not the entire active area that is superconducting. Instead, the area is threaded with a set of superconducting nanowires with a voltage difference V applied across them. This effectively creates an array of small detection regions: an absorbed photon will heat a small region of the active material, including the small portion of the wire passing through that region. Once again, there will be an increase in resistance and drop in current. The change in total R will be proportional to the number of regions that have been heated, which again allows the number of absorbed photons to be determined. If the current in each wire is measured separately, the location of the photon can be determined by which wire loses superconductivity. By having two arrays of nonwires, one above the other, with one set of wires parallel to the x-axis and the other parallel to the y-axis, two-dimensional resolution can be achieved.

SND's have several advantages. They have very low dark count rates and can operate in the visible and infrared and very high repetition rates (on the order of 100 MHz or higher). Uncertainty in photon arrival time can be as low as the picosecond range. They have the disadvantage that they need to be cooled to temperatures well below 10 K in order to superconduct, but they can operate at temperatures higher than a typical edge sensors (as well as being faster). Recent SND designs have achieved quantum efficiencies of near 100%.

22.5   Avalanche Photodiodes

An **avalanche photodiode** (APD) [5] consists of a PN junction maintained at a very high reverse bias. When a photon enters the depletion region, it creates an electron-hole pair. The reverse bias voltage causes the electron and hole to accelerate and increase in energy. If the voltage is high enough, the carriers can gain enough energy to create new electron-hole pairs during collisions with the crystal lattice, which will then also accelerate. The resulting chain reaction leads to an avalanche of charge carriers being generated from a single absorbed photon when the bias voltage V exceeds some minimum level V_0.

Let G be the multiplication factor or current gain, and P be the incoming optical power. Then the mean-squared output photocurrent, assuming quantum efficiency η, is [6]

$$\langle i_c^2 \rangle = 2G^2 \left(\frac{Pe\eta}{hv} \right)^2 . \tag{22.25}$$

Since the output signal power is proportional to i_c^2, the output power also scales proportional to G^2 for fixed input power. The noise will increase with G as well, but an optimal value of G can be found [6] for which the signal to noise ratio is minimized, typically at some value of G on the order of 100. At this optimal value of G, the minimal detectable power is

$$P_{min} = \frac{2hv}{Ge\eta} \sqrt{\frac{kT\Delta v}{R_L}}, \tag{22.26}$$

where T and R_L are the temperature and the load resistance. For large G, there is therefore a large improvement in the sensitivity of the detector, with the dimmest detectable signal scaling inversely with G. This allows detector operation at very low light levels.

Normally, the breakdown voltage V_0 occurs at a few hundred volts. But using specialized doping methods, V_0 values of 1500 or more can be used, allowing sensitivity that reaches the single-photon level. In this high-gain or Geiger-mode operation region, the APD is known as a **single-photon avalanche diode** (SPAD). SPADs can reach gains on the order of $G \sim 10^6$.

APDs can be used over a wide range of photon frequencies, ranging from infrared to x-ray. The material used in the PN junction will depend on the frequency range that is of interest. Germanium is typically used in the range 800–1700 nm. Although having lower multiplication factors, silicon and indium gallium arsenide (InGAs) are also often used: silicon works in the range 190–1100 nm and InGAs in the ranges 950–1550 nm and 1100–1700 nm.

The avalanche photodiode has a very fast response time and high-temperature stability. The chief disadvantage is that very high biasing voltages are required. At very low counting rates, photon counting is possible by counting the number of output pulses detected. However, the nonlinear detector response and the detector recovery time between pulses make photon counting impossible as the beam intensity increases. SPADs are often the detector of choice for quantum optics experiments, as well as being used in applications ranging from fiber optics communications to PET scans and microscopy.

22.6 Homodyne Detection

The detectors in the previous sections are designed to detect the presence or absence of photons, to measure the intensity of an optical beam. But often the *phase* of the light (relative to some reference phase) is also important. This can be achieved using a balanced homodyne detector. This is really a form of intensity interferometry, which allows phases and quadratures to be measured.

The balanced **homodyne detector** looks schematically as shown in Fig. 22.10. One input port contains the signal beam, with amplitude E_s. The other port contains a light beam called the local oscillator, with field amplitude E_{lo}. The local oscillator is a reference beam whose properties are known. We can add a variable phase shift ϕ_{lo} into the path of the local oscillator as shown, so that the field reaching the beam splitter from the right is $E_{lo}e^{i\phi_{lo}}$. ϕ_{lo} is controllable by the experimenter. The term *homodyne* refers to the fact that the local oscillator is taken to have the same frequency as the signal. In a *heterodyne* detector, the local oscillator would have a different frequency from the signal. (Heterodyne circuits are common in radios.)

After the beam splitter, the two input fields mix to form new fields E_1 and E_2 in the two outgoing ports, which are then measured at detectors D_1 and D_2. The photocurrents from the two detectors, i_1 and i_2, are then fed into a circuit that takes the difference of the currents. So the output at the end is the difference current $i = i_1 - i_2$.

The fields leaving the two outgoing sides of the beam splitter are

$$E_1 = \frac{1}{\sqrt{2}}\left(i E_{lo}e^{i\phi_{lo}} + E_s\right) \tag{22.27}$$

$$E_2 = \frac{1}{\sqrt{2}}\left(E_{lo}e^{i\phi_{lo}} + i E_s\right). \tag{22.28}$$

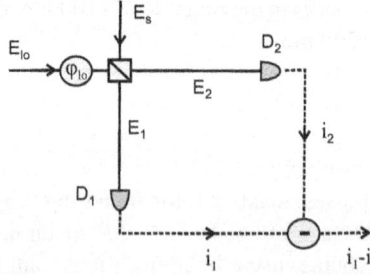

Fig. 22.10 A balanced homodyne detector mixes the signal field E_s with a local oscillator field of phase ϕ_{lo}. After detecting the light exiting both outgoing sides of the beam splitter, the difference of the photocurrents from the two detectors is measured. This difference current can be used to measure either of the two signal field quadratures, depending on the choice of ϕ_{lo}

Up to overall constants (which we will ignore), splitting the signal field into real and imaginary parts amounts to splitting the field into its quadratures (Chap. 5):

$$E_s \sim X + iP. \tag{22.29}$$

Plugging this split into the two output fields, we find

$$E_1 = \frac{1}{\sqrt{2}} \left(i E_{lo} \left(\cos \phi_{lo} + i \sin \phi_{lo} \right) + X + i P \right) \tag{22.30}$$

$$= \frac{1}{\sqrt{2}} \left((X - E_{lo} \sin \phi_{lo}) + i \left(P + E_{lo} \cos \phi_{lo} \right) \right) \tag{22.31}$$

$$E_2 = \frac{1}{\sqrt{2}} \left(i E_{lo} \left(\cos \phi_{lo} + i \sin \phi_{lo} \right) + i X - P \right) \tag{22.32}$$

$$= \frac{1}{\sqrt{2}} \left((E_{lo} \cos \phi_{lo} - P) + i \left(E_{lo} \sin \phi_{lo} \right) + X \right). \tag{22.33}$$

The photocurrent is proportional to the optical intensity I, which is in turn proportional to the square of the field intensity $|E|^2$. So, the output current at the end is (up to overall constants)

$$i_1 - i_2 \sim |E_1|^2 - |E_2|^2 \tag{22.34}$$

$$\sim \frac{1}{2} \Big\{ (X - E_{lo} \sin \phi_{lo})^2 + (P + E_{lo} \cos \phi_{lo})^2 \tag{22.35}$$

$$- (E_{lo} \cos \phi_{lo} - P)^2 - (E_{lo} \sin \phi_{lo})^2 + X \Big\}$$

$$= \frac{1}{2} \Big\{ \Big[X^2 + P^2 + 2 E_{lo} \left(P \cos \phi_{lo} - X \sin \phi_{lo} \right) + E_{lo}^2 \Big] \tag{22.36}$$

$$- \Big[X^2 + P^2 + 2 E_{lo} \left(X \sin \phi_{lo} - P \cos \phi_{lo} \right) + E_{lo}^2 \Big] \Big\}$$

$$= 2 E_{lo} \big(P \cos \phi_{lo} - X \sin \phi_{lo} \big). \tag{22.37}$$

The importance of this result can be seen by setting the phase of the local oscillator to particular values. If $\phi_{lo} = 0$ (or more generally if $\phi_{lo} = n\pi$ for any integer n), then the difference current is

$$i_1 - i_2 \sim E_{lo} P. \tag{22.38}$$

On the other hand, if ϕ_{lo} is set to an odd multiple of π, $\phi_{lo} = \frac{2n+1}{2}\pi$ integer n, then

$$i_1 - i_2 \sim E_{lo} X. \tag{22.39}$$

We see then that by choosing the local oscillator phase appropriately, the output difference current can be used to measure either field quadrature. And once the quadratures are known, the phase (relative to the oscillator) and amplitude of the field can be found:

$$|\alpha| = \sqrt{X^2 + P^2}, \qquad \tan\phi = \frac{P}{X}. \tag{22.40}$$

22.7   Thermal and Low-Coherence Light Sources

Recall that broadband light sources tend to be of low coherence, as measured by the coherence time

$$\tau_c = \frac{2\pi}{\Delta\omega} \tag{22.41}$$

or the longitudinal coherence length

$$L_c = c\tau_c = \frac{\lambda_0{}^2}{\Delta\lambda}, \tag{22.42}$$

where $\Delta\omega$ is the frequency spread, λ_0 is the central frequency of the wavepacket, and the bandwidth $\Delta\lambda$ is related to the frequency spread by

$$\frac{\Delta\lambda}{\lambda_0} = \frac{\Delta\omega}{\omega_0}. \tag{22.43}$$

Most light sources in nature and in everyday life are thermal sources (or *incandescent* sources), in which heat is used to excite electrons in some material into higher energy states, leading to light emission as the electrons drop to back to the ground state. Such thermal sources normally have broad bandwidth and low coherence, with typical coherence times ranging from 10^{-15} s to 10^{-12} s. (This should be compared to lasers, which can have coherence times ranging from about 10^{-9} s to 10^{-4} s or more.)

There are several reasons for the low coherence of thermal light sources. (i) Each electron transition takes only a finite time, so that the resulting emitted wavepacket is of finite duration (10^{-8} to 10^{-9} s). Fourier theory tells us that short-duration wave packets require a broad range of wavelengths. (This defines the *natural* linewidth, via the Heisenberg uncertainty relation.) (ii) Thermal fluctuations in the emitting material mean there is a broad range of energies present in the material, so that ground state electrons can be excited to different higher-energy states. This results in an incoherent superposition of different frequencies in the resulting light. (iii) In addition, each atom emits more or less independently at random times, leading to photons that are randomly superposed in time. (iv) Finally, Doppler and collisional broadening within the material lead to a broadening of emission line widths, with a consequent increase in bandwidth and decrease in coherence. The result is that the output spectrum is continuous and broadband.

The incoherent light emitted by a thermal source at absolute temperature T obeys the standard black body spectrum. The **spectral energy density** $u(T, v)$ or energy per volume per unit frequency is given by the **Planck radiation law**

$$u(T, v) = \frac{du(T)}{dv} = \frac{dP}{dA\, d\Omega dv} dv = \frac{8\pi h v^3}{c^3} \frac{1}{e^{hv/kT} - 1}, \tag{22.44}$$

with the wavelength of peak power emission given by **Wien's displacement law**,

$$\lambda_{peak} = \frac{b_\lambda}{T} \qquad v_{peak} = b_v T, \tag{22.45}$$

where the Wien constants are given by

$$b_\lambda \approx 2.898 \times 10^{-3} \text{ m K}, \quad \text{and} \quad b_v = 5.879 \times 10^{10} \text{ Hz/K}. \tag{22.46}$$

Integrating over frequency, the total intensity (power per area) emitted is given by the **Stefan-Boltzmann law**

$$I = \sigma T^4, \tag{22.47}$$

where the Stefan-Boltzmann constant is $\sigma = \frac{2\pi^5}{15} \frac{k^4}{c^2 h^3} \approx 5.6704 \times 10^{-8} \frac{W}{m^2 K^4}$.

A common incandescent light source consists of tungsten filaments inside an evacuated glass cell. A current running through the filament causes Joule heating, leading to thermal light emission. Tungsten light sources emit light extending from ultraviolet (roughly 300 nm) to near infrared (1400 nm); this range covers the entire visible spectrum and so produces a very white light. One drawback of tungsten filaments is that material from the surface of the filament slowly evaporates and becomes deposited on the glass cell, causing less light to be transmitted as the surface of the glass blackens over time. This can be reduced by filling the glass cell with an inert gas such as krypton or xenon, in order to protect the surface of the tungsten.

A variant of the tungsten lamp is the **halogen lamp**, in which the tungsten filament is in a cell filled with a halogen gas. Halogens are the elements in column VIIA of the periodic table (fluorine, chlorine, bromine, iodine, astatine, tennessine), with bromine and iodine normally used in halogen lamps. Halogen lamps are more compact and efficient than simple tungsten light sources, capable of producing high-intensity white light, with long life spans of 10 years or more.

Fluorescent lights are based on electrical discharge in a dilute gas such as mercury, neon, xenon, or argon. The electrons in an electric current collide with the atoms of the gas, exciting the internal electrons of the atoms to high energy states. In many cases, the excitation of the atoms leads to both visible and UV emissions; commonly the inside of the glass tube is coated with a phosphor, which absorbs the UV and re-emits the energy as visible light. While more costly than incandescent lamps, fluorescent lamps are more efficient, producing greater intensity for lower electrical power input. The spectrum consists of a set of discrete spectral lines, rather than a continuous spectrum. The efficiency is high in part because the gas is chosen to have spectral lines predominantly in the visible range, rather than in the infrared.

Light emitting diodes (LEDs) emit light due to an electric current that causes electrons in a semiconductor to jump across a bandgap. Although the light emission tends to be quasi-monochromatic, the output light is incoherent due to the random emission times and phases of the photons. A typical bandwidth for monochromatic LED's is from 20 to 70 nm, with coherence lengths on the order of microns. White LEDs can be formed using a blue diode coated with a phosphor die that fluoresces over a broader spectral range.

22.8   Coherent Light Sources

The main source of coherent light is lasers. The physics of lasers was discussed in Sect. 5.7; here we discuss specific types of laser light sources and their characteristics.

pn junctions, in addition to being the basis for diodes and transistors, are also the basic building block for semiconductor light sources. Recall that a *pn* junction is formed by a semiconductor which is doped with positively charged (acceptor) impurities on one side of the junction and negatively charged (donor) impurities on the other side. These two sets of impurities provide sources, respectively, of holes and free electrons. The regions with the positive and negative impurities are called the *p* and *n* regions. Without going into detail on the physics of *pn* junctions, the holes and electrons tend to recombine when they encounter each other, releasing energy through both radiative and nonradiative processes. Nonradiative processes include emission of phonons (lattice vibrations) and Auger recombination, in which energy is transferred from one electron to a nearby electron by a short-lived photon. Of more interest to us is radiative decay, in which a photon with energy equal to the electron gap energy is emitted. If the semiconductor is not too large, there is an appreciable chance that the photon can escape the material without absorption, producing a quantum of external light emission. By applying a forward-biased voltage V across the junction (Fig. 22.11a), the result is a light-emitting diode (LED).

The internal radiative efficiency of the LED is defined to be

$$\eta = \frac{W_{rad}}{W_{nonrad}}, \tag{22.48}$$

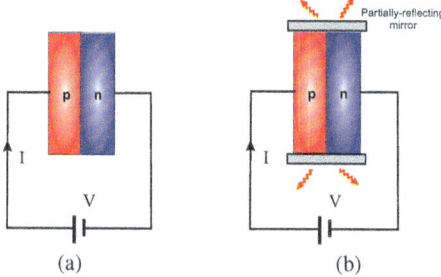

Fig. 22.11 (a) Schematic of an LED: the battery pumps in energy, causing electron-hole pairs to form. When these pairs recombine, a portion of the energy is released as outgoing photons. (b) A laser diode: placing partially reflecting mirrors at the ends of the LED, a resonant cavity is formed. This cavity selects out certain frequencies, leading to more coherent emission

where W_{rad} and W_{nonrad} are the decay rates for the radiative and nonradiative decay modes. The total efficiency is then given by the internal efficiency multiplied by the fraction η_e of photons that escape the material without absorption (the external radiative efficiency)

$$\eta = \eta_e \eta_i. \tag{22.49}$$

Typically, the internal efficiency of an LED is fairly large ($\sim$ 60–95%), but the external efficiency is much lower, typically $\sim$ 5%. When connected to a battery of voltage V producing a current I through the junction, the power the battery supplies is $P_{el} = IV$. The optical power produced as emitted light is then given by

$$P_{opt} = \eta P_{el}. \tag{22.50}$$

As mentioned in the previous section, LED's produce incoherent light. But by adding semitransparent mirrors to the ends of the LED (Fig. 22.11b), a resonant cavity can be formed. As in standard lasers, interference between the waves bouncing back and forth in the cavity lead to destructive interference for most frequencies. The result is a narrowing of the spectral width to a tight band around a single frequency and a corresponding increase in coherence. When operated above a threshold intensity, stimulated emission can also be made to occur. This new device is known as a **laser diode** or **semiconductor laser** and often serves as a convenient coherent light source for photonic integrated circuits or fiber optic systems, with coherence lengths on the order of millimeters to centimeters. By choosing a semiconducting material with an appropriate bandgap, output radiation in the IV, visible, or UV portions of the spectrum can be obtained. The response times of laser diodes to driving current variations are much faster than those of LED's, allowing the output to be switched on and off very rapidly. Intensity modulation frequencies can reach up to 10 GHz, and the maximum intensity output can be as high as 100 mW.

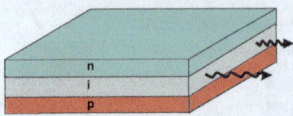

Fig. 22.12 Double heterostructure pin laser diode. An undoped active region *i* is wedged between the *p* and *n* regions. Mirrors (not shown) are on the left and right, with the right-hand mirror partially transparent to allow the laser light to escape. The light beam emitted from the *i* region is roughly elliptical in shape

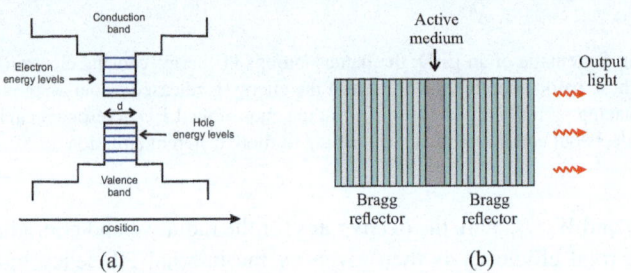

(a)   (b)

Fig. 22.13 (**a**) The energies in a quantum well laser form a containing box with quantized energy levels. The strong confinement leads to improved efficiency. (**b**) Schematic of a VCSEL: two Bragg reflectors form a resonant cavity encompassing the active medium between them. The reflectors serve both as mirrors and spectral filters

Most modern laser diodes have a double heterostructure, with an intrinsic (undoped) region between the *p* and *n* regions (Fig. 22.12). The output beam is elliptical in shape, with a so-called Lambertian distribution, with the intensity in each direction proportional to the cosine of the angle from the axis. This can be a disadvantage when trying to couple the light into an optical fiber with a circular cross section. However, these diodes have high efficiency and produce high-intensity light, due to the fact that electrons and holes mingle together in the intrinsic region, making recombination occur at a high rate. The double heterostructure also increases the stability of the system, making the output less susceptible to temperature and current variations.

If the thickness *d* of the intrinsic region becomes sufficiently small (roughly 10 nm or less), the energy levels of the electrons and holes become quantized (Fig. 22.13a). The electrons and holes can both then be treated as a particle in a box. From introductory quantum physics or modern physics courses, we know that the energy levels of the electrons are then given by

$$E_n = \frac{\hbar^2 \pi^2 n^2}{2md^2},$$

(22.51)

for $n = 0, 1, 2, \ldots$. The hole energy levels are given by the negatives of these values (the zero of energy being taken to be the Fermi level). In this **quantum well laser** structure, the increased density of states inside the wells leads to greatly increased stimulated emission efficiency and a reduced laser threshold current. The quantum well laser is also highly stable with respect to temperature variations. By choosing the well width d when the structure is fabricated, the frequency of the laser can be set to a desired value.

Finally, a **vertical cavity surface-emitting laser (VCSEL)** is shown schematically in Fig. 22.13b. These consist of a several micron-thick active semiconductor material, with one or more quantum wells, wedged between two semiconductor Bragg reflectors. The Bragg reflectors both act as mirrors to form a cavity and serve to filter the output spectrum. A portion of the bean will be transmitted through the end of one reflector, providing the output. Among the advantages of VCSELs are that they are cheap to manufacture in large quantities, they can operate at high temperatures, and they emit at low intensities, reducing the risk of damage to eyes or other sensitive tissues. Arrays of VCSELs have become common for applications such as display screens and computer mice.

22.9   Single-Photon Sources

The ability to produce single photons is necessary for many quantum protocols and quantum devices. Ideally, such a source would produce a single photon (with negligible probability of multiple photon events) on demand at a precisely controlled time, in response to some electrical or optical control signal. In addition, the photon should be produced in a single specified mode, with known direction, energy, helicity, and orbital angular momentum and with high coherence. No known source has all of these properties, but there are a number of sources that can approach an ideal single-photon source to varying degrees of approximation.

For example, a very short, highly attenuated pulsed laser can produce pulses with some small mean number μ of photons per pulse. If a sufficiently strong filter is used to attenuate the beam of sufficiently low amplitude, a mean $\mu \approx 1$ can be arranged. However, the beam is still Poisson-distributed, meaning that the variance is also equal to 1, so that there will still be appreciable probability of more than one photon in the pulse. Similarly, there is a probability of no photons in a given pulse, meaning that even when a single photon is produced, its production time (which pulse it came from) will be probabilistic. Decreasing μ reduces the multi-photon probability, but increases the zero-photon probability and lengthens the data collection time in experiments.

Spontaneous parametric downconversion (Sect. 5.8) produces photons in pairs. It can be arranged to have the two photons in different spatial modes, so that a single photon can be isolated, making a single-photon source. This has the additional advantage that the second photon (the idler) can be used to herald or announce the production of the signal photon, allowing the experimenter to know when a photon is on the way. But on the downside, the production time is still probabilistic, and the signal and idler each on its own has a large bandwidth and low coherence.

Other possibilities are stimulated emission by a single atom, molecule, or quantum dot. For example, a single atom with an appropriate three-level structure (with states labeled $|1\rangle, |2\rangle, |3\rangle$) can be excited by a laser pulse with frequency ω_1 in the transition $|1\rangle \rightarrow |2\rangle$ and then fluoresce rapidly in the decay $|2\rangle \rightarrow |3\rangle$ with frequency ω_2. This produces photons of frequency ω_2 more or less on demand. However, they produce photons emitted in random directions, so extra structure needs to be added to produce anything resembling a "beam" with a fixed direction. Photon production in single-atom sources can be enhanced by placing the atoms in resonant cavities (Chap. 25).

Quantum dots (discussed in more detail in Sect. 26.1) are essentially artificial atoms which can be used in a similar manner as atoms to produce photons, but they have the advantage that their properties are highly controllable [7, 8], and photon production with them can be enhanced by means of so-called optical antennas [9], which can operate at the single-photon level. NV centers (Sect. 26.2) are another solid-state source capable of producing single photons.

In general, the quality of single-photon sources can be characterized in several ways:

• Generation rate: how many photons are generated per second, or per run of the experiment?

• The purity of the photon number distribution: how low is the probability of producing more or less than one photon in a given pulse? This can be quantified using the second-order correlation function $g^{(2)}(\tau)$ discussed in Sect. 10.4: an ideal single-photon source should have $g^{(2)}(0) = 0$.

• Fidelity: generally, the photon that is produced should be in a particular state in order to be useful for a given purpose. The match between the photon and the state can be measured using the quantum fidelity defined in Chap. 3.

• Efficiency: what percent of the photons survive transport from the generation site to the location where they will be used, without loss or degradation of their quantum state? This is an important factor in solid-state sources, especially, where the photon may have to travel some distance through a semiconductor material before being used.

More extensive reviews of single-photon sources may be found in [10–18].

Problems

1. Derive the energy level formula, Eq. 22.51, for a particle in a box. At room temperature, $T \sim 300$ K, estimate how small the box must be in order for the gap between the $n = 0$ and $n = 1$ levels to exceed the thermal fluctuation energy.
2. Suppose a photocell of quantum efficiency η is illuminated with light of frequency ν and optical power P_{opt}.

   (a) Show that the photocurrent, Eq. 22.21, can be written as $i = \frac{P_{opt}}{h\nu}\eta Ge$.

(b) Suppose the light has wavelength 550 nm and power 10 μW. The cell is biased with a voltage of $V_0 = 1.5$ V and has electrode separation $d = 0.2$ mm, with quantum efficiency of $\eta = 70\%$. If the electrons of the semiconductor in the cell have mobility $\mu_e = 3 \times 10^{-2} \frac{m^2}{Vs}$ and lifetime $\tau = 1$ ms, find the gain and the signal current.

3. A particular APD has a responsivity of 25 and efficiency of 80%. If it is illuminated with light of wavelength 800 nm, resulting in a photocurrent of 10 mA, find the current gain and the incident optical power.

4. Show that Eq. 22.43 holds, and then use it to prove Eq. 22.42.

5. Suppose an optical detector requires time Δt to make a measurement. Define the bandwidth $\Delta \nu = \frac{1}{2\Delta t}$ of the detector.

(a) Using the fact that the number of electrons arriving at a detector in a given time period usually obeys Poisson statistics, show that the shot noise in the detector current should have variance $\Delta i^2 = \frac{e\bar{i}}{\Delta t} = 2e\bar{i}\Delta\nu$, where $\bar{i}$ is the mean current and Δt is the averaging time of the detector. So, narrower detector bandwidths or longer integration times give better signal-to-noise ratios.

(b) Denote the mean signal current as i_o and the shot noise current as i_{sh}. Show that the shot noise power across a load resistance R is $P_{sh} = 2e(i_0 + i_{sh})\Delta\nu R$.

6. Suppose a photomultiplier has 15 dynodes and is operated at a voltage of 2000 V, with amplification factor $\delta = 4$. Calculate the charge reaching the anode for one, three, and five input electrons.

7. **Johnson noise** is due to thermal fluctuations in a detection circuit. The energy variations due to these fluctuations are typically $\Delta E_N \sim k_B T$. Suppose the detector circuit has resistance R. Let V_N be the RMS voltage fluctuation in the resistance.

(a) Suppose the detector has bandwidth $\Delta\nu$, as defined in the previous problem. Setting the power variation in the resistance due to ΔE_N equal to the dissipation rate in the resistor due to V_N, find approximate formulas for the RMS Johnson noise voltage and the RMS Johnson noise current in the circuit as functions of $\Delta\nu$.

(b) What effect does increasing the current have on the Johnson noise voltage and current? What effect does increasing bandwidth have? What effect would you expect increases in R and in $\Delta\nu$ to have on the signal-to-noise ratio? (Keep in mind that the signal current is of the form $\bar{i}R^2$, where $\bar{i}$ is average photocurrent.)

8. (For those who have done the previous two problems.) Derive a formula for signal-to-noise ratio $SNR = \frac{P_{signal}}{P_{Johnson}+P_{shot}}$ in terms of detector bandwidth, load resistance, currents, and temperature. You should find that the SNR drops as either the temperature or the bandwidth increases. What are the limits of the SNR as $R \to \infty$ or $R \to 0$?

9. In the treatment of homodyne detection in this chapter, the $e^{i\omega t}$ factors in the fields where ignored, since they were the same for all terms. In heterodyne detection, where the two incoming fields have different frequencies, these factors need to be included. Consider the same setup as in Fig. 22.10, but now with the incident fields having frequencies ω_1 and ω_2. Derive the analog of Eq. 22.37 for this case. Make sure that it reduces to the previous result when $\omega_1 = \omega_2$.

10. (a) In the homodyne detector of Fig. 22.10, suppose the input field is a coherent state, with amplitude $\alpha = |\alpha|e^{i\phi}$. Find the difference current $i_1 - i_2$ as a function of α and ϕ.

   (b) Do the same for the squeezed state with amplitude $\tau = \alpha \cosh \zeta + \alpha^* e^{i\phi} \sinh \zeta$.

References

1. P.G. Datskos, N.V. Lavrik, Detectors: Figures of Merit, in *Encyclopedia of Optical and Photonic Engineering, 2nd ed.*, ed. by C. Hoffman, R. Driggers (CRC Press, Boca Raton, 2016)
2. R.S. Quimby, *Photonics and Lasers: An Introduction* (Wiley-Interscience, New York, 2006)
3. B.E.H Saleh, M.C. Teich, *Fundamentals of Photonics, 3rd ed.* (Wiley, New York, 2019)
4. S.L. Chuang, *Physics of Photonic Devices 2nd ed.*
5. F. Zappa, S. Tisa, A. Tosi, S. Cova, Sensors Actuat. A Phys. **140**, 103 (2007)
6. A. Yariv, *Photonics: Optical Electronics in Modern Communications*, 6th ed. (2007)
7. A.L. Efros, L.E. Brus, ACS Nano **15**, 6192 (2021)
8. T. Heindel, J.H. Kim, N. Gregersen, A. Rastelli, S. Reitzenstein, Adv. Opt. Photon. **15**, 613 (2023)
9. A. Femius Koenderink, ACS Photon. **4**, 710 (2017)
10. B. Lounis, M, Orrit, Rep. Prog. Phys. **68**, 1129 (2005)
11. E. Meyer-Scott, C. Silberhorn, A. Migdall, Rev. Sci. Instrum. **91**, 041101 (2020)
12. M. Oxborrow, A.G. Sinclair, Contemp. Phys. **46**, 173 (2005)
13. S. Scheel, J. Mod. Opt. **56**, 141 (2009)
14. M.D. Eisaman, J. Fan, A. Migdall, S.V. Polyakov, Rev. Sci. Instrum. **82**, 071101 (2011)
15. C.J. Chunnilall, I.P. Degiovanni, S. Kück, I. Müller, A.G. Sinclair, Opt. Eng. **53**, 081910 (2014)
16. I. Aharonovich, D. Englund, M. Toth, Nat. Photon. **10**, 631 (2016)
17. P. Senellart, G. Solomon, A. White, Nat. Nanotechnol. **12**, 1026 (2017)
18. F. Flamini, N. Spagnolo, F. Sciarrino, Rep. Prog. Phys. **82**, 016001 (2019)

Chapter 23
Nuclear Magnetic Resonance

Nuclear magnetic resonance (NMR) uses signals emitted by perturbed nuclear spins to track the distributions of different nuclei within a material sample. Besides having real-world applications ranging from medical diagnostics to art restoration and archaeology, it was one of the first physical platforms used to successfully implement quantum computing operations on a small scale. In the following sections, we will discuss the principles of NMR in general, and then in the final section of the chapter, its application for storing and manipulating qubits will be discussed.

NMR utilizes external magnetic fields applied to the magnetic moments of spin-$\frac{1}{2}$ nuclei in materials such as water, iron, carbon, or nitrogen. The information that can be gained from NMR includes the distribution of particular nuclei in a material, the structure of the molecules in which nuclei reside, and properties of the chemical bonds in which they participate, leading to a broad range of applications in chemistry, physics, and medicine. Probably the most familiar application is **magnetic resonance imaging (MRI)**, which uses NMR to map the distribution of protons in water and in protein and fat molecules for purposes of medical imaging. But over the past few decades, there has also been considerable interest in the use of NMR for quantum information processing. An optical analog of NMR, in which optical pulses are used to entangle the rotational, vibrational, and electronic states of molecules, has also been of intense interest for molecular spectroscopy [1].

The discussion in this chapter presumes a basic background material on dipole moments in magnetic fields (Sect. 4.1.2) and on two-state spin systems (Sect. 6.2).

© The Author(s), under exclusive license to Springer Nature Switzerland AG 2025
D. S. Simon, *Introduction to Quantum Science and Technology*, Undergraduate
Texts in Physics, https://doi.org/10.1007/978-3-031-81315-3_23

23.1 Rabi Oscillations

Consider a spin-$\frac{1}{2}$ nucleus with gyromagnetic ratio γ and magnetic dipole moment $\mu = \gamma S = \frac{\hbar}{2}\gamma\sigma$. For a proton, for example, the gyromagnetic ratio is $\gamma = 5.59\frac{e}{2m_p} = 2.675 \times 10^8 \frac{rad}{s \cdot T}$, with a dipole moment $|\mu| = 1.4 \times 10^{26}\frac{J}{T}$.

Suppose that the nucleus is placed in a strong, uniform magnetic field, pointing along the z-axis, $\boldsymbol{B}_0 = B_0\hat{\boldsymbol{k}}$. If measured along the z-axis, the nucleus is either spin-up (parallel to $\boldsymbol{B}_0$) or spin-down (opposite to $\boldsymbol{B}_0$). In the absence of measurement, the state could be in a superposition of both up and down states. So the most general state can be written in terms of two complex coefficients α and β or in terms of two angles θ and ϕ:

$$|\psi\rangle = \alpha|0\rangle + \beta|1\rangle = e^{-i\phi/2}\cos\frac{\theta}{2}|0\rangle + e^{i\phi/2}\sin\frac{\theta}{2}|1\rangle. \qquad (23.1)$$

The potential energy is $U = -\boldsymbol{\mu} \cdot \boldsymbol{B}_0 = \pm|\mu|B_0$. The higher energy state (opposite to $\boldsymbol{B}_0$) will be denoted $|1\rangle$, with the lower energy state (parallel to $\boldsymbol{B}_0$) denoted $|0\rangle$. At thermal equilibrium, the probabilities of these two states are given by the Boltzmann distribution

$$P(j) = \frac{e^{-\frac{E_j}{kt}}}{\sum_k e^{-\frac{E_j}{k_BT}}}, \qquad (23.2)$$

so that at a given temperature T (in Kelvins)

$$P(0) = \frac{e^{\frac{\mu B_0}{k_BT}}}{e^{\frac{\mu B_0}{k_BT}} + e^{-\frac{\mu B_0}{k_BT}}} = \frac{e^{\frac{\mu B_0}{k_BT}}}{2\cosh\frac{\mu B_0}{k_BT}} \qquad (23.3)$$

$$P(1) = \frac{e^{\frac{-\mu B_0}{k_BT}}}{e^{\frac{\mu B_0}{k_BT}} + e^{-\frac{\mu B_0}{k_BT}}} = \frac{e^{-\frac{\mu B_0}{k_BT}}}{2\cosh\frac{\mu B_0}{k_BT}}, \qquad (23.4)$$

where k_B is Boltzmann's constant. Now imagine a collection of a large number of these spins, either attached to sites of a crystal lattice or free to move in a liquid. At high temperatures and large fields, the spin orientations become randomized, but at room temperature the overwhelming majority of the spins are aligned with each other in the ground state, $|0\rangle$.

Now imagine that in addition to the strong (uniform and constant) field along the z axis there is also a weaker field rotating in the transverse xy-plane at frequency ω. The full external magnetic field is then

$$\boldsymbol{B} = B_0\hat{\boldsymbol{k}} + B_1(\hat{\boldsymbol{i}}\cos\omega t - \hat{\boldsymbol{j}}\sin\omega t). \qquad (23.5)$$

ω is usually a frequency in the radio portion of the spectrum, and so the signal in the electric circuit driving the field rotation is referred to as a radio-frequency signal or **RF signal**. Also define the frequencies

$$\omega_0 = \gamma B_0, \qquad \omega_1 = \gamma B_1. \tag{23.6}$$

ω_0 is the Larmor frequency (Sect. 4.1.2), which gives the precession rate of the spin about the z-axis, while ω_1 is called the **Rabi frequency**. The Rabi frequency describes the frequency of **nutation** or oscillations toward and away from the z-axis. Equivalently, Rabi oscillations are oscillations between the spin-up $|0\rangle$ and spin-down $|1\rangle$ states, as will be seen below.

The Hamiltonian can then be written

$$\hat{H} = -\boldsymbol{\mu} \cdot \boldsymbol{B} = -\frac{\hbar}{2}\omega_0\sigma_z - \frac{\hbar}{2}\omega_1\left(\sigma_x \cos \omega t - \sigma_y \sin \omega t\right). \tag{23.7}$$

Using the explicit forms of the Pauli matrices, this can be put into the matrix form

$$\hat{H} = -\frac{\hbar}{2}\begin{pmatrix} \omega_0 & \omega_1 e^{i\omega t} \\ \omega_1 e^{-i\omega t} & -\omega_0 \end{pmatrix}. \tag{23.8}$$

Let's solve the Schrödinger equation $i\hbar\frac{d}{dt}|\psi(t)\rangle = \hat{H}|\psi(t)\rangle$ for this Hamiltonian. Substituting

$$|\psi(t)\rangle = \alpha(t)|0\rangle + \beta(t)|1\rangle = \begin{pmatrix} \alpha(t) \\ \beta(t) \end{pmatrix} \tag{23.9}$$

into the Schrödinger equation gives two coupled, linear differential equations for $\alpha(t)$ and $\beta(t)$:

$$i\frac{d\alpha(t)}{dt} = -\frac{\omega_0}{2}\alpha(t) - \frac{\omega_1}{2}e^{i\omega t}\beta(t) \tag{23.10}$$

$$i\frac{d\beta(t)}{dt} = -\frac{\omega_1}{2}e^{-i\omega t}\alpha(t) + \frac{\omega_0}{2}\beta(t). \tag{23.11}$$

To solve these equations, first make a change of variables, changing our frame of reference to one that rotates at rate ω_0. Define $\tilde{\alpha}$ and $\tilde{\beta}$ by

$$\alpha(t) = \tilde{\alpha}(t)e^{i\omega_0 t/2} \tag{23.12}$$

$$\beta(t) = \tilde{\beta}(t)e^{-i\omega_0 t/2}. \tag{23.13}$$

Substituting these into the differential equations above results in the new differential equations

$$i\frac{d\tilde{\alpha}(t)}{dt} = -\frac{\omega_1}{2}e^{i\delta t}\tilde{\beta}(t) \tag{23.14}$$

$$i\frac{d\tilde{\beta}(t)}{dt} = -\frac{\omega_1}{2}e^{-i\delta t}\tilde{\alpha}(t), \tag{23.15}$$

where $\delta = \omega - \omega_0$ is the detuning of the RF frequency from the resonant frequency. Taking the derivative of Eq. 23.14 leads to

$$\frac{d^2\tilde{\alpha}(t)}{dt^2} = i\frac{\omega_1}{2}\left(i\delta\tilde{\beta} + \frac{d\tilde{\beta}}{dt}\right)e^{i\delta t} \tag{23.16}$$

$$= i\frac{\omega_1}{2}\left[i\delta\left(-\frac{2i}{\omega_1}e^{-i\delta t}\frac{d\tilde{\alpha}}{dt}\right) + i\frac{\omega_1}{2}e^{-i\delta t}\tilde{\alpha}\right]e^{i\delta t} \tag{23.17}$$

$$= i\delta\frac{d\tilde{\alpha}}{dt} - \left(\frac{\omega_1}{2}\right)^2\tilde{\alpha}, \tag{23.18}$$

where Eq. 23.14 was used in the first term of the second line and Eq. 23.15 was used in the second. So we arrive at a second-order differential equation involving only $\hat{\alpha}$:

$$\frac{d^2\tilde{\alpha}(t)}{dt^2} - i\delta\frac{d\tilde{\alpha}}{dt} + \left(\frac{\omega_1}{2}\right)^2\tilde{\alpha} = 0. \tag{23.19}$$

In a similar manner, an equation for $\tilde{\beta}$ can be found:

$$\frac{d^2\tilde{\beta}(t)}{dt^2} + i\delta\frac{d\tilde{\beta}}{dt} + \left(\frac{\omega_1}{2}\right)^2\tilde{\beta} = 0. \tag{23.20}$$

Now that the equations have been decoupled, they are not hard to solve. It can readily be verified that the general solutions are

$$\tilde{\alpha}(t) = a\,e^{i\Omega_+ t} + b\,e^{i\Omega_- t} \tag{23.21}$$

$$\tilde{\beta}(t) = c\,e^{-i\Omega_+ t} + d\,e^{-i\Omega_- t}, \tag{23.22}$$

where we have defined

$$\Omega_\pm = \frac{1}{2}\left(\delta \pm \Omega\right), \qquad \Omega = \sqrt{\omega_1^2 + \delta^2}, \tag{23.23}$$

and the constants a, b, c, d are determined by the initial conditions. Undoing the change of variables, Eqs. 23.12 and 23.13, gives a solution to the original Schrödinger equation.

Example 23.1.1 As an example, consider the case when $|\psi(0)\rangle = |0\rangle$. Then $\alpha(0) = 1$ and $\beta(0) = 0$. Also, assume that the system is at resonance: $\omega = \omega_0$ and $\delta = 0$. Then the solutions for α and β become

$$\alpha(t) = \cos \frac{\omega_1 t}{2} e^{i\omega_0 t/2} \tag{23.24}$$

$$\beta(t) = i \sin \frac{\omega_1 t}{2} e^{-i\omega_0 t/2}. \tag{23.25}$$

So that (up to an overall phase) the state evolves with time according to

$$|\psi(t)\rangle = \cos \frac{\omega_1 t}{2} |0\rangle + i \sin \frac{\omega_1 t}{2} e^{-i\omega_0 t} |1\rangle. \tag{23.26}$$

Although the state was initially in $|0\rangle$, at most times it will be in a superposition of $|0\rangle$ and $|1\rangle$; at time t, the probabilities of finding it in each of the basis states is

$$P(0 \to 0) = \cos^2 \frac{\omega_1 t}{2} \tag{23.27}$$

$$P(0 \to 1) = \sin^2 \frac{\omega_1 t}{2}, \tag{23.28}$$

so that the system oscillates over time between the ground state and the excited state, as seen in the transition probabilities in Fig. 23.1.

In particular, suppose that the RF signal is on only for a time $t_{\pi/2} = \frac{\pi}{2\omega_1}$. At the end of this time, the state is

$$|\psi(t)\rangle = \frac{1}{\sqrt{2}}\left(|0\rangle + e^{i\frac{\pi\omega_0}{2\omega_1}}|1\rangle\right). \tag{23.29}$$

In other words, after an RF pulse of duration $t_{\pi/2}$ (called a $\frac{\pi}{2}$-**pulse**), the initial state $|0\rangle$ is converted into an equal superposition of $|0\rangle$ and $|1\rangle$. Up to a relative phase, the $\frac{\pi}{2}$ pulse acts like a Hadamard gate or a beam splitter.

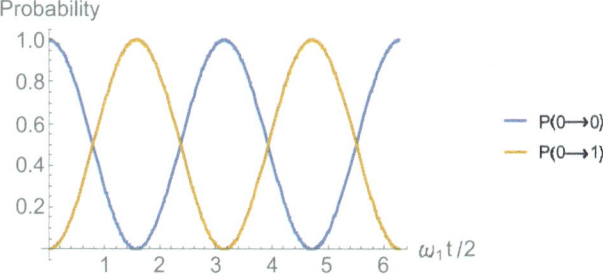

Fig. 23.1 Rabi oscillations. A spin initially in the ground state $|0\rangle$ will oscillate between the states $|0\rangle$ and $|1\rangle$ in response to an RF signal causing the external magnetic field to rotate. Here the probabilities of being found in each state are plotted versus time, with time measured in units of $\frac{2}{\omega_1}$

In a similar manner, if the pulse is of duration $t_\pi = \frac{\pi}{\omega_1}$ (a π-**pulse**), then the initial state $|0\rangle$ is converted to a final state of $|1\rangle$: the π-pulse acts as a NOT gate. ∎

The **Rabi oscillations** of the previous example are an important feature of many physical systems beside NMR. For example, they will also occur in information processing with atomic and ionic systems in Chap. 24. The most important things to remember are that:

- $\frac{\pi}{2}$-*pulses convert computational basis states into superpositions*
- π-*pulses act as NOT gates.*

23.2 Bloch Equations

Rather than looking at the evolution of the quantum states, we can instead look at the evolution of the magnetic moment vector. By doing this, we introduce two important parameters for NMR, the transverse and longitudinal relaxation times.

Recall from Sect. 4.1.2 that in the presence of the static uniform $\boldsymbol{B}_0$ field along the z-axis, the magnetic moment rotates about the z-axis at the Larmor frequency $\omega_0 = \gamma B_0$. Before adding in the transverse field $\boldsymbol{B}_1$, let's re-examine this motion in more detail.

Let $\boldsymbol{M}$ be the magnetization of the material, defined as the expectation value of the sum of all the molecular magnetic dipole moments per volume. The equation of motion for the magnetization is essentially the same as the equation for the individual dipole moment operators:

$$\frac{d\boldsymbol{M}}{dt} = \gamma \, \boldsymbol{M} \times \boldsymbol{B}. \tag{23.30}$$

Given that $\boldsymbol{B} = B_0 \hat{\boldsymbol{k}}$, this leads to the **Bloch equations:**

$$\frac{dM_x}{dt} = \gamma M_y B_0 \tag{23.31}$$

$$\frac{dM_y}{dt} = -\gamma M_x B_0 \tag{23.32}$$

$$\frac{dM_z}{dt} = 0. \tag{23.33}$$

Taking as initial conditions that $M_x(0) = M_\perp$, $M_y(0) = 0$, and $M_z(0) = M_\parallel$, it can be verified that the solution to the Bloch equations is

$$M_x(t) = M_\perp \cos(\omega_0 t) \tag{23.34}$$

$$M_y(t) = -M_\perp \sin(\omega_0 t) \tag{23.35}$$

$$M_z = M_\parallel = \text{constant}, \tag{23.36}$$

which explicitly demonstrates the Larmor precession at frequency ω_0.

When the transverse magnetic field $\boldsymbol{B}_1$ is included, it becomes more convenient to work in a reference frame that rotates at rate ω_0. We won't look at this case here, but details can be found, for example, in [2] or [3].

Now what happens if the motion is perturbed? This could occur through thermal excitations or as a result of an external electromagnetic pulse. We would expect that once the perturbation is ended, the magnetization would return the steady-state motion we've just found. This return to an equilibrium or steady state after perturbation is called a **relaxation process**. In the simplest case, we can assume that the rate of relaxation of each component of $\boldsymbol{M}$ is proportional to the distance the component is from its equilibrium value. So for the z-component, the relaxation equation will be

$$\frac{dM_z}{dt} = \frac{M_\parallel - M_z}{T_1}, \tag{23.37}$$

where T_1 is the characteristic time for M_z to relax toward its equilibrium value $M_\parallel$. T_1 is called the **longitudinal** or **spin-lattice relaxation time**. Similarly, for the transverse components

$$\frac{dM_x}{dt} = -\frac{M_x}{T_2} \tag{23.38}$$

$$\frac{dM_y}{dt} = -\frac{M_y}{T_2}. \tag{23.39}$$

T_2 is the **transverse** or **spin-spin relaxation time** and measures the characteristic time for the transverse part of $\boldsymbol{M}$ to relax back to the Larmor motion. In general, $T_2 < T_1$, and in fact it is usually true that $T_2 << T_1$. This is because neighboring spins may be rotating at slightly different rates (since slight differences in their environment will cause the magnetic field to be a bit different). So the individual spins are getting out of phase with each other in their rotations, causing the *average* value of the transverse magnetization to decay faster than the individual values for each spin separately. This is called **dephasing**.

In the presence of the perturbation, the Bloch equations now become

$$\frac{dM_x}{dt} = \gamma M_y \boldsymbol{B}_0 - \frac{M_x}{T_2} \tag{23.40}$$

$$\frac{dM_y}{dt} = -\gamma M_x \boldsymbol{B}_0 - \frac{M_y}{T_2} \tag{23.41}$$

$$\frac{dM_z}{dt} = \frac{M_\parallel - M_z}{T_1}. \tag{23.42}$$

Starting with a spin initially pointing along the z-axis, $M_{\parallel} = M_0$ and $M_{\perp} = 0$, suppose that it is suddenly perturbed by a rotation into the transverse plane, the Bloch equations are satisfied (as can easily be checked) by

$$M_x = M_0 e^{-t/T_2} \cos(\omega_0 t) \tag{23.43}$$

$$M_y = -M_0 e^{-t/T_2} \sin(\omega_0 t) \tag{23.44}$$

$$M_z(t) = M_0 \left(1 - e^{-t/T_1}\right), \tag{23.45}$$

which describes the tip of the M vector exhibiting a decaying spiral toward its original position on the z-axis.

Although they vary somewhat depending on frequency and on interactions with surroundings, some typical values of T_1 and T_2 are given in the following table.

Material	T_1 (ms)	T_2 (ms)
Water	4000	2000
Muscle	900	50
Fat	250	70
Liver tissue	500	40
Blood	1200	150
Grey matter	800	100
White matter	600	80

Box 23.1 A World of NMR Applications

NMR is ubiquitous in medical diagnostics and chemical spectroscopy, but it has also found a wide range of applications in other areas, both in the lab and in the field. For example, its use has become widespread in the world of art conservation [4–6]. The ability of NMR to determine the concentration of nuclei such as H, C, and P in organic materials allows the identification of specific pigments, varnishes, and primers. The presence of cholesterol, for instance, can be used to distinguish egg tempera binders from acrylic or oil binders. Different types of stains, which require different cleaning methods, can be distinguished from each other, and the state of degradation of an artwork's materials can be determined in a quantitative manner. Further, information about the history of a piece of art can often be extracted using NMR, including the dating of when various components of it were made; an example of this is a Byzantine icon called the Madonna Hodegetria [7]. NMR analysis determined that there were three layers of paint, dating to the 19th, 13th, and 5th centuries, and that the oldest layer had in fact been

(continued)

Box 23.1 (continued)

glued onto the underlying wooden surface after being removed from another surface. When applied to artwork, NMR has the advantage that it can be used in a fairly nondestructive manner; typically, only a minuscule sample needs to be removed from the work. The sample is then often dissolved into a liquid solvent for analysis.

NMR is used in similar manner in archaeology [4], providing analysis of wooden structures in shipwrecks and ancient buildings and paper samples from ancient manuscripts. Analysis of ambergris, resins, and food waste can date the materials and give information about the conditions prevalent when the materials were deposited. For example, NMR spectroscopy has been used to identify the source of the asphalt used in buildings in ancient Babylon [8].

One of the most important areas to which NMR has been applied is environmental science [9]. Soil samples can be analyzed to determine the abundance of heavy metals and radionuclides or to look for minerals that can act as binding sites for pollutants. In an age of climate change, it is vital to understand the sequestering of carbon dioxide in carbonate materials and other minerals, and NMR is an essential tool for determining the CO_2 abundance in these materials.

Given the applications listed above, it is not surprising that a new field of forensic NMR spectroscopy has been developed [10], which is used to detect art forgeries, and to identify explosives at crime scenes and poisons in crime victims, as well as to spot counterfeit drugs, wines with falsified labels, and adulterated foods.

23.3 Magnetic Resonance Imaging

One common application for nuclear magnetic resonance is magnetic resonance imaging (MRI), a form of medical diagnostic imaging used to map the distribution of various nuclei within the human body. Unlike x-rays, MRI can image soft tissue with high resolution and without risk of causing health damage to the patient. It is noninvasive and can distinguish between different types of tissue. Reviews of the physics of MRI can be found in [2, 11].

MRI generally focuses on protons (hydrogen nuclei) inside water or organic molecules. The time required to form an image is $\lesssim 0.1$ s, allowing time-dependent processes, to be followed in real time by stringing large numbers of snapshots together to form a video sequence. This method, called **functional MRI or fMRI**, can be used to track processes such as diffusion of drugs through a system, the beating of a heart, or the functioning of the brain in response to a stimulus.

Fig. 23.2 The free induction signal emitted by an MRI sample. The RF signal (solid curve) oscillates inside an exponentially decaying envelope (dashed curves). The relaxation time can be measured by determining the time for the signal amplitude to decay to $\frac{1}{e}$ (approximately 37%) of its initial amplitude

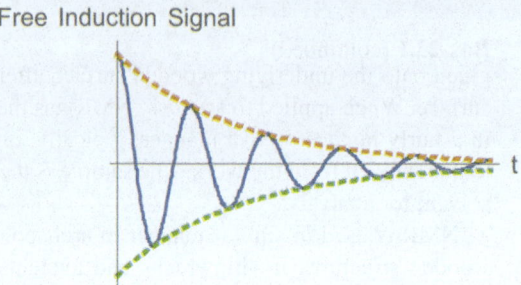

Free Induction Signal

The basic idea is to use an RF pulse to perturb the nuclear spins at or near the resonant frequency, causing them to oscillate and to gradually relax back toward equilibrium. After the initial pulse ends, the oscillation of the spins will produce a new RF signal of its own, called the **free induction signal (FIS)**, which is detected by a radio receiver. The FIS signal will look like the plot in Fig. 23.2 and will soon decay back to zero due to relaxation processes.

Because of interactions between the protons and their surroundings, the resonant frequency of a proton in a particular molecule will be different than the resonant frequency of a free proton. The difference in frequency between a free proton or and a proton in some standard reference molecule is called the **chemical shift**. The chemical shift is usually expressed as a fraction of the resonant frequency of the reference molecule and is typically on the order of parts per million, $\frac{\Delta\nu}{\nu_0} \sim 10^{-6}$; fortunately, a typical NMR apparatus is likely to be able to distinguish fractional shifts on the order of 10^{-7}.

The chemical shift causes protons in different molecules to have different resonant frequencies, allowing the type of molecule to be identified, and the strength of the outgoing signal will be proportional to the density of resonant protons. To produce an image of a single plane in the body, the idea is to produce a field which has a gradient along the axis perpendicular to that plane. As the field varies, the Larmor frequency of the protons will also vary, so that the spins will be resonant only in the vicinity of a thin plane. Tuning the receiver to detect different free induction signals will then allow different planes along the axis to be imaged sequentially. Then, by having gradients along all three axes, stacks of images in three perpendicular directions can be obtained, which can then be pieced together by a computer in order to produce a three-dimensional image.

23.4 NMR as a Quantum Computer

As seen in Sect. 23.1, RF pulses can be used to convert $|0\rangle$ and $|1\rangle$ into linear combinations of each other. The ability to use these pulses to implement single-qubit logic gates should be clear. But for quantum information processing, two-qubit gates are also needed, which means some sort of interaction is needed between the spins. So consider a molecule with two nuclear spins that can be manipulated by NMR. Then the molecular bond between them and the magnetic dipole-dipole forces provide the needed interactions between the spins. This will be the key to engineering controlled logic gates.

Recall from Eq. 23.8 that for a single spin, the interaction with the magnetic field can be written as

$$\hat{H} = -\frac{\hbar}{2} \begin{pmatrix} \omega_0 & \omega_1 e^{i\omega t} \\ \omega_1 e^{-i\omega t} & -\omega_0 \end{pmatrix} \tag{23.46}$$

in the $\{|0\rangle, |1\rangle\}$ basis. ω_0, ω_1, and ω are, respectively, the Larmor, Rabi, and RF frequencies. This can be written in another useful form by defining

$$\sigma_\pm = \frac{1}{2}\left(\sigma_x \pm i\sigma_y\right). \tag{23.47}$$

In matrix form, these operators become

$$\sigma_+ = \begin{pmatrix} 0 & 1 \\ 0 & 0 \end{pmatrix}, \qquad \sigma_- = \begin{pmatrix} 0 & 0 \\ 1 & 0 \end{pmatrix}. \tag{23.48}$$

Effectively σ_+ converts $|0\rangle$ to $|1\rangle$, and σ_- does the reverse, so these can be thought of operators that raise and lower the "spin" (or bit value) of the two-state system. Then Eq. 23.46 can be written as

$$\hat{H} = -\frac{\hbar}{2}\left(\omega_0 \sigma_z + \omega_1 e^{i\omega t}\sigma_+ + \omega_1 e^{-i\omega t}\sigma_-\right). \tag{23.49}$$

Let $|\psi(t)\rangle$ be the state of the system at time t. Recall that the operator $e^{i\omega\sigma_z t/2}$ will rotate a spinor by angle $\omega t/2$ about the z-axis. It is always possible to go to a new frame of reference by defining the state

$$|\tilde{\psi}(t)\rangle = e^{i\omega\sigma_z t/2}|\psi(t)\rangle. \tag{23.50}$$

Our new reference frame is rotating at rate ω, the same rate as the RF field.

In quantum field theory, if $\omega = \omega_0$ is taken to be the free propagation frequency, then the use of the states ψ and $\tilde{\psi}$ in the two reference frames is called the **Schrödinger picture** and **interaction picture**, respectively. In the Schrödinger picture, the time evolution of the state $|\psi\rangle$ is governed by the full Hamiltonian. In the interaction picture, the state $|\tilde{\psi}\rangle$ evolves only according to the free-particle Hamiltonian (all interactions switched off); the time evolution generated by the interactions is swept into the operators, instead of the states. The interaction picture is often useful when a system has complicated interactions of finite time duration (such as in scattering problems), since it allows the interactions to be viewed as causing transitions between well-defined asymptotic input and output states. The input and output states are simply those of the free particle system.

We know that $i\hbar\frac{\partial}{\partial t}|\psi(t)\rangle = \hat{H}|\psi(t)\rangle$ in the original reference frame. Here, when we define $|\tilde{\psi}(x)\rangle$, which defines a reference frame rotating at the RF frequency ω, we know that in new frame there should be a corresponding Hamiltonian $\tilde{H}$ such that $i\hbar\frac{\partial}{\partial t}|\tilde{\psi}(t)\rangle = \tilde{H}|\tilde{\psi}(t)\rangle$. We find the form of this new Hamiltonian as follows:

$$i\hbar\frac{\partial}{\partial t}|\tilde{\psi}(t)\rangle = e^{-i\omega\sigma_z t/2}\left[\frac{\hbar\omega}{2}\sigma_z + e^{i\omega\sigma_z t/2}\hat{H}e^{-i\omega\sigma_z t/2}\right]|\psi(t)\rangle \quad (23.51)$$

$$= \left[\frac{\hbar\omega}{2}\sigma_z + e^{-i\omega\sigma_z t/2}\hat{H}e^{i\omega\sigma_z t/2}\right]|\tilde{\psi}(t)\rangle \quad (23.52)$$

$$\equiv \tilde{H}|\tilde{\psi}(t)\rangle, \quad (23.53)$$

where

$$\tilde{H} = \left[\frac{\hbar\omega}{2}\sigma_z + e^{-i\omega\sigma_z t/2}\hat{H}e^{i\omega\sigma_z t/2}\right] \quad (23.54)$$

First, to simplify notation, we now denote the three Pauli matrices $\sigma_x, \sigma_y, \sigma_z$ simply by $\hat{X}, \hat{Y}, \hat{Z}$. More importantly, the structure of the Hamiltonian can be simplified quite a bit. Start by writing the original Hamiltonian in the form

$$\hat{H} = -\frac{\hbar\omega_0}{2}\hat{Z} - \frac{\hbar\omega_1}{2}\left(\hat{X}\cos\omega t - \hat{Y}\sin\omega t\right). \quad (23.55)$$

Now look at how each term on the right separately transforms into the interaction picture. Since the Z term is unchanged and

$$e^{-i\omega\sigma_z t/2}\hat{X}e^{i\omega\sigma_z t/2} = \hat{X}\cos\omega t + \hat{Y}\sin\omega t \quad (23.56)$$

$$e^{-i\omega\sigma_z t/2}\hat{Y}e^{+i\omega\sigma_z t/2} = \hat{Y}\cos\omega t - \hat{X}\sin\omega t, \quad (23.57)$$

it follows that

$$e^{-i\omega\sigma_z t/2}\hat{H}e^{i\omega\sigma_z t/2} = -\frac{\hbar\omega_0}{2}\hat{Z} - \frac{\hbar\omega_1}{2}\left[\left(\hat{X}\cos\omega t + \hat{Y}\sin\omega t\right)\cos\omega t\right.$$

$$\left. -\left(\hat{Y}\cos\omega t - \hat{X}\sin\omega t\right)\sin\omega t\right]$$

$$= -\frac{\hbar\omega_0}{2}\hat{Z} - \frac{\hbar\omega_1}{2}\hat{X}. \quad (23.58)$$

Therefore, we find that the Hamiltonian in the rotating basis has simplified down to

$$\tilde{H} = \frac{\hbar\delta}{2}\hat{Z} - \frac{\hbar}{2}\omega_1\hat{X}, \quad (23.59)$$

where $\delta = \omega - \omega_0$. Note that, unlike the original Hamiltonian, $\hat{H}$, this new rotating-frame Hamiltonian is time-independent.

Example 23.4.1 The relations of Eqs. 23.56 and 23.57 are easy to demonstrate. For example,

$$
\begin{aligned}
e^{-i\omega\sigma_z t/2}\hat{X}e^{i\omega\sigma_z t/2} &= \left(\cos\frac{\omega t}{2} - i\hat{Z}\sin\frac{\omega t}{2}\right)\hat{X}\left(\cos\frac{\omega t}{2} + i\hat{Z}\sin\frac{\omega t}{2}\right) \\
&= \hat{X}\cos^2\frac{\omega t}{2} + i\sin\frac{\omega t}{2}\cos\frac{\omega t}{2}\left[\hat{X},\hat{Z}\right] + \hat{Z}\hat{X}\hat{Z}\sin^2\frac{\omega t}{2} \\
&= \frac{1}{2}\hat{X}(1 + \cos\omega t) + \frac{i}{2}\left[\hat{X},\hat{Z}\right]\sin\omega t + \frac{1}{2}\hat{Z}\hat{X}\hat{Z}(1 - \cos\omega t) \\
&= \frac{1}{2}\hat{X}(1 + \cos\omega t) + \hat{Y}\sin\omega t - \frac{1}{2}\hat{X}(1 - \cos\omega t) \\
&= \hat{X}\cos\omega t + \hat{Y}\sin\omega t, && (23.60)
\end{aligned}
$$

where the commutation relations of the Pauli matrices have been used, along with half-angle trig identities. The second relation is shown in a similar manner

$$
e^{-i\omega\sigma_z t/2}\hat{Y}e^{+i\omega\sigma_z t/2} = \hat{Y}\cos\omega t - \hat{X}\sin\omega t. \qquad\blacksquare \qquad (23.61)
$$

Now consider two spins on the same molecule. The interaction between them will in general be of the form $\hbar J\hat{Z}_A \otimes \hat{Z}_B$, for some coupling constant J. Here, A and B label the two spins. We will assume that the spin-spin interaction is weak, $|J| << \omega_{1A}, \omega_{1B}$. The full Hamiltonian now involves two copies of the non-interacting Hamiltonian 23.8, or equivalently, 23.49, plus the interaction term

$$
\hat{H} = \hat{H}_1 + \hat{H}_2 + \hbar J\hat{Z}_A \otimes \hat{Z}_B, \qquad (23.62)
$$

where

$$
H_j = -\frac{\hbar}{2}\left(\omega_{0j}\hat{Z}_j + \omega_{1j}e^{i\omega_j t}\sigma_{+j} + \omega_{1j}e^{-i\omega_j t}\sigma_{-j}\right), \qquad (23.63)
$$

where $j = A, B$. Here, we have allowed for the two spins to have different resonance frequencies. We will assume that the detunings are small in the sense that

$$
|\delta_A| = |\omega_A - \omega_{0A}| << \omega_{1A}, \qquad |\delta_B| = |\omega_B - \omega_{0B}| << \omega_{1B}. \qquad (23.64)
$$

Going now to a rotating frame defined by

$$
|\tilde{\psi}_A(t)\rangle \otimes |\tilde{\psi}_B(t)\rangle = e^{-i\omega_A\hat{Z}_A t/2}e^{-i\omega_B\hat{Z}_B t/2}|\psi_A(t)\rangle \otimes |\psi_B(t)\rangle, \qquad (23.65)
$$

leads by the same steps as before to

$$\tilde{H} = \frac{\hbar}{2}\left\{\delta_A \hat{Z}_A + \delta_B \hat{Z}_B - \omega_{1A}\hat{X}_A - \omega_{1B}\hat{X}_B\right\} + \hbar J Z_A \otimes Z_B \qquad (23.66)$$

For the purposes of quantum computing, the active molecule on which the two magnetic dipoles reside is dissolved in a dilute solution. This guarantees that the individual molecules are far enough apart to be non-interacting with each other; only the interactions within the same molecule will be relevant.

So now, suppose that a controlled phase or CZ operation is to be implemented. For $j = A, B$, let $R_{\hat{n},j}(\theta) = e^{-i\theta\sigma\cdot\hat{n}/2}$ be the rotation operator that rotates dipole j by angle θ about the axis along direction $\hat{n}$. So, taking $\hat{n}$ along the z axis,

$$R_{zj}(\frac{\pi}{2}) = \hat{I}\cos\left(\frac{\pi}{4}\right) - i\hat{Z}_j\sin\left(\frac{\pi}{4}\right) = \frac{1}{\sqrt{2}}\left(\hat{I} - i\hat{Z}_j\right). \qquad (23.67)$$

Defining the operator

$$\hat{O}(t) = e^{iJt\left(\hat{Z}_A\otimes\hat{Z}_B\right)}R_{zA}\left(\frac{\pi}{2}\right)\otimes R_{zB}\left(\frac{\pi}{2}\right), \qquad (23.68)$$

then setting $t = \frac{\pi}{4J}$, we have

$$\hat{O}\left(\frac{\pi}{4J}\right) = \left(\frac{1}{\sqrt{2}}\right)^3\left(\hat{I} + iZ_A \otimes Z_B\right)\cdot\left(\hat{I} - \hat{Z}_A \otimes \hat{I}_B\right)\cdot\left(\hat{I} - \hat{Z}_B \otimes \hat{I}_B\right)$$

$$= \frac{1-i}{2\sqrt{2}}\left(\hat{I} + \hat{Z}_A \otimes \hat{I}_B + \hat{I}_A \otimes \hat{Z}_B - \hat{Z}_A \otimes \hat{Z}_B\right). \qquad (23.69)$$

In the basis $|00\rangle, |01\rangle, |10\rangle, |11\rangle$ (where the first entry is A and the second is B), this can be written in matrix form as

$$\hat{O}\left(\frac{\pi}{4J}\right) = e^{i\pi/4}\begin{pmatrix}\hat{I}_A & 0 \\ 0 & \hat{Z}_B\end{pmatrix} = e^{i\pi/4}\begin{pmatrix}1 & 0 & 0 & 0 \\ 0 & 1 & 0 & 0 \\ 0 & 0 & 1 & 0 \\ 0 & 0 & 0 & -1\end{pmatrix} = e^{i\pi/4}\cdot CZ, \qquad (23.70)$$

where CZ is the controlled Z operation.

A Hadamard operation on dipole j is simply a rotation of π about the axis $\hat{n} = \frac{1}{\sqrt{2}}\left(\hat{i} - \hat{k}\right)$. Since this rotation can be implemented by means of an RF pulse as described in Sect. 23.1, the Hadamard operation can be readily implemented experimentally. The ability to target the rotation onto either dipole $j = A$ or $j = B$ depends on the resonant frequencies of the two dipoles being different.

Once the CZ operation and the Hadamard operation can be implemented, then the controlled NOT is simply a combination of these:

$$CNOT = CX = \left(\hat{I}_A \otimes \hat{H}_B \right) \cdot CZ \cdot \left(\hat{I}_A \otimes \hat{H}_B \right) = \begin{pmatrix} \hat{I} & 0 \\ 0 & X \end{pmatrix}. \qquad (23.71)$$

So the NMR system is capable of implementing both single-qubit and two-qubit gates. One problem is that during the action of the gate on one spin, the other spins in the material continue to evolve independently, dephasing from each other: spins in slightly different locations will precess at slightly different phases due to local random effects (imperfections in the magnets creating the field, local thermal fluctuations), eventually moving farther and farther out of phase with each other. This is important because the time required for the interaction term $J \hat{Z}_A \hat{Z}_B$ to flip the spin is much longer (on the order of milliseconds) than the time for the RF field to flip spins (on the order of 10 microseconds). To correct for this, a process called **refocusing** is used. If a π-pulse is applied, then not only will the spins all flip by 180°, but their direction of precession will reverse. Now, the faster spins will be moving faster than the slower spins in the *opposite* direction, undoing the previous dephasing. As the spins all realign themselves, the free induction signal will resurge creating a second copy of itself, called a **spin echo**. However, the resurgence won't be complete since it won't cancel out the T_2 dephasing, so the second copy of the pulse will be of lower amplitude than the first. A sequence of π-pulses will therefore produce a sequence of echo pulses of gradually decreasing amplitudes. In quantum computing, the refocusing is used to keep the system in the desired state long enough to complete the gate operations, while in imaging applications, the spin echo sequence is used to separate T_1 and T_2 effects from each other (see [11]).

One other problem with NMR-based quantum computing is that the qubit is not formed by a single quantum particle, but by a dilute solution containing an enormous number of independent particles. The signal seen during detection is formed collectively by all of these particles together. Maintaining coherence between all of the particles making up one qubit and the particles making up another qubit becomes rapidly more difficult as the number of qubits increases. As a result, in general the signal decreases exponentially with the number of qubits

$$\text{signal strength} \sim 2^{-n},$$

where n is the number of qubits. This makes large-scale computing by this means impractical with current technology. Nonetheless, NMR methods have been used to successfully demonstrate the workings of quantum algorithms on low-qubit-number systems.

Box 23.2 The Quantum Rosetta Stone
It is often the case in physics that systems that seem to be completely unrelated are in fact mathematically identical. One example is the triplet of systems that has been called the **quantum Rosetta stone** [12].

(continued)

Box 23.2 (continued)

The original Rosetta stone was erected in the Egyptian port city of Rosetta (modern-day Rashid) by King Ptolemy V in 196 BC. It was removed during Napoleon's 1798 campaign in Egypt and transported to Cairo. After the British took control of Cairo in 1801, the stone was taken to Plymouth, England. It can be viewed today in the British Museum in London. The Rosetta stone's importance follows from the fact that it has the same text in three languages: ancient Egyptian hieroglyphic and Demotic scripts and Ancient Greek. This allowed Jean-Francois Champollion (1790–1832), a former child prodigy and a protegé of Joseph Fourier, to provide the first decryption of ancient Egyptian hieroglyphics in 1822, beating out Thomas Young (of the Young two-slit experiment), who had previously published a partial decipherment.

Similarly, the triplet of physical systems in the quantum Rosetta stone allows results in one system to immediately by translated into results for the other two. The three systems, shown in Fig. 23.3, are the Mach-Zehnder interferometer, the Ramsey interferometer, and a particular two-qubit quantum gate. In the Mach-Zehnder system, a photon is input to one of the two input ports (A and B); put into a superposition of two paths inside the interferometer, with a phase shift in one path; and then output to two exit ports at the right. Similarly, the **Ramsey interferometer** consists of a two-state system (the two states play the role of the two MZ input ports). An atom in the ground state is put into a superposition of ground and excited states by a $\pi/2$ pulse, which implements a Hadamard transformation. After a phase shift applied to the excited state, another $\pi/2$ pulse is applied, leaving the atom to be read out in either the ground or excited state. Finally, the quantum gate follows a similar procedure. An input qubit, initially in either the $|0\rangle$ or $|1\rangle$ state, is put into a superposition of both states by the Hadamard gate; after a phase shift to the $|0\rangle$ state, a second Hadamard is then applied before reading out the final state.

It is because of equivalences such as these that quantum computing operations can be carried out in many different physical platforms. The quantum Rosetta stone can clearly be extended to find equivalent systems in other physical systems such as cold atoms in optical lattices or quantum dots.

The Ramsey interferometer is used as a component for quantum information processing systems in cavity quantum electrodynamics (Sect. 25.3). It should be mentioned that the Ramsey interferometer is an important quantum device in its own right, appearing in other atom interferometry applications. For example, it can be used to construct an oscillator with an extremely precise and stable oscillation frequency, which in turn forms the basis for precise atomic clocks.

Fig. 23.3 The quantum Rosetta stone—three different physical systems that are mathematically equivalent: the Mach-Zehnder interferometer (top), Ramsey interferometer (middle), and a quantum gate (bottom)

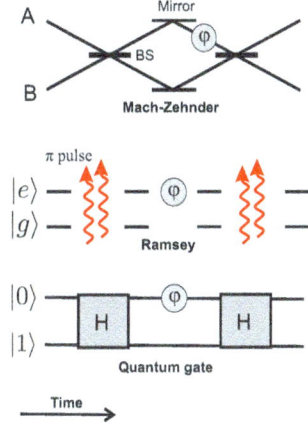

Problems

1. At typical body temperature, and at external fields of 0.5 T and 5.0 T, find the value of $\frac{\mu B_0}{k_B T}$ for electrons, and find the probabilities (Eqs. 23.3 and 23.4) of ground (spin up) and excited (spin down) states for an electron. Do the same for a proton and compare.

2. In thermal equilibrium, the density matrix of a system is $\hat{\rho} = e^{-\beta \hat{H}}/Z$, where Z is the partition function. The free induction signal that is used for readout in an NMR system is proportional to $\text{Tr}(\hat{\rho} H)$, where the Hamiltonian $\hat{H}$ is usually proportional to a linear combination of Pauli matrices. Show for an NMR system with n nuclei that at high temperatures $\hat{\rho} \sim 2^{-n}(1 - \beta \hat{H})$, so that the size of the signal decreases $\sim 2^{-n}$ as the system size increases.

3. (a) Derive Eq. 23.20.
   (b) Explicitly verify that Eqs. 23.21 and 23.22 solve Eqs 23.19 and 23.20.

4. Starting from Eqs. 23.21 and 23.22, undo the change of variables, Eqs. 23.12 and 23.13, to find the general solutions for $\alpha(t)$ and $\beta(t)$.

5. Two hydrogen nuclei are separated by a distance of ~ 1 Å $= 0.1$ nm.

   (a) How large a field gradient is needed to separate their *Larmor* frequencies by 10 Hz? By 100 Hz? Does this seem practical?

   (b) How large a field gradient is needed to separate their *cyclotron* frequencies by 10 Hz? By 100 Hz? Does this seem practical?

6. A proton (or any other magnetic dipole) produces a magnetic field of the form $B_r = \frac{\mu_0}{2\pi} \frac{2\mu}{r^3} \cos\theta$, $B_\theta = \frac{\mu_0}{2\pi} \frac{2\mu}{r^3} \sin\theta$, and $B_\phi = 0$, where r extends from the dipole to the observation point, θ is the angle between r and the dipole direction, and ϕ is the azimuthal angle about the dipole.

   (a) What is the magnitude of the field produced by a proton at a distance of 0.1 nm?

(b) How much does this field change the Larmor and Rabi frequencies of another proton residing at that distance? (Assume the two protons have dipole moments aligned parallel to each other.)

(c) How long does it take for the field of parts (a) and (b) to produce a precession of the second proton's dipole through an angle of π relative to a proton not experiencing that field? Does this tell us anything about the order of magnitude of the coherence times?

7. (a) Equation 23.26 is the solution for Rabi oscillations assuming an initial condition of $|\psi(0)\rangle = |0\rangle$. Find the solution for the case when $|\psi(0)\rangle = |1\rangle$.

(b) Using Eq. 23.26 and the result of part (a), write down 2×2 matrix U that gives the time evolution of the system, $\psi(t) = U|\psi(0)\rangle$.

8. Consider the unperturbed Bloch equations, Eqs. 23.31–23.33. Go to a coordinate system rotating at frequency ω about the z axis by applying the rotation matrix

$$R(t) = \begin{pmatrix} \cos \omega t & \sin \omega t & 0 \\ -\sin \omega t & \cos \omega t & 0 \\ 0 & 0 & 1 \end{pmatrix}.$$

(a) Show that the equations of motion reduce to $\frac{dM}{dt} = \gamma M \times B_{eff}$, with effective magnetic field $B_{eff} = \left(B_0 - \frac{\omega}{\gamma} \right) \hat{k}$.

(b) What do the equations of motion in part (a) reduce to when the detuning vanishes, $\delta = \omega - \omega_0 = 0$?

9. Write the magnetization as a column matrix $M(t)$ with components $M_x(t)$, $M_y(t)$, $M_z(t)$. Write the Bloch equations 23.40–23.42 as a single matrix equation, $\frac{dM}{dt} = NM + v$, and find the elements of the square matrix N and the column vector v.

10. NMR systems are two-level systems driven by RF radiation and therefore should be closely analogous to two-level atoms in optical fields. Analogous to the NMR Bloch equations, Eqs. 23.40–23.42, a set of optical Bloch equations can be written down for the atomic density matrix. Let $\hat{\rho}$ be the density matrix for the electron in the ground state-excited state basis. Define the variables $u = \text{Tr}[\hat{\rho}\sigma_x]$, $v = i\text{Tr}[\hat{\rho}\sigma_y]$, $w = \text{Tr}[\hat{\rho}\sigma_z]$, which are just the expectations of the Pauli operators on the two-dimensional space of electron states. Then the **optical Bloch** or Maxwell-Bloch equations are given by

$$\begin{pmatrix} du/dt \\ dv/dt \\ dw/dt \end{pmatrix} = \begin{pmatrix} -\gamma & \delta & 0 \\ -\delta & -\gamma & -2\Omega \\ 0 & 2\Omega & -2 \end{pmatrix} \begin{pmatrix} u \\ v \\ w \end{pmatrix} - \begin{pmatrix} 0 \\ 0 \\ 2\gamma \end{pmatrix}.$$

(a) Define the vector $\boldsymbol{\sigma}$ with components σ_x, σ_y, σ_z. For no decay ($\gamma = 0$), show that at resonance the vector $\boldsymbol{\sigma}$ simply rotates over time. Explicitly write down the rotation matrix $R(t)$ that carries out the time evolution, $\boldsymbol{\sigma}(t) = R(t)\boldsymbol{\sigma}(0)$.

(b) For no decay, $\gamma = 0$, but for nonzero detuning δ, solve the optical Bloch equations (either analytically or by computer) then use Mathematica or another program to plot the trajectories of the (u, v, w) vector for several initial conditions. Compare to the case of zero detuning.

11. Prove the relation given in Eq. 23.71.

References

1. J.C. Wright, P.C. Chen, Phys. Today **76**(6), 32 (2023)
2. R.K. Hobbie, B.J. Roth, *Intermediate Physics for Medicine and Biology, 5th ed.* (Springer, Berlin, 2015)
3. M. Weissbluth, *Photon-Atom Interactions* (Academic Press, New York, 1989)
4. A. Spyros, Liquid-State NMR in Cultural Heritage and Archaeological Sciences, in *Modern Magnetic Resonance*, ed. by G.A. Webb (Springer Nature, New York, 2018)
5. N. Proietti, D. Capitani, V. Di Tullio, Magnetochemistry **4**, 11 (2018)
6. W.M. Awad, M. Baias, Magn. Reson. Chem. **58**, 792 (2020)
7. N. Proietti, V. Di Tullio, D. Capitani, R. Tomassini, M. Guiso, Appl. Phys. A **113**, 1009 (2013)
8. F. Al-Sammerrai, D. Al-Sammerrai, J. Al-Rawi, Thermochim Acta. **115**(C), 181 (1987)
9. Q. Wang, U.G. Nielsen, Sol. State Nucl. Mag. Res. **110**, 101698 (2020)
10. A.D.C. Santos, L.M. Dutra, L.R.A. Menezes, M.F.C. Santos, A. Barison, TrAC Trends Analyt. Chem. **107**, 31 (2018)
11. S.A. Kane, *Introduction to the Physics in Modern Medicine* (Taylor and Francis, Boca Raton, 2003)
12. H. Lee, P. Kok, J.P. Dowling, J. Mod. Opt. **49**, 2325 (2010)

Chapter 24
Atomic and Ionic Systems

Much of this book has focused on photons, electrons, and solids. But atomic and ionic systems are also of great interest in modern applications of quantum mechanics. Atom interferometry is a well-developed field that is useful for fundamental tests of quantum physics. Atomic and ionic systems allow the implementation of highly stable qubits and the manipulation of those qubits by means of electromagnetic pulses and so are a strong candidate for the implementation of quantum computers, as well as being suitable for many high-precision sensing applications. The atoms and ions are usually used in the form of dilute gases that need to be cooled to low temperatures. So, in the following sections, we first look at methods for trapping and cooling atoms and ions. We then look at how atom-photon interactions can be used to form and process well-defined qubits.

One major advantage for atomic and ionic systems is the existence of both internal (electronic) and external (position and momentum) degrees of freedom, and these can be made to interact with each other. As a result, coupling can be achieved between two sets of qubits, allowing the realizations of controlled logic gates. Trapped atoms also have essentially infinite lifetimes, compared to the brief lifetimes of many photonic and electronic qubits. Another prominent feature is that at sufficiently low temperatures atomic systems in states of integer total spin can form Bose-Einstein condensates, described by a single macroscopic wavefunction; these condensates are similar in many ways to superfluid and superconducting systems.

In addition to computing, atomic systems have found a range of applications such as measurements of accelerations and gravitational fields, atomic clocks, magnetometers, and imaging devices. Recent advances have made it possible to construct compact, low power versions of many of these devices at a scale suitable for incorporation into integrated electronic or photonic chip systems [1].

For more detailed discussion of many of these topics, an excellent introduction is [2]. At a slightly higher level, see also [3, 4], and for more on the theory of photon-atom interactions, see [5]. Ion traps are covered in detail in [6].

© The Author(s), under exclusive license to Springer Nature Switzerland AG 2025
D. S. Simon, *Introduction to Quantum Science and Technology*, Undergraduate
Texts in Physics, https://doi.org/10.1007/978-3-031-81315-3_24

24.1   Laser Cooling

In order to perform operations on single atoms, it will be necessary to trap them
in electromagnetic potential wells. But a prerequisite for this is to first cool the
atoms to the point where thermal fluctuations are too small to allow the atoms to
spontaneously jump out of the wells. Here we will focus mainly on neutral atoms,
although most of this section will apply with little change to charged ions.

A temperature T can be unambiguously associated with a gas of atoms by setting
the average kinetic energy equal to the corresponding thermal energy

$$\frac{3}{2}k_B T = \frac{1}{2}m\langle v^2 \rangle, \tag{24.1}$$

where k_B is Boltzmann's constant. To measure this temperature, a gas of trapped
atoms can be released from the trap; the cloud of atoms then expands into vacuum
in all directions. Lasers bouncing off the cloud provide images of it at several times.
And so, from these images, the expansion rate of the cloud allows the average speed
of the atoms to be determined.

Recall that when light is absorbed or reflected from some surface, it exerts a
force and a pressure on that surface. For the absorbing case, the radiation force and
consequent radiation pressure are

$$F_{rad} = \frac{IA}{c}, \qquad P_{rad} = \frac{I}{c}, \tag{24.2}$$

where I and A are the light intensity and the area of the absorbing or reflecting
surface. For the reflecting case, the change in momentum is doubled, so that the
radiation force and pressure are twice as large.

When dealing with light absorbed or reflected off of individual atoms, the
geometric area is replaced by the interaction cross-sectional σ. σ, which has units of
area, depends on the energy-level structure of the atom, the widths of the absorption
resonances, and the frequency of the incident light. In general, σ is much larger than
the actual geometric cross section of the atom, and it essentially acts as a fictitious
"effective" surface size for the light to strike, in the sense that an interaction radius
r_0 can be defined via $\sigma = \pi r_0^2$, such that light passing within distance r_0 of the
atoms core is likely to interact with the atom, while light passing much farther out
is likely to continue propagating unaffected. Given knowledge of the interaction
cross section, the scattering and absorption rates, R_{sc} and R_{ab}, of the light can be
found. From these rates the net forces per photon on the atom are then $F_{ab} = pR_{ab}$
and $F_{sc} = \Delta p R_{sc}$, where p is the photon momentum and Δp is its change during
scattering.

This radiative pressure can be used to remove kinetic energy from a gas of
warm atoms and thereby to cool the gas. The key to doing this is to have two
counterpropagating lasers of equal frequency striking the atom from opposite
directions and then taking advantage of the Doppler effect.

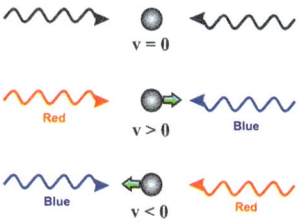

Fig. 24.1 Two counterpropagating lasers of the same frequency are focused on an atom. If the atom is at rest (top), the radiation pressures from the two lasers exactly cancel and the atom stays at rest. If the atom is moving to the right (middle), then it sees the right-hand laser blue-shifted and the left-hand redshifted, leading to a net rightward force, opposite to the velocity. For leftward motion (bottom), the two Doppler shifts are reversed, leading to a net force to the right, again opposite to the velocity

Consider a single dimension first. Suppose the atom is initially moving along the z-axis with speed v as shown in Fig. 24.1, with lasers of frequency ω shining on it from the left and right. The laser light is slightly red-detuned, meaning that its frequency is below the absorption resonance frequency of the atom, $\omega < \omega_0$. For rightward motion ($v > 0$), the atom is moving toward the photons of the right-hand beam and so sees the light blue-shifted to a higher frequency

$$\omega' = \omega + kv, \tag{24.3}$$

where $k = \frac{\omega}{c}$ is the wavenumber of the light. The left-hand beam is shift in the opposite direction: the atom sees it red-shifted to a lower frequency:

$$\omega' = \omega - kv. \tag{24.4}$$

Since the photon momentum is proportional to its frequency, this means that the atom feels a stronger radiation pressure pushing it the left than it feels to the right. As result, there is a net force opposite to its direction of motion. Similarly, if the atom is moving to the left ($v < 0$), then it feels a greater force to the right. In either case, the net effect is to slow the motion of the atom.

The energy that is removed from the atom is radiated away by photons that are emitted during the cooling. As a result, the rate of cooling will depend on the resonance frequency ω_0 for the transition being excited in this process. Since the radiation is emitted in random directions, it carries net average momentum of zero. In some cases, the fluorescence emitted during cooling can be bright enough for individual atoms or ions to be located with the naked eye.

In order to reduce the motion along all three axes, three pairs of counterpropagating beams are needed, one pair along each axis. The net result is that the atom feels a retarding force opposite to its velocity, similar to the viscous force felt by a particle in a fluid

$$F_{ret} = -2\alpha v, \tag{24.5}$$

producing what is known as an **optical molasses**. The drag coefficient is given by

$$\alpha = 2k\frac{\partial F}{\partial \omega}, \tag{24.6}$$

where $\frac{\partial F}{\partial \omega}$ is the slope of the scattering force at the resonant frequency, ω_0. The damping coefficient can be shown [2] to be proportional to $-\delta$, where $\delta = \omega - \omega_0$ is the detuning between the laser frequency ω and the atomic resonant frequency ω_0. So in order achieve damping (positive α), it is necessary that the laser frequency be below the resonant frequency (red-detuning, $\delta < 0$). The damping time required for the cooling to occur is $\tau_{damp} = \frac{m}{2\alpha}$ (see Problem 1).

If the typical transition time and width of the electronic levels in the atom are τ and $\Gamma = \frac{1}{\tau}$, then the minimum temperature that can be achieved by Doppler cooling occurs when the thermal fluctuations are similar in size to the energy uncertainty in the emitted radiation:

$$k_B T = \frac{\hbar \Gamma}{2}. \tag{24.7}$$

This leads to the Doppler cooling limit:

$$T_{min} = \frac{\hbar}{2k_B \tau}. \tag{24.8}$$

This Doppler cooling limit assumes that only two atomic energy levels participate in the transitions. By including additional energy levels, sub-Doppler cooling can be achieved. This is done, for example, through a mechanism known as **sideband cooling**. Sideband cooling takes advantage of the internal states of the atom. Let the initial state of the atom be $|n, m\rangle$, where n is the internal electron state and m labels the external state represented by the translational energy and momentum. States of higher m will be of higher energy, as seen in Fig. 24.2. So suppose that an atom initially in state $|n, m\rangle$ is excited to the state $|n + 1, m - 1\rangle$. When the atom decays

Fig. 24.2 An electron excited to a higher internal atomic state can decay randomly into one of several states of different momentum states. On average, this will lower the energy of the atom and can lead to sub-Doppler cooling

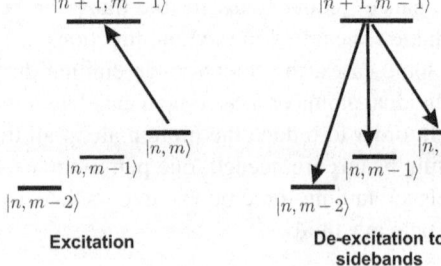

back to internal state n, sometimes it will decay back to the original m value. But it can also decay to nearby sidebands such as $|n, m - 1\rangle$ or $|n, m - 2\rangle$, leading to a lower overall average energy. Through repeated repetitions of this process, the atom can be cooled below the Doppler limit.

24.2 Trapping of Ions

Once cooled, a set of ions or atoms can then be further localized within an electrostatic potential well or by a confining magnetic field. Here, we look at trapping of charged ions; in the next section, we look at electrically neutral atoms. The two most common types of traps are the Paul trap, which uses a combination of static and oscillating electric fields, and the Penning trap, which uses static electric and magnetic fields.

Again, consider a single dimension first, along the z-axis. The obvious way to try to localize a charged atom at the origin is to place two electrodes at positions $z = \pm z_0$ and then to charge them up to the same value (Fig. 24.3a). For example, if the ion is of charge $+q$, the electrodes can both be charged up to some charge $+q_0$. Then both electrodes repel the atom with Coulomb forces, and these two forces become equal when the atom is at the origin. The origin then becomes a stable equilibrium.

The flaw with this procedure is that it only works in one dimension. We can see that by looking at the field lines in the full two-dimensional x-z plane (Fig. 24.3b), instead of just along the axis. We see that the field pushes the atom toward the origin only if it resides on the z axis. If the atom is displaced slightly off-axis in

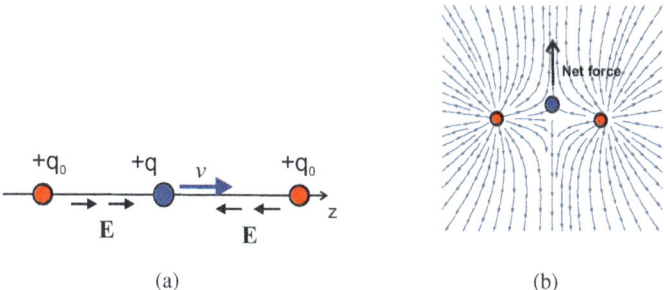

(a)   (b)

Fig. 24.3 (a) In one dimension, a pair of fixed, equally spaced positively charged electrodes (red) push in opposite directions on a mobile positive charge (blue) near the origin. When at the origin, the two forces balance. But if the mobile charge is moving to the right, it comes closer to the right-hand electrode, which will then exert a stronger force on it than the opposing left-hand electrode will. This pushes the charge back toward the origin. If the charge moves to the left, it similarly begins to feel a net rightward force. (b) In two dimensions, the field lines bend away from the z-axis, so the origin is now a saddle point. It remains stable in the z direction, but is unstable in the x direction

the x direction, then the field lines now have components pushing the atom away from the z-axis. So rather than being a stable equilibrium, the origin is now a saddle point: it is stable in the z direction, but unstable in along the x direction.

This is a consequence of **Earnshaw's theorem**, a nineteenth-century result proven in many introductory electromagnetism texts. Recall that in a region free from the sources (electric charges), Gauss' theorem tells us that the electric field must be divergence-free: $\nabla \cdot E = 0$. In integral form, this means that if a closed surface S is drawn around an equilibrium point, for every field line entering S on one side of the point (the right side, say), an equal number of field lines must exit on the left side. So if the equilibrium is stable, with the field pushing the atom toward the equilibrium on one side of S, then there must be points elsewhere on S with the field pushing the atom away from the same point. Therefore:

> **Earnshaw's theorem** - In two or three dimensions, a charged particle cannot be confined at rest at a stable equilibrium by a time-independent electric field.

Confinement therefore requires the addition of additional structures, such as time-varying fields.

So for three dimensions, consider a setup such as shown in Fig. 24.4. The two electrodes along the z axis still provided confinement along the z-axis. But to confine in the x-y plane, four additional electrodes are added in the form of the four wires shown. The desired equilibrium point is at the center of the arrangement, at a distance of r_0 from each of these wires. These electrodes are connected in pairs, as shown, with a time-dependent voltage difference $V_{ac}(t)$ between the pairs. We take the applied voltage to be sinusoidally varying in time

$$V_{ac}(t) = V_0 \cos \Omega t. \tag{24.9}$$

Then, up to a possible constant, the potential in the spatial region between the electrodes is found to be the quadrupole potential

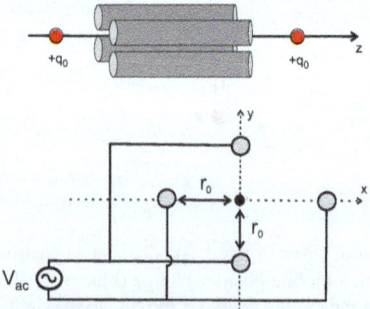

Fig. 24.4 A linear Paul trap. Top: Four rod-like electrodes in a square array are used to contain the particles in the x-y plane. The two equally charged electrons at the left and right continue to provide confinement in the z direction. Bottom: A cross section of the rod electrodes in the x-y plane. They are connected in pairs, with an oscillating voltage V_{ac} applied between the pairs. The confinement region is in the vicinity of the origin

$$\phi(x, y) = \frac{V_0}{2r_0} \cos{(\Omega t)} \left(x^2 - y^2 \right). \tag{24.10}$$

The reader can verify that this satisfies the correct boundary conditions at the electrodes and obeys the Laplace equation $\nabla^2 \phi = 0$ in the region between the electrodes. The resulting electric field in the x-y plane is

$$\boldsymbol{E}(x, y) = -\nabla \phi(x, y) = -\frac{V_0}{r^2} \cos{(\Omega t)} \left(x\hat{\boldsymbol{i}} - y\hat{\boldsymbol{j}} \right). \tag{24.11}$$

So at a given moment, if the force is repulsive along the x-direction, it is attractive in the y direction and vice versa. Because of the oscillatory behavior in time, the attractive and repulsive directions flip periodically. This means that if the flipping frequency is high enough, then the atom cannot get very far away from the axis in any direction before it is pulled back again. As a result, the atom undergoes stable oscillation around the origin, and the distance from the origin remains within a strict upper bound.

The trap described above is called a **linear Paul trap**. At low temperatures, ions placed in the trap will line up in a linear chain, similar to a string of pearls, along the z-axis, with mutual Coulomb repulsions keeping them separated by a fixed distance and linear restoring forces pushing them back to their equilibrium positions when displaced. The two eigenmodes of the ion motion are the in-phase center of mass mode, in which the whole chain oscillates left and right together, or the out-of-phase mode in which adjacent ions oscillate in opposite directions.

The spread of the ground state wavefunction is $\Delta z^2 = \frac{\hbar}{2m\omega}$, where ω is the oscillation frequency along the axis. In the ground state, the distance z_0 between adjacent trapped atoms should be at least as large as this spread: $z_0 \geq \Delta z$. For quantum information processing applications, the system should be controllable with pulses of light, which requires the oscillation amplitudes of the ions to be smaller than the wavelength of the light. Then this is roughly equivalent to the **Lamb-Dicke criterion**, $\eta < 1$, where the **Lamb-Dicke parameter** is defined to be the phase change from one potential minimum to the next:

$$\eta = kz_0 = \frac{2\pi z_0}{\lambda}. \tag{24.12}$$

Rather than a linear geometry, Paul traps can also be constructed in a cylindrical configuration, with the two of the rod-like electrodes replaced by a ring electrode, as shown in Fig. 24.5. The quadrupole field is then created between the ring electrode and the two endcap electrodes above and below it.

Defining a new parameter $Q = \frac{2eV_0}{\Omega^2 M r_0^2}$, where r_0 is the radius of the ring, it can be shown that the ions can be stably trapped for $Q \leq 0.9$. For a trap with size on the order of $r_0 \sim O(1 \text{ mm})$ for $\Omega \sim O(10 \text{ MHz})$, in the radio-frequency range, trapping voltages V_0 for typical ions are therefore usually required to be on the order of a few hundred volts.

Fig. 24.5 A Paul trap in a cylindrically symmetric configuration. The endcaps act as one terminal and the ring electrode as the other terminal, with the AC voltage applied between them

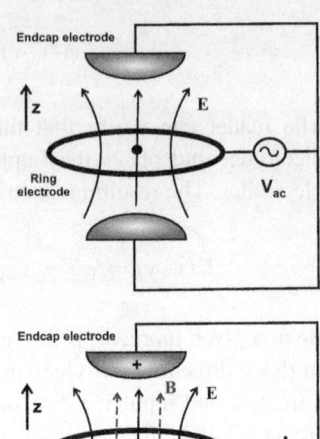

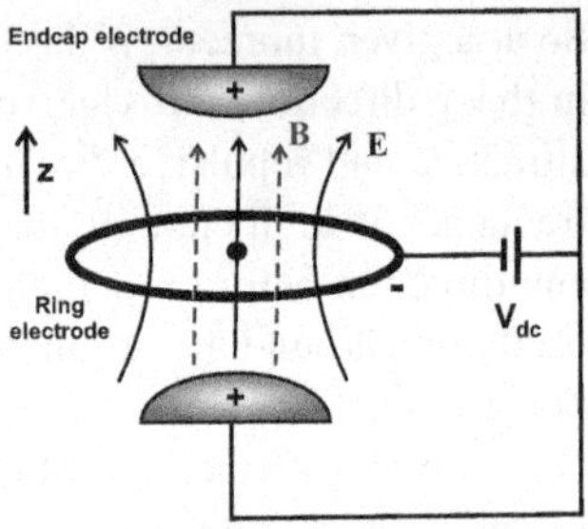

Fig. 24.6 A Penning trap relies only on static fields, but uses two kinds: an electric field (black) and a magnetic field (red, dashed)

An alternative to the Paul trap is the **Penning trap**, which was first built by H. G. Dehmelt, but named after F. M. Penning (1854–1953), whose pioneering work on vacuum tubes inspired the idea for the trapping method. The Penning trap consists of electrodes in the same configuration as the cylindrical Paul trap, but with two changes: (i) the AC voltage is replaced by a DC voltage, and (ii) a uniform magnetic field is applied along the axis of the ring (Fig. 24.6). The magnetic field provides radial confinement, while the quadrupole electric field provides confinement along the axis. Assuming a positively charged ion, the endcaps should be positively charged, and the ring should be negative.

The motion in the axial or z direction is simple harmonic. If the voltage between the electrodes is V_0 and the amplitude of oscillation is z_0, then the oscillation frequency in the axial direction is (Problem 6)

$$\omega_z = \sqrt{\frac{2q V_0}{m z_0^2}}. \tag{24.13}$$

In the transverse, or x-y plane, the motion is a combination of fast cyclotron motion (orbiting around the B field) and slow magnetron motion (from the combined action of the crossed E and B fields). The respective cyclotron and magnetron frequencies are given by

$$\omega_c = \frac{q}{m} B, \qquad \omega_m = \frac{1}{2} \omega_c - \sqrt{\omega_c^2 - 2\omega_z^2}. \tag{24.14}$$

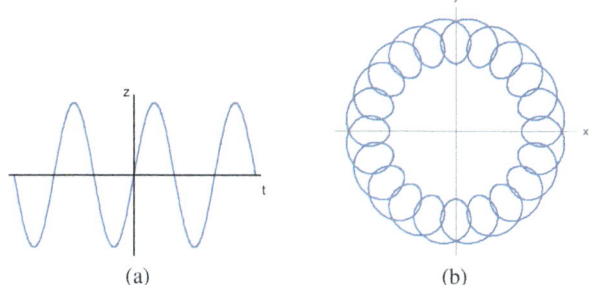

Fig. 24.7 (**a**) The motion of a charged particle in the z direction of a Penning trap is simple harmonic motion. (**b**) The trajectory in the x-y plane is a combination of magnetron motion (the large orbit) and cyclotron motion (the smaller loops)

In general, the three frequencies are very different in size, with $\omega_z \ll \omega_m \ll \omega_c$. Ignoring the slow axial oscillations, the resulting particle trajectories in the transverse plane are similar to the motion of planets in the Ptolemaic model of the solar system: the ions orbit the axis in a large circle while simultaneously tracing out smaller circles (epicycles) around the large orbital circle (Fig. 24.7).

Penning traps have a wide variety of applications. For example, they can be used for mass spectrometry: by measuring the sizes of particle orbits in the trap, the masses of unknown ions can be measured. As another example, the highest precision measurement to date of the proton's magnetic moment was carried out on individual protons isolated in a Penning trap, accurate to three parts in a billion [7].

24.3   Trapping of Neutral Atoms

For neutral atoms, there is no net electric charge to serve as a handle for electric or magnetic fields to grab onto. However, there will still be electric and magnetic dipole moments, and these can also be used for trapping. The couplings of the fields to these moments are generally orders of magnitude smaller than the couplings to the charges, and so the trapping will be less efficient for applied fields of the same size. Recall that dipole forces drop off more rapidly with distance ($\sim r^{-3}$) than Coulomb or Biot-Savart forces ($\sim r^{-2}$).

Consider a small dielectric sphere in the path of a light beam. If the refractive index of the sphere is larger than that of the surrounding medium, then refraction at the surface will bend the paths of light rays inward, as shown in Fig. 24.8. Those light rays represent photon paths, so the dielectric sphere must be exerting a force on the photon to bend its momentum inward. By Newton's third law, the sphere then must feel an equal and opposite force. The vertical components of the forces on the rays of the two sides cancel, leaving a net leftward force on the sphere. The result is that the sphere is pushed toward the focus of the beam, or in other words, toward

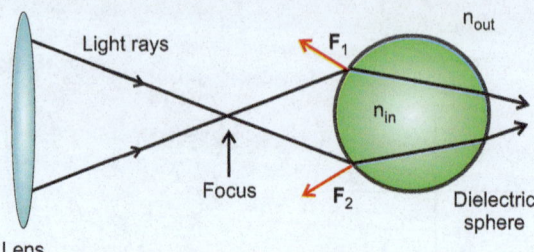

Fig. 24.8 For dielectric sphere with index n_{in} greater than the index of the surroundings, light refracting at the surface will bend inward toward the center of the sphere. The dielectric exerts forces on the photons in order to bend the momentum of the light inward. The equal and opposite forces exerted by the light on the sphere are shown as F_1 and F_2. Summing these forces gives a net force on the sphere toward the point of highest intensity, the focus of the beam

the region of higher intensity. If, on the other hand, the dielectric has index smaller than that of the surrounding medium, then it is pushed away from the focus.

To explain this effect quantitatively, recall that the polarization (the electric dipole moment per volume) is $P = \epsilon_0 \chi E$ for a dielectric of polarizability χ. The potential energy per volume of the dielectric in the presence of the field is then found by integrating the expression for an infinitesimal field

$$du = -P \cdot dE = -\epsilon_0 \chi E \cdot dE, \tag{24.15}$$

to get

$$u = -\frac{1}{2}\epsilon_0 \chi E^2. \tag{24.16}$$

The dipole force per volume is then

$$\frac{dF}{dV} = -\nabla u = +\epsilon_0 \chi E \nabla E = \frac{1}{2}\epsilon_0 \chi \nabla I, \tag{24.17}$$

where $I = |E|^2$ is the intensity. The force is in the direction of the field and intensity gradients, toward the intensity maximum. It thus acts as a restoring force, keeping the particle close to the beam focus.

For an individual atom, rather than a dielectric sphere, the treatment is similar, so if the intensity of the beam is sufficiently high, the end result (see [2] for a derivation) is a force and potential given by

$$F = -\nabla U, \quad \text{where} \quad U \approx \frac{\hbar\Omega^2}{4\delta} = \frac{\hbar\Gamma^2}{8\delta}\frac{I}{I_{sat}}, \tag{24.18}$$

where Ω is the Rabi frequency and the saturation intensity (the intensity at which the upper and lower levels in the electronic transition are equally populated) is given by $I_{sat} = \frac{\pi\hbar c\Gamma}{3\lambda^3}$.

By adjusting the optics in order to change the position of the beam focus, atoms and other microscopic particles can be moved to a desired location as they are dragged along with the focus. This is the basis of **optical tweezers**, which have become a mainstay of nanoscience and microbiology. Optical tweezers are used to manipulate objects as small as a fraction of a nanometer, including individual biological molecules or entire bacteria, and have been used to construct nanodevices such as optical quantum processors containing just a few hundred atoms.

Another option for trapping neutral atoms is a **magneto-optical trap (MOT)**, which combines laser cooling with a quadrupole magnetic field. This uses only static fields, but evades Earnshaw's theorem by making use of the angular momentum structure provided by multiple energy levels. Six lasers are used to form a red-detuned optical molasses. But in this case, the opposing lasers in each of the three pairs should have opposite circular polarization (Fig. 24.9a). Suppose that the atom has a state with $J = m_J = 0$ ground state, $|g\rangle = |J, m_J\rangle = |0, 0\rangle$, where J is total angular momentum, and a $J = 1$ excited state. In the absence of a magnetic field, the energies of the $m_J = 0, \pm 1$ sublevels of the $J = 1$ state would be degenerate. However, here we introduce a pair of coils to produce a quadrupole magnetic field, causing a position-dependent energy level splitting between the three $J = 1$ sublevels. Considering any one of the three axes (z, say), suppose that the atom is moving toward the right, as shown in Fig. 24.9b. Let $\sigma^\pm$ label photons with circular polarizations of angular momentum $\pm \hbar$ along each axis. Then the Doppler shift moves the $\Delta m_J = -1$ transition closer to resonance for the σ^- beam. So

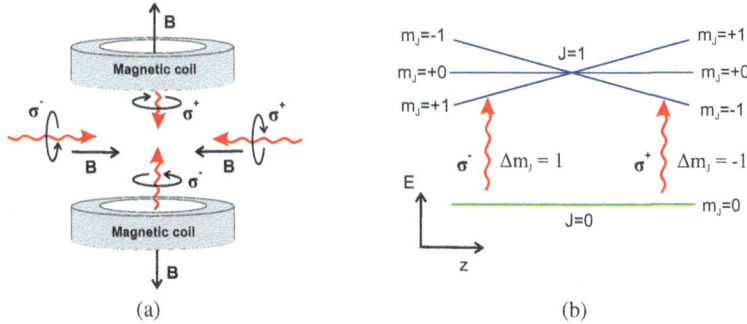

(a) (b)

Fig. 24.9 (**a**) A magneto-optical trap. Three pairs of counterpropagating lasers form an optical molasses (only two are drawn here to avoid clutter). The lasers in each pair are circularly polarized, with opposite angular momentum. The two coils provide a quadrupole magnetic field with a saddle point at the origin, similar the electric field in Fig. 24.3b. (**b**) Inside the MOT, the energy level splitting of the $J = 1$ state depends linearly on position. Selection rules enforce angular momentum conservation, so that for a moving ion, transitions involving one of the circular polarization states is suppressed, while the other is enhanced. the net result is a restoring force keeping the atom trapped near the origin

angular momentum conservation (or equivalently, the atomic selection rules) lead to higher absorption of the σ^- photons while inhibiting absorption of the right-circularly polarized σ^+ photons from the opposite direction. Thus, there is a net linear restoring force pushing the particle back toward the origin. A similar effect, with increased σ^+ absorption and reduced σ^- absorption, occurs if the particle is moving toward the left. The result is confinement along the z-axis. An identical effect leads to an equivalent confinement along the other two axes.

The first MOT trapped 10^6 sodium atoms at a temperature of 500 μK. The atoms were trapped within a region of 0.32 mm, with a containment lifetime of about 2 minutes.

24.4 Optical Lattices

Two counterpropagating laser beams will produce an interference pattern with multiple maxima, similar to the standing wave pattern on a guitar string. For example, two sinusoidal fields propagating in opposite directions along the z-direction lead to a field

$$E(z, t) = E_0 \left[\cos(\omega t - kz) + \cos(\omega t + kz) \right] = 2E_0 \cos kz \cos \omega t, \qquad (24.19)$$

which has maxima separated by distance $\frac{\lambda}{2}$, where $\lambda = \frac{2\pi}{k} = \frac{2\pi c}{\omega}$. Three perpendicular sets of such counterpropagating beams produce a three-dimensional lattice of intensity maxima, with each maximum acting as a potential well. We've seen in the previous sections that atoms feel forces attracting them to intensity maxima of red-detuned laser beams, so if a gas of bosonic atoms is released into this lattice, the atoms will collect at the lattice sites, with confining forces localizing them there. Ideally, if one atom resides at each site (mean occupation number $n_0 \sim 1$), this leads to a particle density $\frac{2}{\lambda}$, with a typical λ on the order of a micron. The potential wells in the lattice are shallow, so the atoms needed to be cooled to temperatures $\leq O(mK)$. These **optical lattices** [3, 8] can serve as analogs of solid-state crystals filled with a gas of bosonic particles in place of the fermionic electron gas in a real crystal. The lattice spacing, $\sim 10^{-7}$ m, is roughly 1000 times larger than in a typical solid, and, unlike in solids, the lattice will be entirely defect-free. Oscillations about the bottom of each well typically occur at a frequency on the order of 10 kHz.

By changing the intensity of the lasers, the depth of the confining wells can be varied, and the separation between the wells can be altered by changing the wavelength. As a result, the rate of tunneling between different lattice sites can be controlled, allowing the interactions to be turned on or off between atoms at neighboring sites.

Let g be the collision rate between atoms within each potential well and $\mathcal{J}$ be the tunneling rate between wells. Then a key parameter, which can be readily controlled by the experimenter, is the dimensionless ratio

$$\xi = \frac{g}{\mathcal{J}}. \tag{24.20}$$

As ξ is varied, a phase transition occurs at some critical value ξ_c [3, 9]. For $n_0 \sim 1$, this critical value is roughly $\xi_c \sim 35$. The two phases are:

- **Superfluid phase ($\xi < \xi_c$).** For small ξ, the tunneling rate dominates and the different lattice sites interact strongly. The system then condenses into a superfluid state, with a single coherent macroscopic quantum state. In this state, there are strong correlations between lattice sites, with a correlation function between creation and annihilation operators at sites i and j being given by

$$\langle \hat{a}_i^\dagger \hat{a}_j \rangle = n_0. \tag{24.21}$$

Notice that this is independent of the distance between the two sites.

- **Insulator phase ($\xi > \xi_c$).** For large ξ, collisions dominate over tunneling. Each atom is largely localized at a single lattice site, within a single potential well, with little tunneling occurring between sites. Even when tunneling does occur, the rapid collisions within each potential well destroy any coherence between the lattice sites. As a result, distinct sites are completely uncorrelated with each other:

$$\langle \hat{a}_i^\dagger \hat{a}_j \rangle = n_0 \delta_{ij}. \tag{24.22}$$

The transition between the superfluid and insulating phases can be seen by illuminating the atoms in the lattice with a laser as they undergo free expansion. The light scattering off the different atoms will form an interference pattern on a screen or camera. In the superfluid phase, the scattered waves are coherent with each other, producing a well-defined diffraction pattern. In the insulating phase, the scattered amplitudes vary randomly in phase (they are incoherent), and so a diffuse blur is formed, with no discernable interference pattern. Such images are shown in Fig. 24.10. From (a) to (h), the depth of the potential well is gradually increasing.

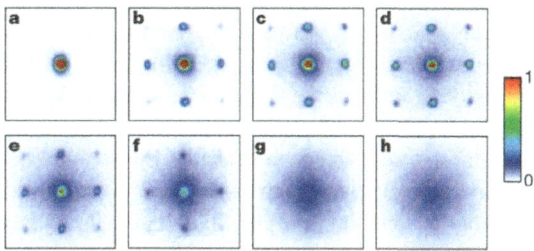

Fig. 24.10 Images of the transition from the superfluid phase (**a**)–(**f**) to the insulating phase (**g**)–(**h**) in an optical lattice. Figure reproduced with permission from M. Greiner, O. Mandel, T. Esslinger, T. W. Hänsch, I. Bloch, Nature (London) **39**, 415 (2002)

This means a decrease in tunneling rates and therefore increasing ξ. The transition between phases can be seen between images (f) and (g), where the value of ξ crosses the critical value.

In the insulating phase, the ability to control the interactions between adjacent sites, combined with the long lifetimes of the trapped states, makes optical lattices of cold atoms a prime candidate for future quantum computing platforms. Optical lattices can also be used to simulate many models used in solid-state physics, including the Hubbard model, and the phase transition between superfluid and insulating phase provides insights into the similar Mott transition between conducting and insulating phases of Mott insulators. Because strong confinement can be achieved at the lattice sites, optical lattices also play a role in nanochemistry, where each site acts as a microscopic test tube within which individual molecules can interact with each other.

24.5 Bose-Einstein Condensate

We know that it is possible for large numbers of bosons to condense into a single quantum state. This is what happens to photons in a laser and to Cooper pairs in a superconductor. In 1924, Einstein predicted that at low temperatures bosonic atoms could also undergo a phase transition so that below some critical temperature, a substantial fraction of the atoms would all be in the ground state. The atoms in this **Bose-Einstein condensate (BEC)** share a single coherent, macroscopic wavefunction, similar to the case in a superconductor.

After Einstein's initial prediction, it took nearly 70 years before the cooling and trapping technology was sufficient to produce such a condensate. The first successful production of a BEC was carried out by the research group led by E. Cornell and C. Weiman, who used a magnet-optical trap, along with an optical molasses and a combination of sisyphus and evaporative cooling (see [2] for descriptions of these cooling methods), in order to produce a BEC of rubidium atoms (^{87}Rb). The atoms were cooled to about 20 nK, and the transition took place at a critical temperature of about $T_c = 170$ nK. The transition was visible by reconstructing the momentum distribution of the atoms; at the transition to the BEC, the distribution suddenly and dramatically narrows as the atoms drop into the ground state. Not long afterward, the group of W. Ketterle observed a BEC in sodium atoms at a much higher transition temperature, $T_c = 2$ μK. Cornell, Weiman, and Ketterle shared the 2001 Nobel in physics for these experiments. Since then, BEC's have been achieved with other atomic species, including lithium and helium, as well as with bosonic quasiparticle excitations such as magnons, polaritons, and excitons.

The thermal de Broglie wavelength of a particle is the de Broglie wavelength at the point when the mean kinetic energy of the particle equals the thermal energy $\frac{1}{2}kT$. For a particle of mass M at absolute temperature T, it is given by

$$\lambda_{dB} = \frac{h}{\sqrt{2\pi MkT}} = \frac{\sqrt{2\pi}\,\hbar}{\sqrt{MkT}}. \tag{24.23}$$

Roughly speaking, the BEC forms when λ_{db} is comparable to the lattice spacing of the material. More precisely, it is shown in many statistical mechanics texts ([10, 11] for instance) that the atom density (number per volume) and critical temperature for the transition to a BEC are related by

$$n = \frac{2.6}{\lambda_{db}^3}, \quad \text{or equivalently,} \quad T_c \approx 3.3125 \frac{\hbar^2 n^{2/3}}{Mk_B}, \tag{24.24}$$

where k_B is Boltzmann's constant and the numerical factors arise from Riemann zeta functions that appear when integrating over the Bose-Einstein distribution function and the density of states.

The fraction of atoms f in the ground state rises quickly to near 100% for temperatures below T_c. For $T < T_c$, this fraction is given approximately by

$$f = 1 - \left(\frac{T}{T_c}\right)^{3/2}, \tag{24.25}$$

as shown in Fig. 24.11.

A number of applications have been proposed or demonstrated for Bose-Einstein condensates. For example, they can be used to produce coherent beams of tightly concentrated atoms, known as **atom lasers**, which are useful for producing high-precision atomic clocks and for lithographic applications such as the fabrication of integrated chips. In 1999, a research group used a BEC to slow light to the pace of a bicycle or a galloping horse ($17\frac{m}{s}$ or 38 mph) [12]. Later experiments showed that the light could actually be brought to a complete stop in the BEC and then restarted. This **slow light** phenomenon has promise for applications in optical information storage and telecommunication systems.

Fig. 24.11 The fraction of atoms in the ground state as a function of temperature. Above the critical temperature, the fraction is close to zero, but below, it is given by $1 - \left(\frac{T}{T_c}\right)^{3/2}$, rising to 100% at $T = 0$

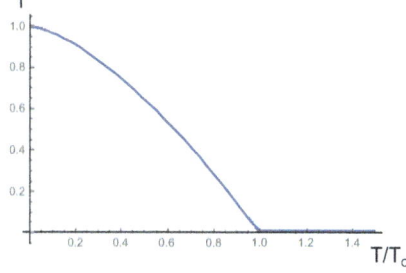

24.6   Photon-Atom Interactions and the Jaynes-Cummings Model

A useful model of atom-field interactions is given by **Jaynes-Cummings model**, introduced in 1963 by Edwin Jaynes and Fred Cummings [13]. The model is a staple of quantum optics, and numerous previous reviews are available, such as in [14–17]. It assumes that a material is made of a large number of identical two-state atoms and ignores any interactions between nearby electron pairs or adjacent atoms.

Suppose an atom in the material is interacting with an external electric field. We assume that the energy per photon $\hbar\omega$ is too low to cause transitions between any of the atomic electron energy levels except for the lowest two, so that the atom can be treated as a two-state system. Label the ground and excited states as $|g\rangle$ and $|e\rangle$, with respective energies E_0 and E_1. These can be taken as the basis for the two-level system and written in matrix form as

$$|g\rangle = \begin{pmatrix} 1 \\ 0 \end{pmatrix}, \qquad |e\rangle = \begin{pmatrix} 0 \\ 1 \end{pmatrix}. \tag{24.26}$$

The resonant frequency of the atom is given by $\hbar\omega_0 = E_1 - E_0$. The detuning of the field from resonance, $\delta = \omega - \omega_0$, is assumed small. Notice that the system will be very similar to the NMR system of the last chapter. The main difference is that here, the electromagnetic field is part of the system, with its own dynamics, while in the NMR system the magnetic field was something imposed from the outside and controlled by the experimenter.

Consider the field in isolation first. We know from Chap. 5 that the field at each point in space can be written in the form of a simple harmonic oscillator. The field Hamiltonian can be written

$$\hat{H}_{em} = \left(\hat{a}^\dagger \hat{a} + \frac{1}{2} \right) \hbar\omega = \hbar\omega \left(\hat{N} + \frac{1}{2} \right), \tag{24.27}$$

where $\hat{a}^\dagger$ and $\hat{a}$ are the photon creation and annihilation operators. The number operator $\hat{N} = \hat{a}^\dagger \hat{a}$ counts the number of photons present. If we adjust the origin of our energy scale for the field appropriately, then the zero-point energy will be shifted to zero, so that we can take

$$\hat{H}_{em} = \hbar\omega\hat{N}, \tag{24.28}$$

This is equivalent to normal ordering of the creation and annihilation operators in the Hamiltonian. (*Normal ordering* means rearranging a product of creation operators to put all of the annihilation operators to the right of the creation operators; this removes any contribution of the vacuum to the quantity being calculated. It is often used in quantum field theory to remove an infinite vacuum contribution to a system's energy.)

Now consider a single two-level atom in isolation. The zero of energy for the atom can also be adjusted to be half-way between the two relevant energy levels E_0 and E_1, so that those levels become $E_0 = -\frac{\hbar\omega_0}{2}$ and $E_1 = +\frac{\hbar\omega_0}{2}$. The atomic Hamiltonian may then be written in the form

$$\hat{H}_{atom} = \frac{\hbar\omega_0}{2}\hat{\sigma}_z. \tag{24.29}$$

For a standard dipole interaction between the field and the atom, the interaction Hamiltonian is $\hat{H}_{int} = -\hat{p} \cdot \hat{E}(r, t)$, where $\hat{p}$ is the electric dipole operator. Since $\hat{p}$ can only connect states of opposite spatial parity, it has no diagonal terms ($|g\rangle\langle g|$ and $|e\rangle\langle e|$). Therefore, making use of the completeness relation, it can be written as

$$\hat{p} = \sum_{i,j=\{e,g\}} |i\rangle\langle i|\hat{p}|j\rangle\langle j| \tag{24.30}$$

$$= |e\rangle\langle e|\hat{p}|g\rangle\langle g| + |g\rangle\langle g|\hat{p}|e\rangle\langle e| \tag{24.31}$$

$$= p\left(\hat{\sigma}_+ + \hat{\sigma}_-\right)\hat{e}, \tag{24.32}$$

where $\hat{e}$ is the direction of $\hat{p}$ and $p = \langle e|\hat{p}|g\rangle = \langle g|\hat{p}|e\rangle$ is the matrix element of $\hat{p}$ between the ground and excited states. Here $\hat{\sigma}_\pm$ are raising and lowering operators for the atomic electron states

$$\hat{\sigma}_- = |g\rangle\langle e| = \begin{pmatrix} 0 & 1 \\ 0 & 0 \end{pmatrix} \tag{24.33}$$

$$\hat{\sigma}_+ = |e\rangle\langle g| = \begin{pmatrix} 0 & 0 \\ 1 & 0 \end{pmatrix}, \tag{24.34}$$

or more compactly

$$\sigma_\pm = \frac{1}{2}\left(\hat{\sigma}_x \pm i\hat{\sigma}_y\right). \tag{24.35}$$

The quantized electric field is of the form (Chap. 5)

$$\hat{E}(r) = \sum_k \kappa_k \left(\hat{a}_k^\dagger e^{ikr} + \hat{a}_k e^{-ikr}\right) \tag{24.36}$$

where $\kappa_k = \sqrt{\frac{\hbar\omega_k}{2\epsilon_0 V}}\epsilon_k$, and we assume the field is polarized along the same direction as the dipole moment. The interaction Hamiltonian then becomes

$$\hat{H}_{int} = \sum_k g_k \left(\hat{a}_k^\dagger e^{ikr} + \hat{a}_k e^{-ikr} \right) \left(\hat{\sigma}_+ + \hat{\sigma}_- \right), \tag{24.37}$$

where $g_k = p\kappa_k$. Expanding out the product leads to four terms

$$\hat{H}_{int} = \sum_k g_k \left(\hat{a}_k^\dagger \hat{\sigma}_+ e^{ikr} + \hat{a}_k^\dagger \hat{\sigma}_- e^{ikr} + \hat{a}_k \hat{\sigma}_+ e^{-ikr} + \hat{a}_k \hat{\sigma}_- e^{-ikr} \right). \tag{24.38}$$

The second and third terms have a clear and sensible meaning: the second term corresponds to an electron dropping from the excited state to the ground state as a new photon is simultaneously created, while the third term corresponds to a photon being absorbed while the electron jumps up to the excited state. The first and last terms, however, correspond to processes that we wouldn't expect to see: the first describes an electron emitting a photon while being excited to a higher state, and the last one describes a photon being emitted while the electron drops down to the ground state. Both of these latter two processes violate energy conservation. So, we can feel confident about dropping them from the Hamiltonian. (An alternative justification for dropping these terms is given in Box 24.1, based on the rotating wave approximation.) So the interaction Hamiltonian reduces to

$$\hat{H}_{int} = \sum_k g_k \left(\hat{a}_k^\dagger \hat{\sigma}_- e^{ikr} + \hat{a}_k \hat{\sigma}_+ e^{-ikr} \right). \tag{24.39}$$

Taking the atom to be at $r = 0$ and assuming there is just a single field mode of wavevector k_0 present further reduces H_{int} to

$$\hat{H}_{int} = g \left(\hat{a}_{k_0}^\dagger \hat{\sigma}_- + \hat{a}_{k_0} \hat{\sigma}_+ \right), \tag{24.40}$$

where $g = g_{k_0}$.

Putting the pieces together, we now have the Hamiltonian for the **Jaynes-Cummings model**:

$$\hat{H} = \hat{H}_{em} + \hat{H}_{atom} + \hat{H}_{int} = \hbar\omega \left(\hat{N} + \frac{1}{2} \right) + \frac{\hbar\omega_0}{2} \hat{Z} + \hbar g \left(\hat{a}_{k_0}^\dagger \hat{\sigma}_- + \hat{a}_{k_0} \hat{\sigma}_+ \right). \tag{24.41}$$

To emphasize again: $\hat{a}^\dagger$ and $\hat{a}$ are raising and lowering operators for the electromagnetic field (creating and annihilating photons), while $\hat{\sigma}_\pm$ are raising and lowering operators for the atom (taking the electrons up and down between the ground and excited state). Again, note that this Hamiltonian is identical to that of the NMR system in Sect. 23.4, except for the inclusion of the term $\hat{H}_{em}$ describing the electromagnetic field.

Box 24.1 Rotating Wave Approximation

An alternative view on why two of the terms can be dropped between Eq. 24.38 and Eq. 24.39 comes from applying the **rotating wave approximation (RWA)**. Consider the interaction picture versions of $\hat{a}$, $\hat{a}^\dagger$, and $\hat{\sigma}_\pm$. In the interaction picture, free particle evolution of a system is swept into the operators, rather than the states, by replacing each time-independent operator O by a time-dependent version

$$\tilde{O}(t) = e^{\frac{i}{\hbar}\hat{H}_0 t} O e^{\frac{i}{\hbar}\hat{H}_0 t}. \tag{24.42}$$

Standard manipulations (Problem 5) lead to the results that

$$\tilde{a}(t) = \hat{a}\, e^{-i\omega t} \qquad \tilde{\sigma}_+(t) = \hat{\sigma}_+ e^{-i\omega_0 t}$$

$$\tilde{a}^\dagger(t) = \hat{a}^\dagger e^{i\omega t} \qquad \tilde{\sigma}_-(t) = \hat{\sigma}_- e^{i\omega_0 t}.$$

Apply this to the interaction picture version of the operator sums and products in Eq. 24.38 (with $r = 0$ for simplicity):

$$\tilde{a}_k^\dagger \tilde{\sigma}_+ + \tilde{a}_k^\dagger \tilde{\sigma}_- + \tilde{a}_k \tilde{\sigma}_+ + \tilde{a}_k \tilde{\sigma}_-$$

$$= \hat{a}_k^\dagger \hat{\sigma}_+ e^{i(\omega+\omega_0)t} + \hat{a}_k^\dagger \hat{\sigma}_- e^{i(\omega-\omega_0)t} + \hat{a}_k \hat{\sigma}_+ e^{i(\omega_0-\omega)t} + \hat{a}_k \hat{\sigma}_- e^{-i(\omega+\omega_0)t}.$$

The factors $e^{\pm i(\omega\pm\omega_0)t}$ cause the terms that they multiply to rotate around the origin in the complex plane. But the terms in which the frequencies add together, $e^{\pm i(\omega+\omega_0)t}$ rotate much faster (they have a higher effective frequency) than the terms in which the frequencies are subtracted from each other, $e^{\pm i(\omega-\omega_0)t} = e^{\pm t\delta}$. Because of the more rapid rotations of the former terms, they circle the origin multiple times before the latter terms have rotated much; as a result, the former terms average to zero (or very close to it) over times in which the latter have not varied much. This is a general phenomenon: rapid oscillations that are superimposed onto much slower motions lead to only small average deviations over the time scale of the slower variations (so-called **secular motions**) and so can often be ignored to leading order.

As a result, the first and last terms in Eq. 24.43 are small compared to the two in the middle, and consequently, they can safely be dropped. The rotating wave approximation is widely used in many applications where two frequencies ω and ω_0 that are not too different from each other both appear in the same system. The RWA is often used in analyzing NMR systems, for example, and will come up again in the next chapter in the context of atoms in resonant cavities.

The states of the coupled atom-field system can be written in the form

$$|n, j\rangle = |n\rangle \otimes |j\rangle, \tag{24.43}$$

where n counts the number of photons present and $j \in \{e, g\}$ represents the state of the atom. In the absence of interactions, these states would be eigenstates (steady states) of the Hamiltonian. But when the interaction term is turned on, it leads to electronic transitions accompanied by energy-conserving photon emissions or absorptions.

In general, the full state of the system is likely to be some superposition of the basis $|n, j\rangle$ states. So denote the full state at time $t = 0$ by $|\psi(0)\rangle$. The goal then is to find the state at some later time t. In the interaction picture (Sect. 23.4 and Box 24.1),

$$|\tilde{\psi}(t)\rangle = e^{\frac{i}{\hbar}\hat{H}_0 t}|\psi(0)\rangle, \tag{24.44}$$

where the tilde denotes interaction picture and $\hat{H}_0$ is the free Hamiltonian obtained by dropping the interaction terms from the Jaynes-Cummings Hamiltonian. The interaction Hamiltonian becomes time-dependent in the interaction picture

$$\tilde{H}_{int}(t) = e^{\frac{i}{\hbar}\hat{H}_0 t}\hat{H}_{int}e^{-\frac{i}{\hbar}\hat{H}_0 t} = g\hbar\left(\hat{a}^\dagger\hat{\sigma}_- e^{+it\delta} + \hat{a}\hat{\sigma}_+ e^{-it\delta}\right), \tag{24.45}$$

where δ is the detuning. The Schrödinger equation then has the form

$$i\hbar\frac{\partial}{\partial t}|\tilde{\psi}(t)\rangle = \tilde{H}_{int}(t)|\tilde{\psi}(t)\rangle. \tag{24.46}$$

An arbitrary state can be expanded in terms of the basis states $|n, j\rangle$, so write

$$|\tilde{\psi}(t)\rangle = \sum_n \left(c_{ng}(t)|n, g\rangle + c_{ne}(t)|n, e\rangle\right). \tag{24.47}$$

Substituting this expansion into the Schrödinger equation leads to two coupled first-order differential equations for the expansion coefficients

$$\frac{dc_{n,e}}{dt} = -ig\sqrt{n+1}\, c_{n+1,g}(t)\, e^{-it\delta} \tag{24.48}$$

$$\frac{dc_{n,g}}{dt} = -ig\sqrt{n}\, c_{n-1,e}(t)\, e^{it\delta} \tag{24.49}$$

These equations can be readily solved (see Problem 4). The general solution is

$$c_{ng}(t) = c_{ng}(0)\cos(g\sqrt{n}t) + \frac{1}{g\sqrt{n}}\frac{dc_{ng}(0)}{dt}\sin(g\sqrt{n}t) \tag{24.50}$$

$$c_{ne}(t) = c_{ne}(0)\cos(g\sqrt{n+1}t) + \frac{1}{g\sqrt{n+1}}\frac{dc_{ne}(0)}{dt}\sin(g\sqrt{n+1}t). \quad (24.51)$$

The probabilities of finding the electron in the ground state and the excited state at time t are, respectively,

$$P_g(t) = \sum_n \left| \langle n, g | \tilde{\psi} \rangle \right|^2 \quad (24.52)$$

$$P_e(t) = \sum_n \left| \langle n, e | \tilde{\psi} \rangle \right|^2. \quad (24.53)$$

So, for example, if an electron is initially in the excited state ($c_{ne}(0) = 1$, $c_{ng}(0) = 0$, with both derivatives initially vanishing), the result is a Rabi oscillation between the ground and excited state

$$P_e(t) = \sum_n \cos^2(g\sqrt{n+1}t), \qquad P_g(t) = \sum_n \sin^2(g\sqrt{n+1}t). \quad (24.54)$$

Notice that even if there is no light initially present (only the c_{1e} nonzero), the probability of a photon being present at later times is nonzero, indicating the presence of spontaneous emission: light spontaneously appears in the material if any electrons are in an excited state and then is repeatedly absorbed and emitted as the light passes through the material. This is a **vacuum Rabi oscillation**.

An interesting effect occurs if the initial state of the field is a coherent state with average photon number $N = |\alpha|^2$

$$|\psi_{em}(0)\rangle = \sum_{n=0}^{\infty} b_n |n\rangle, \quad \text{with} \quad b_n = \frac{\alpha^n}{\sqrt{n!}}e^{-|\alpha|^2/2}, \quad (24.55)$$

and the atom is in an initial superposition state

$$|\psi_{atom}(0)\rangle = c_g |g\rangle + c_e |e\rangle. \quad (24.56)$$

Then the solution is

$$c_{ne}(0)(t) = c_e b_n \cos(g\sqrt{n+1}t) - ic_g b_{n+1}\cos(g\sqrt{n+1}t) \quad (24.57)$$

$$c_{ng}(t) = ic_e b_{n-1}\sin(g\sqrt{n}t) + c_g b_n \cos(g\sqrt{n}t). \quad (24.58)$$

This leads to a decay and resurgence of the light. Defining the population inversion

$$W(t) = P_e(t) - P_g(t) = e^{-N}\sum_{n=0}^{\infty}\frac{N^n}{\sqrt{n!}}\cos(2g\sqrt{n+1}t), \quad (24.59)$$

Fig. 24.12 Decay and recurrence of the population inversion for material exposed to an incident coherent state. The coherent state amplitude used for the plot was $\alpha = \sqrt{15}$

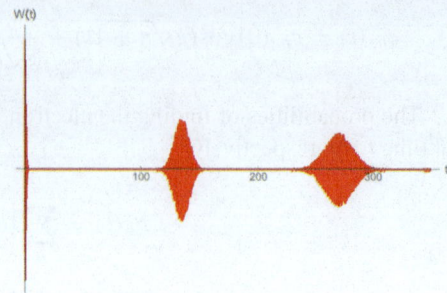

it is seen (Fig. 24.12) that the inversion decays rapidly, seems to have settled down to a value of zero, and then suddenly blows back up to a large amplitude. The cycle of decay and resurrection then repeats periodically with decreasing amplitude at each resurgence.

While the $|n, e\rangle$ and $|n, g\rangle$ states are eigenstates of the free particle system, the reader can easily verify that the eigenstates and eigenvalues of the fully interacting Jaynes-Cummings Hamiltonian are the states

$$|\chi_+\rangle = \cos\frac{\theta}{2}|n, e\rangle + i\sin\frac{\theta}{2}|n+1, g\rangle, \tag{24.60}$$

$$|\chi_-\rangle = \sin\frac{\theta}{2}|n, e\rangle - i\cos\frac{\theta}{2}|n+1, g\rangle, \tag{24.61}$$

$$E_\pm = \left(n + \frac{1}{2}\right)\hbar\omega \pm g\sqrt{n+1}, \tag{24.62}$$

where $\tan\theta \equiv g\sqrt{n+1}/\delta$.

24.7 Trapped Ion Qubits

Recall that we are assuming that only the two lowest energy levels of the atom are playing a role in interactions. So the states of the atom look as in Fig. 24.13a. There are two internal electronic states $|e\rangle$ and $|g\rangle$, often referred to spin states due to their two-level form, and external harmonic-oscillator-like modes from vibrations in space. The full state is then labeled as $|n, m\rangle$, where $n = 0, 1$ (or $n = g, e$) is the internal electron state and m labels the external oscillator state.

Placing the ions in a trap, the internal "spin" states are used as qubits, with the atoms initially in the vibrational ground state, $m = 0$. Using the Jaynes-Cummings model of Sect. 24.6, the states can be manipulated with external fields. For example, tuning the external field to the resonance frequency ω_0 allows internal state rotations on the Bloch sphere to be implemented

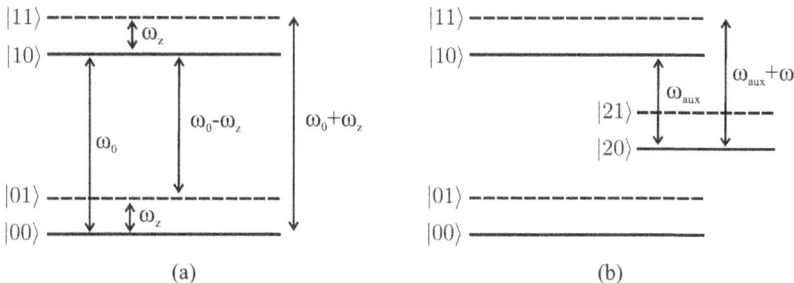

Fig. 24.13 (a) The level structure of ionic qubits. The transition frequency between internal spin states is ω_0, while ω_z is the frequency for transitions between vibrational modes. The solid lines are energy levels for states that are in the vibrational ground state, while dashed lines indicate the first excited vibrational mode. The inclusion of the vibrational modes, along with an external applied field, allows control over transitions between internal modes. (b) Adding transitions to an additional vibrational mode $|2\rangle$, ω_{aux} is the frequency relating vibrational modes $|1\rangle$ and $|2\rangle$

$$R_x(\theta) = e^{-i\theta S_x}, \qquad R_y(\theta) = e^{-i\theta S_y}, \tag{24.63}$$

via the relation

$$R_{\hat{n}}(\theta) = e^{i\theta \hat{n}\cdot\sigma/2}. \tag{24.64}$$

Recalling that any z-rotation can be formed by a composition of x- and y-rotations, then any single-qubit unitary gate can be implemented: any 2×2 unitary transformation can be written as

$$\hat{U} = e^{i\alpha}\, R_z(\beta) R_y(\gamma) R_x(\delta) \tag{24.65}$$

$$= \begin{pmatrix} e^{i\alpha - \frac{\beta}{2} - \frac{\delta}{2}} \cos\frac{\gamma}{2} & -e^{i\alpha - \frac{\beta}{2} + \frac{\delta}{2}} \sin\frac{\gamma}{2} \\ e^{i\alpha + \frac{\beta}{2} - \frac{\delta}{2}} \sin\frac{\gamma}{2} & e^{i\alpha + \frac{\beta}{2} + \frac{\delta}{2}} \cos\frac{\gamma}{2} \end{pmatrix}, \tag{24.66}$$

for some real parameters $\alpha, \beta, \gamma, \delta$.

Implementing two-qubit gates requires extra structure. The simplest way to form a CZ gate, for example, involves the inclusion of a third, auxiliary internal state, $|2\rangle$ (Fig. 24.13b). Let ω_{aux} be the transition frequency between the $|10\rangle$ and $|20\rangle$ levels. Then tuning the laser to $\omega = \omega_{aux} + \omega_z$ causes transitions between the $|1,0\rangle$ and $|2,0\rangle$ levels. A pulse of duration 2π implements a $R_x(2\pi)$ rotation, which leaves all states unchanged except for taking $|11\rangle$ to $-|11\rangle$. Aside from an extra phase that can easily be corrected for, this is a CZ gate.

On the other hand, tuning to $\omega = \omega_{aux} - \omega_z$ and applying a pulse of duration π implements a gate $SWAP'$ which is simply a SWAP gate, up to a phase in one term:

$$SWAP' = \begin{pmatrix} 1 & 0 & 0 & 0 \\ 0 & 0 & 1 & 0 \\ 0 & -1 & 0 & 0 \\ 0 & 0 & 0 & 1 \end{pmatrix}. \tag{24.67}$$

This swaps $|01\rangle$ and $|10\rangle$, leaving other states unchanged.

The reader can then check (Problem 7) that the following sequence leads to a CNOT gate

$$CNOT = H_2 \circ SWAP' \circ CZ \circ SWAP' \circ H_2, \tag{24.68}$$

where H is a Hadamard operation. The implementation of a $CNOT$ gate, together with the single-qubit gates, allows the construction of a universal set of quantum gates.

24.8   Neutral Atom Qubits

Rydberg atoms are atoms in which a valence electron is excited to very high principle quantum numbers (see Box 24.1). Neutral Rydberg atoms form another potential platform for quantum computing. In 2023, a Rydberg neutral-atom computer was demonstrated to carry out an algorithm using 48 logical qubits.

The atoms used are often alkali atoms such as cesium or rubidium, which have a single s-orbital electron in their outer shell, but strontium and ytterbium have also been used. The atoms are cooled to micro-Kelvin temperatures in a vacuum chamber and stored in magneto-optical traps, with either the principal number of the state or the spin used to encode information. Using optical tweezers (see Box 5.2), the atoms can be arranged into spatial arrays that can be rearranged in real time. Interactions can be induced between two atoms by bringing them close together and taking advantage of the fact that the interactions between Rydberg atoms become strong at short distances.

Here, we briefly outline one example of a Rydberg-based neutral atom gate; see [18] for more detail. The levels are as shown in Fig. 24.14: information is stored in the two lower states, $|0\rangle$ and $|1\rangle$, with a Rydberg state $|r\rangle$, $r \gg 1$, used for control. When two such atoms are brought together, external laser pulses can be used to carry out controlled gate operations. For example, consider the **Jaksch sequence** of operations or a pair of atoms:

- First a π pulse is applied at frequency Ω to the control atom. If the control state is initially $|1\rangle$, then this takes the atom $|1\rangle \rightarrow |r\rangle$. An initial state $|0\rangle$ is unaffected.
- Then a 2π pulse is applied to the target atom. This takes the target through the sequence $|1\rangle \rightarrow |r\rangle \rightarrow |1\rangle$. An initial state $|0\rangle$ is again unaffected.
- Finally, a second π pulse takes the control atom back to its initial state.

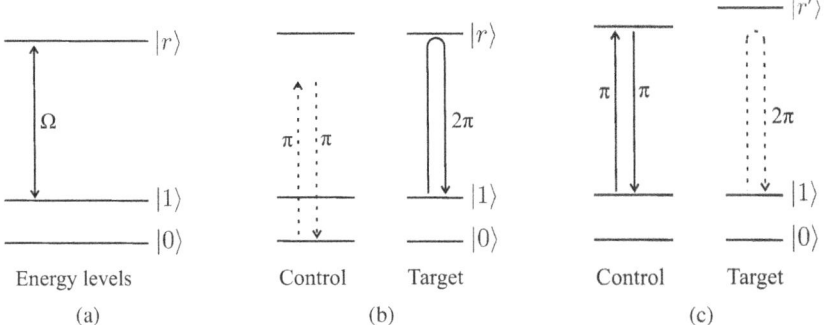

Fig. 24.14 (**a**) Relevant energy levels for neutral atom computing: the ground and first excited state, $|0\rangle$ and $|1\rangle$, plus a highly excited Rydberg state, $|r\rangle$. The frequency separation between $|1\rangle$ and $|r\rangle$ is Ω. (**b**) Two atoms are now brought together. When the control atom is in state $|0\rangle$, the Jaksch sequence causes a phase shift of π in the target qubit. (**c**) When the control atom is in state $|0\rangle$, the first pulse takes the control into the Rydberg state. This causes a Rydberg blockade in the target atom: the original Rydberg state $|r\rangle$ is shifted to $|r'\rangle$ and is now inaccessible via the 2π pulse. The target now no longer accumulates any shift

The result of this sequence depends on the state of the control atom. Let $|n, m\rangle$ denote the state with quantum number n for the control atom and m for the target. The state $|00\rangle$ is decoupled from the interaction Hamiltonian, so nothing happens. The two states $|01\rangle$ and $|10\rangle$ both simply pick up net π phase shifts due to the 2π pulse acting on the $|1\rangle$ state. But when the state is $|11\rangle$, the first pulse puts the control into the $|r\rangle$ state, so that a Rydberg blockade (Box 24.2) prevents the target atom from making a transition during the 2π pulse; in this case, the control gains a π phase and the target gains none. The net result is the unitary transition matrix that is equivalent to a controlled Z gate, up to some phase redefinitions. As was done in the ion case, this can then be converted into a $CNOT$ gate. Together with the single-qubit gates, this again allows the construction of a universal set of quantum gates.

Neutral atom quantum computers have several advantages:

- Single-qubit gates can have quantum fidelities of over 99% and have long coherence times.
- They can operate at or near room temperature, removing the need for complicated cryogenic equipment required for superconducting systems.
- The systems are highly scalable: many atoms can be stored in two-dimensional arrays.
- The ability to move the atoms around with optical tweezers, without doing damage to their quantum state, also means that interactions between specific qubits can be switched on or off simply by moving the atoms closer together or farther apart.

However, the last point also leads to one disadvantage: neutral-atom computers tend to be slow due to the time it takes to physically shuttle the atoms between locations.

Box 24.2 Rydberg Atoms

Rydberg atoms are atoms with very high principle quantum number n. Typical n values could be anywhere from a few dozen up to 1000. The core electrons shield much of the nuclear charge, so that the outer electron sees what looks like the potential from a hydrogen nucleus. As a result, the outermost electron in such an atom obeys the Bohr model to a high degree of accuracy, with the binding energies scale as $E_{bind} \sim \frac{1}{n^2}$ and the atomic radius is $r \sim n^2$. The valence electron is therefore only loosely bound and lives in an orbital far from the nucleus. As such, it is highly responsive to the atom's surroundings. For instance, the energy levels of Rydberg atoms are sensitive to external electric fields, displaying an extremely large Stark effect that allows them to serve as very sensitive field sensors. The exaggerated properties of high-n atoms have made them useful in a number of fields. For example, they have served as a useful lab for studying quantum chaos and many-body systems, and they have been proposed as robust light-atom quantum interfaces. In astrophysics, Rydberg atoms are known to be responsible for some of the radio recombination lines seen in interstellar space and are known to be present in astrophysical plasmas.

Among other unusual properties, Rydberg atoms with high angular momentum quantum numbers tend to have very long lifetimes, due to their low overlap with other orbitals. In addition, collections of Rydberg atoms can condense into long-lived clusters of up to roughly 100 atoms, called **Rydberg matter** [19]. Rydberg matter has been found in interstellar space clouds and the upper atmosphere of some planets. These clusters form a phase of matter high stability, in which the interatomic distance is strongly dependent on the atomic quantum number and which can display properties of conventional liquids, solids, or gases under different circumstances.

Consider two atoms in Rydberg states of quantum number n, separated by a distance r. At short distances, they experience strong dipole-dipole interactions of the form $V = \frac{(n^2 e a_0)^2}{r^3}$, where a_0 is the Bohr radius, while at larger distances their interaction has a van der Waals form, $\sim \frac{1}{r^6}$. Thus, when the quantum numbers are large and the distance are small, the interactions become very strong, scaling $\sim n^4$. For an atom with $n = 100$, the cross-over from van der Waals to dipole-dipole form occurs at around $r \sim 10 \ \mu m$; at this distance the interaction between two Rydberg atoms is about 10^{12} times stronger than the interaction between two ground state ($n = 1$) atoms [20].

(continued)

Box 24.2 (continued)
But at larger distances, the interactions become weak, allowing the atoms to be formed into non-interacting arrays.

Much of the usefulness of Rydberg atoms for quantum information processing is due to the existence of the **Rydberg blockade**. Due to the strong interactions between two Rydberg atoms in close proximity, the energy levels of each atom will experience strong shifts, altering the resonance frequencies. So, suppose a Rydberg atom is in the presence of a radiation field that would normally stimulate a transition to another state. Bringing other Rydberg atoms close to it will alter its energy spectrum and prevent the excitation from occurring. The transition can thus be switched on or off by moving neighboring atoms around; use of this blockade is one method to implement controlled quantum gates.

For more on Rydberg atoms see [20, 21], and for their uses in quantum information processing see [18].

Problems

1. (a) Show that the kinetic energy K in an optical molasses obeys

$$\frac{dK}{dt} = -\frac{K}{\tau_{damp}}$$

with damping time $\tau_{damp} = \frac{m}{2\alpha}$.

   (b) For calcium, the Doppler cooling limit is $T_{min} \approx 240 \, \mu\text{K}$. Find the damping coefficient α and the damping time τ_{damp}. Estimate the most probable atomic speed at the cooling limit.

2. (a) Show that the quadrupole potential $\phi(x, y) = a + b(x^2 - y^2)$ satisfies the Laplace equation $\nabla^2 \phi = 0$ in free space.

   (b) In the linear Paul trap of Fig. 24.4, take the boundary conditions at the electrodes to be $\phi = \phi_0 \pm \frac{V_0}{2} \cos \Omega t$, where the plus sign goes with the electrodes at $x = \pm r_0$ and the minus sign with those at $y = \pm r_0$. Show that the solution of part (a) can satisfy these boundary conditions, and find the required integration constants a and b.

   (c) Find the corresponding electric field in the plane.

   (d) Write down the equations of motion for the x direction, and verify that solutions of the form $x(t) = x_0 \cos\left(c_1 \frac{\Omega t}{2}\right)(1 + c_2 \cos(\Omega t))$ exist. The constants c_1 and c_2 are determined by the initial position and momentum of the ion. (The solutions in the y direction are similar.)

3. Fill in the details to show that the Schrödinger equation Eq. 24.46 reduces to the coupled system of Eqs. 24.48 and 24.49.

4. Explicitly solve Eqs. 24.48 and 24.49, and verify that they are given by 24.50 and 24.51. (Hint: it may help to refer back to Sect. 23.1.)

5. (a) Using the free part of the Jaynes-Cummings Hamiltonian, $\hat{H}_0 = -\frac{\hbar}{2}\omega_0\hat{\sigma}_z + \hbar\omega\hat{a}^\dagger\hat{a}$, find the interaction picture operator $\tilde{a} = e^{\frac{i}{\hbar}\hat{H}_0 t}\hat{a}e^{-\frac{i}{\hbar}\hat{H}_0 t}$ and its conjugate, $\tilde{a}^\dagger$.

   (b) In a similar manner, find the interaction picture operators $\tilde{\sigma}_\pm$. Verify that the results of (a) and (b) match the results given in the text.

6. For an ion in a Penning trap, write down the equation of motion along z-axis for a fixed V_0, and show that the frequency of oscillations along this direction is given by Eq. 24.13. Find the maximum speed of the ion and the amplitude of the oscillation. If the distance to the endcap electrode is z_c, how large can the trapped mass be in order to avoid colliding with the endcap? Given that typically, the ring radius is $r_0 = \sqrt{2}z_c$ what is the radius in terms of this maximum mass?

7. (a) Show that the Hadamard gate H can be implemented using a product of rotations about the x and y axes.

   (b) Let $|mn\rangle \otimes |p\rangle$ denote a two-atom state, with m and n representing the internal "spin" modes of the two atoms and p denoting the common center of mass vibrational mode. By starting with a generic state of the form

   $$\left(a|00\rangle + b|01\rangle + c|10\rangle + d|11\rangle\right) \otimes |0\rangle$$

   and applying the sequence of operations on the right of Eq. 24.68, verify that the identity of Eq. 24.68 is correct.

   (c) Show that the $CNOT$ gate can also be implemented in the form

   $$CNOT = P \circ SWAP_2' \circ CZ \circ SWAP_2'.$$

   Here, the subscript 2 indicates that the second spin mode is being swapped with the vibrational mode, and P is a redefinition of the phase of the second atom's excited vibrational state, $P|m1\rangle \rightarrow = -|m1\rangle$.

8. Suppose a lithium atom in an initial Rydberg state $n = n_i$ is ionized by incident electromagnetic radiation. Using the Rydberg formula of Chap. 4 find the photon frequency required for $n_i = 50$ and $n_i = 100$. What part of the spectrum are these in?

9. The Jaynes-Cummings model Hamiltonian, Eq. 24.41, can be written in matrix form. This matrix is block diagonal, with each 2×2 block representing the states $|g, n\rangle$ and $|e, n-1\rangle$ for a particular value of n. (The exception is the $n = 0$ entry, which is a 1×1 block of its own.)

   (a) Show that the form of the nth block for $n > 1$ is

$$
\begin{pmatrix} n\omega & g\sqrt{n} \\ g\sqrt{n} & (n-1)\omega + (\omega + \delta) \end{pmatrix}.
$$

(b) Diagonalize the block of part (a) to show that the energies of this submatrix are $E_{\pm,n} = \left((n-1)\omega + \omega_0 \pm \sqrt{4ng^2 + \delta^2} \right)$.

References

1. J. Kitchin, Appl. Phys. Revs. **5**, 031302 (2018)
2. C.J. Foot, *Atomic Physics* (Oxford University Press, Oxford, 2005)
3. S. Haroche, J.M. Raimond, *Exploring the Quantum: Atoms, Cavities, and Photons* (Oxford University Press, Oxford, 2006)
4. D.A. Steck, *Quantum and Atom Optics*, available online at http://steck.us/teaching (revision 0.16.1, 16 June 2024)
5. M. Weissbluth, *Photon-Atom Interactions* (Academic Press, New York, 1989)
6. P.K. Gosh, *Ion Traps* (Oxford University Press, Oxford, 1995)
7. G. Schneider, et al., Science **358**, 1081 (2017)
8. L. Guidoni, P. Verkerk, J. Opt. B: Quant. Semiclass. Opt. **1**, R23 (1999)
9. M.P.A. Fisher, P.B. Weichman, G. Grinstein, D.S. Fisher, Phys. Rev. B **40**, 546 (1989)
10. J.P. Sethna, *Statistical Mechanics: Entropy, Order Parameters and Complexity* (Oxford University Press, Oxford, 2006)
11. H. Gould, J. Tobochnik, *Statistical and Thermal Physics: With Computer Applications* (Princeton University Press, Princeton, 2010)
12. L.V. Hau, S.E. Harris, Z. Dutton, C.H. Behrooz, Nature **397**, 594 (1999)
13. E.T. Jaynes, F.W. Cummings, Proc. IEEE **51**, 89 (1963)
14. B.W. Shore, P.L. Knight, J. Mod. Opt. **40**, 1195 (1993)
15. M. Fox, *Quantum optics: An Introduction* (Oxford University Press, Oxford, 2006)
16. C. Gerry, P. Knight, *Introductory Quantum Optics* (Cambridge University Press, Cambridge, 2004)
17. S. Haroche, J.M. Raimond, *Exploring the Quantum: Atoms, Cavities, and Photons* (Oxford University Press, Oxford, 2006)
18. M. Saffman, T.G. Walker, M. Klaus, Rev. Mod. Phys. **82**, 2313 (2010)
19. T. Aasen et al., J. Cluster Sci. **33**, 839 (2022)
20. J.H. Choi, B. Knuffman, T.C. Leibisch, A. Reinhard, G. Raithel, Adv. At. Mol. Opt. Phys **54**, 131 (2007)
21. T.F. Gallagher, *Rydberg Atoms* (Cambridge University Press, Cambridge, 1994)

Chapter 25
Resonant Cavities and Cavity Quantum Electrodynamics

Cavity quantum electrodynamics (CQED), which uses isolated atoms inside resonant electromagnetic cavities, offers another promising physical platform for implementing and manipulating qubits. The cavity itself acts as a spectral filter, only allowing certain frequencies of electromagnetic waves to be present; other frequencies destructively interfere with themselves and rapidly decay away. This allows only single pairs of atomic energy levels, separated by the allowed optical frequency, to participate in interactions. At the same time, photons bouncing back and forth in the cavities have many opportunities to interact with the atoms, effectively amplifying otherwise weak effects. The presence of the mirrors bounding the cavity can also be thought of as altering the vacuum states with which the atom interacts, which will in turn affect properties like atomic energy levels and decay rates of excited states, thus offering additional control over the atom and its interactions. Over the next few sections, we'll look first at the electric field inside a resonant cavity, then we'll add an atom into the cavity. In the last section, we look at implementing quantum information processes using the atom-cavity system.

25.1 Resonant Cavities

In Sect. 5.7, it was seen that a piece of glass placed in a laser cavity acts to form a spectral filter or Fabry-Perot etalon. More generally, two parallel plane mirrors form a **Fabry-Perot cavity**, one of the simplest forms of resonant cavity. Let the complex electric field of a monochromatic optical field in the cavity be

$$E(r, t) = E(r)e^{-i\omega t}. \tag{25.1}$$

Restricting attention to the vicinity of the z-axis, we can ignore the x and y coordinates and seek a solution of the one-dimensional Helmholtz equation along

© The Author(s), under exclusive license to Springer Nature Switzerland AG 2025
D. S. Simon, *Introduction to Quantum Science and Technology*, Undergraduate
Texts in Physics, https://doi.org/10.1007/978-3-031-81315-3_25

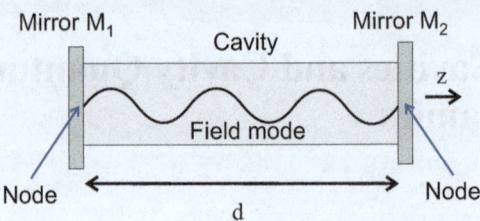

Fig. 25.1 A simple resonant cavity. The field modes must have nodes at the mirrors, which forces the wavelength to satisfy the quantization condition $m\frac{\lambda}{2} = d$, with integer m. This leads to the condition $k_m = \frac{2\pi}{\lambda_m} = \frac{m\pi}{d}$. Modes that do not satisfy this condition interfere destructively with themselves as they move back and forth through the cavity, causing the non-resonant modes to rapidly decay away

the z-axis. A pair of planar mirrors are placed at $z = 0$ and $z = d$, as shown in Fig. 25.1. For the moment, assume that the mirrors are fully reflecting, with no transmission. Their presence will impose boundary conditions on the field: there must be field nodes at each mirror, $E(0) = E(d) = 0$. Assume the light is polarized, so that only one component needs to be considered. We will simply denote this component $E(r)$. The solutions to the Helmholtz equation in the presence of the given boundary conditions are readily found: they are a set of plane wave solutions of the form

$$E_m(r) = E_m(z) = E_{0m} \sin k_m z, \tag{25.2}$$

where the integer $m = 1, 2, 3 \ldots$ labels the quantized wavenumbers

$$k_m = \frac{m\pi}{d}. \tag{25.3}$$

In general, multiple different modes with different m values can coexist, so the general solution is of the form

$$E(z) = \sum_{m=1}^{\infty} E_{0m} \sin k_m z. \tag{25.4}$$

The corresponding resonant frequencies are

$$\nu_m = \frac{c k_m}{2\pi} = \frac{mc}{2d}, \tag{25.5}$$

so that the spacing between resonant levels is

$$\Delta\nu = \nu_{m+1} - \nu_m = \frac{c}{2d}. \tag{25.6}$$

The **mode density**, $M(\nu)$, is defined to be the number of modes per unit frequency, per unit resonator length

$$M(\nu) = 2\left(\frac{1}{d\Delta\nu}\right) = \left(\frac{2}{d}\right) \cdot \left(\frac{2d}{c}\right) = \frac{4}{c}, \tag{25.7}$$

where the extra factor of 2 is coming from the existence of two independent polarization states.

Each photon in the cavity's field can make multiple trips back and forth between the two mirrors. Each round trip consists of two reflections, plus free propagation for a distance $2d$. At each reflection, there is a phase shift of π, while the phase from traveling distance $2d$ is $2dk_m$, so the full phase gain for each round trip is $\Delta\phi = 2dk_m + 2\pi$. The condition for constructive interference is simply that $\Delta\phi$ be equal to a multiple of 2π, which immediately gives back the resonance conditions of Eqs. 25.3 and 25.5. Frequencies that are far from resonance will interfere with themselves in a primarily destructive fashion and will quickly decay away.

Now suppose some of the light is lost on each round trip. Some of that light may absorbed or scattered away from the z-axis. In addition, the mirrors may not be perfectly reflecting, allowing light to escape by being transmitted out the ends of the cavity. All of this can be taken into account by introducing a factor ζ during each round trip: $E(z) \to \zeta E(z)$ during each time around. The transfer factor can be written in polar form, $\zeta = \rho e^{i\phi}$, where $\phi = 2\pi m$ at resonance. The absolute value, $\rho = |\zeta|$, characterizes the loss: $\rho = 1$ for the lossless case, while $\rho < 1$ in the presence of losses.

To find the intensity of the light in the cavity, we can carry out the same sort of argument used in Sect. 4.5 to sum over multiple reflections: each photon can undergo one round trip before being lost, or two round trips, etc. Each of these possibilities needs to be summed, leading to a geometric series:

$$E = E_0 \left(1 + \zeta + \zeta^2 + \zeta^3 + \dots\right) \tag{25.8}$$

$$= \frac{E_0}{1 - \zeta} = \frac{1}{1 - \rho e^{i\phi}}. \tag{25.9}$$

So, given that $I = |E|^2$, the intensity is

$$I = \frac{I_0}{|1 + \rho e^{i\phi}|^2} \tag{25.10}$$

$$= \frac{I_0}{1 + \rho^2 - 2\rho\cos\phi} \tag{25.11}$$

$$= \frac{I_{max}}{1 + \left(\frac{2\mathcal{F}}{\pi}\right)^2 \sin^2\frac{\phi}{2}}, \tag{25.12}$$

where the maximum intensity is the value at resonance ($\phi = 2\pi m$)

$$I_{max} = \frac{I_0}{(1 - \rho)^2},$$ (25.13)

and the cavity **finesse** is defined to be

$$\mathcal{F} = \frac{\pi \sqrt{\rho}}{1 - \rho}.$$ (25.14)

Using the fact that $\phi = 2dk = \frac{4\pi \nu d}{c}$, the intensity can be written in terms of frequency instead of phase

$$I = \frac{I_{max}}{1 + \left(\frac{2\mathcal{F}}{\pi}\right)^2 \sin^2 \frac{\pi \nu}{\Delta \nu}}.$$ (25.15)

In terms of the spacing $\Delta \nu$ between resonant frequencies and the width $\delta \nu$ of each resonant peak, it can also be shown that the finesse can be written as

$$\mathcal{F} = \frac{\Delta \nu}{\delta \nu}.$$ (25.16)

So, for high finesse ($\mathcal{F} \gg 1$), the peaks are far apart and sharply-defined, but as the finesse becomes smaller ($\mathcal{F} \approx 1$ or smaller), the resonant peaks start blurring into each other.

Let r_1 and r_2 be the amplitude for reflection at each mirror. So the intensity reflectances are $\mathcal{R}_1 = |r_1|^2$ and $\mathcal{R}_2 = |r_2|^2$, with corresponding transmittances $\mathcal{T}_j = 1 - \mathcal{R}_j$, for $j = 1, 2$. Loss from absorption and scattering between the mirrors is proportional to propagation distance x, following the well-known **Beer's law**

$$I(x) = I_0 e^{-\alpha_s x},$$ (25.17)

where α_s is the extinction or decay coefficient. In terms of the reflectances, the amplitude for intensity to survive a round trip of length $2d$ is therefore

$$|\rho|^2 = \mathcal{R}_1 \mathcal{R}_2 e^{-2d\alpha_s}.$$ (25.18)

It is common to define a total decay constant α_t that includes losses through the mirrors, as well as the scattering and absorption:

$$|\rho|^2 = e^{-2\alpha_t d}, \quad \alpha_t = \alpha_s - \frac{1}{2d} \ln \mathcal{R}_1 \mathcal{R}_2.$$ (25.19)

The finesse can then be written in terms of α_t as

$$\mathcal{F} = \frac{\pi e^{-\alpha_t d/2}}{1 - e^{-\alpha_t d}}, \tag{25.20}$$

which simplifies for low loss ($\alpha_t d \ll 1$) to

$$\mathcal{F} = \frac{\pi}{\alpha_t d}. \tag{25.21}$$

The average length of time a photon survives in the cavity before it is absorbed, scattered, or lost through the mirrors is called the **photon lifetime**:

$$\tau = \frac{1}{c\alpha_t}. \tag{25.22}$$

Frequency spread and lifetime are always inversely proportional to each other (as in the uncertainty principle or the bandwidth theorem of Fourier analysis), so the width δv of the frequency peaks is given by

$$\delta v = \frac{1}{2\pi \tau} = \frac{c\alpha_t}{2\pi \tau}. \tag{25.23}$$

In addition to the finesse, another common figure of merit for the cavity is the quality factor or **Q factor**, which is defined to be

$$Q = 2\pi \left(\frac{\text{energy stored in cavity}}{\text{energy loss per photon round trip}} \right). \tag{25.24}$$

For the resonant cavity, this becomes

$$Q = \frac{2\pi v_0}{c\alpha_t} = \frac{v_0}{\delta v}\mathcal{F}. \tag{25.25}$$

A cavity of high Q (and therefore high $\mathcal{F}$) is one with low loss, while a leaky cavity is one with low Q.

Cavities with flat mirrors are highly susceptible to alignment problems: if the initial light beam is not perfectly parallel to the axis, then the beam will walk off toward the edge of the mirror and strike the wall of the cavity (Fig. 25.2). A similar problem arises if the two mirrors are not perfectly parallel to each other. So, rather than being flat, the mirrors at the cavity are often concave spherical mirrors, which increase stability by helping to keep the beam focused near the axis.

Additionally, instead of a linear cavity like that shown in Fig. 25.1, other cavity geometries are also common, such as the bow-tie configuration and ring resonator shown in Fig. 25.3b. Ring resonators are of special interest in modern applications and are discussed further in the next section.

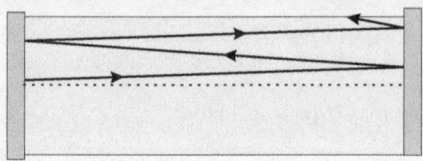

Fig. 25.2 Walk-off in a cavity parallel flat mirrors. Any small misalignments of the beam direction away from the axis, or misalignments of the mirrors so that they are not quite parallel, will lead the light rays to collide with the resonator walls

Fig. 25.3 Alternate resonator geometries: (**a**) A bowtie resonator and (**b**) a ring resonator

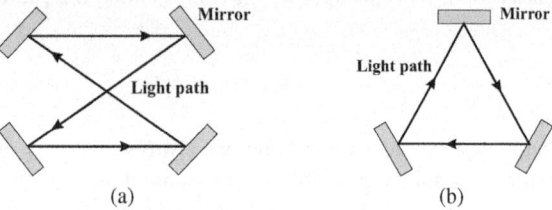

25.2 Ring Resonators and Evanescent Couplers

Optical ring resonators can be formed by looping an optical fiber to form a closed circuit or, more commonly, by constructing an circular or elliptical waveguide on a chip to form an integrated microring resonator. We focus here on such integrated microring resonators, which often have circumferences in the range of a few nanometers. These microrings have a wide range of applications [1–3], including spectral filtering, biosensing, and acting as microring lasers. By varying the distances between two ring resonators, controlled couplings between them can be engineered, making them useful for information processing applications as well. In order to reach small bending radii without losing all of the light out the sides of the waveguide, there has to be high contrast between the refractive indices inside and outside the ring. Often Si ($n \approx 3.48$) and SiO_2 ($n \approx 1.44$) are used for the interior and exterior materials.

Let n be the refractive index in a ring of radius R and circumference $L = 2\pi R$. Then the resonance condition for light in the ring is $kL = 2\pi m$, where $k = nk_0$ (for free space wavenumber $k_0 = \frac{2\pi}{\lambda_0}$) and m is a positive integer. The resonance frequencies are therefore

$$\lambda_m = \frac{nL}{m} = \frac{2\pi Rn}{m}, \qquad m = 1, 2, 3, \ldots. \tag{25.26}$$

The **free spectral range (FSR)** is the spacing between the wavelengths of consecutive modes. Approximating m as a continuous variable for large m, the FSR is

$$\Delta\lambda = |\lambda_{m+1} - \lambda_m| \approx \left|\frac{d\lambda_m}{dm}\right| = \frac{nL}{m^2} = \frac{\lambda_m^2}{nL}. \tag{25.27}$$

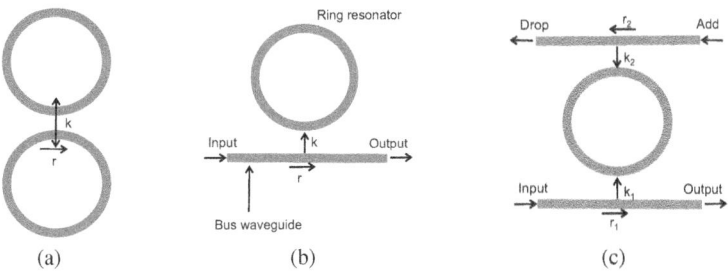

Fig. 25.4 Examples of evanescently coupled systems containing ring resonators. (**a**) Two coupled ring resonators. (**b**) An all pass ring resonator. (**c**) An add drop ring resonator. r and k represent self-coupling and cross-coupling coefficients

Note that small rings will have large FSR, and consequently the resonances will be well-separated from each other.

A ring placed close to another ring (or any other waveguide) will couple evanescently to the second waveguide, allowing the field to leak between nearby rings and waveguides. Recall that in any waveguide, even when total internal reflection occurs there will be an exponentially decaying evanescent field that will extend out beyond the edge of the guide. If a second waveguide is placed close enough to the first one, so that the evanescent field extends into it, amplitude can be exchanged between one waveguide and the other. Common configurations are shown in Fig. 25.4. Figure (a) shows two coupled rings. The configuration in (b), where the ring is coupled to a linear waveguide (the *bus* waveguide), is called an all-pass resonator. The configuration in (c), with the ring coupled to two busses, is called an add-drop resonator.

Consider two waveguides passing close enough for the evanescent waves to provide coupling between them. Two coupling coefficients are defined: r and k are, respectively, the self-coupling and cross-coupling coefficients. Unitarity requires $|r|^2 + |k|^2 = 1$. The input and output amplitudes in the two guides, as defined in Fig. 25.5, are then related by the matrix equation

$$\begin{pmatrix} B_1 \\ B_2 \end{pmatrix} = \begin{pmatrix} r & k^* \\ k & -r^* \end{pmatrix} \begin{pmatrix} A_1 \\ A_2 \end{pmatrix}. \tag{25.28}$$

Notice that the matrix is essentially a beam splitter matrix, compare to Eq. 2.70. Henceforth, we will assume that r and k are real.

Now consider the all-pass configuration. The single-pass phase shift ϕ is defined to be the phase shift gained by light when it circles the ring once, $\phi = kL$. $a = e^{-\alpha L}$ is the amplitude loss for one trip around the loop, due to bending losses, scattering, or absorption. Then it can be shown [2, 4] that the intensity transmission coefficient is

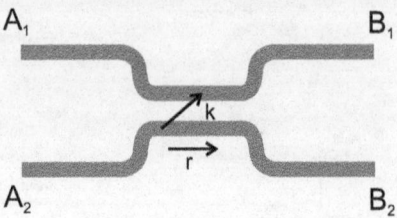

Fig. 25.5 Two evanescently coupled waveguides. The coupling mixes the two input amplitudes A_1 and A_2 to produce outputs B_1 and B_2. The mixing depends on the self-coupling coefficient r and the cross-coupling k, as in Eq. 25.28

$$T = \frac{I_{out}}{I_{in}} = \frac{|A_2|^2}{|A_1|^2} = \frac{r^2 + a^2 - 2ar\cos\phi}{1 + (ar)^2 - 2ar\cos\phi}. \tag{25.29}$$

At or near resonance, the light forms a long-lived standing wave in the ring. In this case, with the light lingering in the ring, there will be less light available to exit at the bus output. So at resonance, there will be a dip in transmission. The full-width at half-maximum of the dip at resonance m will be

$$\text{FWHM} = \frac{(1 - ra)\lambda_m^2}{\pi n_g L \sqrt{ra}}. \tag{25.30}$$

Here, $n_g = n - \lambda_0 \frac{dn}{d\lambda} = \frac{c}{v_g}$, where v_g and n_g are the group velocity and group index. Because of the transmission dip near the resonant frequency, the ring resonator can be used as a spectral filter, removing unwanted transmission frequencies from the output.

The Q-factor and finesse of the all-pass ring resonator are then given by

$$Q = \frac{\pi n_g L \sqrt{ra}}{\lambda_m (1 - ra)} \tag{25.31}$$

$$\mathcal{F} = \frac{\pi \sqrt{ra}}{1 - ra}. \tag{25.32}$$

For the add-drop configuration, there will be two sets of coupling constants: r_1 and k_1 refer to the couplings of the bottom line; r_2 and k_2 are for the upper line. In this case, the Eqs. 25.30, 25.31, 25.32 are altered by replacing $r \rightarrow r_1 r_2$ everywhere. The intensity transmission factors from the input to the pass and drop lines are given by

$$T_p = \frac{r_2^2 a^2 - 2r_1 r_2 a \cos\phi + r_1^2}{1 - 2r_1 r_2 a \cos\phi + (r_1 r_2 a)^2} \tag{25.33}$$

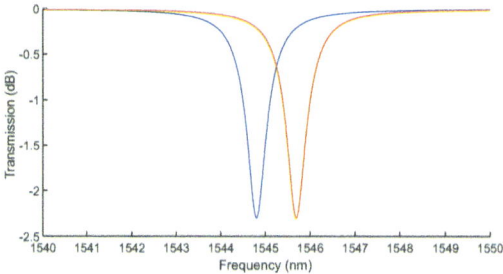

Fig. 25.6 Transmission dip (plotted in decibels) as function of frequency in a ring resonator biosensor. The blue curve is for the buffer solution; red is for test solution. The shift in frequency of the dip provides a measure of the number of bioparticles in the test solution

$$T_d = \frac{a(1-r_1)^2(1-r_2)^2}{1 - 2r_1 r_2 a \cos\phi + (r_1 r_2 a)^2}. \tag{25.34}$$

To see how resonators coupled to buses can be used as sensors, consider a beam of light of constant intensity sent through the all-pass configuration, and suppose the goal is to monitor the density of some type of bioparticle, such as viruses, DNA molecules, or proteins [3]. The ring can be coated with a polymer layer that will trap the particles of interest, leaving them attached to the outside of the ring. This will affect the coupling constants between the ring and the bus lines by an amount proportional to the number of particles attached to the ring. So the procedure is to first immerse the resonator in an aqueous buffer solution in order to provide a reference output. Then the resonator is bathed in the test solution containing the bioparticles. The position of the transmission dip will then be shifted by an amount proportional to the particle density (Fig. 25.6).

Fully integrated on-chip **ring resonator lasers** [5] have become popular for applications in areas such as communication, sensing, and spectroscopy. The ring acts as a spectral filter to achieve narrow output lines. Integrated chip lasers in general have advantages in terms of size and cost and can be part of a complete **photonic integrated circuit (PIC)** on a single chip, but integrated ring resonator lasers in particular have additional advantages including a high level of tunability and narrow line widths.

A schematic depiction of the laser is shown in Fig. 25.7. One or more ring resonators are coupled to the laser on one side of the gain region. In the figure, two loops are shown, of circumferences L_1 and L_2. The circumferences are chosen to be slightly different, in order to achieve more efficient tuning. By varying the temperature, expansion and contraction of the loops mean that the resonant frequencies can be varied in real time. The lengths of the interaction regions (the points where the loops come close to each other) can be used to control the reflectivity of the loops. So the right-hand loop can be arranged to have near complete reflection, while the left region allows some transmission so that an output beam can escape. The region between the two loop mirrors then acts as a Fabry-

Fig. 25.7 Schematic of a ring resonator laser: the rings provide spectral filtering. The loops at the end act as mirrors to form a cavity

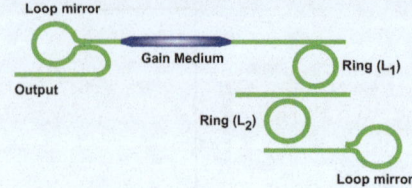

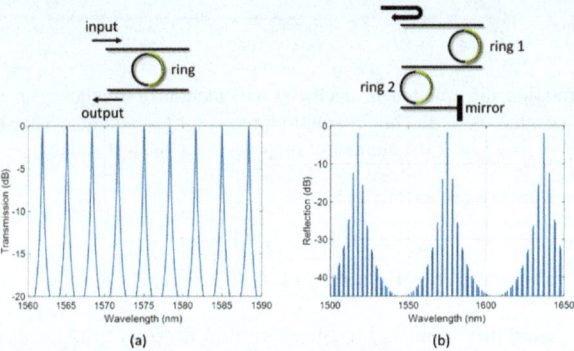

Fig. 25.8 The spectra from ring resonator lasers in (**a**) single ring and (**b**) two-ring configurations. The free spectral range in the two-ring design is given by the distance between the peaks of the envelope (roughly 60 nm here). (Figure reproduced from [5] under Creative Commons Attribution (CC BY), license https://creativecommons.org/licenses/by/4.0/)

Perot cavity, with the loops being used to enhance the frequency filtering and to provide tunability. Typical spectra are shown in Fig. 25.8 for one-ring and two-ring configurations.

The improvement in tuning for two rings, compared to a single ring, is a result of the **optical Vernier effect**. As every physics student learns in their first lab-based physics course, a Vernier caliper uses two different measurement scales to achieve a higher measurement accuracy. Similarly, in optical interferometry, the overlapping response of two interferometers with interference peaks at slightly different frequencies can lead to higher precision in optical measurement (see [6] and references therein for more details). In the optical Vernier effect, the FSR is enhanced by a factor of

$$ M = \left| \frac{FSR_1 \cdot FSR_2}{FSR_1 - FSR_2} \right| = \left| \frac{n_1 L_1}{n_1 L_1 - n_2 L_2} \right|. \tag{25.35} $$

For the ring resonators in the laser, the refractive indices of the rings are usually equal, so this reduces to an enhancement factor of

$$ M = \left| \frac{L_1}{L_1 - L_2} \right|. \tag{25.36} $$

(This assumes that the difference between circumferences is small, $|L_1 - L_2| <<$ L_1, L_2, so that either L_1 or L_2 could be used on the top of the previous equation without making a noticeable difference.)

25.3   Atoms in Cavities

Returning to the parallel-mirror resonator of Fig. 25.1, imagine now that the cavity decreases in length so that d becomes of the same order as the wavelength. Because there must be nodes at the mirrors, no electromagnetic modes exist in the cavity with $\lambda > 2d$. Still assuming that the mirrors at the ends are flat and parallel, consider light polarized parallel to the mirrors (referred to as σ-**polarization**). The field will cause electrons to oscillate in the plane of the mirror, absorbing radiation; as result, there are nodes at the mirrors. For $\lambda > 2d$, there are no allowed modes, since it is impossible to have nodes at both mirrors. So there is a cutoff wavelength, $\lambda_{cut} = 2d$, such that only modes with $\lambda < \lambda_{cut}$ are allowed. For fields below cutoff, the electron excitations in the mirror are long-lived and don't radiate the absorbed radiation back into the cavity. In contrast, for polarizations perpendicular to the mirrors (π-**polarization**), the energy from the cavity simply radiates away into the environment, and so the π-modes decay exponentially in time. As a result, the Q-factor is near zero for π-modes, but can be very high ($Q \sim 10^{11}$ or higher) for σ-modes. σ-polarized light stays in the cavity long enough to interact with any atoms inside the cavity many times. In the high-Q case, any photons emitted by the cavity atoms are rapidly reabsorbed, leading to Rabi oscillations similar to those seen in MRI systems (Sect. 23.1). In this high-Q regime, the excitation probability has the form

$$P_e(t) = \sum_n p(n) \cos^2 \left(\frac{1}{2} \Omega \sqrt{n-1}\, t \right), \tag{25.37}$$

where

$$p(n) = \frac{e^{-\beta n \hbar \omega}}{1 - e^{-\beta \hbar \omega}}, \tag{25.38}$$

the probability of n photons being in the cavity, is given by the Planck distribution. Ω, known as the **Rabi frequency of the vacuum**, is the frequency at which a single mode oscillates in energy. It is given by

$$\Omega = \frac{1}{\hbar} D E_{rms}, \tag{25.39}$$

where E_{rms} is the root-mean-square field strength in the cavity and $D = \langle 1|\boldsymbol{p}|0 \rangle$ is the matrix element of the atomic electric dipole moment $\boldsymbol{p}$ between the ground and excited states. In contrast, in the low-Q regime, above cutoff, the emission rates and resonance width of the cavity atoms are enhanced by a factor $\frac{Q\lambda^3}{V}$. This grows as $\left(\frac{\lambda}{d}\right)^3$ as the wavelength increases.

So the emission rates, Rabi oscillations, and cavity atom interaction rates can all be enhanced or suppressed by controlling the cavity Q-factor. This is the basis of **cavity quantum electrodynamics (CQED)**. CQED is the study of atoms (usually Rydberg atoms) in electromagnetic cavities. The field began emerging in the 1980s and led to the 2012 Nobel Prize being awarded to Serge Haroche and David Wineland for their work in pioneering the field. Among other things, CQED holds promise for creating stable and controllable qubit states, as well as allowing high-precision measurements of atomic properties. Reviews of atoms in cavities and of CQED may be found in [7–13].

The states of the "vacuum" are different inside the cavity than they would be outside of it, forming a discrete set of energy levels, rather than a continuum. Because the atomic electron states interact with the vacuum states via emission and absorption of photons, the states of the electrons will be altered as well. In the standard vacuum (outside of the cavity), the spontaneous decay of an excited electron state would be irreversible, due to the infinite number of available vacuum states and low-lying radiation states. But the cavity reduces the number of radiation states available, which allows atomic transitions to be enhanced or suppressed, or to be made reversible. The atom and the cavity field can exchange energy in periodic manner, producing the previously mentioned Rabi oscillations.

For simplicity, consider an atom inside the cavity with only two energy levels participating in interactions with the field. Label the atomic ground state as $|0\rangle = \begin{pmatrix} 1 \\ 0 \end{pmatrix}$ and excited state as $|1\rangle = \begin{pmatrix} 0 \\ 1 \end{pmatrix}$, with energy levels E_0 and E_1. The resonant frequency of the atom is given by $\hbar\omega_0 = E_1 - E_0$. The zero-point energy can be ignored, which amounts to taking the zero of the energy scale between the two energy levels. So, $E_0 = -\frac{\hbar\omega_0}{2}$ and $E_1 = +\frac{\hbar\omega_0}{2}$. We can then write the atomic Hamiltonian as

$$\hat{H}_{atom} = \frac{\hbar\omega_0}{2}\hat{Z}, \tag{25.40}$$

where $\hat{Z}$ is shorthand for the Pauli matrix $\hat{\sigma}_z$. Assuming the photons in the cavity have frequency ω, the field Hamiltonian can be written

$$\hat{H}_{em} = \hbar\omega\hat{N}, \tag{25.41}$$

where $\hat{N} = \hat{a}^\dagger\hat{a}$ simply counts the number of photons present. For a standard dipole interaction between the field and the atom, the interaction can be put into the form [14]

$$\hat{H}_{int} = -ig\hat{Y}(\hat{a} - \hat{a}^\dagger), \tag{25.42}$$

where g is a coupling constant proportional to the atomic dipole moment and $\hat{Y}$ is the Pauli matrix $\hat{\sigma}_y$ acting on the atomic state. Defining atomic raising and lowering operators,

$$\hat{\sigma}_\pm = \frac{1}{2}\left(\hat{X} \mp i\hat{Y}\right). \tag{25.43}$$

Writing Y in terms of $\hat{\sigma}_\pm$ and applying a rotating wave approximation (Box 24.1), the full Hamiltonian $\hat{H} = \hat{H}_{atom} + \hat{H}_{em} + \hat{H}_{int}$ can be expressed as

$$\hat{H} = \frac{\hbar\omega_0}{2}\hat{Z} + \hbar\omega\hat{N} + g\left(\hat{a}^\dagger\sigma_- + \hat{a}\sigma_+\right) \tag{25.44}$$

which is again the Jaynes-Cummings Hamiltonian (Sect. 24.6). The interaction term has a simple interpretation: the first part (containing $\hat{a}^\dagger$ corresponds to the creation of a photon, with the atomic electron simultaneously being lowered from the $n = 1$ excited state to the $n = 0$ ground state. Similarly, the $\hat{a}$ term describes a photon being absorbed and an atomic electron jumping upward from $n = 0$ to $n = 1$.

Now that the Hamiltonian has been put into Jaynes-Cummings form, it is clear that the construction of qubits and the implementation of quantum gates follows in a similar manner to that of the previous two chapters: π and $\frac{\pi}{2}$ pulses can be used to implement single-qubit gates, while three-level schemes such as that of Sect. 24.7 can be used to implement controlled two-qubit gates.

Box 25.1 The Quantum Vacuum: The Lamb and Casimir Effects

We normally think of the vacuum as a dull and featureless thing: classically, it is simply empty space, with no matter or radiation. But in quantum mechanics, the quantum vacuum is a very different place, full of frenzied activity. The vacuum is akin to a pot of boiling water: just as bubbles of vapor constantly appear and disappear in the boiling water, virtual photons and other particles are constantly appearing and disappearing from the vacuum. These are responsible for the electromagnetic field fluctuations discussed in Chap. 5. Not only can these vacuum fluctuations affect charged particles in atoms and solids, but the presence or absence of matter can also affect the vacuum. These two facts are clearly demonstrated by the Lamb shift and the Casimir effect, respectively.

The first tangible effect of vacuum fluctuations to be discovered was the **Lamb shift**, first seen in an experiment carried out by Lamb and Retherford in 1947. The Dirac equation predicted that the $2S_{\frac{1}{2}}$ and $2P_{\frac{1}{2}}$ energy levels of hydrogen should be degenerate. But experiment show them to differ by a tiny amount: $\Delta E \approx 4 \times 10^{-6}$eV (or equivalently, $\frac{\Delta E}{\hbar} \approx 1057.86$ MHz) out of an expected value of $|E| = 13.6$ eV. This is a shift of only about .00003%.

(continued)

Box 25.1 (continued)

Within months of its discovery, Hans Bethe had given an explanation of this difference: electrons in the two energy levels interact by different amounts with virtual photons appearing out of the vacuum, or in other words, the electron interacts with the vacuum fluctuations. Bethe's original calculations came close to the experimentally observed value, giving an expression for the magnitude of the shift that can be expressed in the form

$$\Delta E = -\frac{4\alpha^2}{15m^2}|\psi(0)|^2,$$

where m is the electron mass, α is the fine-structure constant, and $\psi(0)$ is the electron wavefunction at the center of the atom. For a quantitative derivation, see [15]. More sophisticated calculations done since Bethe's time give even better agreement between theory and experiment.

The converse effect, of matter affecting the vacuum, is demonstrated by the **Casimir effect**, whose existence was predicted by Hendrick Casimir in 1948. Consider two parallel metal plates separated by some distance d. The electric field between the plates has to have nodes on the conductors, so only wavelengths satisfying $\frac{\lambda}{2}n = d$ can be supported between the plates, where n is any positive integer. As a result, there is a discrete, countable set of quantized field modes between the plates. But outside the plates, there is an uncountable continuum of unquantized field modes. So, it is qualitatively clear that with more vacuum field modes outside the plates than inside, there are more virtual photons scattering off the outer sides of the plate than the inner sides. Consequently, there should be a net force pulling the plates toward each other. Quantitative calculations back up this qualitative conclusion (see, e.g., [16]), with the force given by

$$F = -\frac{\hbar c\pi^2}{240d^4};$$

the $\frac{1}{d^4}$ dependence guarantees that the force will be exceedingly weak at macroscopic scales and difficult to measure. Although the existence of the Casimir effect was demonstrated earlier, its magnitude was not measured quantitatively until 1997, when it was shown to match the theoretical expectation to within 5% [17].

Problems

1. Suppose a vacuum-filled ($n = 1.0$) resonator cavity of length $d = 60$ cm and $Q = 10^6$. The resonator mirrors have reflectances of 0.98 and 0.99.

(a) Find the overall loss coefficient, α_r and finesse $\mathcal{F} = 80$.

(b) Find the ratio $\frac{\delta\nu}{\nu_0}$. For central frequency of $\nu_0 = 800$ nm, find the width $\delta\nu$ of the resonant peaks and the spacing $\Delta\nu$ between them.

(b) How long does it for half of the initial energy in the cavity to decay away?

2. A microring resonator has radius of 4 µm and operates at the telecom wavelength $\lambda_0 = 1550$ nm. The ring is of silicon ($n = 3.48$) and is surrounded by silicon dioxide ($n = 1.44$).

(a) What is the shortest wavelength that can be sustained in the ring?

(b) What is the free spectral range at the shortest wavelength? How much does the FSR change for higher-order excitations, say around $m \sim 10$? How easily can λ_{10} and λ_{11} be separated, compared to λ_1 and λ_2?

3. (a) The FWHM, the Q-factor, and finesse for a ring resonator in the all-pass configuration are given in Eqs. 25.30 25.31, 25.32. By analogy to these, find the FWHM, Q, and F for a ring resonator in the add-drop configuration.

(b) For an all-pass ring resonator, find a formula for the width $\delta\nu_m$ of the mth resonance peak in terms of R, r, a, and L.

4. (a) Consider an arbitrary 2×2 matrix, $U = \begin{pmatrix} a & b \\ c & d \end{pmatrix}$, where the entries may all be complex. Show that requiring U to be unitary gives a set of four constraint equations on the coefficients a, b, c, d, and that the waveguide coupler matrix, Eq. 25.28, satisfies all of these constraints.

(b) Do the constraints of part (a) uniquely fix U to the form of Eq. 25.28? What are some other forms allowed by unitarity?

5. The density operator of a Rydberg atom in a resonator obeys a set of single-photon Bloch equations of the form [12]

$$\frac{d\hat{\rho}_{11}}{dt} = \frac{i}{2}\Omega_0 V \qquad\qquad \frac{d\hat{\rho}_{11}}{dt} = -\frac{\omega_0}{Q}\rho_{22} - \frac{i}{2}\Omega_0 V$$

$$\frac{d\hat{\rho}_{33}}{dt} = \frac{\omega_0}{Q}\rho_{22} \qquad\qquad \frac{dV}{dt} = i\Omega_0 W - \frac{1}{2}\left(\frac{\omega_0}{Q}\right)V,$$

where $V = \hat{\rho}_{12} - \hat{\rho}_{21}$ and $W = \hat{\rho}_{11} - \hat{\rho}_{22}$. Here, ω_0 is the atom's resonant frequency, and Ω_0 is the vacuum Rabi frequency introduced in Chap. 24. Taking $\hat{\rho}_{11}$, $\hat{\rho}_{22}$, and V as the independent variables, these equations can be written in the form

$$\frac{d}{dt}\begin{pmatrix} \hat{\rho}_{11} \\ \hat{\rho}_{22} \\ V \end{pmatrix} = M \begin{pmatrix} \hat{\rho}_{11} \\ \hat{\rho}_{22} \\ V \end{pmatrix}.$$

(a) Write out the matrix M explicitly. Find the eigenvalues and label them Λ_0, $\Lambda_\pm$, where $\Lambda_+ > \Lambda_0 > \Lambda_-$.

(b) Show that when $\frac{\omega_0}{Q} < 2\Omega_0$ the eigenvalues $\Lambda_\pm$ are complex and that the probability of an excited state remaining excited, $P_{e \to e}(t)$, will be a damped oscillation over time.

(c) Similarly, for $\frac{\omega_0}{Q} > 2\Omega_0$ show that $\Lambda_\pm$ are real, so that the excited state decays irreversibly over time. (Comparing the decay rate to the decay rate outside the cavity, it can be shown that the spontaneous decay rate is increased by a **Purcell enhancement** factor $\sim Q\lambda_0^3/n^3 V$. See [13] for more details.)

6. Suppose a resonant cavity contains a definite but unknown number of photons n. Further, suppose a Rydberg atom in the cavity is prepared in state $\frac{1}{\sqrt{2}}|\psi_0\rangle \big(|e\rangle + |g\rangle\big)$, where $|e\rangle$ and $|g\rangle$ represent the ground and excited states of the atom and $|\psi_0\rangle$ is the state of the cavity field. A $\frac{\pi}{2}$ pulse is used applied so that the state of the atoms-plus-cavity system at time t is [12, 18]

$$|\Psi(t)\rangle = \frac{1}{2}\Big[\big(e^{-i\chi t(n+1)} - e^{+i\chi tn}\big)|e\rangle + \big(e^{-i\chi t(n+1)} + e^{+i\chi tn}\big)|g\rangle\Big]|n\rangle,$$

where χ is a coupling constant, where χ is a coupling constant.

(a) Calculate the probabilities of finding the atom in the ground and excited states at time t, $P_e(t)$ and $P_g(t)$. Show that measuring the oscillation rate provides a means of measuring the photon number n in the cavity. (This is a QND measurement: n doesn't change in the process of measurement.)

(b) Suppose that the measurement takes time T. What is the largest value of n that can be measured by this method?

7. (Entangling atoms) Suppose two atoms are sent through a cavity. Initially, the cavity is in the ground state (no photons) $|0\rangle_{cav}$, while one atom is in its excited state and the other in its ground state: $|e\rangle_1, |g\rangle_2$. Let t_1 and t_2 be times that are required for each atom to cross the cavity and exit. If λ is a coupling constant representing the atom-field interaction strength, the state of the first atom and the cavity will oscillate since the atom can emit and reabsorb photons. The state at the output will be

$$|e\rangle_1|0\rangle_{cav} \to \frac{1}{\sqrt{2}}\Big(\cos \lambda t_1 |e\rangle_1|0\rangle_{cav} - i \sin \lambda t_1 |g\rangle_1|1\rangle_{cav}\Big).$$

(a) At what time t_1 will the system have equal likelihood of having 0 photons and of having 1 photon? What state has $|e\rangle_1|0\rangle_{cav}$ evolved into at this time?

(b) Now send in the second atom. If there happens to be a photon in the cavity $|1\rangle$ then what will the state $|g\rangle_2|1\rangle_{cav}$ evolve into at time t_2?

(c) Use the result of part (b) in the final state of part (a) to find the full state that $|e\rangle_1, |g\rangle_2|0\rangle_{cav}$ evolves into. Show that if t_2 is chosen appropriately the pair of atoms will end up in a Ψ^+ Bell state, so that the state of the full system is $|\Psi^+\rangle|0\rangle_{cav}$. At what time t_2 does this occur? Is there a time at which the atom pair is in the $|\Psi^+\rangle$ Bell state; if so at what time does this happen?

(d) Is it possible to use a variation on this procedure to produce the other two Bell states, $|\Phi^\pm\rangle$? Give an explanation of why or why not.

References

1. W. Bogaerts, et al., Laser Photonics Rev. **6**, 47 (2012)
2. J. Heebner, R. Grover, T. Ibrahim, *Optical Microresonators: Theory, Fabrication, and Applications* (Springer, New York, 2008)
3. A.K. Sarkaleh, B.V. Lahijani, H. Saberkari, A. Esmaeeli, J. Med. Signals Sensors **7**, 185 (2017)
4. A. Yariv, P. Yeh, *Photonics: Optical Electronics in Modern Communication, 6th ed.* (Oxford University Press, Oxford, 2006)
5. T. Komljenovic, et al., Appl. Sci. **7**, 732 (2017)
6. A.D. Gomes, Sensors **19**, 5431 (2019)
7. R. Miller, T.E. Northup, K.M. Binbaum, A. Boca, A.D. Boozer, H.J. Kimble, J. Phys. B: At. Mol. Opt. Phys. **38**, 5551 (2005)
8. H. Walther, B.T.H. Varcoe, B.G. Englert, T. Becker, Rep. Mod. Phys. **69**, 1325 (2006)
9. S. Haroche, J.M. Raimon, *Exploring the Quantum: Atoms, Cavities, and Photons* (Oxford University Press, Oxford, 2006)
10. S. Haroche, D. Kleppner, Sci. Am. **42**(1), 24 (1989)
11. S.M. Girvin, Circuit QED: superconducting qubits coupled to microwave photons, in *Quantum Machines: Measurement and Control of Engineered Quantum Systems,* ed. by M. Devoret, B. Huard, R. Schoelkopf, L.F. Cugliandolo, *Lecture Notes of the Les Houches Summer School,* Vol. 96, July 2011 (Oxford University Press, Oxford, 2014)
12. C.G. Gerry, P.L. Knight, *Introductory Quantum Optics* (Cambridge University Press, Cambridge, 2005)
13. M. Fox, *Quantum Optics* (Oxford Press, Oxford, 2006)
14. M.A. Nielsen, I.L. Chuang, *Quantum Computation and Quantum Information: 10th Anniversary Edition* (Cambridge University Press, Cambridge, 2011)
15. W. Greiner, J. Reinhart, *Quantum Electrodynamics, 3rd ed.* (Springer, Heidelberg, 1994)
16. P.W. Milonni, *The Quantum Vacuum - An Introduction to Quantum Electrodynamics* (Academic Press, New York, 1994)
17. S.K. Lamoreaux, Phys. Rev. Lett. **78**, 5 (1997)
18. M. Brune et al., Phys. Rev. Lett. **72**, 3339 (1994)

Chapter 26
Solid-State Qubits

In addition to optical, atomic, and ionic systems, information processing can also be carried out with qubits formed within solid-state systems. In this chapter, brief introductions are given to several of these solid-state platforms, including quantum dots, superconducting qubits, and nitrogen-vacancy (NV) centers.

Each of the solid-state systems discussed below has its advantages and disadvantages as platforms for constructing and manipulating qubits. Quantum dots and NV centers, for example, can operate at room temperature, and systems using them can be easily miniaturized. In contrast, superconducting systems require extremely low temperature and are highly sensitive to environmental disturbances, but they can operate at high speed and with high-precision control. All of these platforms seem to be highly scalable to large numbers of qubits.

26.1 Quantum Dots

A **quantum dot** is a structure, usually in a semiconductor, that confines electrons within a small three-dimensional volume. In other words, it is a three dimensional potential well. Due to the confinement, the electrons have a discrete energy level structure, similar to the energy levels in an atom. Because of this, quantum dots are often called **artificial atoms**. By choosing an appropriate volume size and shape, the spacing of the energy levels can be tailored to fit the desired application. Typical dot sizes may range from tens to hundreds of nanometers. These dots can be attached to a semiconductor substrate or can be floating in a colloidal suspension (Fig. 26.1).

Quantum dots have an extraordinarily broad range of applications. Since the emission and absorption spectra are controlled by the dot size, quantum dots are useful to create fluorescence at desired frequencies in biomedical applications. They can also be mixed into semiconductor ink and applied onto the surface of other materials to produce cheap and flexible solar cells. Several companies now use

© The Author(s), under exclusive license to Springer Nature Switzerland AG 2025
D. S. Simon, *Introduction to Quantum Science and Technology*, Undergraduate
Texts in Physics, https://doi.org/10.1007/978-3-031-81315-3_26

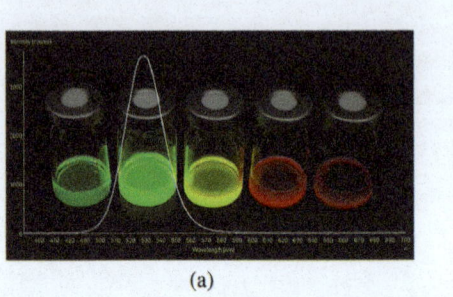 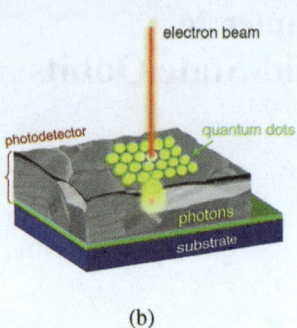

(a)   (b)

Fig. 26.1 (**a**) Quantum dots of different sizes in solution will fluoresce at different frequencies, allowing optical sources of precisely controlled wavelength. (Image source: NASA) (**b**) Quantum dots can also be embedded on a semiconductor substrate for use as optical detectors, electron detectors, or solar cells. Here, the dots are used to detect electrons; the dots absorb energy from the electrons in the beam and convert it into visible light that can be detected by an adjacent photodetector (Image source: Dill/NIST)

quantum dots in television screens, and graphene quantum dots can be used to create nanoscale transistors. As a result of their rapid growth in technological importance, M. Bawendi, L. Brus, and A. Ekimov shared the 2023 Nobel Prize for Chemistry for the discovery and synthesis of quantum dots.

Quantum dots can also be used as qubits. There are two main schemes for doing this. One uses excitons (electron-hole pairs) in a pair of neighboring weakly coupled quantum dots. So a set of quantum states can be defined:

$$|00\rangle = \text{state with exciton in dot 1}$$

$$|01\rangle = \text{state with the electron in dot 2 and the hole in dot 1}$$

$$|10\rangle = \text{state with the electron in dot 1 and the hole in dot 2}$$

$$|11\rangle = \text{state with the exciton in dot 2}$$

The $|01\rangle$ and $|10\rangle$ states are degenerate and will form themselves into symmetric and antisymmetric superpositions if they are close enough (within a few nanometers) for tunneling to occur. In general, the excitons have short decoherence times, on the order of tens of picoseconds. As in NMR systems, the exciton states in the two quantum wells can be manipulated by using π-pulses from a laser, allowing the formation of CNOT and SWAP gates.

A second quantum dot qubit scheme can be carried out using the spin of the electrons in the dot. These spins have longer lifetimes than excitons, with decoherence times on the order of μs. Electron-based qubits also have the advantage that they can be transported through a network. Because of their lighter mass, electron spins can be rotated more quickly than proton spins and therefore lead to faster gates than can be formed with NMR systems.

Another important application of quantum dots is their use as single-photon sources [1]. A semiconductor quantum dot is embedded in another semiconducting material with a much larger bandgap. The surrounding material then provides a barrier confining electrons inside the dot, since the contained electrons don't have the required energy to reach the conduction band of the outside material. A laser pulse can then excite an electron inside the well, creating electron-hole pairs. These pairs, being tightly confined in a small region, will form an exciton. After a short time, the electron and hole will recombine to produce a single photon of known energy. Single-photon emission can be induced in this manner in the infrared, visible, and ultraviolet portions of the spectrum, spanning wavelengths roughly from 1.5 μm to 280 nm. In particular, photons can be created in the infrared telecom bands. Since the photons are emitted randomly in all directions, directional emission requires extra structures to be added, such as waveguides and cavities, in order to enhance the probability of photon arrival in the desired direction. Early approaches to single-photon emission from quantum dots required low temperatures in order to minimize phonon noise, but the allowed temperatures have been rising over time.

The energy levels between the ground and first excited states can be easily found. When an electron is promoted to a higher energy band, the resulting electron-hole pair will have a reduced mass $\frac{1}{m^*} = \frac{1}{m_e} + \frac{1}{m_h}$, where m_e and m_h are the effective masses of the electron and hole. Then there is an energy term from the confinement of the particles in the quantum well of diameter d, $E_{confine} = \frac{\pi^2 \hbar^2 n^2}{2m^* d^2} = \frac{n^2 h^2}{8m^* d^2}$, for $n = 1, 2, \ldots$. In addition, there is Coulomb energy from the attraction between the electron and resulting hole. This takes the form of $E_{coul} = -\frac{1.8e^2}{2\pi \epsilon_r \epsilon_0 d}$, where the 1.8 factor comes from numerical calculation of the average distance between the electron and hole, given their orbital structure. If the quantum dot is made from a material that has an intrinsic bulk bandgap of E_g, then the bandgap between the two lowest-lying energy bands is found by adding the previous two energy terms (with $n = 1$) to the intrinsic material bandgap: $E_{g,dot} = E_g + \frac{h^2}{8m^* d^2} - \frac{1.8e^2}{2\pi \epsilon_r \epsilon_0 d}$. Frequencies of photons absorbed or emitted during transitions between energy levels will be determined by the differences between the confinement energies of the levels.

One interesting effect that can be seen in quantum dots is the Coulomb blockade. A **tunnel diode** is a device where a thin insulating film is placed between two electrodes. If the film is thin enough, then electrons are capable of tunneling between the electrodes to produce a current. Tunnel diodes are low power devices that were often used in the past in triggering circuits for oscilloscopes and tuning circuits for televisions. Suppose such a diode is altered by wedging a quantum dot between two thin conducting films, with electrodes on the outside (Fig. 26.2a). Suppose one electron tunnels through the dot. Then a charge $q = \pm e$ accumulates on each electrode. This requires a voltage difference of $V = \frac{e}{C}$, where the capacitance of a spherical quantum dot of radius r and dielectric constant ϵ_r is given by $C = 4\pi \epsilon_r \epsilon_0 r$. But this voltage will create a uniform electric field $E = \frac{V}{2r}$ inside the quantum dot, which will require a potential energy $U = \frac{1}{2}CV^2 = \frac{e^2}{2C}$. So, if the voltage applied is less than the threshold voltage $V_{th} = \frac{e^2}{2C}$, then there will not be enough energy to charge the capacitor: no electron will cross the dot, and no

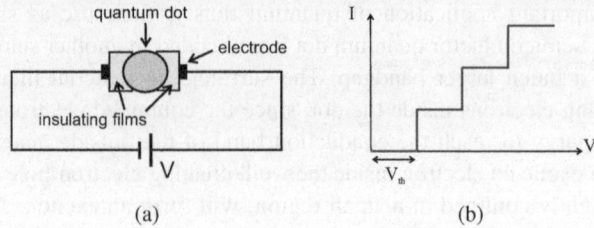

Fig. 26.2 (**a**) Electrons can tunnel through a quantum dot with thin insulating films on each side, producing an electric current. (**b**) The current shows a step-like structure due to the Coulomb blockade: each increase in current requires a finite increase in voltage to support the increased electrostatic energy that each transported electron produces in the quantum dot

current will flow through the circuit. This is called the **Coulomb blockade**. If the voltage starts at $V = 0$ and is slowly increased, there will be a sudden jump in current at $V = V_{Th}$. If the voltage continues to increase, the current will remain constant until the voltage reaches $V = 2V_{Th}$. At this point, the energy supplied can support the electric field created by a charge of $\pm 2e$ on the two sides of the capacitor. Continuing this way, the current will show a series of equally spaced plateaus (Fig. 26.2b) forming a staircase structure, similar to that of the quantum Hall effect. These plateaus are due to the quantized value of the electron charge, and the width of the steps is determined by the size and material of the quantum dot.

A recent review of the properties of quantum dots can be found in [2]. Reviews of biomedical applications can be found in [3, 4], while applications in quantum information technology are reviewed in [5].

26.2   Lattice Defects and NV Centers

In the search for practical physical implementations of qubit states, trade-offs have to be made. Platforms like ions and neutral atoms can be well isolated from noise (and so have long coherence times), and they can be easily controlled and manipulated with electromagnetic pulses. But it is hard to make them interact with each other, meaning that implementations of controlled gates are difficult.

One promising option that allows long coherence lengths, ease of control, and inter-qubit interactions is the **nitrogen vacancy (NV) center** in diamond. These are defects in the crystal lattice that form atom-like structures, with electrons that are highly localized near the defect and that have a set of energy levels that allow a sequence of readily controllable transitions. They have the advantage of being able to implement controlled gates. Like superconducting junctions (next section), they can serve as both qubits and as field sensors, but in contrast to superconducting systems, they can be used at room temperature. More extensive reviews of NV centers can be found in [6, 7].

Diamond forms a lattice with each carbon atom bound to its four nearest neighbors to form a tetrahedral structure. Of all known materials, diamond has the greatest density of atoms per volume. Together with the strong covalent bonds of the atoms and the rigidity of the tetrahedral lattice, this leads to diamond being the hardest and least compressible known material. In addition, it has very high mass density and high thermal conductivity. Diamonds are highly transparent in and near the visible region of the spectrum.

Like any crystal lattice, diamond will typically contain some density of impurities. For example, diamond may host **color centers**, a type of impurity or defect that has an electronic structure capable of fluorescence and which can be readily identified by its emission and absorption spectrum.

An NV center in diamond is a particular type of color center, consisting of a missing carbon atom (the vacancy) sitting next to a nitrogen impurity. The NV center can come in three charge states (positively charged, negatively charged, or neutral). Of the three, the negatively charged state NV^- is the most useful and will be the only one considered here. The NV^- state contains six electrons: two from the N impurity, three from the dangling C bonds, and one captured from the surrounding lattice. The electrons at the dangling bonds introduce states in the middle of an indirect bandgap, making them energetically isolated from other states, contributing to the long coherence times.

The relevant set of states is shown in Fig. 26.3. There is a ground state $|g\rangle$ separated from an excited state $|e\rangle$. Each of these is split into a spin triplet of states with $m_s = 0, \pm1$. In the absence of an external field, the $m_s = \pm1$ states are degenerate with each other. When an external magnetic field is turned on, they gain an energy splitting $\Delta E = E_{+1} - E_{-1}$ proportional to the size of B. The spacing between the $m_s = 0$ state and the $m_s = \pm1$ states leads to resonant frequencies in the microwave region: $\nu = \Delta E/\hbar = 2.87$ GHz for $|g\rangle$ and 1.42 GHz for $|e\rangle$. These correspond to wavelengths of $\lambda = 10$ cm and $\lambda = 21$ cm, respectively. In contrast, the transition between $|e\rangle$ and $|g\rangle$ is at 637 nm, giving a resonant frequency of 4.7×10^{14} Hz, which falls in the visible (red) portion of the spectrum. So optical light beams can be used to drive $|g\rangle \leftrightarrow |e\rangle$ transitions, while microwave pulses can stimulate transitions within each spin triplet.

The density of NV centers has to be low enough so that they can be identified individually by a confocal optical microscope and addressed individually by laser

Fig. 26.3 Energy levels of the negatively charged NV center. Transitions between $|e\rangle$ and $|g\rangle$ occur with a resonant frequency in the visible range, while transitions within the same spin triplet are in the microwave

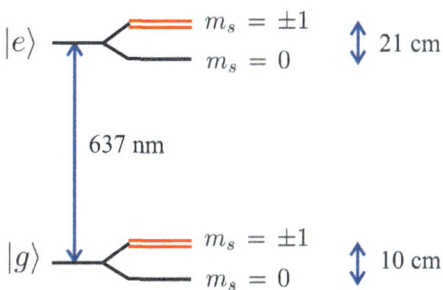

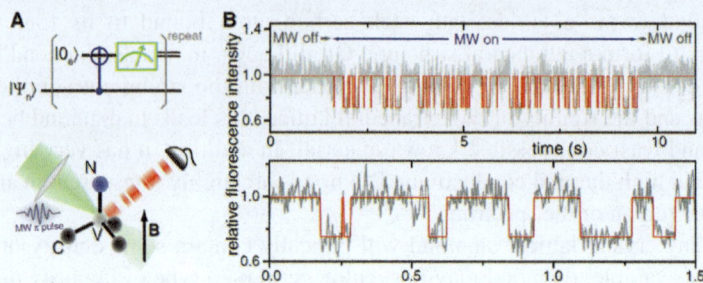

Fig. 26.4 A CNOT gate based on the NV^- center in diamond. (**a**) The schematic of the gate is shown at the top. Below, the π-pulse control mechanism is shown. (**b**) The fluorescence emissions from the electrons are shown on two different time scales as microwave pulses are applied. The jumps in the red fluorescence intensity curve are due to ^{14}N spin transitions. The black trace in the foreground shows the microwave pulse turning off and on. When the microwave pulses are turned off, the fluorescence rate remains high since the electron states are then uncorrelated with the nuclear spin, which will be in a mixed state. (Figure reproduced with permission from [8])

pulses. These centers can be in a single sample of bulk diamond or separated onto different nanodiamond crystals. Depending on how the material was prepared, the average separation between NV centers can range from nm to 10's of μm.

The coherence time is temperature dependent and can be as large as milliseconds at room temperature. If the NV center is being operated as a quantum logic gate, this is long enough for millions of operations to be carried out, since transitions can be induced on nanosecond time scales. For temperatures below about 80 K, the coherence length can be even longer, on the order of seconds.

For use as a qubit, first note that nuclear spins on their own are normally highly isolated from their surroundings, with coherence times that can be on the order of hours. But in an NV center, the electron states at the impurity can act as an intermediary, coupling external control and measurement pulses to the spins of the nitrogen and surrounding carbon nuclei. Within the electronic spin triplets shown in Fig. 26.3, there are hyperfine splittings with transition frequencies between them that depend on the states of the C and N nuclei. A narrowband microwave π-pulse flips the electron spin in a manner that is conditional on the state of the nuclei. The state can then be measured from the emitted fluorescence trace. This can be seen in Fig. 26.4. The NV center can then be viewed as a controlled-NOT gate, with the nuclear state as the control and the electron as the target. Note that the electron fluorescence allows the nuclear states to be measured without perturbing those states, or in other words, it serves as a quantum nondemolition (QND) measurement of nuclear states.

Since the resonant frequencies of the nuclei depend on the strengths of external fields, the NV center can also serve as a field sensor. In this capacity, the NV center should be at the surface of the diamond in order to be brought within a few nanometers of the sample. Then Zeeman and Stark shifts in the resonance frequencies of the electron levels allow both electric and magnetic fields to be

Fig. 26.5 An example of using NV-based sensors to image individual cells. (**a**) and (**b**): The fluorescence traces from two different points. (**c**) The reconstructed image from the traces emitted over the entire sample. (Figure reproduced with permission from [9])

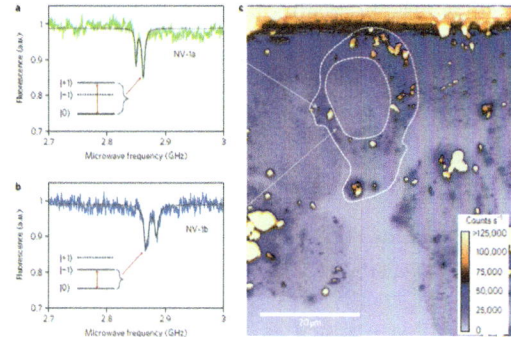

measured via the spectrum of emitted fluorescence. This can provide information about the location and density of electric and magnetic dipoles in a material, which has applications for sensing and imaging in biology, down to the molecular level. An example is shown in Fig. 26.5 where nanodiamond sensors, stimulated by microwave pulses, are used as probes in an individual living cell, with the emitted fluorescence allowing imaging of the state. For a review of biological applications, see [10].

26.3 Superconducting Qubits and Circuit QED

One extremely promising road to forming practical quantum information processing systems is the use of Josephson junctions. Recall from Chap. 7 that a Josephson junction is a pair of superconductors separated by a small insulating gap or barrier. Despite the gap, current flows across the junction as Cooper pairs tunnel from one superconductor to the other. On the two sides of the junction, there are two macroscopic quantum states, with a phase difference between their wavefunctions.

The Josephson junction allows qubits to be encoded into macroscopic variables in a highly controllable manner. These qubits can be put into superposition states that remain coherent for several microseconds or longer, which is sufficient time for hundreds of quantum logic operations to be carried out. The qubits can be connected electromagnetically, and the size of the energy level spacings allow the qubits to couple to each other at microwave frequencies (tens of GHz). The qubits themselves can be encoded into physical variables in several ways; these various types of qubits are known as **charge qubits** (or **Cooper pair boxes**), **phase qubits**, **flux qubits**, and **transmons**. In the following two subsections we will discuss charge qubits and transmons in detail. For flux and charge qubits, see the references. Good reviews of superconducting qubits include [11–14].

First, briefly recall a few results from Sect. 7.3. A Josephson junction is a thin insulating gap between two superconductors. The current across the junction is due to tunneling of Cooper pairs from one superconducting region to the other and has the form

$$I(\phi) = I_c \sin \phi, \tag{26.1}$$

where ϕ is the phase difference between the two macroscopic wavefunctions in the two superconducting regions. The phase evolves at a rate proportional to the voltage across the junction

$$\frac{\partial \phi}{\partial t} = \frac{2eV}{\hbar}. \tag{26.2}$$

It can be seen then that the junction acts as a nonlinear inductor. Since the voltage across an inductor of inductance L (ignoring a sign from Lenz' law) is

$$V = L\frac{dI}{dt}, \tag{26.3}$$

we find that the inductance of the junction varies nonlinearly with phase

$$L = \frac{V}{\partial I/\partial dt} = \frac{\hbar}{2I_c e \cos \phi} = \frac{L_J}{\cos \phi}, \tag{26.4}$$

where $L_J = \frac{\hbar}{2eI_c}$.

An insulating gap such as in the Josephson junction will also have a capacitance. The charge stored on each side of the junction (the "plates" of the capacitor) will be proportional to the number of Cooper pairs that have tunneled across the junction, with each Cooper pair carrying a charge of $2e$. So the energy stored in the junction has a capacitive (electric) term and an inductive (magnetic) term. The energy takes the form

$$\hat{H} = 4E_C \hat{n}^2 - E_J \cos \hat{\phi}, \tag{26.5}$$

where

$$E_J = L_J I_c^2, \qquad E_C = \frac{e^2}{2C}. \tag{26.6}$$

Typically, the capacitance of the junction is likely to be in the femto-Farad (fF) to pico-Farad (pF) range, with inductance in the nano- to micro-Henry range. By appropriately engineering the parameters of the junction, the ratio of Josephson energy to charging energy, E_J/E_C, can vary over seven orders of magnitude from about 10^{-1} to 10^6.

Note that the Hamiltonian, number of Cooper pairs, and phase are all now being treated as quantum operators. The number and phase variables are conjugate operators, $[\hat{\phi}, \hat{n}] = i$, which leads to an uncertainty relation $\Delta\phi\Delta n \geq \frac{1}{2}$.

In order to be useful, a qubit needs to couple in some way to other qubits, to gate operations, and to a readout device. In the case of superconducting qubits, because of the spacing between the qubit energy levels, this coupling is generally achieved by using microwave signals (wavelengths roughly 1 to 10 cm) transmitted from one qubit to another. A system of transmon qubits connected by microwave transmission lines is called a **circuit QED** system and operates to some extent in analogy to cavity QED (CQED) systems. Circuit QED offers the possibility of conducting quantum optics experiments on integrated chips; however, being dependent on superconducting Josephson junctions, such systems must be maintained at mK temperatures or lower. Extensive reviews of circuit QED and of analogies with CQED may be found in [15–19].

26.3.1 Charge Qubits

Before discussing charge qubits or Cooper-pair boxes, first recall the standard parallel LC circuit shown in Fig. 26.6. The impedances of the inductor and capacitor are given by $Z_L = i\omega L$ and $Z_C = \frac{1}{i\omega C}$. At resonance, these impedances cancel, $Z_L + Z_C = 0$, leaving the parallel LC combination looking like a simple conducting wire. This happens at the resonance frequency $\omega_0 = \frac{1}{\sqrt{LC}}$. So the voltage source supplying a signal of frequency ω to the circuit will produce maximum current at $\omega = \omega_0$, and the current will decay as ω moves away from the resonance value, ω_0.

So now consider the **Cooper pair box (CPB)** or **charge qubit** circuit in Fig. 26.7a. It consists of a Josephson junction (the box with the cross) and a capacitor C_g connected to a voltage source V_g. Recall that the Josephson junction acts like a parallel inductor and capacitor with values C_J and L_J, so this is similar to the LC circuit above, with an additional series capacitor added. The two main differences between the Cooper pair box and the LC circuit are that the inductor is nonlinear and that the shaded segment of the diagram (known as the **island**) is electrically isolated from the rest of the circuit (the **reservoir**). Classically, no charge can flow between the island and the reservoir; any charge exchanged between

Fig. 26.6 A parallel LC circuit, with resonant frequency $\omega_0 = \frac{1}{\sqrt{LC}}$

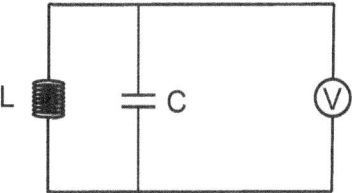

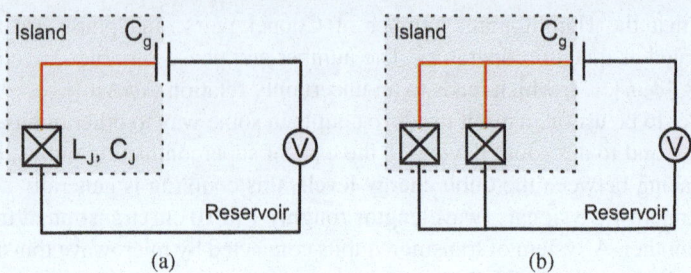

Fig. 26.7 (a) A Cooper pair box (CPB) or charge qubit is formed by Cooper pairs tunneling through a Josephson junction, leaving the two sides of the junction with opposite sign charges. (b) A split CPB replaces the single junction with a pair of Josephson junctions, so that the ratio E_J/E_C can be controlled by varying the flux in the superconducting loop formed by the junctions

them is due to quantum tunneling. The reservoir may be connected to additional electronics, but the island should stay isolated.

Often, the single junction is replaced by a parallel combination of junctions (in other words, by a SQUID structure), as in Fig. 26.7b. This is called a **split CPB**. This operates in essentially the same manner as the ordinary CPB, but it allows the value of E_J to be adjusted by tuning the flux Φ inside the loop formed by the SQUID:

$E_J(\Phi) = 2\,E_J|_{\Phi=0}\cos\left(\frac{\pi\Phi}{\Phi_0}\right)$ (see Sect. 7.3.2). Here $\Phi_0 = \frac{h}{2e}$ is the flux quantum.

By adjusting the voltage V_g, an initial number of Cooper pairs, n_g, can be preset to be exchanged across the junction, producing a fixed initial charge on the junction plates. This defines the operating point of the circuit, taken to be the zero of energy for the n-dependent part of the Hamiltonian. So the Hamiltonian will now be taken to be of the form

$$\hat{H} = 4E_C\left(\hat{n} - n_g\right)^2 - E_J\cos\hat{\phi}. \tag{26.7}$$

For the CPB, we will always assume that the ratio of inductive and capacitive energies obeys $E_J/E_C < 1$. It can be shown (Problem 5) that the Hamiltonian can be rewritten in the form

$$\hat{H} = 4E_C\left(n - n_g\right)^2 |N\rangle\langle N| + \frac{1}{2}E_J \sum_n \left(|n\rangle\langle n+1| + |n+1\rangle\langle n|\right), \tag{26.8}$$

where $|n\rangle$ is the eigenstate of the number operator, with eigenvalue n: $\hat{n}|n\rangle = n|n\rangle$. The first term in the Hamiltonian represents a potential energy term, measuring the energy stored on the junction for a given number of Cooper pairs n. The second term is a transition energy: it tells us how much energy is required to add or remove a pair from the junction.

The eigenvalues of $\hat{H}$ are plotted in Fig. 26.8 as a function of the operating point n_g. When n_g is an integer, the energy difference between the first two levels $E_{01} = E_1 - E_0$ is large, but the difference between the second and third levels $E_{12} = E_2 -$

Fig. 26.8 The first three energy levels of the Hamiltonian 26.7. When n_g is a half integer, the first two levels are well separated from the next level and are capable of forming a qubit. Figure reproduced from [12]

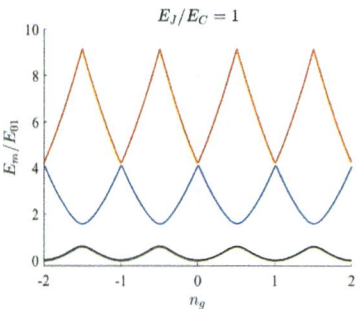

E_1 is small. So these values are unacceptable for constructing qubit states, since the lower two levels (acting as the 0 and 1 states) are not sufficiently separated from the third level. Half-integer values of n_g are more promising. Not only is $E_1 - E_0 <<$ $E_2 - E_1$, allowing well-defined qubit state at low energies, but the minima of the two lowest-lying levels are nearly flat at those values, which helps improve coherence times.

So the Josephson junction is capable of admitting a qubit in the form of the quantized charge on the two sides of the junction. Unfortunately, this qubit is too sensitive to charge noise, noise due to movement of charge near defects in the junction or in the surrounding material. It increases as E_J/E_C decreases and can cause decoherence of the qubit. To address this issue, additional structure can be added to the circuit, leading to the transmon of the next subsection.

26.3.2 Transmons

To improve on the CPB qubit, start with the split CPB of Fig. 26.7b, and add an additional shunt capacitance C_B, as in Fig. 26.9. This will reduce the value of E_C, which will now be given by $E_C = \frac{e^2}{2C_\Sigma}$, where $C_\Sigma = C_J + C_B + C_g$. So, the addition of C_B will alter the ratio E_J/E_C, changing the energy eigenvalues. In particular, if $C_B >> C_g$, then the sensitivity to charge noise will be greatly reduced; in fact, the charge noise and the tunneling probability through the junction will both decrease exponentially with E_J/E_C. This reduction in noise can increase the coherence time of the qubit by an order of magnitude or more. The charge qubit formed by this system is called a **transmon**.

An explicit expression for the energy eigenvalues can be given [20], involving a special function called the Mathieu characteristic value. We won't give it here, but instead we examine plots of the energy levels. The effect on the energies can be seen in Fig. 26.10, where the energy levels are plotted versus n_g for different values of E_J/E_C. It can be seen that as E_J/E_C increases the separation between E_2 and E_1 increases and the level E_2 becomes flatter, but of which improve the quality of the qubit formed by the two-level (E_0 and E_1) system. The drawback is that as the ratio

Fig. 26.9 Adding a large shunt capacitance C_B to the split CPB circuit yields a transmon circuit. The charge qubit, or transmon, that is produced is less sensitive to charge noise and has a longer coherence time than the CPB

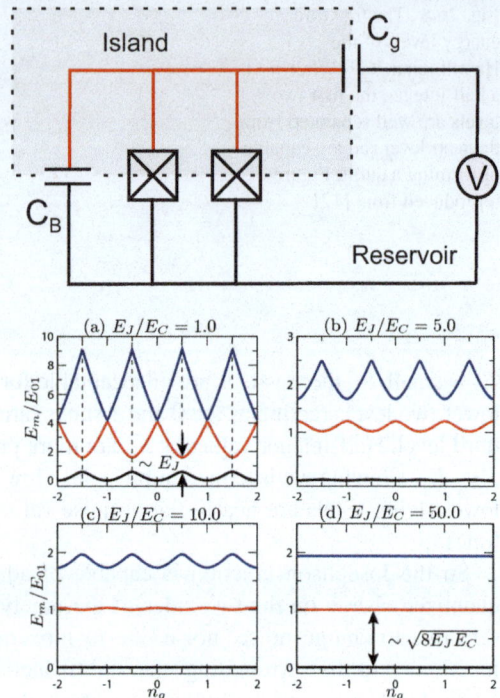

Fig. 26.10 The first three energy levels of the transmon Hamiltonian 26.7 for different values of the ratio E_J/E_C. Figure reproduced with permission from [20]

becomes too large, the energy levels become harmonic (i.e., they look like those of the simple harmonic oscillator): the difference between the first two levels starts becoming equal to the difference between the second pair of levels. As the level differences and transition frequencies between the pairs become equal, interference can occur between the different transitions, making the system unusable for qubit storage. So a balance between low-charge (high E_J/E_C) and anharmonicity (low E_J/E_C) has to be struck.

Typically, the operating frequency of the transmon qubit, $\omega = (E_1 - E_0)/\hbar$, is on the order of 1–10 GHz.

Other types of superconducting qubits are possible, in which the information is encoded into flux or phase variables. For more about these, see [13] for discussion of flux qubits and [11] for phase qubits.

Problems

1. Imagine a spherical quantum dot of radius r at room temperature $T \sim 300$ K. The dot is made of silicon, which has relative permittivity $\epsilon_r = 11.5$. In order for a Coulomb blockade effect to be visible, the energy change at each step has

to be well above the average thermal energy. How small must the dot's radius be in order to observe the Coulomb blockade?

2. A particular quantum dot is made from a material with bulk bandgap $E_g = 1.4$ eV. The effective masses for electrons and holes in this material are $m_h = 0.1$ m and $m_e = 0.5$ m, where m is the free electron mass.

   (a) What is the energy gap for a quantum dot of radius $r = 2$ nm? For a radius of $r = 100$ nm?

   (b) In what part of the optical spectrum are the photons emitted by low-lying excitations when $r = 2$ nm?

3. Consider a quantum dot made of $CdSe$, which has a bulk bandgap of $E_g = 1.74$ eV and dielectric constant $\epsilon_r = 4.8$. The effective masses of electrons and holes are $m_h = 0.13$ m and $m_e = 0.45$ m, where m is the free electron mass. Find the frequencies and wavelengths of the two lowest-frequency photons that can be emitted by the dot.

4. The total current through a Josephson junction can be written as $I = I_J + I_R + I_C = I_J + \frac{V}{R} + C\frac{dV}{dt}$. From Chap. 7 we know that the junction current and voltage are described by the Josephson equations $I(t) = I_c \sin\Theta(t)$ and $V = \frac{\Phi_0}{2\pi}\frac{d\Phi}{dt}$, where the flux quantum is $\Phi_0 = \frac{h}{2e}$ and $\Theta(t)$ is the phase difference between the two superconductors.

   (a) Show that these relations can be arranged into the form of the differential equation $0 = m_{eff}\frac{d^2\Theta}{dt^2} + D\frac{d\Theta}{dt} + \frac{\partial}{\partial\Theta}U(\Theta)$ with potential

   $$U(\Theta) = -E_J\left(\frac{I}{I_c}\Theta + \cos\Theta\right).$$

   The phase difference therefore behaves like the position variable for a particle in potential U. Find the damping coefficient D and effective mass m_{eff}.

   (b) Plot $U(\Theta)/E_J$ versus Θ for several values of $\frac{I}{I_c}$ between 0 and 2. Plot for Θ spanning a range of at least 0 to 6π. For small values of $\frac{I}{I_c}$, you should be able to see why $U(\Theta)$ is sometimes called the washboard potential.

   (c) Show that the potential has a sequence of local minima for $\frac{I}{I_c} < 1$. Since the system is trapped near one of these minima, the phase is constant and no voltage occurs (DC Josephson effect).

   (d) Show that the minima vanish for $\frac{I}{I_c} > 1$ and that the potential has negative slope. Therefore the system evolves over time toward increasing Θ, leading (via the second Josephson equation) to a voltage across the junction (AC Josephson effect).

5. Let $\hat{\phi}$ and $\hat{n}$ be the operators representing the phase difference between the two superconductors in a Josephson junction and the number of Cooper pairs.

The quantization of the junction variables requires the canonical commutation relation $\left[\hat{\phi}, \hat{n}\right] = i$ to hold.

(a) Show that $e^{\pm i\hat{\phi}}$ act as raising and lowering operators. Do this by first showing that $\left[\hat{n}, e^{\pm i\hat{\phi}}\right] = \pm e^{\pm i\hat{\phi}}$ and then that $e^{\pm i\hat{\phi}}|n\rangle = |n \pm 1\rangle$. (Since the charge operator is proportional to the number of Cooper pairs, $e^{\pm i\hat{\phi}}$ also act to raise and lower charge.)

(b) Show that $e^{\pm i\hat{\phi}} = \sum_n |n \pm 1\rangle\langle n|$ and that therefore $\cos\hat{\phi} = \frac{1}{2}\sum_n \left(|n + 1\rangle\langle n| + |n - 1\rangle\langle n|\right)$. This shows that Eq. 26.7 is equivalent to Eq. 26.8.

6. In close analogy to cavity QED, the circuit QED Hamiltonian can again be put into Jaynes-Cummings form, with the role of the cavity and the atom being played by the transition line and the superconducting qubit. Let n be the number of Cooper pairs present, and let g and e denote ground and excited states of the qubit. Then the two excited states $|e, n\rangle$ and $|g, n + 1\rangle$ tend to mix, producing a pair of "dressed" excited states

$$|+, n\rangle = \cos\theta|e, n\rangle + \sin\theta|g, n+1\rangle, \qquad |-, n\rangle = -\sin\theta|e, n\rangle + \cos\theta|g, n+1\rangle,$$

with mixing angle $\theta = \frac{1}{2}\tan^{-1}\left(\frac{2g\sqrt{n+1}}{\delta}\right)$, where $\delta = \omega - \omega_0$ is the detuning and g is a coupling constant. (The corresponding energies are given in Problem 9 of Chap. 24.)

(a) Draw a right triangle with 2θ as one angle and adjacent side of length δ. Fill in the lengths of the other two sides, and then use the triangle to find $\cos\theta$ and $\sin\theta$ as functions of δ, g, and n.

(b) If $\delta \ll g$, then Rabi oscillations occur between $|e, n\rangle$ and $|g, n + 1\rangle$, as in Sect. 23.1. But if $\delta \gg g$ (the dispersive regime), show that

$$|+, n\rangle \approx |e, n\rangle + \frac{g\sqrt{n + 1}}{\delta}|g, n+1\rangle, \quad |-, n\rangle \approx -\frac{g\sqrt{n + 1}}{\delta}|e, n\rangle + |g, n+1\rangle,$$

which reduces back to the unmixed states for $g \to 0$ or $\Delta \to \infty$, as they should.

References

1. Y. Arakawa, M.J. Holmes, Appl. Phys. Rev. **7**, 021309 (2020)
2. A.L. Efros, L.E. Brus, ACS Nano **15**, 6192 (2021)
3. A.A.H. Abellatif, et al., Int. J. Nanomed. **17**, 1951 (2022)
4. A.M. Wagner, J.M. Knipe, G. Orive, N.A. Peppas, Acta Biomater. **94**, 44 (2019)

5. T. Heindel, J.H. Kim, N. Gregersen, A. Rastelli, S. Reitzenstein, Adv. Opt. Photon. **15**, 613 (2023)
6. R. Schirhagl, K. Chang, M. Loretz, C.L. Degen, Annu. Rev. Phys. Chem. **65**, 83 (2013)
7. L. Childress, R. Walsworth, M. Lukin, Phys. Today **67**, 38 (2014)
8. P. Neumann et al., Science **329**, 542 (2010)
9. L.P. McGuinness et al., Nat. Nanotech. **6**, 358 (2011)
10. T. Zhang, et al., ACS Sensors **6**, 2077 (2021)
11. W.D. Oliver, Superconducting Qubits, in *Quantum Information Processing: Lecture Notes of 44th IFF Spring School*, ed. by D.P. DiVincenzo (2013). https://inis.iaea.org/collection/NCLCollectionStore/Public/46/123/46123072.pdf
12. T. Roth, R. Ma, W.C. Chew, The transmon qubit for electromagnetics engineers: an introduction, in *IEEE Antennas and Propagation Magazine, Jun 7, 2022* (2022), p. 2
13. P. Krantz, M. Kjaergaard, F. Yan, T.P. Orlando, S. Gustavsson, W.D. Oliver, Appl. Phys. Rev. **6**, 021318 (2019)
14. S. Bravyi, O. Dial, J.M. Gambetta, D. Gil, Z. Nazarioa, J. Appl. Phys. **132**, 160902 (2022)
15. R.J. Schoelkopf, S.M. Girvin, Nature **451**, 664 (2008)
16. S.M. Girvin, Circuit QED: superconducting qubits coupled to microwave photons in *Quantum Machines: Measurement and Control of Engineered Quantum Systems*, ed. by M. Devoret, et al. Lecture Notes of the Les Houches Summer School, vol. 96 (2011) (Oxford Academic, Oxford, 2014). https://doi.org/10.1093/acprof:oso/9780199681181.003.0003
17. A. Blais, A.L. Grimsmo, S.M. Girvin, A. Wallraff, Rev. Mod. Phys. **93**, 025005 (2021)
18. A. Blais, S.M. Girvin, W.D. Oliver, Nat. Phys. **16**, 247 (2020)
19. S. Haroche, M. Brune, J.M. Raimond, Nat. Phys. **16**, 243 (2020)
20. J. Koch et al., Phys. Rev. A **76**, 042319 (2007)

Chapter 27
Future Prospects and Guide to Additional Topics

Quantum science and quantum technology are broad, rapidly growing fields, so it is impossible to cover all aspects in a comprehensive manner in a single book. In this final chapter, brief descriptions are given of a few of the topics that have not been covered in the previous chapters, with references to books and review articles that contain more detailed coverage.

27.1 Quantum Memories

All computers require memory registers that can store results until they are needed for further computations or for readout. In a classical computer, there are a variety of ways of storing information, including MOS transistors, the spin states of ferromagnetic materials, and optical disks, among others.

Quantum computers also need to be able to store information, so a prerequisite for a fully functioning quantum computer is a **quantum memory**. Essentially, this is a means of absorbing a qubit (usually the state of a photon), storing it in an atom or collection of atoms, and then re-emitting it in a controlled manner on demand.

There have long been two main approaches to quantum memory, as well as hybrid systems that incorporate aspects of both. These consist of optically controlled memories and photon echo memories. An **optically controlled memory** uses an atomic system in which a strong control pulse stimulates absorption or controlled emission of a weak signal pulse, allowing the optical signal to be stored and then released after a controllable delay. These approaches usually make use of a three-state atomic system (a so-called Λ-structure). For example, the absorbed photon may cause an electron to be excited from the ground state $|g\rangle$ to the excited state $|e\rangle$, while a separate, lower-frequency control pulse caused the electron to drop to a stable auxiliary state $|s\rangle$ for storage. When the qubit is to be released, another control pulse causes $|s\rangle$ to be excited back to $|e\rangle$, which quickly decays back to

© The Author(s), under exclusive license to Springer Nature Switzerland AG 2025
D. S. Simon, *Introduction to Quantum Science and Technology*, Undergraduate
Texts in Physics, https://doi.org/10.1007/978-3-031-81315-3_27

$|g\rangle$, releasing a new photon in the original state. This approach can be viewed as slowing the optical pulse speed to zero during storage, using a phenomenon known as electromagnetically induced transparency. The **photon echo** or **Hahn echo** protocol is similar to the spin echo of NMR systems: a large ensemble of atoms is rephased in order for interference of their output emissions to produce a replica of the original input signal. These memory protocols can be implemented in number of different physical platforms, such as rare-earth ion-doped solids, spin systems, NV centers in diamonds, and alkali vapors.

A number of reviews exist on the subject of quantum memories; these include [1–7].

27.2   Quantum Walks

Most introductory statistical mechanics courses cover classical random walks, in which (in one dimension) a particle chooses randomly to move to the left or to the right. The probability of motion in each direction is fixed, and the steps are assumed to be independent of each other. The position x at a given time is then a random variable, distributed according to a binomial distribution. One important characteristic of the classical walk is that the particle motion is **diffusive**: the distance Δx traveled in time t is proportional to the *square root* of time

$$\Delta x \sim \sqrt{t}. \tag{27.1}$$

Since the 1990s, there has been intense interest in **quantum walks** [8], the quantum mechanical analog of the classical random walk. In the quantum case, rather than moving either to the left *or* to the right, the particle has an amplitude both to move left *and* to move right. In other words, if the positions are sites on a discrete lattice labeled by integer m, then a particle initially localized at site m_0, $|\psi(0)\rangle = |m_0\rangle$, will be in a state

$$|\psi(1)\rangle = \alpha|m_0 - 1\rangle + \beta|m_0 + 1\rangle \tag{27.2}$$

after one step, where $|\alpha|^2 + |\beta|^2 = 1$. After two time steps, the particle will be a superposition of position states $|m_0\rangle$, $|m_0 - 2\rangle$, and $|m_0 + 2\rangle$, and so on. As a result, over time the particle wavefunction becomes spread over many lattice sites, and the amplitudes arriving at each site by different paths can interfere with each other, leading to quantum effects not present in the classical walk. In particular, the diffusive motion of the classical random walk becomes **ballistic motion** in the quantum case: the expectation value of the distance traveled in time t is now *directly proportional* to time:

$$\langle \Delta x \rangle \sim t. \tag{27.3}$$

Because of this ballistic motion, the quantum walk can interact with a larger range of sites in the same amount of time, allowing a speed-up of search algorithms and transport processes. In fact, it can be shown that quantum walks can be used to model *any* quantum algorithm [9] and that use of quantum walks can, in some cases, cause an exponential speed up compared to classical algorithms. In addition, quantum walks allow the possibility of perfect quantum state transfer [10], which can be used to transport quantum states without alteration to desired locations in a quantum computer or quantum network. Quantum walks can even be used to implement systems with distinct topological phases [11].

Quantum walks can be arranged to occur in many different degrees of freedom; the walk does not necessarily have to occur in position space, but can take place, for example, in momentum space or in the space of spin or orbital angular momentum. See [12, 13] for introductory reviews on the physics of quantum walks, and [14] for a survey of results and applications with a comprehensive bibliography.

27.3 Quantum Internet

The Internet is essentially just a network of connections linking many computers together, along with a set of protocols for how they communicate with each other. One of the current dreams of researchers is the construction of a **quantum Internet**: a network of linked quantum computers, communicating with each other via single photons or entangled photon pairs. Such a quantum network would have a number of advantages over the classical internet. For example, the use of quantum cryptographic methods allows enhanced security, and the quantum links between the individual computers would provide a way of scaling quantum computing power up to larger numbers of qubits, using distributed quantum computing with multiple computers working on different parts of a single calculation. More generally, quantum networks could also be used to transmit quantum states to desired locations for quantum sensing or microscopy.

There are a number of technical issues that need to be resolved before a quantum Internet becomes a reality. Just to list a few:

- A means has to be arranged to reliably direct the quantum state through the network to its desired location. Optical switches are one possibility for doing this; another method is to use constructive and destructive interference between entangled photon pairs [15].
- Loss has to be kept very low within the network. Loss is even more damaging in a quantum network than in a classical one, since the signal may only contain one or two photons to begin with. Decoherence also has to be suppressed in order to maintain the quantum nature of the signal.
- In a large-scale quantum network, the individual nodes would most likely be connected by optical fibers. Current optical fibers are only low-loss in the infrared telecom windows. So nonlinear optical devices may be needed to reduce the

frequency of downconverted photons from the optical range into the infrared. Newer technology for this is also becoming available, such as ensembles of cold ^{85}Rb atoms [16].

- A means is needed to connect flying qubits (photons) traveling between nodes of the network to stationary qubits at the node site.
- Current single-photon sources are probabilistic, and most have a chance of producing more than one photon in a given pulse. High-quality, deterministic sources that produce a single photon on demand are needed. A great deal of progress has been made in this direction using stimulated emission from single atoms or quantum dots.

Potential solutions for most of these issues are already in the works, so that practically useful quantum networks may start to become a reality within a few years. Maintaining photonic quantum states over long distances has also been achieved, making it feasible to eventually build up large-scale networks; for example, since 2022 researchers at Argonne National Lab and the University of Chicago have transmitted high-quality quantum states over a fiber-optic network containing six nodes and 125 miles of optical fiber. Similarly, scientists from Brookhaven National Laboratory, Stony Brook University, and the U.S. Department of Energy's Energy Sciences Network have transmitted entangled states over distances of 11 miles or more and are now working with a quantum network containing over 80 miles of fiber. Needless to say, companies like IBM and Google have also begun taking an active role in this research.

For further information, see [17–21] and references therein.

27.4   Quantum Batteries

In a standard rechargeable electrical storage battery, a material at one end of the battery is gradually oxidized, losing electrons which then reduce (add electrons to) a different material at the other end. This continues until the first material is completely oxidized and has no more spare electrons to release. By applying an external voltage, the battery can be recharged by forcing the electrons back into the original material. Normally, the larger a battery is, the longer it takes to become fully charged. But a new category of **quantum batteries** [22] has recently been developed, which exhibit a property called **superabsorption** [23], in which the charging time actually *decreases* as the battery becomes larger. These batteries make use of a photosensitive dye called Lumogen-F Orange [24], in which the molecules can be put into coherent superpositions or entangled states.

In a situation reminiscent of the resolution improvement found in entangled sensors, there is evidence that entangled molecules can charge up to N times faster than N independent molecules [25]. This opens up the possibilities for charging electric vehicles in a matter of seconds instead of hours or of developing vastly improved solar cells for large-scale power generation. This work is still in an early stage, and practical limitations will probably prevent reaching the full theoretical N-fold improvement rate, but there are hopes it will eventually lead to commercially available quantum batteries.

27.5 Quantum Biology

Biological systems generally operate at or near room temperature, and they are constantly exchanging matter (food, oxygen, etc.) and energy (light and heat) with their environment. Normally, such systems that are warm, open, and strongly coupled to their environment will exhibit very rapid decoherence, so that coherent quantum states are next to impossible to observe. But there have been a number of recent studies that indicate quantum-coherent states and quantum processes may nevertheless play a role in some biological processes. For example, the photon-absorbing bioantennas that carry out the light-harvesting stage of photosynthesis operate with near 100% quantum efficiency [26], a number of studies have indicated that biomolecules can exhibit spatial coherence over several nanometers and temporal coherences of up to tenths of picoseconds [27–29], and there is evidence that environmentally induced interactions can increase transport efficiency of quantum states [30, 31]. All of this adds up to evidence that purely quantum processes, which cannot be described with a classical picture, may be essential in some aspects of biology. In addition, it has been suggested that quantum effects play important roles in bird navigation [32, 33] and the olfactory sense [34, 35].

Conversely, there have been suggestions that biological molecules may be useful for carrying out human-designed quantum processes. For example, Ref. [36] suggests that photon antibunching statistics and non-classical photon states for interferometry experiments may be generated using proteins, and [37] proposes the possibility that quantum information processing algorithms can be carried out with nucleotides. Further, a number of groups have proposed methods of carrying out photosynthesis using human-engineered systems [38–42]. Fueling further interest in such ideas is a relatively recent experiment [43] demonstrating entanglement between the state of a bacterium and a photon.

Extensive reviews of quantum biology may be found in [44–46].

References

1. M. Afzelius, N. Gisin, H. de Riedmatten, Phys. Today **68**(12), 42 (2015)
2. A.I. Lvovsky, B.C. Sanders, W. Tittel, Nat. Photonics **3**, 706 (2009)
3. C. Simon et al., Eur. Phys. J. D - At. Mol. Opt. Plasma Phys. **58**, 1 (2010)
4. W. Tittel et al., Laser Photonics Rev. **4**, 244 (2010)
5. F. Bussiéres et al., Mod. Opt. **60**, 1519 (2013)
6. K. Heshami et al., J. Mod. Opt. **63**, 2005 (2016)
7. L. Ma, O. Slattery, X. Tang, J Opt. **19**, 043001 (2017)

8. Y. Aharonov, L. Davidovich, N. Zagury, Phys. Rev. A **48**, 1687 (1993)
9. A.C. Childs, Phys. Rev. Lett. **102**, 180501 (2009)
10. V.M. Kendon, C. Tamon, J. Comp. Theor. Nanoscience **8**, 1 (2010)
11. T. Kitagawa, Quantum Inf. Process **11**, 1107 (2012)
12. J. Kempe, Contemp. Phys. **44**, 307 (2003)
13. V. Kendon, Where to quantum walk, arXiv:1107.3795v1 [quant-ph] 19 Jul 2011
14. K. Kadian, S. Garhwal, A. Kumar, Comp. Sci. Rev. **41**, 100419 (2021)
15. S. Osawa, D.S. Simon, A.V. Sergienko, Phys. Rev. A **104**, 012617 (2021)
16. A. Kumar, A. Suleymanzade, M. Stone, L. Taneja, A. Anferov, D.I. Schuster, J. Simon, Nature **615**, 614 (2023)
17. S. Wehner, D. Elkouss, R. Hanson, Sciene **362**, 303 (2018)
18. D. Castelvecchi, Nature **554**, 289 (2018)
19. A. Thoss, *Photonics Spectra* (2023), p. 52
20. H.J. Kimble, Nature **453**, 1023 (2019)
21. S. Ritter, Nature **484**, 195 (2012)
22. J.Q. Quach, G. Cerullo, T. Virgili, Joule **7**, 2195 (2023)
23. R. Alicki, M. Fannes, Phys. Rev. E **87**, 042123 (2013)
24. J.Q. Quach et al., Sci. Adv. **8**, eabk3160 (2022)
25. S. Juliá-Farré, T. Salamon, A. Riera, M. N. Bera, and M. Lewenstein, Phys. Rev. Res. **2**, 023113 (2020)
26. R. van Grondelle, V. Novoderezhkin, Phys. Chem. Chem Phys. **8**, 793 (2006)
27. M. Chachisvilis, T. Pullerits, M.R. Jones, C.N. Hunter, V. Sundström, Chem. Phys. Lett. **224**, 345 (1994)
28. M. Vos, M.R. Jones, C.N. Hunter, J. Breton, J.C. Lambry, J.L. Martin, Proc. Natl Acad. Sci. USA **91**, 701 (1994)
29. R. Kumble, S. Palese, R.W. Visschers, P.L. Dutton, R.M. Hochstrasser, Chem. Phys. Lett. **261**, 396 (1996)
30. M. Mohseni, P. Rebentrost, S. Lloyd, A. Aspuru-Guzik, J. Chem. Phys. **129**, 174106 (2008)
31. A. Shabani, M. Mohseni, H. Rabitz, S. Lloyd, Phys. Rev. E **89**, 042706 (2014)
32. K. Schulten, C.E. Swenberg, A. Weller, Z. Phys. Chem. **111**, 1 (1978)
33. Y. Zhang, G.P. Berman, S. Kais, Int. J. Quant. Chem. **115**, 1327 (2015)
34. G.M. Dyson, Perfumery Essent. Oil Rec. **19**, 456 (1928)
35. L. Turin, Chem. Senses **21**, 773 (1996)
36. G. Sanchez-Mosteiro et al., Chem. Phys. Chem. **5**, 1782 (2004)
37. J. Jones, M. Mosca, Phys. Rev. Lett. **83**, 1050 (1999)
38. G.D. Scholes, G.R. Fleming, A. Olaya-Castro, R. van Grondelle, Nat. Chem. **3**, 763 (2011)
39. G.D. Scholes, T. Mirkovic, D.B. Turner, F. Fassioli, A. Buchleitner, Energy Environ. Sci. **5**, 9374 (2012)
40. A. Olaya-Castro, A. Nazir, G.R. Fleming, Phil. Trans. R. Soc. A **370**, 3613 (2013)
41. C. Creatore, M.A. Parker, S. Emmott, A.W. Chin, Phys. Rev. Lett. **111**, 253601 (2013)
42. M. Sarovar, K.B. Whaley, New J. Phys. **15**, 013030 2013
43. C. Marletto, D.M. Coles, T. Farrow, V. Vedral, J. Phys. Commun. **2**, 101001 (2018)
44. M. Arndt, T. Juffmann, V. Vedral, HFSP J. **3**, 386 (2009)
45. A. Marais et al., J. R. Soc. Interface **15**, 20180640 (2018)
46. J. Cao et al., Sci. Adv. **6**(14), eaaz4888 (2020)

Appendix A
Math Review

In this appendix, a quick review is given to some areas of mathematics that are needed for the main text. Useful formulas such as those for Gaussian integrals are also given for easy reference.

A.1 Complex Numbers

Although the idea of taking square roots of negative numbers was mentioned by Hero of Alexandria nearly 2000 years ago, the term *imaginary numbers* didn't occur until it was used by Descartes in the seventeenth century. The very term implied their disreputable nature, and it would be another century or more before they began to gain some acceptance among mathematicians. Despite the fact that they still strike fear into schoolchildren (and their parents) today, they have become vital in areas ranging from electromagnetic theory and electrical engineering to quantum mechanics and abstract algebra.

Defining the imaginary unit $i = \sqrt{-1}$, then a complex number is defined to be an object composed from a sum of real and imaginary numbers: $z = x + i\,y$, where x and y are real numbers, known respectively as the real and imaginary parts of z: $x = Re(x)$, $y = Im(z)$. The complex conjugate is obtained by flipping the sign of the imaginary part: $z^* = x - iy$.

It is common to think of x and y as coordinates in a plane, as shown in Fig. A.1a, with z representing the point at those coordinates. If a line is drawn from the origin to z, then the length of that line, written $|z|$, is called the magnitude of the complex number. The angle of the line from the x axis (called the real axis in this context) to z is called the complex phase angle, θ. Basic trigonometry applied to the figure below then gives the following relations:

$$|z|^2 = x^2 + y^2 \qquad \text{and} \qquad \tan\theta = \frac{y}{x} \qquad (A.1)$$

© The Author(s), under exclusive license to Springer Nature Switzerland AG 2025
D. S. Simon, *Introduction to Quantum Science and Technology*, Undergraduate
Texts in Physics, https://doi.org/10.1007/978-3-031-81315-3

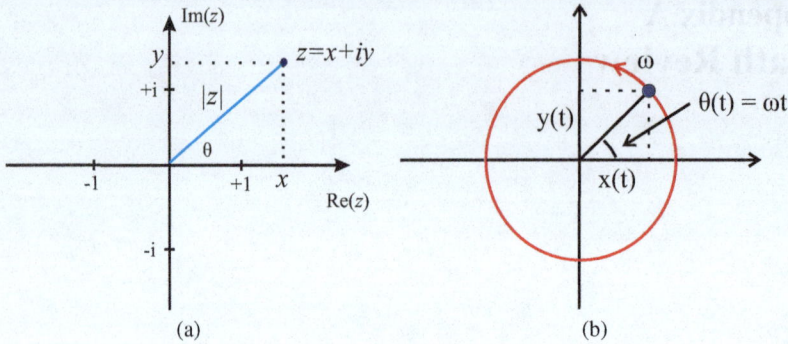

Fig. A.1 (**a**) Using the real and imaginary parts as coordinates in the complex plane, a complex number z can be viewed as a point in the plane. The right triangle formed by the blue line and the dotted lines normal to the axes can be used to relate x, y, $|z|$, and θ by means of the Pythagorean theorem and the definitions of the standard trig functions. (**b**) Euler's theorem: the position of a particle on the unit circle in the complex plane can be used to describe circular motion at rate ω

Notice that the magnitude can also be obtained by multiplying z with its complex conjugate: $|z|^2 = z^*z$. Given the expression for $|z|^2$ in the previous paragraph this is trivial to show:

$$z^*z = (x - iy)(x + iy) = x^2 + ixy - ixy + y^2 = x^2 + y^2 = |z|^2. \qquad (A.2)$$

One of the reasons complex numbers are so useful is that they can be exponentiated to describe periodic motions. This is a consequence of **Euler's theorem**:

$$e^{i\theta} = \cos\theta + i\sin\theta. \qquad (A.3)$$

The most straightforward way to prove this is simply to expand both sides in Taylor series and see that the two series are equal. To see the connection to periodic motion, imagine that a particle is traveling at constant angular speed ω counterclockwise around the unit circle in the complex plane (see Fig. A.1b). Then the angle from the real axis at time t is $\theta(t) = \omega t$. Applying Euler's theorem, we find

$$e^{i\omega t} = \cos\omega t + i\sin\omega t. \qquad (A.4)$$

So the real and imaginary parts of the complex exponential describe simultaneous oscillations (90° out of phase with each other) in the real and imaginary directions: $x(t) = \cos\omega t$ and $y(t) = \sin\omega t$. Mathematically, the bouncing motion of a spring can simply be viewed as the shadow of a circular motion, projected onto a line (one of the axes, or any other line through the origin).

Complex numbers are often used to describe mechanical or electromagnetic waves or quantum wavefunctions. They are preferable to using sines and cosines, because exponentials are much easier to work with algebraically than trigonometric functions, as can be seen from the next example. If the variable undergoing the wave motion is a physically measurable quantity like an electric field or the height of a water wave, then the usual procedure is to allow the variable to be complex while carrying out any algebraic manipulations, then take the real part at the end to get back the real, physical variable.

Example A.1.1 Use complex exponentials to derive the sum formula for $\cos(\alpha + \beta)$.

Start by looking at the exponential instead of the cosine, then use Euler's theorem:

$$e^{i(\alpha+\beta)} = \cos(\alpha + \beta) + i \sin(\alpha + \beta). \tag{A.5}$$

But we can also write the same exponential as

$$e^{i\alpha} e^{i\beta} = (\cos\alpha + i\sin\alpha)(\cos\beta + i\sin\beta) \tag{A.6}$$

$$= (\cos\alpha \cos\beta - \sin\alpha \sin\beta) + i(\cos\alpha \sin\beta + \sin\alpha \cos\beta). \tag{A.7}$$

Setting the real parts of expressions in Eqs. A.5 and A.7 equal to each other, we find the summation formula for the cosine:

$$\cos(\alpha + \beta) = \cos\alpha \cos\beta - \sin\alpha \sin\beta. \tag{A.8}$$

As a bonus, we also get the summation formula for sines without any additional work, simply by setting the imaginary parts equal:

$$\sin(\alpha + \beta) = \cos\alpha \sin\beta + \sin\alpha \cos\beta. \qquad \blacksquare \tag{A.9}$$

In particular, complex exponentials describe propagating waves in a simple manner. The function $\psi_R(x,t) = Ae^{-i(kx-\omega t)}$ describes a wave of amplitude A, angular frequency ω, and wave number k traveling toward the right, while $\psi_L(x,t) = Ae^{-i(kx+\omega t)}$ describes a similar wave moving toward the left.

A.2   Matrices

A **matrix** is a rectangular array of real or complex numbers, such as the following:

$$\begin{pmatrix} 3 & 2 & 1 \\ 0 & -7 & 13 \end{pmatrix}, \quad \begin{pmatrix} 0 & i & 0 & -1 \\ -i & 0 & -1 & 0 \\ 0 & -1 & i & 0 \\ -1 & 0 & 0 & -1 \end{pmatrix}, \quad (-4\ 7\ 9), \quad \begin{pmatrix} 1 \\ 0 \\ 1 \end{pmatrix}.$$

Given a matrix A, the entry in the ith row and jth column is often denoted by a pair of indices: A_{ij}. So for the matrix $A = \begin{pmatrix} 0 & -i \\ i & 0 \end{pmatrix}$, the entries are $A_{11} = A_{22} = 0$, $A_{12} = -i$, $A_{21} = i$. A matrix with n rows and m columns is referred to as an $(n \times m)$ matrix.

The product of two matrices $C = AB$ is well-defined as long the number of columns of A equals the number of rows of B. Then the entries of C are given by

$$C_{ij} = \sum_k A_{ik} B_{kj}. \qquad (A.10)$$

Roughly speaking this equation tells us to multiply the elements of each row of A times the elements of each column of B, as in the following example.

Example A.2.1 Consider the matrices

$$A = \begin{pmatrix} 1 & 2 \\ -3 & 4 \\ 6 & -7 \end{pmatrix}, \qquad \begin{pmatrix} 5 & 3 \\ 0 & 1 \end{pmatrix}.$$

The product AB is

$$AB = \begin{pmatrix} 1 & 2 \\ -3 & 4 \\ 6 & -7 \end{pmatrix} \begin{pmatrix} 5 & 3 \\ 0 & 1 \end{pmatrix}$$

$$= \begin{pmatrix} (1)(5) + (2)(0) & (1)(3) + (2)(1) \\ (-3)(5) + (4)(0) & (-3)(3) + (4)(1) \\ (6)(5) + (-7)(0) & (6)(3) + (-7)(1) \end{pmatrix}$$

$$= \begin{pmatrix} 5 & 5 \\ -15 & -5 \\ 30 & 11 \end{pmatrix}.$$

In contrast, the product BA is not defined, since the number of columns of B does not match the number of rows of A. ∎

The **transpose** A^T of a matrix A is obtained by flipping the matrix across the main descending diagonal or equivalently by interchanging the rows with the columns; for example, if $M = \begin{pmatrix} 1 & 2i \\ 3 - 4i & 4i \end{pmatrix}$, then the transpose is $M^T = \begin{pmatrix} 1 & 3 - 4i \\ 2i & 4i \end{pmatrix}$. If the elements of A are A_{ij}, then the elements of A^T are found simply by interchanging the indices: $\left(A^T \right)_{ji} = A_{ij}$.

The **Hermitian adjoint** or Hermitian transpose $A^\dagger$ is the complex conjugate of the transpose: $A^\dagger = (A^T)^*$ has elements $(A^\dagger)_{ji} = A_{ij}^*$. So for the matrix M defined above, the adjoint is $M^\dagger = \begin{pmatrix} 1 & 3+4i \\ -2i & -4i \end{pmatrix}$.

Consider a square $n \times n$ matrix S. Two important numbers associated with this matrix are the determinant and the trace. The **determinant** of matrix A is denoted by either $|A|$ or Det A and is defined by

$$Det\ A = \sum_{i_1=1}^{n} \sum_{i_2=1}^{n} \cdots \sum_{i_n=1}^{n} \epsilon_{i_1 i_2 \ldots i_n} A_{1i_1} A_{2i_2} \ldots A_{ni_n}, \qquad (A.11)$$

where $\epsilon_{ijk\ldots}$ is the antisymmetric Levi-Civita tensor defined in Appendix C. (Eq. C.15 or C.16). For $n = 2$ or $n = 3$, the determinant is easy to find:

$$A = \begin{pmatrix} a & b \\ c & d \end{pmatrix} \longrightarrow Det\ A = ad - bc$$

$$B = \begin{pmatrix} a & b & c \\ d & e & f \\ g & h & i \end{pmatrix} \longrightarrow Det\ B = aei + bfg + cdh - ceg - bdi - afh.$$

The **trace** is simply the sum of the diagonal elements: if

$$A = \begin{pmatrix} a_{11} & a_{12} & a_{13} & \cdots & a_{1n} \\ a_{21} & a_{22} & a_{23} & \cdots & a_{2n} \\ & & \ddots & & \\ a_{n1} & a_{n2} & a_{n3} & \cdots & a_{nn} \end{pmatrix} \qquad (A.12)$$

then

$$Tr\ A = \sum_{i=1}^{n} a_{nn} = a_{11} + a_{22} + a_{33} + \cdots + a_{nn}. \qquad (A.13)$$

Column matrices are often used to represent vectors. For example, the three-dimensional vector $\boldsymbol{v} = v_x \hat{\boldsymbol{i}} + v_y \hat{\boldsymbol{j}} + v_z \hat{\boldsymbol{k}}$ would be represented by the column matrix

$$\tilde{v} = \begin{pmatrix} v_x \\ v_y \\ v_z \end{pmatrix}.$$

The dot product between two vectors would then simply be the inner product of one column matrix with the transpose of the other:

$$\boldsymbol{v} \cdot \boldsymbol{w} = \tilde{\boldsymbol{v}}^T \tilde{\boldsymbol{w}} = \begin{pmatrix} v_x & v_y & v_z \end{pmatrix} \begin{pmatrix} w_x \\ w_y \\ w_z \end{pmatrix} = v_x w_x + v_y w_y + v_z w_z. \qquad (A.14)$$

The generalization to vectors of arbitrary dimension is then obvious. The one extra feature that is present in quantum mechanics is that states are represented as vectors with complex entries. In this case, the inner product involves a Hermitian adjoint in place of the transpose:

$$\boldsymbol{v} \cdot \boldsymbol{w} = \tilde{\boldsymbol{v}}^\dagger \tilde{\boldsymbol{w}} = \begin{pmatrix} v_x^* & v_y^* & v_z^* \end{pmatrix} \begin{pmatrix} w_x \\ w_y \\ w_z \end{pmatrix} = v_x^* w_x + v_y^* w_y + v_z^* w_z. \qquad (A.15)$$

The identity matrix I has zeros everywhere, except ones along the main diagonal:

$$I = \begin{pmatrix} 1\,0\,0\,0\ldots \\ 0\,1\,0\,0 \\ 0\,0\,1\,0 \\ 0\,0\,0\,1 \\ \vdots \qquad \ddots \end{pmatrix}, \qquad (A.16)$$

obeys $IM = MI = M$ for any matrix M. This allows inverses of matrices to be defined, so that $M^{-1}M = MM^{-1} = I$.

For a square $n \times n$ matrix M, any n-dimensional vector (or column matrix) $\boldsymbol{v}$ such that

$$M\boldsymbol{v} = \lambda \boldsymbol{v} \qquad (A.17)$$

is called an **eigenvector** of M. The number λ is complex in general and is called an **eigenvalue** of M. In the special case that M is Hermitian, λ will be real, and if M is unitary, then the eigenvalues will obey $|\lambda| = 1$. If the matrix is representing a quantum mechanical observable, then the eigenvalues represent the allowed values of that variable; for example, the eigenvalues of the Hamiltonian operator will be the allowed energy levels of the system. In most cases encountered in practice, an $n \times n$ matrix will have n linearly independent eigenvectors, and so any state in the space the matrix acts on can be expanded in terms of these n eigenvectors. The eigenvalues are found by solving the equation

$$\mathrm{Det}(M - \lambda I) = 0. \qquad (A.18)$$

Math software packages such as Matlab and Mathematica have built-in operations that can be used to easily find eigenvectors, eigenvalues, and inverses of matrices.

A matrix M with eigenvalues λ_j can be diagonalized. In other words, there is some invertible matrix V acting on M as a similarity transformation, $D = V^{-1}MV$ such that the matrix D is zero everywhere except on the main diagonal. The diagonal entries are simply the eigenvalues of M. So the transformation by V simply rotates the coordinate system so that the basis vectors point along the eigenvectors of the matrix. The matrix V is readily found if the eigenvectors are known: the columns of V are the normalized eigenvectors of M. Notice that V is not unique, since the eigenvectors can be permuted among themselves. Two matrices are simultaneously diagonalizable (i.e., they can be diagonalized by the same transformation V) if and only if they commute, $[A, B] = 0$.

In their most basic form, matrices are simply a useful way to deal with systems described by sets of linear equations. Consider the following, for example.

Example A.2.2 Imagine that we wish to solve for the currents in an electric circuit, such as that of Fig. A.2. The Kirchhoff loop and junction rules give us a set of three linear equations for the currents:

$$I_2 + I_3 = I_1 \tag{A.19}$$

$$R_1 I_1 + R_2 I_2 = V_0 \tag{A.20}$$

$$R_1 I_1 + R_3 I_3 = V_0. \tag{A.21}$$

This set of simultaneous equations can be conveniently written in the form of a single matrix equation

$$RI = V, \tag{A.22}$$

where we have defined

$$I = \begin{pmatrix} I_1 \\ I_2 \\ I_3 \end{pmatrix}, \qquad V = \begin{pmatrix} 0 \\ V_0 \\ V_0 \end{pmatrix}, \qquad R = \begin{pmatrix} 1 & -1 & -1 \\ R_1 & R_2 & 0 \\ R_1 & 0 & R_3 \end{pmatrix}. \tag{A.23}$$

Solving for the unknown currents then simply amounts to inverting the matrix equation

Fig. A.2 A simple electric circuit whose Kirchhoff analysis results in Eqs. A.19–A.21

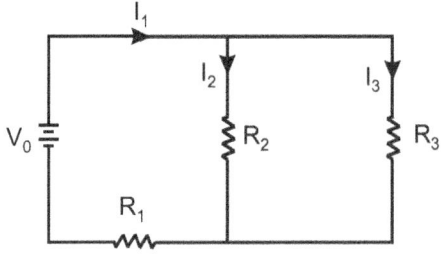

$$I = R^{-1}V, \tag{A.24}$$

where it can be readily verified that

$$R^{-1} = \frac{1}{\Delta} \begin{pmatrix} R_2 R_3 & R_3 & R_2 \\ R_1 R_3 & R_1 + R_3 & R_1 \\ R_1 R_2 & R_1 & R_1 + R_2 \end{pmatrix}, \tag{A.25}$$

with

$$\Delta = \mathrm{Det}\, R = R_1 R_2 + R_1 R_3 + R_2 R_3. \quad \blacksquare \tag{A.26}$$

Problems

1. (a) Find the products AB, AC, and BC, where

$$A = \begin{pmatrix} 1 & 1 \\ 1 & -1 \end{pmatrix}, \quad B = \begin{pmatrix} 2 & 0 \\ 1 & -2 \end{pmatrix} \quad C = \begin{pmatrix} -3 \\ 3 \end{pmatrix}. \tag{A.27}$$

   (b) Can the matrices A and B be simultaneously diagonalized?

   (c) Can the matrices $D = \begin{pmatrix} 0 & 1 \\ 1 & 0 \end{pmatrix}$ and $F = \begin{pmatrix} 1 & 1 \\ 1 & 1 \end{pmatrix}$ be simultaneously diagonalized? If so, find a similarity matrix V that diagonalizes them both.

2. (a) Find the eigenvalues and eigenvectors of the matrix $M = \begin{pmatrix} 1 & 1 \\ 1 & -1 \end{pmatrix}$.

   (b) Find the eigenvalues of R in Eq. A.23.

3. Consider the three-dimensional Grover coin, a matrix that arises in the study quantum walks:

$$G = \frac{1}{3} \begin{pmatrix} -1 & 2 & 2 \\ 2 & -1 & 2 \\ 2 & 2 & -1 \end{pmatrix}. \tag{A.28}$$

   (a) Find the determinant and trace of G.
   (b) Find the eigenvalues and eigenvectors of G.
   (c) Verify that the determinant and trace equal, respectively, the sum and the product of the eigenvalues.

A.3   Probability Distributions

Probability distributions are ubiquitous in physics and engineering, as well as in the social sciences and many other areas. Here we give a brief review of probability distributions in general and of two of the most important distributions in physics.

Consider some random variable x. This is some variable whose value we can predict ahead of time and which may have different values each time it is measured. The lack of predictability may come from a classical source of randomness (rolling of dice, thermal fluctuations, electromagnetic static in a circuit, etc.) or may be of a fundamental nature (due to Heisenberg uncertainty).

First consider the case where x is a discrete variable; in other words, its possible values x_n can be labeled by a set of integers, $n = 1, 2, 3 \ldots, N$, where N could be finite or infinite. A **discrete probability distribution** describing x is a collection of all of the x_n and their probabilities, p_n. The distribution can be described by a table giving the x_n and p_n values (if N is finite) or by a histogram; see Fig. A.3. These probabilities have to obey some fairly obvious constraints:

$$0 \le p_n \le 1, \quad \text{for all } n \tag{A.29}$$

$$\sum_{n=0}^{\infty} p_n = 1. \tag{A.30}$$

The **population mean** μ and the **population standard deviation** σ_p describe the center and width of the probability histogram that would be found if you had perfect knowledge of the distribution. These are defined by

$$\mu = \sum_{n=0}^{N} p_n x_n \tag{A.31}$$

$$\sigma_p = \sqrt{\frac{\sum_{n=0}^{N} p_n (x_n - \mu)^2}{N}}. \tag{A.32}$$

Fig. A.3 A simple discrete probability distribution $p_n = \frac{9-(n-3)^2}{35}$ for $n = 0, 1, \ldots, 6$. The distribution is expressed as a table of probabilities and as a histogram

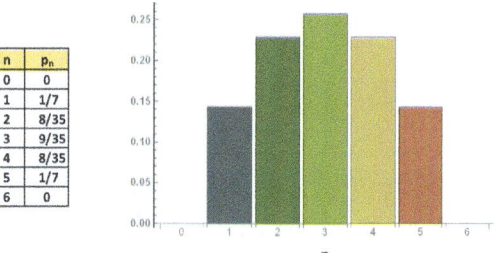

n	p_n
0	0
1	1/7
2	8/35
3	9/35
4	8/35
5	1/7
6	0

The **variance** is simply the square of the standard deviation, $\text{var} = \sigma_p^2$. It is easy to show (Problem 3) that the variance can also be written

$$\text{var} = \sigma_p^2 = \overline{x^2} - (\bar{x})^2 = \langle x^2 \rangle - \langle x \rangle^2, \tag{A.33}$$

where in the last expression we have used brackets $\langle\ \rangle$ to denote means instead of bars to make the order of operations clearer.

In the real world, we never know the exact values of μ and σ. Instead, we have estimates of them. To obtain these estimates, we take a **sample**: we take a set of N randomly chosen measurements of x, where N is known as the data size. If the values of x we obtain from measurements on this sample are $x_1, x_2, \ldots x_N$, then the values of μ and σ_p are estimated by the sample mean $\bar{x}$ and the sample standard deviation σ. These are defined by

$$\bar{x} = \frac{1}{N} \sum_{j=0}^{N} x_j \tag{A.34}$$

$$\sigma = \sqrt{\frac{\sum_{j=0}^{N}(x_j - \bar{x})^2}{N-1}}. \tag{A.35}$$

Often $N \gg 1$, so that the $N - 1$ on the bottom of σ can be approximated by N; then σ and σ_p have similar forms. Henceforth, we won't distinguish between sample and population when we discuss means and standard deviations, unless the distinction is important, and we will drop the subscript from σ_p.

If x is a continuous variable, then the corresponding **continuous probability distribution** is given by a continuous function, $p(x)$. Unlike the p_n in the discrete case, $p(x)$ is not a probability, but a **probability density** or probability of unit x: the probability of being in small range x centered at x is $p(x)dx$. More generally, the probability of x being found in the range $a < x < b$ is given by

$$P(a < x < b) = \int_a^b p(x)dx. \tag{A.36}$$

The two constraints of Eqs. A.29 and A.30 in the continuous case become

$$0 \le p(x), \quad \text{for all } x \tag{A.37}$$

$$\int_{-\infty}^{\infty} p(x)dx = 1. \tag{A.38}$$

Note that in the continuous case, there is no upper limit on the size of $p(x)$. The expressions for means and standard deviations in the continuous case are the same as for the discrete case, but with integrals over x instead of sums over n.

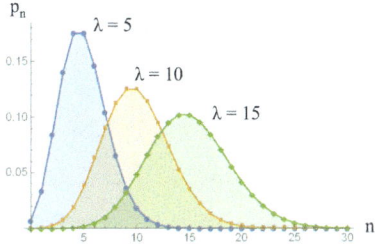

Fig. A.4 The Poisson distribution for $\lambda = 5$, $\lambda = 10$, and $\lambda = 15$. In each case, the mean is at $\mu = \lambda$ and the standard deviation is $\sigma = \sqrt{\lambda}$. Notice that as λ increases, the distribution becomes more symmetric, and it starts to look more like a normal distribution

For the types of applications discussed in this book, the two most important probability distributions are the discrete Poisson distribution and the continuous normal distribution.

Let x be a random variable taking on non-negative integer values: $x = 0, 1, 2, \ldots$, and let λ be some real, positive parameter. Then the **Poisson distribution** is of the form

$$p_n \equiv P(x_n) = \frac{\lambda^n}{n!} e^{-\lambda}. \tag{A.39}$$

Clearly, the p_n are all positive. The normalization constraint A.30 is also readily shown to hold:

$$\sum_{n=0}^{\infty} p_n = \sum_{n=0}^{\infty} \frac{\lambda^n}{n!} e^{-\lambda} = e^{-\lambda} \sum_{n=0}^{\infty} \frac{\lambda^n}{n!} = e^{-\lambda} e^{\lambda} = 1. \tag{A.40}$$

The mean and variance of the Poisson distribution are given by (Problem 4):

$$\mu = \sigma^2 = \lambda. \tag{A.41}$$

Note that everything about the distribution is determined by the single number λ. Plots of the Poisson distribution for several values of λ are shown in Fig. A.4.

The Poisson distribution describes situations in which a random event occurs at a constant average rate, and one wants to know the probability of n such events occurring in a given time period. For example:

- Suppose that the average number of calls per day, λ, to your business, Quantum Industries, Inc., is constant. Then the probability p_n of exactly n calls coming in on a randomly chosen business day is given by the Poisson distributed.
- Similarly, if λ is the average decay rate of a radioactive nucleus (in decays per second), then the number of decays in a randomly chosen 1 s interval is Poisson-

Fig. A.5 The Gaussian or
normal distribution for $\sigma = 1$,
$\sigma = 2$, and $\sigma = 3$. In each
case, the mean is at $\mu = 10$

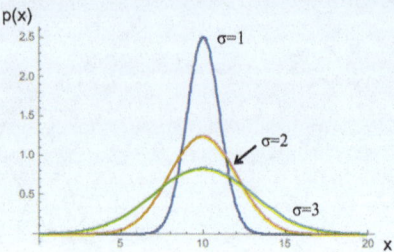

Fig. A.5 The Gaussian or
normal distribution for $\sigma = 1$,
$\sigma = 2$, and $\sigma = 3$. In each
case, the mean is at $\mu = 10$

distributed. (This assumes that the decay isn't so fast that the decay rate changes
noticeably over the one second interval.)

- If a pulsed laser has an average number λ of photons per pulse, then the Poisson
  distribution gives the probability of n photons in a given pulse.
- If a stream of electrons in an electrical current or a stream of photons in a light
  ray are striking some detector surface, then the number n of particles detected in
  a given microsecond will be Poisson distributed, where λ is the average number
  per microsecond. The noise in the detector reading from fluctuations of n around
  the mean are then referred to as shot noise.

Now suppose x is a continuous variable (not necessarily positive). Then the
Gaussian or normal distribution with mean μ and σ is given by

$$p(x) = \frac{1}{\sqrt{2\pi}\,\sigma} e^{-\frac{(x-\mu)^2}{2\sigma^2}} . \tag{A.42}$$

Note that everything there is to know about the distribution is determined by the pair
of numbers μ and σ (Fig. A.5).

The Gaussian distribution arises in every field of the physical and social sciences.
This is a result of the **central limit theorem**, which roughly says the following:
Let $p(x)$ be any sufficiently well-behaved continuous probability distribution (*not*
necessarily normal) with mean μ and standard deviation σ. Take a sequence of
samples of size n and measure the means of each sample. Then as n increases,
the fluctuation of the sample mean $\bar{x}$ away from the population mean μ gets
proportionally smaller compared to μ (i.e., the signal to noise ratio increases), and
for sufficiently large n the sample means $\bar{x}$ themselves become normally distributed:
the average value of $\bar{x}$ is a random variable with mean μ and standard deviation $\frac{\sigma}{\sqrt{n}}$.
So for most continuous distributions, the normal distribution arises naturally in the
limit of large sample sizes. Note also that the Poisson distribution looks more and
more like the normal distribution as λ increases; this can be clearly seen in Fig. A.4.

To find probabilities from the normal probability density, $p(x)$ must be integrated
over, as in Eq. A.36. The resulting Gaussian integrals are discussed in the next
section.

Problems

1. Prove that the values μ and σ in Eq. A.42 are the population mean and the standard deviation of the distribution.
2. (a) Show that the mean of the continuous exponential distribution,

$$p(x) = \frac{1}{\sqrt{2\pi\sigma^2}} e^{-\frac{(x-\mu)^2}{2\sigma^2}},$$

   is μ. Find the variance.
   (b) Suppose a continuous variable is uniformly distributed over the interval $a \le x \le b$. Write down the probability density for x, then find the mean and variance of the distribution.
3. Show that the variance can be written in the form of Eq. A.33.
4. Show that the Poisson mean and variance are as given in Eq. A.41

A.4 Gaussian Integrals and Their Relatives

Most of the problems that can be exactly solved in quantum mechanics without resort to approximations are problems that can be reduced to computing a Gaussian integral or some other integral that has some simple relation to a Gaussian integral. Here we give some common Gaussian-related integrals.

The basic Gaussian integral in one dimension is

$$\int_{-\infty}^{\infty} e^{-x^2} dx = 2 \int_{0}^{\infty} e^{-x^2} dx = \sqrt{\pi}. \tag{A.43}$$

Shifting and rescaling the integration variable, we have (for real constants a and b):

$$\int_{-\infty}^{\infty} e^{-a(x+b)^2} dx = \sqrt{\frac{\pi}{a}}. \tag{A.44}$$

In d dimensions, these integrals generalize to

$$\int e^{-x^2} d^d x = (\pi)^{d/2} \quad \text{and} \quad \int e^{-a(x+\cdot b)^2} d^d x = \left(\frac{\pi}{a}\right)^{d/2}. \tag{A.45}$$

Adding linear and constant terms in the exponent, completion of a square followed by use of the formulas above leads to the result

$$\int_{-\infty}^{\infty} e^{-ax^2+bx+c} dx = \sqrt{\frac{\pi}{a}} e^{b^2/4a+c}. \tag{A.46}$$

Gaussian-like integrals with powers in front of the exponential follow easily from the results above. For example,

$$\int_0^\infty x^{2n} e^{-ax^2} dx = \frac{(2n-1)!!}{a^n 2^{n+1}} \sqrt{\frac{\pi}{a}} \tag{A.47}$$

$$\int_0^\infty x^{2n+1} e^{-ax^2} dx = \frac{n!}{2a^{n+1}}, \tag{A.48}$$

where the double factorial is defined by $(2n - 1)!! = (2n - 2)(2n - 3)(2n - 5)\ldots(3)(1)$. Or, more generally,

$$\int_0^\infty x^\beta e^{-ax^2} dx = \frac{\Gamma\left(\frac{\beta+1}{2}\right)}{2a^{(\beta+1)/2}}, \tag{A.49}$$

where β is not necessarily an integer and Γ is the Euler gamma function. (If the argument of the gamma function is a positive integer, then it reduces to a factorial function: $\Gamma(n) = (n - 1)!$.) If the integration range is extended from $0, \infty$) to $(-\infty, \infty)$, the integral of $x^{2n} e^{-ax^2}$ is doubled (due to the even integrand), while the integral of $x^{2n+1} e^{-ax^2}$ vanishes because the integrand is odd. For even powers, these integrals can be derived by repeatedly taking derivatives of Eq. A.43; for example,

$$\int_0^\infty x^2 e^{-ax^2} dx = -\frac{d}{da} \int_0^\infty e^{-ax^2} dx \tag{A.50}$$

$$= -\frac{d}{da} \frac{1}{2} \sqrt{\frac{\pi}{a}} \tag{A.51}$$

$$= \frac{1}{4} \sqrt{\frac{\pi}{a^3}}. \tag{A.52}$$

The results above can also be generalized by making x into a d-dimensional vector and replacing a by a $d \times d$ matrix A:

$$\int_{-\infty}^\infty e^{-\frac{1}{2} \sum_{i,j=1}^d x_i A_{ij} x_j} d^d x = \int_{-\infty}^\infty e^{-\frac{1}{2} x^T A x} d^d x = \sqrt{\frac{(2\pi)^d}{Det(A)}}. \tag{A.53}$$

Take A to be a d-dimensional unit matrix; this reduces back to Eq. A.45.

A.5   Fourier Transforms and Delta Functions

Fourier transforms take a function and decompose it as a sum of periodic signals or plane waves of different frequencies or wave vectors. Any well-behaved wave packet or wave pulse of finite size can be built as a sum or integral of harmonic plane waves, and the Fourier transform simply gives a weighting to how much each different plane wave contributes to the total. The narrower the original function, the broader the Fourier transform (i.e., the more plane waves that need to be included). The extreme limit of this is the Dirac delta function—it is infinitely narrow, so the Fourier function must be infinitely broad; in fact the Fourier transform is constant (see below).

For a function $f(x)$ in d dimensions, the **Fourier transform** $\mathcal{F}: f(x) \rightarrow \tilde{f}(k)$ is defined as

$$\tilde{f}(k) = \mathcal{F}[f(x)] = \left(\frac{1}{2\pi}\right)^{d/2} \int f(x) \, e^{-ik \cdot x} \, d^d x, \tag{A.54}$$

with the inverse transform given as

$$f(x) = \mathcal{F}^{-1}[\tilde{f}(k)] = \left(\frac{1}{2\pi}\right)^{d/2} \int \tilde{f}(k) \, e^{+ik \cdot x} \, d^d k. \tag{A.55}$$

Here, the integrals go from $-\infty$ to $+\infty$ along all coordinate directions. (Be warned that there are varying conventions, and some authors distribute the powers of 2π differently in front of these definitions.)

When x represents the physical position in three-dimensional space, then the Fourier partner variable k is the wave vector, with magnitude given by the wavenumber $k = \frac{2\pi}{\lambda}$. A common one-dimensional case is when x represents the time, t, so that k is actually the angular frequency, ω:

$$\tilde{f}(\omega) = \frac{1}{\sqrt{2\pi}} \int_{-\infty}^{\infty} f(t) \, e^{-i\omega t} \, dt. \tag{A.56}$$

Among the most important properties of the Fourier transform are the **shift theorem**

$$\mathcal{F}[f(x+a)] = e^{ik \cdot a} \mathcal{F}[f(x)], \tag{A.57}$$

and the action of derivatives on the transform

$$\mathcal{F}\left[\frac{\partial}{\partial x_j} f(x)\right] = ik_j \mathcal{F}[f(x)]. \tag{A.58}$$

A useful binary operation on a pair of functions $f(x)$ and $g(x)$ is their **convolution**, defined by

$$f(x) * g(x) = \int f(y)g(x-y) d^d y. \tag{A.59}$$

The convolution often appears in areas such as optics and signal processing, and it has a very simple behavior under Fourier transforms—the transform of a convolution is simply the product of the Fourier transforms:

$$\mathcal{F}[f(x) * g(x)] = \mathcal{F}[f(x)]\mathcal{F}[g(x)]. \tag{A.60}$$

For a function that depends only on the magnitude $r = |x|$, not on angles, the **radial Fourier transform** is

$$\mathcal{F}[f(x)] = \tilde{f}(k) = 4\pi \int_0^\infty f(r)\frac{\sin kr}{kr}r^2 \, dr = \frac{4\pi}{k} \int_0^\infty r \, \sin(kr)f(r) \, dr, \tag{A.61}$$

where $k = |k|$ and the factors of 4π come from the angular integrations.

The Fourier transform pair $f(x)$ and $\tilde{f}(k)$ defined above assumes that f is a continuous function. Often in applications, Fourier transforms are also defined for functions of a discrete variable. For example, in solids the Hamiltonian is often a sum over terms labeled by the discrete integer n, which labels the locations of the various lattice sites of the crystal. Assume there are N lattice sites, $1 \le n \le N$. Then the Fourier transform to the corresponding discrete momentum variable k in one dimension is

$$\tilde{c}_k = \frac{1}{\sqrt{N}} \sum_{n=1}^N e^{-ikn} c_n, \tag{A.62}$$

with inverse

$$c_n = \frac{1}{\sqrt{N}} \sum_{k=1}^N e^{+ikn} \tilde{c}_k. \tag{A.63}$$

The d-dimensional version simply makes k and n into vectors and sums over all components; for example,

$$\tilde{c}_k = \frac{1}{\sqrt{N}} \sum_{n_1=1}^N \cdots \sum_{n_d=1}^N e^{-ik\cdot n} c_n \tag{A.64}$$

The **Dirac delta function** is an example of a distribution (a structure that is really only defined when embedded inside an integral) and plays an important role in quantum mechanics and many other areas of physics. Recall that the delta function is only nonzero at one point, with an infinitely narrow, infinitely tall spike at that point. Formally, the delta function (in one dimension) is defined by the relations

$$\int_a^b f(x)\delta(x - x_0) = \begin{cases} 1, & \text{if } a < x_0 < b \\ 0, & \text{otherwise} \end{cases} \tag{A.65}$$

$$\int_{-\infty}^{\infty} \delta(x - x_0) = 1. \tag{A.66}$$

The second property simply gives the normalization of the delta function, while the first relation is the **sifting property**: the delta function picks out the value of function at the location of the delta function spike. The generalization to d dimensions is straightforward.

The Dirac delta function can be defined as the limit of a sequence of ordinary functions in a number of ways. But in practice, the most useful representation of the delta function is as the Fourier transform of a constant function:

$$\delta(x - x_0) = \frac{1}{\sqrt{2\pi}} \int_{-\infty}^{\infty} e^{-ik(x-x_0)} dx. \tag{A.67}$$

The inverse transform, of course, gives the result:

$$\frac{1}{\sqrt{2\pi}} \int_{-\infty}^{\infty} \delta(x - x_0) e^{+ik(x-x_0)} dx = \frac{1}{\sqrt{2\pi}}. \tag{A.68}$$

Similarly, the discrete Fourier transform can be used to define the **Kronecker delta**:

$$\delta_{jk} = \begin{cases} 1, & \text{if } j = k \\ 0, & \text{if } j \neq k \end{cases} \tag{A.69}$$

$$= \frac{1}{N} \sum_{n=1}^{N} e^{\frac{2\pi i n}{N}(j-k)}. \tag{A.70}$$

One particularly useful and simple Fourier transform is that of a Gaussian function. The Fourier transform of a Gaussian with width σ is simply a Gaussian with width $\frac{1}{2\pi\sigma}$:

$$f(x) = \frac{1}{\sqrt{2\pi\sigma^2}} e^{-\frac{x^2}{2\sigma^2}} \qquad \longrightarrow \qquad \tilde{f}(k) = \sqrt{2\pi\sigma^2} e^{-2\pi^2\sigma^2 k^2}. \tag{A.71}$$

A.6   The Geometric Series

The infinite geometric series (and its truncation to a finite number of terms) commonly arise in many areas of math, science, and engineering and make several appearances in this book. Here, we give two basic results and briefly derive them.

The geometric series is the infinite series defined by:

$$s \equiv \sum_{n=0}^{\infty} x^n, \tag{A.72}$$

where x may be a complex number. The series converges to a finite result if $|x| < 1$ and diverges for $|x| \geq 1$. In the case of convergence, the sum is given by

$$s = \sum_{n=0}^{\infty} x^n = \frac{1}{1-x}. \tag{A.73}$$

Proof: Writing out the terms explicitly, we have

$$s = 1 + x + x^2 + x^3 + x^4 + \ldots \tag{A.74}$$

$$= 1 + x(1 + x + x^2 + x^3 + \ldots) \tag{A.75}$$

$$= 1 + xs. \tag{A.76}$$

So, $(1 - x)s = 1$. Solving for s gives the result. •

The truncated version of the geometric series, consisting of just the terms up to finite order N, is also useful:

$$\sum_{n=0}^{N} x^n = \frac{1 - x^{N+1}}{1-x} \tag{A.77}$$

Proof:

$$\sum_{n=0}^{N} x^n = \sum_{n=0}^{\infty} x^n - \sum_{n=N+1}^{\infty} x^n = \sum_{n=0}^{\infty} x^n - x^{N+1} \sum_{n=0}^{\infty} x^n \tag{A.78}$$

$$= \left(1 - x^{N+1}\right) \sum_{n=0}^{\infty} x^n = \frac{1 - x^{N+1}}{1-x}. \quad • \tag{A.79}$$

Problems

1. Compute the finite sum

$$\sum_{n=M}^{N} x^n.$$

2. Making use of your knowledge of the geometric series, compute the infinite series

$$\sum_{n=0}^{\infty} nx^n.$$

Appendix B
The Fresnel Equations

Recall that if light is incident on the interface between two media with different indices of refraction, then in general part of the light will reflect back into the first medium, while part will refract into the second material. Snell's law and the law of reflection tell us the directions that the refracted and reflected light will travel, but they don't tell us *how much* of the light (what fraction of the incident intensity) will reflect and how much will be transmitted into the second material. To determine the reflected and transmitted intensities, the **Fresnel equations** are needed.

The Fresnel equations follow from the boundary conditions that the electric fields have to obey at the material interface. Derivations can be found in any electromagnetism textbook; here we will just state the results. The form of the Fresnel equations will depend on the polarization of the light. Recall that the **incident plane** of the light is the plane formed by the rays of the incident and reflected beams (Fig. B.1). If light is polarized in the incident plane, then it is said to be **p-polarized** (p for parallel) or **TM-polarized**. TM stands for *transverse magnetic*, since the magnetic field must be perpendicular to the plane of incidence if the electric field is contained in the incident plane. On the other hand, if the light is polarized perpendicular to the incident plane, then it is called **s-polarized** or **TE-polarized**. TE stands for *tranverse electric*, while s is for *senkrecht*, the German word for perpendicular.

Let E_i, E_r, and E_t by the electric field amplitudes in the incident, reflected, and transmitted (refracted) light, respectively. Then the (dimensionless) **reflection and transmission amplitudes** are given by the ratios of the reflected and transmitted fields to the incident field:

$$r = \frac{E_r}{E_i}, \qquad t = \frac{E_t}{E_i}. \tag{B.1}$$

The **reflectivity** and the **transmissivity** are the corresponding ratios of intensities:

© The Author(s), under exclusive license to Springer Nature Switzerland AG 2025
D. S. Simon, *Introduction to Quantum Science and Technology*, Undergraduate
Texts in Physics, https://doi.org/10.1007/978-3-031-81315-3

Fig. B.1 The plane of incidence is the plane containing the incident and reflected rays. Light polarizes parallel to this plane is s-polarized (TE), while light polarized in the plane is p-polarized (TM)

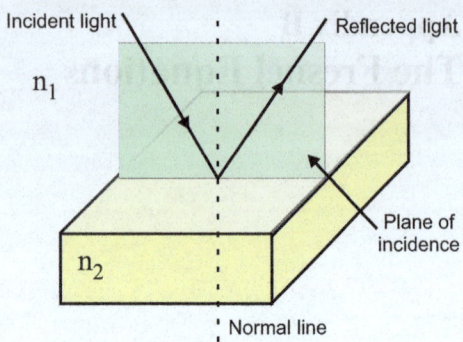

Incident light

Reflected light

n_1

Plane of incidence

n_2

Normal line

$$R = |r|^2 = \left| \frac{E_r}{E_i} \right|^2 = \frac{I_r}{I_i} \tag{B.2}$$

$$T = |t|^2 = \left| \frac{E_t}{E_i} \right|^2 = \frac{I_t}{I_i}. \tag{B.3}$$

Notice that for energy to be conserved, it must be true that $R + T = 1$.

Let θ_1 and θ_2 be the angles from the normal line of the incident and refracted beams. Then the Fresnel equations for s-polarized light are

$$r = \frac{n_1 \cos\theta_1 - n_2 \cos\theta_2}{n_1 \cos\theta_1 + n_2 \cos\theta_2} = -\frac{\sin(\theta_1 - \theta_2)}{\sin(\theta_1 + \theta_2)} \tag{B.4}$$

$$t = \frac{2n_1 \cos\theta_1}{n_1 \cos\theta_1 + n_2 \cos\theta_2} = +\frac{2 \sin\theta_2 \cos\theta_1}{\sin(\theta_1 + \theta_2)}, \tag{B.5}$$

where standard trig identities can be used to show the second equality in each line.

For p-polarized light, the corresponding equations are

$$r = \frac{n_2 \cos\theta_1 - n_1 \cos\theta_2}{n_1 \cos\theta_2 + n_2 \cos\theta_1} = +\frac{\tan(\theta_1 - \theta_2)}{\tan(\theta_1 + \theta_2)} \tag{B.6}$$

$$t = \frac{2n_1 \cos\theta_1}{n_1 \cos\theta_2 + n_2 \cos\theta_1} = +\frac{2 \sin\theta_2 \cos\theta_1}{\sin(\theta_1 + \theta_2)\cos(\theta_1 - \theta_2)}. \tag{B.7}$$

Appendix C
From Vectors to Hilbert Space

At an algebraic level, quantum mechanics describes the behavior of vectors in an infinite-dimensional Hilbert space. These Hilbert spaces are really just straightforward generalizations of the space of three dimensional vectors students learn about in freshman physics classes. In this appendix, the Hilbert spaces of quantum mechanics are described in analogy to three dimensional vector spaces.

C.1 N-Dimensional Vectors

First, we recall three-dimensional vectors and introduce some notation that makes them easier to generalize. Since quantum mechanics is uses states and wavefunctions that are inherently complex, we also need to generalize from real to complex vector spaces.

A vector $\boldsymbol{v}$ is simply an arrow in space. It can always be written in three dimensions in the form

$$\boldsymbol{v} = v_x \hat{\boldsymbol{i}} + v_y \hat{\boldsymbol{j}} + v_z \hat{\boldsymbol{k}}, \tag{C.1}$$

where $\hat{\boldsymbol{i}}$, $\hat{\boldsymbol{j}}$, and $\hat{\boldsymbol{k}}$ are unit vectors in the x-, y-, and z-directions, respectively. These unit vectors are required to be orthonormal or in other words

$$|\hat{\boldsymbol{i}}| = |\hat{\boldsymbol{j}}| = |\hat{\boldsymbol{k}}| = 1 \qquad \text{(normalized)} \tag{C.2}$$

$$\hat{\boldsymbol{i}} \cdot \hat{\boldsymbol{j}} = \hat{\boldsymbol{i}} \cdot \hat{\boldsymbol{k}} = \hat{\boldsymbol{j}} \cdot \hat{\boldsymbol{k}} \qquad \text{(mutually orthogonal)} \tag{C.3}$$

For a pair of vectors $\boldsymbol{v}$ and $\boldsymbol{w}$, the **inner product** (the dot product) and the **magnitude** are defined according to

$$\boldsymbol{v} \cdot \boldsymbol{w} = \boldsymbol{w} \cdot \boldsymbol{v} = v_x w_x + v_y w_y + v_z w_z = |\boldsymbol{v}||\boldsymbol{w}| \cos\theta \tag{C.4}$$

© The Author(s), under exclusive license to Springer Nature Switzerland AG 2025
D. S. Simon, *Introduction to Quantum Science and Technology*, Undergraduate
Texts in Physics, https://doi.org/10.1007/978-3-031-81315-3

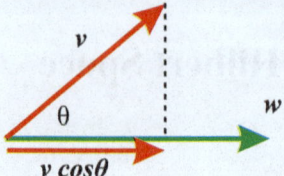

Fig. C.1 Projecting v onto the direction of w leads to a new vector of length $v \cos \theta$, pointing in the direction of w. Multiplying the length of this projection times the length of w gives the dot product. The process is symmetric, in the sense that we could instead have projected w onto v and ended up with the same result

and

$$|v| = \sqrt{v \cdot v} = v_x^2 + v_y^2 + v_z^2, \tag{C.5}$$

where θ is the angle between the two vectors. In other words, the dot product projects one vector onto the direction of the other and multiplies the portions of the two vectors that point in the same direction, dropping the parts that are perpendicular to each other. The dot product is completely symmetric: the result of the product is independent of the order of multiplication (or equivalently, independent of which vector is projected onto which) (Fig. C.1).

Similarly, the cross-product multiplies the parts of the vectors that are perpendicular, and it is antisymmetric:

$$v \times w = -w \times v = \begin{vmatrix} \hat{i} & \hat{j} & \hat{k} \\ v_x & v_y & v_z \\ w_x & w_y & w_z \end{vmatrix}. \tag{C.6}$$

In order to generalize the above definitions to an arbitrary number of dimensions, we need to introduce additional notation. Let N be the number of dimensions. Then the three unit vectors $\hat{i}$, $\hat{j}$, and $\hat{k}$ are replaced by a set of n unit vectors. In the **ket notation** introduced by Dirac, these vectors are written as

$$|1\rangle, |2\rangle, |3\rangle, \ldots |N\rangle. \tag{C.7}$$

A generic unit vector will be denoted $|j\rangle$, where $j \in \{1, 2, \ldots N\}$. (An alternative notation often used is $\hat{e}_j$.) The generic component expression for a vector given in Eq. C.1 would then become

$$v = \sum_{i=1}^{N} v_i |i\rangle. \tag{C.8}$$

The unit vectors can most easily be thought of as column vectors of length N

$$|1\rangle = \begin{pmatrix} 1 \\ 0 \\ 0 \\ \vdots \end{pmatrix}, \qquad |2\rangle = \begin{pmatrix} 0 \\ 1 \\ 0 \\ \vdots \end{pmatrix}, \qquad |3\rangle = \begin{pmatrix} 0 \\ 0 \\ 1 \\ \vdots \end{pmatrix}, \ldots, \qquad (C.9)$$

so a generic vector will be written

$$|v\rangle = \begin{pmatrix} v_1 \\ v_2 \\ v_3 \\ \vdots \end{pmatrix}, \ldots \qquad (C.10)$$

We now allow the components v_j to be complex, so that we can then define the **adjoint** or **conjugate vectors**; if $|v\rangle$ is the ket defined above, then the corresponding adjoint vector, also called a **bra vector**, is the complex-conjugated row vector

$$\langle v| = \begin{pmatrix} v_1^* & v_2^* & v_3^* & \ldots \end{pmatrix} \qquad (C.11)$$

where $*$ represents complex conjugation. The inner product which generalizes the dot product of Eq. C.4 now involves multiplying the row-vector bra corresponding to one vector with the column-vector ket corresponding to the other (thereby getting a bra-ket or bracket):

$$\langle v|w\rangle = \sum_{j=1}^{N} v_j^* w_j. \qquad (C.12)$$

Notice switching the order of the vectors now introduces a complex conjugation: $\langle w|v\rangle = \langle v|w\rangle^*$. The magnitude of a vector, as in Eq. C.5 then become

$$|v| = \sqrt{\langle v|v\rangle} = \sum_{j=1} v_j^* v_j = \sum_{j=1} |v_j|^2. \qquad (C.13)$$

The orthonormality conditions for the unit vectors can now be written compactly as

$$\langle i|j\rangle = \delta_{ij}, \qquad (C.14)$$

where δ_{ij} is the Kronecker delta.

The definitions above satisfy the formal definition of a **linear vector space**: a vector space is a set V of objects over some field F (usually the real or complex numbers) that has two operations defined on it:

Fig. C.2 Cycling through the numbers 1, 2, 3 clockwise gives a plus sign to the antisymmetric tensor (left), while cycling through them counter-clockwise leads to a minus sign (right)

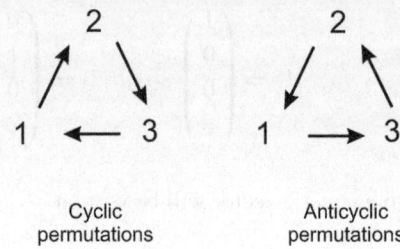

Cyclic permutations

Anticyclic permutations

(i) A vector addition operation taking the Cartesian product $V \times V \to V$: $(v, w) \to v + w$,
(ii) Scalar multiplication taking $F \times V \to V$, $(c, v) \to cv$, for any $c \in F$ and $v \in V$.

The usefulness of vector spaces is that they generalize Cartesian space in a way that allows the use of the standard machinery of linear algebra to be used. In the case of interest to us, F will be the field of complex numbers, $F = \mathbb{C}$, and V will represent a set of quantum states.

For completeness, we should also discuss the generalization of the cross-product. To do this we first introduce the completely antisymmetric **Levi-Civita tensor**. This is an array of numbers $\epsilon_{ijk...}$, with N integer indices in N dimensions. In three dimensions, it is given by

$$\epsilon_{ijk} = \begin{cases} +1 \text{ , if } \{i, j, k\} \text{ are a cyclic permutation of } \{123\} \\ 0 \text{ , if any of } \{i, j, k\} \text{ are equal} \\ -1 \text{ , if } \{i, j, k\} \text{ are an anticyclic permutation of } \{123\} \end{cases} \tag{C.15}$$

Here, 123, 231, and 312 are cyclic permutations, while 132, 321, and 213 are anticyclic (see Fig. C.2). This tensor is completely antisymmetric in the sense that interchanging any of the two indices will introduce a sign change. For example, $\epsilon_{231} = -\epsilon_{321}$. This definition readily generalizes to N dimensions. The tensor will now have N indices (each with potential values running from 1 to N): $\epsilon_{i_1, i_2, i_3, ..., i_N}$, such that

$$\epsilon_{i_1, i_2, i_3, ..., i_N} = \begin{cases} +1 \text{ , if } i_1, ..., i_N \text{ are a cyclic permutation of } 1, 2 ..., N \\ 0 \text{ , if any of } i_1, ..., i_N \text{ are equal} \\ -1 \text{ , if } i_1, ..., i_N \text{ are an anticyclic permutation of } 1, 2 ..., N \end{cases} \tag{C.16}$$

With the Levi-Civita tensor in hand, the three-dimensional cross-product can be written as

$$v \times w = \sum_{i,j,k} \epsilon_{ijk} v_i w_j |k\rangle, \tag{C.17}$$

where i, j, k run from 1 to 3. This generalizes in an obvious manner to antisymmetric products of N vectors in N dimensions:

$$\{A, B, C, \ldots, Z\} \quad \rightarrow \quad \sum_{i_1, \ldots i_N} \epsilon_{i_1, i_2, i_3, \ldots, i_N} A_{i_1} B_{i_2} C_{i_3} \ldots Z_{i_N}. \tag{C.18}$$

These antisymmetric products often come up in the description of angular momentum and of axial vector forces (such as magnetic forces) that don't change sign under parity (spatial reflection) transformations.

C.2   Hilbert Spaces and Non-Countable Dimensions

In the previous section, the basis vectors formed either a finite set, or a countably infinite set. In either case, each vector can be described by a discrete set of labels. However, in quantum mechanics we will often want to use vectors (such as $|\psi\rangle$ or $|\phi\rangle$) that describe the state of a quantum system and that are labeled by a *continuous* variable such as position x, time t, or momentum p. Examples might be of forms such as $|\psi(x, t)\rangle$ or $|\phi(p)\rangle$.

Consider position x as an example. In the discrete case, each component is labeled by an integer and represents one entry in the column vector of Eq. C.10. Basis vectors are states of definite position, $|x\rangle$. But there are an infinite number of x values, forming a continuum. So if we think of vectors like $|\psi(x)\rangle$ as column vectors, they will have an uncountably infinite number of rows: each x value labels a different row in the column vector. Clearly, we can't explicitly write down the vector any more, so we need some indirect way to extract information from it. There are two main ways of doing this:

(i) We can operate on the vector with a Hermitian operator in order to extract the value of some physical observable.
(ii) Or, we can project onto a lower-dimensional subspace. For example, we can take a state vector $|\psi\rangle$ and project it onto a particular value of position, x_0. The result is the wavefunction at x_0:

$$\psi(x_0) = \langle x_0|\psi\rangle. \tag{C.19}$$

This amounts to taking the component of the infinite-dimensional state vector in the "direction" x_0. It is worth stressing again that here each position value in real space, x_0, x_1, etc. corresponds to a different direction in the infinite directional vector space.

To generalize inner products and magnitudes from the previous section, we simply replace discrete labels like i, j, k by continuous labels such as x, t, p, replace sums over the discrete labels by integrals over continuous ones, and replace Kronecker deltas by Dirac delta functions. Thus, inner products now have the form

(in one spatial dimension)

$$\langle \phi | \psi \rangle = \int \phi(x)^* \psi(x) \, dx \qquad (C.20)$$

and the magnitude is given by

$$|\psi|^2 = \langle \psi | \psi \rangle = \int \psi^*(x) \psi(x) dx, \qquad (C.21)$$

where integrals are over all allowed x values and the functions $\phi(x)$ and $\psi(x)$ are related to state vectors as in Eq. C.19. We normalize according to $\langle \psi | \psi \rangle = 1$.

One remarkable fact should be noted at this point: the choice of projecting onto the position variable x to get a position space wavefunction $\psi(x)$ is not unique. We could instead have written the wavefunctions over any continuous variable (such as momentum) and integrated over this new variable to get the inner product. The remarkable thing about this is that the inner products and magnitudes would still come out to the same values. The transition from describing the system in terms of x to describing it in terms of p is simply a change of basis: the values of all physically measurable quantities should remain unchanged.

Appendix D
Lightning Review of Special Relativity

In four-dimensional space-time, a vector will have a single time component t and three space components (x, y, z). These are conveniently combined into a single four-vector x with components x^μ, where μ is an index that runs from 0 to 4: $x^0 = ct$, $x^1 = x$, $x^2 = y$, $x^3 = z$. Similarly, the energy scalar and the momentum vector can be combined into a single energy-momentum four-vector, p, with components $p^\mu = (\frac{E}{c}, p^1, p^2, p^3)$. Notice that three-vectors are in bold print, while four-vectors are not; so for example, we can write $p = (\frac{E}{c}, \boldsymbol{p})$.

Products in four dimensions are similar to the dot product in three dimensions, except that the space components come in with opposite sign to the time component. There are different conventions, but we will use the convention where the space components gain the minus sign. (This is the common convention in particle physics; in general relativity, the convention where the time components gain a minus sign is more common.) So, for example:

$$x^2 \equiv x \cdot x = c^2 t^2 - x^2 - y^2 - z^2 = c^2 t^2 - \boldsymbol{x}^2, \tag{D.1}$$

$$p^2 \equiv p \cdot p = c^2 \left(\frac{E}{c}\right)^2 - \boldsymbol{p} \cdot \boldsymbol{p}, \tag{D.2}$$

$$x \cdot p = Et - \boldsymbol{p} \cdot \boldsymbol{x}. \tag{D.3}$$

Notation can be made more compact by defining the flat-space **metric**, a four-by-four matrix,

$$\eta = \eta^{-1} = \begin{pmatrix} 1 & 0 & 0 & 0 \\ 0 & -1 & 0 & 0 \\ 0 & 0 & -1 & 0 \\ 0 & 0 & 0 & -1 \end{pmatrix}, \tag{D.4}$$

© The Author(s), under exclusive license to Springer Nature Switzerland AG 2025
D. S. Simon, *Introduction to Quantum Science and Technology*, Undergraduate
Texts in Physics, https://doi.org/10.1007/978-3-031-81315-3

with components denoted $\eta_{\mu\nu} = \eta^{\mu\nu}$, where μ and ν run from 0 to 3. Then four-vectors can be thought of as column matrices and contracted with η to get the inner product. For example,

$$p \cdot x = p^T \eta x = \sum_{\mu=0}^{3} p^\mu \eta_{\mu\nu} x^\nu = p^\mu \eta_{\mu\nu} x^\nu, \tag{D.5}$$

where in the last equality, we have introduced the **Einstein summation convention**: any index that is repeated twice (once up and once down) is assumed to be summed over.

The metric tensor can also be used to raise and lower indices:

$$x_\mu = \eta_{\mu\nu} x^\nu, \qquad x^\mu = \eta^{\mu\nu} x_\nu, \tag{D.6}$$

so that inner products can be written (still using the summation convention) in the additional, more compact forms

$$p \cdot x = p_\mu x^\mu = p^\mu x_\mu. \tag{D.7}$$

Because of the minus signs in the metric, we have

$$x_0 = x^0, \; x_1 = -x^1, \; x_2 = -x^2, \; x_3 = -x^3, \tag{D.8}$$

Four-dimensional dot products are relativistically invariant scalars: they have the same value in every inertial reference frame. These invariants often have simple meaning. For example, the product of the four-momentum with itself is proportional to the particle's rest mass: $p^2 = m^2 c^2$.

The Schrödinger equation is nonrelativistic: the space and time derivatives enter into it in different ways, and so it will not be covariant under Lorentz transformations. As a result, it is not useful for describing relativistic particles. One possible relativistic generalization is obtained by simply taking the relativistic mass shell condition $p^2 c^2 + m^2 c^4 = E^2$, multiplying it by a relativistic wavefunction ϕ, and substituting the quantum operators $E \to I\hbar\partial_t$ and $p \to -i\hbar\nabla$ into it. The result is the **Klein-Gordon equation**:

$$\left(\frac{1}{c^2}\partial_t^2 - \nabla^2 + \frac{\hbar^2}{c^2}m^2\right)\phi(x) = 0. \tag{D.9}$$

For a particle of speed v, the Schrödinger equation re-emerges as an approximation to the Klein-Gordon equation for $\frac{v}{c} \ll 1$.

The Klein-Gordon equation has to be satisfied by all free-particles, regardless of their spin. However, for spin-$\frac{1}{2}$ particles, the Klein-Gordon equation, while necessary, is not sufficient for a full description of the particle's behavior. To include the effects of spin, it is necessary to use the **Dirac equation**, which is discussed briefly in Sect. 21.6.

Appendix E
Brief Introduction to Groups

A **discrete group** G of order N is a collection of mathematical objects $\{g_1, g_2, \ldots, g_N\}$ (where N may be infinite) and a product or composition operation $g_i \cdot g_j$ such that the following conditions hold:

1. **Closure property:** If g_i and g_j are any two elements of G, then so is $g_i \cdot g_j$. (More compactly, $g_i, g_j \in G$ implies $g_i \cdot g_j \in G$.)
2. **Associativity:** If $g_i, g_j, g_k \in G$, then $g_i \cdot (g_j \cdot g_k) = (g_i \cdot g_j) \cdot g_k$.
3. **Identity element:** there exists an element I in G such that $I \cdot g_j = g_j \cdot I = g_j$, for all j.
4. **Inverses:** For every g_j there is an element g_j^{-1} of the group such that $g_j \cdot g_j^{-1} = g_j^{-1} \cdot g_j = I$.

Most often the group operation is not explicitly written, so that $g_i \cdot g_j$ is simply written as $g_i g_j$. If the order N of the group is finite, then N repetitions of the same operation will always give back the identity: $g_j^N = I$ for all j.

In physics, the mathematical objects making up the group are usually operators of some sort (spin operators, translation and rotation operators, etc.). If all of the elements of the group commute ($g_i g_j = g_j g_i$ for all i and j), then the group is called **Abelian**. If even one pair fails to commute, then it is **non-Abelian**.

Example E.0.1 Examples of groups are easy to find. To give just a few:

(1) $\mathbb{Z}$: The integers form an Abelian group under ordinary addition, with the identity element being the number 0 and the inverses of the positive integers given by the negative integers.
(2) $\mathbb{R}$: The real numbers form an Abelian group under ordinary addition, with the identity element being the number 0 and the inverses of the positive real numbers given by the negative real numbers.
(3) $\mathbb{R}^+$: The positive real numbers also form an Abelian group under multiplication with identity element 1 and the inverse of element b given by $\frac{1}{b}$.
(4) $\mathbb{Z}_2$: The set $\{0, 1\}$ is a group under addition (mod 2), with identity element 0.

© The Author(s), under exclusive license to Springer Nature Switzerland AG 2025
D. S. Simon, *Introduction to Quantum Science and Technology*, Undergraduate Texts in Physics, https://doi.org/10.1007/978-3-031-81315-3

(5) $SO(2)$: Rotations in two-dimensional space form an Abelian group, where the group multiplication simply consists of performing one rotation followed by the other.

(6) $SO(3)$: Rotations in three-dimensional space form a non-Abelian group. The reader can easily verify that this is non-Abelian: picture an object rotated by $\frac{\pi}{2}$ about the z-axis followed by a $\frac{\pi}{2}$ rotation about the x-axis, and then picture the same object after the rotations are carried out in opposite order. The object will be oriented differently in each case.

A group may also be a **continuous group**, if instead of having a discrete label like the i, j, k above, its elements are labeled by a continuous variable, $g(x)$. Examples of continuous groups are the translation and rotation groups. If the elements $g(x)$ form a smooth, differentiable manifold (a sphere, a plane, etc.) as x is varied, then the group is a **Lie group**. The rotation group and translation groups are therefore Lie groups.

Suppose that G has a subset $S = \{s_1, s_2, \ldots, s_m\}$ of elements such that all elements of G can be formed from products of powers of S elements, $g = s_1^{n_1} s_2^{n_2} \ldots s_m^{n_m}$, for some set of integers $n_1, \ldots n_m$. Then the elements of G are called the **generators** of the group. For example, the group of rotational symmetries of a regular octagon consists of *eight* elements (the rotations through angles $\theta_n = \frac{2\pi n}{8}$ for $n = 0, 1, \ldots, +6, +7$), but these are all powers of a single generator: the rotation by θ_1. This one generator generates the rest of the group. In general, the generators determine the structure of the entire group, and often groups of infinite rank are generated by a finite set of elements.

Physical operators that make up a group have to act on something. The space of objects or states that the operators act on is the *representation space*. The action of each group element on this space can be represented as a matrix. The set of matrices, which have to obey all of the group requirements listed above (with matrix multiplication replacing the group operation) are called a **representation** of the group. These representations can come in different sizes: for a **d-dimensional representation**, the matrices are $d \times d$ and the representation space is spanned by d basis vectors.

For example, the angular momentum operators have many representations of different dimensions. The spin one-half representation is two-dimensional and is given by the Pauli matrices acting on a space spanned by the spin-up and spin-down basis states. The spin-one representation is three-dimensional; it is given by three 3×3 matrices acting on a space spanned by basis states with $L_z = +1, 0, -1$. Notice that although these two representations have different dimensions (2 for spin-1/2 and 3 for spin-1), the order is 3 for both (there are three matrices in both cases). The order is a property of the group and is representation-independent, but the dimension depends on the representation.

A subset of states in the representation space that mix only with each other under the group transformations and don't get transformed into elements outside the subset are called an **irreducible representation**. If a representation isn't irreducible, then it is called reducible. Bases can be found in which reducible representations are block-diagonalized, with each block representing an irreducible representation. In other words, reducible representations are direct sums of irreducible ones, so for many purposes it suffices to know the irreducible representations. In particle physics, the states of fundamental particles like an electron, quark, or photon form irreducible representations of the symmetry groups. The states of composite particles like a proton (which is made from three quarks) or a helium nucleus form reducible representations.

Groups are central to physics because they describe symmetries of a system: if a system is unchanged (i.e. symmetric or invariant) under a set of transformations, those transformations will always form a group. Especially important is the case when a physical system is unchanged by a continuous group of symmetry transformations: **Noether's theorem** implies that every continuous symmetry leads to a conservation law (see any mechanics book that covers Lagrangian methods, such as [1], or any book on particle physics or quantum fields, such as [2]). For example, linear momentum conservation, angular momentum conservation, and energy conservation arise as a result, respectively, of symmetry under spatial translations, spatial rotations, and time translations.

An important example of a group in n dimensions is the **orthogonal group** $O(n)$. Any element $\hat{R}$ of the group must satisfy the relation $\hat{R}^T \hat{R} = \hat{R}\hat{R}^T = \hat{I}$, where $\hat{I}$ is the identity element; this orthogonality constraint guarantees that the rotations do not change the length of the vectors on which they act. $O(n)$ is the group of rotations and reflections acting on real vectors. Reflections will have determinant -1, while rotations have determinant $+1$. The rotation subgroup is the special orthogonal group, $SO(n)$, of orthogonal matrices with determinant $+1$.

In quantum mechanics, the vectors are complex, so the orthogonal group has to be generalized to the **unitary group**: $U(n)$ is the group whose elements obey $\hat{U}^\dagger \hat{U} = \hat{U}\hat{U}^\dagger = I$. The determinant of a unitary matrix can lie anywhere on the unit circle in the complex plane. By restricting to those with determinant equal to 1, we get the special unitary group, $SU(n)$, which is of special importance in particle physics. $SU(n)$ matrices can be thought of as rotations that act on complex vector spaces, while $U(n)$ transformations rotate as well as multiplying by an overall phase.

For quantum information processing, the important group is $U(n)$: the action of any gate that acts on a set of m qubits can viewed as a unitary operation in the group $U(2^m)$. A universal set of quantum logic gates is any set that can generate the full $U(n)$ group. In particular, the single qubit transformation $U(2)$ can be written as products of phase factors and $SU(2)$ elements. The $SU(2)$ group in turn is spanned by the Pauli matrices and the identity matrix.

For more detailed treatments of group theory and its applications in physics, see [2–5]

References

1. H. Goldstein, C.P. Poole, J. Safko, *Classical Mechanics*, 3rd edn. (Pearson, New York, 2011)
2. A. Zee, *Quantum Field Theory in a Nutshell*, 2nd edn. (Princeton University Press, Princeton, 2010)
3. M. Hamermesh, *Group Theory and Its Application to Physical Problems* (Dover Publications, New York, 1989)
4. R. Gilmore, *Lie Groups, Physics, and Geometry: An Introduction for Physicists, Engineers and Chemists* (Cambridge University, Cambridge, 2008)
5. M. Tinkham, *Group Theory and Quantum Mechanics* (Dover Publications, New York, 2003)

Appendix F
A Bit of Number Theory

Number theory studies the properties of non-negative integers. Here we give a brief description of the results needed to understand the RSA encryption scheme in Chap. 18.

Let N be a positive number. Two integers x and y are said to be equal modulo N, written $x = y \pmod{N}$, if they differ by an integer multiple of N: $x = y + kN$, where $k \in \mathbb{Z}$.

Let G_N be the set of positive integers $\leq N$ which share no common factors with N. It is easy to show that G_N forms a group under multiplication mod N:

1. First note that if q and p are in G_N, then their product pq can have no common factors with N, and so pq is also therefore an element of G_N. G_N is thus closed under mod N multiplication.
2. There is an identity element, which is simply the number 1.
3. Modular multiplication is associative: $a(bc) \pmod{N} = (ab)c \pmod{N}$.
4. Inverses exist. To see this, let $a, b, c \in G_N$ be integers such that $ab = ac \pmod{N}$. Then, $a(b - c)$ is a multiple of N. But since a has no common factors with N it follows that $b - c$ must be a multiple of N, or equivalently that $b = c \pmod{N}$. So if $b \neq c$ then $ab \neq ac$. Multiplication of the elements of G_N by a fixed element a is therefore one-to-one and simply permutes the elements of G_N. As a result, for each b there has to be an a that takes b to $1 \pmod{N}$ under the permutation. Thus, for each b, there is an inverse $a = b^{-1}$ under mod N multiplication.

G_N is therefore an Abelian group under mod N multiplication. Let K be the **order of the group** G_N (i.e., the number of elements in the group). For each individual element of G_N, we define a separate number called the **order of the particular group element**: the order k of element $a \in G_N$ is the smallest integer such that $a^k = 1 \pmod{N}$. For each $a \in G_N$, k divides into K.

Suppose now that N is prime. Then by definition, no integer less than N shares any common factors with N. So the order of the group G_N is $K = N - 1$. For any

© The Author(s), under exclusive license to Springer Nature Switzerland AG 2025
D. S. Simon, *Introduction to Quantum Science and Technology*, Undergraduate
Texts in Physics, https://doi.org/10.1007/978-3-031-81315-3

a, $N - 1$ must be a multiple of the order of a, so this then implies **Fermat's little theorem** (see [1] or appendix E of [2]):

> The order of any group element divides evenly into the order of the group, so for any $a \in G_N$, $a^{N-1} = 1$ (mod N).

Fermat's little theorem has an extension due to Euler. Suppose that p and q are primes and that neither of them divides into a, and define $r = (q - 1)(p - 1)$. Then no powers of p or q are divisible into a. Since a^{q-1} is not divisible by p and a^{p-1} is not divisible by q, then Fermat's little theorem implies that

$$\left(a^{q-1} \right)^{p-1} = 1 \text{ (mod } p) \tag{F.1}$$

$$\left(a^{p-1} \right)^{q-1} = 1 \text{ (mod } q). \tag{F.2}$$

Combining the last two lines, we find that

$$a^{(q-1)(p-1)} = 1 \text{ (mod } pq), \tag{F.3}$$

or equivalently

$$a^{1+r} = a \text{ (mod } pq). \tag{F.4}$$

More generally

$$a^{1+kr} = a \text{ (mod } pq), \tag{F.5}$$

for any positive integer k.

Now, given an integer c that has no common factors with r. Since c is an element of the group G_r, it must have a mod r inverse $d = c^{-1} \in G_r$, such that $cd = 1$ (mod r). This implies that there must be an integer k such that

$$cd = 1 + kr = 1 + k(q - 1)(p - 1). \tag{F.6}$$

From Eqs. F.5 and F.6, it follows that any integer a must satisfy $a^{cd} = a$ (mod pq). Thus, if $b = a^c$ (mod pq), then

$$b^d = a^{cd} \text{ (mod } pq), \tag{F.7}$$

or

$$b^d = a \text{ (mod } N). \tag{F.8}$$

The latter result is the desired **Euler extension of Fermat's little theorem**. It forms the basis of the RSA encryption scheme (Chap. 18).

References

1. N.D. Mermin, *Quantum Computer Science: An Introduction* (Cambridge University, Cambridge, 2007)
2. S.M. Barnett, *Quantum Information* (Oxford University, Oxford, 2009)

Index

© The Editor(s) (if applicable) and The Author(s), under exclusive license to
Springer Nature Switzerland AG 2025
D. S. Simon, *Introduction to Quantum Science and Technology*, Undergraduate
Texts in Physics, https://doi.org/10.1007/978-3-031-81315-3

The manufacturer's authorised representative in the EU is Springer
Nature Customer Service Centre GmbH, Europaplatz 3, 69115 Heidelberg,
Germany. If you have any concerns regarding our products, please
contact ProductSafety@springernature.com

Printed and bound by CPI Group (UK) Ltd, Croydon, CR0 4YY

27/04/2026

02097573-0011